COLLEGE ALGEBRA AND TRIGONOMETRY

Lawrence G. Gilligan
Robert B. Nenno

GOODYEAR PUBLISHING COMPANY, INC.
Santa Monica, California

Library of Congress Cataloging in Publication Data

Gilligan, Lawrence G.
College algebra and trigonometry.

Includes index.
1. Algebra. 2. Trigonometry. I. Nenno, Robert B., joint author. II. Title.
QA154.2.G533 512'.13 80-25418
ISBN 0-8302-1992-7

Y-1992-0

ISBN: 0-8302-1992-7

Photo credits:
Photo of Descartes on p. 187 from Culver Pictures, Inc.
Photo of W. Leontief on p. 426 from Arthur Grace; Stock, Boston.

Current Printing (last digit):

10 9 8 7 6 5 4 3 2 1

Supervising Production Editor: Pam Price
Compositor and Text Designer: Sandy Bennett
Cover Design: Marshall Licht, The Design Element
Art: Michael M. Fujita

Printed in the United States of America

To our buddy

Nelson G. Rich

Contents

CHAPTER 1
Basic Concepts of Algebra 1

1.1 The Real Number System 2
1.2 Order Relations and Absolute Value 10
1.3 Integral Exponents 16
1.4 Polynomials 24
1.5 Factoring 34
1.6 Fractional Expressions 39
1.7 Radical Expressions 48
1.8 Rational Exponents 55
1.9 Complex Numbers 60

Chapter 1 in Review 66
Summary 66
Vocabulary/Symbol True-False Quiz 67
Chapter Test 67
Student Research Project 68

CHAPTER 2
Equations, Inequalities, and Applications 69

2.1 Linear Equations 70
2.2 Applied Problems and Formulas 75
2.3 Linear and Absolute Value Inequalities 89
2.4 Quadratic Equations and Applications 97
2.5 Additional Types of Equations 106
2.6 Quadratic and Fractional Inequalities 114

Chapter 2 in Review 122
Summary 122
Vocabulary/Symbol True-False Quiz 122
Chapter Test 123
Student Research Project 124

CHAPTER 3
Relations, Functions, and Graphs 125

3.1 The Cartesian Coordinate System 126
3.2 Relations and Functions 132
3.3 Graphing Linear Functions 145
3.4 Graphing Quadratic Functions 153
3.5 Other Linear and Quadratic Relations 160
3.6 The Inverse of a Relation 168
3.7 Applications—Variation 174

Chapter 3 in Review 183
Summary 183
Vocabulary/Symbol True-False Quiz 183
Chapter Test 184
Student Research Project 186

CHAPTER 4
Topics in Analytic Geometry 187

4.1 Piecewise Functions 188
4.2 Algebraic Clues to Graphing 195
4.3 Graphing Rational Functions 204
4.4 The Conic Sections 210

Chapter 4 in Review 222
Summary 222
Vocabulary/Symbol True-False Quiz 222
Chapter Test 223
Student Research Project 223

CHAPTER 5
Exponential and Logarithmic Functions 224

5.1 Exponential Functions 225
5.2 Logarithmic Functions 233
5.3 Properties of Logarithms 241
5.4 More About Common Logarithms 246
5.5 More About Natural Logarithms 252
5.6 Exponential and Logarithmic Equations 255
5.7 Applications 260

Chapter 5 in Review 270
Summary 270
Vocabulary/Symbol True-False Quiz 271
Chapter Test 271
Student Research Project 272

CHAPTER 6
The Trigonometric Functions 273

6.1 Angles and Their Measurement 274
6.2 Triangles and Trigonometric Ratios 282
6.3 Trigonometric Functions of a Real Number 294
6.4 Evaluating Trigonometric Functions 306
6.5 More Graphs of the Trigonometric Functions 315
6.6 Applications 325

Chapter 6 in Review 335
Summary 335
Vocabulary/Symbol True-False Quiz 336
Chapter Test 336
Student Research Project 337

CHAPTER 7
More About Trigonometry 338

7.1 Law of Sines and Law of Cosines 339
7.2 Trigonometric Identities 346
7.3 Trigonometric Identities and Equations 355
7.4 Inverse Trigonometric Functions 363
7.5 The Polar Coordinate System 369
7.6 Vectors 375
7.7 DeMoivre's Theorem 384

Chapter 7 in Review 392
Summary 392
Vocabulary/Symbol True-False Quiz 393
Chapter Test 393
Student Research Project 394

CHAPTER 8
Systems of Equations and Inequalities 395

8.1 Systems of Linear Equations: 2-by-2 Systems 396
8.2 Systems of Three Linear Equations in Three Variables 403
8.3 Nonlinear Systems 409
8.4 Systems of Inequalities and Linear Programming 414

Chapter 8 in Review 423
Summary 423
Vocabulary/Symbol True-False Quiz 423
Chapter Test 424
Student Research Project 425

CHAPTER 9
Matrices and Systems of Linear Equations 426

9.1 Matrix Solutions to Systems of Linear Equations 427
9.2 Matrix Algebra 436
9.3 The Matrix Inverse 444
9.4 Determinants and Their Properties 451
9.5 Cramer's Rule 458

Chapter 9 in Review 462
Summary 462
Vocabulary/Symbol True-False Quiz 463
Chapter Test 464
Student Research Project 465

CHAPTER 10
Polynomials 466

10.1 The Remainder and Factor Theorems 467
10.2 The Rational Root Theorem 472
10.3 Approximating Irrational Zeros of Polynomials 478
10.4 Partial Fractions 484

Chapter 10 in Review 490
Summary 490
Vocabulary/Symbol True-False Quiz 490
Chapter Test 490
Student Research Project 491

CHAPTER 11
Sequences and Series 492

11.1 Introduction 493
11.2 Arithmetic Sequences 500
11.3 Geometric Sequences 506
11.4 Mathematical Induction 513
11.5 The Binomial Theorem and the Binomial Series 518

Chapter 11 in Review 525
Summary 525
Vocabulary/Symbol True-False Quiz 525
Chapter Test 526
Student Research Project 526

Appendices 528

Appendix 1 Sets 528
Appendix 2 The Metric System 530

Tables 533

Table 1 Common Logarithms 533
Table 2 Natural Logarithms 535
Table 3 Trigonometric Values 537

Answers to Selected Problems 547

Index 579

Preface

TO THE INSTRUCTOR

Through the use of almost 400 completely solved examples and over 3700 exercises, we believe we have accomplished our goal of presenting the elementary functions of algebra and trigonometry in a truly readable text. Other features that we feel enhance this book are calculator exercises, computer exercises, a wide range of useful applications, and pertinent chapter openers. Each chapter opener consists of an application of mathematics to another discipline. These chapter openers are designed to give students a chance to think about the practical applications of mathematics. At the end of each chapter there is a review section that includes a summary, vocabulary true/false quiz, chapter test, and student research project. Research projects often include a reading in the history of the mathematics presented in that chapter.

The prerequisite for the material in this book is a second course in high school algebra or a college course in intermediate algebra. The book is designed for people preparing for calculus or other college-level mathematics and it can be used for a one- or two-semester course that meets three or four times per week. It should be noted that most of the material in Chapter 1 is review and the time spent in that chapter should be minimal.

TO THE STUDENT

This book was written for you—so use it! There are almost 400 completely solved examples that should be read carefully *before* attempting to complete assignments. Most exercises are keyed directly to the examples. When studying for examinations, try taking the chapter tests under examlike conditions. (In other words, limit your time to the usual exam time allotment, don't look back searching for answers, etc.) The vocabulary true/false quizzes will also be of benefit to you.

Finally, we have learned that the best critic is the student. Thus, should you care to offer criticisms or opinions, or if you have any questions, please feel free to contact us; we promise to reply.

Lawrence G. Gilligan
Robert B. Nenno
c/o Goodyear Publishing Company, Inc.
1640 Fifth Street
Santa Monica, CA 90401

Acknowledgements

We wish to thank Rod Gauvin of GRW College Marketing Group, without whose initial encouragement and motivation, this project would never have begun. Our meetings at Greenstreets will always be remembered.

We also want to thank several people at Goodyear Publishing Company.

To our longtime friend, Jack Pritchard, we want you to know that your faith in us often kept us going.

To our mathematics editor, Pete Marshall, we say "thanks" for coming up with answers to our questions when we needed them and for always keeping us on the right track.

To our supervising production editor, Pam Tully Price, who must have had enough work for three people, we give our thanks. Following long hours of proofreading, checking and rechecking artwork, and occasional moments of panic, it was always possible to look forward to a phone call and never having to "hang ten" for too long.

To others who worked on the production or other facets of this book, including Sandy Bennett, Mike Holmes, Pamela Sullivan, and Susan Caney-Peterson, we want you to know that we are extremely pleased with your work.

To the reviewers, Donald LaTorre, Clemson University; William R. Wilson, Central Piedmont Community College; John Burns, Mt. San Antonio College; and Vincent Motto, Asnontuck Community College, we give our sincerest thanks for your advice and guidance.

Susan Gilligan has to be the neatest mathematics typist in "el mundo." Sue, we don't know how you do it, but you have every right to be proud of it. Thanks!

Finally, to all the students who have in many ways helped to develop this material and our style, we extend our appreciation.

Lawrence G. Gilligan
Robert B. Nenno

CHAPTER **1**

Basic Concepts of Algebra

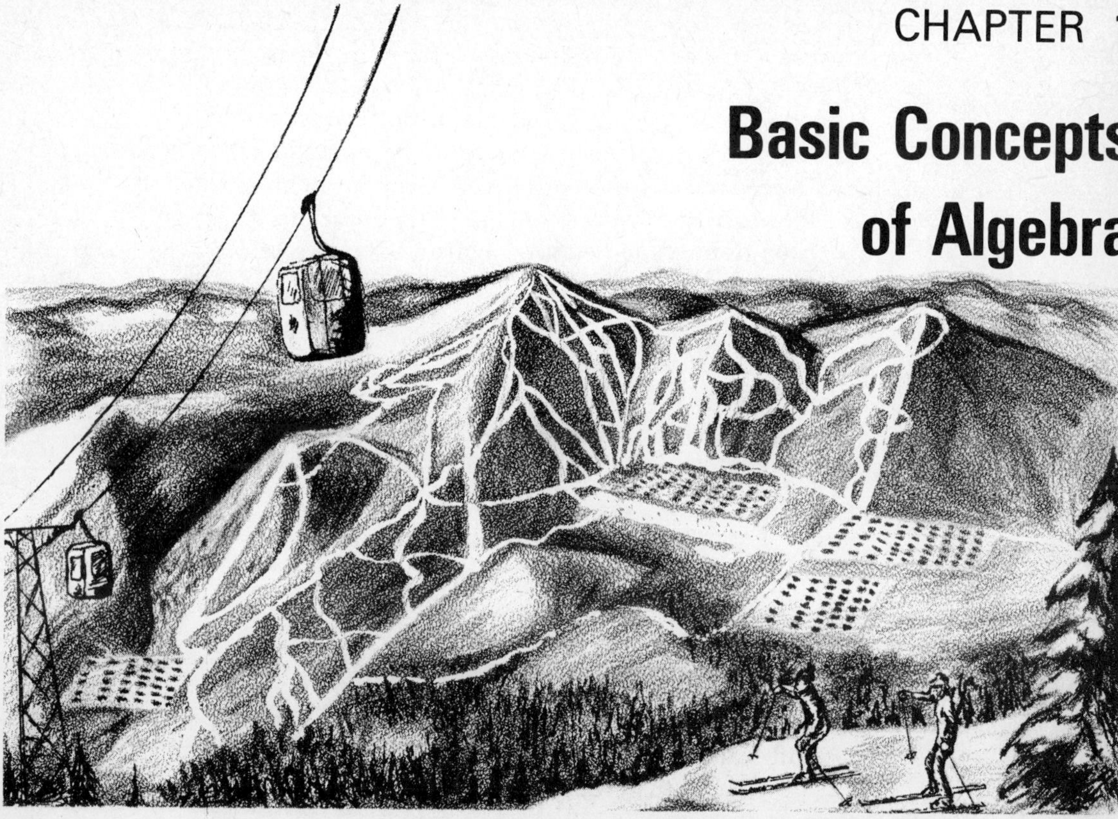

How to ski Killington's four mountains

Formula of K = 4M + 4PL + 7SA
produces unbeatable skiing combinations

The formula for Killington is: K = 4M + 4PL + 7SA. Puzzling? It needn't be. M stands for *mountains,* PL stands for *parking lots*, and SA stands for *skiing areas.* In other words, Killington consists of four separate mountains plus four separate parking lots plus seven separate skiing areas—and every single element is connected to every other element.

Consequently no matter where you park, no matter which mountain you choose to start skiing on and no matter which particular ski area you try first, *you can get to every other parking lot, mountain or ski area without taking off your skis* (except, naturally, to ride up in the gondola).

Reprinted from The Killington Ski Scene Winter 1978–79, *Sherburne Corporation, 1978.*

1.1 THE REAL NUMBER SYSTEM

Development of the number system began about 5000 years ago with the notion of counting. Since then various types of numbers have been defined and classified into different sets. We now describe these sets.

The Natural Numbers or Counting Numbers

$N = \{1, 2, 3, \ldots\}$

The Integers

$$I = \{\underbrace{\ldots, -3, -2, -1}_{\text{negative integers}}, \underbrace{0}_{\text{zero}}, \underbrace{1, 2, 3, \ldots}_{\text{positive integers (or natural numbers)}}\}$$

The Rational Numbers

Q is the set of all numbers of the form $\frac{a}{b}$, such that a is an integer and b is a nonzero integer.*

Notice that the set of rational numbers includes the set of integers (if $b = 1$, $\frac{a}{b}$ is an integer). Furthermore, division by zero is not defined.

Another way to characterize a rational number is through its decimal representation. A decimal that terminates or repeats is *always* a rational number.

Examples of rational numbers:

$$\frac{17}{5} = 3.4 \qquad \text{(terminating)}$$

$$\frac{-2}{3} = -.666\ldots \qquad \text{(repeating)}$$

$$\frac{229}{90} = 2.5444\ldots \qquad \text{(repeating)}$$

*This can be written as $Q = \left\{\frac{a}{b} \,\middle|\, a \in I, b \in I, b \neq 0\right\}$. The reader who is unfamiliar with this set notation should read Appendix 1.

A repeating decimal such as 2.5444 . . . is often written with a bar over the repeating part. Thus, $2.544\ldots = 2.5\overline{4}$.

Any number that cannot be expressed as a quotient (division) of integers, (or equivalently, its decimal representation is nonterminating and nonrepeating) is *not* a rational number. The set of all such numbers is called the **set of irrational numbers,** T.

The Irrational Numbers

T is the set of all numbers that cannot be expressed as a quotient of integers (the decimal representation of such a number is nonterminating and nonrepeating).

Examples of irrational numbers:

$$\left.\begin{aligned} \sqrt{2} &= 1.41421356\ldots \\ \pi &= 3.14159265\ldots \\ -\frac{\pi}{3} &= -1.04719755\ldots \\ -\sqrt[3]{5} &= -1.7099759\ldots \end{aligned}\right\} \text{nonterminating and nonrepeating}$$

We are now in a position to state the definition of a **real number.**

DEFINITION

> A **real number** is any number that is either rational or irrational.

The Real Numbers

$\mathcal{R}$, the **set of real numbers,** consists of the set of rational numbers together with the set of irrational numbers.*

Examples of real numbers: $-3, -1, 6, -\sqrt{\frac{1}{2}}, 2\pi, \frac{5}{2}, 2018.91$.

The set of real numbers is often displayed on a **real number line**:

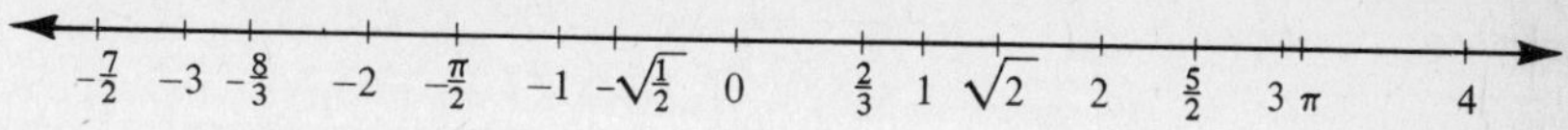

EXAMPLE 1 Answer true or false.

a. Every integer is a rational number.
b. Every integer is a natural number.
c. A repeating decimal is an irrational number.
d. Every rational number is a real number.
e. Every real number is a rational number.
f. $\sqrt{8}$ is a real number.
g. $\sqrt{8}$ is a rational number.
h. $\sqrt{8}$ is an irrational number.

---

*This can be written as $\mathcal{R} = Q \cup T$.

Solution a. True b. False c. False d. True
e. False f. True g. False h. True

Most of the work in this book is with real numbers or expressions that represent real numbers. As a result one must be very careful to use real numbers properly. The following list of eleven axioms pertains to the real number system: a, b, and c represent any real numbers.

Closure Properties

1. $a + b$ is a real number
2. ab is a real number

Commutative Properties

3. $a + b = b + a$
4. $ab = ba$

Associative Properties

5. $a + (b + c) = (a + b) + c$
6. $a(bc) = (ab)c$

Identity Properties

7. $a + 0 = a$ (0: additive identity)
8. $a \cdot 1 = a$ (1: multiplicative identity)

Inverse Properties

9. For every real number a, there exists a real number $-a$ such that $a + (-a) = 0$ ($-a$: additive inverse of a)
10. For every real number $a \neq 0$, there exists a real number $\frac{1}{a}$ such that $a \cdot \frac{1}{a} = 1$ ($\frac{1}{a}$: multiplicative inverse (or reciprocal) of a)

Distributive Property

11. $a(b + c) = ab + ac$

These eleven axioms apply to addition and multiplication. The operations of subtraction and division are now defined.

DEFINITION

The subtraction of b from a, written $a - b$, is defined as:

$$a - b = a + (-b)$$

That is, to subtract b from a, *add* a and the additive inverse of b.

DEFINITION

> The division of a by the nonzero number b, written $a \div b$ or $\frac{a}{b}$, is defined as:
>
> $$\frac{a}{b} = a \cdot \frac{1}{b}$$

That is, in order to divide a by b, *multiply* a and the reciprocal of b.

One of the most common uses of the eleven axioms of real numbers is in simplifying algebraic expressions containing one or more variables. A variable is a letter that is used to represent any number in some set, usually $\mathcal{R}$. Some examples follow.

EXAMPLE 2 What properties are used in going from one step to the next in the simplification process shown below?

$$\begin{array}{l} 5x + (2 + 3x) \\ 5x + 3x + 2 \\ x \cdot 5 + x \cdot 3 + 2 \\ x(5 + 3) + 2 \\ x \cdot 8 + 2 \\ 8x + 2 \end{array}$$

Solution

$5x + (2 + 3x)$	
	Commutative and Associative Properties for Addition
$5x + 3x + 2$	
	Commutative Property for Multiplication
$x \cdot 5 + x \cdot 3 + 2$	
	Distributive Property
$x(5 + 3) + 2$	
	Adding 5 and 3
$x \cdot 8 + 2$	
	Commutative Property for Multiplication
$8x + 2$	

We can write $5x + (2 + 3x) = 8x + 2$.

EXAMPLE 3 What properties are used in going from one step to the next in the simplification process shown below?

$$2(2x - 3y) - 4(3x - 2) + 1$$
$$4x - 6y - 12x + 8 + 1$$
$$4x + (-6y) + (-12x) + 9$$
$$4x + (-12x) + (-6y) + 9$$
$$x \cdot 4 + x(-12) + (-6y) + 9$$
$$x[4 + (-12)] + (-6y) + 9$$
$$x \cdot (-8) + (-6y) + 9$$
$$-8x + (-6y) + 9$$
$$-8x - 6y + 9$$

Solution

$$2(2x - 3y) - 4(3x - 2) + 1$$
Distributive Property
$$4x - 6y - 12x + 8 + 1$$
Definition of Subtraction
Also $8 + 1 = 9$
$$4x + (-6y) + (-12x) + 9$$
Commutative Property for Addition
$$4x + (-12x) + (-6y) + 9$$
Commutative Property for Multiplication
$$x \cdot 4 + x(-12) + (-6y) + 9$$
Distributive Property
$$x[4 + (-12)] + (-6y) + 9$$
$4 + (-12) = -8$
$$x \cdot (-8) + (-6y) + 9$$
Commutative Property for Multiplication
$$-8x + (-6y) + 9$$
Definition of Subtraction
$$-8x - 6y + 9$$

We can write $2(2x - 3y) - 4(3x - 2) + 1 = -8x - 6y + 9$.

The two previous examples emphasize that the simplification of algebraic expressions involves a careful application of various properties. In practice, we shorten the time needed to do this by removing parentheses, called *expanding*, and then *collecting like terms* (terms are algebraic or arithmetic expressions that are added to or subtracted from other expressions). In the previous example, the *like terms* $4x$ and $-12x$ are added to give $-8x$. Similarly, the like terms 8 and 1 are added to give 9. More commonly, the simplification in Example 3 will be done as:

$$2(2x - 3y) - 4(3x - 2) + 1$$
Expand
$$4x - 6y - 12x + 8 + 1$$
Collect like terms
$$-8x - 6y + 9$$

In the next example we simplify by expanding and collecting like terms. We use the **subscripted variables** x_1 and x_2; the **subscripts** 1 and 2 are used to denote the two *different* variables x_1 and x_2.

EXAMPLE 4 Simplify: $5x_1 - 3[x_2 + 2(x_1 + 2)] + 12$.

Solution

$5x_1 - 3[x_2 + 2(x_1 + 2)] + 12$	
	Expand innermost grouping symbols
$5x_1 - 3[x_2 + 2x_1 + 4] + 12$	
	Expand
$5x_1 - 3x_2 - 6x_1 - 12 + 12$	
	Collect like terms
$-x_1 - 3x_2$	

The axioms of the real numbers and definitions are facts that can be used to prove new statements, called **theorems**. A statement about the real numbers that is not true can be shown to be false by exhibiting a **counterexample**. For example, the statement

$$a + (bc) = (a + b) \cdot (a + c)$$

is *false* because when $a = 3$, $b = 4$, and $c = 5$ we have

$$3 + (4 \cdot 5) = (3 + 4) \cdot (3 + 5)$$
$$23 = 56$$

Thus, $a + (bc) \neq (a + b)(a + c)$.

Now, we direct our attention to the number zero which "behaves" in ways unlike any other real number.

Laws about Zero

1. For any real number a, $a + 0 = a$. This is the Identity Property for Addition.
2. For any real number a, $a - 0 = a$.
3. For any real number a, $a \cdot 0 = 0$.
4. If a and b are real numbers and $ab = 0$, then $a = 0$ or $b = 0$ or both.
5. If $a \neq 0$, $\dfrac{0}{a} = 0$.
6. If $a \neq 0$, $\dfrac{a}{0}$ is undefined.
7. $\dfrac{0}{0}$ is **indeterminate**. Indeterminate expressions are studied in calculus. As a result, $\dfrac{0}{0}$ will not be discussed here.

There are also many special ways that quotients behave. We conclude this section by stating certain laws concerning quotients (all denominators are nonzero real numbers).

Laws about Quotients

1. $\dfrac{a}{b} = \dfrac{c}{d}$ if and only if $ad = bc$
2. $\dfrac{ac}{bc} = \dfrac{a}{b}$
3. $\dfrac{a}{b} \pm \dfrac{c}{b} = \dfrac{a \pm c}{b}$
4. $\dfrac{a}{b} \pm \dfrac{c}{d} = \dfrac{ad + bc}{bd}$
5. $\dfrac{a}{b} \cdot \dfrac{c}{d} = \dfrac{ac}{bd}$
6. $\dfrac{a}{b} \div \dfrac{c}{d} = \dfrac{a}{b} \cdot \dfrac{d}{c} = \dfrac{ad}{bc}$

EXERCISES FOR SECTION 1.1

In Exercises 1 through 18, answer true or false. (See Example 1.)

1. Every integer is a rational number.
2. Every rational number is an integer.
3. Every integer is a natural number.
4. Every natural number is an integer.
5. Every rational number is a real number.
6. Every real number is a rational number.
7. Every real number is an irrational number.
8. Every irrational number is a real number.
9. .12 is a rational number.
10. $.\overline{12}$ is a rational number.
11. .12 is a real number.
12. $.\overline{12}$ is a real number.
13. $\sqrt{4}$ is an irrational number.
14. $\sqrt{4}$ is a rational number.
15. $\sqrt{4}$ is a real number.
16. $\sqrt{5}$ is an irrational number.
17. 6 is a real number.
18. 6 is an integer.

In 19 through 22, state the property used in going from each step to the next. (See Examples 2 and 3.)

19.
$6x - (2x + 4)$
 a.
$6x - 1(2x + 4)$
 b.
$6x - 2x - 4$
 c.
$6x + (-2x) + (-4)$
 d.
$x \cdot 6 + x(-2) + (-4)$
 e.
$x[6 + (-2)] + (-4)$
 f.
$x \cdot 4 + (-4)$
 g.
$4x - 4$

20.
$1 + 2(x - 4)$
 a.
$1 + 2x - 8$
 b.
$1 + 2x + (-8)$
 c.
$2x + 1 + (-8)$
 d.
$2x + (-7)$
 e.
$2x - 7$

21. $(3x + 2) + (6x - 2)$
a.
$(3x + 2) + [6x + (-2)]$
b.
$[(3x + 2) + 6x] + (-2)$
c.
$[(2 + 3x) + 6x] + (-2)$
d.
$[2 + (3x + 6x)] + (-2)$
e.
$[2 + 9x] + (-2)$
f.
$[9x + 2] + (-2)$
g.
$9x + [2 + (-2)]$
h.
$9x + 0$
i.
$9x$

22. $(w + x)(y + z)$
a.
$(w + x)y + (w + x)z$
b.
$y(w + x) + z(w + x)$
c.
$yw + yx + zw + zx$
d.
$yw + zw + yx + zx$

In 23 through 36, simplify each expression by expanding and collecting like terms. (See Example 4.)

23. $2(x + 4) + 3$
24. $2(x + 4) + 3x$
25. $(3x + 4) + (4x - 2)$
26. $3(3x + 4) + 2(4x - 2)$
27. $4y - (4y - 5)$
28. $3x - 2[2y + (2x - 1)]$
29. $(3w + 1) - (w + 1)$
30. $x_1 - 5(x_1 + 2) + 11(x_1 - 2)$
31. $x_1 - 5(x_1 + x_2) + 11(x_1 - x_2)$
32. $(x^2 + 3x + 2) - (x^2 - 2x + 4)$
33. $2(x^2 + 7x + 6) - 3(x^2 - 2x + 1)$
34. $(2x + 1)(3x - 2)$
35. $(4y_1 - y_2)(2y_1 + y_2)$
36. $6(x - 2y) + 4[x - 3(x - 2y)] + 1$

In 37 through 53, answer true or false. If true, state why. If false, exhibit a counterexample or state why. (Consult the various laws presented in this section.)

37. $6(x + 3) = 6x + 18$
38. $3(2 - x) = 3[2 + (-x)]$
39. $a + (b \cdot c) = (a + b)(a + c)$
40. $a - (b + c) = (a - b) + c$
41. $(a + b) \div c = a + (b \div c)$
42. $2 - (x + 1) = 2 - 1(x + 1)$
43. $(a + b) \div c = (a + b) \cdot \frac{1}{c}$
44. If $ab = 0$ and $a \neq 0$, then $b = 0$.
45. $\frac{a}{b} + \frac{c}{d} = \frac{a + c}{b + d}$
46. $\frac{7}{0} = 0$
47. $\frac{a}{b} \cdot \frac{c}{b} = \frac{ac}{b}$
48. $\frac{x + 2}{y + 2} = \frac{x}{y}$
49. $\frac{x + 1}{x} = 1$
50. $\frac{2}{11} + \frac{5}{8} = \frac{71}{88}$
51. $\frac{2}{11} - \frac{5}{8} = -\frac{39}{88}$
52. $\frac{2}{3} \cdot \frac{4}{7} = \frac{8}{21}$
53. $\frac{2}{3} \div \frac{4}{7} = \frac{7}{6}$

Calculator Exercise

54. a. As stated in this section, division by zero is *not* defined. On a calculator, follow the sequence:

[8] [÷] [0] [=]

What is the result?

b. Divide 8 by the number "nearest" to zero that your calculator will accept. What is the result?

Computer Exercise

55. In the BASIC* computer language, the operations +, −, ·, ÷ are represented, respectively, by the symbols +, −, *, /.

a. Consider the following program:

```
10 LET X = 5 + 6 * (7/2)
20 PRINT X.
30 STOP
```

What value of X will be outputted?

b. The symbols follow a *priority of operation* which means that * and / get performed (from left to right) before + and − but, as in algebra, any expression in parentheses is evaluated first. What is the value of X3 on output for the program below?

```
100 LET X1 = 6
120 LET X2 = 2
130 LET X3 = 2 * X1 - 5 * (X2 - X1) / X1
140 PRINT "X3 ="; X3
150 STOP
```

c. What is the danger of replacing statement number 1ØØ with 1ØØ LET X1 = Ø?

1.2 Order Relations and Absolute Value

In the previous section, the real number line was seen to be a pictorial way of displaying the real numbers. Each point on the line corresponds to one and only one real number. On the number line if a real

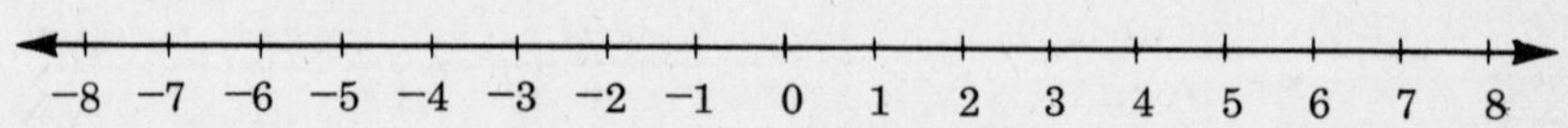

number, a, is to the left of another real number, b, we write $a < b$ and say "a is less than b." Also, notice that the difference, $b - a$, must be a positive number. We have the following definition:

DEFINITION

> Given that a and b are any two real numbers, **a is less than b** ($a < b$) if and only if $b - a$ is positive.

Also, $b > a$ (read "b is greater than a") is equivalent to $a < b$. For example, $13 < 15$, $-8 < -7$, and $1 > 0$ are all true statements.

*BASIC is a registered trademark of the Trustees of Dartmouth College. BASIC stands for Beginner's All-purpose Symbolic Instruction Code. In this textbook, when the word "BASIC" is used, it refers to the language in the generic sense.

The relations "$<$" and "$>$" are called **order relations.** Statements such as $3 < 12$ and $-2 > -8$ are **statements of inequality** (or simply, **inequalities**). Two additional order relations are "$\leqslant$" (less than or equal to) and "$\geqslant$" (greater than or equal to). Thus, $a \leqslant b$ means either $a < b$ or $a = b$.

EXAMPLE 1 Place the symbol $\leqslant$ or $>$ between each pair of numbers to create an equality that is true.

a. 10 3 b. −2 −5 c. −8 6 d. 4 4

Solution a. $10 > 3$ b. $-2 > -5$ c. $-8 \leqslant 6$ d. $4 \leqslant 4$

A glance at the real number line discloses the fact that any real number and its additive inverse are the same distance from 0. For example, the number 6 is situated 6 units to the *right* of the **origin** (the point representing 0), and the number −6 is situated 6 units to the *left* of the origin.

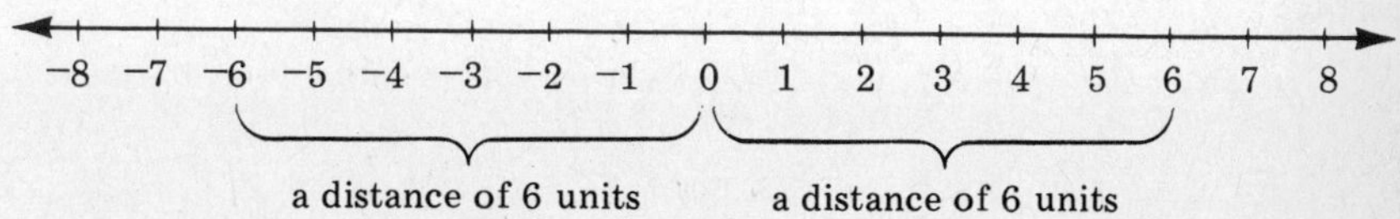

The distance that a real number, x, is from the origin is written in symbols as $|x|$ and is known as **the absolute value of x.** Thus an absolute value represents distance and cannot be negative.

EXAMPLE 2 Evaluate the following absolute values.

a. $|-12|$ b. $|2|$ c. $|0|$ d. $|7 + (-4)|$ e. $|7| + |-4|$

Solution a. $|-12| = 12$ b. $|2| = 2$ c. $|0| = 0$

d. $|7 + (-4)| = |3| = 3$ e. $|7| + |-4| = 7 + 4 = 11$

Thus far, absolute value has been represented as distance on the real number line. We now state a more formal definition.

DEFINITION

The **absolute value** of the real number x, symbolized as $|x|$ is

$$|x| = \begin{cases} x \text{ provided } x > 0 \\ 0 \text{ provided } x = 0 \\ -x \text{ provided } x < 0 \end{cases}$$

Note that as a result of this three-part definition, it is again seen that an absolute value cannot be negative, because

when $x > 0$, $|x| = \boxed{x}$ ← $\boxed{\text{This}}$ is nonnegative.

when $x = 0$, $|x| = \boxed{0}$ ← $\boxed{\text{This}}$ is nonnegative.

when $x < 0$, $|x| = \boxed{-x}$ ← $\boxed{\text{This}}$ is nonnegative since x itself is negative.

Thus the statement $|x| = -6$ is impossible, because there is no real number whose absolute value (distance from 0) is -6. On the other hand the statement $|x| = 6$ *is* possible, because there are two real numbers whose absolute value (distance from 0) is 6, namely $+6$ or -6. We have illustrated the theorem stated below.

THEOREM 1 Suppose b is a positive real number. Then: if $|a| = b$, then $a = \pm b$ and conversely.

This theorem is used in Chapter 2 as well as in the two examples that follow.

EXAMPLE 3 Use Theorem 1 in order to draw a conclusion about the real number x.

a. $|x| = 5$ b. $|x| = 4$ c. $|x| = 12.5$

Solution a. $|x| = 5$ leads to $x = \pm 5$. b. $|x| = 4$ leads to $x = \pm 4$.

c. $|x| = 12.5$ leads to $x = \pm 12.5$.

EXAMPLE 4 Use Theorem 1 in order to draw a conclusion about $|x|$.

a. $x = \pm 2$ b. $x = \pm 1$ c. $x = \pm \pi$

Solution a. $x = \pm 2$ leads to $|x| = 2$. b. $x = \pm 1$ leads to $|x| = 1$.

c. $x = \pm \pi$ leads to $|x| = \pi$.

Next we use the notion of absolute value to discuss the distance between any two points on the real number line.

EXAMPLE 5 Evaluate $|x_1 - x_2|$ and $|x_2 - x_1|$ for each pair of values of x_1 and x_2 given below. In each case, graph the points on a real number line.

a. $x_1 = 5,\ x_2 = -2$ b. $x_1 = 6,\ x_2 = 10$ c. $x_1 = -7,\ x_2 = -2$

Solution a. $|x_1 - x_2| = |5 - (-2)| = |7| = 7$

$|x_2 - x_1| = |-2 - 5| = |-7| = 7$

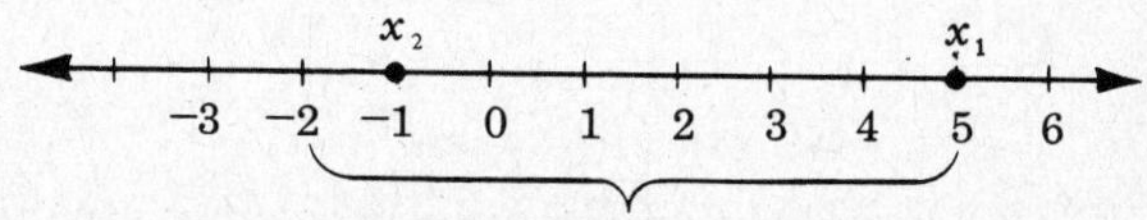

x_1 and x_2 are 7 units apart

b. $|x_1 - x_2| = |6 - 10| = |-4| = 4$

$|x_2 - x_1| = |10 - 6| = |4| = 4$

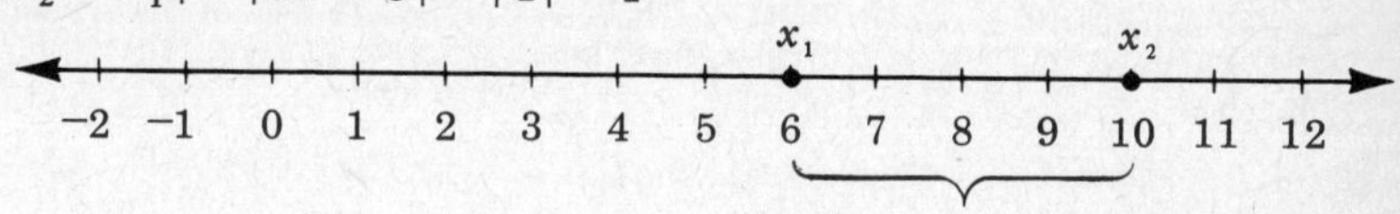

x_1 and x_2 are 4 units apart

c. $|x_1 - x_2| = |-7 - (-2)| = |-5| = 5$

$|x_2 - x_1| = |-2 - (-7)| = |5| = 5$

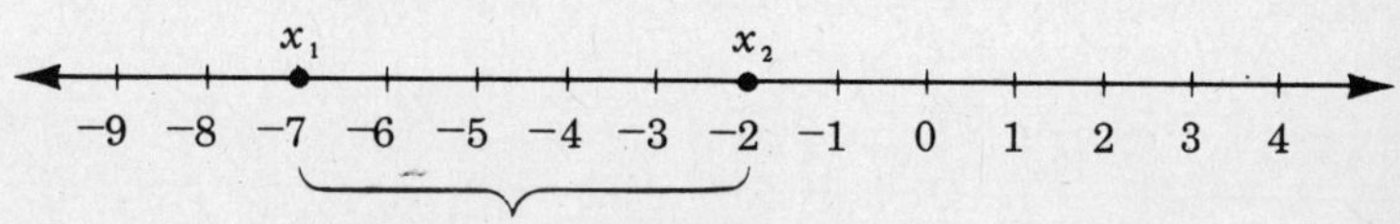

x_1 and x_2 are 5 units apart

Example 5 illustrates that:

> The distance between two real numbers, x_1 and x_2, on the real number line is equal to $|x_1 - x_2|$ or $|x_2 - x_1|$. Thus,
>
> $$|x_1 - x_2| = |x_2 - x_1|.$$

We close this section with an example of another property of absolute value.

EXAMPLE 6 Evaluate $|x|^2$ and x^2 for each value of x given below.

a. $x = -7$ b. $x = 0$ c. $x = 4$

Solution a. $|x|^2 = |-7|^2 = 7^2 = 49$

$x^2 = (-7)^2 = 49$

b. $|x|^2 = |0|^2 = 0^2 = 0$

$x^2 = (0)^2 = 0$

c. $|x|^2 = |4|^2 = 4^2 = 16$

$x^2 = (4)^2 = 16$

In each part of Example 6, it is seen that $|x|^2 = x^2$. In general, it can be shown that:

> If x is any real number, then $|x|^2 = x^2$.

EXERCISES FOR SECTION 1.2

In Exercises 1 through 10, place the symbol "$\leqslant$" or "$>$" between each pair of numbers to create an inequality that is true. (See Example 1.)

1. 2 8
2. 8 2
3. −7 −10
4. −10 −7
5. 7 −3
6. $-\frac{1}{2}$ $-\frac{1}{4}$
7. 8 −|24|
8. 4 −6
9. $\frac{22}{7}$ π
10. $0.\overline{3}$ 0.3

In Exercises 11 through 20, place the symbol "$\geqslant$" or "$<$" between each pair of numbers to make an inequality that is true. (See Example 1.)

11. 2 0
12. 0 2
13. −2 0
14. 0 −2
15. 4 −4
16. −4 4
17. −100 2
18. $\frac{\pi}{2}$ $\frac{\pi}{4}$
19. 1.6 1.63
20. $0.\overline{6}$ 0.6

In Exercises 21 through 30, evaluate each of the absolute values. (See Examples 2 and 6.)

21. $|-10|$
22. $|6|^2$
23. $|9-4|$
24. $|9|-|4|$
25. $|9-4|^2$
26. $|9|^2-|4|^2$
27. $|6+(-8)|$
28. $|6|+|-8|$
29. $|4-7|$
30. $|7-4|$

In Exercises 31 through 36, use Theorem 1 to draw a conclusion about the real number x. (See Example 3.)

31. $|x| = 10$
32. $|x| = 6.2$
33. $|x| = 1$
34. $|x| = 0.02$
35. $|x| = \pi$
36. $7 = |x|$

In Exercises 37 through 42, use Theorem 1 in order to draw a conclusion about $|x|$. (See Example 4.)

37. $x = \pm 3$
38. $x = \pm \frac{\pi}{2}$
39. $x = \pm 6$
40. $x = \pm 9.2$
41. $x = \pm 2\pi$
42. $x = \pm .001$

In Exercises 43 through 48, use the absolute value concept to find the distance between each pair of numbers, x_1 and x_2. (See Example 5.)

43. $x_1 = 2,\ x_2 = 5$
44. $x_1 = -2,\ x_2 = 6$
45. $x_1 = 0,\ x_2 = 4$
46. $x_1 = 0,\ x_2 = -4$
47. $x_1 = 10,\ x_2 = -10$
48. $x_1 = 5,\ x_2 = -5$

49. If the distance between the numbers p and q on the real number line is 6 and p is the real number 2, what are the possible values for q?
50. If $|p - q| = 8$, and $q = 1$, what are the possible values for p?
51. If $|x - y| = 6$, and $x = 2$, what are the possible values for y?
52. If $|x_2 - x_1| = 1$, and $x_1 = 0$, what are the possible values for x_2?
53. If $|x_1 - x_2| = 0$, and $x_1 = 0$, what are the possible values for x_2?
54. It can be shown that the absolute value of the product of any two real numbers is given by

$$|xy| = |x| \cdot |y|$$

Verify this law for the values of x and y given below.

a. $x = 4, y = 2$ b. $x = 4, y = -2$ c. $x = -4, y = 2$
d. $x = -4, y = -2$ e. $x = 4, y = 0$ f. $x = 0, y = 0$

55. It can be shown that

$$|x + y| \leqslant |x| + |y|$$

Verify this law, called the **triangle inequality**, for the values of x and y given below.

a. $x = 5, y = 6$ b. $x = 5, y = -6$ c. $x = -5, y = 6$
d. $x = -5, y = -6$ e. $x = 5, y = 0$ f. $x = 0, y = 0$

56. True or false: $\left|\frac{x}{y}\right| = \frac{|x|}{|y|}$. Before answering, pick various real values of x and y to support your answer.
57. True or false: $|x - y| \leqslant |x| + |y|$. Before answering, pick various real values of x and y to support your answer.

Computer Exercises

58. *Branching* to various parts of a computer program is often desirable. One way of determining *which* branch to go to, is by using the order relations. For example, consider BASIC statements 1Ø and 2Ø below:

```
 ⋮
1Ø  IF X < 10    THEN GO TO 5ØØ
2Ø  IF X > 60    THEN LET X = 4 * Y.
 ⋮
```

If X has been assigned a value less than 10, the program will go to statement number 500. If X has been assigned a value greater than 60, X is reassigned a value—this time equal to 4Y.

a. Determine what happens during execution of the following program if A = 6, B = 2, C = −3:

```
100  LET D = B*B - 4*A*C
110  IF D<0  THEN GO TO 999
120  IF D = 0  THEN PRINT "ONE ROOT"
130  IF D>0  THEN PRINT XI
999  STOP
```

b. Rework exercise (a) above if A = 1, B = −6, and C = 9.

c. Rework exercise (a) above if A = 4, B = −1, and C = 3.

59. The $|x|$ notation is usually written as ABS(X) in BASIC.

a. Determine the output of the program below if P = −3.1416.

```
200  LET Q = 6.9
210  LET R = Q*P/180
220  LET S = ABS(R)
230  IF S< ABS(P/2) THEN PRINT "S = " ; S
240  IF S>= ABS(P/2) THEN PRINT "OOPS"
250  STOP
```

b. Change the value of Q in statement 200 to −150 and run that program.

1.3 INTEGRAL EXPONENTS

When the real number x is multiplied by itself a number of times as in $x \cdot x \cdot x$, each x in the indicated product is called a **factor** and it is convenient to use the notation

$$x^2 = \underbrace{x \cdot x}_{\text{2 factors}}, \qquad x^3 = \underbrace{x \cdot x \cdot x}_{\text{3 factors}}, \qquad x^4 = \underbrace{x \cdot x \cdot x \cdot x}_{\text{4 factors}}, \text{ etc.}$$

In general, we have the following

DEFINITION

> If n is a positive integer,
>
> $$x^n = \underbrace{x \cdot x \cdot x \cdot \ldots \cdot x}_{n \text{ factors}}$$

We read x^n as "x to the nth power" or "x to the n"; n is called the **exponent** and x is called the **base**. Usually x^2 is read "x squared" and

x^3 is read "x cubed." When another factor multiplies a power of x, that factor is called a **coefficient**. Thus, with $4x^5$ (or equivalently $4 \cdot x \cdot x \cdot x \cdot x \cdot x$), 4 is the *coefficient* of x^5. Similarly, with yx^3 (or equivalently $y \cdot x \cdot x \cdot x$), y is the *coefficient* of x^3.

In this section expressions involving exponents will be simplified. Presently, we will work with **monomials** (one-term expressions).

EXAMPLE 1
a. Rewrite the product of monomials: $x^4 \cdot x^3$.
b. Rewrite the power of the monomial: $(x^4)^3$.
c. Rewrite the power of the product of monomials: $(xy)^4$.

Solution
a. $x^4 \cdot x^3 = (x \cdot x \cdot x \cdot x)(x \cdot x \cdot x)$
$x^4 \cdot x^3 = x^7$

b. $(x^4)^3 = x^4 \cdot x^4 \cdot x^4$
$(x^4)^3 = (x \cdot x \cdot x \cdot x)(x \cdot x \cdot x \cdot x)(x \cdot x \cdot x \cdot x)$
$(x^4)^3 = x^{12}$

c. $(xy)^4 = (xy)(xy)(xy)(xy)$
$(xy)^4 = (x \cdot x \cdot x \cdot x)(y \cdot y \cdot y \cdot y)$
$(xy)^4 = x^4y^4$

The solutions in Example 1 illustrate, respectively, the three laws of exponents that are now stated.

THEOREM 2

Laws of Exponents

If x and y are real numbers and a and b are positive integers, then

#1 $x^a \cdot x^b = x^{a+b}$
#2 $(x^a)^b = x^{ab}$
#3 $(xy)^a = x^a y^a$

The proofs for Theorem 2 can be structured by using a technique known as Mathematical Induction (see Chapter 11). Moreover, these laws can be extended to $x^a \cdot x^b \cdot x^c = x^{a+b+c}$, $[(x^a)^b]^c = x^{abc}$, $(xyz)^a = x^a y^a z^a$, etc.

EXAMPLE 2 Rewrite each of the following by using Theorem 2.

a. $x^4 \cdot x^2$ b. $y^2 \cdot y^3 \cdot 4y^3$ c. $(4y)^3(2y^2)^3$

d. $(5x^2y)(3x^3y^5)$ e. $(ab^2c^3)^4$

Solution

a. $x^4 \cdot x^2$

x^6 — Exponent Law #1

b. $y^2 \cdot y^3 \cdot 4y^3$

$4 \cdot y^2 \cdot y^3 \cdot y^3$ — Commutative and Associative Properties for Multiplication

$4y^8$ — Exponent Law #1

c. $(4y)^3(2y^2)^3$

$(4y \cdot 2y^2)^3$ — Exponent Law #3

$(8y^3)^3$ — Commutative and Associative Properties for Multiplication and also Exponent Rule #1

$8^3(y^3)^3$ — Exponent Law #3

$512y^9$ — Exponent Law #2

d. $(5x^2y)(3x^3y^5)$

$5 \cdot 3 \cdot x^2 \cdot x^3 \cdot y \cdot y^5$ — Commutative and Associative Properties for Multiplication

$15x^5y^6$ — Exponent Law #1

e. $(ab^2c^3)^4$

$a^4b^8c^{12}$ — Exponent Law #2

At this stage we extend our knowledge of exponents by defining a *zero* exponent and a *negative* integer exponent.

DEFINITION

Zero Exponent *

$$x^0 = 1 \qquad (x \neq 0)$$

*This definition avoids discussing the expression 0^0. The reason is because 0^0 is indeterminate. Indeterminate expressions are investigated in calculus.

DEFINITION

Negative Integer Exponent

If a is a positive integer, then

$$x^{-a} = \frac{1}{x^a} \qquad (x \neq 0)$$

We mention here that the three laws of exponents apply to zero and negative exponents as well. For example,

$$(x^3)^{-4} = \frac{1}{(x^3)^4} = \frac{1}{x^{12}} = x^{-12}$$

It is possible to prove two additional laws of exponents that are now stated (the proofs are outlined in Exercise 57).

THEOREM 3

Laws of Exponents

If x and y are real numbers and a and b are integers, then

#4 $\dfrac{x^a}{x^b} = x^{a-b} \qquad (x \neq 0)$

#5 $\left(\dfrac{x}{y}\right)^a = \dfrac{x^a}{y^a} \qquad (y \neq 0)$

EXAMPLE 3 Rewrite each of the following. Express each answer using only positive exponents.

a. $\dfrac{1}{3^{-2}}$ b. $\dfrac{x^{-6}}{x^{-2}}$ c. $\dfrac{a^0}{a^{-4}}$

d. $\left(\dfrac{1}{x^3}\right)^4$ e. $\left(\dfrac{xy^2}{3z^4}\right)^3$ f. $\dfrac{2p^2q^{-2}r^3}{8p^{-3}q^4r^{-1}}$

Solution a. $\dfrac{1}{3^{-2}} = \dfrac{1}{\frac{1}{3^2}} = 3^2 = 9$

b. By using Exponent Law #4, $\dfrac{x^{-6}}{x^{-2}} = x^{-6-(-2)} = x^{-4} = \dfrac{1}{x^4}$

c. $\dfrac{a^0}{a^{-4}}$ — Exponent Law #4

$a^{0-(-4)}$

a^4

d. $\left(\dfrac{1}{x^3}\right)^4$ — Exponent Law #5

$\dfrac{1^4}{(x^3)^4}$ — Exponent Law #2

$\dfrac{1}{x^{12}}$

e. $\left(\dfrac{xy^2}{3z^4}\right)^3$ — Exponent Law #5

$\dfrac{(xy^2)^3}{(3z^4)^3}$ — Exponent Law #3

$\dfrac{x^3(y^2)^3}{3^3(z^4)^3}$ — Exponent Law #2

$\dfrac{x^3y^6}{27z^{12}}$

f. $\dfrac{2p^2q^{-2}r^3}{8p^{-3}q^4r^{-1}}$ — Law #5 about Quotients (Section 1.1)

$\dfrac{2}{8} \cdot \dfrac{p^2}{p^{-3}} \cdot \dfrac{q^{-2}}{q^4} \cdot \dfrac{r^3}{r^{-1}}$ — Exponent Law #4

$\dfrac{1}{4}p^5 \cdot q^{-6} \cdot r^4$ — Definition of a negative integer exponent

$\dfrac{p^5r^4}{4q^6}$

One practical use of exponents is with rewriting numbers in **scientific notation**. For example, the mass of an atom of uranium is 0.000000000000000000000395 grams.* Such a number is awkward to read, write, and use in calculations. As an alternative, scientific notation is used. To write a number in scientific notation, it is written as a product of two numbers: the first is a number between 1 and 10 and the second is a power of 10.

*Metric and English units as well as conversions between them are discussed in Appendix 2.

EXAMPLE 4 Write each of the following numbers in scientific notation.

a. 33,400 b. 0.021 c. 0.0000000798

Solution

a. $33{,}400 = 3.34 \cdot 10{,}000$

$33{,}400 = 3.34 \cdot 10^4$

b. $0.021 = 2.1 \cdot \dfrac{1}{100}$

$0.021 = 2.1 \cdot 10^{-2}$

c. $0.0000000798 = 7.98 \cdot \dfrac{1}{100{,}000{,}000}$

$0.0000000798 = 7.98 \cdot 10^{-8}$

In effect, the positive exponent "4" on the "10" in part a signifies that the decimal point would need to be moved 4 places to the *right* in order to obtain the number in its original form. The negative exponent "−2" on the "10" in part b signifies that the decimal point would need to be moved 2 places to the *left* in order to obtain the number in its original form. In the next example we show how scientific notation can be used as an aid when calculating with large or small numbers.

EXAMPLE 5 Use scientific notation to calculate

a. $992{,}000{,}000 \cdot 40{,}000$ b. $\dfrac{(0.0000002)^3}{0.000054}$

Solution

a. $992{,}000{,}000 \cdot 40{,}000$

Write each number in scientific notation

$9.92 \cdot 10^8 \cdot 4 \cdot 10^4$

Exponent Law #1

$39.68 \cdot 10^{12}$

Write 39.68 in scientific notation

$3.968 \cdot 10^{13}$

b. $\dfrac{(0.0000002)^3}{0.000054}$

Write each number in scientific notation

$\dfrac{(2 \cdot 10^{-7})^3}{5.4 \cdot 10^{-5}}$

Exponent Laws #2 and 3

$\dfrac{8 \cdot 10^{-21}}{5.4 \cdot 10^{-5}}$

Exponent Law #4

$1.4815 \cdot 10^{-16}$

Use can be made of an electronic calculator in performing computations such as those in Example 5. Some calculators allow the entry of a number in scientific notation and some automatically convert "big" and "small" numbers into scientific notation. In any case, if you have a calculator that employs scientific notation, the answers in Example 5 would appear as:

	Display	
a.	3.968	13
	the number between 1 and 10	the exponent on 10
b.	1.4815	−16
	the number between 1 and 10	the exponent on 10

EXERCISES FOR SECTION 1.3

In Exercises 1 through 42, rewrite each expression using the laws of exponents. Express each answer using only positive exponents. (See Examples 1 and 3.)

1. $(x^6)^2$
2. $(x^2)^{-3}$
3. $(x^2y)^5$
4. $(x^2y)^{-5}$
5. $x^2\cdot x^5$
6. $x^2\cdot x^{-5}$
7. $x^{-2}\cdot x^{-5}$
8. $\dfrac{x^2}{x^5}$
9. $\dfrac{x^5}{x^2}$
10. $\dfrac{x^2}{x^{-5}}$
11. $\dfrac{x^{-2}}{x^{-5}}$
12. $\dfrac{1}{3^{-1}x^{-2}y^{-2}}$
13. $\dfrac{(368)^0}{y^{-2}}$
14. $2z^2\cdot z^3\cdot z^{-5}$
15. $2z^2\cdot 3z^{-4}$
16. $(2z)^2\cdot(3z)^{-4}$
17. $(w^2)(-2w^3)\left(\dfrac{1}{2}w\right)$
18. $\left(\dfrac{p^{-1}q^{-2}r^{-1}}{2p^2q^3r^4}\right)^0$
19. $[(x^{-2})^3]^4$
20. $[(x^{-2})^3]^{-4}$
21. $\dfrac{(x+y)^5}{(x+y)^6}$
22. $\dfrac{1}{x^{-2}y^{-3}z^{-4}}$
23. $(xy^{-2})^{-2}$
24. $(-x)(-y)^{-2}$
25. $\left(\dfrac{x}{y}\right)^{-1}\cdot\dfrac{x}{y}$
26. $\left(\dfrac{x}{y}\right)^{-2}\cdot\dfrac{y}{x}$
27. $\dfrac{x^{-2}y^{-2}}{(xy)^{-2}}$
28. $\dfrac{2a^2b^3}{ab}\cdot\dfrac{ab^4}{(ab)^2}$
29. $\dfrac{(-2a^3)^2}{a}$
30. $\dfrac{(a+b)^0}{a+c}$
31. $\dfrac{pq}{p^2q^3}$
32. $\left(\dfrac{pq}{p^2q^3}\right)^{-1}$
33. $\dfrac{3x^4y^3z^{-2}}{81x^2y^{-3}z^{-2}}$
34. $[(x^2)^4(x^4)^2]^0\cdot\left(\dfrac{2}{x}\right)^{-1}$
35. $[(x^{-1})^{-2}]^{-3}$
36. $\dfrac{(UV)^3}{(UV)^4}$
37. $\dfrac{3\cdot 4^{-1}}{3^{-1}\cdot 4}$

38. $\left(\frac{x}{y}\right)^2 \div \frac{x}{y^2}$

39. $\frac{2}{x^{-3}y^{-1}}$

40. $\frac{(2x)^3 \cdot (3y)^2}{x^3 \cdot y^2}$

41. $\frac{2x^3 \cdot 3y^2}{x^3 \cdot y^2}$

42. $\frac{p^2q^3}{q^{-2}r} \div \frac{r^2q}{2p^3q^5}$

In Exercises 43 through 48, write each number in scientific notation. (See Example 4.)

43. 23,900,000,000
44. 0.04
45. 0.4
46. 4.0
47. 0.000000093
48. 463,000

In Exercises 49 through 56, perform each calculation using scientific notation. (See Example 5.)

49. $5{,}600 \cdot 4{,}920{,}000{,}000$

50. $5{,}600 \cdot 0.000000492$

51. $\frac{(400{,}000)}{(0.0002)}$

52. $\frac{0.000002}{56{,}000{,}000}$

53. $\frac{(400{,}000)^5}{(0.0002)^3}$

54. $\frac{(0.0000002)^4}{56{,}000{,}000}$

55. $(0.6)^4 \cdot (22.4 \cdot 10^{23})$

56. $(93{,}000{,}000)(0.006)^3$

57. a. Complete the reasons for each step in the proof of #2 of Theorem 2.

Statement	Reason
1. $x^a = \underbrace{x \cdot x \cdot x \cdot \ldots \cdot x}_{a \text{ factors of } x}$	1.
2. $(x^a)^b = \underbrace{(x^a) \cdot (x^a) \cdot \ldots \cdot (x^a)}_{b \text{ factors of } x^a}$	2.
3. $(x^a)^b = x^{ab}$	3.

b. Complete the reasons for each step in the proof of #4 of Theorem 3.

Statement	Reason
1. $\frac{x^a}{x^b} = \frac{(x^{a-b}) \cdot x^b}{x^b}$	1.
2. $\frac{x^a}{x^b} = \frac{x^{a-b}}{1} \cdot \frac{x^b}{x^b}$	2.
3. $\frac{x^a}{x^b} = x^{a-b} \cdot 1$	3.
4. $\frac{x^a}{x^b} = x^{a-b}$	4.

c. Complete the reasons for each step in the proof of #5 of Theorem 3.

Statement	Reason
1. $\left(\dfrac{x}{y}\right)^a = \underbrace{\dfrac{x}{y} \cdot \dfrac{x}{y} \cdot \ldots \cdot \dfrac{x}{y}}_{a \text{ factors of } \frac{x}{y}}$	1.
2. $\left(\dfrac{x}{y}\right)^a = \dfrac{\overbrace{x \cdot x \cdot \ldots \cdot x}^{a \text{ factors of } x}}{\underbrace{y \cdot y \cdot \ldots \cdot y}_{a \text{ factors of } y}}$	2.
3. $\left(\dfrac{x}{y}\right)^a = \dfrac{x^a}{y^a}$	3.

Computer Exercise

58. In the BASIC programming language, exponentiation is symbolized by ↑. (The symbol ↑ is probably the most common symbol; some computers use ∧ and others use **.) Thus, X ↑ 3 means x^3.

a. Run the following program.

```
1Ø  PRINT "X", "X ↑ 2"
2Ø  FOR X = 1 TO 10
3Ø  LET Y = X ↑ 2
4Ø  PRINT X, Y
5Ø  NEXT X
6Ø  STOP
```

b. Refine the program to print a table of squares and cubes of the first twenty positive integers.

1.4 POLYNOMIALS

Several laws of algebra have been examined thus far. In this section many of these laws will be used in order to add, subtract, multiply and divide **polynomials**. Loosely speaking, a polynomial is an algebraic expression consisting of a finite number of terms (added or subtracted). Furthermore, each term consists of a finite number of factors of real numbers and/or variables, like 2, x, y, etc. Expressions such as $4x^5 - 3x^3 + 2x^2 + x + 3$, $\frac{1}{2}y^2 - x^2y^2 + 1$, $4 - 7x$, and $6xyz$ are polynomials, but $x^{-3} + 2x^2 - x + xy$ and $2x^3 - \frac{1}{x^2} + 1$ are *not* because polynomials never have variables in a denominator. In order to be more concise we state the following definition.

DEFINITION A **polynomial** in the single variable x is an expression of the form,

$$a_n x^n + a_{n-1} x^{n-1} + a_{n-2} x^{n-2} + \ldots + a_3 x^3 + a_2 x^2 + a_1 x + a_0,$$

where each coefficient a_i is a real number and n is a **nonnegative integer.**

The real number a_n is called the **leading coefficient** and the real number a_0 is called the **constant term.** If $a_n \neq 0$, the polynomial is said to be of **degree** n. A polynomial in more than one variable can also be defined. For instance, $x^2y^3z - xyz$ is a polynomial in the three variables x,y,z.

A polynomial having three terms is called a **trinomial**, a polynomial having two terms is called a **binomial**, and a polynomial having one term is called a **monomial**. Thus, $x^2 + 5x + 4$, $3x - 2y + z$, and $x^2y^2 - 2xy + 1$ are trinomials; $2x + 3$, $x^2 - 4$, and $a^2 + b^2$ are binomials; y^2, $-2xyz$, and 7 are monomials. Several examples are now presented.

EXAMPLE 1 In parts a and b, add the given polynomials.

a. $2x^4 - 3x^3 + x^2 + 7x - 4$, $4x^4 + x^3 - 3x^2 - 6x - 1$

b. $4x^2 + 4xy - 9y^2 + 2x - 1$, $6xy + 9y^2 + 4x + 6$

Solution

a. $(2x^4 - 3x^3 + x^2 + 7x - 4) + (4x^4 + x^3 - 3x^2 - 6x - 1)$ — Expand and collect like terms

$6x^4 - 2x^3 - 2x^2 + x - 5$

b. $(4x^2 + 4xy - 9y^2 + 2x - 1) + (6xy + 9y^2 + 4x + 6)$ — Expand and collect like terms

$4x^2 + 10xy + 6x + 5$

EXAMPLE 2 In parts a and b, perform the indicated subtraction of the polynomials from Example 1.

a. $(2x^4 - 3x^3 + x^2 + 7x - 4) - (4x^4 + x^3 - 3x^2 - 6x - 1)$

b. $(4x^2 + 4xy - 9y^2 + 2x - 1) - (6xy + 9y^2 + 4x + 6)$

Solution a. $(2x^4 - 3x^3 + x^2 + 7x - 4) - (4x^4 + x^3 - 3x^2 - 6x - 1)$ — Expand

$2x^4 - 3x^3 + x^2 + 7x - 4 - 4x^4 - x^3 + 3x^2 + 6x + 1$ — Collect like terms

$-2x^4 - 4x^3 + 4x^2 + 13x - 3$

b. $(4x^2 + 4xy - 9y^2 + 2x - 1) - (6xy + 9y^2 + 4x + 6)$ — Expand

$4x^2 + 4xy - 9y^2 + 2x - 1 - 6xy - 9y^2 - 4x - 6$ — Collect like terms

$4x^2 - 2xy - 18y^2 - 2x - 7$

EXAMPLE 3 In parts a and b, perform the indicated multiplication of polynomials.

a. $x^2(2x - 7)$ b. $(x^3 + 2x)(x - 2)$

Solution a. $x^2(2x - 7)$ — Expand

$x^2 \cdot 2x - x^2 \cdot 7$

$2x^3 - 7x^2$

b. $(x^3 + 2x)(x - 2)$ — Expand

$(x^3 + 2x)(x) - (x^3 + 2x)(2)$ — Expand

$x^4 + 2x^2 - 2x^3 - 4x$ — Rearrange terms*

$x^4 - 2x^3 + 2x^2 - 4x$

Part b of Example 3 is a product of two binomials. Such products occur commonly in algebra and calculus so we point out that the answer is a sum of the **f**irst, **o**uter, **i**nner, and **l**ast terms. This pattern (called **foil**) is useful for mentally multiplying two binomials. To multiply $(a + b)(c + d)$:

f	product of **f**irst terms:	ac	
o	product of **o**uter terms:	ad	$ac + ad + bc + bd$
i	product of **i**nner terms:	bc	
l	product of **l**ast terms:	bd	

In the next example, products of trinomials are encountered.

*It is somewhat traditional, in algebra, to write polynomials in one variable in descending powers of the variable. We retain that tradition here.

EXAMPLE 4 In parts a and b, perform the indicated multiplication of polynomials.

a. $(x^2 - 2x + 3)(2x^2 + x - 4)$ b. $(x^2 - 2x + 3)^2$

Solution a.

$$(x^2 - 2x + 3)(2x^2 + x - 4)$$

Expand

$$(x^2 - 2x + 3)2x^2 + (x^2 - 2x + 3)x - (x^2 - 2x + 3)4$$

Commutative Law for Multiplication

$$2x^2(x^2 - 2x + 3) + x(x^2 - 2x + 3) - 4(x^2 - 2x + 3)$$

Expand

$$2x^4 - 4x^3 + 6x^2 + x^3 - 2x^2 + 3x - 4x^2 + 8x - 12$$

Collect like terms

$$2x^4 - 3x^3 + 11x - 12$$

b.

$$(x^2 - 2x + 3)^2$$

Definition of a positive integral exponent

$$(x^2 - 2x + 3)(x^2 - 2x + 3)$$

Expand

$$x^2(x^2 - 2x + 3) - 2x(x^2 - 2x + 3) + 3(x^2 - 2x + 3)$$

Expand

$$x^4 - 2x^3 + 3x^2 - 2x^3 + 4x^2 - 6x + 3x^2 - 6x + 9$$

Collect like terms

$$x^4 - 4x^3 + 10x^2 - 12x + 9$$

Below we list some special products that occur frequently in algebra and calculus. It is strongly recommended that the reader verify each of these by performing the indicated multiplications.

Special Products

1. $(a + b)(c + d) = ac + ad + bc + bd$
2. $(a + b)(a - b) = a^2 - b^2$
3. $(a + b)^2 = a^2 + 2ab + b^2$
4. $(a - b)^2 = a^2 - 2ab + b^2$
5. $(a + b)^3 = a^3 + 3a^2b + 3ab^2 + b^3$
6. $(a - b)^3 = a^3 - 3a^2b + 3ab^2 - b^3$

Commiting these to memory can be a great time saver as the next example shows.

EXAMPLE 5 a. Expand $(3x - y)^3$. b. Expand $\left(x^2 + \dfrac{2}{x^2}\right)^3$.

Solution a. Use special product #6, with $a = 3x$, $b = y$.

$$(3x - y)^3 = (3x)^3 - 3(3x)^2(y) + 3(3x)(y)^2 - (y)^3$$
$$(3x - y)^3 = 27x^3 - 27x^2y + 9xy^2 - y^3$$

b. Even though $x^2 + \dfrac{2}{x^2}$ is not a polynomial, we can still use special product #5 with $a = x^2$ and $b = \dfrac{2}{x^2}$.

$$\left(x^2 + \frac{2}{x^2}\right)^3 = (x^2)^3 + 3(x^2)^2\left(\frac{2}{x^2}\right) + 3(x^2)\left(\frac{2}{x^2}\right)^2 + \left(\frac{2}{x^2}\right)^3$$
$$\left(x^2 + \frac{2}{x^2}\right)^3 = x^6 + 6x^2 + \frac{12}{x^2} + \frac{8}{x^6}$$

The examples thus far illustrate that the sum, difference, or product of polynomials is another polynomial—such is always the case. Howevery, with division, things change.

Division of a polynomial by a monomial is an application of previous rules, as the next example shows.

EXAMPLE 6 a. Divide $2x^6 - 4x^4 + 3x^2$ by $2x^2$.

b. Divide $4x^2y^3 - 3x^4y + 2x^5 - 4y$ by x^2y.

Solution a.

$$\frac{2x^6 - 4x^4 + 3x^2}{2x^2} = \frac{2x^6}{2x^2} - \frac{4x^4}{2x^2} + \frac{3x^2}{2x^2}$$
$$\frac{2x^6 - 4x^4 + 3x^2}{2x^2} = x^4 - 2x^2 + \frac{3}{2}$$

b.

$$\frac{4x^2y^3 - 3x^4y + 2x^5 - 4y}{x^2y} = \frac{4x^2y^3}{x^2y} - \frac{3x^4y}{x^2y} + \frac{2x^5}{x^2y} - \frac{4y}{x^2y}$$
$$\frac{4x^2y^3 - 3x^4y + 2x^5 - 4y}{x^2y} = 4y^2 - 3x^2 + \frac{2x^3}{y} - \frac{4}{x^2}$$

Notice in part a of Example 6 that the answer *is* a polynomial whereas in part b the answer *is not* a polynomial.

Before proceeding any further with polynomial division, we introduce **polynomial notation**. Frequently, a polynomial in one variable, say x, is denoted by the symbol $P(x)$, $Q(x)$, etc, where the x in parentheses denotes the variable. We read $P(x)$ as "P of x" or "P at

x."* For example, we can write $P(x) = x^3 - 5x + 4$, $Q(x) = 6x + 7$, $R(y) = y^3 - y + 3$, etc.

EXAMPLE 7 Given $P(x) = x^3 - 5x + 4$, determine

a. $P(2)$ b. $P(0)$ c. $P(-1)$

Solution a. The symbol $P(2)$ denotes the value of $P(x)$ when $x = 2$.

$$\begin{aligned} P(x) &= x^3 - 5x + 4 \\ P(2) &= 2^3 - 5(2) + 4 \\ P(2) &= 8 - 10 + 4 \\ P(2) &= 2 \end{aligned}$$

b. $$\begin{aligned} P(x) &= x^3 - 5x + 4 \\ P(0) &= 0^3 - 5(0) + 4 \\ P(0) &= 4 \end{aligned}$$

c. $$\begin{aligned} P(x) &= x^3 - 5x + 4 \\ P(-1) &= (-1)^3 - 5(-1) + 4 \\ P(-1) &= 8 \end{aligned}$$

Next we review some terminology associated with division of real numbers. Consider the division $67 \div 4$.

$$\begin{array}{rl} & 16 \leftarrow \text{Quotient } (Q) \\ \text{Divisor} \rightarrow & 4\,\overline{)\,67} \leftarrow \text{Dividend } (P) \\ (D) & \underline{4} \\ & 27 \\ & \underline{24} \\ & 3 \leftarrow \text{Remainder } (R) \end{array}$$

Observe: $$\begin{array}{ccccc} \dfrac{67}{4} & = & 16 & + & \dfrac{3}{4} \\ \downarrow & & \downarrow & & \downarrow \\ \dfrac{P}{D} & = & Q & + & \dfrac{R}{D} \end{array}$$

and

$$\begin{array}{ccccccc} 67 & = & 4 & \cdot & 16 & + & 3 \\ \downarrow & & \downarrow & & \downarrow & & \downarrow \\ P & = & D & \cdot & Q & + & R \end{array}$$

The following laws have been illustrated.

Division Law Form I

$$\boxed{\frac{P}{D} = Q + \frac{R}{D}}$$

*$P(x)$ *does not* mean P multiplied by x.

Division Law Form II

$$\boxed{P = D \cdot Q + R}$$

When dividing two polynomials, we will denote the dividend by $P(x)$, the divisor by $D(x)$, the quotient by $Q(x)$, and the remainder by $R(x)$. The division laws will be written

$$\frac{P(x)}{D(x)} = Q(x) + \frac{R(x)}{D(x)} \quad \leftarrow \text{Form I}$$

$$P(x) = D(x) \cdot Q(x) + R(x) \quad \leftarrow \text{Form II}$$

In the rest of this section we discuss polynomial division where the divisor, $D(x)$, has the form $x - c$, where c is a real number.

EXAMPLE 8 Divide $2x^4 - 3x^2 - 15x + 1$ by $x - 2$.

Solution Long division (as with real numbers) is shown below.

$$\begin{array}{r l}
 & 2x^3 + 4x^2 + 5x \;-\; 5 \\
x - 2 \,\big) & 2x^4 + 0x^3 - 3x^2 - 15x + 1 \\
 & \underline{2x^4 - 4x^3} \\
 & \qquad 4x^3 - 3x^2 \\
 & \qquad \underline{4x^3 - 8x^2} \\
 & \qquad\qquad 5x^2 - 15x \\
 & \qquad\qquad \underline{5x^2 - 10x} \\
 & \qquad\qquad\quad - 5x + 1 \\
 & \qquad\qquad\quad \underline{- 5x + 10} \\
 & \qquad\qquad\qquad\quad - 9
\end{array}$$

← In the dividend we "fill in" $0x^3$ for the missing x^3 term.

← Always continue long division until the remainder is 0 *or* until the remainder has a smaller degree than the divisor.

Using Form I we write the answer as:

$$\underbrace{\frac{2x^4 - 3x^2 - 15x + 1}{x - 2}}_{\frac{P(x)}{D(x)}} = \underbrace{2x^3 + 4x^2 + 5x - 5}_{Q(x)} + \underbrace{\frac{-9}{x - 2}}_{\frac{R(x)}{D(x)}}$$

In Exercise 87 the reader is asked to "check" this division by using Form II.

A shorter way of performing the division of a polynomial by a divisor of the form $x - c$ is with the method of **synthetic division** explained below. Let us rewrite the division in Example 8 with only the coefficients. It looks like:

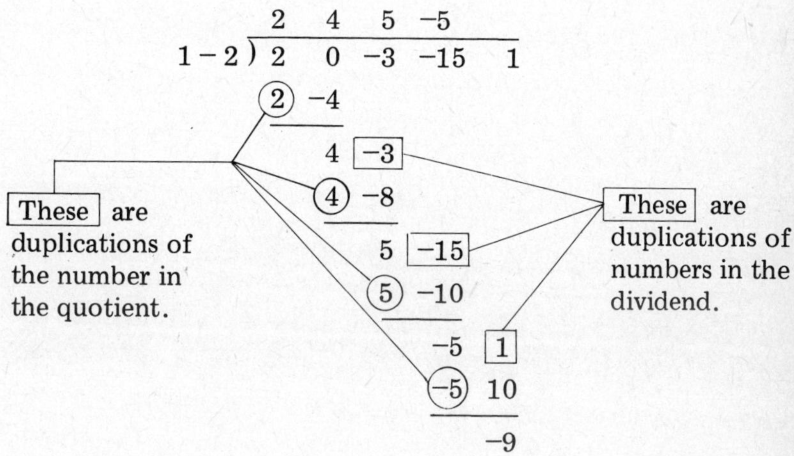

If we eliminate all duplications and "compress" the other numbers, the division looks like:

$$\begin{array}{r|rrrrr} & 2 & 4 & 5 & -5 & \\ \hline 1-2\,) & 2 & 0 & -3 & -15 & 1 \\ & & -4 & -8 & -10 & 10 \\ \hline & & & & & -9 \end{array}$$

Next, we omit the 1 in the divisor and relocate the coefficients of the quotient on the bottom row.

$$\begin{array}{r|rrrrr} -2 & 2 & 0 & -3 & -15 & 1 \\ & & -4 & -8 & -10 & 10 \\ \hline & 2 & 4 & 5 & -5 & -9 \end{array}$$

Each number in this row comes from multiplying by −2 and afterward a subtraction is performed.

Multiplying by −2 and subtracting is equivalent to multiplying by 2 and adding. This is depicted below.

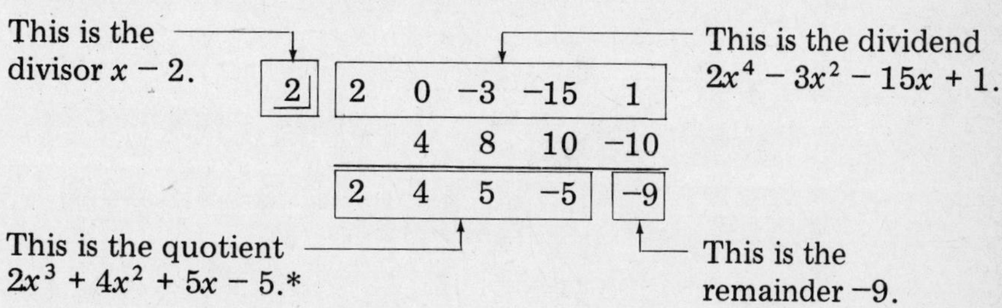

*The degree of $Q(x)$ is always one less than the degree of $P(x)$ when dividing by $x - c$.

Each number in the second row results from the product of the "divisor number" with the previous number in the third row. Each number in the third row is the sum of the numbers above it.

EXAMPLE 9 Perform each division by using synthetic division.

a. Divide $x^3 - 3x^2 + 2x + 1$ by $x - 3$.

b. Divide $4x^5 + x^2 + 6$ by $x + 1$.

Solution a.

3	1	−3	2	1
		3	0	6
	1	0	2	7

$Q(x) = x^2 + 2$

$R(x) = 7$

Thus, $$\frac{x^3 - 3x^2 + 2x + 1}{x - 3} = x^2 + 2 + \frac{7}{x - 3}$$

b.

−1	4	0	0	1	0	6
		−4	4	−4	3	−3
	4	−4	4	−3	3	3

$Q(x) = 4x^4 - 4x^3 + 4x^2 - 3x + 3$

$R(x) = 3$

$$\frac{4x^5 + x^2 + 6}{x + 1} = 4x^4 - 4x^3 + 4x^2 - 3x + 3 + \frac{3}{x + 1}$$

In Exercise 88 the reader is asked to check each division by using Form II.

EXERCISES FOR SECTION 1.4

In Exercises 1 through 10, identify each expression that is a polynomial.

1. $3x^6 - x + 1$
2. $2x^4 - x^9 + x^8 - 4$
3. $x^2 + \dfrac{x}{y} - 2$
4. $x^2 + x^{-2} + 2$
5. $x^3 - 3x^2 + 2x - 9$
6. $x^2 + 5x + 4$
7. $\dfrac{1}{y} + \dfrac{1}{x} + 1$
8. $x^2y^3 - y^3x^2 + \dfrac{1}{xyz}$
9. $x^5 - 2x^3 + x^2 + 1$
10. $xy + x^2 + y^2 + x + y + 1$

In Exercises 11 through 20, perform the indicated operation. (See Examples 1 and 2.)

11. $(x^2 - 3x + 4) + (x^3 + 3x - 4)$
12. $(x^2 - 3x + 4) - (x^3 + 3x - 4)$
13. $(5y^3 - 2y^2 - 1) - (y^2 - 2y - 3)$
14. $(2r^2 + r) + (5r^3 + 7)$
15. $(6x^4 - x + 1) + (x^5 - x^3 + 6)$

16. $2(x^3 - 3x^2 + 7x - 1) + 4(x^3 - 2x + 1)$
17. $3y(y^2 - 2y - 5) + y(2y^2 - 5y + 3)$
18. $y^2(2y^2 - 5y + 3) - 3(y^2 - 2y - 5)$
19. $(4x^4 + 2x^3 - 3x^2 + x + 6) - (4x^4 + 3x^2)$
20. $(a^6 + a^4 + 1) - (a^6 + 2a^4 + a^3 - a^2 + 1)$

In Exercises 21 through 54, perform the indicated operation. (See Examples 3, 4, and 5.)

21. $x^3(x^2 - 3x)$
22. $2x^2(-x^2 - 2x)$
23. $-2x^4(3x + 1)$
24. $(a - b)(2a + b)$
25. $(a - b)(c - d)$
26. $(3x^2 - 1)(x + 2)$
27. $(2x - 1)(4x + 5)$
28. $(6x - 1)(2x + 1)$
29. $(x - 1)(x + 1)$
30. $(x - 2)(x + 2)$
31. $(x - 2)(x + y + 2)$
32. $(x + y - 2)(x + y + 2)$
33. $(a + b + c)(d + e + f)$
34. $(2y^2 + y - 1)(y^2 + 2y + 1)$
35. $(3x^2 - 2x + 1)(x - 4)$
36. $(x^2 + y^2 + yx + 1)(x - 4)$
37. $(x^2 - 2)(x^2 + 2)$
38. $(r - 2r^2)(2r + 1)^2$
39. $(x - 3)(x + 4)(x - 1)$
40. $(x - 3)(x + 4)(x - 1)^2$
41. $(x + 1)^3$
42. $(x - 1)^3$
43. $(2y + 1)^3$
44. $(2y - 1)^3$
45. $(2x + 3)^3$
46. $(2x - 3)^3$
47. $(x + 1)^4$
48. $(x - 1)^4$
49. $(x + 1)^5$
50. $(x - 1)^5$
51. $\left(x - \dfrac{1}{x}\right)^3$
52. $(x + x^{-2})^3$
53. $\left(2x + \dfrac{1}{x}\right)^3$
54. $\left(\dfrac{1}{y} + \dfrac{1}{y^2}\right)^2$

In Exercises 55 through 64, perform the indicated operation. (See Example 6.)

55. $\dfrac{2x^7 - x^5 + 4x}{x}$
56. $\dfrac{2x^7 - x^5 + 4x}{x^2}$
57. $\dfrac{3x^5 - 2x^2 + 2}{3x}$
58. $\dfrac{x^2y^3 - xy}{xy}$
59. $\dfrac{w^3 - 3w^2 + 1}{w^3}$
60. $\dfrac{x^2y^2z^3 - xyz}{xyz}$
61. $\dfrac{3xy + 1}{xy}$
62. $\dfrac{x^2 - 5x + 4}{x^2}$
63. $\dfrac{x^3 - 5x^2 + 4}{x^2}$
64. $\dfrac{x^3 - 5x^2 + 4}{x^2y}$

In Exercises 65 through 72, evaluate as directed. (See Example 7.)

65. If $P(x) = x^3 - 3x^2 + 1$, find $P(0)$.
66. If $P(x) = 2x^4 - 3x^2 + x - 1$, find $P(1)$.

67. If $Q(x) = x^2 - 7x + 12$, find $Q(4)$.
68. If $Q(x) = x^2 - 7x + 12$, find $Q(3)$.
69. If $R(x) = x^2 - 2$, find $R(-2)$.
70. If $P(x) = x^4 + x^3 + x^2 + x + 1$, find $P(-1)$.
71. If $W(y) = y^4 - 2y + 3$, find $W(0)$.
72. If $W(r) = r^4 - 2r + 3$, find $W(0)$.

In Exercises 73 through 76, perform the indicated long division and express the answer using Form I of the division law. (See Example 8.)

73. $x - 1 \overline{)\, 2x^3 - 3x^2 + 2x - 1}$

74. $x + 1 \overline{)\, 2x^3 - 3x^2 + 2x - 1}$

75. $x + 3 \overline{)\, x^4 + 0x^3 - x^2 + 20x + 1}$

76. $x - 1 \overline{)\, x^5 + 0x^4 + 0x^3 + 0x^2 + 0x - 1}$

In Exercises 77 through 86, perform each division by using synthetic division and write the answer using Form I of the division law. (See Example 9.)

77. $x^2 - 7x + 12 \div x - 5$
78. $x^2 - 7x + 12 \div x - 4$
79. $x^3 - 1 \div x - 1$
80. $2x^3 + 2x^2 - 3x + 4 \div x - 2$
81. $2x^3 + 2x^2 - 3x + 4 \div x + 2$
82. $x^4 - 1 \div x - 1$
83. $5x^3 - 14x^2 + 4x - 21 \div x - 3$
84. $3x^3 - 11x - 1 \div x + 2$
85. $2x^5 + x^4 - 14x^3 + x + 2 \div x + 3$
86. $x^6 + 2x^2 + 5 \div x + 1$
87. Use Form II of the division law to check the answer to Example 8. *Hint:* Form II gives $2x^4 - 3x^2 - 15x + 1 = (x - 2)(2x^3 + 4x^2 + 5x - 5) - 9$. Now simplify the right-hand side and see if it is identical to the left-hand side.
88. Use Form II of the division law to check each answer in Example 9.

In Exercises 89 through 94, answer true or false. If true, state why. If false, exhibit a counterexample.

89. $(a + b)^2 = a^2 + b^2$
90. $(a - b)^2 = a^2 - b^2$
91. $\dfrac{a + b}{a} = 1 + b$
92. $\dfrac{a + b}{a} = b$
93. $\dfrac{a + b}{a} = 1 + \dfrac{b}{a}$
94. $(a + b)^3 = a^3 + b^3$

1.5 FACTORING

When an algebraic expression such as $x^2 + 8x + 7$ is written as a product of other expressions as in $x^2 + 8x + 7 = (x + 7)(x + 1)$*, the process is called **factoring**. Also, both $x + 7$ and $x + 1$ are **factors** of the original expression $x^2 + 8x + 7$. In this section factoring of polynomials is discussed. The ability to factor a polynomial is important later on and often spells the difference between success and frustration with algebra.

*The reader can verify this by applying the foil method to $(x + 7)(x + 1)$.

The first factoring process discussed is the extraction of a **common factor**. For instance, consider $6x^2y^3 + 3xy^2 - 3x^3y$. Individually examine each of the terms $6x^2y^3$, $3xy^2$ and $3x^3y$ for *any and all* common factors. The result is $3xy$. Thus we write:

$$6x^2y^3 + 3xy^2 - 3x^3y$$
$$3xy(2xy^2 + y - x^2)$$
Distributive Property

The factor $3xy$ is called the **greatest common factor** or GCF. This factoring or any other can be checked by expanding the answer.

EXAMPLE 1 Factor each expression by extracting the GCF.

a. $x^4 + 3x^3 - 11x^2$

b. $4w^3z^3 - 2w^3z^2 + 12w^2z$

Solution a. $x^4 + 3x^3 - 11x^2$
$x^2(x^2 + 3x - 11)$ Distributive Property

b. $4w^3z^3 - 2w^3z^2 + 12w^2z$
$2w^2z(2wz^2 - wz + 6)$

In Example 1, the GCF in each case was a monomial. It can also be the case that the GCF is a binomial, as we now see.

EXAMPLE 2 Factor each expression by extracting the GCF.

a. $a^2(b - 1) + b^2(b - 1)$

b. $x^2 + xy + bx + by$

Solution a. Comparing the two terms $a^2(b - 1)$ and $b^2(b - 1)$ there is the GCF $b - 1$. Thus,

$$a^2(b - 1) + b^2(b - 1)$$
$$(b - 1)(a^2 + b^2)$$

b. Comparing the four terms, x^2, xy, bx, and by, there is no GCF (other than 1). However, by *grouping* the terms as shown, we can accomplish a factorization.

$$(x^2 + xy) + (bx + by)$$
$$x(x + y) + b(x + y)$$
Distributive Property (used twice)
$$(x + y)(x + b)$$

The method of factoring in part b of Example 2 is known as **factoring by grouping.**

At the end of the previous section there were several exercises involving the product of two binomials $(ax + b)(cx + d)$ where a, b, c, d are real numbers. Multiplying shows:

$$(ax + b)(cx + d) = acx^2 + (ad + bc)x + bd$$

This result suggests that second degree polynomials in a single variable may be factorable as the product of two binomials. Furthermore, the product of first terms ax and cx gives the second degree term acx^2 and the product of last terms b and d gives the constant term bd. These observations will now be used to factor such polynomials.

EXAMPLE 3 Factor each second degree polynomial.

a. $x^2 - 2x - 8$ b. $6x^2 - x - 2$

Solution a. Write $x^2 - 2x - 8 = (ax + b)(cx + d)$.

The product of a and c must be 1. The product of b and d must be -8. Trial and error shows

$$x^2 - 2x - 8 = (x - 4)(x + 2)$$

b. $6x^2 - x - 2 = (ax + b)(cx + d)$

The product of a and c must be 6. The product of b and d must be -2. Trial and error shows

$$(6x^2 - x - 2) = (3x - 2)(2x + 1)$$

Often the trial and error method of Example 3 is adaptable to expressions that are not second degree polynomials in a single variable, as the next example shows.

EXAMPLE 4 Factor each expression.

a. $6x^2 + 11xy - 2y^2$ b. $6x^2y^2 + 15xy^2 - 9y^2$

Solution a. Expressions of the form $Ax^2 + Bxy + Cy^2$ are often factorable as the product of two binomials $(ax + by)(cx + dy)$. Proceeding, we write $6x^2 + 11xy - 2y^2 = (ax + by)(cx + dy)$ and note that the product of a and c must be 6 and the product of b and d must be -2. Trial and error produces the correct result

$$6x^2 + 11xy - 2y^2 = (x + 2y)(6x - y)$$

b. $6x^2y^2 + 15xy^2 - 9y^2$
$3y^2(2x^2 + 5x - 3)$ — Factor the GCF
$3y^2(2x - 1)(x + 3)$

Often expressions can be factored by using a certain formula. Three such formulas are listed below. The reader is asked to verify each of these in Exercise 67.

Factoring Formulas

1. $a^2 - b^2 = (a - b)(a + b)$	**Difference of Two Squares**
2. $a^3 - b^3 = (a - b)(a^2 + ab + b^2)$	**Difference of Two Cubes**
3. $a^3 + b^3 = (a + b)(a^2 - ab + b^2)$	**Sum of Two Cubes**

We use these formulas in the next two examples.

EXAMPLE 5 Factor each expression.

a. $4x^2 - 25y^2$ b. $x^4 - 16y^4$ c. $8x^3y - 18xy^3$

Solution a. We use the difference of two squares formula with $a = 2x$ and $b = 5y$.

$$4x^2 - 25y^2 = (2x - 5y)(2x + 5y)$$

b. We use the difference of two squares formula with $a = x^2$ and $b = 4y^2$.

$$x^4 - 16y^4 = \underbrace{(x^2 - 4y^2)}(x^2 + 4y^2)$$

This factor can be factored again as the difference of two squares.

$$x^4 - 16y^4 = (x - 2y)(x + 2y)(x^2 + 4y^2)$$

c. $8x^3y - 18xy^3$
$2xy(4x^2 - 9y^2)$ — Factor the GCF
$2xy(2x - 3y)(2x + 3y)$ — Factor the difference of two squares

EXAMPLE 6 Factor each expression.

a. $x^3 - 8y^3$ b. $64r^3 + 27s^6$ c. $2(x - y)^2 - 2$

Solution a. Use the difference of two cubes formula with $a = x$ and $b = 2y$.

$$x^3 - 8y^3 = (x - 2y)[x^2 + x \cdot 2y + (2y)^2]$$
$$x^3 - 8y^3 = (x - 2y)(x^2 + 2xy + 4y^2)$$

b. Use the sum of two cubes formula with $a = 4r$ and $b = 3s^2$.

$$64r^3 + 27s^6 = (4r + 3s^2)[(4r)^2 - (4r)(3s^2) + (3s^2)^2]$$
$$64r^3 + 27s^6 = (4r + 3s^3)(16r^4 - 12rs^2 + 9s^4)$$

c. $2(x - y)^2 - 2$
$2[(x - y)^2 - 1]$ — Factor the GCF
$2[(x - y) - 1][(x - y) + 1]$ — Difference of two squares with $a = x - y$ and $b = 1$
$2(x - y - 1)(x - y + 1)$

EXERCISES FOR SECTION 1.5

In Exercises 1 through 66 factor each expression. (See Examples 1 through 6.)

1. $x^4 + 2x^2$
2. $2x^5 + wx^3$
3. $x^2y - xy + y^2$
4. $3xrz^2 - 3yrz^2 + 3rz^3$
5. $2xyz + 2xy$
6. $-2x^2 - 8xz$
7. $-2z^2x^2 - 2xyz + 2zxy^2$
8. $4a^2b^2 + 4ab^3 + 8ab^2$
9. $3a^4b^4 + 3a^2b^6$
10. $x^2yz^2 - 2xy^2z^2 + 4xyz^2$
11. $ax + ay + bx + by$
12. $ax + ay - bx - by$
13. $ax - ay + bx - by$
14. $2ax - 4ay - bx + 2by$
15. $2cx + 3cy + 2ax + 3ay$
16. $x^2 + 7x + 12$
17. $x^2 - 7x + 12$
18. $6x^2 + x - 1$
19. $6x^2 + 7x + 1$
20. $x^2 + 3xy + 2y^2$
21. $2x^2 + xy - 3y^2$
22. $15x^2 + 42xy + 24y^2$
23. $8x^2 + 6xy - 20y^2$
24. $x^2 + 7x + 10$
25. $3x^2 - 19x + 20$
26. $20x^2 - 19x + 3$
27. $x^2 - 15x + 56$
28. $2x^2 - x - 10$
29. $4x^2 - 16xy - 20y^2$
30. $3x^2 + 6x - 144$

31. $x^3 - 4x^2 - 12x$
32. $x^4 - 4x^3 - 12x^2$
33. $6x^5 - x^4 - 2x^3$
34. $8x^3 - 8x^2 + 2x$
35. $9x^3y + 9x^2y^2 + 2xy^3$
36. $xz^3 + 3xz^2 - 4xz$
37. $2y^3 + 4y^2 + 2y$
38. $3x^2 + 18x - 21$
39. $4x^2 + 12x + 9$
40. $12w^4 + 16w^3 - 16w^2$
41. $x^2 - y^2$
42. $4x^2 - y^2$
43. $x^3 - y^3$
44. $2x^3 + 2y^3$
45. $x^6 - y^6$
46. $16x^2 - 81$
47. $x^3 - 1$
48. $64x^2 - 1$
49. $64x^3 - 16x^2 + x$
50. $x^4 - y^2$
51. $x^4 - 16y^4$
52. $x^5 - 16xy^4$
53. $8x^3 + 1$
54. $125x^3 - 1$
55. $125x^3 + y^3$
56. $x^3 - xy^2$
57. $(x^2 + 1)^2 - 4$
58. $x(b^2 - 1) + y(b^2 - 1)$
59. $(x + y)a^2 + (x + y)2a + (x + y)$
60. $ax + x + ay + y$
61. $10x^2 + 10y^2$
62. $w(w^2 - 1) + (w^2 - 1)$
63. $x^8 - 1$
64. $(a + b)^2 - (x + y)^2$
65. $a^4 - (a + 1)^2$
66. $(a - b)^2 - (x - y)^2$
67. Verify each formula below by multiplying.
    a. $(a^2 - b^2) = (a - b)(a + b)$
    b. $(a^3 - b^3) = (a - b)(a^2 + ab + b^2)$
    c. $(a^3 + b^3) = (a + b)(a^2 - ab + b^2)$
68. True or False: $a^2 + b^2 = (a + b)(a + b)$

1.6 FRACTIONAL EXPRESSIONS

In this section we will discuss simplifying, multiplying, dividing, adding, and subtracting fractions whose numerators and/or denominators contain algebraic expressions. Many of the examples will be concerned with quotients of polynomials—so called **rational expressions**—because they occur so frequently in mathematics and related subjects.

The techniques employed when manipulating fractions will primarily involve the Laws about Quotients presented in Section 1.1. One of those laws is $\frac{ac}{bc} = \frac{a}{b}$. Translated into words this says that a factor that is *common* to both the numerator and denominator of a given fraction can be removed without affecting the value of that fraction. The word "cancellation" is sometimes used to describe this process. Notice, however, that $\frac{ac}{bc}$ can be written as $\frac{a}{b} \cdot \frac{c}{c}$ by using Law #5 about Quotients. This , in turn, becomes $\frac{a}{b} \cdot 1$ or just $\frac{a}{b}$. Thus, the "disappearance" or "cancellation" of c is, in reality, a division of c by itself. Once all common factors have been removed from the numerator and denominator of a given fraction we say that the fraction has been **reduced to lowest terms.**

EXAMPLE 1 Reduce $\dfrac{x^2 - 3x - 4}{x^2 - 6x + 8}$ to lowest terms.

Solution In order to reduce any fraction to lowest terms the numerator and denominator should be factored and then common factors should be cancelled. Observe:

$$\frac{x^2 - 3x - 4}{x^2 - 6x + 8}$$

Factor

$$\frac{(x+1)(x-4)}{(x-2)(x-4)}$$

Cancel

$$\frac{x+1}{x-2}$$

Some comments are in order: first, do not make the common mistake of cancelling x's in the answer above. This is *not* correct algebra because the x's are *not* factors, they are terms. Second, the answer can be checked by using Law #1 about Quotients. The reader is asked to do this in Exercise 1. Third, because we have variables in the denominator, it should be said that the reduction $\dfrac{x^2 - 3x - 4}{x^2 - 6x + 8} = \dfrac{x+1}{x-2}$ is correct provided $x \neq 2, x \neq 4$. In the examples that follow we will assume that denominators are not zero.

EXAMPLE 2 Reduce $\dfrac{2x^3 - 3x^2 - 2x}{4 - x^2}$ to lowest terms.

Solution

$$\frac{2x^3 - 3x^2 - 2x}{4 - x^2}$$

Factor

$$\frac{x(2x+1)(x-2)}{(2+x)(2-x)}$$

Rewrite $x - 2$ as $-1(2 - x)$

$$\frac{x(2x+1)(-1)(2-x)}{(2+x)(2-x)}$$

Cancel

$$\frac{-x(2x+1)}{x+2}$$

Multiplication of fractions is accomplished by using Law #5 about quotients, namely $\frac{a}{b} \cdot \frac{c}{d} = \frac{ac}{bd}$. Translated into words this says that to multiply two fractions, multiply the numerators and divide the result by the product of the denominators.

EXAMPLE 3 Multiply $\frac{x^2 - 4x + 4}{6x + 3} \cdot \frac{2x^2 - 5x - 3}{4x^2 - 7x - 2}$ and reduce the answer to lowest terms.

Solution

$$\frac{x^2 - 4x + 4}{6x + 3} \cdot \frac{2x^2 - 5x - 3}{4x^2 - 7x - 2}$$

Factor

$$\frac{(x-2)^2}{3(2x+1)} \cdot \frac{(2x+1)(x-3)}{(4x+1)(x-2)}$$

Multiply and cancel

$$\frac{(x-2)(x-3)}{3(4x+1)}$$

Division of fractions is performed by using Law #6 about Quotients, namely $\frac{a}{b} \div \frac{c}{d} = \frac{a}{b} \cdot \frac{d}{c}$. Translated into words this says that to divide two fractions, invert the divisor and multiply.

EXAMPLE 4 Divide $\frac{x^3 - x^2}{y^2 - 1} \div \frac{x^3}{y - 1}$ and reduce the answer to lowest terms.

Solution

$$\frac{x^3 - x^2}{y^2 - 1} \div \frac{x^3}{y - 1}$$

Convert to multiplication

$$\frac{x^3 - x^2}{y^2 - 1} \cdot \frac{y - 1}{x^3}$$

Factor

$$\frac{x^2(x-1)}{(y-1)(y+1)} \cdot \frac{y-1}{x^3}$$

Multiply and cancel

$$\frac{x-1}{x(y+1)}$$

When adding or subtracting fractions we will use Law #3 about Quotients, namely $\frac{a}{b} \pm \frac{c}{b} = \frac{a \pm c}{b}$. We call b the **common denomi-**

nator and to add $\frac{2x+1}{x} + \frac{3x-5}{x}$, we merely add the numerators and retain the denominator x:

$$\frac{2x+1}{x} + \frac{3x-5}{x} = \frac{5x-4}{x}$$

If the fractions do not have a common denominator, one will be constructed using the **least common multiple (LCM)*** of the denominators. We illustrate with numerical fractions. To add $\frac{1}{20} + \frac{5}{24}$, we first determine the LCM of the denominators. Factor each denominator:

$$20 = 2 \cdot 2 \cdot 5$$
$$24 = 2 \cdot 2 \cdot 2 \cdot 3$$

The LCM is formed by using *each* factor the *greatest* number of times it appears in any one of the original denominators. The LCM of 20 and 24 is $2 \cdot 2 \cdot 2 \cdot 3 \cdot 5 = 120$. Next, each fraction is "built up" so that its denominator is 120:

$$\frac{1}{20} + \frac{5}{24}$$

$$\frac{1}{2 \cdot 2 \cdot 5} + \frac{5}{2 \cdot 2 \cdot 2 \cdot 3}$$

$$\frac{1 \cdot 2 \cdot 3}{2 \cdot 2 \cdot 5 \cdot 2 \cdot 3} + \frac{5 \cdot 5}{2 \cdot 2 \cdot 2 \cdot 3 \cdot 5}$$

Multiply the numerator and denominator of 1/20 by $2 \cdot 3$

Multiply the numerator and denominator of 5/24 by 5

$$\frac{6}{120} + \frac{25}{120}$$

$$\frac{31}{120}$$

EXAMPLE 5 Perform the indicated operation and express the answer in lowest terms.

$$\frac{5(x-1)}{x^3yz} + \frac{2}{xz^3}$$

*The LCM is the smallest or least common denominator available. Its use makes the calculation easier.

Solution With the denominators x^3yz and xz^3, the LCM $= x^3yz^3$. Thus the numerator and denominator of the first fraction are multiplied by z^2 and the numerator and denominator of the second are multiplied by x^2y.

$$\frac{5(x-1)}{x^3yz} + \frac{2}{xz^3} = \frac{5(x-1)\cdot z^2}{x^3yz\cdot z^2} + \frac{2\cdot x^2y}{xz^3\cdot x^2y}$$

$$= \underbrace{\frac{5xz^2 - 5z^2 + 2x^2y}{x^3yz^3}}_{\text{This result is in lowest terms.}}$$

EXAMPLE 6 Perform the indicated operation and express the answer in lowest terms.

$$\frac{3x}{x^2+5x+6} - \frac{2x-1}{x^2+6x+9}$$

Solution

$$\frac{3x}{x^2+5x+6} - \frac{2x-1}{x^2+6x+9} = \frac{3x}{(x+2)(x+3)} - \frac{(2x-1)}{(x+3)(x+3)}$$

Now that the denominators have been factored, it is seen that the LCM $= (x+2)(x+3)^2$. Continuing,

$$\frac{3x}{x^2+5x+6} - \frac{2x-1}{x^2+6x+9} = \frac{3x(x+3)}{(x+2)(x+3)(x+3)}$$

$$- \frac{(2x-1)(x+2)}{(x+3)(x+3)(x+2)}$$

$$= \frac{3x(x+3) - (2x-1)(x+2)}{(x+2)(x+3)^2}$$

$$= \frac{3x^2 + 9x - 2x^2 - 3x + 2}{(x+2)(x+3)^2}$$

$$= \underbrace{\frac{x^2+6x+2}{(x+2)(x+3)^2}}_{\text{This result is in lowest terms.}}$$

EXAMPLE 7 Perform the indicated operations and simplify.

$$2x - \frac{1}{x+1} + \frac{1}{(x+1)^2}$$

Solution The given expression can be written as $\frac{2x}{1} - \frac{1}{x+1} + \frac{1}{(x+1)^2}$. The three denominators 1, $x + 1$, $(x + 1)^2$ have an LCM $= (x + 1)^2$. Continuing,

$$\begin{aligned}
2x - \frac{1}{x+1} + \frac{1}{(x+1)^2} &= \frac{2x \cdot (x+1)^2}{1 \cdot (x+1)^2} - \frac{1 \cdot (x+1)}{(x+1)(x+1)} + \frac{1}{(x+1)^2} \\
&= \frac{2x(x+1)^2 - 1(x+1) + 1}{(x+1)^2} \\
&= \frac{2x^3 + 4x^2 + 2x - x - 1 + 1}{(x+1)^2} \\
&= \frac{x(2x^2 + 4x + 1)}{(x+1)^2}
\end{aligned}$$

When there is a fraction within the numerator or denominator of another fraction, the resulting expression is a **complex fraction.** It is usually desirable to simplify complex fractions and the next two examples show how to do that.

EXAMPLE 8 Simplify $\dfrac{\dfrac{x^2}{y^2} - 1}{\dfrac{x}{y} - 1}$.

Solution We show two solutions below. In I, the subtractions are performed first within the numerator and denominator, followed by the division. In II, the numerator is factored as the difference of two squares, followed by simplification.

Method I

$$\frac{\dfrac{x^2}{y^2} - 1}{\dfrac{x}{y} - 1} = \frac{\dfrac{x^2}{y^2} - \dfrac{y^2}{y^2}}{\dfrac{x}{y} - \dfrac{y}{y}}$$

Method II

$$\frac{\dfrac{x^2}{y^2} - 1}{\dfrac{x}{y} - 1} = \frac{\left(\dfrac{x}{y} - 1\right)\left(\dfrac{x}{y} + 1\right)}{\dfrac{x}{y} - 1}$$

$$= \frac{\frac{x^2 - y^2}{y^2}}{\frac{x - y}{y}}$$

$$= \frac{x^2 - y^2}{y^2} \cdot \frac{y}{x - y}$$

$$= \frac{(x - y)(x + y)y}{y^2(x - y)}$$

$$= \frac{x + y}{y}$$

$$= \frac{x}{y} + 1$$

$$= \frac{x}{y} + \frac{y}{y}$$

$$= \frac{x + y}{y}$$

EXAMPLE 9 Simplify $\dfrac{x - x^{-1}}{y - y^{-2}}$.

Solution

$$\frac{x - x^{-1}}{y - y^{-2}} = \frac{x - \frac{1}{x}}{y - \frac{1}{y^2}}$$

Next, multiply numerator and denominator of the complex fraction by xy^2.*

$$\frac{x - x^{-1}}{y - y^{-2}} = \frac{xy^2\left(x - \frac{1}{x}\right)}{xy^2\left(y - \frac{1}{y^2}\right)}$$

$$= \frac{x^2y^2 - y^2}{xy^3 - x}$$

$$= \frac{y^2(x^2 - 1)}{x(y^3 - 1)}$$

---

*xy^2 is the LCM of the denominators of the fractions $\frac{1}{x}$ and $\frac{1}{y^2}$.

EXERCISES FOR SECTION 1.6

1. Law #1 about Quotients is: $\frac{a}{b} = \frac{c}{d}$ if and only if $ad = bc$. Use this law to check the answer to Example 1.

In Exercises 2 through 10, reduce each fraction to lowest terms. (See Examples 1 and 2.)

2. $\dfrac{x+2}{4x+8}$
3. $\dfrac{x-2}{2-x}$
4. $\dfrac{x-y}{y-x}$
5. $\dfrac{x^2+2x-8}{x^2-4}$
6. $\dfrac{2x+2}{3x^2+2x-1}$
7. $\dfrac{x^3+27}{x^2+7x+12}$
8. $\dfrac{x-2x^2}{1-5x+6x^2}$
9. $\dfrac{5x^2-8x-4}{10x^2+19x+6}$
10. $\dfrac{x^3-8}{x^2-4x+4}$

In Exercises 11 through 30, perform the indicated operation and reduce the answer to lowest terms. (See Examples 3 and 4.)

11. $\dfrac{6x-2}{x+2} \cdot \dfrac{x-1}{(3x-1)(x+1)}$
12. $\dfrac{2x+4}{x-1} \cdot \dfrac{3x-3}{x^2-4}$
13. $\dfrac{y^2-x^2}{2} \cdot \dfrac{x+y}{x-y}$
14. $\dfrac{2x+5}{4x^2+25} \cdot \dfrac{2x-5}{4x^2-25}$
15. $\dfrac{xy}{8-2x-x^2} \cdot \dfrac{4+12x-7x^2}{xy^3}$
16. $\dfrac{x+1}{x^2+1} \cdot \dfrac{x+1}{x^3+1}$
17. $\dfrac{x}{y-1} \div \dfrac{y}{y-1}$
18. $\dfrac{1}{x(y-1)} \cdot \dfrac{x^3-x}{y} \cdot \dfrac{y^2-y}{2x+2}$
19. $\dfrac{x^3-x^2}{2y^3} \div \dfrac{x^3-x}{y^2}$
20. $\dfrac{a^3-64}{a-4} \div \dfrac{a^2+4a+16}{a^2-8a+16}$
21. $\dfrac{ax+ay+bx+by}{a-b} \cdot \dfrac{a^2-b^2}{x^2-y^2}$
22. $\dfrac{x-3}{2x+1} \div \dfrac{9-x^2}{x+3}$
23. $\dfrac{2x^2-11x-6}{2x^2+3x+1} \div \dfrac{x^2-6x+9}{x^2-5x-6}$
24. $\dfrac{1}{x^2} \cdot \dfrac{x^3y}{2} \div \dfrac{x}{y}$
25. $\dfrac{x}{y}\left(\dfrac{x}{y}-1\right) \div \left(\dfrac{x}{y}-1\right)^2$
26. $\dfrac{4t^2-4t+1}{2t^2+t-1} \div \dfrac{4t^2-1}{4t+2}$
27. $\dfrac{a^2-8ab+16b^2}{a+4b} \cdot \dfrac{a^2+3ab-4b^2}{a^2-3ab-4b^2}$
28. $\dfrac{r^2-25}{r-5} \div \dfrac{r^2+6r+5}{r+6}$
29. $\dfrac{r+w}{r^2+w^2} \cdot \dfrac{w^4}{r^2-w^2}$
30. $\dfrac{2ax-bx-2ay+by}{(2a-b)^2} \div \dfrac{x-y}{2a^2+ab-b^2}$

In Exercises 31 through 50, perform the indicated operation and reduce the answer to lowest terms. (See Examples 5, 6, and 7.)

31. $\dfrac{2}{xy^2} + \dfrac{4}{xyz^3}$

32. $\dfrac{1}{x} - \dfrac{1}{xy}$

33. $\dfrac{1}{x} + \dfrac{1}{y} + \dfrac{1}{z}$

34. $\dfrac{1}{x} + \dfrac{1}{y} - \dfrac{1}{z}$

35. $\dfrac{2}{x-3} + \dfrac{x}{x+4}$

36. $\dfrac{1}{x(x-1)} + \dfrac{2}{(x-1)^2}$

37. $\dfrac{3x+3}{x^2-1} - \dfrac{1}{x-1}$

38. $\dfrac{2}{x^2+3x+2} + \dfrac{1}{x^2+2x+1}$

39. $\dfrac{x-1}{x^2+3x+2} + \dfrac{1}{x^2+2x+1}$

40. $\dfrac{1}{a} + \dfrac{1}{a^2} - \dfrac{1}{a^3}$

41. $\dfrac{1}{a-1} + \dfrac{1}{(a-1)^2} - \dfrac{1}{(a-1)^3}$

42. $\dfrac{1}{s^2-2s} + \dfrac{1}{4-2s}$

43. $\dfrac{2}{a(a+1)} - \dfrac{1}{a^2} + \dfrac{1}{a+1}$

44. $w + \dfrac{1}{w} + \dfrac{1}{w^2} + \dfrac{1}{w^3}$

45. $1 - \dfrac{2}{k+2}$

46. $\dfrac{2x}{2x^2-3x+1} + \dfrac{3x}{x^2-1}$

47. $\dfrac{2a+3b}{a^2+3ab+2b^2} + \dfrac{a+b}{2a^2+7ab+6b^2}$

48. $\dfrac{2a+3b}{a^2+3ab+2b^2} - \dfrac{a+b}{2a^2+7ab+6b^2}$

49. $\dfrac{1}{a^2-9b^2} + \dfrac{2}{a^2+6ab+9b^2} - \dfrac{3}{a-3b}$

50. $\dfrac{a}{a^2-9b^2} + \dfrac{2}{a^2+6ab+9b^2} - \dfrac{3}{a-3b}$

In Exercises 51 through 60, simplify each complex fraction. (See Examples 8 and 9.)

51. $\dfrac{1+\dfrac{1}{x}}{x}$

52. $\dfrac{1+\dfrac{1}{x}}{1+\dfrac{1}{a}}$

53. $\dfrac{x+\dfrac{1}{2}}{x-\dfrac{1}{4}}$

54. $\dfrac{\dfrac{a^2}{b^2}-1}{\dfrac{a}{b}+1}$

55. $\dfrac{x}{\dfrac{1}{x} - \dfrac{1}{x^2}}$ 56. $\dfrac{\dfrac{1}{x+h} - \dfrac{1}{x}}{h}$ 57. $\dfrac{r + r^{-1}}{r + r^{-2}}$

58. $\dfrac{\dfrac{1}{x} + \dfrac{1}{y}}{(xy)^{-1}}$ 59. $\dfrac{(xy)^{-1}}{\dfrac{1}{x} + \dfrac{1}{y}}$ 60. $\dfrac{\dfrac{1}{x+h+1} - \dfrac{1}{x+1}}{h}$

In Exercises 61 through 64, answer true or false. If true, state why. If false, exhibit a counterexample.

61. $\dfrac{1}{p} + \dfrac{1}{q} = \dfrac{1}{p+q}$ 62. $\dfrac{1}{p} + \dfrac{1}{q} = \dfrac{2}{p+q}$

63. $\dfrac{p+q}{r} = \dfrac{p}{r} + \dfrac{q}{r}$ 64. $\dfrac{1}{p} \cdot \dfrac{1}{q} = \dfrac{1}{pq}$

1.7 RADICAL EXPRESSIONS

The number 49 is the square of 7; also 49 is the square of -7. We say that 7 is a **square root** of 49 and also that -7 is a square root of 49. Similarly, -64 is the cube of -4, which leads us to say that -4 is the **cube root** of -64. To be more precise we state the following:

DEFINITION

> Let n be a positive integer ($n \geqslant 2$) and a be a real number. The number a is called **an *n*th root** of b provided $a^n = b$.

In light of the fact that 49 has two square roots, it is seen that the nth root of a real number is *not unique.* However, in the following definition the **principal *n*th root** of a real number is defined and *is* unique.

DEFINITION

> The **principal *n*th root** of a real number b, denoted by $\sqrt[n]{b}$, is defined as:*
>
> $\sqrt[n]{b}$ is the positive nth root of b, if n is odd and $b > 0$;
> $\sqrt[n]{b}$ is the negative nth root of b, if n is odd and $b < 0$;
> $\sqrt[n]{b}$ is the positive nth root of b, if n is even and $b > 0$.

*The case where n is even and $b < 0$ is not covered by this definition, but is discussed in Section 1.9.

In the case of the square root, $\sqrt[2]{b}$ is written as $\sqrt{b}$. In the expression $\sqrt[n]{b}$, n is called the **index**, b is called the **radicand**, and $\sqrt[n]{\ }$ is called the **radical sign**. As a result of the previous definitions we have the following theorem.

THEOREM 4

The principal nth root of a real number b must satisfy:

$$(\sqrt[n]{b})^n = b.$$

EXAMPLE 1 Determine each of the following.

a. $\sqrt[3]{27}$ b. $\sqrt[5]{-32}$ c. $\sqrt{49}$ d. $\sqrt[4]{10{,}000}$

Solution

a. $\sqrt[3]{27} = 3$ because $(\sqrt[3]{27})^3 = 3^3 = 27$.
b. $\sqrt[5]{-32} = -2$ because $(\sqrt[5]{-32})^5 = (-2)^5 = -32$.
c. $\sqrt{49} = 7$ because $(\sqrt{49})^2 = (7)^2 = 49$.
d. $\sqrt[4]{10{,}000} = 10$ because $(\sqrt[4]{10{,}000})^4 = (10)^4 = 10{,}000$.

The following rules for radicals with the indicated restrictions can be proven.

Rules for Radicals

1. If $x \geqslant 0$, $\sqrt[n]{x^n} = x$.
2. If $x \geqslant 0$ and $y \geqslant 0$, $\sqrt[n]{xy} = \sqrt[n]{x} \cdot \sqrt[n]{y}$
3. If $x \geqslant 0$ and $y > 0$, $\sqrt[n]{\dfrac{x}{y}} = \dfrac{\sqrt[n]{x}}{\sqrt[n]{y}}$
4. If $x \geqslant 0$, $\sqrt[m]{\sqrt[n]{x}} = \sqrt[mn]{x}$

We now use these rules to *simplify* radical expressions. Thus, in the examples that follow we assume that the variables are nonnegative.

EXAMPLE 2 Simplify

a. $\sqrt{27}$ b. $\sqrt[3]{16ab^5}$

Solution By using Radical Rules #1 and #2,

a. $\sqrt{27} = \sqrt{3^2 \cdot 3} = \sqrt{3^2} \cdot \sqrt{3} = 3\sqrt{3}$

b. $\sqrt[3]{16ab^5} = \sqrt[3]{8b^3 \cdot 2ab^2} = \sqrt[3]{(2b)^3 \cdot 2ab^2} = 2b\sqrt[3]{2ab^2}$

In part a we "removed" the perfect square 3^2 from the radicand and in part b the perfect cube $(2b)^3$ was "removed" from the radicand. One meaning then of *simplifying a radical* of index n is to remove all perfect nth powers from the radicand. In the next example another meaning of simplify is explored.

EXAMPLE 3 Simplify

a. $\dfrac{\sqrt{2x}}{\sqrt{5}}$ b. $\sqrt[3]{\dfrac{a}{b}}$

Solution a. $\dfrac{\sqrt{2x}}{\sqrt{5}} = \dfrac{\sqrt{2x} \cdot \sqrt{5}}{\sqrt{5} \cdot \sqrt{5}}$ by multiplying numerator and denominator by $\sqrt{5}$

$= \dfrac{\sqrt{10x}}{\sqrt{5^2}}$ by using Radical Rules #1 and #3

$= \dfrac{\sqrt{10x}}{5}$

b. $\sqrt[3]{\dfrac{a}{b}} = \sqrt[3]{\dfrac{a \cdot b^2}{b \cdot b^2}}$ by multiplying numerator and denominator of the radicand by b^2

$= \sqrt[3]{\dfrac{ab^2}{b^3}} = \dfrac{\sqrt[3]{ab^2}}{\sqrt[3]{b^3}}$ by using Radical Rules #1 and #3

$= \dfrac{\sqrt[3]{ab^2}}{b}$

In part a the original expression was rewritten without any radical in the denominator. This is called **rationalizing the denominator.** In part b the original expression was rewritten without any fractions under the radical sign. Both of these techniques are considered to be part of *simplifying a radical*. In the next example one more simplification technique is demonstrated.

EXAMPLE 4 Simplify.

a. $\sqrt[6]{27}$ b. $\sqrt{\sqrt[6]{x^3}}$

Solution a. $\sqrt[6]{27} = \sqrt{\sqrt[3]{27}}$ by using Radical Rule #4

$\sqrt[6]{27} = \sqrt{3}$

b. $\sqrt{\sqrt[6]{x^3}} = \sqrt[12]{x^3} = \sqrt[4]{\sqrt[3]{x^3}}$ by using Radical Rule #4

$\sqrt{\sqrt[6]{x^3}} = \sqrt[4]{x}$

In parts a and b the size of the index was made smaller by suitable manipulation. Thus, simplifying radicals means several things: removal of perfect nth powers from the radicand of a radical whose index is n, rationalizing denominators, eliminating fractions from radicands, and making indices as small as possible.

The basic arithmetic operations (multiplying, dividing, adding, and subtracting) can be performed on radical expressions as the next example shows.

EXAMPLE 5 Perform the indicated operation and simplify.

a. $\sqrt[3]{2z^2w} \cdot \sqrt[3]{4z^4w^5}$ b. $\dfrac{\sqrt{56xy^3}}{\sqrt{2x^2y}}$ c. $\sqrt{8x} - \sqrt{72xy^2}$

Solution a. $\sqrt[3]{2z^2w} \cdot \sqrt[3]{4z^4w^5} = \sqrt[3]{8z^6w^6}$

$= \sqrt[3]{(2z^2w^2)^3}$

$= 2z^2w^2$

b. $$\frac{\sqrt{56xy^3}}{\sqrt{2x^2y}} = \sqrt{\frac{56xy^3}{2x^2y}} = \sqrt{\frac{28y^2}{x}} = \sqrt{\frac{4y^2 \cdot 7 \cdot x}{x \cdot x}}$$

$$= \sqrt{\frac{(2y)^2 \cdot 7x}{x^2}} = \sqrt{\left(\frac{2y}{x}\right)^2 \cdot 7x} = \sqrt{\left(\frac{2y}{x}\right)^2} \cdot \sqrt{7x}$$

$$= \frac{2y}{x}\sqrt{7x}$$

c. $\sqrt{8x} - \sqrt{72xy^2}$ must be restructured since we have no rule for subtracting radicals.

$$\sqrt{8x} - \sqrt{72xy^2}$$
$$\sqrt{4 \cdot 2x} - \sqrt{36y^2 \cdot 2x}$$
$$\sqrt{2^2 \cdot 2x} - \sqrt{(6y)^2 \cdot 2x}$$
$$\sqrt{2^2} \cdot \sqrt{2x} - \sqrt{(6y)^2} \cdot \sqrt{2x}$$
$$2\sqrt{2x} - 6y\sqrt{2x}$$
$$(2 - 6y)\sqrt{2x}$$

Extract the common factor $\sqrt{2x}$

It was mentioned earlier that the radical rules can be used if the variables represent nonnegative real numbers. However, sometimes variables are negative. Let us consider the expression $\sqrt{x^2}$. Radical Rule #1 assures us that $\sqrt{x^2} = x$, if $x \geqslant 0$. But what if $x < 0$? Then $\sqrt{x^2} = \sqrt{(-x)^2}$, since $x^2 = (-x)^2$. Furthermore, $\sqrt{(-x)^2} = -x$ by Radical Rule #1. Thus we have:

$$\sqrt{x^2} = x \quad \text{if } x \geqslant 0$$
$$\sqrt{x^2} = -x \quad \text{if } x < 0$$

The two previous statements can be combined to read $\sqrt{x^2} = |x|$. We have proven:

THEOREM 5

If x is any real number, $\sqrt{x^2} = |x|$.

This theorem is used in Chapter 2 to obtain certain results. Also, it is used from time to time in the study of calculus and related areas. Moreover, it provides us with a relationship between radicals and absolute value. In simplifying algebraic expressions where all variables represent nonnegative real numbers, we can write

$$\sqrt{3x^2z} = \sqrt{x^2 \cdot 3z} = \sqrt{x^2} \cdot \sqrt{3z} = x\sqrt{3z}$$

But if variables can be negative, we must write

$$\sqrt{3x^2z} = \sqrt{x^2 \cdot 3z} = \sqrt{x^2} \cdot \sqrt{3z} = |x|\sqrt{3z}.$$

Much of the time, when working with radicals, we will be able to treat all variables as nonnegative real numbers. When that is not the case, it will be noted.

Another aspect worthy of noting here is that if a statement of equality is known to be true, then both sides of that equality can be raised to the same positive integral power and the equality sign will be preserved. For example, since $4 = 4$ we can *square both sides* and obtain $4^2 = 4^2$ or $16 = 16$. We can *cube both sides* and write $4^3 = 4^3$ or $64 = 64$. We can *raise both sides to the 4th power* and write $4^4 = 4^4$ or $256 = 256$.

It is also possible to take a principal root of both sides of an equality and the equality sign will be preserved. For example, since $9 = 9$, taking the *principal square root of both sides* yields $\sqrt{9} = \sqrt{9}$ or $3 = 3$. Starting with $8 = 8$ and taking the *principal cube root of both sides* yields $\sqrt[3]{8} = \sqrt[3]{8}$ or $2 = 2$, etc. We have been illustrating a rule or principle that will be discussed further as the need arises. We now state that principle.

Principle of Exponentiation

A statement of equality can be rewritten by raising both sides to the same positive integral power or by taking the principal nth root of both sides.

Many of the examples in this section involve principal roots of conveniently chosen numbers, such as $\sqrt[3]{27} = 3$, $\sqrt[5]{32} = 2$, etc. However, the reader who has access to an electronic calculator with a y^x key can determine the principal nth root of any positive real number to several decimal places. This will be examined further in the next section.

EXERCISES FOR SECTION 1.7

In Exercises 1 through 8, determine the principal root. (See Example 1.)

1. $\sqrt{25}$
2. $\sqrt{36}$
3. $\sqrt[3]{27}$
4. $\sqrt[3]{-27}$
5. $\sqrt[4]{16}$
6. $\sqrt[6]{64}$
7. $\sqrt[8]{1}$
8. $\sqrt[5]{243}$

In Exercises 9 through 30, simplify each expression. Assume all variables are nonnegative. (See Examples 2, 3, and 4.)

9. $\sqrt{12}$
10. $\sqrt[3]{54}$
11. $\sqrt[3]{16}$
12. $\sqrt[4]{32}$
13. $\sqrt[5]{128}$
14. $\sqrt[5]{96}$
15. $\sqrt{8x^4y^5z^6}$
16. $\sqrt[3]{125ab^2c^7}$
17. $\dfrac{\sqrt{5x}}{\sqrt{7y}}$

18. $\sqrt[3]{\dfrac{a^2}{b}}$ 19. $\sqrt{\dfrac{5x^2y}{2x^3y^4}}$ 20. $\dfrac{\sqrt[4]{32h^3d}}{\sqrt[4]{d^3}}$

21. $\sqrt{\sqrt{6}}$ 22. $\sqrt[3]{\sqrt{64}}$ 23. $\sqrt{x\sqrt{x^2}}$

24. $\sqrt[3]{x\sqrt{4x^2}}$ 25. $\sqrt[4]{4}$ 26. $\sqrt[6]{8}$

27. $\sqrt[4]{4a^8b^4}$ 28. $\sqrt[8]{16}$ 29. $\sqrt[4]{4x^4y^8}$

30. $\dfrac{\sqrt[3]{27ab^2}}{\sqrt[3]{b^2}}$

In Exercises 31 through 58, perform the indicated operation and simplify. Assume all variables are nonnegative. (See Example 5.)

31. $\sqrt{2x} \cdot \sqrt{4x}$ 32. $\sqrt[3]{2x} \cdot \sqrt[3]{4x}$

33. $\sqrt{3x^2yz^4} \cdot \sqrt{4xy^3z^5}$ 34. $\sqrt{(2x)^3} \cdot \sqrt{2xy^3}$

35. $\sqrt{5ab} \cdot \sqrt{10a^5b^6}$ 36. $\dfrac{\sqrt{10a^5b^6}}{\sqrt{5ab}}$

37. $\dfrac{\sqrt[3]{36w^5z^2}}{\sqrt[3]{4wz^{-1}}}$ 38. $\sqrt{\dfrac{x^2}{y}} \cdot \sqrt{\dfrac{x}{y^2}}$

39. $\sqrt{(x+y)^2}$ 40. $\sqrt[3]{(x+y)^3}$

41. $\sqrt{(x+y)^3} \cdot \sqrt{(x+y)^2}$ 42. $(\sqrt[3]{xyz})^3$

43. $\sqrt[5]{xy^6} \cdot \sqrt[5]{\dfrac{x^4}{y}}$ 44. $\dfrac{1}{\sqrt{2x^2y}}$

45. $\sqrt{x^2y^5z^{-2}} \cdot \sqrt{x^3y^2z^2}$ 46. $\sqrt{2a^3b^4c} \cdot \sqrt{ab} \cdot \sqrt{ac}$

47. $\sqrt[4]{a^2b^2c^4} \cdot \sqrt[4]{abc^4} \cdot \sqrt[4]{a^4bc^5}$ 48. $\dfrac{1}{\sqrt{x}} \cdot \sqrt{y}$

49. $(\sqrt[3]{xy} \cdot \sqrt[3]{x^2y^4})^2$ 50. $\sqrt{8} + \sqrt{2}$

51. $\sqrt{25x^3} - \sqrt{4x^3}$ 52. $\sqrt{25x^3} - \sqrt{4x}$

53. $1 - \sqrt{16xy^2} - \sqrt{xy^2}$ 54. $\sqrt{4a} + \sqrt{9a} + \sqrt{16b} + \sqrt{25b}$

55. $\sqrt{4a} + \sqrt{9a} - (\sqrt{16b} - \sqrt{25b})$ 56. $\sqrt[3]{a^3b} + \sqrt[3]{8b}$

57. $\sqrt{2}(\sqrt{2} - \sqrt{2x})$ 58. $\sqrt{\dfrac{x}{y}}\left(\sqrt{\dfrac{x}{y}} - 1\right)$

59. a. Rationalize the denominator of $\dfrac{2}{\sqrt{2} - \sqrt{3}}$ by multiplying numerator and denominator by $\sqrt{2} + \sqrt{3}$.

b. Rationalize the denominator of $\dfrac{4x}{\sqrt{x} + \sqrt{y}}$ by multiplying numerator and denominator by $\sqrt{x} - \sqrt{y}$.

c. Rationalize the denominator of $\dfrac{\sqrt{3} + \sqrt{5}}{\sqrt{5} - \sqrt{3}}$.

60. Rationalize the *numerator* of $\dfrac{\sqrt{x+h} - \sqrt{x}}{h}$ by multiplying numerator and denominator by $\sqrt{x+h} + \sqrt{x}$.

61. Rationalize the *numerator* of $\dfrac{\dfrac{1}{\sqrt{x+h}} - \dfrac{1}{\sqrt{x}}}{h}$ by multiplying numerator and denominator by $\dfrac{1}{\sqrt{x+h}} + \dfrac{1}{\sqrt{x}}$.

In Exercises 62 through 65, answer true or false. If true, state why. If false, exhibit a counterexample.

62. If $a \geqslant 0$ and $b \geqslant 0$, $\sqrt{a+b} = \sqrt{a} + \sqrt{b}$.

63. If $a \geqslant 0$ and $b \geqslant 0$, $\sqrt{a-b} = \sqrt{a} - \sqrt{b}$.

64. If $a \geqslant 0$ and $b \geqslant 0$, then $\sqrt{ab} = \sqrt{a} \cdot \sqrt{b}$.

65. If $a \geqslant 0$ and $b > 0$, then $\sqrt{\dfrac{a}{b}} = \dfrac{\sqrt{a}}{\sqrt{b}}$.

Computer Exercises

66. The symbol $\sqrt{x}$ is written as SQR(X) in many versions of BASIC. Run the following program:

```
1Ø FOR X = 1 TO 20
2Ø LET Y = SQR(X)
3Ø PRINT X, Y
4Ø NEXT X
5Ø STOP
```

67. Refine the program above to create a program that will print a table of numbers from 1 to 50, their squares, and their principal square roots.

68. Write a program that calculates Y = SQR(X ↑ 2) for any inputted value of x. How does this relate to Theorem 5?

1.8 RATIONAL EXPONENTS

The Principle of Exponentiation is concerned with powers and roots of real numbers. It is possible to unify these two concepts by defining the notion of a rational number exponent. In this section expressions such as $x^{1/2}$, $(3x)^{2/3}$, $(4y)^{-3/4}$ will be defined. This will be done in a manner such that the exponent laws of Section 1.3 remain true. To begin with, consider the expression $x^{1/2}$. Using the exponent law $(a^m)^n = a^{mn}$, we can write: $(x^{1/2})^2 = x^1 = x$. But we already know that $(\sqrt{x}\,)^2 = x$. Thus $x^{1/2}$ is defined to be $\sqrt{x}$. Similarly, consider $x^{1/n}$. Again using the exponent rule $(a^m)^n = a^{mn}$, $(x^{1/n})^n = x^1 = x$, but also $(\sqrt[n]{x}\,)^n = x$. Thus we have the following:

DEFINITION

$$x^{1/n} = \sqrt[n]{x} \quad *$$

Next, consider $x^{m/n}$, where m/n is a positive rational number.

$x^{m/n}$ can be written as $(x^{1/n})^m = (\sqrt[n]{x}\,)^m$
$x^{m/n}$ can be written as $(x^m)^{1/n} = \sqrt[n]{x^m}$

Thus we have

DEFINITION

If m/n is a positive rational number, then

$$x^{m/n} = (\sqrt[n]{x}\,)^m = \sqrt[n]{x^m}$$

Negative rational exponents are handled in the same manner as negative integral exponents.

DEFINITION

If m/n is a positive rational number, then

$$x^{-m/n} = \frac{1}{x^{m/n}}$$

EXAMPLE 1 Simplify.

a. $49^{1/2}$ b. $49^{-1/2}$ c. $125^{2/3}$ d. $(-32)^{-2/5}$

Solution a. $49^{1/2} = \sqrt{49} = 7$

b. $49^{-1/2} = \dfrac{1}{\sqrt{49}} = \dfrac{1}{7}$

c. $125^{2/3} = (\sqrt[3]{125}\,)^2 = (5)^2 = 25$

d. $(-32)^{-2/5} = \dfrac{1}{(-32)^{2/5}} = \dfrac{1}{(\sqrt[5]{-32}\,)^2} = \dfrac{1}{(-2)^2} = \dfrac{1}{4}$

Next, algebraic expressions are simplified.

*In this section, whenever we write $\sqrt[n]{x}$ it will be assumed that n and x are such that $\sqrt[n]{x}$ exists.

EXAMPLE 2 Simplify.

a. $(x^{2/3}y^{1/3}z)^2$ b. $(2x^{1/3})(3x^{1/4})$ c. $(3a^{2/3})^3(b^{-2}c^{-1})^{1/2}$

Solution a. $(x^{2/3}y^{1/3}z)^2 = (x^{2/3})^2(y^{1/3})^2(z)^2 = x^{4/3}y^{2/3}z^2$

b. $(2x^{1/3})(3x^{1/4}) = 2 \cdot 3 \cdot x^{1/3} \cdot x^{1/4} = 6x^{4/12} \cdot x^{3/12} = 6x^{7/12}$

c. $$\begin{aligned}(3a^{2/3})^3(b^{-2}c^{-1})^{1/2} &= 27(a^{2/3})^3(b^{-2})^{1/2}(c^{-1})^{1/2}\\ &= 27a^2b^{-1}c^{-1/2} = \frac{27a^2}{b\sqrt{c}}\\ &= \frac{27a^2\sqrt{c}}{bc}\end{aligned}$$

Rational exponents provide us with an alternative method for performing multiplications and divisions with radicals as the next example shows. Note in part b that the radicals have different indices and thus no rule of radicals is directly applicable.

EXAMPLE 3 Simplify.

a. $\sqrt{x^2y} \cdot \sqrt{xy}$ b. $\dfrac{\sqrt[3]{x^2}}{\sqrt{x}}$

Solution a. $$\begin{aligned}\sqrt{x^2y} \cdot \sqrt{xy} &= (x^2y)^{1/2} \cdot (xy)^{1/2} = x \cdot y^{1/2} \cdot x^{1/2} \cdot y^{1/2}\\ &= x \cdot y^{1/2} \cdot y^{1/2} \cdot x^{1/2} = x \cdot y \cdot x^{1/2}\\ &= xy\sqrt{x}\end{aligned}$$

b. $$\frac{\sqrt[3]{x^2}}{\sqrt{x}} = \frac{x^{2/3}}{x^{1/2}} = \frac{x^{4/6}}{x^{3/6}} = x^{1/6} = \sqrt[6]{x}$$

Now that we have a knowledge of fractional exponents we shall return to some previous notions of this chapter and provide some additional insights.

EXAMPLE 4 Use scientific notation to calculate $(\sqrt{.00000081})^3$

Solution $$\begin{aligned}(\sqrt{.00000081})^3 &= (.00000081)^{3/2}\\ &= (81 \cdot 10^{-8})^{3/2}\\ &= 81^{3/2} \cdot (10^{-8})^{3/2}\\ &= (\sqrt{81})^3 \cdot 10^{-12}\\ &= 729 \cdot 10^{-12} = 7.29 \cdot 10^{-10}\end{aligned}$$

EXAMPLE 5 Factor each expression.

a. $x^2 - 3$ b. $x^2 - y$

Solution a. $x^2 - 3 = x^2 - (3^{1/2})^2$
$= (x - 3^{1/2})(x + 3^{1/2})$
$= (x - \sqrt{3})(x + \sqrt{3})$

b. $x^2 - y = (x - \sqrt{y})(x + \sqrt{y})$

EXAMPLE 6 Perform the indicated operations and simplify.

a. $(x^2 - \sqrt{y})^2$ b. $(\sqrt{x} - \sqrt{y})(\sqrt{x} + \sqrt{y})$

Solution a. $(x^2 - \sqrt{y})^2 = (x^2)^2 - 2(x^2)(\sqrt{y}) + (\sqrt{y})^2$
$= x^4 - 2x^2\sqrt{y} + y$

b. $(\sqrt{x} - \sqrt{y})(\sqrt{x} + \sqrt{y}) = \sqrt{x^2} + \sqrt{xy} - \sqrt{xy} - \sqrt{y^2}$
$= x - y$

In the final example of this section a method of factoring known as **completing the square** is examined.

EXAMPLE 7 Factor each expression.

a. $x^2 + 8x + 7$ b. $x^2 - 2x - 5$

Solution a. First consider $(x + a)^2 = x^2 + 2ax + a^2$ and notice that one-half of the coefficient of x, when squared, produces the last term a^2. Now consider $x^2 + 8x + 7$. One-half of the coefficient of x, when squared, is 4^2 or 16. So we write:

$$\begin{aligned} x^2 + 8x + 7 &= x^2 + 8x + 16 - 16 + 7 \\ &= x^2 + 8x + 16 - 9 \\ &= (x + 4)^2 - 9 \\ &= [(x + 4) - 3][(x + 4) + 3] \\ &= (x + 1)(x + 7) \end{aligned}$$

b. $$\begin{aligned} x^2 - 2x - 5 &= x^2 - 2x + 1 - 1 - 5 \\ &= x^2 - 2x + 1 - 6 \\ &= (x - 1)^2 - 6 \\ &= [(x - 1) - \sqrt{6}][(x - 1) + \sqrt{6}] \end{aligned}$$

EXERCISES FOR SECTION 1.8

In Exercises 1 through 20, simplify each number. (See Example 1.)

1. $9^{1/2}$	2. $9^{-1/2}$	3. $8^{1/3}$	4. $8^{-1/3}$
5. $8^{2/3}$	6. $8^{-2/3}$	7. $8^{4/3}$	8. $8^{-4/3}$
9. $32^{1/5}$	10. $32^{2/5}$	11. $32^{3/5}$	12. $32^{4/5}$
13. $32^{6/5}$	14. $32^{-3/5}$	15. $81^{1/4}$	16. $81^{3/4}$
17. $81^{-2/4}$	18. $25^{3/2}$	19. $25^{-5/2}$	20. $(100{,}000)^{1/5}$

In Exercises 21 through 50, simplify each expression. (See Examples 2 and 3.)

21. $x^{1/3} \cdot x^{1/3}$
22. $x^{1/2} \cdot x^{1/3}$
23. $a^{2/3} \cdot a^{4/3}$
24. $\dfrac{x^{5/6}}{x^{1/3}}$
25. $(x^2y^{1/3})^3$
26. $(3x^{1/2})^2$
27. $\dfrac{(x^{1/2}y^{2/3})^6}{x^3y^3}$
28. $(x^{2/3})(6x^{1/3}y)$
29. $(x^{2/3})(6x^{1/3}y^{1/4})^2$
30. $(x^{1/3}y^{1/4}z^{2/5})^{1/2}$
31. $[(x^{1/2})^{1/2}]^{1/2}$
32. $\left[\left(\dfrac{1}{x}\right)^{-1/2}\right]^{1/2}$
33. $x^{1/2}(x^{1/2} - y)$
34. $(p^{1/3}q^{2/3})^{-1} \div p^{1/3}q^{2/3}$
35. $\dfrac{(a^{-1/2})^{-2}}{(b^{-1})^{-1/2}}$
36. $(r^{-1/2})^2(w^{3/4})^{4/3}$
37. $(x^{1/2}y^{2/3}z^{-2})^{-1}$
38. $(8x^3)^{2/3} \cdot (32x)^{2/5}$
39. $(x^{3/8}y^{3/4})^{-4}$
40. $(a^{3/2}b^{3/4})^{-1/3}$
41. $\sqrt{x} \cdot \sqrt[3]{x}$
42. $\sqrt[3]{a^2} \cdot \sqrt[4]{a^3}$
43. $\dfrac{\sqrt[4]{xy}}{\sqrt[3]{xy}}$
44. $\sqrt{2x^2y^3} \cdot \sqrt{3x^3y^3}$
45. $\sqrt{\sqrt{x}}$
46. $\sqrt[4]{\sqrt[3]{\sqrt{x}}}$
47. $\sqrt{4a^2b} \cdot \sqrt[4]{16a^4b}$
48. $\dfrac{\sqrt{2pqr^3}}{\sqrt[3]{2pqr^3}}$
49. $\sqrt{x+y} \cdot \sqrt[3]{x+y}$
50. $\dfrac{\sqrt{(x+y)^2}}{\sqrt{(x+y)^0}}$

In Exercises 51 through 54, use scientific notation to evaluate each number. (See Example 4.)

51. $\sqrt{0.000009}$
52. $\sqrt{16 \cdot 10^{-6}}$
53. $\sqrt[5]{32 \cdot 10^{-5}}$
54. $(\sqrt[3]{0.027})^4$

In Exercises 55 through 60, factor each expression. (See Example 5.)

55. $x^2 - 5$	56. $8 - x^2$	57. $a - b^2$
58. $2x - y^2$	59. $x - 4y^2$	60. $3x^2 - 2y^2$

In Exercises 61 through 68, perform the indicated operations and simplify. (See Example 6.)

61. $(\sqrt{x} + \sqrt{y})^2$
62. $(\sqrt{x} - \sqrt{y})^2$
63. $(\sqrt{a} - \sqrt{b})(\sqrt{a} + \sqrt{b})$
64. $(\sqrt{2a} - \sqrt{2b})(\sqrt{2a} + \sqrt{2b})$
65. $(x^{1/2} + 5)^2$
66. $\left(1 + \dfrac{1}{\sqrt{x}}\right)^2$
67. $\left(\sqrt{x} - \dfrac{1}{\sqrt{x}}\right)^2$
68. $(1 + \sqrt{x})^3$

In Exercises 69 through 77, factor each expression by completing the square. (See Example 7.)

69. $x^2 - 2x - 4$
70. $x^2 - 2x - 1$
71. $x^2 - 2x - 3$
72. $x^2 + 4x + 1$
73. $x^2 + 4x + 2$
74. $x^2 + 4x + 3$
75. $x^2 + 4x$
76. $y^2 - 5y + 1$
77. $y^2 - 5y + \frac{1}{4}$

Calculator Exercises

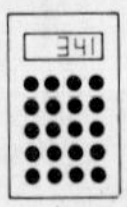

Using the y^x key evaluate each number to four decimal places.

78. $(943)^{1/10}$
79. $(.00006742)^{1/25}$
80. $(8)^{2.06}$
81. $(2)^{1/3}$
82. $(67)^{2/5}$
83. $(100)^{3/4}$
84. $\sqrt{47}$
85. $\sqrt{747}$
86. $\sqrt[5]{47}$
87. $\sqrt[10]{100}$
88. $\sqrt[3]{29}$
89. $\sqrt[4]{16.78}$
90. Many calculators are not as "smart" as we would like them to be. For instance, we know $\sqrt[3]{-8} = (-8)^{1/3} = -2$.
    a. How does your calculator perform when you "ask" for $\sqrt[3]{-8}$?
    b. We know $\sqrt[4]{81} - 3$ is equal to zero. What does your calculator get?

1.9 COMPLEX NUMBERS

Until now, the various operations and rules that have been developed have pertained to real numbers. In fact, most of the work in the following chapters centers around real numbers. However, there are occasions when we need to look deeper into the meaning of "number." We begin to do that in this section. When the principal nth root of a real number, $\sqrt[n]{x}$, was defined, the definition did not cover the case where n is even and $x < 0$. Thus, expressions like $\sqrt{-1}$, $\sqrt{-5}$, etc., have not been defined. There is *no* way to define $\sqrt{-1}$ to be a real number because if $\sqrt{-1}$ were real, then its square $(\sqrt{-1})^2$ would be equal to -1. But squares of real numbers are not negative. Since $\sqrt{-1}$ is *not* real, it might at first seem to be impossible to work with this symbol. However, we will use it to generate a new set of numbers called the set of **imaginary numbers**, denoted by J.

The Imaginary Numbers

J is the set of all numbers of the form $b\sqrt{-1}$, where b is a nonzero real number.

Thus an imaginary number is any nonzero real number multiple of $\sqrt{-1}$. Examples are $2\sqrt{-1}, -3\sqrt{-1}, \frac{1}{2}\sqrt{-1}$, etc. The name *imaginary* was first used in the 1800's, but such numbers were known of for hundreds of years before that. It is an unfortunate name because it seems to imply that such numbers are "fake" or "fictitious." Such is not the case, and in fact, we will show how to add, subtract, multiply, and divide imaginary numbers in this section.

Since $\sqrt{-1}$ is of basic importance, the symbol i is often used to represent it. That is, $i = \sqrt{-1}$. As a result, it is correct to say that an imaginary number is any number of the form bi, where b is a nonzero real number and $i = \sqrt{-1}$. The number i has the property that $i^2 = -1$. In the following example other powers of i are calculated by using the basic fact that $i^2 = -1$.

EXAMPLE 1 Find:

a. i^4 b. i^3 c. i^{20}

Solution

a. $i^4 = (i^2)^2 = (-1)^2 = 1$

b. $i^3 = i^2 \cdot i = -1 \cdot i = -i$

c. $i^{20} = (i^4)^5 = (1)^5 = 1$

The square root of *any* negative real number can be written in the *bi* form as is now demonstrated.

EXAMPLE 2 Express each of the following numbers in the *bi* form.

a. $\sqrt{-25}$ b. $\sqrt{-6}$ c. $\sqrt{-12}$

Solution

a. $\sqrt{-25} = \sqrt{25} \cdot \sqrt{-1} = 5i$

b. $\sqrt{-6} = \sqrt{6} \cdot \sqrt{-1} = \sqrt{6}\,i$

c. $\sqrt{-12} = \sqrt{4} \cdot \sqrt{3} \cdot \sqrt{-1} = 2\sqrt{3}\,i$

Consider a real number, a, added to the imaginary number bi. The result, $a + bi$, is neither real nor imaginary. We are led to the following.

The Complex Numbers

The set of **complex numbers,** $\mathbb{C}$, is the set of all such numbers of the form $a + bi$, where a and b are real numbers and $i = \sqrt{-1}$.

In the complex number $a + bi$, a is called the **real part** and bi is called the **imaginary part**. Examples of complex numbers are $3 + \sqrt{2}\,i$, $\sqrt{2} - i$, $-\frac{1}{2} + i$, $4 + 0i = 4$, $0 - 3i = -3i$, etc. We can view the system of real numbers, $\mathcal{R}$, as being included in the system of complex numbes, $\mathbb{C}$, because any real number a is a complex number of the form $a + 0i$. We can also view the system of imaginary numbers, J, as being included in the system of complex numbers because any imaginary number is a complex number of the form $0 + bi$, $b \neq 0$. Reflection over the discussions about numbers in Section 1.1 as well as the present section produces the following "number tree."

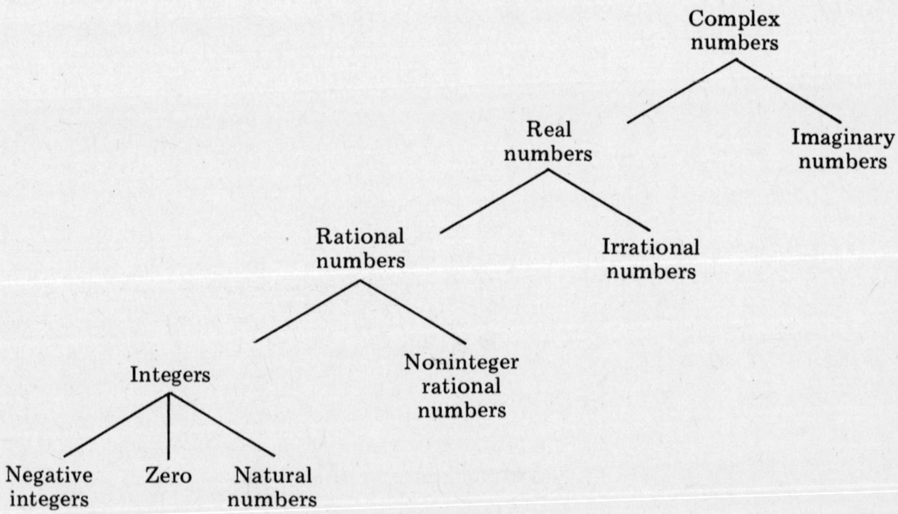

Next, more detail about the system of complex numbers is considered.

DEFINITION

The two complex numbers $a + bi$ and $c + di$ are said to be equal if and only if $a = c$ and $b = d$.

EXAMPLE 3 For what values of x and y are the given complex numbers equal?

a. $x + yi, -2 + 4i$ b. $2 + xi, 1 + x - yi$

Solution a. $x + yi$ will equal $-2 + 4i$ when, as prescribed by the definition above,

$$x = -2$$
$$y = 4$$

b. $2 + xi$ will equal $1 + x - yi$ when

$$2 = 1 + x$$
$$x = -y$$

Thus, by inspection, $x = 1, y = -1$.

In order for the complex numbers to be usable, it is necessary to define addition, subtraction, multiplication, and division.

DEFINITION

$$(a + bi) + (c + di) = (a + c) + (b + d)i$$
$$(a + bi) - (c + di) = (a - c) + (b - d)i$$
$$(a + bi)(c + di) = (ac - bd) + (ad + bc)i$$
$$\frac{a + bi}{c + di} = \frac{ac + bd}{c^2 + d^2} + \frac{bc - ad}{c^2 + d^2}i$$

Although we will not prove it here, the structure of the definitions above leads to the fact that the complex numbers obey the eleven axioms of the real number system that were discussed in Section 1.1. Furthermore, as the reader is asked to show in Exercise 60, the definitions are such that, when $b = 0$ and $d = 0$ (that is, imaginary parts become equal to zero) each definition gives a correct result about real numbers.

The definitions for addition and subtraction translate into words as: adding or subtracting complex numbers is performed by using the usual rules for removing parentheses and collecting like terms.

EXAMPLE 4 Perform the indicated operation.

a. $(2 - 4i) + (-1 + 6i)$ b. $(2 - 4i) - (-1 + 6i)$

Solution a. $(2 - 4i) + (-1 + 6i) = 2 - 4i - 1 + 6i$

$$= 2 - 1 - 4i + 6i$$
$$= 1 + 2i$$

b. $(2 - 4i) - (-1 + 6i) = 2 - 4i + 1 - 6i$
$= 3 - 10i$

Multiplication of complex numbers is easy to remember because, as the next example shows, it can be performed by treating $a + bi$ and $c + di$ as a product of binomials.

EXAMPLE 5 Perform the indicated operation.

a. $(a + bi)(c + di)$ b. $(6 - 2i)(\frac{1}{2} + 3i)$

Solution a. $(a + bi)(c + di) = ac + adi + bci + bdi^2$ by using foil

$$= ac + adi + bci - bd$$
$$= ac - bd + adi + bci$$
$$= (ac - bd) + (ad + bc)i$$

b. $(6 - 2i)(\frac{1}{2} + 3i) = 3 + 18i - i - 6i^2$

$$= 3 + 17i + 6$$
$$= 9 + 17i$$

Division of complex numbers is easy to remember because, as the next example shows, $\frac{a + bi}{c + di}$ can be performed by multiplying numerator and denominator by $c - di$. The complex number $c - di$ is called the **complex conjugate** of the complex number $c + di$ and vice versa.

EXAMPLE 6 Perform the indicated operation.

a. $\frac{a + bi}{c + di}$ b. $\frac{3 - 2i}{-1 + 4i}$

Solution a. $\frac{a + bi}{c + di} = \frac{(a + bi)(c - di)}{(c + di)(c - di)} = \frac{ac - adi + bci - bdi^2}{c^2 - cdi + cdi - d^2i^2}$

$$= \frac{ac - adi + bci + bd}{c^2 + d^2}$$
$$= \frac{ac + bd}{c^2 + d^2} + \frac{bc - ad}{c^2 + d^2}i$$

b. $$\frac{3-2i}{-1+4i} = \frac{(3-2i)(-1-4i)}{(-1+4i)(-1-4i)} = \frac{-3-12i+2i+8i^2}{1+4i-4i-16i^2}$$
$$= \frac{-3-10i-8}{1+16} = -\frac{11}{17} - \frac{5}{8}i$$

EXERCISES FOR SECTION 1.9

In Exercises 1 through 10, determine each power of i. (See Example 1.)

1. i^2 2. i^3 3. i^4 4. i^5 5. i^6
6. i^7 7. i^8 8. i^9 9. i^{10} 10. i^{70}

In Exercises 11 through 20, express each number in the bi form. (See Example 2.)

11. $\sqrt{-36}$ 12. $\sqrt{-49}$ 13. $\sqrt{-100}$ 14. $\sqrt{-2}$ 15. $\sqrt{-8}$
16. $\sqrt{-16}$ 17. $\sqrt{-20}$ 18. $\sqrt{-21}$ 19. $\sqrt{-28}$ 20. $\sqrt{-30}$

In Exercises 21 through 26, determine the values of x and y that will make the given complex numbers equal. (See Example 3.)

21. $\frac{1}{2} - 4i,\ x + yi$ 22. $\frac{1}{2} - 4i,\ x - yi$
23. $x + 2i,\ 3 + yi$ 24. $-x - 2i,\ 3 - yi$
25. $1 + x - 2i,\ 3 - 2yi$ 26. $x - y + i,\ 1 + yi$

In Exercises 27 through 52, perform the indicated operation. Write your answer in the $a + bi$ form. (See Examples 4, 5, and 6.)

27. $(3 + 2i) - (2 + i)$ 28. $(2 - i) + (4 + 5i)$
29. $\frac{1+i}{1-i}$ 30. $(3 + 4i) + (-2 + i)$
31. $(3 + i) - (3 + i)$ 32. $(4 + i) - (3 + i)$
33. $(2 + 0i)(4 - i)$ 34. $2(2 + i) + 3(-2 + 4i)$
35. $4(2 + i)$ 36. $(5 - \frac{1}{2}i) - (\frac{1}{2} + 3i)$
37. $i(2 + i)$ 38. $\frac{2+i}{6-i}$
39. $(4 + i)(2 + i)$ 40. $i(2 + i)(3 - 4i)$
41. $\frac{4+i}{2+i}$ 42. $\frac{a+bi}{2-i}$
43. $(3 - 2i)^2$ 44. $\left(1 + \frac{1}{\sqrt{2}}i\right)^2$
45. $(3i)(4i)$ 46. $(3i)(1 + 4i)$
47. $(1 + i)^3$ 48. $(1 + i)^4$
49. $(x + yi)(x - yi)$ 50. $(2 + 3i) - 6i$
51. $(\sqrt{2} + \sqrt{3}\,i)(\sqrt{2} - \sqrt{3}\,i)$ 52. $\frac{\sqrt{2} + \sqrt{3}\,i}{\sqrt{2} - \sqrt{3}\,i}$

In Exercises 53 through 59, answer true or false.

53. Every real number is a complex number.
54. Every complex number is a real number.
55. Every complex number is an imaginary number.
56. Every integer is a complex number.
57. A real number is a complex number whose imaginary part is zero.
58. An imaginary number is a complex number whose real part is zero.
59. The product of any complex number and its complex conjugate is a real number.
60. In the definitions for adding, subtracting, multiplying, and dividing complex numbers let $b = 0$ and $d = 0$ and show that each becomes a correct result for real numbers.

CHAPTER 1 IN REVIEW

I. Summary

This chapter began with a discussion of different number systems, the counting numbers, the integers, the rational numbers, the irrational numbers, and the real numbers. There are eleven axioms for the system of real numbers upon which much of algebra is built.

Order relations and absolute value were discussed in Section 1.2 and we saw that the distance between the two real numbers, x_1 and x_2, on the real number line is equal to $|x_1 - x_2| = |x_2 - x_1|$.

The laws of exponents, as they apply to integers, were the topic of Section 1.3 and we used those laws in conjunction with scientific notation. Positive integral exponents were used to state the definition of a polynomial in Section 1.4, after which addition, subtraction, multiplication, and division of such expressions were performed. When a polynomial in the single variable x is divided by a factor of the form $x - c$, synthetic division is a great time saver.

In Section 1.5 various techniques of factoring were viewed: removal of a common factor, factoring by grouping, factoring polynomials of the form $ax^2 + bx + c$, factoring the difference of two squares, factoring the sum and difference of two cubes. These techniques were all seen to be useful in finding the LCM which was needed to add and subtract fractions in Section 1.6.

The laws of radicals were investigated in Section 1.7 and the investigation continued in Section 1.8, where rational exponents were defined and used.

In Section 1.9 we returned to an investigation of number systems —this time, imaginary and complex numbers. The basic arithmetic operations of adding, subtracting, multiplying, and dividing were defined for complex numbers.

II. Vocabulary/Symbol True–False Quiz

If the statement is always true, write "TRUE." However, if the statement can be false, replace the underlined word or phrase with a different word or phrase that will make the new statement true.

1. Every real number is a rational number.
2. Every real number is a complex number.
3. The expression $\frac{x}{x-1}$ is undefined if $x = \underline{0}$.
4. $0.\overline{41}$ is a rational number.
5. For any real number x, $\sqrt{x^2} = \underline{x}$.
6. $3^4 \underline{\leqslant} 3^4$.
7. $(x + 3)^2 = \underline{x^2 + 9}$.
8. $\sqrt{x^2 + 6x + 9} = \underline{|x + 3|}$.
9. $\sqrt{-100} = \underline{10i}$.
10. $\frac{1}{b} + \frac{2}{d} = \underline{\frac{3}{b+d}}$.

III. Chapter Test

1. Write 0.00084 in scientific notation.
2. Evaluate: a. $|3 - 14|$ b. $|3| - |14|$
3. Simplify: a. $3 - [x - 2(x - 4)]$ b. $14 - x[x - 2(x + 1)]$

In questions 4 through 6 perform the indicated operation.

4. $x(2x - 1)(x + 4)$
5. $x^2 - 3x + 4 - 2(x - 1) + x(x + 2)$
6. $\dfrac{2y^7 + 3y^6 + y^2}{y^2}$
7. If $P(x) = x^3 - 3x^2 + 1$, find a. $P(-3)$ b. $P(0)$

In questions 8 through 12, factor each expression, if possible.

8. $x^2 + 7x + 12$
9. $t^8 - 1$
10. $3x^2y^2 - 3y^2$
11. $4x^2 + 12x + 9$
12. $4x^2 - 9y^2$

In questions 13 and 14, perform the indicated operation and express the answer in lowest terms.

13. $\dfrac{x}{x^2 - 7x + 12} - \dfrac{4x}{x^2 - 9}$
14. $\dfrac{\frac{1}{x} - \frac{1}{y}}{\frac{1}{x} + y}$
15. Use synthetic division to find the quotient and remainder:

$$3x^3 - x + 4 \div x + 1$$

16. Perform the indicated operation and simplify:

$$\sqrt{8x} + \sqrt{32x} - \sqrt{128x}$$

17. Simplify $\left(\dfrac{x^{1/3}y^{1/2}}{z}\right)^6$

18. Simplify $i^4 + i^2 + i^{102}$

In questions 19 and 20, perform the indicated operation and simplify.

19. $(3 + 2i)(4 - i)$

20. $\dfrac{2 - i}{2 + i}$

IV. Student Research Project

One irrational number, π (first seen on page 3), has a rather interesting history. Although proven to be irrational in 1767, an attempt was made by one state to legislate the value of π. (See below.)

1. Find ten circular objects and measure the diameter and circumference of each. What pattern do you find?
2. Prepare a report of about 100 words on the history of pi. You may want to consult the following works:

Eves, Howard. *An Introduction to the History of Mathematics*, 4th ed. Holt, Rinehart and Winston, New York, 1976.

Schepler, H. C. "The Chronology of π." *Mathematics Magazine*, January/February, March/April, May/June, 1950, pp. 165-170, 279-283, 216-228.

"House Bill Number 246, Indiana State Legislature, 1897," Proceedings of the Indiana Academy of Science, volume 45 (1935), pp. 206-210.

A table of π to 4,000 decimal places appears in *The Lore of Large Numbers* by P. J. Davis, Random House, New York, 1961.

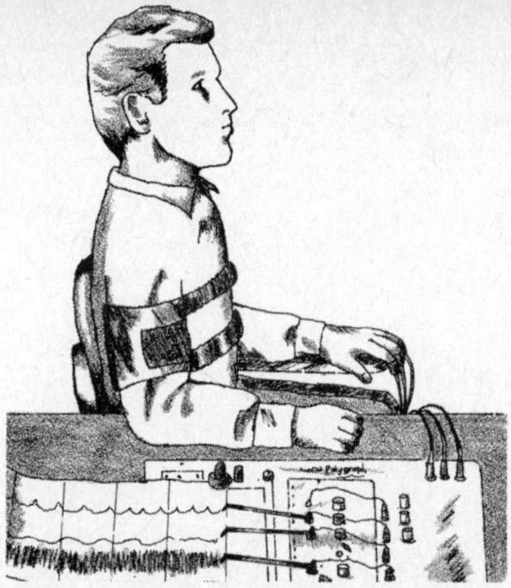

CHAPTER **2**

Equations, Inequalities, and Applications

When someone takes a lie-detector test, the changes in his breathing, heart rate and skin resistance tell the polygrapher when the suspect is lying. But, as David Lykken points out, innocent people also may appear guilty when doubt, fear and lack of confidence in the test increase their autonomic responses to incriminating questions. These emotions can generate shifts that make anyone look guilty. Lykken has devised a formula to explain how this works:

$$R = LF\left(\frac{1-C}{2}\right) + M.$$

An innocent person's autonomic response (R) is determined by (L) his basic emotional reactivity, (F) his fear of being found guilty, (C) his level of confidence in the test (ranging from −1, no confidence, to +1, complete confidence), and (M), his average response to irrelevant questions.

When an innocent suspect tells the truth, and is not afraid ($F = 0$) or generally reacts minimally ($L = 0$), then $LF = 0$, and

$$0\left(\frac{1-C}{2}\right) = 0$$

so that $R = M$. The innocent person will respond to an incriminating question just as he will to irrelevant queries. And if he is completely confident that the test will exonerate him, then $C = 1$ so

$$\frac{1-C}{2} = \frac{0}{2} = 0.$$

Here we are also left with R equaling M, and the innocent person won't respond dramatically to incriminating questions.

In both of these cases the polygrapher probably will report "deception not indicated," and the suspect will be cleared.

But if an innocent person is afraid, or he reacts emotionally to questions, or he is not confident that the test will exonerate him, his autonomic responses to incriminating questions will be greater (larger than M) than those to trivial queries. A polygrapher in this situation would be inclined to report "deception indicated."

The moral to the story is simple. If you're ever asked to prove your innocence by taking a lie-detector test, have faith and stay cool.

Margie Casady, "Formula for Fibbing," reprinted from Psychology Today, *March 1975, p. 59.*

2.1 LINEAR EQUATIONS

A statement of equality, an **equation**, is a way of expressing a *balanced* relationship between two expressions. Thus

$$4 + 9 = 13 \qquad x - 9 = 15 \qquad \text{and} \qquad x^2 = \sqrt{x + 1}$$

are equations; the latter two are called *equations in the variable* x. In the equation $x - 9 = 15$, notice that if the number 10 is substituted for x the statement of equality is false. However, if 24 is substituted for x, the statement is true—thus 24 is said to be a **solution** to $x - 9 = 15$. The collection of all solutions to an equation is called the **solution set.**

Notice that the equation $3y + 2 = 2 + 3y$ is true *regardless* of what value is substituted for y. Such an equation is called an **identity**. On the other hand, a **conditional equation** is an equation that is not true for every possible replacement of the variable.

When we speak of *solving* a conditional equation in algebra, we imply the determination of its solution set. This is often accomplished by using the following rules:

Rules for Solving Conditional Equations

1. Any number or expression can be added to or subtracted from *both sides* of an equation. (In symbols: $a = b$ if and only if $a + c = b + c$.)
2. Any expression that represents a nonzero quantity can be multiplied or divided on *both sides* of an equation. (In symbols: if $c \neq 0$, $a = b$ if and only if $ac = bc$.)

Applying these rules to an equation will always yield or produce an **equivalent equation.*** Generally, our goal is to obtain *simpler* equivalent equations as is the case in the examples that follow.

EXAMPLE 1 Solve $3x - 1 = 11$.

*Equivalent equations having the same solution set.

Solution The reasons for each step are provided at the right:

$$3x - 1 = 11$$
$$3x - 1 + 1 = 11 + 1 \qquad \text{Add 1 to both sides}$$
$$3x = 12$$
$$\tfrac{1}{3}(3x) = \tfrac{1}{3}(12)$$
$$x = 4 \qquad \text{Multiply both sides by } \tfrac{1}{3}$$

The solution to the last equation is quite obviously 4. Since each equation in the sequence is equivalent to it, 4 is the only solution to the original equation $3x - 1 = 11$. We can also say that the solution set of $3x - 1 = 11$ is $\{4\}$.

Finally as a safeguard, we *check* the solution:

Check:
$$3x - 1 = 11$$
$$3(4) - 1 = 11$$
$$12 - 1 = 11$$
$$11 = 11 \ \checkmark$$

EXAMPLE 2 Solve $2(3x - 1) = 4x + 5$.

Solution The reader should supply the reason for each step:

$$2(3x - 1) = 4x + 5$$
$$6x - 2 = 4x + 5$$
$$6x - 4x = 5 + 2$$
$$2x = 7$$
$$x = \tfrac{7}{2}$$

Check:
$$2(3x - 1) = 4x + 5$$
$$2(3 \cdot \tfrac{7}{2} - 1) = 4 \cdot \tfrac{7}{2} + 5$$
$$2(\tfrac{21}{2} - 1) = 14 + 5$$
$$19 = 19 \ \checkmark$$

Thus, the solution is $\frac{7}{2}$.

Each of the preceding examples involves solving a **linear equation.** By definition, a linear equation in x is any equation that can be written as

General Linear Equation

$$ax + b = 0 \qquad (a \neq 0) \qquad (1)$$

Employing the technique we have used of isolating the variable, the solution to $ax + b = 0$ is found in the following sequence of steps:

$$ax + b = 0$$
$$ax = -b$$
$$\frac{1}{a}(ax) = \frac{1}{a}(-b)$$
$$x = \frac{-b}{a} \qquad (2)$$

Equation (1) is referred to as the **standard form of a linear equation in a single variable.** Equation (2) is a formula for its solution. The letters a and b represent real numbers.

EXAMPLE 3 Use formula (2) to solve $3x + 4 = 5$.

Solution First we rewrite $3x + 4 = 5$ in the standard form $ax + b = 0$.

$$3x + 4 = 5$$
$$3x + 4 - 5 = 5 - 5$$
$$3x - 1 = 0$$

With $a = 3$ and $b = -1$ we obtain

$$x = \frac{-b}{a}$$
$$x = \frac{-(-1)}{3}$$
$$x = \tfrac{1}{3}$$

Check:

$$3x + 4 = 5$$
$$3(\tfrac{1}{3}) + 4 = 5$$
$$5 = 5 \ \checkmark$$

Hence, the solution is $\frac{1}{3}$.

We continue this section with examples that illustrate certain equations can have infinitely many solutions and other equations can have no solutions.

EXAMPLE 4 Use the technique of isolating the variable and try to solve

$$3x + 9 = 3(x - 4)$$

Solution

$$3x + 9 = 3(x - 4)$$

Expand

$$3x + 9 = 3x - 12$$

Subtract $3x$ from both sides

$$9 = -12$$

Since $9 = -12$ is NEVER true, its equivalent equation $3x + 9 = 3(x - 4)$ is *never true* (regardless of any real value substitution for x). We say that $3x + 9 = 3(x - 4)$ *has no solution* (or equivalently its solution set is the empty set, $\emptyset$.

EXAMPLE 5 Solve $8 - 2(y - 11) = y - 3(y - 10)$.

Solution

$$8 - 2(y - 11) = y - 3(y - 10)$$

Expand

$$8 - 2y + 22 = y - 3y + 30$$

Collect like terms

$$-2y + 30 = -2y + 30$$

Add $2y$ to both sides

$$30 = 30$$

We have $30 = 30$ being equivalent to the original equation. But since $30 = 30$ is ALWAYS true (regardless of any real number replacement for y), the solution set is $\mathcal{R}$, the set of all real numbers. Thus, any real number solves the equation $8 - 2(y - 11) = y - 3(y - 10)$; it is an identity.

We conclude this section with an equation that involves absolute value.

EXAMPLE 6 Solve the equation $|x - 1| = 7$.

Solution Recalling the definition of absolute value from Chapter 1, $|x - 1| = 7$ becomes equivalent to

$$x - 1 = 7 \quad \text{or} \quad x - 1 = -7$$
$$x = 8 \quad \text{or} \quad x = -6$$

Thus, $|x - 1| = 7$ is an equation equivalent to a *pair* of linear equations, whose solutions are respectively 8 and -6. So, the solution set for $|x - 1| = 7$ is $\{8, -6\}$.

The reader should verify that these solutions check in the original equation.

EXERCISES FOR SECTION 2.1

In Exercises 1 through 30, solve each equation. (See Examples 1, 2, 4, and 5.)

1. $3x - 1 = 14$
2. $2x - 7 = 9$
3. $5x - 12 = 13$
4. $7x + 9 = 30$
5. $3y + 13 = 4$
6. $2x + 11 = 3$
7. $2x - 8 = 7$
8. $9z - 5 = 8$
9. $4 - x = x + 1$
10. $3 - 2x = x$
11. $8 - 5x = 2x + 15$
12. $3t - 1 = \frac{1}{4}$
13. $\frac{2}{7}s - \frac{1}{7} = \frac{6}{7}$
14. $\frac{3}{4}x - 5 = \frac{1}{4}$
15. $8(2x - 1) = -8$
16. $2(x - 1) = x + 4$
17. $4 - 2(x + 1) = -8 - x$
18. $9 - 2x = 7(x + 41) + 1$
19. $0.3x - 0.2 = 2(x - 0.4)$
20. $0.7x - 1 = 0.3(x - 0.5)$
21. $5(3y - 1) = 2y + 34$
22. $1 - 5(x - 12) = -2 + 9(x - 7)$
23. $3(z - 1) = 3z - 1$
24. $8H - 4 = 8(H - 4)$
25. $2 - 3(x - 1) = 5 - 3x$
26. $10 - 4(x + 11) = x - 5(x + 7) + 1$
27. $5 - 2(x + 14) = x - 3(x + 10)$
28. $5 - 2(x + 14) = x - 3(x + 10) + 7$
29. $4 - 5[3 - 2(x - 1)] = 2x + 1$
30. $6 - 4[10 - 3(x + 7)] + x = \dfrac{2x}{3}$

In Exercises 31 through 40, use the formula on page 72 to solve each equation. (See Example 3.)

31. $3x + 7 = 13$
32. $4q + 5 = 1$
33. $-2t + 4 = 11$
34. $3z - 11 = 42$
35. $2x - 6 = x + 1$
36. $10t - 4 = 7t + 18$
37. $19y - 81 = 3y + 157$
38. $5(z + 1) = 35$
39. $5 - 2x = 9$
40. $5 - 2x = -9$

In Exercises 41 through 48, determine the solution set for each absolute value equation. (See Example 6.)

41. $|x - 4| = 11$
42. $|y - 7| = 9$
43. $|2x - 8| = 5$
44. $|3 - 2x| = 1$
45. $|7 - 2x| = 4$
46. $|7 - 2t| = 0$
47. $|3x + 8| = -4$
48. $|x - 1| = -7$

Hint: for 47 and 48 refer to the comments *following* the definition of absolute value in Chapter 1.

For questions 49 through 52, answer true or false.

49. If two equations are equivalent, then they have the same solution set.
50. An identity is an equation that is true for every possible replacement of the variable.

51. Any equation that can be written in the standard form $ax + b = 0$ is a linear equation ($a \neq 0$).
52. It is possible that the solution set of a conditional equation is $\emptyset$.

Computer Exercises

The General Linear Equation (1), $ax + b = 0$, can be solved using the following program:

```
 5 PRINT "INPUT A, B"
1Ø INPUT A, B
2Ø LET C = -B/A
3Ø PRINT "THE SOLUTION TO";A;"X + ";B;"=Ø"
4Ø PRINT "  IS ";C
5Ø STOP
```

53. Use the program to solve the following equations:
    a. $3x + 5 = 0$ b. $\frac{1}{5}x - \frac{2}{3} = 0$ c. $0.17x - 0.46 = 0$
54. Refine the above program to provide a "check" to see if A = Ø before statement 2Ø.

2.2 APPLIED PROBLEMS AND FORMULAS

The mechanics of solving equations are relatively abstract and to some people seem a bit "cut and dried." However, the real utility of algebra is in solving problems. Frequently, a quantitative problem can be symbolized and translated into an equation. Once the equation is solved, we have a solution to the problem.

In this text, problem solving is dissected into three main steps. They are:

Step 1. **Problem translation**—symbolize the unknown quantities and translate the problem into an equation (that is, an **algebraic model**).
Step 2. **Equation solution**—solve the equation.
Step 3. **Check**—make sure the answer obtained from step 2 checks in the *original statement* of the problem.

The applied problems in this section fall into three basic categories: business problems, motion problems, and mixture problems.

Business Problems

In the example that follows, the concept of **markup** is discussed. The retail price of an item, P, is the cost (to the dealer), C, plus a markup, M. In symbols,

$$P = C + M.$$

EXAMPLE 1 The cost of an AM–FM car radio is \$42. If the markup is 40% of the retail price what is the retail price?

Solution

STEP 1 Problem translation

Let P represent the retail price (in dollars). Then the markup is $.4P$. Since

$$P = C + M, \text{ we obtain}$$
$$P = 42 + 0.4P \leftarrow \text{This is our algebraic model.}$$

STEP 2 Equation solution

$$P = 42 + 0.4P$$
$$P - 0.4P = 42$$
$$0.6P = 42$$
$$P = \frac{42}{0.6}$$
$$P = 70$$

Thus the retail price is \$70.

STEP 3 Check

The retail price is \$70 and 40% of the retail price is \$28. The cost added to the markup (\$42 + \$28) *does* total the retail price (\$70). ✓

EXAMPLE 2 Thirty members of a fraternity attended a concert where two types of tickets were sold—"turf" and "bleachers." Turf tickets sold for \$6.50 and bleacher tickets sold for \$5.00. If the total cost for the thirty members was \$187.50, how many purchased bleacher tickets and how many purchased turf tickets?

Solution

STEP 1 Problem translation

First, we must decide what the *unknown* is, that is what we are asked to find. The unknown is shaded above—it is the number of tickets of each type of seat.

Let x be the number of turf tickets.

Since there are 30 in all, $30 - x$ represents the number of bleacher tickets purchased. Thus,

$$\text{total for turf tickets} = 6.5 \cdot x$$
$$\text{total for bleacher tickets} = 5(30 - x)$$

$$\underbrace{\text{Cost of turf tickets}}_{6.5x} + \underbrace{\text{Cost of bleacher tickets}}_{5(30-x)} = \underbrace{\text{Total Cost (in dollars)}}_{187.5} \leftarrow \text{This is our algebraic model.}$$

STEP 2
Equation solution

$$6.5x + 5(30 - x) = 187.5$$
$$6.5x + 150 - 5x = 187.5$$
$$1.5x = 37.5$$
$$x = 25$$

Hence, the fraternity purchased 25 turf tickets and $30 - x = 5$ bleacher tickets.

STEP 3
Check

25 turf tickets at \$6.50 each =	\$162.50
5 bleacher tickets at \$5.00 each =	25.00
Total for 30 tickets =	\$187.50 ✓

The next example of a business problem involves the concept of **simple interest**. The simple interest, I, for an investment (or principal) of P dollars at an annual rate r for t years is

$$I = Prt$$

For example, if \$600 is invested at an annual rate of 5%, the amount of simple interest earned after two years is

$$I = Prt$$
$$I = \$600 \cdot 0.05 \cdot 2$$
$$I = \$60$$

EXAMPLE 3 Betty Jo invested \$3000 for one year, part at 6% and part at 5%. If the simple interest for one year from the 6% investment exceeded the simple interest for one year from the 5% investment by \$92, how much did she invest at the 6% rate?

Solution

STEP 1
Problem translation

Let x represent the amount (in dollars) invested at 6%. Then $3000 - x$ is the amount invested at 5%. If we denote by I_6 the simple interest for one year from the 6% investment, then

$$I_6 = 0.06x$$

Similarly,

$$I_5 = 0.05(3000 - x)$$

Also, we have

$$I_6 - I_5 = 92$$

because the interest from the 6% investment exceeded the interest from the 5% investment by $92.

$$\begin{array}{ccccc} I_6 & - & I_5 & = & 92 \\ \downarrow & & \downarrow & & \\ 0.06x & - & 0.05(3000-x) & = & 92 \end{array} \leftarrow \text{This is our algebraic model.}$$

STEP 2
Equation solution

$$\begin{aligned} 0.06x - 0.05(3000-x) &= 92 \\ 0.06x - 150 + 0.05x &= 92 \\ 0.11x &= 242 \\ x &= 2200 \end{aligned}$$

Thus, Betty Jo invested $2200 at 6%.

STEP 3
Check

$2200 at 6% yields $132 simple interest for one year. The remainder, $800 at 5% yields $40. $132 exceeds $40 by $92. ✓

Motion Problems

The equation

$$d = rt$$

is a formula for the distance, d, travelled by an object in straight-line (or *rectilinear*) motion where r is the constant rate of travel and t represents the time of travel. Typical units of measure for d, r, and t appear below:*

	Metric			*English*		
distance	*centimeters (cm)*	*meters (m)*	*kilometers (km)*	*feet (ft)*	*inches (in)*	*miles (mi)*
rate	cm/sec	m/sec	km/hr	ft/sec	in/sec	mi/hr
time	sec	sec	hr	sec	sec	hr

*Metric and English units as well as conversions between them are discussed in Appendix 2.

EXAMPLE 4 Two long-distance runners, Hal and Corine, run the same course with Hal travelling 2 mi/hr slower than Corine. If Corine finishes the course in 2 hours and 24 minutes and Hal takes 3 hours to finish, what is the speed of each runner and the length of the course?

Solution

STEP 1 Problem translation

Let r represent Corine's rate. Then $r - 2$ will represent Hal's rate. Denoting Corine's distance by d_C and Hal's distance by d_H, we have

$$d_C = r \cdot t = r \cdot 2.4 = 2.4r \qquad (2 \text{ hrs } 24 \text{ min} = 2.4 \text{ hr})$$
$$d_H = (r-2)t = (r-2)3 = 3r - 6$$

Also, $d_C = d_H$ (they run the same course):

$$2.4r = 3r - 6 \leftarrow \text{This is our algebraic model.}$$

STEP 2 Equation solution

$$2.4r = 3r - 6$$
$$-0.6r = -6 \qquad \text{Add } -3r \text{ to both sides.}$$
$$r = 10$$

Thus, Corine's rate is 10 mi/hr, Hal's rate is 8 mi/hr and the length of the course is

$$d_C = rt = 10 \cdot 2.4 = 24$$

Hence, the course length is 24 miles.

STEP 3 Check

Hal's rate of 8 mi/hr is 2 mi/hr slower than Corine's. Hal's distance is $8 \cdot 3 = 24$ miles; Corine's is $10 \cdot 2.4 = 24$ miles. Everything checks. ✓

The next example of a distance problem shows how a sketch can be utilized as an aid in translating a problem.

EXAMPLE 5 Two backpackers, Andy and Jorge, leave their house at noon. Andy travels due east at a uniform rate that is 3 km/hr faster than Jorge, who is travelling west. After 4 hours, they are 60 kilometers apart. Find the rate of each backpacker.

Solution

STEP 1 Problem translation

The following sketch provides a *pictorial model* and helps to summarize the given information. We use d_A for Andy's distance and d_J for Jorge's:

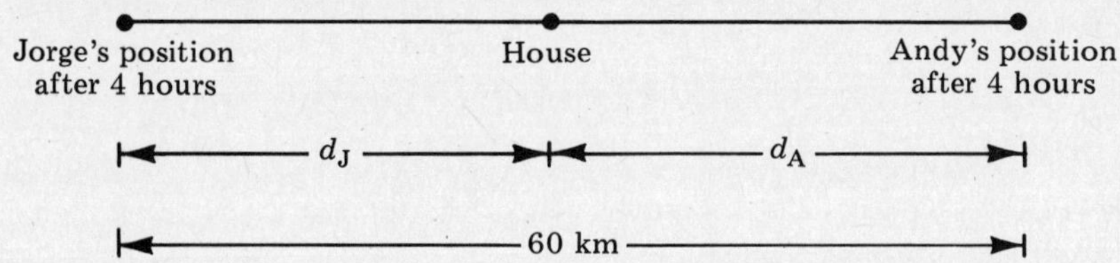

Let r represent Andy's rate. Then, $r - 3$ is Jorge's rate. Now, since the *sum* of Andy's and Jorge's distances must be 60, we have

$$d_A + d_J = 60$$

$$r \cdot 4 + (r - 3)4 = 60 \leftarrow \text{This is our algebraic model.}$$

STEP 2 Equation solution

$$4r + (r - 3)4 = 60$$
$$4r + 4r - 12 = 60$$
$$8r - 12 = 60$$
$$8r = 72$$
$$r = 9$$

Thus, r, Andy's rate, is 9 km/hr and Jorge's rate, $r - 3$, is 6 km/hr.

STEP 3 Check

Andy, hiking at the rate of 9 km/hr for 4 hours, travels 36 km. Jorge travels $6 \cdot 4 = 24$ km. Hence, after 4 hours they are $36 + 24 = 60$ km. apart. ✓

Mixture Problems

The two examples that follow involve the notion of *mixing* or blending two types of products to produce a third type of product of intermediate value. We will use

$$\underset{\text{value of one unit}}{V} \cdot \underset{\text{number of units}}{n} = \underset{\text{total value}}{T}$$

EXAMPLE 6 A nutritionist wishes to mix two types of nuts as a trailmix for hikers. One type sells for \$1 per pound and the more expensive type sells for \$1.25 per pound. She wants to combine the two to create a mixture that will sell for \$1.15 per pound. How much of the \$1.25 nuts must be added to 10 pounds of the cheaper nuts so that the value of the new mixture is equal to the total value of the other two?

Solution

STEP 1 Problem translation

Let x represent the number of pounds of the \$1.25 nuts needed. Now, we convert the information to table form:

	Value of one pound ·	*Number of pounds* =	*Total value*
Cheaper nuts	1.00	10	10
More expensive nuts	1.25	x	
Mixture	1.15		

Since there are 10 pounds of the \$1 nuts and x pounds of the \$1.25 nuts, there will be $x + 10$ pounds of the mixture. Also, notice that we multiply the first two columns to get the third:

	V ·	n =	T
Cheaper nuts	1.00	10	10
More expensive nuts	1.25	x	$1.25x$
Mixture	1.15	$x + 10$	$1.15(x + 10)$

We want the value of the mixture to be equal to the sum of the values of the other two. Therefore:

$$10 + 1.25x = 1.15(x + 10) \leftarrow \text{This is our algebraic model.}$$

STEP 2 Equation solution

$$10 + 1.25x = 1.15x + 11.5$$
$$.10x = 1.5$$
$$x = 15$$

Thus, 15 pounds of the \$1.25 nuts are needed.

STEP 3 Check

The 10 pounds of the \$1 nuts would sell for a total of \$10.00.

The 15 pounds of the \$1.25 nuts would sell for a total of \$18.75.

The 25 pounds (10 + 15) of mixture at \$1.15 per pound would sell for \$28.75.

$10 + 18.75 = 28.75$ ✓

EXAMPLE 7 Diammonium phosphate is a chemical that is added to water to make a fire retardant solution used in fighting forest fires. Suppose a 100-gallon solution that is 15% diammonium phosphate must be made into a 35% solution. How much pure diammonium phosphate must be combined with the weaker solution?

Solution

STEP 1 Problem translation

Let x = amount of diammonium phosphate (in gallons) needed. We complete the table below with "value" being interpreted as percent of diammonium phosphate in each solution.

	V ·	n =	T
	Percent of diammonium phosphate (in decimal form) ·	*Number of gallons* =	*Total gallons of diammonium phosphate*
Original solution	0.15	100	15
Pure diammonium phosphate	1.00	x	x
New solution	0.35	$100 + x$	$0.35(100 + x)$

Thus, our equation is

$$15 + x = 0.35(100 + x) \leftarrow \text{This is our algebraic model.}$$

STEP 2 Equation solution

$$15 + x = 0.35(100 + x)$$

$$15 + x = 35 + .35x$$

$$0.65x = 20$$

$$x = \frac{400}{13} \approx 30.77$$

Therefore, to get the desired 35% solution, we should combine 30.77 gallons of pure diammonium phosphate with the 100 gallons of 15% solution.

STEP 3 Check

After combining the 30.77 gallons of pure diammonium phosphate and 100 gallons of the 15% solution, there is 130.77 gallons of which 45.77(15 + 30.77) is diammonium phosphate. $\frac{45.77}{130.77} \approx 35\%$. ✓

In the remainder of this section we turn our attention to **literal equations**. By a literal equation we mean an equation involving two or more variables. A **formula** is a literal equation stating a general fact, rule, or principle symbolically. The reader is probably familiar with some of these formulas from the study of geometry:

$A = lw$	(area of a rectangle)
$C = 2\pi r$	(circumference of a circle)
$A = \frac{1}{2}bh$	(area of a triangle)
$P = 2l + 2w$	(perimeter of a rectangle)
$A = \pi r^2$	(area of a circle)

These will be used in various places as the need arises. Two other formulas, $d = rt$ and $I = Prt$ have been used already in this section. In virtually every field there exist algebraic models or formulas, as representations of various phenomena.

The next example illustrates how we can use formulas and rewrite them by *solving explicitly* for one of the variables.

EXAMPLE 8 Symmetrically tapered pieces of wood are often used for furniture legs, arms, etc. A formula used to determine the taper, T (in inches per foot) of each side is $T = \dfrac{6(W - X)}{L}$, where W, X, and L (measured in inches) are indicated in the diagram below.

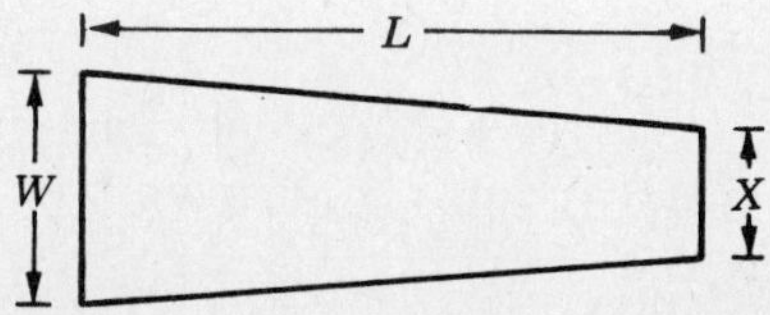

a. Determine the taper if $W = 5$ inches, $X = 2$ inches, and $L = 24$ inches.
b. Solve the formula for X.
c. Determine X, if $W = 4$ inches, $L = 36$ inches and $T = \frac{1}{4}$ inch per foot.

Solution a. With $W = 5$, $X = 2$, and $L = 24$, we substitute into the formula:

$$T = \frac{6(W - X)}{L}$$

$$T = \frac{6(5 - 2)}{24}$$

$$T = \frac{3}{4}$$

Thus, the taper is $\frac{3}{4}$ inches per foot.

b.

$$T = \frac{6(W - X)}{L}$$

Multiply both sides by $\frac{L}{6}$

$$\frac{TL}{6} = W - X$$

Add X to both sides

$$X + \frac{TL}{6} = W$$

Subtract $\frac{TL}{6}$ from both sides

$$X = W - \frac{TL}{6}$$

c. Here, with $W = 4$, $L = 36$ and $T = \frac{1}{4}$, we now use the formula developed in part b.

$$X = W - \frac{TL}{6}$$

$$X = 4 - \frac{(\frac{1}{4})(36)}{6}$$

$$X = 4 - \tfrac{3}{2}$$

$$X = 2\tfrac{1}{2}$$

Hence, the distance X is equal to $2\frac{1}{2}$ inches.

The final example in this section illustrates *how* to construct a formula from certain facts.

EXAMPLE 9 At Friendly Felix Auto Sales, salespersons are salaried and also work on a commission basis. The regular salary is \$4 per hour and they also receive 1% of all gross sales. Let S represent the total income of a salesperson, h represent the number of hours worked, and g represent the gross amount of sales. Construct a formula for the total income of a salesperson.

Solution

Total income	=	salary			+	commission	
		hourly rate	·	number of hours		1% of gross amount of sales	
↓		↓		↓		↓	↓
S	=	4	·	h	+	0.01 ·	g

Hence, $S = 4h + 0.01g$ is the desired formula.

EXERCISES FOR SECTION 2.2

1. The cost of a guitar is $80. If the markup is 60% of the retail price, what is the retail price?
2. If the markup is 48% of the retail price, what is the retail price for a CB radio that costs $61.62?
3. If the markup is 35% of the cost, what is the retail price of a pair of jeans that costs $10?
4. A small airline offers first-class and economy seats. For a certain flight, the first-class seats cost $190 each and the economy seats cost $155 each. If the total receipts for 13 tickets were $2120, how many of each type of seat were occupied?
5. Twenty members of a fraternity attended a concert where two types of tickets were sold—"orchestra" and "mezzanine." Orchestra tickets sold for $5.50 each and mezzanine tickets sold for $4.00 each. If the total cost for the twenty members was $89.00, how many sat in the orchestra seats?
6. Jack invested $2800 for one year, part at 6% and part at 5%. If the simple annual interest from the 6% investment exceeded the interest from the 5% investment by $80.00, how much was invested at the 6% rate?
7. Herbie invested $30,000 in two stocks for one year. One yielded annual interest at a rate of 11% and the other yielded annual interest at a rate of 5.5%. Find the amounts invested at each rate if:
   a. The amounts of annual income from each stock were equal.
   b. The amount of annual income from the 11% stock was four times that of the annual income from the 5.5% stock.
8. Two long-distance runners, Alice and Sean, run the same course with Sean travelling 2 mi/hr slower than Alice. Alice finishes the course in 3 hours, while it takes Sean 4 hours to finish. What is the speed of each runner and the length of the course?
9. Two bicyclists, Katie and Andy, leave their house at the same time. Katie travels south at a uniform rate that is 10 km/hr faster than Andy who is travelling north. After 5 hours they are 160 kilometers apart. Find the rate of travel for each cyclist.
10. A truck leaves San Francisco for Dixon, Illinois at 3 p.m., travelling at an average rate of 40 mi/hr. Two hours later another truck leaves, following the same route as the first, and travels at an average rate of 60 mi/hr. When will the second truck overtake the first?
11. Two horseback riders, on the same straight path 38.5 miles apart, begin to travel at noon toward each other at uniform speeds of 4.5 mi/hr and 6.5 mi/hr. When will they meet?
12. The owner of a coffee and spice shop wants to blend two types of coffee; the more expensive brand sells for $7.50 per pound and the cheaper brand sells for $6.00 per pound. If she wants a mixture of 15 pounds to sell for $6.50 per pound, how many pounds of each should she use?
13. A dietician desires to prepare a special diet containing 40% protein. He has two foods available, one containing 20% protein and the other containing 50% protein. The desired 40% mixture will be made by adding a certain amount of the 20% protein food to 400 grams of the 50% protein food. How many grams of the 20% protein food must be added?

14. Al notices that the 3-gallon solution in his radiator is only 20% antifreeze. How much pure antifreeze must be added so that the radiator will contain a 50% solution?
15. Suppose, after Al has added the amount of antifreeze called for in problem 14, he also desires to fill the radiator to its capacity of 5 gallons. Now, what should he do?
16. Harry had \$2.20 in nickels, dimes, and quarters. He had one more quarter than he had dimes and he had one more nickel than twice the number of quarters. How many coins of each type did he have?
17. Two backpackers left from the same point and travelled in the same direction on the Appalachian Trail. The first backpacker left one hour earlier than the second and travelled at 2 mi/hr. The second backpacker travelled at an average rate of 4 mi/hr. By 3:00 p.m., the second backpacker had overtaken the first and was 7 miles ahead. At what time did the first backpacker leave the starting point?
18. If the markup on a pair of jogging shorts is 25% of the retail price, what is the retail price for a pair of jogging shorts that costs \$8.50?
19. The retail price of a 4 x 8 sheet of plywood is \$20.95. The markup is 30% of the cost. How much is the cost?
20. How much pure gold must be added to a 30-kilogram alloy that is 15% gold in order to raise the percent of gold to 20%?
21. Rosanna has three types of bills in her wallet—singles, fives, and tens. She has three more five-dollar bills than ten-dollar bills and three times as many singles as five-dollar bills. How many bills of each type does she have if the total amount of money in her wallet is \$42?
22. Juan invested \$3,000 in two bank accounts; one paid interest at a 6% annual rate and the other at a $7\frac{1}{4}\%$ annual rate. If the annual interest from the 6% account exceeded the interest Juan made on the $7\frac{1}{4}\%$ account by \$15.70, how much did he invest in each account?
23. Two trains leave a station at the same time and travel in opposite directions. After 6 hours they are 720 miles apart. The faster train travels at an average rate that is 20 mi/hr more than the slower train. Find the average rate of each train.
24. When Sam invested his savings of \$10,000 in two businesses (call them A and B), he *lost* at a 4% annual rate on A, but *gained* at a 7% annual rate on B.
    a. If he invested an equal amount in each business, how much income did he have from his investment after one year?
    b. If he invested \$1,000 more in business A than business B, how much income did he have from his investment after one year?
25. Ms. Lombardi purchased \$13 worth of stamps. She bought 20 more 13¢ stamps than 9¢ stamps and twice as many 15¢ stamps as 9¢ stamps. How many of each denomination did she buy?
26. The Gonzales family attended the movies one evening with the cost of an adult ticket ("adult" means "over 12 years of age") being \$3.50, while a child's ticket cost \$1.75. If there are 6 people in the Gonzales family, and the total cost for all was \$19.25, how many adults and how many children are in the Gonzales family?

27. Melinda's boat can travel downstream in 1 hour and it takes her 2 hours to travel upstream the same distance. If the rate of the river is 7 mi/hr, how fast can Melinda travel in still water?
28. A pharmacist needs a 15% boric acid solution. How much pure boric acid solution must be added to 2 liters of a solution, that is presently 6.5% boric acid, to get the desired concentration?
29. If milk containing 1% butterfat is to be added to cream containing 35% butterfat to obtain 510 gallons of 2% butterfat milk, how much of each will be needed?
30. A florist sells daffodils for \$3 per dozen, petunias for \$3 per dozen, and roses for \$15 per dozen. He wishes to create 15 bouquets (consisting of 12 flowers each) made up of all three of these flowers, to sell for \$7 each. If each bouquet is to have the same number of daffodils as roses, how many dozen of each type of flower will be needed to make sure that the bouquets sell for the same total amount as the flowers would have sold for separately?
31. Harry Homeowner uses a solution that is 60% insecticide for his trees and shrubs. How much water and how much 60% solution should be mixed together to create 5 gallons of a 20% solution that can be used on his grass?
32. Kim is working in her lab. She has two solutions of hydrochloric acid—one is a 60% concentration, the other is a 20% concentration.
    a. How much of the 20% solution must be added to 10 liters of the 60% solution so that a 45% solution results?
    b. How many liters of the 45% solution will she now have?
33. A 16% acid solution is to be diluted by combining it with water. How much water must be combined with 100 ml of it to make a 12% solution?
34. In the field of fire-fighting hydraulics, the equation $h = \frac{1}{2}p + 26$ is a formula for the horizontal range, h (measured in feet), of water coming from a $\frac{3}{4}$-inch nozzle with nozzle pressure, p (measured in pounds per square inch).
    a. What is the horizontal range of a $\frac{3}{4}$-inch nozzle with a 60-pound-per-square-inch nozzle pressure?
    b. Solve the equation for p.
    c. Compute the nozzle pressure (p) for a $\frac{3}{4}$-inch nozzle whose horizontal range is 72 feet.
35. In the sport of scuba diving, the equation $P = 0.45D + 14.7$ is a formula for the pressure, P (measured in pounds per square inch) on a diver at a depth, D (measured in feet) in sea water.
    a. What is the pressure on a diver at a depth of 80 feet in sea water?
    b. Solve the equation for D.
    c. Compute the depth that a diver is at, if the pressure is 100 pounds per square inch.
36. Solid top decks on houses are generally built with a taper to allow for water run-off. A formula that gives the taper, T (in inches per foot) of the top is $T = \dfrac{12(W-X)}{L}$, where W, X, and L (measured in inches) appear in the diagram at the right.

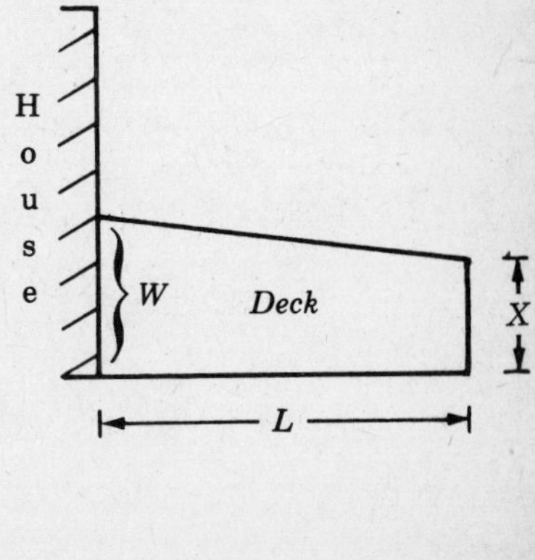

a. Determine the taper if $W = 24$ inches, $X = 22$ inches, and $L = 96$ inches.
b. Solve the formula for W.
c. Determine W if $X = 20$ inches, $L = 100$ inches, and $T = \frac{1}{4}$ inch per foot.

37. Solve $A = lw$ for w.
38. Solve $P = 2l + 2w$ for l.
39. Solve $d = rt$ for r.
40. Solve $A = \frac{1}{2}bh$ for h.
41. Solve $C = \pi d$ for d.
42. Solve $V = lwh$ for w.
43. Solve $v_2 = v_1 + at$ for t.
44. Solve $l = a + (n - 1)d$ for d.
45. Solve $l = a + (n - 1)d$ for n.
46. Solve $3x + 4y = 7$ for y.
47. Solve $ax + by = c$ for y. (Assume $b \neq 0$.)
48. Solve $ax + by = c$ for x. (Assume $a \neq 0$.)
49. A certain pediatrician believes that the number of hours, h, a child of age a (in years), should sleep each night is represented by the formula

$$h = 18 - 1.1a \qquad \text{(provided } a \leqslant 10\text{)}$$

a. Solve this formula for a.
b. Find a if $h = 8$.

50. The equation $E = 8LWH^2$ is an algebraic model for the energy, E (in foot-pounds), that can be expected from sea waves, where:

L is the length (in feet) of the wave (distance between successive crests),
H is the height (in feet) of the wave (vertical distance from trough to crest),
W is the width (in feet) of the wave.

a. Solve this equation for W.
b. If a certain wave that is 50 feet long and 8 feet high has 400,000 foot-pounds of energy, how wide is the wave?

51. A formula for converting liters, L, to quarts, q, is $L = 0.943q$.
a. Solve the formula for q. (The answer is a formula for converting quarts to liters!)
b. How many quarts are there in 2 liters?

52. A formula for converting kilometers, K, to miles, M, is $M = 0.621K$.
a. Solve the formula for K. (The answer is a formula for converting miles to kilometers!)
b. How many kilometers are there in 3 miles?

53. The relationship betweeen Fahrenheit (F) and Celsius (C) temperatures is given by the formula $5F - 9C = 160$.
a. Solve the formula for F.
b. Solve the formula for C.
c. Using the appropriate formula (either part a or b above), find the Fahrenheit equivalent to $100° C$.
d. Using the appropriate formula (either part a or b above), find the Celsius equivalent of $-10° F$.

54. The energy that an object has due to its motion is called **kinetic energy**. The equation $K = \frac{1}{2}mv^2$ is an algebraic model for the kinetic energy, K (in foot-pounds), that an object of mass, m (measured in pounds), has when it is travelling at a speed, v (measured in feet/second).
a. Solve this equation for m.
b. What is the mass of an object that has a kinetic energy of 200 foot-pounds and a speed of 9 ft/sec?

55. Solid state digital watches use a vibrating quartz crystal to regulate the recording of time. The formula $s = \dfrac{v}{32{,}768}$ is an algebraic model that gives the number of seconds elapsed, s, after v vibrations of the quartz crystal.
   a. After the quartz crystal vibrates 1,048,576 times, how many seconds have elapsed?
   b. Solve the formula for v.
   c. How many vibrations does the quartz crystal make in 1 second?
   d. How many vibrations does the quartz crystal make in 5 seconds?

56. Bob is a woodworker specializing in custom-made tables and chairs. He charges \$920 for each table and \$85 for each chair. If t represents the number of tables he makes and c represents the number of chairs, construct a formula for his earnings, E (in dollars).

57. a. The Ace Trucking Company pays apprentice loaders \$30 a day and journeymen loaders \$50 a day. Let P represent the total money paid out on a given day when there were a apprentice loaders working and j journeymen loaders working. Construct a formula for the total money paid out.
   b. Determine how much money is paid out by the Ace Trucking Company, if on a particular day they employ 15 apprentice loaders and 5 journeymen loaders.

58. The Mark IV Construction Company owns an apartment complex with one- and two-bedroom apartments. One-bedroom apartments will rent for \$295 per month and two-bedroom apartments will rent for \$345 per month. Let M represent the total monthly income from x one-bedroom apartments and y two-bedroom apartments. Construct a formula for the total monthly income.

59. For question 58, construct a formula for the *yearly* income, K, from the apartments.

60. The Car-U-Drive Auto Rental Agency charges \$8 per day plus 5¢ per mile for renting a Gremlin. Using d for number of days and m for number of miles, a formula for finding the cost of renting a Gremlin, C_G (in dollars), is:

$$C_G = 8d + 0.05m$$

For a Toyota, the company charges \$5 per day plus 20¢ per mile and the formula for C_T (cost of renting the Toyota) is:

$$C_T = 5d + 0.20m$$

   a. Suppose you'll need a car for 3 days to drive 500 miles. Which car has a better rental deal?
   b. How much will you save by taking the better deal?

2.3 LINEAR AND ABSOLUTE VALUE INEQUALITIES

In Chapter 1, we saw that the real number line can be used to display the real numbers since each point on the line corresponds to one and only one real number. The order relations $<, \leqslant, >, \geqslant$ were also defined and investigated. Recall,

$a < b$ means a is less than b

$a \leqslant b$ means a is less than or equal to b

$a > b$ means a is greater than b

$a \geqslant b$ means a is greater than or equal to b.

In the remainder of this section we will be concerned with solving and graphing inequalities. The basic process for solving an inequality is the same as for solving an equation—it involves constructing equivalent inequalities (inequalities having the same solution set) by using some basic principles. Those principles are highlighted in the following theorem:

THEOREM 1

Given that a, b, and c are real numbers, then:

1. if $a < b$, then $a + c < b + c$, and conversely
2. if $a < b$, then $ac < bc$, provided $c > 0$, and conversely
3. if $a < b$, then $ac > bc$, provided $c < 0$, and conversely

The proof of this theorem is outlined in Exercise 51. The reader is cautioned that part 3 of the theorem means that if both sides of an inequality are multiplied by a *negative* real number, the *sense of the inequality must be reversed.* Thus

$$-2 < 5 \text{ is equivalent to } 2 > -5$$

and

$$-x < 3 \text{ is equivalent to } x > -3$$

Before proceeding to some examples, we point out that the theorem remains true if *any* of the other order relations are substituted for "<." For example, if $a \geqslant b$, then $ac \geqslant bc$, provided $c > 0$, and conversely.

EXAMPLE 1 Solve the inequality: $2 - 3x \leqslant 14$.

Solution

$$2 - 3x \leqslant 14$$

Add -2 to both sides

$$-3x \leqslant 12$$

Multiply both sides by $-\frac{1}{3}$

$$x \geqslant -4$$

The solution set consists of *infinitely many* numbers—all those real numbers that are greater than or equal to -4. We graph that solution set below:

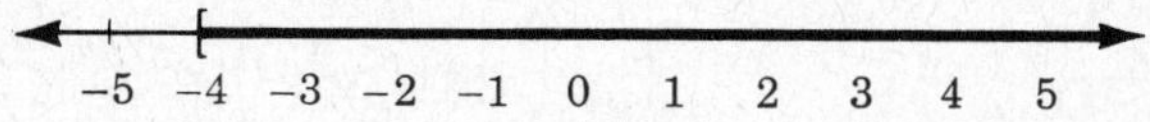

We use a bracket, [, to indicate that -4 is included in the solution set; if an end point is not included, we use a parenthesis, (.

For convenience, inequalities are sometimes written in **compound form**. For example,

$$-4 \leqslant 3x - 1 < 8$$

means

$$-4 \leqslant 3x - 1 \quad \text{and} \quad 3x - 1 < 8.$$

In the next example, we solve a compound inequality.

EXAMPLE 2 Solve $-4 \leqslant 3x - 1 < 8$.

Solution

$$-4 \leqslant 3x - 1 < 8$$

Add 1 to each member of the compound inequality

$$-3 \leqslant 3x < 9$$

Divide each member of the compound inequality by 3

$$-1 \leqslant x < 3$$

The graph of the solution set is:

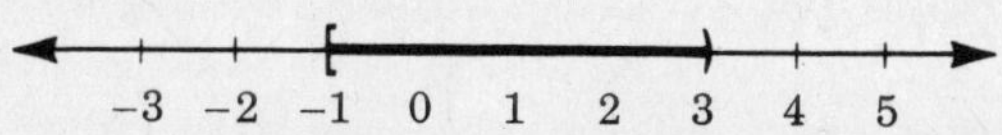

We use **interval notation** as a notational aid, occasionally. The above solution set, $\{x \mid -1 \leqslant x < 3\}$, can be rewritten in compact form as $[-1,3)$. Other variations appear below:

Interval Notation	Set Notation	Graph
$[a,b]$	$\{x \mid a \leqslant x \leqslant b\}$	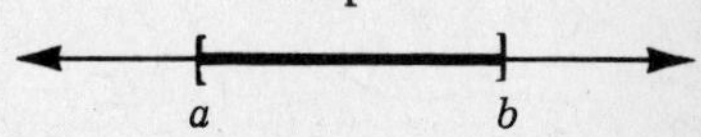
$(a,b]$	$\{x \mid a < x \leqslant b\}$	
(a,b)	$\{x \mid a < x < b\}$	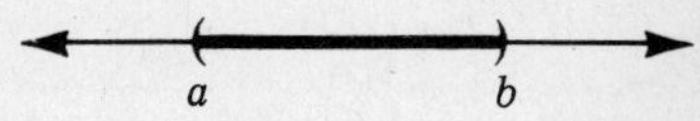

Also, we can have **unbounded** intervals:

$[a, \infty)$	$\{x \mid x \geqslant a\}$	
(a, ∞)	$\{x \mid x > a\}$	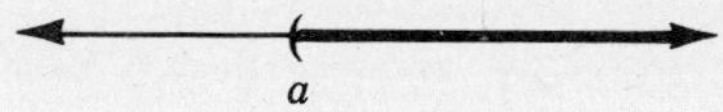
$(-\infty, b]$	$\{x \mid x \leqslant b\}$	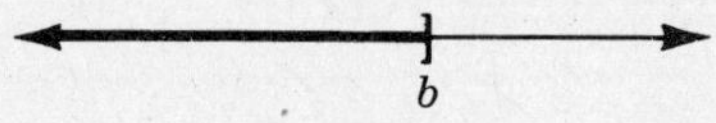
$(-\infty, b)$	$\{x \mid x < b\}$	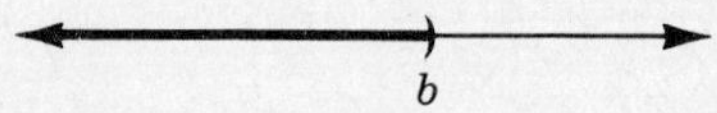

EXAMPLE 3 Solve $-4 < \dfrac{3 - 2x}{5} \leqslant 9$ and represent the solution using interval notation.

Solution

$$-4 < \frac{3-2x}{5} \leqslant 9$$

$$-20 < 3 - 2x \leqslant 45$$

$$-23 < -2x \leqslant 42$$

$$\frac{23}{2} > x \geqslant -21$$

or

$$-21 \leqslant x < \frac{23}{2}$$

Thus, the set of solutions is $[-21, \frac{23}{2})$.

The next example is an illustration of how inequalities can be applied in the business world. In order for a business to make a profit, its total revenue, R, must be greater than its total costs, C; that is, $R > C$. Often, after a business has researched its marketing and manufacturing processes, it may have equations on hand that relate revenue and cost to the number of items sold. When this is the case, it is possible to determine the minimum number of items that must be sold in order to make a profit.

EXAMPLE 4 At Sue's Geranium Villa the revenue equation is $R = 1.8x$, where x represents the number of geraniums sold per season. Also, the cost equation is $C = 120 + 1.5x$. How many geraniums must be sold per season in order to realize a profit?

Solution We must have $R > C$. Hence,

$$1.8x > 120 + 1.5x$$
$$0.3x > 120$$
$$x > 400$$

Thus, in order to make a profit, Sue's Geranium Villa must sell more than 400 geraniums per season.

Now, we shall examine inequalities involving absolute value. Recall from Chapter 1 that $|y| = 4$ can be graphically interpreted as those real numbers which are 4 units from 0. Similarly, $|y| \leqslant 4$ means that values of y must lie *within* 4 units of zero (inclusively) or equivalently, the distance, y, from 0 is *less than or equal* to 4 units:

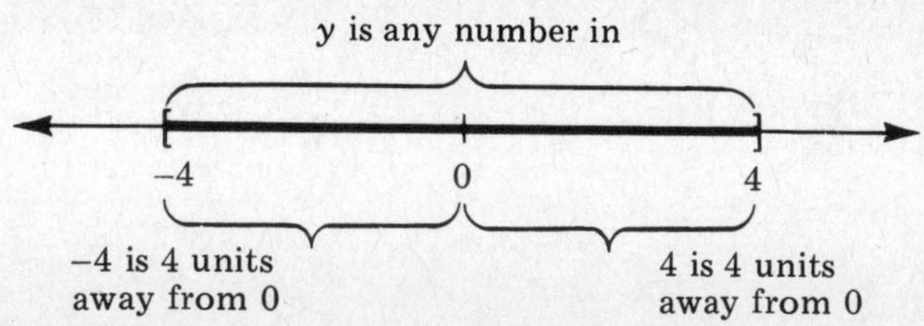

Thus, $|y| \leqslant 4$ means

$$-4 \leqslant y \leqslant 4.$$

In general, we have the following:

$$|a| \leqslant b \text{ if and only if } -b \leqslant a \leqslant b \qquad \text{(provided } b \geqslant 0) \qquad (1)$$

Also, since $|y| > 4$ represents all numbers on the real number line whose distance from 0 is *greater than* 4 units,

$$|y| > 4 \text{ means } y > 4 \text{ or } y < -4:$$

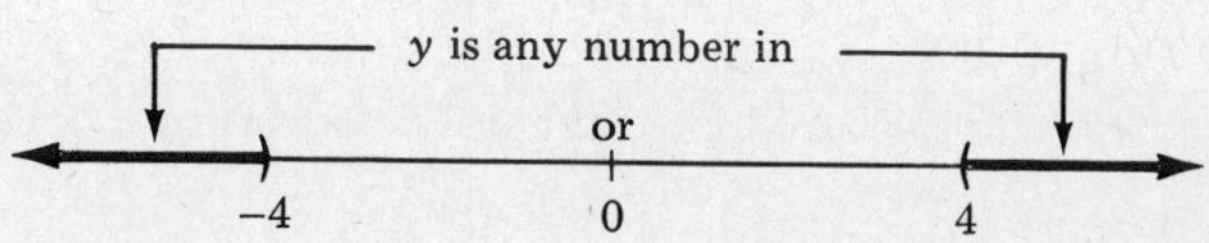

We have:

$$|a| > b \text{ if and only if } a > b \text{ or } a < -b \quad \text{(provided } b \geqslant 0) \qquad (2)$$

Other variations are:

$$|a| < b \text{ if and only if } -b < a < b \quad (\text{provided } b > 0) \tag{3}$$

$$|a| \geqslant b \text{ if and only if } a \geqslant b \text{ or } a \leqslant -b \quad (\text{provided } b \geqslant 0) \tag{4}$$

Some examples follow.

EXAMPLE 5 Solve $|2x - 5| < 9$.

Solution According to (3) we can rewrite $|2x - 5| < 9$ as

$$-9 < 2x - 5 < 9$$
$$-4 < 2x < 14$$
$$-2 < x < 7$$

In interval notation, we have: $(-2,7)$.

EXAMPLE 6 Solve $|3x + 4| \geqslant 19$.

Solution According to (4) we can rewrite $|3x + 4| \geqslant 19$ as

$$3x + 4 \geqslant 19 \quad \text{or} \quad 3x + 4 \leqslant -19$$
$$3x \geqslant 15 \quad \text{or} \quad 3x \leqslant -23$$
$$x \geqslant 5 \quad \text{or} \quad x \leqslant -\tfrac{23}{3}$$

Thus any real number greater than or equal to 5 or any real number less than or equal to $-\frac{23}{3}$ is a solution. Graphically, we have

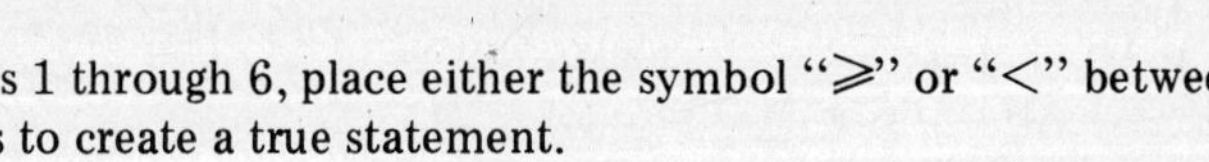

In interval notation we can write, $[5,\infty) \cup (-\infty, -\frac{23}{3}]$.

EXERCISES FOR SECTION 2.3

In Exercises 1 through 6, place either the symbol "$\geqslant$" or "$<$" between each pair of numbers to create a true statement.

1. 3 9
2. 2 −1
3. −2 1
4. 7 |−8|
5. |−3| |−5|
6. π 3.14

In Exercises 7 through 45: a) solve each inequality and b) express the set of solutions in interval notation. (See Examples 1, 2, 3, 5, and 6.)

7. $3 - 2x \leqslant 7$
8. $2x - 3 \leqslant 7$
9. $1 - x > 0$
10. $1 - x \leqslant 0$
11. $-x \geqslant 0$
12. $3y - 14 \leqslant 7y + 14$
13. $6 - x < 3x + 10$
14. $-1 - 2x > -3 - x$
15. $2(3x + 1) \geqslant -10$
16. $-2(5x + 7) \leqslant 3x - 14$
17. $3(x - 2) > -2(1 - x)$
18. $3 < x + 1 < 4$
19. $-2 < \dfrac{3x + 1}{6} \leqslant 19$
20. $0 \leqslant \dfrac{4z + 6}{2} \leqslant 1$
21. $0 \leqslant 3x + 1 < 10$
22. $-20 < 4x + 8 < -12$
23. $-5 < 1 - x < -2$
24. $-5 < x - 1 < -2$
25. $-7.3 \leqslant 2x + 3.1 < 14.8$
26. $|y - 1| < 2$
27. $|y - 1| \geqslant 2$
28. $|2 - x| \leqslant 7$
29. $|2 - x| > 7$
30. $|2x + 1| < 4$
31. $|2x - 5| < 9$
32. $|3t + 8| \geqslant 20$
33. $|-3t - 8| \geqslant 20$
34. $|3 - 2y| > 7$
35. $|z + 4| \geqslant 2$
36. $|2x + 3| < \sqrt{5}$
37. $\left|\dfrac{1}{4} - 3x\right| < 1$
38. $\left|\dfrac{2 - 7x}{4}\right| \geqslant 3$
39. $\left|\dfrac{x - 1}{6}\right| < \dfrac{1}{2}$
40. $\left|\dfrac{3 - x}{2}\right| < 0.2$
41. $|x + 10| \leqslant -1$
42. $|3x - 11| < -6$
43. $|3x - 11| \geqslant -6$
44. $|x + 1| \leqslant 0$
45. $|x + 1| < 0$

In questions 46 through 48, solve each problem. (See Example 4.)

46. The Home Grown Music Company manufactures records and knows that its revenue equation is $R = 3.2x$, where x is the number of records sold in a month. Also, their cost equation is $C = 180 + 2x$. How many records must be sold in order to make a profit?
47. At Shirley's Chrysanthemum Farm the revenue equation is $R = 2x$, where x is the number of chrysanthemums sold per season and the cost equation is $C = 100 + 1.5x$. How many chrysanthemums must be sold per season in order to make a profit?
48. A family-operated company makes custom blue jeans to sell in its own boutique shop. The revenue equation is $R = 10x$, where x is the number of pairs of jeans sold per week. The cost equation is $C = 210 + \frac{5}{4}x$. How many pairs of jeans must be sold in order to make a profit?
49. The relationship between Fahrenheit (F) and Celsius (C) temperatures is given by $F = \frac{9}{5}C + 32$. If the daily range of temperatures in Greely, Colorado was $-13^\circ F$ to $23^\circ F$, what was the corresponding Celsius temperature range?
50. A student made grades of 58, 61, 77, and 88 on the first four hourly exams. What grade must he obtain on the fifth hourly exam so that his average will be between 70 and 80?

51. a. To prove the first part of Theorem 1 on page 90, the following sequence of steps might be used:

Statement	Reason
1. a and b are real numbers and $a < b$	1. Given
2. $b - a$ is positive	2. Definition of $a < b$
3. $(b + c) - (a + c) = b - a$	3. The associative, distributive, and commutative laws of addition can be applied to $(b + c) - (a + c)$
4. $(b + c) - (a + c)$ is positive	4. Statements 2 and 3
5. Hence, $a + c < b + c$	5. Definition of "<"

Supply the reasons in the proof of the converse of the above: if a, b, c are real numbers and $a + c < b + c$ then $a < b$.

Statement	Reason
1. a, b, and c are real numbers and $a + c < b + c$	1. Given
2. $(b + c) - (a + c)$ is positive	2. Definition of "<"
3. $(b + c) - (a + c) = b - a$	3.
4. $b - a$ is positive	4.
5. Hence $a < b$	5.

b. Complete the reasons in the proof below of part 2 of Theorem 1 on page 90:

Statement	Reason
1. a, b, and c are real numbers $a < b$ and $c > 0$	1. Given
2. c is positive	2.
3. $b - a$ is positive	3.
4. $(b - a)c$ is positive	4.
5. $(b - a)c = bc - ac$	5.
6. $bc - ac$ is positive	6.
7. Hence, $ac < bc$	7.

c. Complete the reasons in the proof below of the converse of part 2 of Theorem 1 on page 90:

Statement	Reason
1. a, b, and c are real numbers and $ac < bc$ and $c > 0$	1. Given
2. $bc - ac$ is positive	2. Definition of "<"
3. $bc - ac = (b - a)c$	3.

4. $(b-a)c$ is positive 4.
5. c is positive 5.
6. $b-a$ is positive 6.
7. Hence, $a < b$ 7.

d. Prove part 3 of Theorem 1 on page 90.
e. Prove part 3 (converse) of Theorem 1 on page 90.

2.4 QUADRATIC EQUATIONS AND APPLICATIONS

In this section, we will examine two methods for solving equations of the form

$$ax^2 + bx + c = 0 \qquad (a \neq 0). \tag{1}$$

Such equations are called **quadratic equations.** The a, b, and c represent real numbers, and equation (1) is referred to as the **standard form of a quadratic equation in a single variable.** In the first two examples that follow, we employ factoring (see Section 1.5).

EXAMPLE 1 Solve $x^2 - 9x - 10 = 0$.

Solution First, we factor the left-hand side of the equation:

$$x^2 - 9x - 10 = 0$$
$$(x-10)(x+1) = 0$$

Here, we use a fact about real numbers: if $ab = 0$, then $a = 0$ or $b = 0$.* Thus, the original equation will be solved when either factor is equal to 0:

$$(x-10)(x+1) = 0$$
$$x - 10 = 0 \quad \mid \quad x + 1 = 0$$
$$x = 10 \quad \mid \quad x = -1$$

Notice, there are *two* solutions (or **roots**); the solution set, verified by the check on the following page, is $\{10, -1\}$.

*See Laws about Zero in Section 1.1.

Check for $x = 10$:

$$x^2 - 9x - 10 = 0$$
$$(10)^2 - 9(10) - 10 = 0$$
$$100 - 90 - 10 = 0$$
$$0 = 0 \checkmark$$

Check for $x = -1$:

$$x^2 - 9x - 10 = 0$$
$$(-1)^2 - 9(-1) - 10 = 0$$
$$1 + 9 - 10 = 0$$
$$0 = 0 \checkmark$$

EXAMPLE 2 Solve $4x^2 = 4x - 1$.

Solution As we saw in the previous example, zero plays an important part in the factoring method of solving quadratic equations; so, first, we write the equation in standard form:

$$4x^2 = 4x - 1$$
Rewrite in standard form
$$4x^2 - 4x + 1 = 0$$
Factor
$$(2x - 1)(2x - 1) = 0$$
$$2x - 1 = 0 \quad | \quad 2x - 1 = 0$$
$$x = \tfrac{1}{2} \quad | \quad x = \tfrac{1}{2}$$

Here, $\frac{1}{2}$ is called a **root of multiplicity two** or simply a **double root**.

Check:
$$4x^2 = 4x - 1$$
$$4(\tfrac{1}{2})^2 = 4(\tfrac{1}{2}) - 1$$
$$1 = 2 - 1$$
$$1 = 1 \checkmark$$

Each of the previous examples involved factoring one side of the equation. A more general method, using a *formula* to solve the quadratic equation is sometimes less time consuming. We develop the formula below:

$$ax^2 + bx + c = 0$$
Subtract c from both sides
$$ax^2 + bx = -c$$
Multiply both sides by $4a$
$$4a^2x^2 + 4abx = -4ac$$
Add b^2 to both sides
$$4a^2x^2 + 4abx + b^2 = b^2 - 4ac$$
Factor the left-hand side
$$(2ax + b)^2 = b^2 - 4ac$$
Take the principal square root of both sides
$$\sqrt{(2ax + b)^2} = \sqrt{b^2 - 4ac}$$

$$\sqrt{(2ax+b)^2} = \sqrt{b^2-4ac}$$

Recall $\sqrt{Q^2} = |Q|$

$$|2ax+b| = \sqrt{b^2-4ac}$$

If $|Q| = P$, then $Q = \pm P$ (see p. 11)

$$2ax + b = \pm\sqrt{b^2-4ac}$$

Subtract b from both sides

$$2ax = -b \pm \sqrt{b^2-4ac}$$

Divide both sides by $2a$

$$x = \frac{-b \pm \sqrt{b^2-4ac}}{2a} \qquad (2)$$

Equation (2) is **the quadratic formula** for the two solutions to

$$ax^2 + bx + c = 0.$$

Those two solutions are:

$$x_1 = \frac{-b + \sqrt{b^2-4ac}}{2a} \quad \text{and} \quad x_2 = \frac{-b - \sqrt{b^2-4ac}}{2a}$$

The radicand, $b^2 - 4ac$, called the **discriminant**, affects the nature of the two roots listed above. If $b^2 - 4ac$ is positive, the two solutions are real and distinct and are given by (2). If $b^2 - 4ac$ is zero, the two solutions are real and equal; $x_1 = x_2 = \frac{-b}{2a}$. If $b^2 - 4ac$ is negative, the solutions are complex numbers and are given by (2). Some examples follow.

EXAMPLE 3 Solve $x^2 = 1 - 2x$.

Solution First we rewrite the equation in standard form:

$$x^2 + 2x - 1 = 0$$

Using (2) with $a = 1$, $b = 2$, and $c = -1$, we have:

$$x = \frac{-2 \pm \sqrt{(2)^2 - 4(1)(-1)}}{2(1)}$$

$$x = \frac{-2 \pm \sqrt{8}}{2}$$

$$x = \frac{-2 \pm 2\sqrt{2}}{2}$$

The two solutions are:

$$x = -1 + \sqrt{2} \approx 0.414 \quad \text{and} \quad -1 - \sqrt{2} \approx -2.414$$

We leave the check for these solutions to the reader in Exercise 47.

If the discriminant, $b^2 - 4ac$, is zero, then a double root occurs as the next example illustrates.

EXAMPLE 4 Solve $16x^2 - 24x + 9 = 0$.

Solution Here, $a = 16$, $b = -24$, and $c = 9$.

$$x = \frac{-b \pm \sqrt{b^2 - 4ac}}{2a}$$

$$x = \frac{24 \pm \sqrt{576 - 4(16)(9)}}{2(16)}$$

$$x = \frac{24 \pm \sqrt{0}}{32}$$

$$x = \tfrac{3}{4}$$

Thus, there is exactly one unique solution, $x = \frac{3}{4}$.

Check:

$$16x^2 - 24x + 9 = 0$$
$$16(\tfrac{3}{4})^2 - 24(\tfrac{3}{4}) + 9 = 0$$
$$9 - 18 + 9 = 0$$
$$0 = 0 \checkmark$$

EXAMPLE 5 Solve $5x^2 + 4x = -1$.

Solution First we rewrite the equation in standard form:

$$5x^2 + 4x + 1 = 0$$

With $a = 5$, $b = 4$, $c = 1$ we obtain:

$$x = \frac{-b \pm \sqrt{b^2 - 4ac}}{2a}$$

$$x = \frac{-4 \pm \sqrt{16 - 4(5)(1)}}{10}$$

$$x = \frac{-4 \pm \sqrt{-4}}{10}$$

$$x = \frac{-4 \pm 2i}{10} \text{ or } \frac{2}{5} \pm \frac{1}{5}i$$

The check is left to the reader in Exercise 55.

In Section 2.2 we saw that, given a literal equation, we might sometimes want to solve it explicitly for a particular variable. The next example illustrates how we can do that if one of the variables is in a quadratic expression.

EXAMPLE 6 Solve for x: $x^2 - 3x - 18 = y$.

Solution First, we subtract y from both sides:

$$x^2 - 3x - 18 - y = 0 \leftarrow \text{This equation is in standard form.}$$

Now with $a = 1$, $b = -3$, and $c = -18 - y$ we use (2):

$$x = \frac{-b \pm \sqrt{b^2 - 4ac}}{2a}$$

$$x = \frac{3 \pm \sqrt{9 - 4(-18 - y)}}{2}$$

$$x = \frac{3 \pm \sqrt{4y + 81}}{2}$$

We conclude this section with two examples of applied problems that involve quadratic equations.

EXAMPLE 7 A page being designed for a book is to have $1\frac{1}{2}$-inch margins on top, right, and bottom and a $2\frac{1}{2}$-inch left-hand margin. The height of the page is 2 inches more than the width and the area of the printed matter is 40 square inches. Find the height and width of the entire page.

Solution

STEP 1 Problem translation

Let x = the width of the page. We first create a pictorial model:

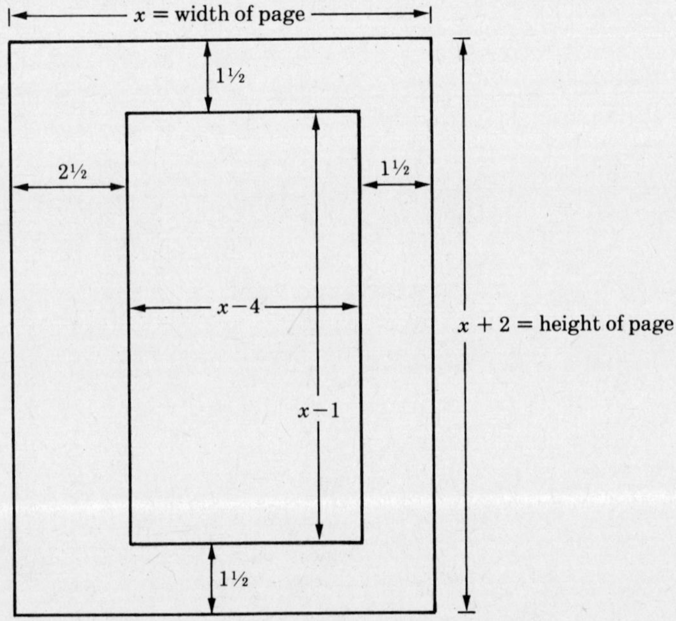

Since the area of the printed matter is 40, we have:

$(x - 4)(x - 1) = 40$ ← This is our algebraic model.

STEP 2 Equation solution

$$(x - 4)(x - 1) = 40$$
$$x^2 - 5x + 4 = 40$$
$$x^2 - 5x - 36 = 0$$

with $a = 1, b = -5, c = -36$:

$$x = \frac{5 \pm \sqrt{25 - 4(1)(-36)}}{2}$$

$$x = \frac{5 \pm \sqrt{169}}{2}$$

$$x = \frac{5 + 13}{2} = 9 \quad \text{or} \quad x = \frac{5 - 13}{2} = \boxed{-3}$$

We reject -3 because x is a page width.

Thus, the width is 9 inches and the height is 11 inches.

STEP 3 Check

Recall the pictorial model:

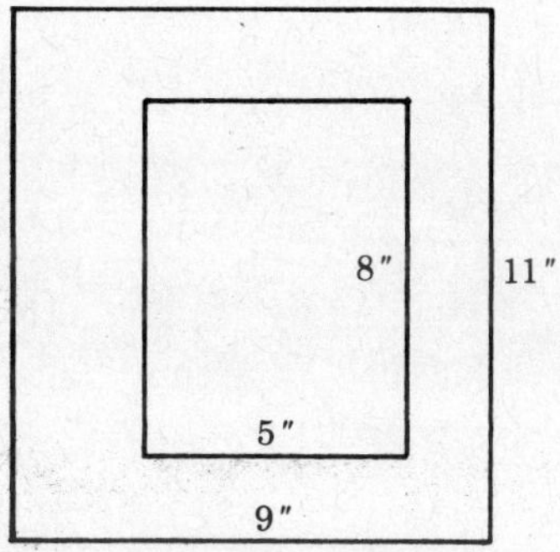

The area of the printed matter is

$$8 \cdot 5 = 40 \checkmark$$

Also, the height *is* 2 inches more than the width.

EXAMPLE 8 The Pyric Security Company manufactures smoke detectors for the home. The profit in dollars, P, they make by selling d smoke detectors is given by the formula

$$P = 2d^2 + 9d - 800$$

How many smoke detectors must they sell to realize a profit of \$1270?

Solution With $P = 1270$ we have:

$$1270 = 2d^2 + 9d - 800$$

or

$$2d^2 + 9d - 2070 = 0$$

$$d = \frac{-9 \pm \sqrt{81 - 4(2)(-2070)}}{2(2)}$$

$$d = \frac{-9 \pm \sqrt{16641}}{4}$$

$$d = \frac{-9 \pm 129}{4}$$

$$d = \frac{-9 + 129}{4} \quad \text{or} \quad d = \frac{-9 - 129}{4}$$

$$d = 30 \qquad\qquad d = \boxed{-34.5}$$

We reject −34.5 since d represents a number of smoke detectors and cannot be negative.

Thus, if they sell 30 smoke detectors, the company will realize a profit of $1270.

EXERCISES FOR SECTION 2.4

In Exercises 1 through 10, use factoring to solve each quadratic equation. (See Examples 1 and 2.)

1. $x^2 - 7x - 18 = 0$
2. $x^2 - 9x + 20 = 0$
3. $x^2 + 13x + 12 = 0$
4. $y^2 - 10y + 24 = 0$
5. $y^2 - 10y - 24 = 0$
6. $x^2 = 36$
7. $4x^2 + 20x + 25 = 0$
8. $9x^2 = 6x - 1$
9. $2x^2 - 7x - 9 = 0$
10. $6x^2 + 13x = 5$

In Exercises 11 through 40, use the quadratic formula to solve each quadratic equation. (See Examples 3, 4, and 5.)

11. $x^2 - 5x + 6 = 0$
12. $x^2 + 5x - 24 = 0$
13. $x^2 + 4 = 0$
14. $x^2 - 4x + 5 = 0$
15. $2x^2 + 5x = 3$
16. $2y^2 - 3 = y$
17. $4x^2 - 16x + 15 = 0$
18. $3z^2 + 7z + 2 = 0$
19. $4t^2 + 17t + 4 = 0$
20. $3x^2 + 2x = 1$
21. $3x^2 = 2x + 1$
22. $3x^2 + 1 = 4x$
23. $3x^2 + 4x + 1 = 0$
24. $r^2 + 2r + 2 = 0$
25. $r^2 - 2r + 2 = 0$
26. $x^2 + 3x - 7 = 0$
27. $y^2 - 3y - 2 = 0$
28. $-y^2 + 3y + 2 = 0$
29. $4x^2 - 24x + 37 = 0$
30. $4x^2 + 12x + 9 = 0$
31. $4x^2 - 12x + 9 = 0$
32. $5y^2 - 7y - 1 = 0$
33. $5y^2 - 7y + 1 = 0$
34. $5y^2 + 7y - 1 = 0$
35. $x^2 + \sqrt{18}x + 4 = 0$
36. $x^2 + x + 1 = 0$
37. $2x^2 - 3x - 11 = 0$
38. $2x^2 - 4x + \frac{1}{4} = 0$
39. $t^2 + \sqrt{8}\,t + 2 = 0$
40. $-\frac{1}{3}x^2 - \frac{2}{3}x = \frac{1}{3}$

In Exercises 41 through 46, solve each equation for x. (See Example 6.)

41. $x^2 - 7x + 12 = y$
42. $y = x^2 - 9x + 2$
43. $s = -16x^2 + 64x$
44. $yx^2 + 2x + 1 = y$
45. $y = 2x^2 + 3x + 4$
46. $y = ax^2 + bx + c$
47. Check the solutions obtained in Example 3.

Solve each applied problem in Exercises 48 through 53. (See Examples 7 and 8.)

48. A page being designed for a math textbook is to have $1\frac{1}{2}$-inch margins on all four sides. The height of the page is to be 4 inches more than the width. The area of the printed matter is to be 45 square inches. Find the height and width of the entire page.
49. A carpenter has been hired by a supermarket's manager to install a wooden fence to enclose a trash pickup area behind the store. One side is bounded by a concrete wall and needs no fencing. The other three sides need fencing. If there is 100 feet of fencing available, and the enclosed area is to be 1200 square feet, what will be the dimensions of the trash pickup area?
50. A square piece of cardboard is to be formed into an open-top box by cutting 2-centimeter squares out of each corner and folding up the sides as indicated below.

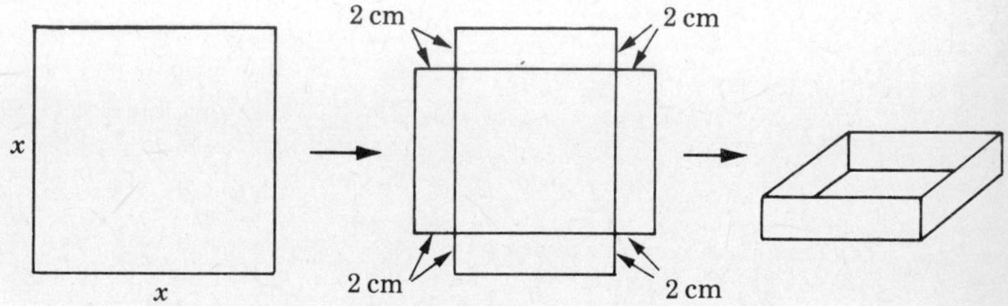

If the volume of the box is to be 968 cubic centimeters, what should the dimensions of the original square piece of cardboard be?

51. If an object is thrown upward from the top of a building 160 feet high, with an initial speed of 80 ft/sec, the formula

$$h = -16t^2 + 80t + 160$$

gives the height of the object, h, above the ground after t seconds have elapsed (h is measured in feet and t is measured in seconds).

a. How high is the object after 2 seconds?
b. After how many seconds will the object be 260 feet above the ground?
c. After how many seconds will the object be 196 feet above the ground?

52. The equation $d = \frac{2}{9}t^2 - \frac{40}{3}t$ gives the distance, d (in feet), that a computer-controlled diving bell will descend in t seconds. (The process starts at the surface level of the water when $t = 0$.)

a. How deep is the diving bell after 15 seconds? (*Note:* Negative values of d are distances *below* the surface.)
b. After how many seconds will the diving bell be 102 feet below the surface?

53. The equation $C = 42s^2 - 177s + 396$ is an algebraic model for the calories per hour, C, consumed by a person walking at a speed of s miles per hour $(2 \leqslant s \leqslant 5)$. Solve this equation for s.

54. Prove the following:

The quadratic *expression* $ax^2 + bx + c$ is factorable as

$$a(x - r_1)(x - r_2)$$

where r_1 and r_2 are the solutions (or roots) to the quadratic *equation* $ax^2 + bx + c = 0$. (Assume $a \neq 0$ and $b^2 - 4ac \geqslant 0$.)

55. Check each of the solutions obtained in Example 5.

Computer Exercises

The following program was written by a student to solve quadratic equations:

```
10  PRINT "THIS PROGRAM SOLVES THE QUADRATIC"
20  PRINT "EQUATION AX ↑ 2 + BX + C = Ø"
25  PRINT "PROVIDED A IS NOT ZERO AND THE SOLUTIONS ARE"
28  PRINT "REAL"
3Ø  PRINT "INPUT THE VALUES A, B, C"
4Ø  INPUT A, B, C
5Ø  IF A = Ø THEN GOTO 99Ø
6Ø  LET D = B ↑ 2 – 4*A*C
7Ø  IF D >= Ø THEN GOTO 1ØØ
8Ø  PRINT "THE ROOTS ARE NOT REAL"
9Ø  GOTO 999
1ØØ LET X1 = (D ↑ .5 – B)/(2*A)
11Ø LET X2 = (–(D ↑ .5) – B)/(2*A)
12Ø PRINT "THE TWO SOLUTIONS ARE: "; X1; "AND"; X2
9ØØ GOTO 999
99Ø PRINT "A = Ø. THIS IS NOT A QUADRATIC EQUATION"
999 STOP
```

Use the program to solve the following equations:

56. $2x^2 + 5x - 3 = 0$

57. $4x^2 - 8x + 1 = 0$

58. $2x^2 + x + 10 = 0$

59. $0x^2 + 3x - 4 = 0$

60. $2x^2 + 4x - 11 = 0$

2.5 ADDITIONAL TYPES OF EQUATIONS

In this section we examine methods for solving three additional types of equations: equations quadratic in form, fractional equations, and equations involving radical expressions. There are several examples of each type.

Equations that are Quadratic in Form

An equation in x is said to be **quadratic in form** if it can be written as

$$ap^2 + bp + c = 0 \qquad (a \neq 0) \tag{1}$$

where p is an expression in x. The technique used is a *change of variable* method illustrated in the next example.

EXAMPLE 1 Solve the equation: $x^4 - 7x^2 + 12 = 0$.

Solution Since $x^4 - 7x^2 + 12 = 0$ can be rewritten as $(x^2)^2 - 7x^2 + 12 = 0$, we make a change of variable by letting $p = x^2$. We have:

$$p^2 - 7p + 12 = 0 \tag{2}$$

Notice that equation (2) is a quadratic equation in p and in the form of (1). We solve for p by using the quadratic formula with $a = 1$, $b = -7$, $c = 12$:

$$p = \frac{-b \pm \sqrt{b^2 - 4ac}}{2a}$$

$$p = \frac{7 \pm \sqrt{49 - 48}}{2}$$

$$p = 4 \quad \text{or} \quad p = 3$$

To find x, recall $p = x^2$, thus,

$$x^2 = p = 4 \quad \text{or} \quad x^2 = p = 3$$
$$x = \pm 2 \quad \text{or} \quad x = \pm\sqrt{3}$$

The solutions are: $2, -2, \sqrt{3}, -\sqrt{3}$.

The reader should verify that each of these four solutions checks in the original equation.

EXAMPLE 2 Solve: $x^{2/3} + 2x^{1/3} - 35 = 0$.

Solution $x^{2/3} + 2x^{1/3} - 35 = 0$ can be written as $(x^{1/3})^2 + 2(x^{1/3}) - 35 = 0$. This suggests the substitution $p = x^{1/3}$. Thus we have

$$p^2 + 2p - 35 = 0 \tag{3}$$

Solving equation (3) by factoring yields:

$$(p + 7)(p - 5) = 0$$
$$p + 7 = 0 \quad | \quad p - 5 = 0$$
$$p = -7 \quad | \quad p = 5$$

Thus $$x^{1/3} = -7 \quad | \quad x^{1/3} = 5$$

Now, we find x by cubing both sides of each equation:

$$(x^{1/3})^3 = (-7)^3 \qquad\qquad (x^{1/3})^3 = (5)^3$$
$$x = -343 \qquad\qquad x = 125$$

Check for $x = -343$:

$$x^{2/3} + 2x^{1/3} - 35 = 0$$
$$(-343)^{2/3} + 2(-343)^{1/3} - 35 = 0$$
$$49 - 14 - 35 = 0$$
$$0 = 0 \checkmark$$

Check for $x = 125$:

$$x^{2/3} + 2x^{1/3} - 35 = 0$$
$$(125)^{2/3} + 2(125)^{1/3} - 35 = 0$$
$$25 + 10 - 35 = 0$$
$$0 = 0 \checkmark$$

Thus, the solution set is $\{-343, 125\}$.

EXAMPLE 3 Solve: $3x + 8x^{1/2} = 3$.

Solution Rewriting as $3x + 8x^{1/2} - 3 = 0$ and letting $p = x^{1/2}$ we obtain

$$3p^2 + 8p - 3 = 0,$$

a quadratic equation in p. Solving by factoring yields:

$$(3p - 1)(p + 3) = 0$$
$$3p - 1 = 0 \quad\Big|\quad p + 3 = 0$$
$$p = \tfrac{1}{3} \quad\Big|\quad p = -3$$

Since $x = p^2$, the two corresponding values of x are:

$$x = p^2 = (\tfrac{1}{3})^2 = \tfrac{1}{9} \qquad \text{and} \qquad x = p^2 = (-3)^2 = 9$$

Check for $x = \frac{1}{9}$:

$$3x + 8x^{1/2} = 3$$
$$3(\tfrac{1}{9}) + 8(\tfrac{1}{9})^{1/2} = 3$$
$$\tfrac{1}{3} + \tfrac{8}{3} = 3$$
$$3 = 3 \checkmark$$

Check for $x = 9$:

$$3x + 8x^{1/2} = 3$$
$$3(9) + 8(9)^{1/2} = 3$$
$$27 + 24 = 3$$
$$51 = 3 \times$$

The *only* solution to $3x + 8x^{1/2} = 3$ is $x = \frac{1}{9}$.

We caution the reader by pointing out that it is necessary to return to the *original equation* when performing a check. In the preceding example, the result $p = -3$, although it is certainly a solution to $3p^2 + 8p - 3 = 0$, does not yield a solution to the original equation.* Such solutions (ones that do not solve an original equation but *do* solve a newer, derived equation) are called **extraneous solutions.**

Fractional Equations By a **fractional equation** we mean any equation involving a variable in at least one fraction's denominator. The method of solving a fractional equation that we shall employ here is multiplying both sides of the equation by the LCM of all the denominators (see Section 1.6).

EXAMPLE 4 Solve: $\dfrac{3y}{y+1} + \dfrac{9}{y^2-1} = \dfrac{5}{y-1}$. (4)

Solution Multiply both sides of (4) by the LCM of the three denominators, $y+1$, y^2-1, and $y-1$ which is $(y+1)(y-1)$:

$$\frac{3y}{y+1} + \frac{9}{y^2-1} = \frac{5}{y-1}$$

Multiply both sides by $(y+1)(y-1)$

$$3y(y-1) + 9 = 5(y+1)$$

$$3y^2 - 3y + 9 = 5y + 5$$

$$3y^2 - 8y + 4 = 0$$

Factor the left-hand side**

$$(3y-2)(y-2) = 0$$

$$3y - 2 = 0 \quad \Big| \quad y - 2 = 0$$

$$y = \tfrac{2}{3} \quad \Big| \quad y = 2$$

Check for $y = \frac{2}{3}$:

$$\frac{3y}{y+1} + \frac{9}{y^2-1} = \frac{5}{y-1}$$

$$\frac{2}{\frac{2}{3}+1} + \frac{9}{\frac{4}{9}-1} = \frac{5}{\frac{2}{3}-1}$$

$$\frac{6}{5} - \frac{81}{5} = -15$$

$$-\frac{75}{5} = -15$$

$$-15 = -15 \checkmark$$

Check for $y = 2$:

$$\frac{3y}{y+1} + \frac{9}{y^2-1} = \frac{5}{y-1}$$

$$\frac{6}{3} + \frac{9}{4-1} = \frac{5}{2-1}$$

$$2 + 3 = 5$$

$$5 = 5 \checkmark$$

Thus, the solutions are $y = \frac{2}{3}$ and $y = 2$.

---

*The reason is that the substitution $x = p^2$ and the substitution $x^{1/2} = p$ are only equivalent if $p \geqslant 0$.

**Also, we could have used the quadratic formula.

EXAMPLE 5 Solve: $\dfrac{x}{x-5} - \dfrac{5}{x+5} = \dfrac{10x}{x^2-25}$.

Solution Multiply both sides by $(x+5)(x-5)$:

$$\frac{x}{x-5} - \frac{5}{x+5} = \frac{10x}{x^2-25}$$
$$x(x+5) - 5(x-5) = 10x$$
$$x^2 + 5x - 5x + 25 = 10x$$
$$x^2 - 10x + 25 = 0$$
$$(x-5)(x-5) = 0$$
$$x = 5$$

Check:

$$\frac{x}{x-5} - \frac{5}{x+5} = \frac{10x}{x^2-25}$$
$$\frac{5}{5-5} - \frac{5}{5+5} = \frac{10(5)}{5^2-25}$$

Here we have a zero denominator—an impossibility. Therefore, $x = 5$ is *not* a solution to the original equation; it is extraneous. The equation $\dfrac{x}{x-5} - \dfrac{5}{x+5} = \dfrac{10x}{x^2-25}$ has *no solution.*

Equations Involving Radical Expressions We shall proceed to solve equations in which a variable occurs within a radical symbol. In order to do so, we use the Principle of Exponentiation (see Section 1.7).

EXAMPLE 6 Solve for x: $\sqrt{2x} + x = 0$.

Solution After "isolating" $\sqrt{2x}$, we square both sides:

$$\sqrt{2x} = -x$$
$$2x = x^2$$
$$x^2 - 2x = 0$$
$$x(x-2) = 0$$
$$x = 0 \mid x = 2$$

Check for $x = 0$:

$$\sqrt{2x} + x = 0$$
$$\sqrt{0} + 0 = 0$$
$$0 = 0 \checkmark$$

Check for $x = 2$:

$$\sqrt{2x} + x = 0$$
$$\sqrt{4} + 2 = 0$$
$$4 = 0 \times$$

The only solution to $\sqrt{2x} + x = 0$ is $x = 0$.

EXAMPLE 7 Solve for x: $\sqrt{x-2} + \sqrt{2x+3} = 8$.

Solution Since there are two radical symbols, the technique is to "isolate" one of them, square, and then repeat:

$$\sqrt{x-2} + \sqrt{2x+3} = 8$$
$$\sqrt{x-2} = 8 - \sqrt{2x+3} \quad \text{Square both sides}$$
$$x - 2 = 64 - 16\sqrt{2x+3} + 2x + 3 \quad \text{"Isolate" } \sqrt{2x+3}$$
$$16\sqrt{2x+3} = x + 69 \quad \text{Square both sides}$$
$$256(2x+3) = x^2 + 138x + 4761$$
$$512x + 768 = x^2 + 138x + 4761$$
$$x^2 - 374x + 3993 = 0^*$$
$$(x-11)(x-363) = 0$$
$$x - 11 = 0 \quad | \quad x - 363 = 0$$
$$x = 11 \quad | \quad x = 363$$

Check for $x = 11$:

$$\sqrt{x-2} + \sqrt{2x+3} = 8$$
$$\sqrt{11-2} + \sqrt{22+3} = 8$$
$$3 + 5 = 8$$
$$8 = 8 \checkmark$$

Check for $x = 363$:

$$\sqrt{x-2} + \sqrt{2x+3} = 8$$
$$\sqrt{361} + \sqrt{729} = 8$$
$$19 + 27 = 8$$
$$46 = 8 \times$$

Thus, the only solution to $\sqrt{x-2} + \sqrt{2x+3} = 8$ is $x = 11$.

We conclude this section by returning to problems with an applied setting that involve equations of the type we have recently seen.

---

*We could also have used the quadratic formula with $a = 1$, $b = -374$, and $c = 3993$.

EXAMPLE 8 A department store buyer bought some calculators for \$320. Then, she sold all but 10 of them. On each calculator sold, a profit of \$4 (over initial cost) was made. Furthermore, the calculators sold brought in \$360. How many calculators did the store buy initially?

Solution

STEP 1 Problem translation

Let x = the number of calculators bought initially.

Then, $x - 10$ = the number sold.

We use the fact that:

$$\underbrace{\text{Number of calculators sold}}_{(x-10)} \cdot \underbrace{\text{Selling price of one calculator}}_{\left(\frac{320}{x}+4\right)} = \underbrace{\text{Total income from sales}}_{360}$$

This is our algebraic model.

Thus, we have a fractional equation.

STEP 2 Equation solution

$$(x-10)\left(\frac{320}{x}+4\right) = 360$$

Expand

$$320 + 4x - \frac{3200}{x} - 40 = 360$$

Multiply both sides by x

$$320x + 4x^2 - 3200 - 40x = 360x$$

$$4x^2 - 80x - 3200 = 0$$

$$x^2 - 20x - 800 = 0$$

$$(x-40)(x+20) = 0$$

$$x - 40 = 0 \quad | \quad x + 20 = 0$$

$$x = 40 \quad | \quad \boxed{x = -20} \leftarrow \text{reject}$$

Thus, 40 calculators were initially bought.

STEP 3 Check

If 40 calculators were bought for \$320, each one cost \$8. If all but 10 (i.e., 30) were sold for a \$4 profit (that's \$12 each), the total from sales would be \$360. ✓

EXERCISES FOR SECTION 2.5

In Exercises 1 through 21, solve for x. (See Examples 1, 2, and 3.)

1. $x^4 - 9x^2 + 20 = 0$
2. $x^4 - 10x^2 + 9 = 0$
3. $x^4 - 6x^2 + 8 = 0$
4. $x^4 - 12x^2 + 20 = 0$
5. $x^5 - 12x^3 + 20x = 0$
6. $x^4 + 144 = 25x^2$
7. $3x^4 - 13x^2 + 4 = 0$
8. $5x^4 - 16x^2 + 11 = 0$
9. $10x^4 - 32x^2 + 22 = 0$
10. $x^{2/3} + 7x^{1/3} + 12 = 0$
11. $3x + 5\sqrt{x} - 2 = 0$
12. $2x - 7x^{1/2} + 6 = 0$
13. $x^{2/3} - 3x^{1/3} + 2 = 0$
14. $x^{2/3} - 5x^{1/3} + 6 = 0$
15. $x^{1/3} - 3x^{1/6} + 2 = 0$
16. $x^4 - 4x^2 + 1 = 0$
17. $x^{5/3} - 5x^{4/3} + 6x = 0$
18. $2x - 8\sqrt{x} + 6 = 0$
19. $2x^{-2} + 3x^{-1} - 20 = 0$
20. $x^2 - 6x^{3/2} + 9x = 0$
21. $2x^{2/3} + 7x^{1/3} + 6 = 0$

In Exercises 22 through 33, solve each fractional equation for x. (See Examples 4 and 5.)

22. $\dfrac{1}{x} + \dfrac{4}{x} = 3$
23. $\dfrac{2}{5} = \dfrac{3}{x}$
24. $\dfrac{5x}{x-1} = 2$
25. $\dfrac{5x}{x-1} = 5$
26. $\dfrac{2x}{x-1} + \dfrac{1}{2} = 1$
27. $\dfrac{x}{2x+7} + \dfrac{1}{x} = \dfrac{6}{2x^2+7x}$
28. $\dfrac{1}{2x+5} - \dfrac{1}{2x-5} = \dfrac{2x}{4x^2-25}$
29. $\dfrac{1}{2x+5} - \dfrac{1}{2x-5} = \dfrac{2}{4x^2-25}$
30. $\dfrac{1}{3x+2} + \dfrac{2}{x-4} = \dfrac{3}{3x^2-10x-8}$
31. $\dfrac{1}{3x+2} - \dfrac{2}{x-4} = \dfrac{3}{3x^2-10x-8}$
32. $\dfrac{x+2}{x} + \dfrac{1}{x+4} = \dfrac{-2x-6}{x(x+4)}$
33. $\dfrac{x+2}{x} - \dfrac{1}{x+4} = \dfrac{x+1}{x+4}$

In Exercises 34 through 45, solve each radical equation for x. (See Examples 6 and 7.)

34. $\sqrt{3x} + x = 0$
35. $\sqrt{3x} = 20$
36. $3 - \sqrt{2x} = 1$
37. $3 + \sqrt{2x} = 1$
38. $x - \sqrt{x} = 2$
39. $\sqrt{x+1} = \sqrt{3x+1} - 2$
40. $\sqrt{x+3} + 1 = \sqrt{2-x}$
41. $\sqrt{x+10} + \sqrt{13-3x} = 7$
42. $\sqrt{x} - \sqrt{x-12} = 2$
43. $\sqrt{-3x} - \sqrt{x+3} = 0$
44. $4 + \sqrt[5]{5x-1} = 4 + \sqrt[5]{19}$
45. $\sqrt[3]{6x-1} = 3$

Solve each applied problem in Exercises 46 and 47. (See Example 8.)

46. A real estate agent bought some land in the country for \$66,000. Then she sold part of the land, keeping 40 acres. On each acre sold, a profit of \$400 (over initial cost) was made. Furthermore, the portion of land sold brought in \$70,000. How many acres did the real estate agent buy initially?

47. A music store owner bought a shipment of folk guitars for \$3,000. One was damaged in shipment beyond the point of repair. Each of the others was marked up \$35 and sold. The income from these sales was \$4655.
    a. How many guitars were purchased initially?
    b. What was the initial cost of each guitar?
    c. What was the selling price of each guitar?
48. The equation $\frac{1}{d} = \frac{pq}{p+q}$ is used in optometry. Solve this equation for q.
49. The equation $d^2 = \frac{m^2}{4(1-A)}$ is used by geologists in seismic prospecting of the earth. Solve this equation for A.
50. The time, T, it takes a pendulum of length l to swing through one complete oscillation is given in the algebraic model

$$T = 2\pi \sqrt{\frac{l}{9.8}}$$

    a. Solve this equation for l.
    b. If T is measured in seconds and l in meters, how long would a pendulum have to be in order to have an oscillation time of 1 second?
51. Many stores offer purchases on the installment plan, whereby the purchaser agrees to pay a certain finance charge. When the loan is paid off early, the purchaser does not have to pay the entire finance charge. The formula

$$C = \frac{k(k+1)}{n(n+1)} F,$$

gives the amount of finance charge that you *do not* have to pay, C, where

F = original finance charge
n = original number of payments intended
k = number of payments remaining when the loan is paid off.

    a. If the original finance charge on a 36-month loan was \$400, determine C if the loan is paid off after 12 months.
    b. Solve this formula for F.
    c. Solve this formula for k.

2.6 QUADRATIC AND FRACTIONAL INEQUALITIES

A **quadratic inequality in x** is an inequality having one of the forms:

$$ax^2 + bx + c < 0$$
$$ax^2 + bx + c \leq 0$$
$$ax^2 + bx + c > 0$$
$$ax^2 + bx + c \geq 0$$

where a, b, and c are real numbers ($a \neq 0$).

In order to solve such an inequality we will use a graphical method that can also be extended to solve other inequalities as well. The basic preliminary idea is the **sign graph** of a binomial. For example, the sign of the real number $x - 3$ has three distinct possibilities:

negative	zero	positive
$-$	0	$+$
$x - 3 < 0$	$x - 3 = 0$	$x - 3 > 0$

These possibilities are projected on the number line below:

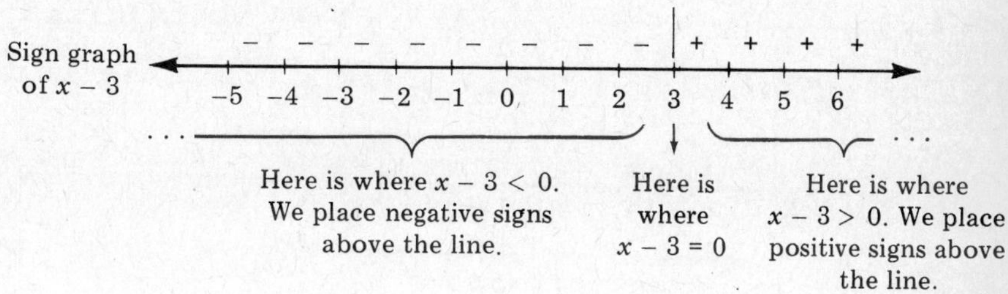

To solve a quadratic inequality, we will construct sign graphs for each factor. For instance, to solve $x^2 - x - 6 < 0$, we first factor to obtain:

$$(x - 3)(x + 2) < 0 \tag{1}$$

Next, sign graphs for each binomial factor $x - 3$ and $x + 2$ are constructed:

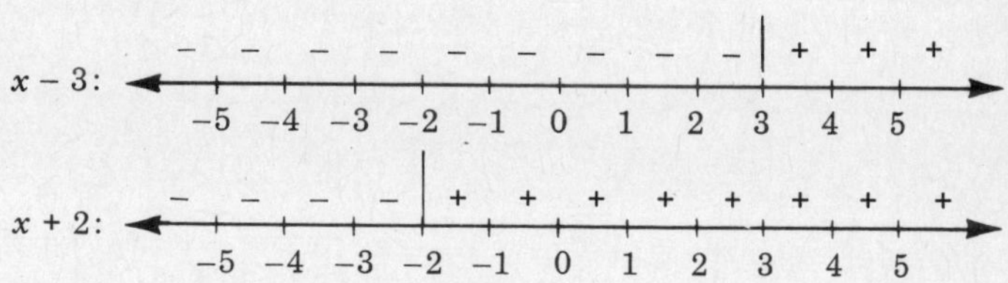

Equation (1) dictates that the product of factors be negative. Thus, we look for territory on the sign graphs with *opposite* signs. We shade that portion as follows:

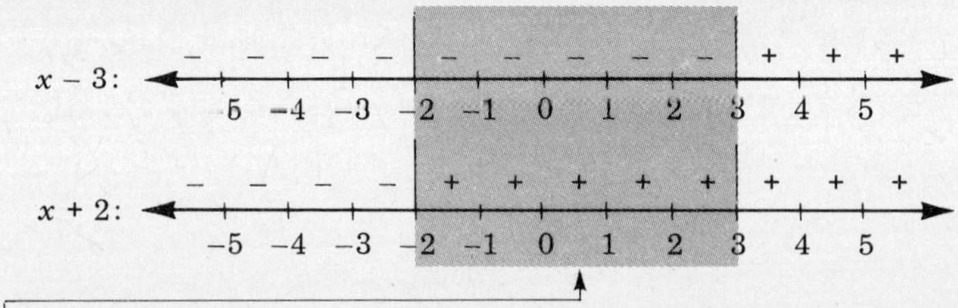

In this shaded region, $x - 3$ and $x + 2$ have *opposite* signs and thus their product is negative. Thus, the solution set of $x^2 - x - 6 < 0$ is $-2 < x < 3$. A graph can be drawn as

EXAMPLE 1 Solve the quadratic inequality: $x^2 + 3x - 4 \geqslant 0$.

Solution First realize that $x^2 + 3x - 4$ is factored as $(x + 4)(x - 1)$ and we will be interested in constructing sign graphs for $x + 4$ and $x - 1$. Also, the "$\geqslant$" sign *includes equality* so the numbers -4 and 1 will be included in our answer. The "$>$" part is solved below:

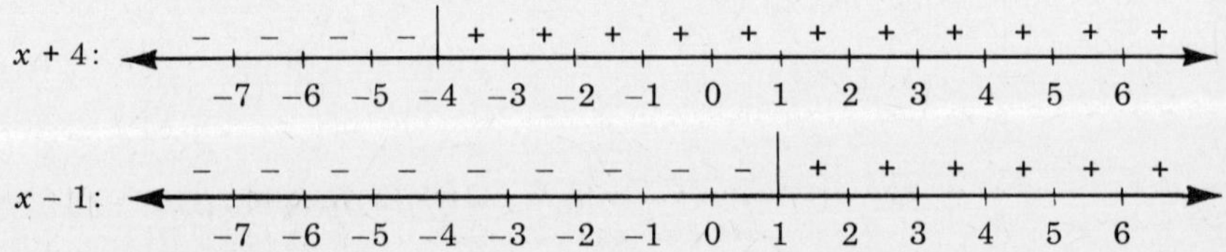

We want the product $(x + 4)(x - 1)$ to be *positive* (greater than 0) so now we shade the territory where the factors $x + 4$ and $x - 1$ have the same signs:

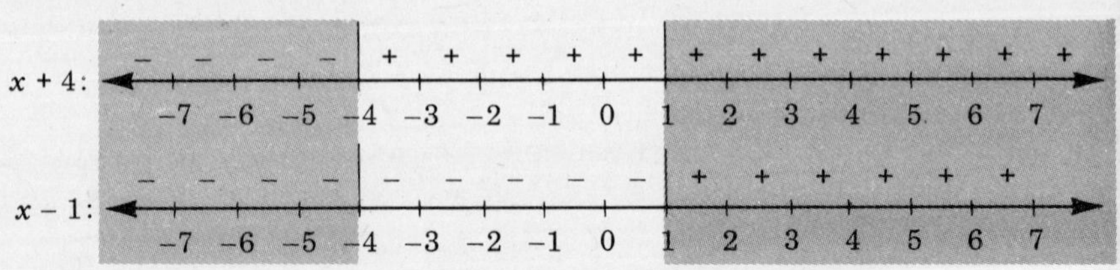

Thus, the solution is $x \leqslant -4$ or $x \geqslant 1$. In interval notation, we write $(-\infty, -4] \cup [1, \infty)$ and the graph of the solution appears below.

EXAMPLE 2 Solve: $6 \geqslant t(t-1)$.

Solution

$$6 \geqslant t(t-1)$$
$$6 \geqslant t^2 - t$$
$$t^2 - t - 6 \leqslant 0^*$$
$$(t-3)(t+2) \leqslant 0$$

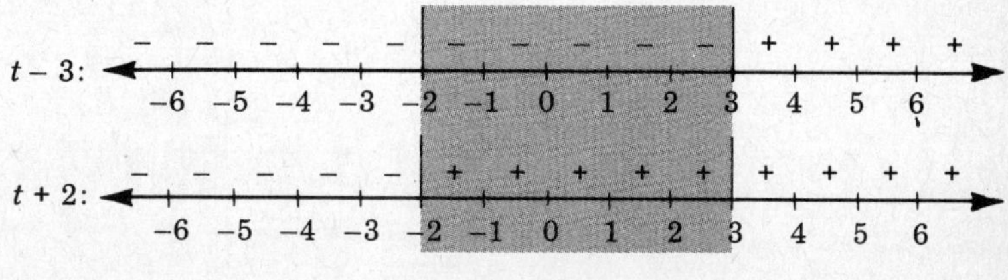

The shaded area is where the factors have opposite signs, thus ensuring their product is negative. The solution is $-2 \leqslant t \leqslant 3$ and in interval notation, $[-2, 3]$.

EXAMPLE 3 Solve: $x^2 + 2x - 1 > 0$.

Solution Although $x^2 + 2x - 1$ is not factorable using only integers, we can factor it as:

$$[x - (-1 + \sqrt{2})][x - (-1 - \sqrt{2})] > 0^{**}$$

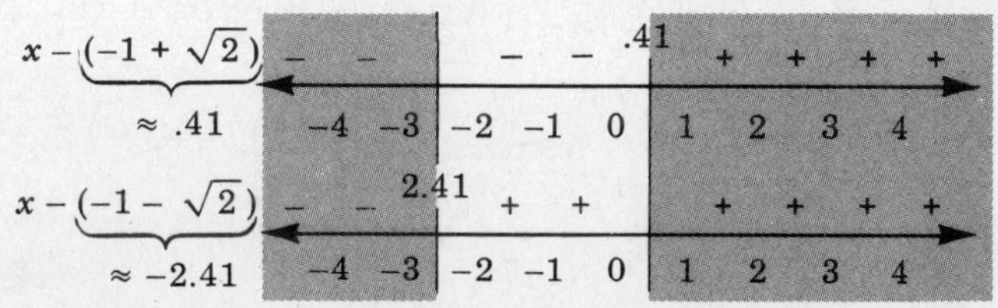

We want that territory where $[x - (-1 + \sqrt{2})]$ and $[x - (-1 - \sqrt{2})]$ have the same sign. Thus, we have

$$(-\infty, -1 - \sqrt{2}) \cup (-1 + \sqrt{2}, \infty)$$

The concepts employed thus far can be extended to expressions involving three or more linear factors, as the next example illustrates.

---

*Note that the logic of the sign graph method discussed in this section requires that one side of the inequality be equal to zero.

**The reader can verify this by Example 3 on page 99.

EXAMPLE 4 Solve $(x-1)(3-x)(x+4) \leqslant 0$.

Solution We construct sign graphs for each factor:

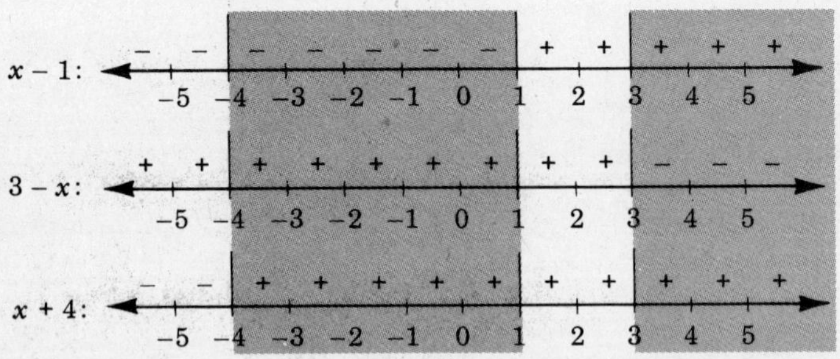

The solution includes those numbers for which the product $(x-1)(3-x)(x+4)$ is negative or zero. Hence, the solution is

$$-4 \leqslant x \leqslant 1 \quad \text{or} \quad x \geqslant 3$$

In interval notation, $[-4,1] \cup [3,\infty)$.

To solve an inequality involving a fraction, a similar technique is employed, but first care must be taken to obtain zero on one side of the inequality.

EXAMPLE 5 Solve $\dfrac{x}{x-2} \geqslant 0$.

Solution We construct the sign graphs below:

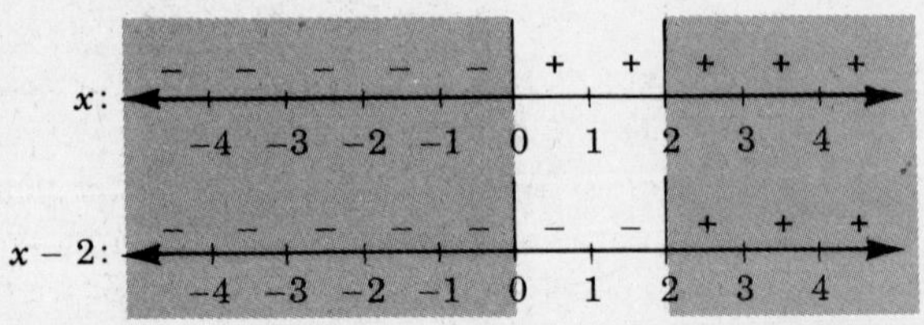

We want the *quotient* $\dfrac{x}{x-2}$ to be equal to or greater than zero. The fraction $\dfrac{x}{x-2}$ will equal zero only when $x = 0$. The quotient will be positive at places where x and $x-2$ have the same sign. Thus, the solution is

$$x \leqslant 0 \quad \text{or} \quad x > 2$$

or equivalently, $(-\infty,0] \cup (2,\infty)$.

EXAMPLE 6 Solve $\frac{x+3}{x-4} - \frac{10}{3} \leqslant 0.$

Solution First, we must combine the terms on the left in order to obtain the fraction

$$\frac{3(x+3)}{3(x-4)} - \frac{10(x-4)}{3(x-4)} \leqslant 0$$

$$\frac{3x+9-10x+40}{3(x-4)} \leqslant 0$$

$$\frac{-7x+49}{3(x-4)} \leqslant 0$$

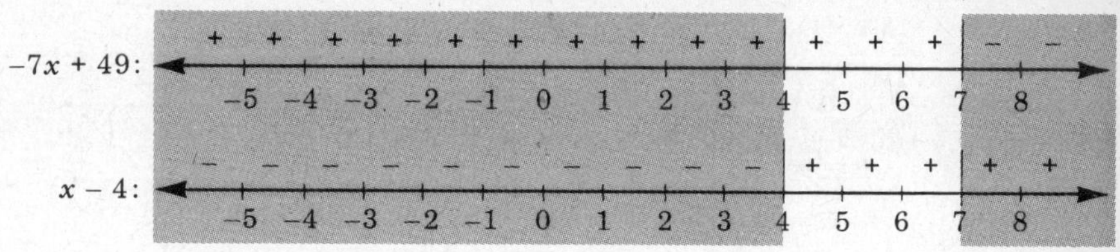

The shaded territory is where the fraction $\frac{-7x+49}{3(x-4)}$ is negative. The solution set is $(-\infty, 4) \cup [7, \infty)$.

We conclude this section with an application to business and economics by examining **supply equations**. A supply equation is an equation that relates the two variables q_S and p, where

q_S = number of items supplied by manufacturer

p = price charged per item by manufacturer

We will make use of the following principle:

Number of items sold	·	Profit per item	=	Total profit from sales

EXAMPLE 7 It costs a machine shop \$2 to produce a certain type of case-hardened bolt for use in a speed shifter. The supply equation is $q_S = 40p + 12$. What price should be charged per bolt so that a profit of at least \$1460 is realized?

Solution

STEP 1
Problem translation

Since p is the selling price of each bolt,

$$p - 2 = \text{profit per bolt}$$

$$\underbrace{\text{Number of bolts sold}} \cdot \underbrace{\text{Profit per bolt}} \geqslant 1460$$

$$(40p + 12) \cdot (p - 2) \geqslant 1460$$ ← This is our algebraic model

STEP 2
Solving the inequality

$$(40p + 12)(p - 2) \geqslant 1460$$
$$40p^2 - 80p + 12p - 24 \geqslant 1460$$
$$40p^2 - 68p - 1484 \geqslant 0$$
$$10p^2 - 17p - 371 \geqslant 0$$
$$(p - 7)(10p + 53) \geqslant 0$$

We construct the sign graphs:

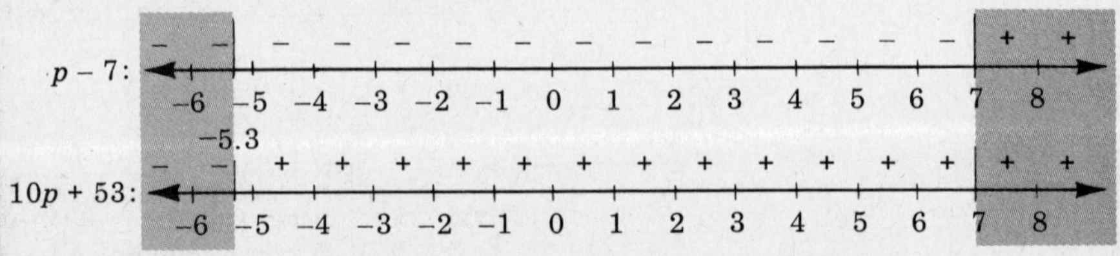

The shaded area represents numbers for which $(p - 7)(10p + 53)$ is positive. Thus, the inequality's solution is:

$$p \leqslant -5.3 \quad \text{or} \quad p \geqslant 7$$

However, we reject this part of the solution because the selling price must be positive. Thus, a price, p, of at least \$7 per bolt will result in a total profit of at least \$1460.

STEP 3
Check

Because of the infinite number of solutions, we perform a partial check.

a. When $p = \$7$, the profit per bolt is \$5 and $q_S = 40p + 12 = 292$.

Total profit = 292 · 5 = \$1460 ✓

b. When $p > \$7$, say \$8, the profit per bolt is \$6 and $q_S = 40p + 12 = 332$.

Total profit = 332 · 6 = \$1992 ✓

c. When $p < \$7$, say \$6, the profit per bolt is \$4 and $q_S = 40p + 12 = 252$.

Total profit = 252 · 4 = \$1008 ✓

EXERCISES FOR SECTION 2.6

In Exercises 1 through 30, solve each quadratic inequality. (See Examples 1, 2, and 3.)

1. $x^2 - 3x - 10 < 0$
2. $x^2 - 3x - 10 \geqslant 0$
3. $x^2 - 3x - 10 \leqslant 0$
4. $x^2 - 3x - 10 > 0$
5. $x^2 - 6x + 8 \geqslant 0$
6. $2x^2 - 12x + 16 \geqslant 0$
7. $2x^2 - 3x - 5 < 0$
8. $2x^2 + 9x - 5 \leqslant 0$
9. $3x^2 - 10x > 48$
10. $-3x^2 + 10x < -48$
11. $9x^2 + 2x - 7 \geqslant 0$
12. $x^2 > x + 2$
13. $x^2 \leqslant x$
14. $-10x^2 < x$
15. $2x^2 - 5x \geqslant 12$
16. $3x^2 + 13x - 30 > 0$
17. $4x^2 + 13x + 3 < 0$
18. $2x^2 - 9x + 7 \leqslant 0$
19. $8 - 12x^2 \leqslant 29x$
20. $2t^2 - 4t - 10 < 0$
21. $y^2 > 3$
22. $z^2 \leqslant 4z + 3$
23. $x^2 - 6x + 6 \geqslant 0$
24. $x^2 - 6x + 3 \leqslant -3$
25. $3x^2 + 6x \leqslant 30$
26. $x^2 - 6x + 9 \leqslant 0$
27. $x^2 > 0$
28. $x^2 \geqslant 0$
29. $4x^2 - 20x + 25 > 0$
30. $4x^2 - 20x + 25 \geqslant 0$

In Exercises 31 through 36, solve each inequality. (See Example 4.)

31. $(x - 1)(x - 3)(x + 4) \leqslant 0$
32. $(x + 1)(x - 3)(x + 4) > 0$
33. $(2x + 1)(5 - x)(x + 2) \geqslant 0$
34. $(8x + 3)(2x - 1)(x + 7) < 0$
35. $-5(x - 1)(x - 2)(x - 3) \leqslant 0$
36. $(1 - x)(2 - x)(3 - x) > 0$

In Exercises 37 through 48, solve each fractional inequality. (See Examples 5 and 6.)

37. $\dfrac{x + 1}{x - 2} < 0$
38. $\dfrac{x + 1}{x - 2} \leqslant 0$
39. $\dfrac{x - 2}{x + 1} \leqslant 0$
40. $\dfrac{x^2 + 3x + 2}{x - 4} > 0$
41. $\dfrac{x^2 + 3x + 2}{x - 4} \geqslant 15$
42. $\dfrac{x^2 + 5x - 6}{x - 2} \leqslant 0$
43. $\dfrac{x^2 + 5x - 6}{x - 2} < 0$
44. $\dfrac{2x + 1}{x - 6} \geqslant 0$
45. $\dfrac{1}{x} + \dfrac{3}{2x - 1} < 4$
46. $\dfrac{1}{x} + \dfrac{3}{5(2x + 3)} - \dfrac{16}{5(2x + 3)} < 0$
47. $\dfrac{1}{x} + \dfrac{3}{5(2x + 3)} - \dfrac{16}{5(2x + 3)} \leqslant 0$
48. $\dfrac{(x^2 - 5)^2}{x^2 + 4x + 4} \geqslant 0$

In Exercises 49 and 50, solve each applied problem. (See Example 7.)

49. It costs a bakery \$1 to make a certain type of cake. The supply equation is $q_S = 50p + 10$. What price should be charged per cake to realize a profit of at least \$80?

50. It costs a clothing factory \$4 to make a short-sleeve shirt. The supply equation is $q_S = 50p + 20$.
    a. What price should be charged per shirt in order for the company to make a total profit of at least \$3120?
    b. How many shirts should be made to sell at the price determined in part a?

CHAPTER 2 IN REVIEW

I. Summary

In this chapter we examined methods for solving several different types of equations and inequalities. In Section 2.1, we solved linear equations in one variable by applying certain principles in order to construct simpler, equivalent equations. These equations were then applied, in Section 2.2, to several settings (business, physics, etc.). Also, we studied literal equations and saw the advantages of being able to solve explicitly for one of the variables. In Section 2.3, linear and absolute value inequalities in one variable were solved by using certain principles in order to construct simpler equivalent inequalities.

Quadratic equations were solved using two methods: factoring and the quadratic formula. Also, we examined quadratic equations (and inequalities) in several applied settings. In Section 2.5, some other types of equations solved were fractional equations, radical equations, and equations quadratic in form. It was at this point in the chapter that we encountered extraneous solutions. Remember, an extraneous solution is *not* a solution to the original equation, it is only a solution to a derived equation that is not equivalent to the original one. Hence, any extraneous solution should *not* be included in the solution set for the original equation.

II. Vocabulary/Symbol True–False Quiz

If the statement is true, write "True." However, if the statement is false, replace the underlined word or phrase with a different word or phrase that will make the new statement true.

1. An equation that is true regardless of the value substituted for the variable, is called an identity.
2. It is possible that the solution set of a conditional equation is $\emptyset$.
3. Two equations are equivalent if they have the same solution sets.
4. A formula is a literal equation stating a general rule, fact, or principle symbolically.

5. The solution set of $|x + 1| = 8$ is $\underline{\{7\}}$.
6. If $a < b$, then $\underline{ac < bc}$, provided $c < 0$.
7. The interval (3,4) contains <u>two</u> integers.
8. If $ax^2 + bx + c = 0$ and $a \neq 0$, then

$$\underline{x = \frac{-b \pm \sqrt{b^2 - 4ac}}{2a}}$$

9. The discriminant of $ax^2 + bx + c = 0$ is $\underline{\sqrt{b^2 - 4ac}}$.
10. If the real number x solves an equation that was derived from an original equation but does not solve that original equation, x is called an <u>extraneous</u> solution to the original equation.

III. Chapter Test

1. Solve for x: $9x - 6 = 12$.
2. Solve for x: $2(1 - 3x) = x + 7$.
3. Solve for x: $ax + b = c$.
4. Two different qualities of hamburger meat sell for \$2.38 per pound and \$1.38 per pound. A butcher decides to mix the two together to create a third, intermediate grade hamburger to sell for \$1.98 per pound. If he has 400 pounds of the \$1.38 hamburger, how much of the \$2.38 hamburger must be added so that the value of the \$1.98-per-pound mixture is equal to the total value of the other two?

In questions 5 through 10, solve for x.

5. $2x^2 - 13x - 7 = 0$
6. $9x^2 + 6x + 1 = 0$
7. $3x^2 + 2x - 2 = 0$
8. $\sqrt{3x - 2} - \sqrt{x - 5} = 3$
9. $x^4 - 8x^2 + 15 = 0$
10. $\dfrac{x}{x + 2} + \dfrac{3x}{x + 4} = \dfrac{13x + 10}{x^2 + 6x + 8}$

In questions 11 through 20 a) solve each inequality and b) express your answer in interval notation.

11. $4x - 1 < 7$
12. $3(1 - x) \geqslant 8$
13. $|x - 7| < 3$
14. $|2x - 1| > 11$
15. $\left|\dfrac{1 - 6x}{2}\right| \leqslant 4$
16. $x^2 - 7x + 12 \geqslant 0$
17. $6x^2 + 19x < 7$
18. $2x^2 + 7x - 1 \geqslant 0$
19. $(1 - x)(x + 3)(x - 5) < 0$
20. $\dfrac{(x - 1)(3x + 11)}{x - 4} \leqslant 0$

IV. Student Research Project

The history and evolution of algebraic symbols (in particular, the use of the equals symbol) is an interesting study. Prepare a brief report on this topic. The following sources may prove useful:

Bergamini, David. *Mathematics*, Life-Science Library, New York, 1969, Chapter 3.

"Image of Computation" in *Readings for Mathematics: A Humanistic Approach*, edited by G. Joseph Wimbish. Wadsworth, Belmont, California, 1972, pp. 172–191.

Eves, Howard. *An Introduction to the History of Mathematics*, 4th ed., Holt, Rinehart and Winston, New York, 1976, Chapter 8.

Jourdain, Philip E. B. "The Nature of Mathematics" in *The World of Mathematics*, edited by James R. Newman. Simon and Schuster, New York, 1966.

CHAPTER 3

Relations, Functions and Graphs

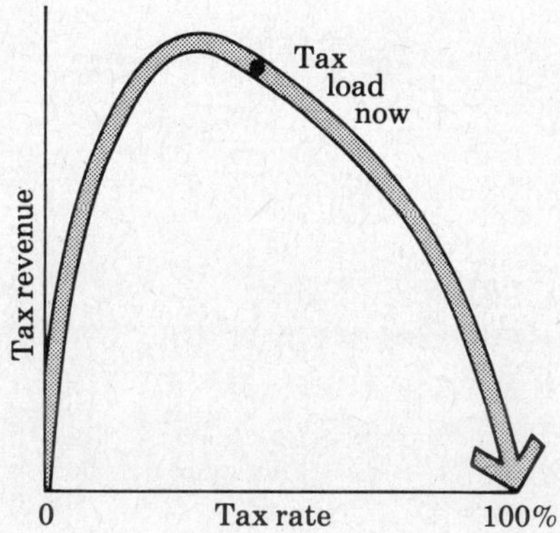

Behind the "Laffer Curve"—

Q Will you describe the "Laffer curve" that stimulated your tax-cut idea?

A Arthur Laffer teaches business economics at the University of Southern California, and his curve amounts to a chart illustrating that tax revenue obeys the law of diminishing returns. There are two points on the curve at which taxes do not gain any revenue: zero and 100 percent tax rates. At 100 percent, production ceases. This is nearly what's happening to some people and businesses in some countries in Europe, where the No. 1 sport is not soccer but escaping the tax collectors.

Of course, at zero taxes you get zero revenue. As you raise taxes from zero, you gain revenue because production and industry are only slightly discouraged and taxes will be paid. And as you lower taxes from 100 percent to 99 or 85 or whatever, you get a pickup in production. Increased revenues are then fed into the treasuries of government at the federal, state and local levels.

The thrust of the Laffer curve is to help find the equilibrium point between zero and 100 percent tax rates where people are willing to maximize their output and still pay their taxes. I think that is somewhere around 25 percent in peacetime. When tax rates go beyond 25 percent—as they have in America and particularly in New York state—revenues are lost, since production diminishes and people are laid off. Layoffs lead to more government spending for assistance payments, and so you have more spending and thus even higher taxes in the future.

In America, we tax work, savings, thrift, investment and entrepreneurship—and we subsidize nonwork, unemployment, leisure, debt, spending and mediocrity.

3.1 THE CARTESIAN COORDINATE SYSTEM

In Chapter 2, the majority of equations we encountered were equations that involved a single variable. In this chapter, we will examine relationships between *two* variables, such as $x + y = 5$, $y = x^2 - 4$, and $3r - 4s = 17$.

Unlike the one-value solution for most single-variable equations, equations with two variables usually have infinitely many solutions. Consider the equation $x + y = 5$. This equation states an algebraic relationship between a *pair* of numbers (namely, x and y): that is, their sum is 5. The solution set of $x + y = 5$ is, therefore, made up of *pairs* of numbers—an x and a corresponding y value. Thus, we are led to the following definition.

DEFINITION

The **Cartesian product** of sets A and B, denoted by $A \times B$, is the set consisting of all pairs such that the first member in the pair is an element of A and the second member in the pair an element of B. That is,

$$A \times B = \{(x,y) \mid x \in A \text{ and } y \in B\}$$

For our purposes $\mathcal{R}$, the set of real numbers, will play the role of both A and B in the preceding definition. Hence,

$$\mathcal{R} \times \mathcal{R} = \{(x,y) \mid x \in \mathcal{R} \text{ and } y \in \mathcal{R}\}.$$

In words, $\mathcal{R} \times \mathcal{R}$ is an infinite set consisting of pairs of numbers where the first number, x, is real and the second number, y, is real. Solution sets, then, to equations in two variables, will be subsets of $\mathcal{R} \times \mathcal{R}$ and, as stated previously, will usually be infinitely large. Hence, the solution set to $x + y = 5$ is the subset of $\mathcal{R} \times \mathcal{R}$ where the sum of x and y equals 5.

Solutions to equations in two variables will often be *graphed* (displayed pictorially) and the system that we will use to graph these solutions is called the **Cartesian** (or **rectangular**) **coordinate system**. To construct it, we intersect two real number lines perpendicularly at their zero positions on a flat surface (called a **plane**); the point where these lines intersect is called the **origin**. The system is depicted on the following page.

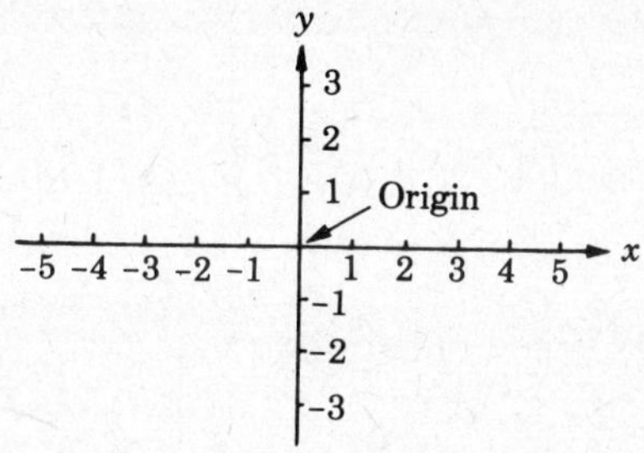

The horizontal real number line is labeled with an "x" and is called the x**-axis**. The vertical real number line is the y**-axis**. The arrows on the axes indicate positive direction.

Every point in the plane can now be associated with an **ordered pair** of numbers (x,y) in $\mathbb{R} \times \mathbb{R}$. The first component, x, represents the distance travelled horizontally from the origin and the second component, y, represents the distance travelled vertically to get to that point. For example, the point labeled $(4,-3)$ is arrived at by starting at the origin and travelling 4 units to the right and then 3 units downward (the negative sign in -3 indicates negative direction for y). The point $(4,-3)$ as well as three others are "plotted" (that is, graphed) below:

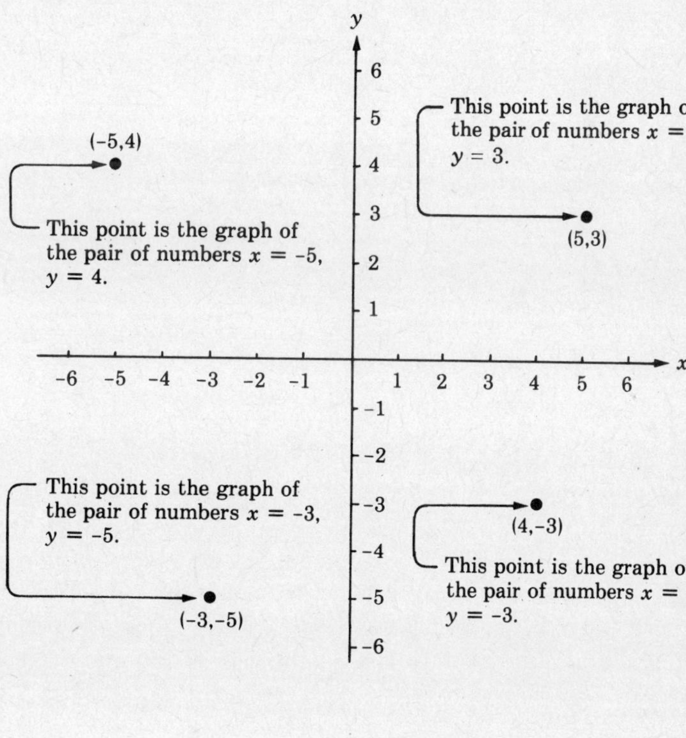

Before we proceed with examples, some definitions are in order. For the ordered pair, (x, y), the x value is called the **abscissa** and the y value is called the **ordinate**. Collectively, the abscissa and the ordinate are referred to as the **coordinates** of a point. Finally, we note that the two axes separate the plane into four regions, called **quadrants**. We label these quadrants I, II, III, and IV according to the following diagram:

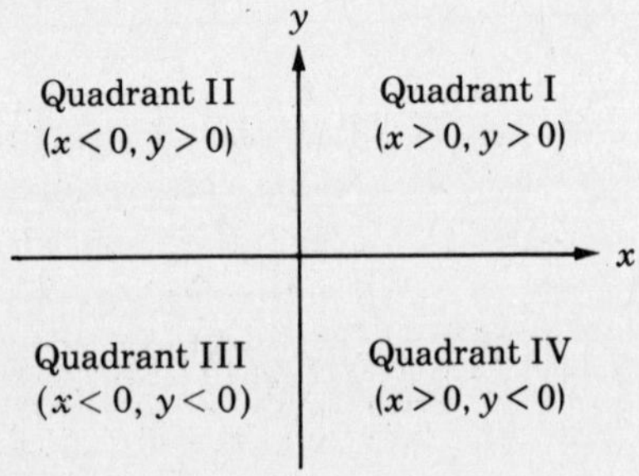

EXAMPLE 1 Graph the points $(-3,0)$, $(4,0)$, $(-2,0)$, $(0,0)$, $(1,0)$, and $(\pi,0)$. What can be said about any point whose ordinate is 0?

Solution

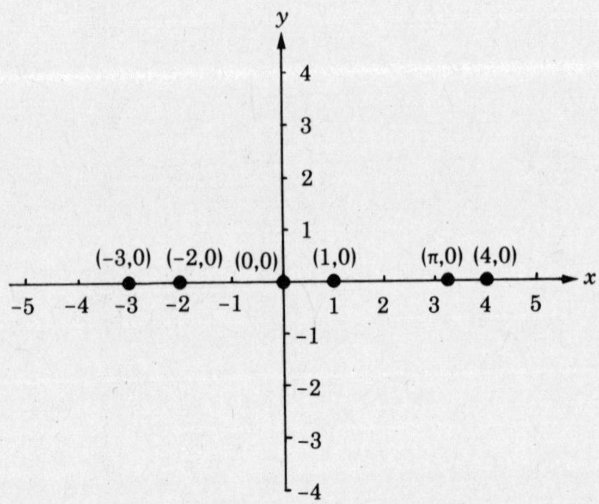

Points whose ordinates are zero must lie on the x-axis.

In future chapters, as well as the study of calculus, it will be necessary to find the *distance* between two points. The discussion that follows is the establishment of the **distance formula**.

We begin with two points, P_1 and P_2. Let P_1 be (x_1, y_1) and P_2 be (x_2, y_2). We depict them below along with the point Q, whose coordinates are (x_2, y_1):

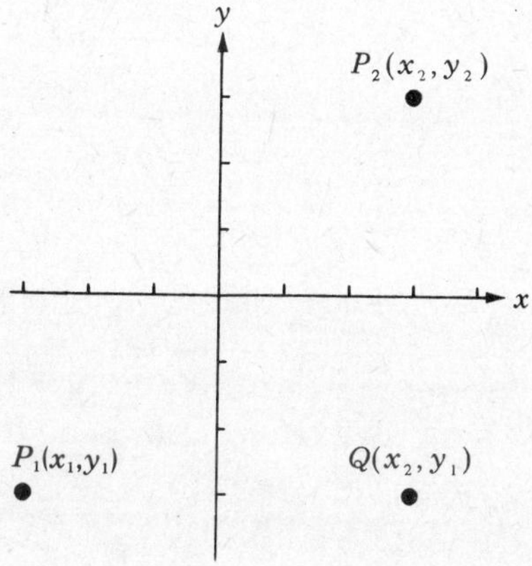

From our discussion in Chapter 1, we know that the distance between P_1 and Q is $|x_1 - x_2|$. Similarly, the distance between P_2 and Q is $|y_1 - y_2|$. Now, if we join P_1 to P_2, P_2 to Q, and Q to P_1 with three line segments, we can use the Pythagorean formula* to find the desired distance, d, between P_1 and P_2:

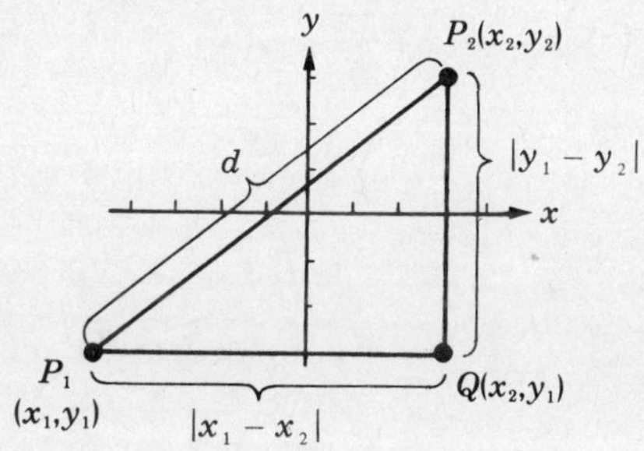

Thus,

$$d^2 = |x_1 - x_2|^2 + |y_1 - y_2|^2$$

For any real number, $|Z|^2 = Z^2$

$$d^2 = (x_1 - x_2)^2 + (y_1 - y_2)^2$$

Take the principal square root of both sides

$$d = \sqrt{(x_1 - x_2)^2 + (y_1 - y_2)^2}$$

---

*Recall from geometry, the Pythagorean formula: For the right triangle 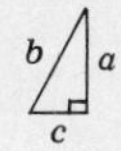 it is always true that $c^2 = a^2 + b^2$.

This proves the following theorem:

THEOREM 1

Distance Formula. If (x_1, y_1) and (x_2, y_2) are any two points in the plane, then the distance, d, between them is given by:

$$d = \sqrt{(x_1 - x_2)^2 + (y_1 - y_2)^2}$$

EXAMPLE 2 Show that the points $A(-4,4)$, $B(4,5)$, and $C(0,-2)$ form the vertices of an isosceles triangle.

Solution We let $\overline{AB}$ represent the length of the line segment from A to B. We must now show that two of the three quantities $\overline{AB}$, $\overline{BC}$, $\overline{AC}$ are equal.

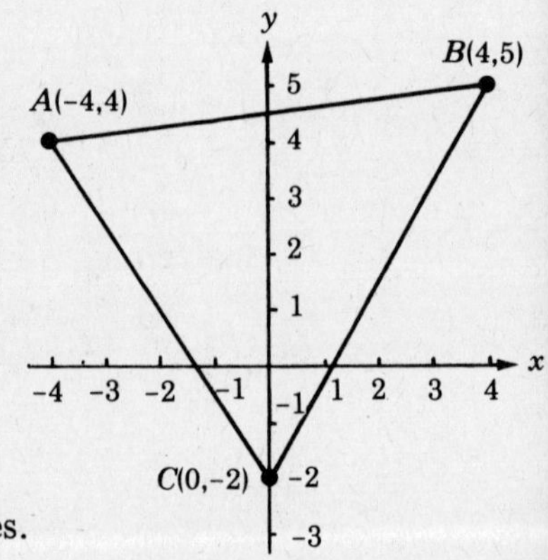

$$\overline{AB} = \sqrt{(-4-4)^2 + (4-5)^2} = \sqrt{65}$$

$$\overline{AC} = \sqrt{(-4-0)^2 + (4+2)^2} = \sqrt{52}$$

$$\overline{BC} = \sqrt{(4-0)^2 + (5+2)^2} = \sqrt{65}$$

Since $\overline{AB} = \overline{BC}$, triangle ABC is isosceles.

The proof of the following theorem is left to the reader (see Exercise 35).

THEOREM 2

Midpoint Formula. If $P_1(x_1, y_1)$ and $P_2(x_2, y_2)$ are any two points in the plane, then the coordinates of the midpoint of the line segment from P_1 to P_2 are given by:

$$\left(\frac{x_1 + x_2}{2}, \frac{y_1 + y_2}{2}\right)$$

EXAMPLE 3 Find the coordinates of the midpoint, M, of the line segment joining $A(-5,-2)$ to $B(3,6)$.

Solution Using the previous theorem,

$$M = \left(\frac{-5+3}{2}, \frac{-2+6}{2}\right) = (-1,2)$$

EXERCISES FOR SECTION 3.1

In Exercises 1 and 2, plot the given points on a Cartesian coordinate system. (See illustration preceding Example 1 and also Example 1.)

1. $A(4,5), B(-4,-5), C(-3,2), D(-1,0), E(3,0), F(2,-3), G(-3,-2)$.
2. a. $A(0,4), B(0,2), C(0,0), D(0,-2), E(0,-4), F(0,-\pi)$.
   b. What can be said about the collection of all points whose abscissa is zero?
3. Find the coordinates of the points labeled below:

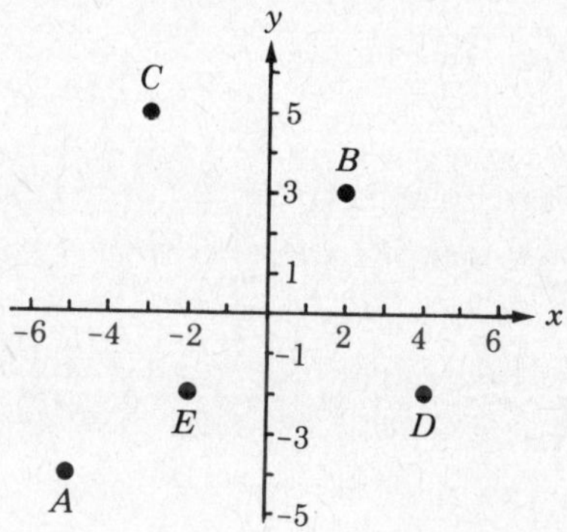

In Exercises 4 through 13, find the distance between P_1 and P_2. (Use the Distance Formula.)

4. $P_1(4,6); P_2(1,2)$
5. $P_1(6,12); P_2(1,0)$
6. $P_1(-2,7); P_2(-14,12)$
7. $P_1(-4,-11); P_2(2,-3)$
8. $P_1(3,-4); P_2(7,-1)$
9. $P_1(-1,-2); P_2(-3,-4)$
10. $P_1(4,6); P_2(4,6)$
11. $P_1(-1,1.5); P_2(3,-6.5)$
12. $P_1(1,7); P_2(-7,8)$
13. $P_1(\sqrt{2},\pi); P_2(-\sqrt{2},\pi)$
14. [True or False] Quadrant II = $\{(x,y) \mid x < 0 \text{ and } y > 0\}$.
15. [True or False] Quadrant IV = $\{(x,y) \mid x > 0 \text{ and } y < 0\}$.

In Exercises 16 through 21, a) graph the points, b) find the coordinates of the midpoint of the line segment joining them. (See Example 3.)

16. $P_1(3,5); P_2(1,1)$
17. $P_1(4,7); P_2(3,5)$
18. $P_1(-1,6); P_2(3,-8)$
19. $P_1(-1,7); P_2(6,-4)$
20. $P_1(\sqrt{2},3); P_2(\sqrt{2},4)$
21. $P_1(-\pi,7); P_2(\pi,-2)$

In Exercises 22 through 27, follow arguments similar to those explained in Example 2.

22. Show that $(-1,4)$, $(7,5)$, and $(3,-2)$ form the vertices of an isosceles triangle.
23. Show that A, B, and C are collinear (fall on the same straight line) by showing that $\overline{AB} + \overline{AC} = \overline{BC}$. In each case, graph the points.

    a. $A(4,2); B(1,-2); C(10,10)$
    b. $A(-2,-2); B(-3,-5); C(-1,1)$
24. Show that the points $(6,6)$, $(-1,-1)$, and $(-5,3)$ form the vertices of a right triangle.

25. Show that the points (−4,−1), (5,−9), and (1,1) form the vertices of a right triangle.
26. Show that the points (8,−10), (9,−11), (8,−12), and (7,−11) form the vertices of a square.
27. Show that the points (0,−2), (0,−4), (−1,−3), and (1,−3) form the vertices of a square.
28. Determine the value of x such that $(x,-1)$ is 26 units from (4, 23).
29. Determine the value of y such that $(3,y)$ is 6 units from (3,7).
30. Find x such that the point $(x,4)$ is 10 units from (4,−2).
31. Find y such that the points $(5,y)$ and (−5,19) are $10\sqrt{5}$ units apart.
32. Find the point on the y-axis that is equidistant from the points (1,−2) and (5,−1).
33. Find the point on the x-axis that is equidistant from the points (−4,7) and (3,8).
34. a. Graph 10 points whose distance from the origin is 3 units.
    b. What equation represents the set of *all* points (x,y) whose distance from (0,0) is 3 units?
    c. Find the (generalized) equation of a circle of radius r whose center is the origin.
35. The proof of Theorem 2 requires two parts:
    a. Show that $\overline{P_1M} = \overline{MP_2}$.
    b. Show that $\overline{P_1M} + \overline{MP_2} = \overline{P_1P_2}$ (that is, the points are collinear).
36. a. Plot the following points: (6,6), (4,4), (2,2), (1,1), (0,0), (−1,−1), (−3,−3), (−5,−5).
    b. What do you conjecture about the graph of $\{(x,y) \mid x = y\}$?

Computer Exercise

37. Write a BASIC program that will find the distance between any two points in the plane.

3.2 RELATIONS AND FUNCTIONS

In this section we continue to make use of the notion of ordered pairs. Any set of ordered pairs of real numbers is called a **relation**. There are several ways of describing relations:

1. We can list ordered pairs. That is,

$$R = \{(1,4), (-3,3), (-2,-4), (5,0), (3,-4)\}$$

   is a relation.

2. We can describe a relation by means of its graph. For example, the relation R is depicted in the diagram on the following page.

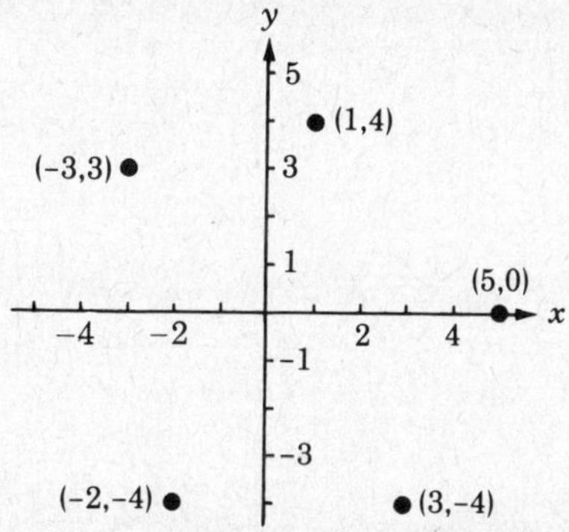

3. A relation can also be described by an equation (or inequality). For example, $S = \{(x, y) \mid y = x + 2\}$ (or simply $y = x + 2$) describes a relation that consists of an infinitude of ordered pairs of numbers.

The set of all first components (the abscissas) of the ordered pairs in a relation is called the **domain** of that relation; the set of all second components (the ordinates) of the ordered pairs in a relation is called the **range** of that relation. For example, if $T = \{(1,4), (-3,3), (-2,-4), (5,0), (3,-4)\}$ the domain of T is $\{1,-3,-2,5,3\}$ and the range of T is $\{4,3,-4,0\}$. Some examples follow.

EXAMPLE 1 Determine the domain and range of the relation graphed below.

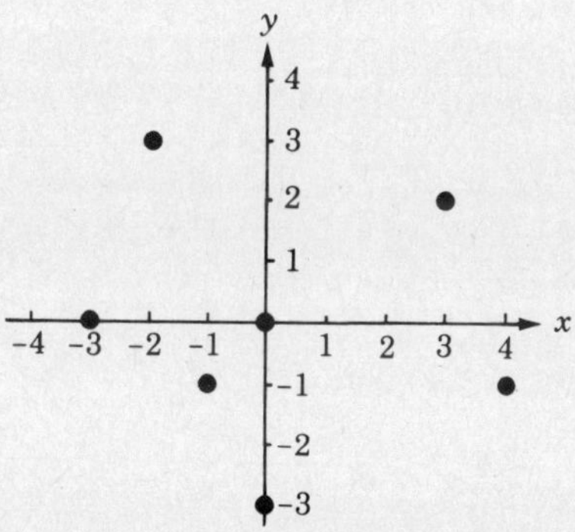

Solution The relation is $\{(-3,0), (-2,3), (-1,-1), (0,0), (0,-3), (3,2), (4,-1)\}$. The domain is $\{-3,-2,-1,0,3,4\}$; the range is $\{0,3,-1,-3,2\}$.

EXAMPLE 2 Determine the domain and range of the relation graphed below.

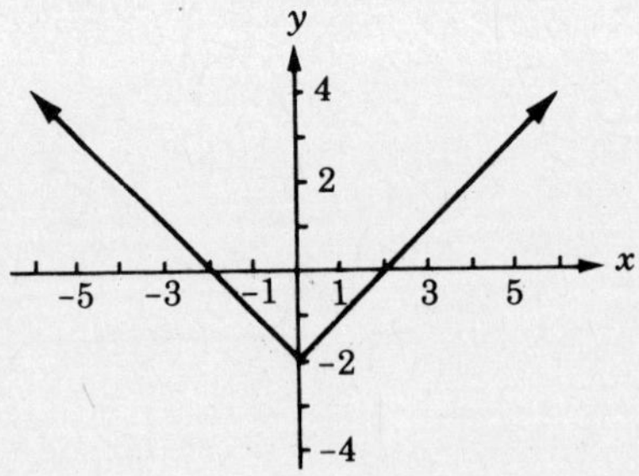

Solution The relation described by the graph contains an infinitude of ordered pairs. The arrows indicate that the graph continues indefinitely both to the left and to the right; the domain is therefore the set of real numbers $\mathcal{R}$. Since the graph does not extend lower than the point $(0,-2)$ but rises indefinitely, the range is $\{y \mid y \geqslant -2\}$.

EXAMPLE 3 Find the domain and range of the relation $\{(x,y) \mid y = x^2 + 1\}$.

Solution There is an infinitude of ordered pairs in this relation and we choose to graph the relation in order to determine its domain and range. The graph of $\{(x,y) \mid y = x^2 + 1\}$ or equivalently, the graph of $y = x^2 + 1$ can be found by listing "many" x values with their corresponding y values as shown in the table below:

x	$y = x^2 + 1$
−3	10
−2	5
−1	2
0	1
1	2
2	5
3	10
−1.5	3.25
2.5	7.25

Next, we plot these points and draw a smooth curve through them:

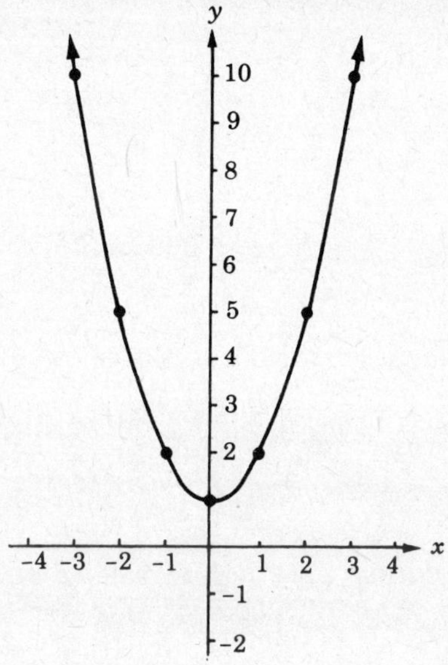

We place the arrows on the graph to indicate that for "extreme" values of x, the y values increase. Finally, the domain is $\mathcal{R}$ and the range is $\{y \mid y \geqslant 1\}$.

A relation can also be thought of as a *correspondence* (or **map**) between one set of numbers (the domain) and another set of numbers (the range). For example, the relation $W = \{(1,2), (2,-4), (3,-4)\}$ *maps* 1 to 2, *maps* 2 to -4, and *maps* 3 to -4. This can be represented in the following diagram, called a **correspondence diagram** or **map**:

W:

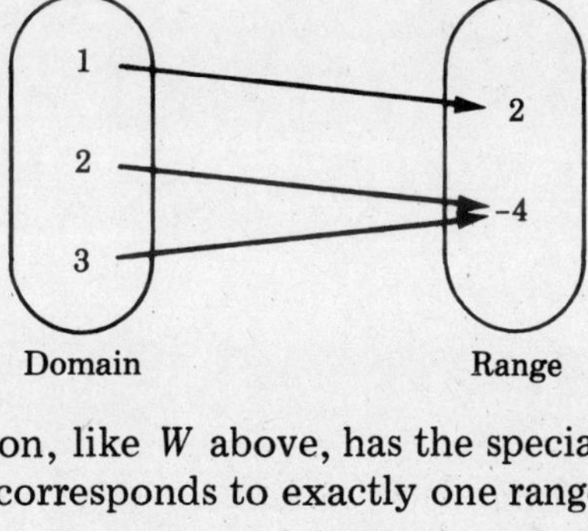

When a relation, like W above, has the special property that each domain element corresponds to exactly one range element it is called a **function**.

DEFINITION

> A function is a relation having the special property that each element in the domain corresponds to *exactly one* range element.

Some examples follow.

EXAMPLE 4 Which of the following relations are functions?

a. $f = \{(1,2), (3,4), (5,4)\}$
b. $g = \{(2,3), (3,2)\}$
c. $h = \{(1,\sqrt{2}), (-1,3), (-1,\pi)\}$

Solution

a. f is a function because each element in its domain corresponds to exactly one element in its range.
b. g is a function because each element in its domain corresponds to exactly one element in its range.
c. h is *not* a function because the domain element -1 corresponds to *two* range elements, 3 and π.

EXAMPLE 5 Is the relation described by the graph below a function? What is the relation's domain? What is the relation's range?

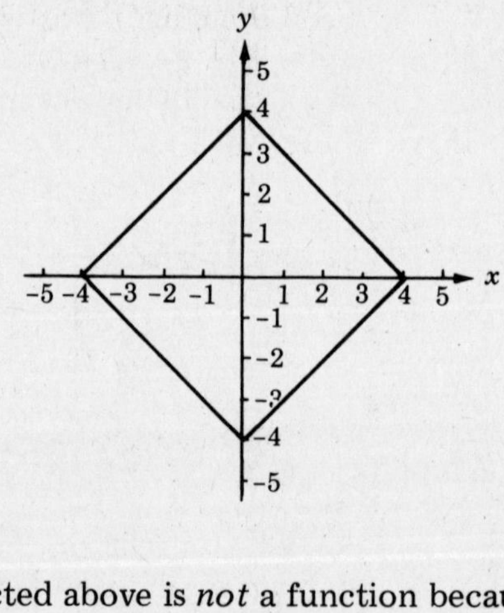

Solution The relation depicted above is *not* a function because the domain element $x = 1$, for example, corresponds to *two* elements in the range, 3 and -3:

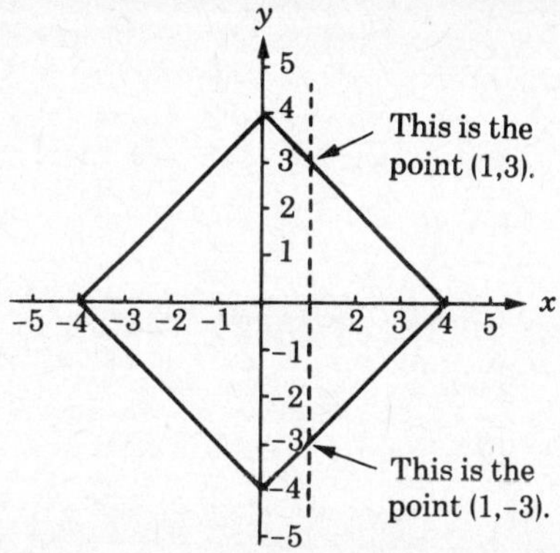

The domain is $\{x \mid -4 \leqslant x \leqslant 4\}$; the range is $\{y \mid -4 \leqslant y \leqslant 4\}$.

The example above illustrates the *vertical line test* for a function:

> **Vertical Line Test for a Function**
>
> A relation is a function if and only if a vertical line placed anywhere on the graph intersects that graph in one and only one point.

It was mentioned earlier in this section that one way of describing a relation is by means of an equation. The equation $y = 3x - 5$ states a correspondence between x and y. If x is 4, y is 7. If x is 1, y is -2. Further, notice that $y = 3x - 5$ describes a function because, given any x value, there will be one and only one value of y. When an equation (such as $y = 3x - 5$) represents a function, we often replace y by the symbol $f(x)$, $g(x)$, or $h(x)$. We read $f(x)$ as "f of x" or "f at x." Thus, $y = 3x - 5$ can be written as:

$$f(x) = 3x - 5$$

This **function notation** is used in the same way that polynomial notation was used in Chapter 1. The next two examples illustrate function notation.

EXAMPLE 6 Given $f(x) = 3x - 5$, find:

a. $f(-2)$ b. $f(0)$ c. $f(4)$ d. $f(\frac{5}{3})$

Solution a. $f(-2)$ means evaluate $3x - 5$ for $x = -2$:

$$f(x) = 3x - 5$$
$$f(-2) = 3(-2) - 5$$
$$f(-2) = -11$$

Graphically, this corresponds to the point $(-2,-11)$.

b.

$$f(x) = 3x - 5$$
$$f(0) = 3(0) - 5$$
$$f(0) = -5$$

Graphically, this corresponds to the point $(0,-5)$. At this point the graph of $f(x) = 3x - 5$ meets the y-axis. Such a point is called a **y-intercept** of the function.

c.

$$f(x) = 3x - 5$$
$$f(4) = 3(4) - 5$$
$$f(4) = 7$$

Graphically, this corresponds to the point $(4,7)$.

d.

$$f(x) = 3x - 5$$
$$f(\tfrac{5}{3}) = 3(\tfrac{5}{3}) - 5$$
$$f(\tfrac{5}{3}) = 0$$

Graphically, this corresponds to the point $(\frac{5}{3},0)$. At this point the graph of $f(x) = 3x - 5$ meets the x-axis. Such a point is called an **x-intercept** of the function.

A number a having the property that $f(a) = 0$ is called a **zero of the function** f. The graph of the function $f(x) = 3x - 5$ is shown on the following page.

EXAMPLE 7 Given the function g described by $g(x) = 2x^2 - 5x - 3$, find:

a. $g(-1)$ b. $g(\frac{1}{2})$ c. $g(a + 2)$ d. Zeros of $g(x)$

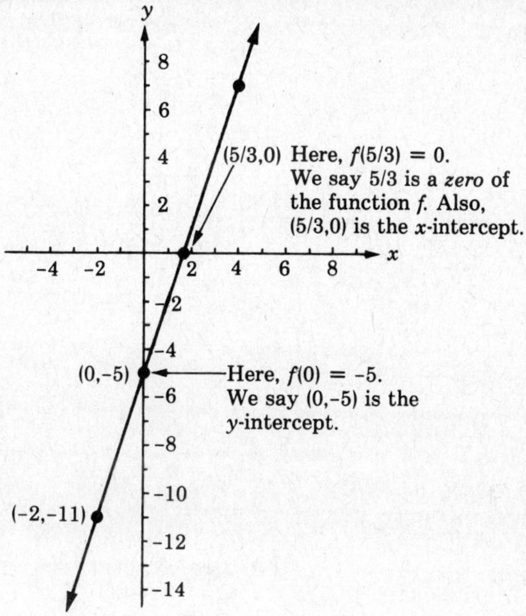

Solution a. $g(x) = 2x^2 - 5x - 3$

$g(-1) = 2(-1)^2 - 5(-1) - 3$

$g(-1) = 4$

b. $g(x) = 2x^2 - 5x - 3$

$g(\tfrac{1}{2}) = 2(\tfrac{1}{2})^2 - 5(\tfrac{1}{2}) - 3$

$g(\tfrac{1}{2}) = -5$

c. $g(x) = 2x^2 - 5x - 3$

$g(a + 2) = 2(a + 2)^2 - 5(a + 2) - 3$

$g(a + 2) = 2a^2 + 3a - 5$

d. To obtain zeros of $g(x)$ we let $g(x) = 0$. Thus,

$$2x^2 - 5x - 3 = 0$$
$$(2x + 1)(x - 3) = 0$$
$$x = -\tfrac{1}{2},\ x = 3$$

The function concept is fundamental to that branch of mathematics called calculus. In particular, the quantity

$$\frac{f(x + h) - f(x)}{h} \qquad (h \neq 0)$$

is one that must frequently be found. The next example illustrates how to determine $\dfrac{f(x + h) - f(x)}{h}$.

EXAMPLE 8 If $f(x) = x^3 - 7x + 5$, find $\dfrac{f(x + h) - f(x)}{h}$.

Solution First, we determine $f(x + h)$:

$$f(x) = x^3 - 7x + 5$$
$$f(x + h) = (x + h)^3 - 7(x + h) + 5$$
$$f(x + h) = x^3 + 3x^2h + 3xh^2 + h^3 - 7x - 7h + 5$$

Now, since $f(x) = x^3 - 7x + 5$, then

$$f(x + h) - f(x) = 3x^2h + 3xh^2 + h^3 - 7h.$$

Thus, $\dfrac{f(x + h) - f(x)}{h} = 3x^2 + 3xh + h^2 - 7,\ h \neq 0.$

Another notation used in calculus that we mention here is the **composite function.**

DEFINITION Given two functions $g(x)$ and $h(x)$, the composite function $g \circ h(x)$ is defined by

$$g \circ h(x) = g(h(x)).$$

EXAMPLE 9 Given $f(x) = x^2 + 1$ and $g(x) = 5x - 1$,

a. Determine $f \circ g(2)$
b. Determine $g \circ f(2)$
c. Determine $f \circ g(x)$
d. Determine $g \circ f(x)$

Solution

a. $f \circ g(x) = f(g(x))$
$f \circ g(2) = f(g(2))$
$f \circ g(2) = f(9)$
$f \circ g(2) = 82$

b. $g \circ f(x) = g(f(x))$
$g \circ f(2) = g(f(2))$
$g \circ f(2) = g(5)$
$g \circ f(2) = 24$

c. $f \circ g(x) = f(g(x))$
$f \circ g(x) = f(5x - 1)$
$f \circ g(x) = (5x - 1)^2 + 1$
$f \circ g(x) = 25x^2 - 10x + 2$

d. $g \circ f(x) = g(f(x))$
$g \circ f(x) = g(x^2 + 1)$
$g \circ f(x) = 5(x^2 + 1) - 1$
$g \circ f(x) = 5x^2 + 4$

Comparing the results in the example above shows that, in general, $f \circ g(x) \neq g \circ f(x)$.

EXAMPLE 10 The volume of a spherical balloon of radius r is given by $V = f(r) = \frac{4}{3}\pi r^3$. Suppose air is being pumped into the balloon causing its radius to increase at the constant rate of 4 in/sec, that is, $r = g(t) = 4t$.

a. Determine $f \circ g(t)$.
b. Determine volume as a function of time.
c. What is the relationship between parts a and b?

Solution a. $f \circ g(t) = f(g(t))$
$f \circ g(t) = f(4t)$
$f \circ g(t) = \frac{4}{3}\pi(4t)^3$
$f \circ g(t) = \frac{256}{3}\pi t^3$

b. To find V as a function of t we write $V = \frac{4}{3}\pi r^3$ and then replace r by its equal, $4t$. Thus,

$$V = \frac{4}{3}\pi(4t)^3 = \frac{256}{3}\pi t^3.$$

c. $V = f \circ g(t)$. In other words, the notion of composite function can be used to eliminate a variable, like r, between two equations.

EXERCISES FOR SECTION 3.2

For each of the relations in Exercises 1 through 22: a) find the domain, b) find the range, c) determine whether or not the relation is a function. (See Examples 1, 2, 4, and 5.)

1. The relation represented by:

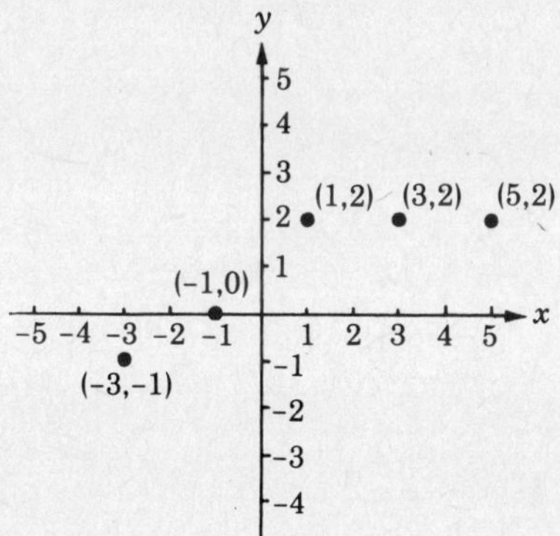

2. The relation represented by:

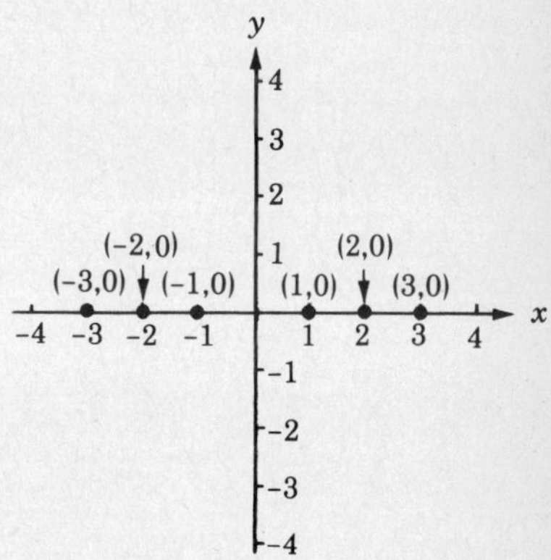

3. The relation represented by:

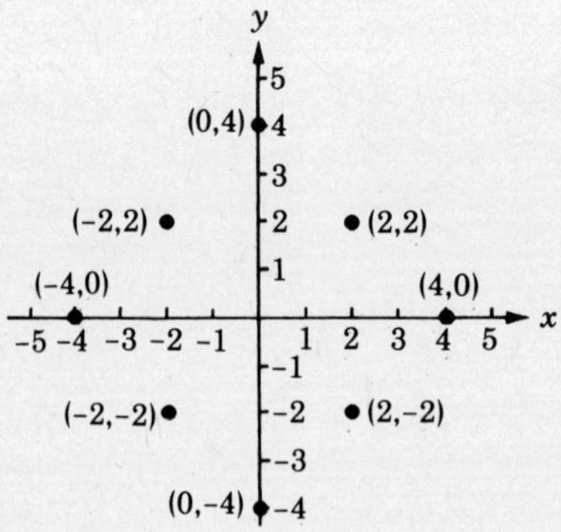

4. The relation represented by:

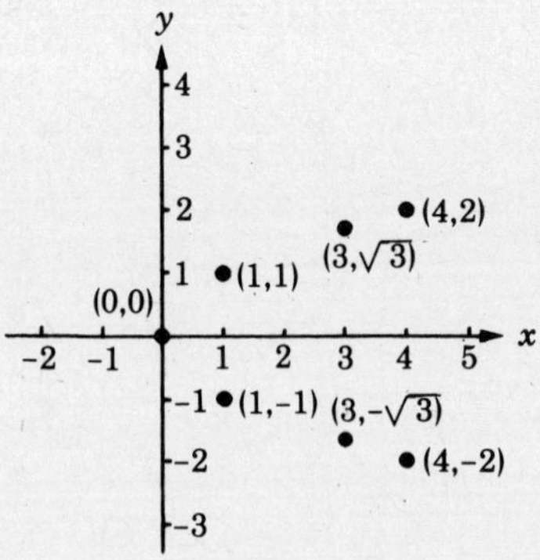

5. The relation represented by:

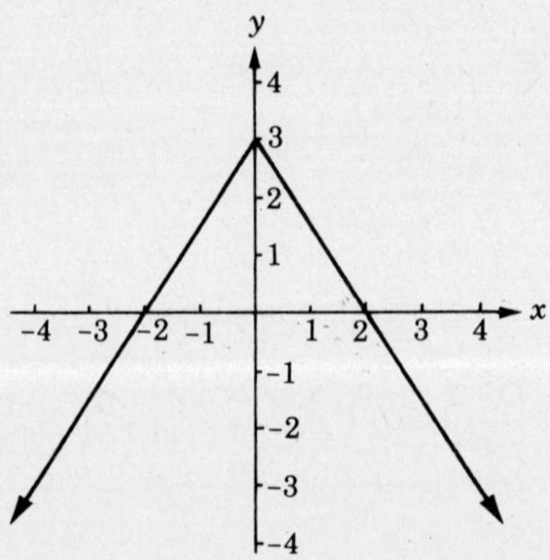

6. The relation represented by:

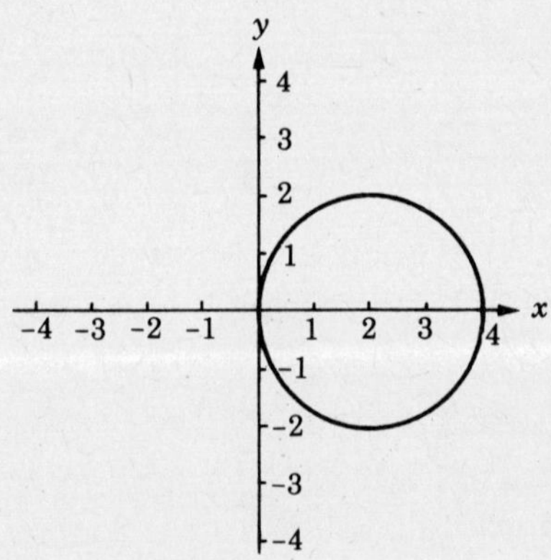

7. The relation represented by:

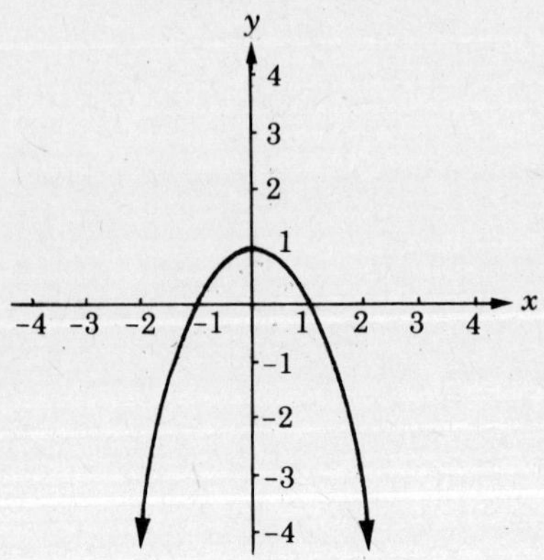

8. The relation represented by:

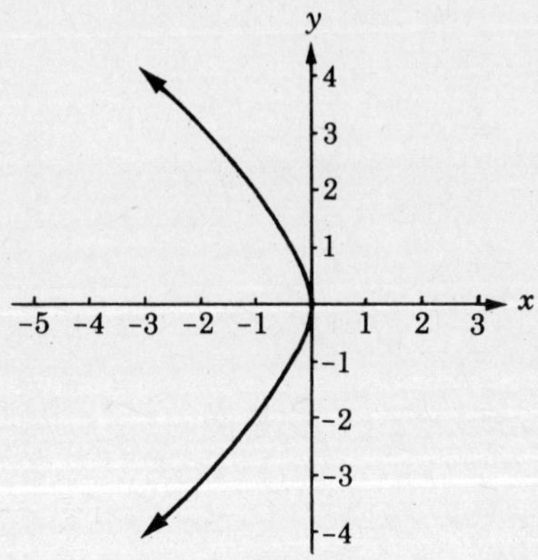

9. The relation represented by:

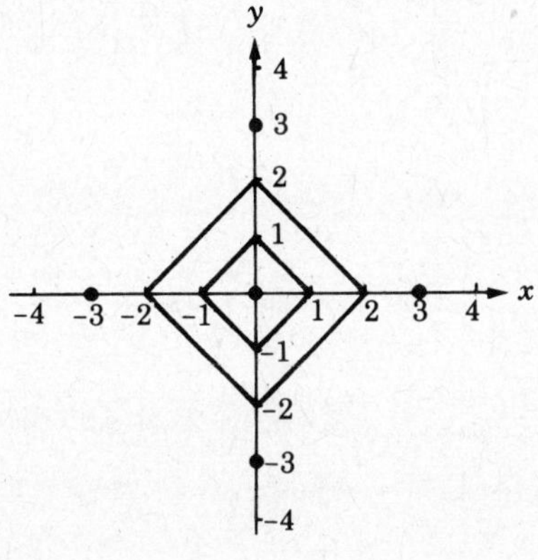

10. The relation represented by:

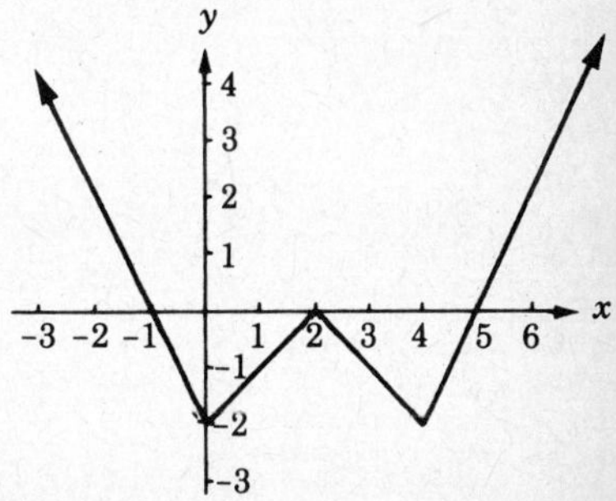

11. $R_1 = \{(6,0), (0,6)\}$

12. $R_2 = \{(1,4), (1,5), (-2,-3)\}$

13. $R_3 = \{(-\sqrt{2},0), (\sqrt{2},0\}$

14. $R_4 = \{(2,3), (-2,-3), (4,1), (1,4)\}$

15. $R_5 = \{(3,12)\}$

16. $R_6 = \{(4,0), (5,0), (6,0), (0,0)\}$

17. $R_7 = \{(12,3), (12,6), (12,9), (12,11)\}$

18. $R_8 = \{(0,0), (10,0), (-10,0), (0,-10)\}$

19.

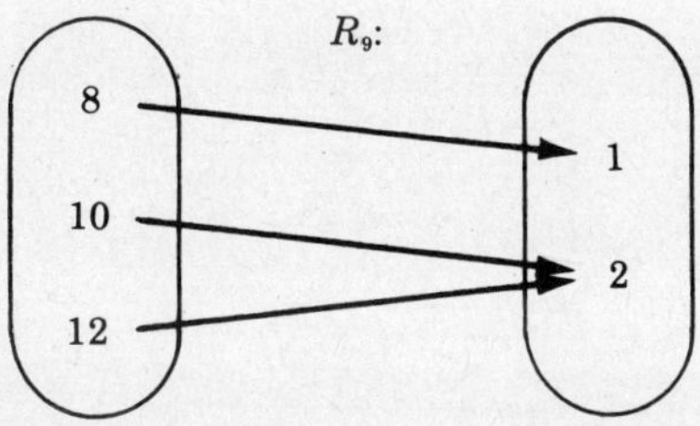

20.

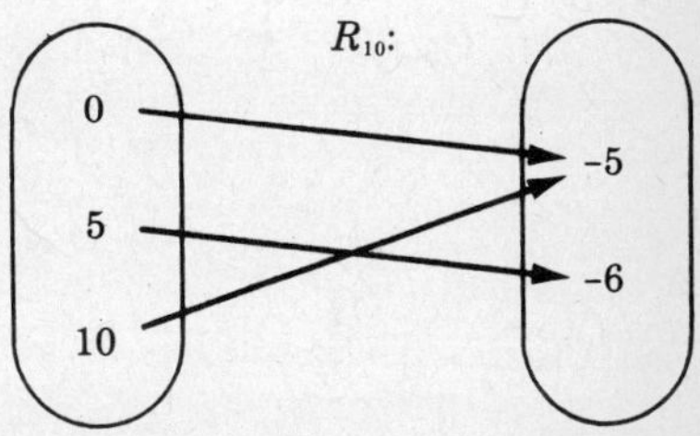

21.

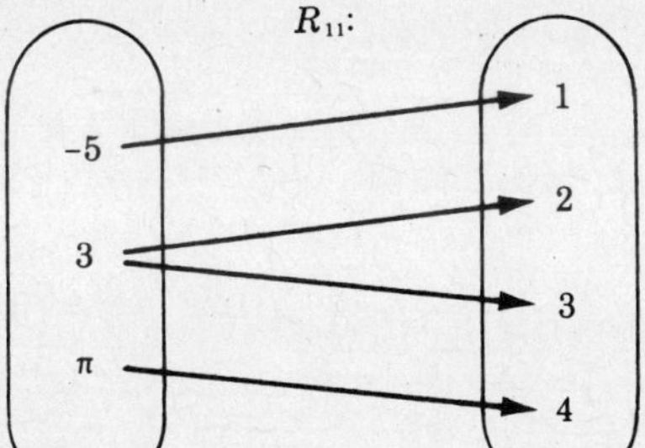

22.

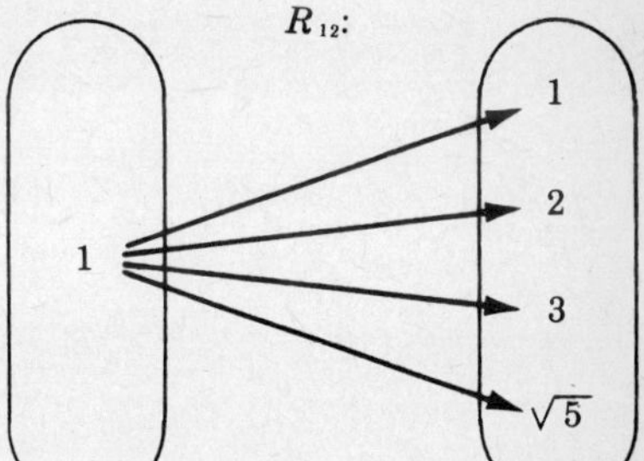

In Exercises 23 through 26: a) sketch the graph by determining at least seven ordered pairs that are elements of the given relation, b) determine the domain, c) determine the range, d) determine whether or not the relation is a function. (See Example 3.)

23. $\{(x,y) \mid y = 4x + 1\}$
24. $\{(x,y) \mid y = x^2\}$
25. $\{(x,y) \mid y = x^2 - 4\}$
26. $\{(x,y) \mid x = y^2 + 1\}$

In Exercises 27 through 35 find: a) $f(0)$ b) $f(-3)$ c) $f(2)$
d) $\dfrac{f(x+h) - f(x)}{h}$ $(h \neq 0)$ (See Examples 6, 7, and 8.)

27. $f(x) = 2x + 1$
28. $f(x) = x^2 + 7$
29. $f(x) = \dfrac{x-1}{2}$
30. $f(x) = 3x^3 + 2x - 4$
31. $f(x) = \dfrac{2}{x-2}$
32. $f(x) = \dfrac{x}{x-1}$
33. $f(x) = \sqrt{x+3}$ (*Hint:* for part d), rationalize the numerator.)
34. $f(x) = 7$
35. $f(x) = mx + b$

In Exercises 36 through 41, find a) the y-intercept and b) the zeros of each function. (See Example 6 and 7.)

36. $f(x) = 3x - 2$
37. $f(x) = 3x^2 - 8x - 3$
38. $g(x) = x^2 - 11x + 30$
39. $g(x) = 2x^2 - x - 4$
40. $h(x) = 2 + x$
41. $h(x) = mx + b$

In Exercises 42 through 45, find a) $f \circ g(x)$, b) $g \circ f(x)$ (See Example 9.)

42. $f(x) = 3x - 1$; $g(x) = 1 - 3x$
43. $f(x) = \dfrac{1}{x^2}$; $g(x) = -x + 3$
44. $f(x) = x^2 - 2$; $g(x) = 2x + 1$
45. $f(x) = 2$; $g(x) = x^2 + 2x$
46. The area of a circle is $A = f(r) = \pi r^2$. Suppose the radius r is increasing according to the relationship $r = g(t) = t + 2$, where t is time.
    a. Determine $f \circ g(t)$.
    b. Determine A as a function of time.
    c. Compare parts a and b.
47. A manufacturer of tennis racquets has projected that her cost, C, of making x tennis racquets is given by $C = f(x) = 200 + 10x$. Furthermore, the income, S, from selling all the racquets is given by $S = g(C) = 1.4C$.
    a. Determine $g \circ f(x)$.
    b. Determine S as a function of x.
    c. Compare parts a and b.

A function f is called a **one-to-one function** if and only if every range element is associated with *exactly one* domain element. In Exercises 48 through 57, determine whether or not the given functions are one-to-one functions.

48. $\{(2,7), (3,8), (9,9)\}$
49. $\{(2,7), (3,8), (9,8)\}$
50. $\{(0,0), (5,5), (3,5)\}$
51. $\{(4,6), (6,4)\}$

52.

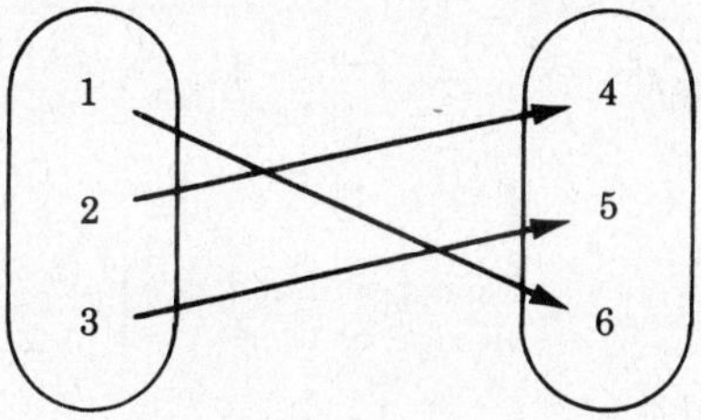

53.

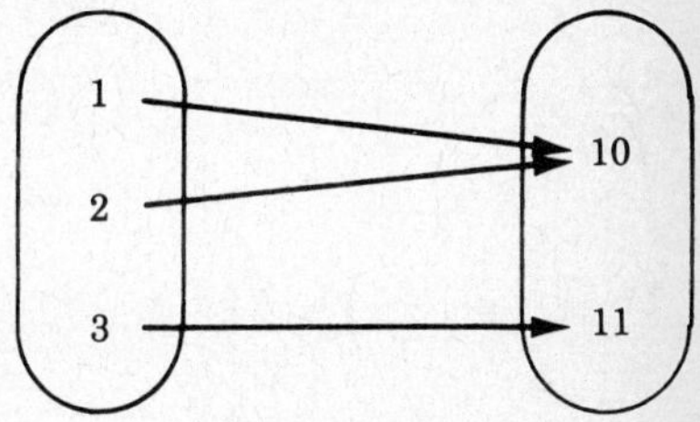

54. $\{(x, y) \mid y = x^2\}$

55. $\{(x, y) \mid y = x + 5\}$

56. $f(x) = \dfrac{1}{x}$

57. $g(x) = \dfrac{1}{x^2}$

58. The function, $f(x) = x$, is called the **identity function.** Notice that each element *corresponds* (or "is mapped") to *itself* by evaluating the following:
a. $f(5)$ b. $f(-6)$ c. $f(0)$ d. $f(1)$
e. $f(a^2 + 7)$ f. $\dfrac{f(x+h) - f(x)}{h}$ $(h \neq 0)$

Computer Exercises

A function is defined in BASIC as follows:

```
10 DEF FNF(X) = X↑2
```

Statement 10, for example, defines $f(x)$ to be x^2. Write a BASIC program to find $f(0)$, $f(1)$, and $f(2)$ for each of the following functions:

59. $f(x) = x^2$

60. $f(x) = \dfrac{1}{x - 2}$

61. $f(x) = \sqrt{x^2 + 1}$

62. $f(x) = \sqrt{1 - x^2}$

3.3 GRAPHING LINEAR FUNCTIONS

In the previous section, we observed that one way of describing a relation in two variables was by means of an equation. In the remainder of this chapter, we will be concerned with graphing relations when they are described by an equation. For example, consider the equation $y = 2x + 3$. As pointed out previously, each solution to $y = 2x + 3$ is an ordered pair of numbers, each of which can be graphed as a point in the Cartesian coordinate system. Since for *any* real number x there will be a corresponding y value, there are infinitely many solutions to $y = 2x + 3$. We list six of them in the following table.

x	y
−4	−5
−2	−1
−1	1
0	3
1	5
1.5	6

These six solutions are graphed below.

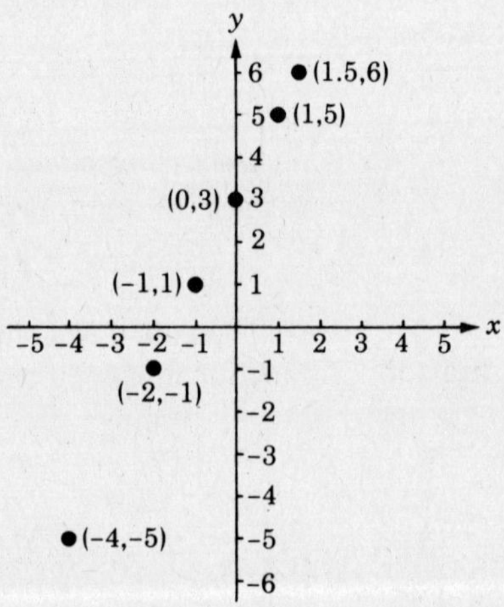

The graph above gives us a good idea as to where the other solutions to $y = 2x + 3$ are located: namely, on the straight line connecting the six points. The complete graph of $y = 2x + 3$ follows.

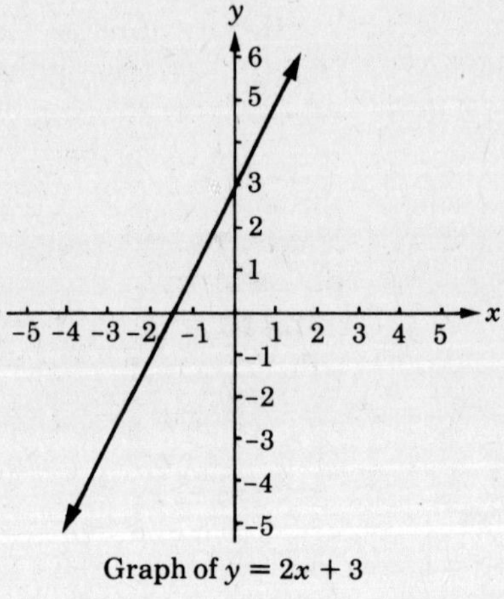

Graph of $y = 2x + 3$

The following observations can be made from the graph above:

1. The line, extending indefinitely in both directions (represented by the arrows), represents *all* solutions to $y = 2x + 3$.
2. The equation $y = 2x + 3$ describes a function. (The Vertical Line Test verifies this.) We might also write $f(x) = 2x + 3$.
3. The domain of this function is $\mathcal{R}$ because the graph extends in- indefinitely to the left and right.
4. The range of this function is $\mathcal{R}$ because the graph extends indefinitely upward and downward.

We have the following:

DEFINITION Let m and b be real numbers. The function $f(x) = mx + b$ is called a **linear function.**

In the previous illustration, $y = 2x + 3$, notice that m is 2 and b is 3. Some examples follow.

EXAMPLE 1
a. Graph $f(x) = -2x + 1$.
b. What are the values of m and b?
c. State the domain and range of f.

Solution a. We choose three values for x and find the corresponding values of $f(x)$ and list them with the graph below.

x	$y = f(x)$
2	−3
0	1
−2	5

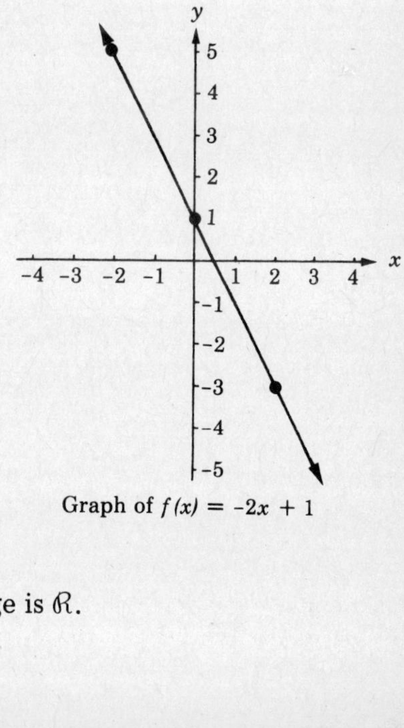

Graph of $f(x) = -2x + 1$

b. $m = -2$, $b = 1$

c. The domain is $\mathcal{R}$ and the range is $\mathcal{R}$.

EXAMPLE 2
a. Graph $f(x) = 2$.
b. What are the values of m and b?
c. State the domain and range of f.

Solution
a. Regardless of what x value we choose, the corresponding y value *must* be 2. The graph is the horizontal line shown below.

x	y
−3	2
0	2
3	2

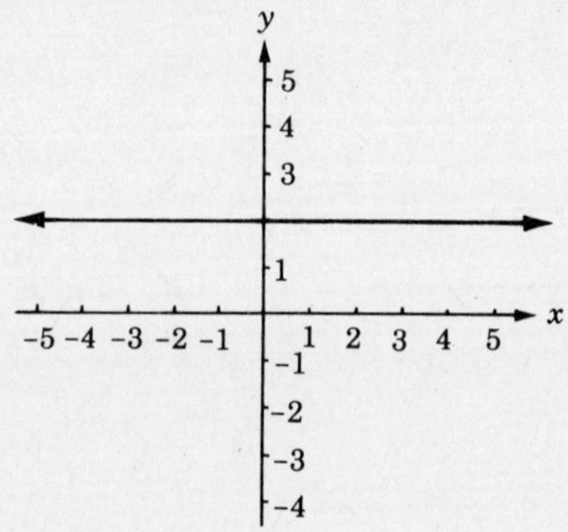

b. m is 0; b is 2.

c. The domain is $\mathcal{R}$; the range is $\{y \mid y = 2\}$.

The previous examples can be summarized by the following facts:

1. The graph of $y = mx + b$ is a straight line.
2. The graph passes through $(0, b)$.
3. The domain of $y = mx + b$ is $\mathcal{R}$.
4. If $m \neq 0$, the range of $y = mx + b$ is $\mathcal{R}$.
5. If $m = 0$, the graph of $y = mx + b$ is a horizontal line (through $(0,b)$) with range $\{y \mid y = b\}$.

Since the graph of $y = mx + b$ passes through the y-intercept $(0,b)$, we often refer to b, itself, as the y-intercept. What follows is an interpretation of m in the linear function $y = mx + b$.

Suppose we have two points, call them (x_1, y_1) and (x_2, y_2), that lie on the line $y = mx + b$. Since they lie on the line, both (x_1, y_1) and (x_2, y_2) satisfy the equation $y = mx + b$. That is,

$$y_1 = mx_1 + b \tag{1}$$

$$y_2 = mx_2 + b \tag{2}$$

If equation (2) is subtracted from equation (1), we obtain:

$$y_1 - y_2 = mx_1 + b - (mx_2 + b)$$
$$y_1 - y_2 = mx_1 - mx_2$$
$$m = \frac{y_1 - y_2}{x_1 - x_2}$$

Solve for m

Thus, m is the difference (or change) of the ordinates divided by the difference (or change) of the abscissas, for *any two points on the line.* The ratio $m = \frac{y_1 - y_2}{x_1 - x_2}$ is called the **slope** of the line and, as the graph below illustrates, it is a measure of the line's inclination.

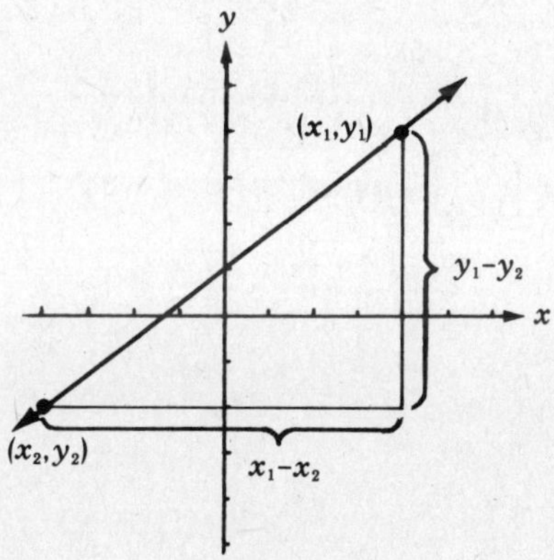

EXAMPLE 3 Find the equation of the line whose slope is 4 and whose y-intercept is −3.

Solution Since $m = 4$ and $b = -3$, the equation of the line is $y = 4x - 3$.

EXAMPLE 4 Find the slope of the straight line connecting the points (−4,2) and (−3,0). Graph the line.

Solution Let (x_1, y_1) be (−4,2) and (x_2, y_2) be (−3,0). Then

$$m = \frac{y_1 - y_2}{x_1 - x_2} = \frac{2 - 0}{-4 - (-3)} = -2$$

The slope is −2. The graph appears as follows:

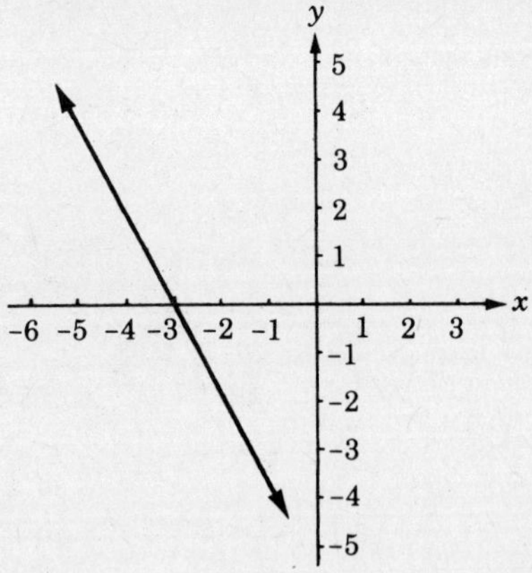

The equation $y = mx + b$ is referred to as the **slope-intercept form** of a straight line. If we are given any point (x_1, y_1) on a line and the line's slope m, we can derive another form of a stright line:

$$y = mx + b$$
Since (x_1, y_1) is on the line
$$y_1 = mx_1 + b$$
Solve for b
$$b = y_1 - mx_1$$
Substitute for b in $y = mx + b$
$$y = mx + y_1 - mx_1$$
$$y = m(x - x_1) + y_1$$

Point-Slope Form of a Straight Line

The straight line that has slope m and passes through the point (x_1, y_1) is described by the equation

$$y = m(x - x_1) + y_1.$$

EXAMPLE 5 Find the equation of the line that has a slope of 2 and passes through (3,5).

Solution Using the point-slope form, we have:

$$y = m(x - x_1) + y_1$$
$$y = 2(x - 3) + 5$$
$$y = 2x - 1.$$

EXAMPLE 6 a. Find the equation of the line passing through the points $(-4,2)$ and $(-3,0)$.
b. What is the y-intercept?

Solution a. First we find the slope:

$$m = \frac{2-0}{-4-(-3)} = -2$$

Now, we use the point-slope form:

$$y = m(x - x_1) + y_1$$
$$y = -2(x - x_1) + y_1$$
$$y = -2(x + 4) + 2$$
$$y = -2x - 6$$

We use $(-4,2)$ for (x_1, y_1); $(-3,0)$ would also work

b. $b = -6$. Incidentally, this line was graphed in Example 4.

In the next example we use the fact that if two lines are parallel, their slopes are equal (see Exercise 40).

EXAMPLE 7 Determine the equation of the line that passes through the point $(1,-3)$ and is parallel to the line $2x + y = 4$.

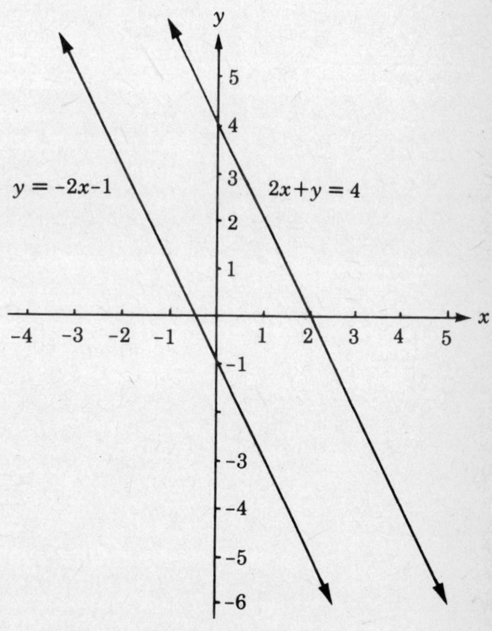

Solution With $m = -2$ (why?) we use the point-slope form:

$$y = m(x - x_1) + y_1$$
$$y = -2(x - 1) - 3$$
$$y = -2x - 1$$

The two lines are graphed on the right.

EXERCISES FOR SECTION 3.3

In Exercises 1 through 12: a) graph the function, b) find the slope and y-intercept, c) state the domain and range. (See Examples 1 and 2.)

1. $f(x) = -3x + 2$
2. $f(x) = 4x - 1$
3. $y = x + 1$
4. $f(x) = 6$
5. $f(x) = -3$
6. $y = 2$
7. $y = -x + \frac{1}{2}$
8. $g(x) = 0$
9. $h(x) = x$
10. $f(x) = -2 - x$
11. $f(x) = 1 - x$
12. $f(x) = \frac{1 - 2x}{2}$

In Exercises 13 through 21, find the slope of the straight line passing through the given points. (See Example 4.)

13. $(-4,3)$ and $(-3,0)$
14. $(6,7)$ and $(5,2)$
15. $(2,5)$ and $(0,0)$
16. $(2,-7)$ and $(4,1)$
17. $(-1,-6)$ and $(-3,-3)$
18. $(2,-4)$ and $(-1,-5)$
19. $(1,1+\sqrt{3})$ and $(2,1-\sqrt{3})$
20. $(-3,2)$ and $(-4,2)$
21. $(-3,2)$ and $(-3,4)$

In Exercises 22 through 39, determine the equation of the line described. (See Examples 3, 5, 6, and 7.)

22. slope = 2; y-intercept = 1
23. slope = -2; y-intercept = 4
24. slope = 0; y-intercept = 3
25. slope = 3; y-intercept = 0
26. passes through $(3,4)$; slope = 1
27. passes through $(-3,4)$; slope = 1
28. passes through $(-3,4)$; slope = -1
29. passes through $(-3,-4)$; slope = -1
30. passes through $(-1,5)$; slope = 2
31. passes through $(-6,-1)$; slope = -1
32. passes through $(3,-5)$; slope = -3
33. passes through $(2,5)$ and $(-1,-4)$
34. passes through $(0,0)$ and $(1,5)$
35. passes through $(1,4)$ and $(-3,4)$
36. passes through $(2,9)$ and $(1,5)$
37. passes through $(-2,4)$ and $(3,9)$
38. passes through $(1,-7)$ and $(-2,-3)$
39. passes through $(-3,8)$ and $(2,8)$

40. Show that if two lines are parallel, their slopes are equal. Start by letting the first line be $y = m_1x + b_1$ and the other be $y = m_2x + b_2$ $(b_1 \neq b_2)$ and show that if (x_1,y_1) and (x_2,y_2) are on the first line then $(x_1,y_1-(b_1-b_2))$ and $(x_2,y_2-(b_1-b_2))$ must be on the second line and by direct calculation, $m_2 = m_1$. See the following diagram.

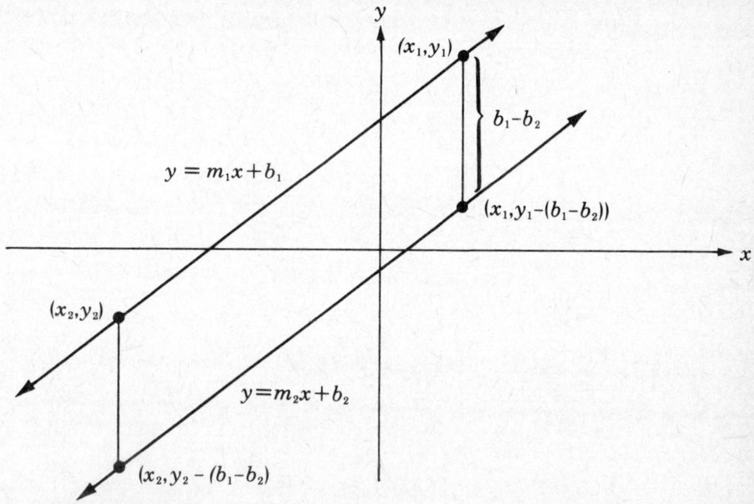

41. What is the equation of the horizontal line passing through (4, 3)?
42. What is the equation of the x-axis?
43. If two lines, $y = m_1x + b_1$ and $y = m_2x + b_2$ are perpendicular, then $m_1 \cdot m_2 = -1$ (provided neither line is horizontal).
    a. What is the equation of the line perpendicular to $y = 2x + 1$ and passing through (3, 5)?
    b. What is the equation of the line perpendicular to $2x + 3y = 6$ and passing through the origin?
44. The point-slope form $y = m(x - x_1) + y_1$ can be rewritten if we substitute $\dfrac{y_1 - y_2}{x_1 - x_2}$ for m:

$$y = \frac{y_1 - y_2}{x_1 - x_2}(x - x_1) + y_1$$

This resulting equation is referred to as the **two-point form** of a linear equation. Use this form to determine the equation of the line passing through each of the following pairs of points:
    a. (6, 7) and (5, 5)   b. (4, −4) and (2, −3)
45. Show that the linear function $f(x) = mx + b$ is a one-to-one function provided $m \neq 0$. (See explanation of one-to-one function preceding Exercise 48 in Section 3.2.) (*Hint:* show that if $x_1 \neq x_2$, then $f(x_1) \neq f(x_2)$.)

3.4 GRAPHING QUADRATIC FUNCTIONS

In Chapter 2, we examined quadratic equations that contained one variable. Recall, the general quadratic equation in one variable is

$$ax^2 + bx + c = 0 \qquad (a \neq 0)$$

In this section, we will be concerned with graphing quadratic equations in two variables whose general equation is:

$$y = ax^2 + bx + c \qquad (a \neq 0)$$

Such an equation was graphed in Example 3 of Section 3.2; there the graph of $y = x^2 + 1$ was found by plotting a large number of ordered pairs. The examples that follow use a more analytical method for graphing quadratic equations in two variables.

EXAMPLE 1 Graph $y = x^2 - 2x - 3$.

Solution Points of particular assistance in graphing are the *intercepts*—the x-intercepts and the y-intercepts. The x-intercepts (or *zeros*) are found by setting y equal to zero.

$$x^2 - 2x - 3 = y$$

$$x^2 - 2x - 3 = 0$$

To find x-intercepts, set y equal to 0

$$(x - 3)(x + 1) = 0$$

$$x - 3 = 0 \quad \Big| \quad x + 1 = 0$$

$$x = 3 \quad \Big| \quad x = -1$$

← These are the zeros of $y = x^2 - 2x - 3$.

Thus, the two x-intercepts are $(3,0)$ and $(-1,0)$. The y-intercepts are found by setting x equal to zero:

$$y = x^2 - 2x - 3$$

$$y = 0^2 - 0 - 3$$

$$y = -3$$

The y-intercept is $(0,-3)$. We list the three intercepts with four other points in table form below.

x	y
3	0
−1	0
0	−3
−2	5
1	−4
2	−3
4	5

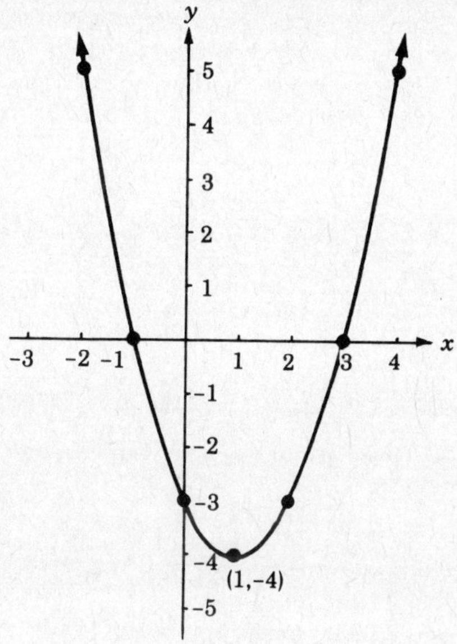

Graph of $y=x^2-2x-3$

The following observations should be made from studying the previous example:

1. The equation $y = x^2 - 2x - 3$ represents a function (the Vertical Line Test verifies this fact). In general, $f(x) = ax^2 + bx + c$ is a function; it is called the **quadratic function.**
2. The x-intercepts (or zeros) are found by setting $y = 0$ in the equation $y = ax^2 + bx + c$ and then solving for x.
3. The y-intercept is found by setting $x = 0$ in the equation $y = ax^2 + bx + c$ and solving for y.

The graph presented in the previous example is called a **parabola.** Although the shape and position of a parabola with respect to the axes change as a, b, and c change, the quadratic equation $y = ax^2 + bx + c$ always represents a parabola. In the case of $y = x^2 - 2x - 3$ above, we say the parabola *opens upward*. Also, the point $(1,-4)$ is the "turning point" of this parabola; it is called the **vertex.** In the case of the upward opening parabola $y = x^2 - 2x - 3$, the vertex $(1,-4)$ represents the parabola's *minimum point*. Finally, notice that the domain of $y = x^2 - 2x - 3$ is $\mathcal{R}$; the range is $\{y \mid y \geqslant -4\}$. The next example illustrates a parabola that *opens downward.*

EXAMPLE 2 Graph $y = -x^2 + 6x - 5$.

Solution The x-intercepts are $(1,0)$ and $(5,0)$ and the y-intercept is $(0,-5)$. Other points are listed below.

x	y
1	0
5	0
0	−5
2	3
3	4
4	3
6	−5

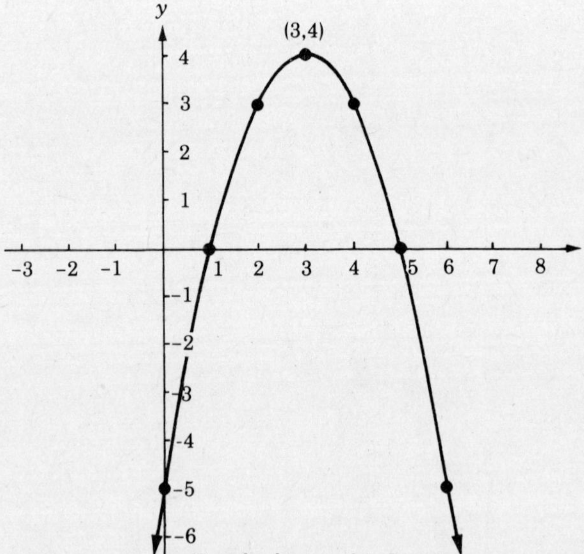

Graph of $y = -x^2 + 6x - 5$

Notice that the domain is $\Re$, the range is $\{y \mid y \leqslant 4\}$, and the vertex (*a maximum point*) is $(3,4)$. Also, the parabola opens downward.

A vertical line through the vertex of a parabola whose equation is $f(x) = ax^2 + bx + c$ is called the *axis of symmetry* because the left- and right-hand parts of the parabola are symmetrical.

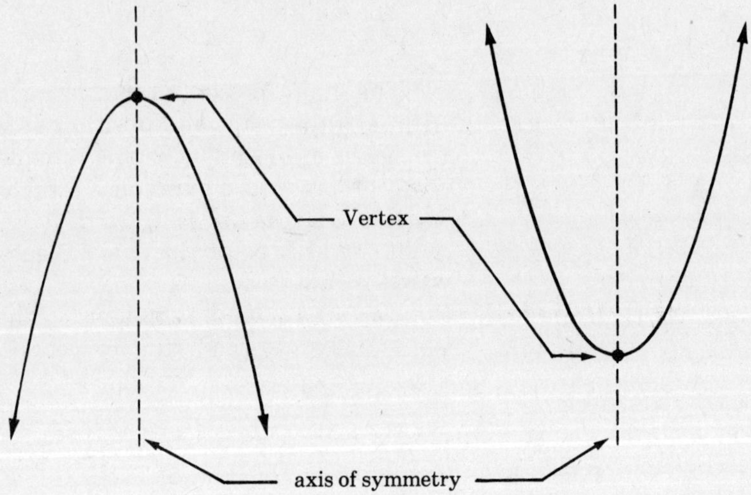

We now use the symmetry of the parabola to derive an important result. The graph of $f(x) = ax^2 + bx + c$ is a parabola that opens either upward or downward. For discussion purposes, let us consider the following diagram.

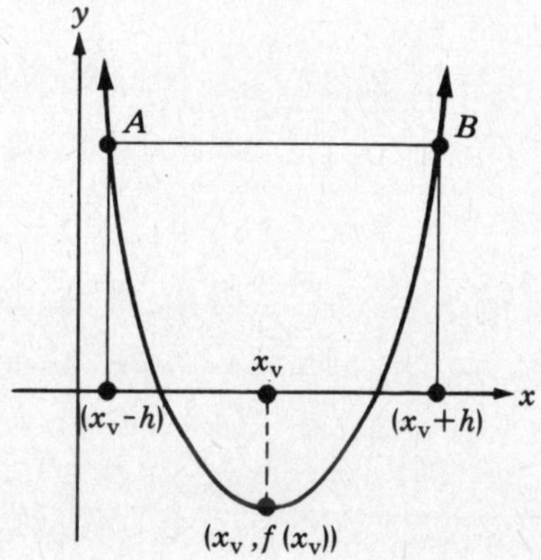

Let x_v be the abscissa of the vertex. The distance h is measured from each side of x_v. A's coordinates are $(x_v - h, f(x_v - h))$; B's coordinates are $(x_v + h, f(x_v + h))$. Due to the symmetry of the parabola,

$$f(x_v - h) = f(x_v + h)$$

$$a(x_v - h)^2 + b(x_v - h) + c = a(x_v + h)^2 + b(x_v + h) + c \quad \text{Expand and simplify}$$

$$-2ahx_v - bh = 2ahx_v + bh$$

$$-4ahx_v = 2bh$$

$$x_v = \frac{-2bh}{4ah} = -\frac{b}{2a} \quad \leftarrow \text{This is a formula for the abscissa of the vertex.}$$

The ordinate of the vertex is obtained by evaluating $f(x_v) = f\left(-\frac{b}{2a}\right)$. The result is $\frac{-b^2 + 4ac}{4a}$ (see Exercise 29). We are led to the following theorem.

THEOREM 3 The coordinates of the vertex of the parabola whose equation is $f(x) = ax^2 + bx + c$ are

$$\left(-\frac{b}{2a}, \frac{-b^2 + 4ac}{4a}\right).$$

The important features of the two previous examples and Theorem 3 are summarized below.

Graph of $y = ax^2 + bx + c$

1. The graph of $y = ax^2 + bx + c$ is a parabola opening upward if $a > 0$ and opening downward if $a < 0$.
2. The x-intercepts, if they exist*, are given by

$$\left(\frac{-b + \sqrt{b^2 - 4ac}}{2a}, 0\right) \quad \text{and} \quad \left(\frac{-b - \sqrt{b^2 - 4ac}}{2a}, 0\right).$$

3. The y-intercept of $y = ax^2 + bx + c$ is $(0,c)$.
4. The coordinates of the vertex are $\left(\frac{-b}{2a}, \frac{-b^2 + 4ac}{4a}\right)$.

We conclude this section with another example.

EXAMPLE 3 a. Does the parabola represented by $f(x) = 2x^2 - 4x - 1$ open upward or downward?
b. Find the intercepts.
c. Find the coordinates of the vertex.
d. Graph $y = 2x^2 - 4x - 1$.
e. What are the domain and range of this quadratic function?

Solution a. Since $a = 2 > 0$, the parabola opens upward.

b. The x-intercepts are $\left(\frac{2 + \sqrt{6}}{2}, 0\right)$ and $\left(\frac{2 - \sqrt{6}}{2}, 0\right)$ or approximately $(2.2,0)$ and $(-0.2,0)$. The y-intercept is $(0,-1)$.

c. $-\frac{b}{2a} = -\frac{-4}{2(2)} = 1$ is the abscissa of the vertex. The ordinate of the vertex is -3. Thus $(1,-3)$ is the vertex.

*If $b^2 - 4ac < 0$, the graph will not have any x-intercepts (see Example 3, Section 3.2 and Exercise 30 of this section.) When $b^2 - 4ac < 0$, the zeros of $f(x) = ax^2 + bx + c$ are not real.

d.

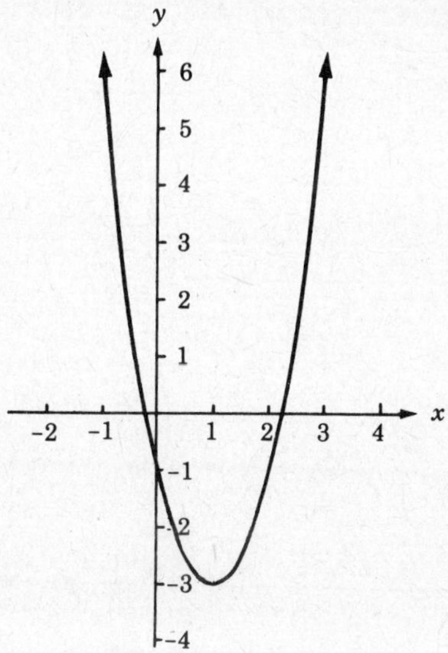

Graph of $y = 2x^2 - 4x - 1$.

e. The domain is $\mathcal{R}$. The range is $\{y \mid y \geqslant -3\}$.

EXERCISES FOR SECTION 3.4

In Exercises 1 through 28: a) determine whether the parabola opens upward or downward, b) find all intercepts, c) find the coordinates of the vertex, d) graph the parabola, and e) from the graph, state the domain and range. (See Examples 1, 2, and 3.)

1. $y = x^2 + x - 6$
2. $y = x^2 - x - 6$
3. $y = -x^2 - x + 6$
4. $y = -x^2 + x + 6$
5. $y = x^2 + 2x - 8$
6. $y = x^2 - 2x - 8$
7. $y = x^2 - 2x + 2$
8. $y = x^2 + 2x + 2$
9. $y = -3x^2 + 2x + 8$
10. $y = 3x^2 + 2x - 8$
11. $y = 3x^2 - 14x + 8$
12. $3x^2 + y = 14x - 8$
13. $y = x^2 + 4$
14. $y = x^2 - 4$
15. $y = 4 - x^2$
16. $y = x^2 + x + 1$
17. $y = 4x^2 - 24x + 37$
18. $y = x^2 - 4x + 5$
19. $y = 4x^2 - 4x + 1$
20. $y = 3x^2 - 6x + 4$
21. $2x^2 = -y - 8x - 8$
22. $y = \dfrac{6x - x^2 - 9}{2}$
23. $y = (2x - 3)^2$
24. $y = -(2x - 3)^2$
25. $y = x^2 + 2x + 1$
26. $y = 2x^2 + 5x + 3$
27. $y = 5 - 4x + 4x^2$
28. $y = x^2 + 4x + 13$

29. Using $f(x) = ax^2 + bx + c$, show that

$$f\left(\frac{-b}{2a}\right) = \frac{-b^2 + 4ac}{4a}$$

30. Show that, in the quadratic function $f(x) = ax^2 + bx + c$:
   a. if $b^2 - 4ac < 0$, there are no x-intercepts,
   b. if $b^2 - 4ac = 0$, there is exactly one intercept,
   c. if $b^2 - 4ac > 0$, there are two distinct intercepts.

31. Show that the equation $y - \frac{-b^2 + 4ac}{4a} = a\left(x - \frac{-b}{2a}\right)^2$ is equivalent to $y = ax^2 + bx + c$. $(a \neq 0)$
(Although this equation appears more awkward, it is sometimes preferred. If you think of the vertex as (h, k) this new form is

$$y - k = a(x - h)^2$$

and is sometimes referred to as the **vertex-form** of a quadratic equation.)

32. Consider the quadratic function $f(x) = x^2 - 5x + 4$. Let $(x, f(x))$ be any point on the graph of this function and let $(x + h, f(x + h))$ be any *other* point on the graph. See diagram on the right.
   a. Show that the slope of line $\overline{PQ}$ is given by $\frac{f(x+h)-f(x)}{h}$
   b. Determine $\frac{f(x + h) - f(x)}{h}$ by using $f(x) = x^2 - 5x + 4$.

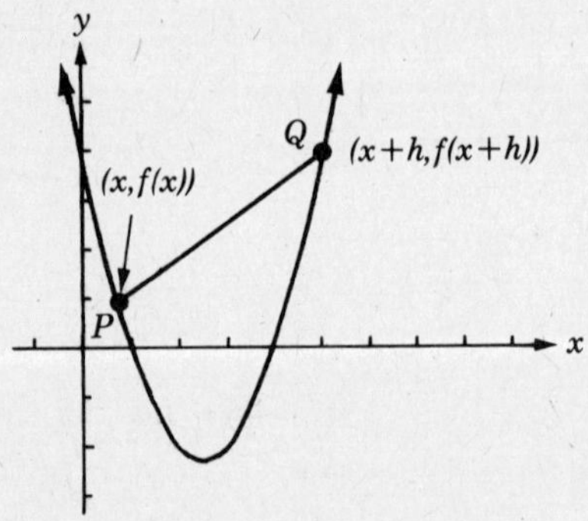

3.5 OTHER LINEAR AND QUADRATIC RELATIONS

In the previous two sections, we examined the graphs of certain functions. Particularly, we have seen that the graph of $y = mx + b$ is a straight line and the graph of $y = ax^2 + bx + c$ is a parabola. In this section we continue to graph lines and parabolas but they will no longer represent *functions.*

The next example illustrates that a *vertical* straight line is represented by the general relation $x = a$ (where a is any real number).

EXAMPLE 1 a. Graph the relation $x = 3$.
b. What are the domain and range of this relation?

Solution a. Since the abscissa *must* equal 3, regardless of the ordinate, the points (3,4), (3,−1), and (3,−3) are points on the graph of $x = 3$.

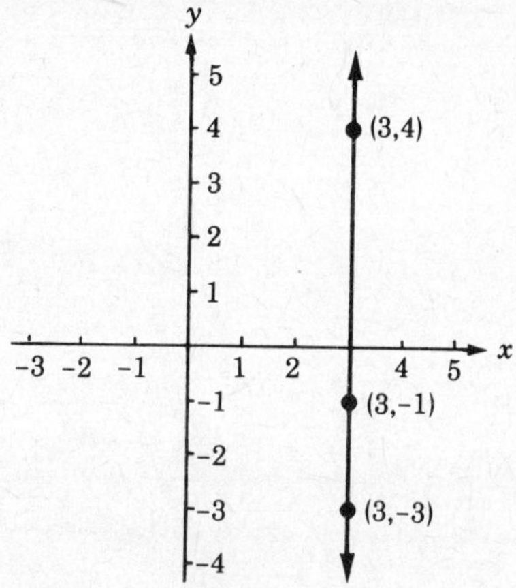

Graph of $x = 3$

b. From the graph, the domain is $\{x \mid x = 3\}$; the range is $\mathcal{R}$.

In the example above, notice that the slope of the vertical line is *undefined*. For the points (3,4) and (3,−1) we have:

$$m = \frac{y_1 - y_2}{x_1 - x_2} = \frac{4-(-1)}{3-3} = \frac{5}{0} \qquad \text{which is } \textit{undefined}.$$

In the general case with $x = a$ the slope formula gives

$$m = \frac{y_1 - y_2}{x_1 - x_2} = \frac{y_1 - y_2}{a - a} = \frac{y_1 - y_2}{0}.$$

Thus, *for vertical lines, the slope does not exist.*

Another linear relation is a **linear inequality in two variables**. For example, consider

$$y \geqslant 2x - 1$$

The graph of $y \geqslant 2x - 1$ includes all points on the line $y = 2x - 1$ as well as those points satisfying $y > 2x - 1$. Notice that the graph of $y = 2x - 1$ separates the plane into three distinct parts:

1. the points *on* $y = 2x - 1$ (the line itself);
2. the points *above* the line (called a **half-plane**);
3. the points in the half-plane *below* the line.

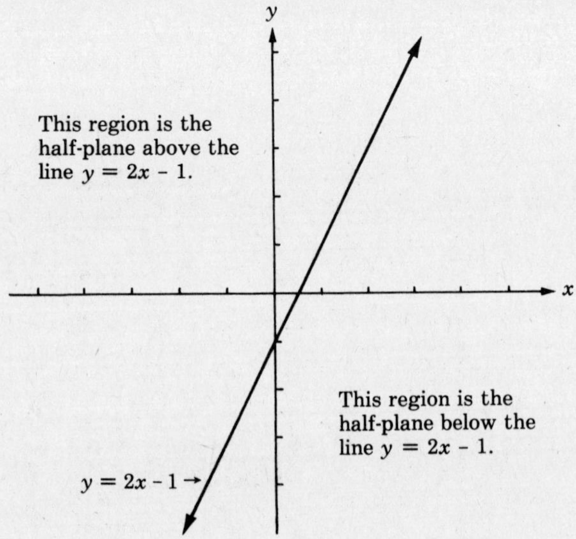

In order to determine which half-plane is the one that represents the desired relation $y > 2x - 1$, we choose a *test point* on each side of the line to see whether or not it satisfies $y > 2x - 1$:

Test point below $y = 2x - 1$:

Does $(3,-4)$ satisfy $y > 2x - 1$?

$-4 > 2(3) - 1$?

$-4 > 6 - 1$?

$-4 > 5$?

No. Thus, the point $(3,-4)$ is not in the desired region.

Test point above $y = 2x - 1$:

Does $(-1,2)$ satisfy $y > 2x - 1$?

$2 > 2(-1) - 1$?

$2 > -2 - 1$?

$2 > -3$?

Yes. The point $(-1,2)$ is in the desired region.

The half-plane above the line, along with the line, represent the solutions to $y \geqslant 2x - 1$. The graph appears as follows.

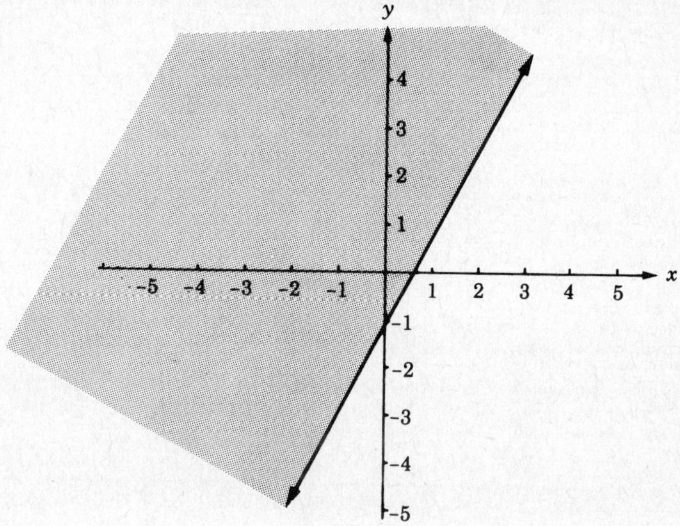

Graph of $y \geqslant 2x - 1$

EXAMPLE 2 Graph the linear relation $y < -2x + 3$.

Solution We graphically represent the line $y = -2x + 3$ as a dashed line to indicate equality does not hold.

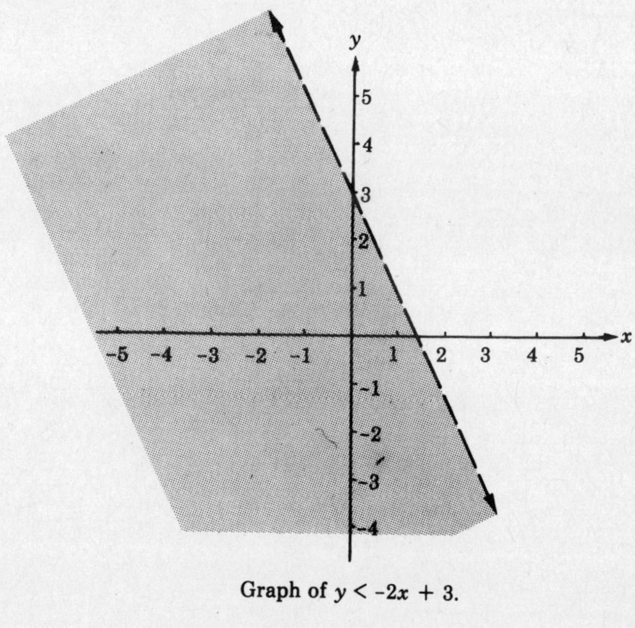

Graph of $y < -2x + 3$.

Next, we try test points:

Test point above $y = -2x + 3$:
Does $(4,4)$ satisfy $y < -2x + 3$?

$$4 < -2(4) + 3 \quad ?$$
$$4 < -5 \quad ?$$

No. Thus $(4,4)$ is not in the desired half-plane.

Test point below $y = -2x + 3$:
Does $(0,0)$ satisfy $y < -2x + 3$?

$$0 < -2(0) + 3 \quad ?$$
$$0 < 3 \quad ?$$

Yes. Hence, the half-plane below the line is shaded.

Next, we graph a **quadratic inequality in two variables.**

EXAMPLE 3 a. Graph $y \leqslant x^2 + 2x - 3$.
b. What are the domain and range of this relation?

Solution a. We know that the equality $y = x^2 + 2x - 3$ is graphically represented as a parabola. The inequality $y < x^2 + 2x - 3$ will be represented either by the points "inside" or the points "outside" the parabola.

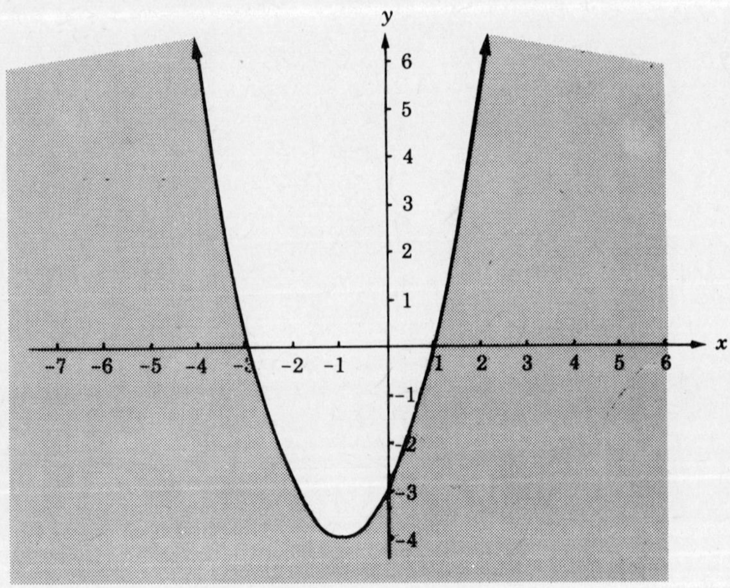

Graph of $y \leqslant x^2 + 2x - 3$

Again, we use test points:

Test point outside $y = x^2 + 2x - 3$:
Does $(5,-2)$ satisfy $y < x^2 + 2x - 3$?

$$-2 < (5)^2 + 2(5) - 3 \quad ?$$
$$-2 < 32 \quad ?$$

Yes. Thus, we shade those points "outside" the parabola.

Test point inside $y = x^2 + 2x - 3$:
Does $(0,0)$ satisfy $y < x^2 + 2x - 3$?

$$0 < 0^2 + 2(0) - 3 \quad ?$$
$$0 < -3 \quad ?$$

No. Thus, the points "inside" the parabola are not shaded.

Another type of relation we shall examine in this section is a relation of the form:

$$x = ay^2 + by + c \qquad (a \neq 0).$$

An equation in this form represents a parabola opening to the left or the right. The following properties, similar to those presented in the previous section, apply:

Graph of $x = ay^2 + by + c$

1. The graph of $x = ay^2 + by + c$ is a parabola opening to the right if $a > 0$ and opening to the left if $a < 0$.
2. The y-intercepts, if they exist, are given by

$$\left(0, \frac{-b + \sqrt{b^2 - 4ac}}{2a}\right) \quad \text{and} \quad \left(0, \frac{-b - \sqrt{b^2 - 4ac}}{2a}\right)$$

3. The x-intercept of $x = ay^2 + by + c$ is $(c,0)$.
4. The coordinates of the vertex are $\left(\frac{-b^2 + 4ac}{4a}, \frac{-b}{2a}\right)$.

EXAMPLE 4
a. Does the parabola $x = -y^2 + 6y - 5$ open to the left or right?
b. Find the intercepts of $x = -y^2 + 6y - 5$.
c. What are the coordinates of the vertex?
d. Graph the relation $x = -y^2 + 6y - 5$.
e. What are the domain and range of this relation?

Solution
a. Since $a = -1$, $a < 0$ and the parabola opens to the left.
b. The intercepts are $(0,1)$, $(0,5)$, and $(-5,0)$.
c. The vertex is $(4,3)$.
d.

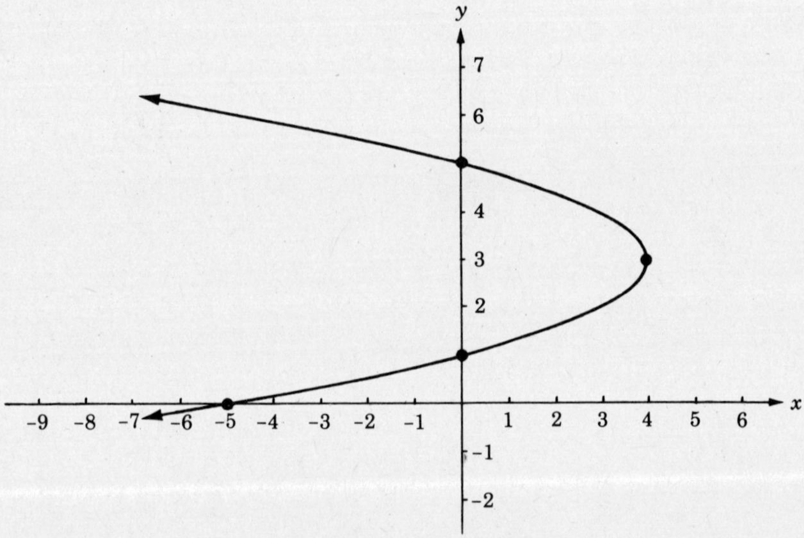

Graph of $x = -y^2 + 6y - 5$

e. From the graph, the domain is $\{x \mid x \leqslant 4\}$ and the range is $\mathcal{R}$.

The reader is urged to compare the graph above with the graph of $y = -x^2 + 6x - 5$ in Example 2, Section 3.4.

We conclude this section with another example of graphing a quadratic inequality.

EXAMPLE 5 Graph $x < 2y^2 + 3y - 2$.

Solution The parabola $x = 2y^2 + 3y - 2$ opens to the right; has intercepts $(0,\frac{1}{2})$, $(0,-2)$, and $(-2,0)$; and vertex $(-\frac{25}{8},-\frac{3}{4})$. To determine whether we graph the points inside or outside the parabola, we choose two test points:

Test point inside $x = 2y^2 + 3y - 2$:
Does $(0,0)$ satisfy $x < 2y^2 + 3y - 2$?

$$0 < 2(0)^2 + 3(0) - 2 \quad ?$$
$$0 < -2 \quad ?$$

No. Thus, the region inside the parabola is not shaded.

Test point outside $x = 2y^2 + 3y - 2$:
Does $(0,6)$ satisfy $x < 2y^2 + 3y - 2$?

$$0 < 2(6)^2 + 3(6) - 2 \quad ?$$
$$0 < 88 \quad ?$$

Yes. Thus, the desired region is outside the parabola.

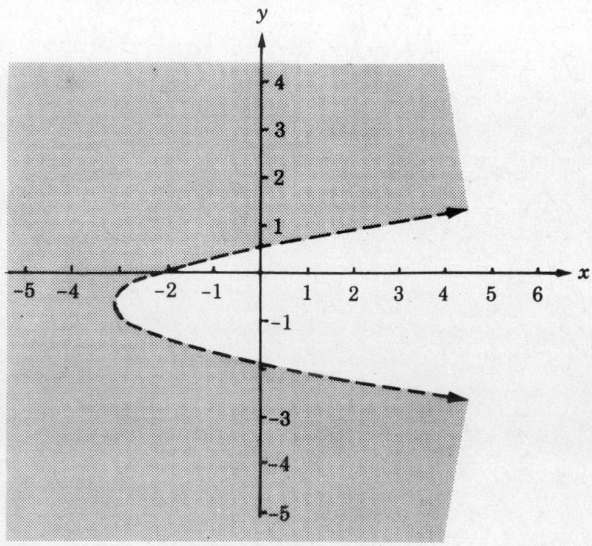

Graph of $x < 2y^2 + 3y - 2$

The reader is urged to verify that none of the relations graphed in this section represent functions (use the Vertical Line Test).

EXERCISES FOR SECTION 3.5

In Exercises 1 through 6: a) graph the given relation and b) state its domain and range. (See Example 1.)

1. $x = 2$	2. $x = -2$	3. $x = 4$
4. $x = 0$	5. $x + 1 = 2$	6. $\dfrac{2x+1}{3} = 4$

In Exercises 7 through 24: a) graph the given relation and b) state its domain and range. (See Examples 2 and 3.)

7. $y \geqslant 2x + 3$	8. $y < x - 2$	9. $y \leqslant 2x - 3$
10. $x + y < 3$	11. $x + y > 8$	12. $y \leqslant -2x$
13. $y \geqslant 1$	14. $x < -\frac{1}{2}$	15. $x + y > 0$
16. $y \leqslant x^2 - 3x - 4$	17. $y > x^2 + 3x - 4$	18. $y \geqslant 4 - x^2$
19. $y < x^2 + 1$	20. $y \leqslant 2x^2 - 5x + 3$	21. $y > 2x^2 - 5x - 3$
22. $y \geqslant x^2$	23. $y < x^2 + x + 1$	24. $1 - x^2 > y$

In Exercises 25 through 33: a) determine which way the parabola opens, b) find all intercepts, c) find the vertex, d) graph the relation, and e) state its domain and range. (See Example 4.)

25. $x = y^2 + 2y - 8$	26. $x = y^2 - 6y + 5$	27. $x = -y^2 + 6y - 5$
28. $x = 2y^2 - y + 1$	29. $x = 4 - y^2$	30. $x = y^2 - 4$
31. $x = 4y^2 - 4y + 1$	32. $x = 2y^2 + y - 2$	33. $x = 2 - \dfrac{y^2}{2}$

In Exercises 34 through 42, graph the given relation. (See Example 5.)

34. $x \leqslant y^2 + 2y - 8$	35. $x \geqslant y^2 - 6y + 5$	36. $x < -y^2 + 6y - 5$
37. $x > 2y^2 - y + 1$	38. $x \geqslant 4 - y^2$	39. $x < y^2 - 4$
40. $x > 4y^2 - 4y + 1$	41. $x \leqslant 2y^2 + y - 2$	42. $x > 2 - \dfrac{y^2}{2}$

3.6 THE INVERSE OF A RELATION

Thus far in this chapter we have described relations (and functions) in four ways:

1. by listing the ordered pairs in set form
2. by using a correspondence diagram
3. by means of a graph
4. by means of an equation

If R is some relation, the **inverse** of R is the relation obtained by interchanging the components in each ordered pair of R. More formally, we have:

DEFINITION

Let R be a relation. The **inverse** of R, denoted by R^{-1}, is given by

$$R^{-1} = \{(b,a) \mid (a,b) \in R\}*$$

Thus, if $R = \{(1,3), (-2,6), (4,-1)\}$, then $R^{-1} = \{(3,1), (6,-2), (-1,4)\}$.

EXAMPLE 1

a. Given $g = \{(3,-2), (4,0), (2,1)\}$, what are the domain and range of g?
b. Find g^{-1}.
c. Determine the domain and range of g^{-1}.
d. Graph g and g^{-1} on the same set of axes.

Solution

a. The domain of $g = \{3,4,2\}$; the range of $g = \{-2,0,1\}$
b. $g^{-1} = \{(-2,3), (0,4), (1,2)\}$
c. The domain of $g^{-1} = \{-2,0,1\}$; the range of $g^{-1} = \{3,4,2\}$
d.

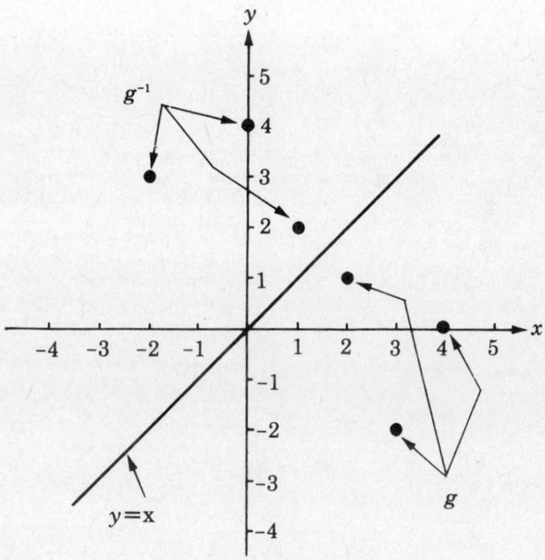

Graphs of g and g^{-1}

*The symbol R^{-1} is read "the inverse of R" or simply "R inverse" and is not to be confused with exponent notation.

Notice three things from the example above that are true for *any* relation g:

1. the domain of g is the range of g^{-1}
2. the range of g is the domain of g^{-1}
3. the graph of g^{-1} is the "mirror image" of g about the line $y = x$

The next two examples illustrate the technique for finding the inverse relation when the relation is described by an equation.

EXAMPLE 2 Consider the function $f = \{(x, y) \mid y = 2x - 3\}$.

a. Find f^{-1}.
b. Graph both f and f^{-1} on the same set of axes.

Solution a. The inverse is found by interchanging the two variables in the equation $y = 2x - 3$. We have

$$f^{-1} = \{(x, y) \mid x = 2y - 3\}$$

We prefer to rewrite $x = 2y - 3$ and solve for y:

$$x = 2y - 3$$
$$2y = x + 3$$
$$y = \frac{x}{2} + \frac{3}{2}$$

Equivalently, we can write $f^{-1}(x) = \dfrac{x}{2} + \dfrac{3}{2}$.

b.

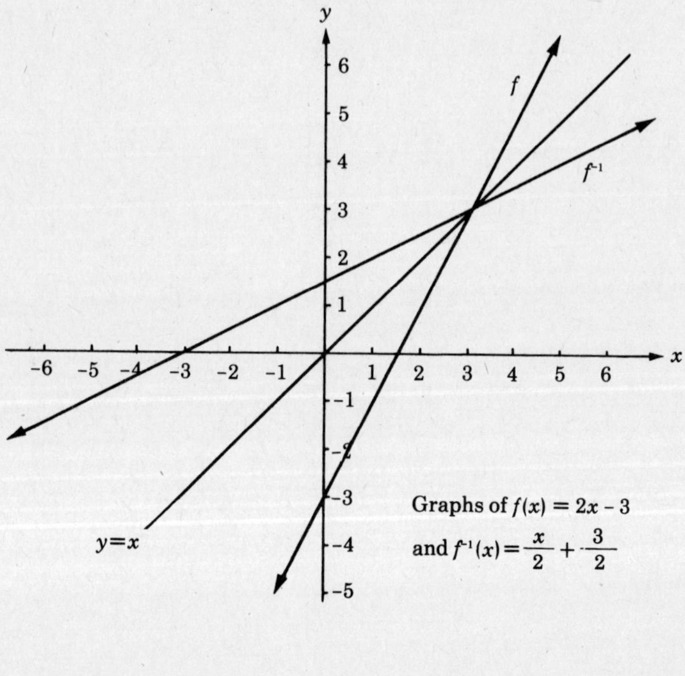

Graphs of $f(x) = 2x - 3$ and $f^{-1}(x) = \frac{x}{2} + \frac{3}{2}$

Notice that both f and f^{-1} *are* functions (the Vertical Line Test verifies this).

EXAMPLE 3 a. Find the inverse of the relation defined by $y = x^2 + 4$.
b. Graph both relations on the same set of axes.

Solution a. To find the inverse, we interchange x and y:

$y = x^2 + 4$ ← This is the given relation.
$x = y^2 + 4$ ← This is the inverse relation.

Although it is not necessary, we solve for y:

$$\left.\begin{aligned} x &= y^2 + 4 \\ y^2 &= x - 4 \\ y &= \pm\sqrt{x-4} \end{aligned}\right\} \text{These are all equivalent equations.}$$

Hence, the inverse is described by the equation $y = \pm\sqrt{x-4}$.

b. We know $y = x^2 + 4$ is a parabola opening upward with vertex at $(0,4)$ and $y = \pm\sqrt{x-4}$ (or its equivalent, $x = y^2 + 4$) is a parabola opening to the right with vertex at $(4,0)$:

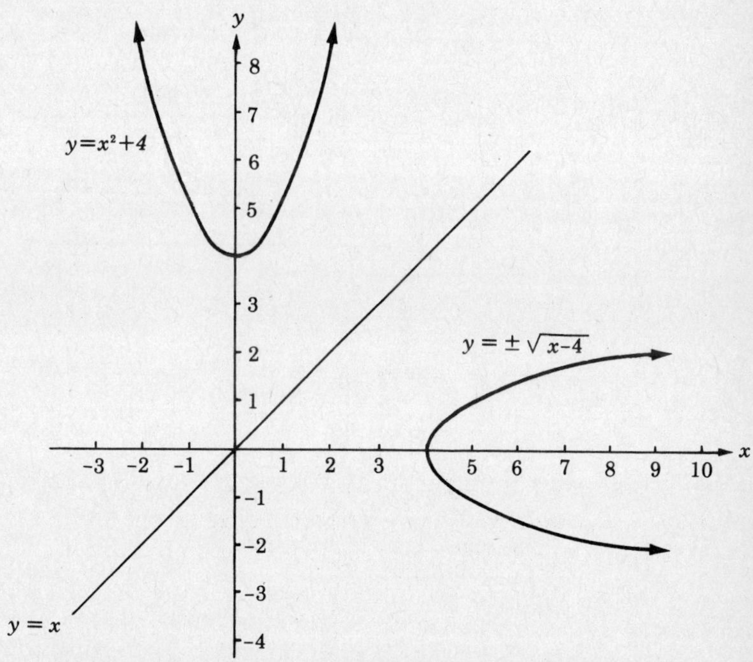

Graphs of the relation $y = x^2 + 4$ and its inverse $y = \pm\sqrt{x-4}$.

Notice that the given relation, $y = x^2 + 4$ *is* a function but its inverse is *not*.

In summary, then, we have seen that the inverse of a relation might or might not, itself, be a function.

When a relation, f, and its inverse, f^{-1}, are *both* functions, it can be shown that

$$f(f^{-1}(x)) = x$$

and

$$f^{-1}(f(x)) = x$$

To illustrate this, consider $f(x) = 2x - 3$ (Example 2):

$$f(x) = 2x - 3 \qquad\qquad f^{-1}(x) = \frac{x}{2} + \frac{3}{2}$$

$$f(f^{-1}(x)) = f\left(\frac{x}{2} + \frac{3}{2}\right) = 2\left(\frac{x}{2} + \frac{3}{2}\right) - 3 = x + 3 - 3 = x$$

and

$$f^{-1}(f(x)) = f^{-1}(2x - 3) = \frac{2x - 3}{2} + \frac{3}{2} = \frac{2x}{2} = x.$$

For g and g^{-1} of Example 1, we have the following:

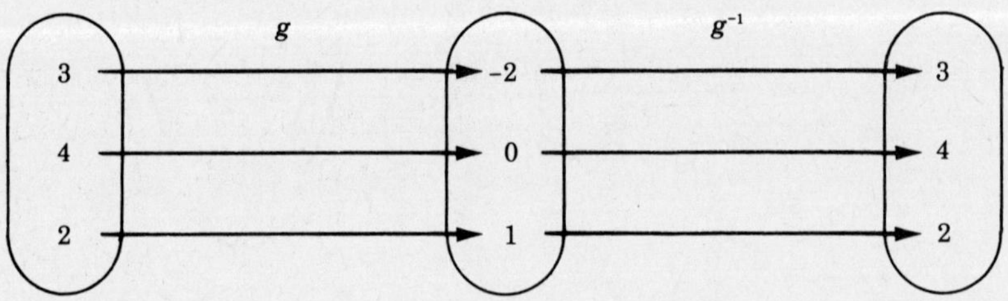

EXAMPLE 4 Let $f(x) = x^3 + 1$.

a. Find $f^{-1}(x)$.
b. Is $f^{-1}(x)$ a function?
c. Verify that $f(f^{-1}(x)) = x$.
d. Verify that $f^{-1}(f(x)) = x$.

Solution a. To find the inverse of $y = x^3 + 1$, we interchange x and y:

$$x = y^3 + 1$$

We solve for y:

$$y^3 = x - 1$$
$$y = \sqrt[3]{x - 1}$$

or

$$f^{-1}(x) = \sqrt[3]{x-1}$$

b. Yes, $f^{-1}(x)$ is a function (any x in the domain, $\mathcal{R}$, yields *exactly* one value of $f^{-1}(x)$).

c. $f(x) = x^3 + 1$ and $f^{-1}(x) = \sqrt[3]{x-1}$
$f(f^{-1}(x)) = f(\sqrt[3]{x-1}) = (\sqrt[3]{x-1})^3 + 1 = x - 1 + 1 = x.$

d. $f^{-1}(f(x)) = f^{-1}(x^3+1) = \sqrt[3]{(x^3+1)-1} = \sqrt[3]{x^3} = x.$

EXERCISES FOR SECTION 3.6

In Exercises 1 through 23: a) find the inverse of the given relation, b) graph both the relation and its inverse on the same set of axes, c) determine whether or not the given relation and its inverse are functions. (See Examples 1, 2, and 3.)

1. $R = \{(1,2), (3,4)\}$
2. $R = \{(-1,2), (3,5), (4,2)\}$
3. $R = \{(-1,5), (-1,7)\}$
4. $R = \{(0,1), (1,0)\}$
5. $R = \{(x,y) \mid y = x + 3\}$
6. $R = \{(x,y) \mid y = 2x\}$
7. $R = \{(x,y) \mid y = 2x + 3\}$
8. $R = \{(x,y) \mid y = \frac{1}{2}x + 1\}$
9. $R = \{(x,y) \mid x + y = 2\}$
10. $R = \{(x,y) \mid y = x\}$
11. $R = \{(x,y) \mid y = 2\}$
12. $R = \{(x,y) \mid x = 3\}$
13. $R = \{(x,y) \mid y = x^2 + 1\}$
14. $R = \{(x,y) \mid y = x^2\}$
15. $R = \{(x,y) \mid x = y^2 - 2\}$
16. $R = \left\{(x,y) \mid y = \dfrac{x^2}{2}\right\}$
17. $R = \{(x,y) \mid y = x^2 + 4x + 3\}$
18. $R = \{(x,y) \mid y = x^2 - 2x - 8\}$
19. $R = \{(x,y) \mid x = y^2 + y - 6\}$
20. $R = \{(x,y) \mid x = -y^2\}$
21. $R = \{(x,y) \mid y = 9 - x^2\}$
22. $R = \{(x,y) \mid y = x^3\}$
23. $R = \{(x,y) \mid y = (x+1)^3\}$

In Exercises 24 through 30: a) find $f^{-1}(x)$ and b) verify $f(f^{-1}(x)) = f^{-1}(f(x)) = x$. (See Example 4.)

24. $f(x) = 3x$
25. $f(x) = 2x - 5$
26. $f(x) = 9x + 11$
27. $f(x) = \dfrac{x+7}{6}$
28. $f(x) = \dfrac{1}{x+1}$
29. $f(x) = \dfrac{1}{x^3}$
30. $f(x) = x^5$
31. We have the Vertical Line Test to determine whether or not a certain relation is a function. How could a *Horizontal Line Test* be used on $y = f(x)$ to determine whether or not $f^{-1}(x)$ is a function? Explain.
32. Prove that if f is a one-to-one function, then f^{-1} must be a function. (See the explanation of a one-to-one function preceding Exercise 48 in Section 3.2.)
33. Let $f(x) = mx + b$. a) Find $f^{-1}(x)$. b) What restrictions (if any) must be made on m and b to insure that $f^{-1}(x)$ is a function?
34. Let $f(x) = ax^2 + bx + c$ $(a \neq 0)$. a) Find $f^{-1}(x)$. b) Is f^{-1} a function?

3.7 APPLICATIONS—VARIATION

The techniques of graphing acquired in this chapter will be applied to graphs of a technical or applied nature in the present section. The applications involve linear, quadratic, and other functions.

Natural Science Problems

EXAMPLE 1 In an experiment in physics lab, Corky has measured voltage and current in a certain electrical circuit. The following data were obtained:

Current I (in amperes)	*Voltage V (in volts)*
0	12
100	10
150	9
200	8

a. Graph the points listed above, using I for the horizontal axis and V for the vertical axis.
b. Construct an equation that could serve as an algebraic representation (model) for this experiment.
c. Graph the equation in part b.

Solution a. Notice that since both current and voltage will always be positive, we only need to be concerned with the first quadrant. Also, observe that 3 units of voltage on the vertical axis are represented by the same length as 25 units of current on the horizontal axis. We say the graph exhibits a 3 to 25 **scale ratio**.

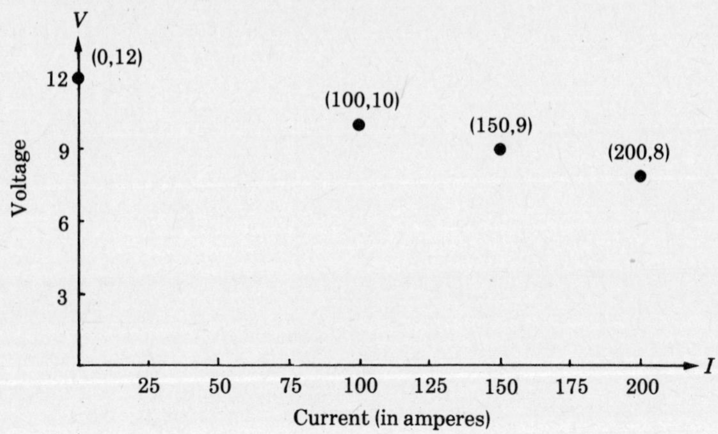

b. With slope $= \dfrac{12-10}{0-100} = -.02$ and intercept $(0,12)$ we have

$$V = -.02I + 12$$

c. Notice that the equation "fits" the data:

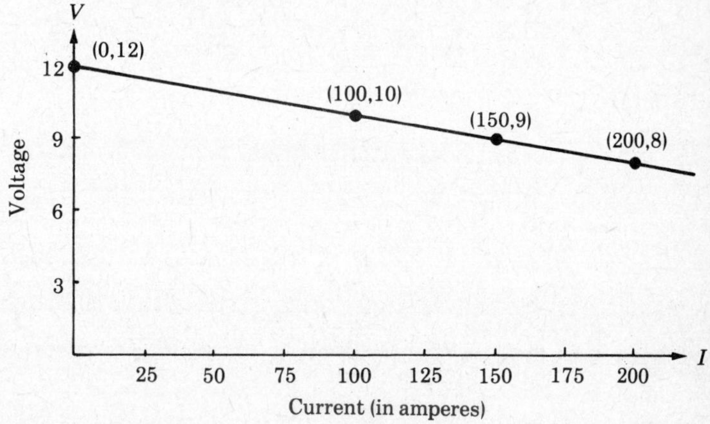

The next example involves an application of a quadratic equation.

EXAMPLE 2 The relation $D = 0.05v^2 + 1.1v$ gives the minimum safe stopping distance, D (in feet), for a motor vehicle travelling at a velocity v (in mi/hr).

a. Graph the relation $D = 0.05v^2 + 1.1v$.
b. Adjust the graph in part a to account only for the physical aspects of the problem.
c. What is the minimum safe stopping distance for a motor vehicle travelling at a speed of 40 mi/hr?

Solution a. The graph of $D = 0.05v^2 + 1.1v$ is a parabola opening upward with vertex $(-11, -6.05)$ and intercepts at $(0,0)$ and $(-22,0)$ (we choose to use a 40 to 10 scale ratio).

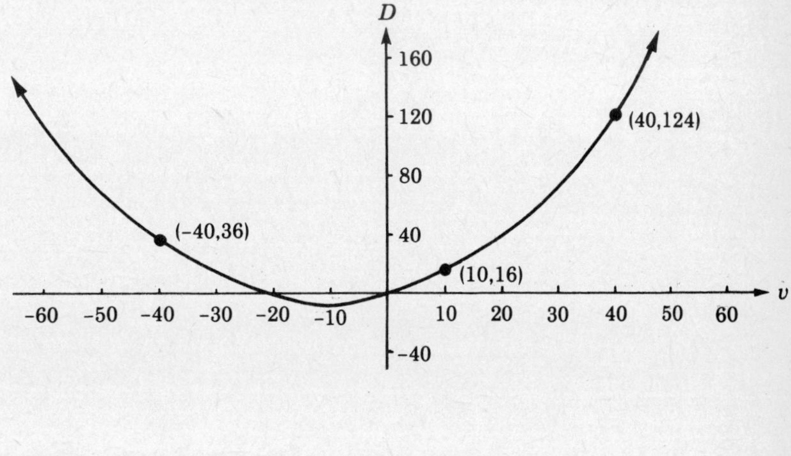

b. Since $v \geqslant 0$, we have:

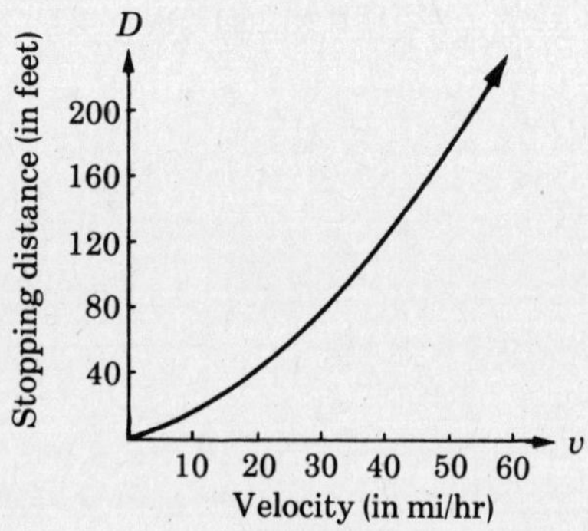

c. $D = 0.05v^2 + 1.1v$

$D = 0.05(40)^2 + 1.1(40)$

$D = 80 + 44$

$D = 124$ feet

Business Problem

EXAMPLE 3 A company assumes that its sales revenue, R, is solely a function of how many units of an item, x, it sells. Further, its marketing research has determined:

$$R = 100x - x^2$$

(R is in units of \$100, x is number of units sold.)

How many units of the product must be sold in order to maximize sales revenue?

Solution We graph the function below and point out that R is maximized at the vertex. Thus the maximum sales revenue is \$250,000 and is obobtained when 50 units are sold:

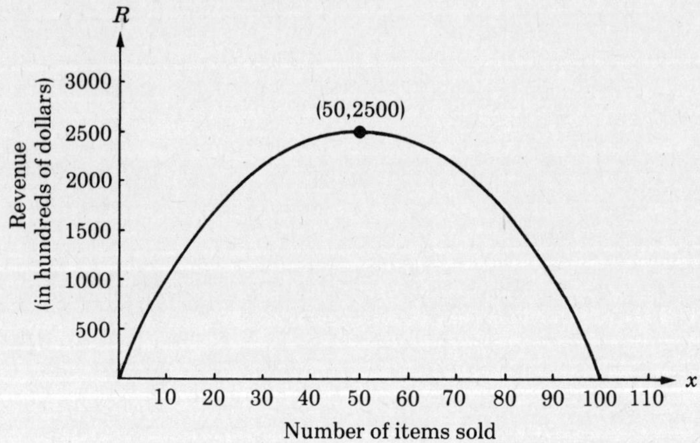

Variation Problems

The applications in this portion of the section are from different areas and rely upon the following definitions.

DEFINITION

P varies directly as the nth power of Q means that

$$P = kQ^n,$$

where k is some real number and $n > 0$.*

DEFINITION

P varies inversely as the nth power of Q means that

$$P = \frac{k}{Q^n},$$

where k is a real number and $n > 0$.**

DEFINITION

P varies jointly as the mth power of Q and the nth power of S means that

$$P = kQ^m S^n,$$

where k is a real number and $m > 0$, $n > 0$.***

In these definitions, k is called the **proportionality constant.**

EXAMPLE 4

a. If y varies directly as the first power of x, express this fact as an equation.
b. If y varies directly as the second power of x, express this fact as an equation.
c. If y varies inversely as the third power of x, express this fact as an equation.

Solution

a. $y = kx$. Here we say that y is *directly proportional to* x.
b. $y = kx^2$. Here we say that y is *directly proportional to* the square of x.
c. $y = \frac{k}{x^3}$. Here we say that y is *inversely proportional to* the cube of x.

*Direct variation is often referred to as *direct proportion.*
**Inverse variation is often referred to as *inverse proportion.*
***Joint variation is often referred to as *joint proportion.*

EXAMPLE 5 Express the following in equation form.

a. The circumference, C, of a circle is directly proportional to its diameter, d.
b. The area, A, of a circle is directly proportional to the square of its radius, r.
c. The volume, V, of a cone varies jointly as the square of the base radius, r, and the first power of the height, h.

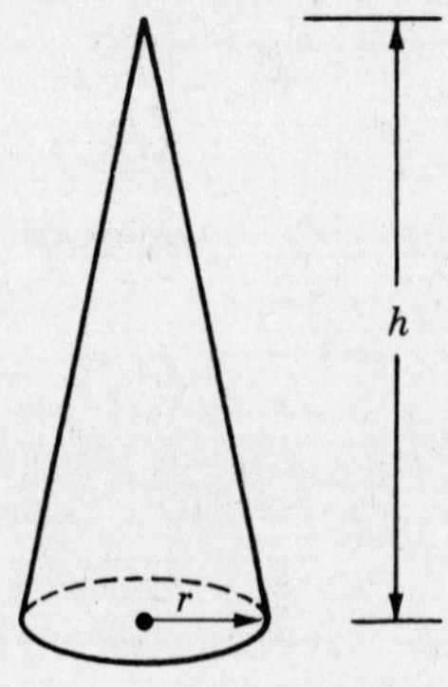

Solution a. $C = kd$. However, from geometry it is known that $k = \pi$. Thus,

$$C = \pi d.$$

b. $A = kr^2$. From geometry it is known that $k = \pi$. Thus,

$$A = \pi r^2.$$

c. $V = kr^2h$. From geometry it is known that $k = \frac{1}{3}\pi$. Thus,

$$V = \tfrac{1}{3}\pi r^2 h.$$

In the next examples we show a method for determining k.

EXAMPLE 6 Suppose y varies directly as the square of x and inversely as the square root of w.

a. Express this in equation form.
b. Determine the proportionality constant knowing that $y = 10$ when $x = 3$ and $w = 16$.
c. Combine the results of parts a and b.

Solution a. We have both direct and inverse variation in this problem. Thus,

$$y = k\frac{x^2}{\sqrt{w}}.$$

b. Since $y = 10$ when $x = 3$ and $w = 16$, substitute these values into the answer to a.

$$10 = k\frac{(3)^2}{\sqrt{16}}, k = \frac{40}{9}.$$

c. Now we can write,

$$y = \frac{40}{9} \cdot \frac{x^2}{\sqrt{w}}.$$

EXAMPLE 7 Under perfect conditions, the horsepower, H, of an automobile engine varies jointly as the square of a piston diameter, D, and the number of pistons, N.

a. Express this information in equation form.
b. What happens to the horsepower if piston diameter doubles?
c. What happens to the horsepower if piston diameter is cut in half and the number of pistons doubles?

Solution a. $H = kD^2N$

b. Use the subscript "o" for "original values."
Use the subscript "n" for "new values."

$$H_o = kD_o^2N_o \text{ and } H_n = kD_n^2N_n.$$

Also $D_n = 2D_o$ and $N_n = N_o$.
Thus, $H_n = k(2D_o)^2N_o = 4kD_o^2N_o$
$H_n = 4H_o$

Hence, doubling piston diameter causes horsepower to quadruple.

c. $H_o = kD_o^2N_o$ and $H_n = kD_n^2N_n$.
Also $D_n = \frac{1}{2}D_o$ and $N_n = 2N_o$.
Thus, $H_n = k(\frac{1}{2}D_o)^2(2N_o) = \frac{1}{2}kD_o^2N_o$
$H_n = \frac{1}{2}H_o$

Hence, cutting piston diameter in half and doubling the number of pistons will cause the horsepower to be cut in half.

EXERCISES FOR SECTION 3.7

1. It is known from physics that when an object moves with a constant velocity, the relationship between distance travelled and time elapsed is a linear function. Suppose the following numbers pertain to a particular experiment involving an object in motion at constant velocity.

Time t (in seconds)	*Distance d (in meters)*
0	0
2	25
4	50
6	75

   a. Graph the points listed above using t for the horizontal axis, d for the vertical axis, and a 25 to 2 scale ratio.
   b. Construct the (linear) equation that serves as an algebraic model for this experiment and graph that relationship.
   c. Use the result of part b to find the distance travelled after 7 seconds.

2. One indicator used by cosmetologists in studying healthy human hair, is its breaking strength when stretched. It is known that, for healthy hair, the relationship between the force required to break a strand of hair and its diameter is a linear function. An experiment yielded the following data:

Diameter d (in microns)	*Breaking force F (in grams)*
55	26
65	34
75	42

   a. Graph the points listed above. Use d for the horizontal axis and F for the vertical axis.
   b. Construct the (linear) equation that serves as an algebraic model for this experiment and graph that relationship.
   c. Use the result in part b to determine F when $d = 58$.

3. Scientists use a temperature scale known as the **Kelvin** temperature scale. The lowest possible temperature (called absolute zero) is 0° K (K denotes Kelvin degrees). The relationship between Kelvin temperature (T_K) and Celsius temperature (T_C) is

$$T_K = T_C + 273$$

   a. Graph $T_K = T_C + 273$ with T_K on the vertical axis.
   b. A comfortable room temperature is 20° C. What is the corresponding Kelvin temperature?

4. *Hooke's Law* states that the relationship between the length x (in inches) that a spring stretches (increases in length) and the force F (in pounds) applied to the spring, is a linear relationship. In symbols, $F = kx$, where k is the proportionality constant (its value depends upon the particular spring used).
   a. If a spring stretches 4 inches when a force of 10 pounds is applied, what is k?

b. On the same set of axes graph $f = kx$ for $k = 1, 2, 4$, and 6. (You need only use the first quadrant.)

5. The equation $d = \frac{2}{9}t^2 - \frac{40}{3}t$ gives the distance d (in feet) that a computer-controlled diving bell will descend in t seconds. The process starts at the surface level of the water when $t = 0$. (*Note:* the distance *below* the surface is represented by negative values of d in this model.)
   a. Using a 100:10 scale ratio, graph $d = \frac{2}{9}t^2 - \frac{40}{3}t$. Place d on the vertical axis.
   b. What part of the graph pertains to the physical aspects of the problem?
   c. What is the greatest depth reached?
   d. What is the time of the entire diving experiment?
6. The boiling point of water is 100° C or 212° F; the freezing point of water is 0° C or 32° F. The relationship between Celsius and Fahrenheit temperatures is linear.
   a. Construct an algebraic model for the linear relationship between C and F.
   b. Graph the equation from part a using C for the horizontal axis variable.
7. A farmer desires to enclose a rectangular region using 900 feet of fencing. He is undecided on the area of the rectangle to enclose.
   a. Express the area, A, of the rectangle in terms of its dimensions.

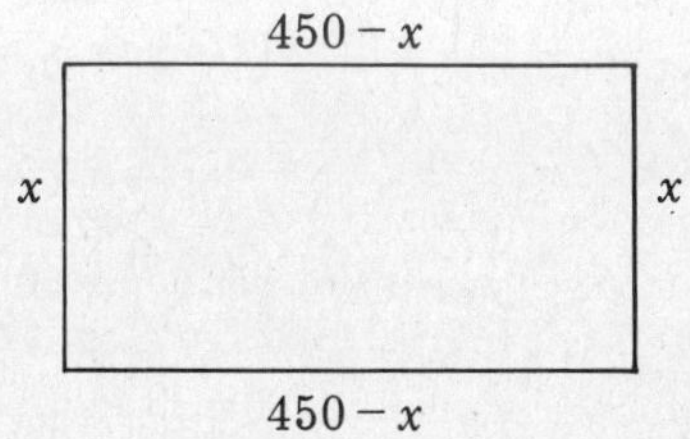

   b. On a sheet of graph paper, graph the equation from part a. (Depict A on the vertical axis.)
   c. What part of the graph pertains to the physical aspects of this problem?
   d. What dimensions should be chosen for the field to attain a maximum area?
8. In a test for the concentration of a certain drug in a person's blood, it was found that the amount of the drug present, D (in parts per million), was given by the equation, $D = -0.2t^2 + 0.4t$, where t is measured in hours ($t = 0$ is the moment the experiment begins).
   a. Graph this equation on a sheet of graph paper. (Depict D on the vertical axis.)
   b. What part of the graph pertains to the physical aspects of this problem?
   c. At what time, t, is the concentration a maximum?
9. The cost, C (in dollars), of producing x FM stereo receivers is given by $C = 200 + 10x + \frac{1}{4}x^2$.
   a. Graph this equation on a sheet of graph paper. (Depict C on the vertical axis.)
   b. What part of the graph pertains to the monetary aspects of this problem?

10. An object is thrown upward from the top of a building that is 100 feet high. The equation $d = -16t^2 + 32t + 100$ gives the object's height, d (in feet), above the ground at any time, t.
    a. Graph this equation on a sheet of graph paper. (Depict d on the vertical axis.)
    b. What part of the graph pertains to the physical aspects of this problem?
    c. At what time, t, does the object hit the ground?
    d. At what time, t, is the object's height above the ground a maximum?
11. If w varies directly as the square of z, express this fact as an equation.
12. If y varies inversely as the square root of x, express this fact as an equation.
13. If z varies jointly as the cube of x and the square of y, express this fact as an equation.
14. If F varies jointly as the square of x and the square root of y, express this fact as an equation.
15. If F is directly proportional to the product of M with m, and inversely proportional to the square of r, express this fact as an equation.
16. The area of a square, A, is directly proportional to the square of its side, s.
    a. Express this as an equation.
    b. What is the value of the constant of proportionality?
17. The perimeter of a square, P, varies directly as the first power of its side, s.
    a. Express this as an equation.
    b. What is the value of the constant of proportionality?
18. Suppose y is directly proportional to the fourth power of x.
    a. Express this as an equation.
    b. If $y = 5$ when $x = 2$, determine the proportionality constant.
    c. Combine the results of parts a and b.
19. Suppose Q is inversely proportional to the square of r.
    a. Express this as an equation.
    b. If $Q = .002$ when $r = 4$, determine the proportionality constant.
    c. Combine the results of parts a and b.
    d. Determine Q, when $r = 3$.
    e. Determine Q, when $r = 6$.
20. The resistance to flow, R, in a tube is directly proportional to the length, L, of the tube and inversely proportional to the fourth power of the radius, r.
    a. Express this as an equation.
    b. What happens to the resistance if the length of the tube doubles?
    c. What happens to the resistance if the radius of the tube doubles?
    d. What happens to the resistance if the length of the tube quadruples and the radius is halved?
21. The time required for a complete oscillation, T, of a pendulum is proportional to the square root of its length, L.
    a. Express this as an equation.
    b. If the length quadruples, what happens to T?
    c. If the length increases by a factor of 1.1, what happens to T?
22. The distance travelled by a freely falling body, s, is directly proportional to the square of the time, t.
    a. Express this as an equation.
    b. It has been found that $s = 64$ ft when $t = 2$ sec. Determine the constant of proportionality.

c. Find s when $t = 1$ sec.
d. Find s when $t = 3$ sec.

23. The energy from a windmill, E, is directly proportional to the cube of the wind velocity, v.
a. Express this as an equation.
b. If velocity doubles, what happens to E?
c. If velocity triples, what happens to E?

CHAPTER 3 IN REVIEW

I. Summary

This chapter began with a discussion of the relationship that exists between the points in the plane (on a Cartesian coordinate system) and the pairs of numbers that solve a given equation in two variables. Fundamentally, the points form a graph that pictorially represents a given equation.

The notions of relation and function were defined in Section 3.2. We saw there were four ways to represent a relation:

1. by an equation in two variables
2. by a graph in the Cartesian coordinate system
3. by a set of ordered pairs
4. by a correspondence diagram

In Section 3.3 we examined the linear function $y = mx + b$ where m is the slope and b is the y-intercept and in Section 3.4 saw that the quadratic equation $y = ax^2 + bx + c$ can be represented by a parabola whose vertex is $\left(\frac{-b}{2a}, \frac{-b^2 + 4ac}{4a}\right)$.

In Section 3.5 we saw that vertical lines have the algebraic representation $x = a$ and that parabolas opening to the right or left have the form $x = ay^2 + by + c$. We also graphed linear and quadratic inequalities.

Section 3.6 introduced the notion of the inverse of a relation. If R is a relation, then its inverse, R^{-1}, is a collection of points from R with abscissas and ordinates interchanged. Finally, in Section 3.7, both linear and quadratic equations were seen in applied settings. Direct, inverse, and joint variation in various applied settings were also explored.

II. Vocabulary/Symbol True-False Quiz

If the statement is true, write "TRUE." However, if the statement is false, replace the underlined word or phrase with a different word or phrase that will make the new statement true.

1. The distance between (2,−4) and (−3,2) is $\underline{\sqrt{37}}$.

2. The domain of the function described below is $\mathcal{R}$.

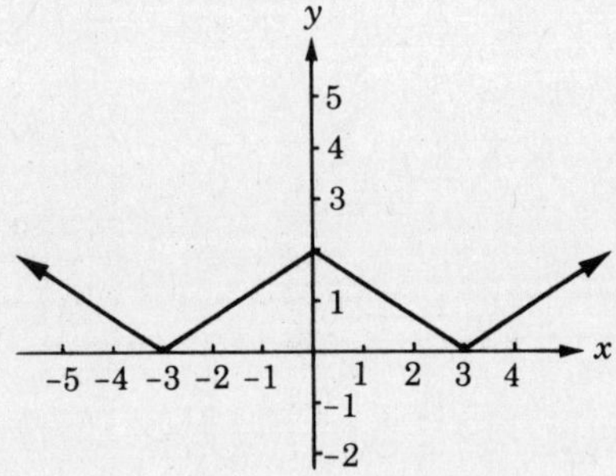

3. The equation of the straight line passing through $(3,5)$ and $(-1,1)$ is $y = x + 2$.
4. The slope of $x = 3$ is 0.
5. The parabola $y = 4 - x^2$ opens upward and has vertex $(0,4)$.
6. The x-intercepts of $y = x^2 - 7x + 12$ are $(0,3)$ and $(0,4)$.
7. The graph of $y \geqslant x^2$ is given below.

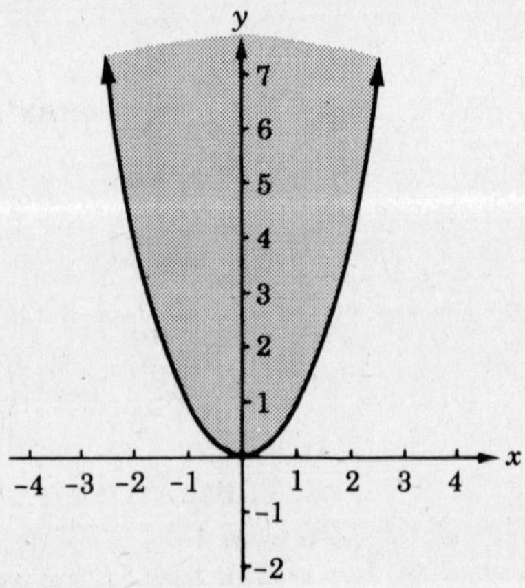

8. If R is a relation, then the graph of R^{-1} is the mirror image of the graph of R reflected about the line $y = x$.
9. If $f(x) = x^2$ and $g(x) = \sqrt{x}$ then $f(g(x)) = x$.
10. If $f(x) = 3x - 1$ then $f^{-1}(x) = \dfrac{x+1}{3}$.

III. Chapter Test

1. Define what is meant by function.
2. Find the distance between the points $(2,4)$ and $(-1,8)$.

3. Consider the relation R graphed below.

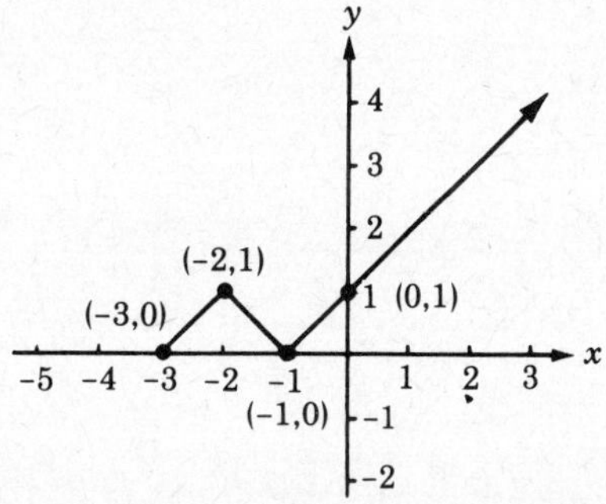

   a. Is R a function? b. State the domain and range of R.
4. Given $f(x) = x - 5x^2$, find
   a. $f(-1)$ b. $\dfrac{f(x+h) - f(x)}{h}$ $(h \neq 0)$
5. Determine the equation of the straight line passing through $(2,-3)$ and $(6,5)$.
6. Determine the slope of the straight line $x + y = 0$.
7. Determine the equation of the line that passes through $(-3,4)$ and has slope equal to 1.
8. Graph $y = 2x - 2$.
9. Graph $x = 4$.
10. Graph $y \leqslant 3x - 1$.
11. Given $y = 2x^2 + 4x + 2$
    a. Determine which way it opens, its intercepts, and its vertex.
    b. Graph $y = 2x^2 + 4x + 2$.
    c. State the domain and range.
12. a. Graph $y > 4 - x^2$.
    b. State the domain and range.
13. a. Graph $x = y^2 - 6y + 9$.
    b. State the domain and range.
14. Given $f(x) = 2x + 3$
    a. Find $f^{-1}(x)$.
    b. Graph f and f^{-1} on the same set of axes.
    c. Find $f(f^{-1}(x))$.
    d. Find $f^{-1}(f(x))$.
15. If $g(x) = x^2 + 1$
    a. Find $g^{-1}(x)$.
    b. Is $g^{-1}(x)$ a function?
16. Scuba divers experience an increase in pressure on their bodies (above the normal atmospheric pressure of 14.7 lb/in^2) as they descend to greater depths in sea water. It is known that the rela-

tionship between pressure and depth is a linear relationship. The equation

$$P = .45D + 14.7$$

serves as an algebraic model relating the total pressure P (in lb/in^2) on a person at a depth D feet in sea water. (*Note:* the distance *below* the surface is represented by positive values of D in this model.)

a. Graph this relation with D on the horizontal axis and P on the vertical axis for values of D in $[0,200]$. Use a 20 to 40 scale ratio.
b. Determine the pressure on a scuba diver who is at a depth of 180 feet.

IV. Student Research Project

1. The Cartesian coordinate system owes its name to the person who first used it, René Descartes. Prepare a biography of Descartes. In addition to encyclopedias, the following classic sources may be useful:

   Bell, E. T. *Men of Mathematics*, Simon and Schuster, New York, 1937.

   Turnbull, H. W. *The Great Mathematicians*, New York University Press, New York, 1961.

   Descartes, René. "The Geometry" in *The World of Mathematics* edited by James R. Newman, Simon and Schuster, New York, 1966.

2. The coordinate plane can be used as a geometrical tool to "prove" what we usually consider to be algebraic relationships. One such example is the quadratic formula. Consider the equation $x^2 - 7x + 12 = 0$. If we a) plot the points $B(0,1)$ and $Q(7,12)$; b) draw a circle with BQ as its diameter; c) we find the circle cuts the x-axis at the roots of $x^2 - 7x + 12 = 0$. This is called the **Method of Carlyle.**
   a. Perform steps a), b), and c) for $x^2 - 7x + 12 = 0$. What are the roots?
   b. The method above is generalized for $x^2 - gx + h = 0$ in *An Introduction to the History of Mathematics*, 4th ed., by Howard Eves, Holt, Rinehart and Winston, New York, 1976, p. 80. Geometrically, approximate the roots of $x^2 - 4x + 1 = 0$. Use graph paper, ruler, and compass.

CHAPTER **4**

Topics in Analytic Geometry

René Descartes (1596–1650)

René Descartes was a French philosopher and scientist. He earned a law degree, served with the military, and in 1628 settled in Holland, where he did most of his writing. Descartes improved the symbolism of algebra and is often regarded as the founder of analytic geometry. In *Discourse on Method* he made analytic geometry known to his contemporaries and his writing on this subject led to the present-day methods that you will study in this chapter.

4.1 PIECEWISE FUNCTIONS

Most of the functions graphed thus far have had relatively simple algebraic representations. For example, linear and quadratic functions can be expressed as:

$$\text{Linear function:} \qquad f(x) = mx + b$$

$$\text{Quadratic function:} \quad f(x) = ax^2 + bx + c \qquad (a \neq 0)$$

In this section we will examine functions, called **piecewise functions**, that are defined differently for various portions of their domain. For example,

$$f(x) = \begin{cases} x^2 - 3x + 2 & \text{if } x \leqslant 3 \\ -x + 5 & \text{if } x > 3 \end{cases}$$

is a quadratic function for some values of x (namely, $x \leqslant 3$) and is a linear function for other values of x ($x > 3$) in its domain. We graph this function and others in the examples that follow.

EXAMPLE 1 a. Graph $f(x) = \begin{cases} x^2 - 3x + 2 & \text{if } x \leqslant 3 \\ -x + 5 & \text{if } x > 3 \end{cases}$ b. Evaluate $f(3)$.

Solution We graph $y = x^2 - 3x + 2$ and $y = -x + 5$ below.

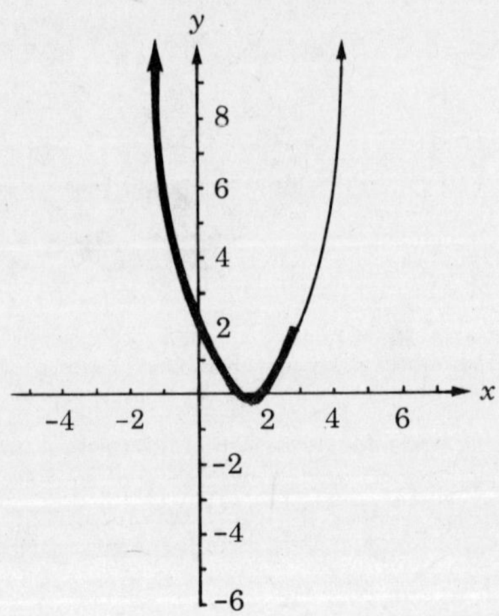

$y = x^2 - 3x + 2$

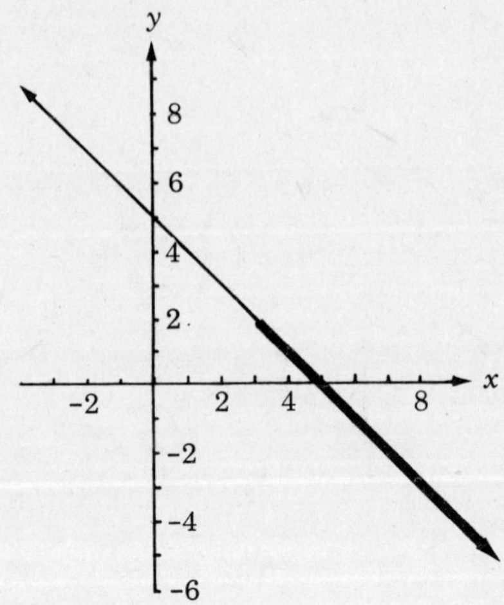

$y = -x + 5$

Notice that we are only interested in the darkened portions of each graph above. Combining these, we have the graph $y = f(x)$ below:

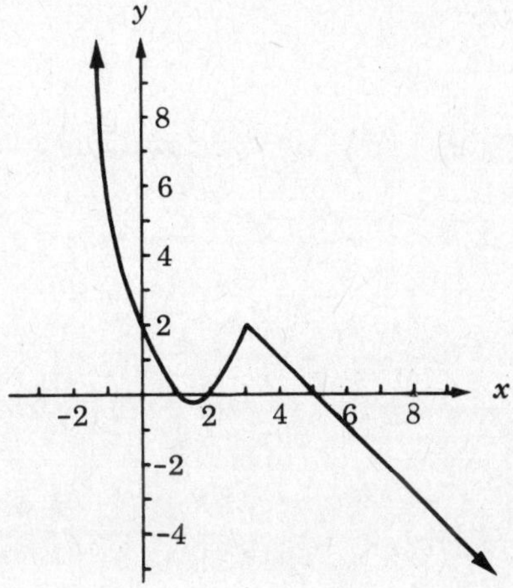

Graph of $y = f(x) = \begin{cases} x^2 - 3x + 2 & x \leqslant 3 \\ -x + 5 & x > 3 \end{cases}$

b. $f(3) = 3^2 - 3(3) + 2 = 2$

In the example above, notice that the entire graph of $y = f(x)$ can be traced without lifting the pencil off the page. We say f is a **continuous function.*** All linear and quadratic functions are continuous but the function graphed in Example 2 is not—it is called a **discontinuous** function.

EXAMPLE 2 Graph $g(x) = \begin{cases} x^2 - 4 & \text{if } x \leqslant 1 \\ 4 - x^2 & \text{if } x > 1 \end{cases}$

Solution Notice that $y = g(x)$ has a "jump" at $x = 1$ and $g(1) = -3$. The function g is said to be *discontinuous* at $x = 1$.

*The definition of continuous function must necessarily be left to the study of calculus. Now, we attempt only to give the reader a very basic introduction to the notion of continuity.

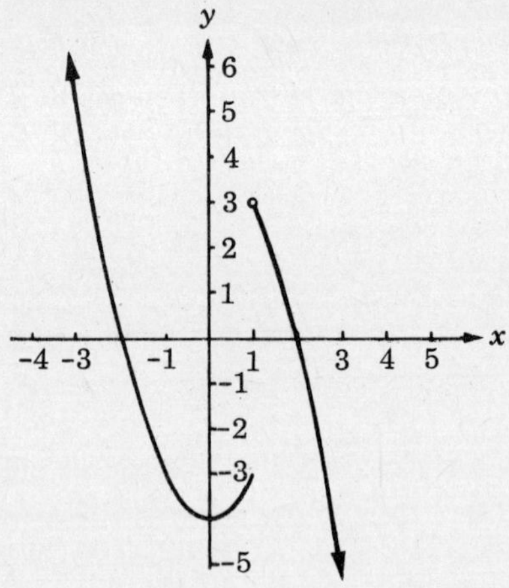

Graph of $g(x) = \begin{cases} x^2 - 4 & \text{if } x \leqslant 1 \\ 4 - x^2 & \text{if } x > 1 \end{cases}$

The function $f(x) = |x|$ can be expressed as a piecewise function, using the definition of absolute value in Section 1.2.

$$f(x) = |x| = \begin{cases} -x & \text{if } x < 0 \\ x & \text{if } x \geqslant 0 \end{cases}$$

Notice that the two "pieces" ($y = x$ and $y = -x$) are each linear functions; we graph the desired parts of those lines to form the graph of $y = |x|$ on the right.

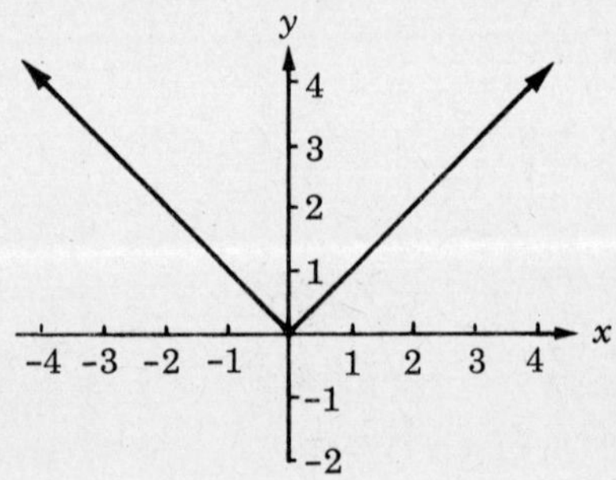

Graph of $y = |x|$

EXAMPLE 3 Graph $y = |2x - 1|$.

Solution Using the definition of absolute value, we can write

$$y = \begin{cases} -(2x - 1) & \text{if } 2x - 1 < 0 \\ 2x - 1 & \text{if } 2x - 1 \geqslant 0 \end{cases}$$

or equivalently,

$$y = \begin{cases} -2x + 1 & \text{if } x < \frac{1}{2} \\ 2x - 1 & \text{if } x \geqslant \frac{1}{2} \end{cases}$$

We have the following continuous function:

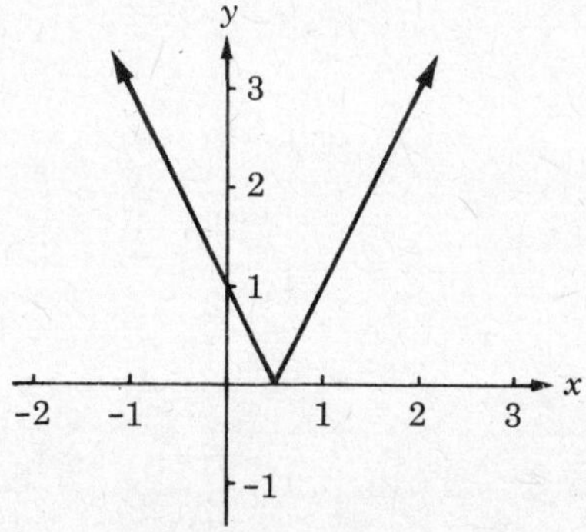

Graph of $y = |x|$

EXAMPLE 4 Graph $y = |1 - x^2|$.

Solution Using the definition of absolute value,

$$y = \begin{cases} -(1 - x^2) & \text{if } 1 - x^2 < 0 \\ 1 - x^2 & \text{if } 1 - x^2 \geqslant 0 \end{cases}$$

or equivalently*

$$y = \begin{cases} -1 + x^2 & \text{if } x < -1 \\ 1 - x^2 & \text{if } -1 \leqslant x \leqslant 1 \\ -1 + x^2 & \text{if } x > 1 \end{cases}$$

The graph consists of portions of two parabolas as shown on the right.

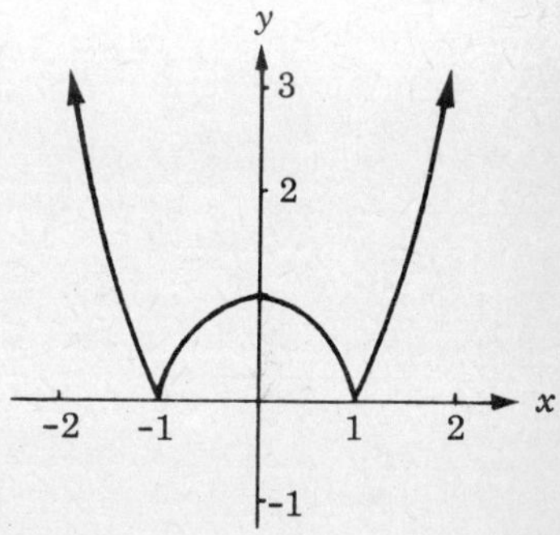

Graph of $y = |1 - x^2|$

One special piecewise function, known as the **greatest integer function** is described by:

"the set of all ordered pairs where the ordinate is the greatest integer that is less than or equal to the abscissa."

*Recall, from Chapter 2, $1 - x^2 < 0$ is equivalent to $x < -1$ or $x > 1$, and $1 - x^2 \geqslant 0$ is equivalent to $-1 \leqslant x \leqslant 1$.

That can be interpreted as:

$$y = \begin{cases} \vdots \\ -1 & \text{if } -1 \leqslant x < 0 \\ 0 & \text{if } 0 \leqslant x < 1 \\ 1 & \text{if } 1 \leqslant x < 2 \\ 2 & \text{if } 2 \leqslant x < 3 \\ \vdots \end{cases}$$

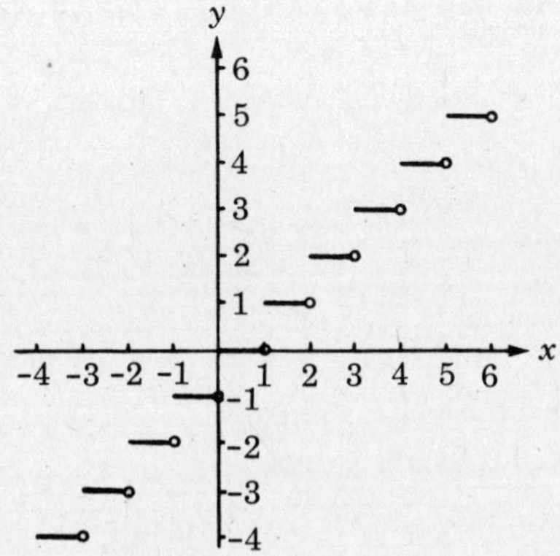

Graph of $y = [\![x]\!]$

We use the notation

$$f(x) = [\![x]\!]$$

to represent the greatest integer function. Its graph appears on the right. For obvious reasons, functions that resemble $y = [\![x]\!]$ are referred to as **step functions**. Another example of a step function follows.

EXAMPLE 5 On 1979 New York State income tax returns, the computation of "Household Credit" is given by:

Gross income		Household credit
at least	*but less than*	
0	\$ 5000	\$65
\$ 5000	6000	50
6000	7000	40
7000	25000	35
25000 and over		0

Graph household credit, y, as a function of gross income, x.

Solution Notice the 1:200 scale ratio in the graph below.

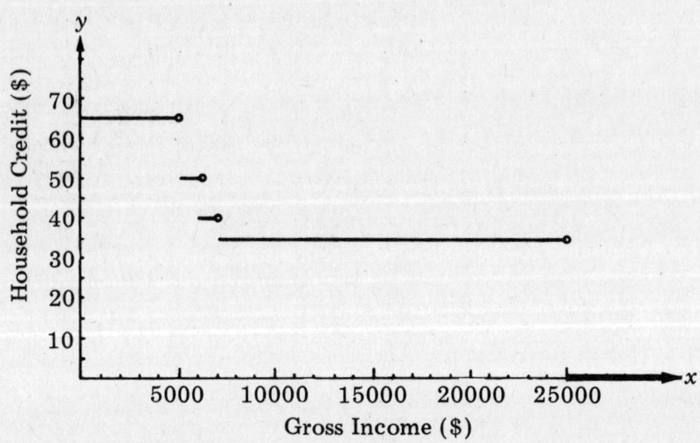

EXERCISES FOR SECTION 4.1

In Exercises 1 through 25, a) graph $y = f(x)$ and b) determine whether or not the function is continuous. (See the illustration preceding Example 1 and Examples 1, 2, 3, and 4.)

1. $f(x) = \begin{cases} x^2 - 4x - 5 & \text{if } x \leq 0 \\ -2x - 5 & \text{if } x > 0 \end{cases}$

2. $f(x) = \begin{cases} 2x + 1 & \text{if } x \leq 2 \\ 9 - x^2 & \text{if } x > 2 \end{cases}$

3. $f(x) = \begin{cases} x^2 + 2x - 8 & \text{if } x \leq 3 \\ 2x - 3 & \text{if } x > 3 \end{cases}$

4. $f(x) = \begin{cases} x^2 + 2x - 8 & \text{if } x < 3 \\ 2x - 3 & \text{if } x \geq 3 \end{cases}$

5. $f(x) = \begin{cases} x - 1 & \text{if } x < -2 \\ 1 - x & \text{if } x \geq -2 \end{cases}$

6. $f(x) = \begin{cases} 4 - 3x & \text{if } x \leq -\frac{1}{3} \\ 5 & \text{if } -\frac{1}{3} < x < 3 \\ 2 - x & \text{if } x \geq 3 \end{cases}$

7. $f(x) = \begin{cases} 3 & \text{if } x < -1 \\ 4 - x^2 & \text{if } -1 \leq x \leq 1 \\ 3 & \text{if } x > 1 \end{cases}$

8. $f(x) = \begin{cases} -2x - 6 & \text{if } x < -3 \\ x^2 - 9 & \text{if } -3 \leq x \leq 3 \\ 2x - 6 & \text{if } x > 3 \end{cases}$

9. $f(x) = \begin{cases} 0 & \text{if } x < -1 \\ x & \text{if } -1 \leq x \leq 1 \\ 1 & \text{if } x > 1 \end{cases}$

10. $f(x) = \begin{cases} 0 & \text{if } x \leq -1 \\ x & \text{if } -1 < x \leq 1 \\ 1 & \text{if } x > 1 \end{cases}$

11. $f(x) = \begin{cases} -1 & \text{if } x \leq -1 \\ x & \text{if } -1 < x < 1 \\ 1 & \text{if } x > 1 \end{cases}$

12. $f(x) = \begin{cases} -\sqrt{-x} & \text{if } x < -1 \\ x & \text{if } -1 \leq x \leq 1 \\ \sqrt{x} & \text{if } x > 1 \end{cases}$

13. $f(x) = \begin{cases} -\sqrt{-x} & \text{if } x < -1 \\ x & \text{if } -1 < x < 1 \\ \sqrt{x} & \text{if } x > 1 \end{cases}$

14. $f(x) = \begin{cases} -4 & \text{if } x \leq -2 \\ 4x + 4 & \text{if } -2 < x \leq -1 \\ x^2 - 1 & \text{if } -1 < x \leq 1 \\ -4x + 4 & \text{if } 1 < x \leq 2 \\ -4 & \text{if } x > 2 \end{cases}$

15. $f(x) = |2x|$
16. $f(x) = |x - 3|$
17. $f(x) = |2x + 1|$
18. $f(x) = |3 - x|$
19. $f(x) = |1 - 2x|$
20. $f(x) = -|x|$
21. $f(x) = |4 - x^2|$
22. $f(x) = |x^2 - 4|$
23. $f(x) = |x^2 + 4|$
24. $f(x) = |x^2 + 2x - 8|$
25. $f(x) = \frac{1}{2}|x^2 + 3x - 4|$
26. Consider the relation

$$g(x) = \begin{cases} x & x \leq -1 \\ 1 - x^2 & -1 \leq x \leq 1 \\ 0 & x \geq 1 \end{cases}$$

Is $g(x)$ a function? Explain.

27. a. What is the domain of f in Exercise 7?
    b. What is the range of f in Exercise 7?
28. a. What is the domain of f in Exercise 13?
    b. What is the range of f in Exercise 13?
29. a. What is the domain of g in Exercise 26?
    b. What is the range of g in Exercise 26?

In Exercises 30 through 32, graph the given function. (See Example 5 and the illustration preceding it.)

30. $y = [\![x + 3]\!]$ 31. $y = [\![x]\!] + 3$ 32. $y = [\![2x]\!]$

33. Consider the first-class postage function.

Cost of postage y (in cents)	*Weight of package x (in ounces)* at least	but less than
15	0	1
28	1	2
41	2	3
54	3	4
⋮	⋮	⋮

a. Graph the function.
b. State its domain and range.

34. (True or False) The graph of the postage function in Exercise 33 is identical to the graph of

$$y = 13[\![x]\!] + 15 \qquad (x \geqslant 0)$$

35. The notion of piecewise functions can be used to analyze and graph more complicated functions such as

$$y = |x + 1| + |x - 1|$$

by first considering significant "breaks" in the domain. For example, $|x + 1|$ should be "broken" into the parts for $x < -1$ and $x \geqslant -1$. We know $|x + 1| = -(x + 1)$ for $x < -1$ and $|x + 1| = x + 1$ for $x \geqslant -1$. Similarly, $|x - 1|$ should be considered as $-(x - 1)$ for $x < 1$ and as $x - 1$ for $x \geqslant 1$. *Collectively*, we have

$$y = \begin{cases} -(x + 1) - (x - 1) & x < -1 \\ x + 1 - (x - 1) & -1 \leqslant x < 1 \\ x + 1 + x - 1 & x \geqslant 1 \end{cases}$$

or equivalently

$$y = \begin{cases} -2x & x < -1 \\ 2 & -1 \leqslant x < 1 \\ 2x & x \geqslant 1 \end{cases}$$

Graph $y = |x + 1| + |x - 1|$.

36. a. Use a development that parallels that of Exercise 35 to graph

$$h(x) = |x| + |2x - 1|$$

b. What is the domain of h?
c. What is the range of h?

37. Graph $y = |x + 2| - |x|$. (See Exercise 35.)

38. a. Graph $y = |x^2 - 4| - |3x|$. (See Exercise 35.)
b. Is this relation a function?
c. Is this relation a continuous function?
d. State the domain and range of this relation.

4.2 ALGEBRAIC CLUES TO GRAPHING

As the algebraic structure of a function becomes more complicated, usually so does its graph. A so-called **rational function** is a function that can be expressed as $\frac{P(x)}{Q(x)}$ where $P(x)$ and $Q(x)$ are polynomials. We will graph such functions in Section 4.3. The complicated nature of these graphs necessitates that we use *analytical* methods to help in their construction. In short, plotting six points and "connecting the dots" will not work!

One analytical tool we shall employ is the concept of symmetry.

Symmetry

Consider each of the three graphs below.

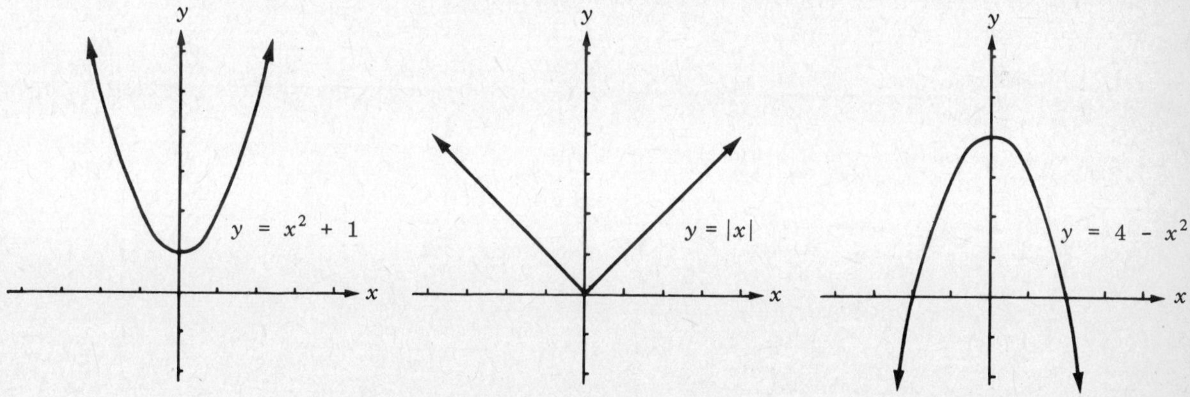

Each of the graphs is *symmetric with respect to the y-axis*. That is, for any number x in the domain of the function, both points (x, y) and $(-x, y)$ lie on the graph:

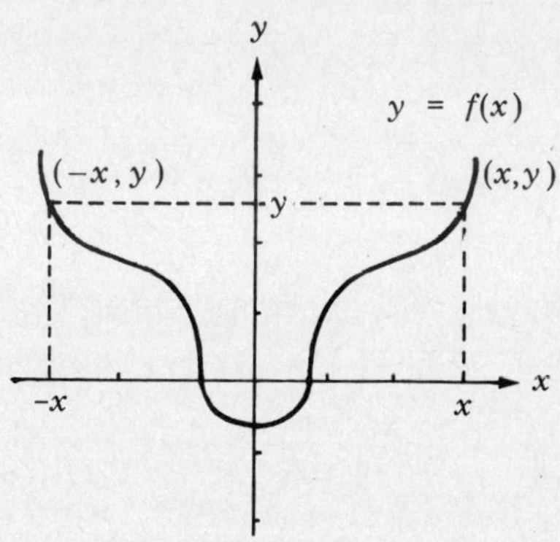

A function symmetric with respect to the y-axis.

We define y-axis symmetry in more formal terms below.

DEFINITION

y-axis Symmetry

The graph of $y = f(x)$ is symmetric with respect to the y-axis if and only if $f(-x) = f(x)$. Equivalently, the graph of $y = f(x)$ is symmetric with respect to the y-axis if and only if replacing x by $-x$ produces the same equation.

EXAMPLE 1 Determine whether the graph of $f(x) = \dfrac{1}{x^2}$ is symmetric with respect to the y-axis.

Solution Yes, the graph of $f(x) = \dfrac{1}{x^2}$ is symmetric with respect to the y-axis because $f(-x) = f(x)$.

$$f(-x) = \frac{1}{(-x)^2} = \frac{1}{x^2} = f(x).$$

EXAMPLE 2 Determine whether the graph of $f(x) = \dfrac{1}{x^2 + 7x + 12}$ is symmetric with respect to the y-axis.

Solution No, it is not. We have:

$$f(-x) = \frac{1}{(-x)^2 + 7(-x) + 12} = \frac{1}{x^2 - 7x + 12} \neq f(x)$$

Another type of symmetry that the graph of a function might exhibit is **symmetry with respect to the origin.** Such symmetry is displayed by the graph below.

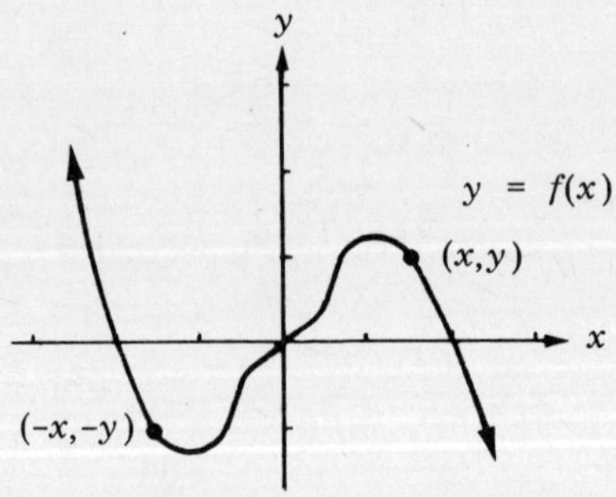

A function symmetric with respect to the origin.

The formal definition of origin symmetry follows.

DEFINITION

Origin Symmetry

The graph of $y = f(x)$ is symmetric with respect to the origin if and only if $f(-x) = -f(x)$. Equivalently, the graph of $y = f(x)$ is symmetric with respect to the origin if and only if replacing x by $-x$ and simultaneously replacing y by $-y$ produces the same equation.

EXAMPLE 3 Determine whether the graph of $f(x) = \frac{4}{x^3}$ is symmetric with respect to the origin.

Solution Yes, because $f(-x) = \frac{4}{(-x)^3} = \frac{4}{-x^3} = -\frac{4}{x^3} = -f(x)$.

EXAMPLE 4 Is the graph of $y = \frac{1}{x+1}$ symmetric with respect to the origin?

Solution Taking the original equation $y = \frac{1}{x+1}$ and simultaneously replacing x by $-x$ and y by $-y$ gives:

$$-y = \frac{1}{-x+1} \quad \text{or} \quad y = \frac{1}{x-1}$$

However, this latter equation is not the same as the original. Thus, there is *no origin symmetry*.

EXAMPLE 5 Consider the relation $x^2 + y^2 = 9$. Discuss the graph's symmetry.

Solution First, check for y-axis symmetry. Does substituting $-x$ for x result in the same equation?

$$(-x)^2 + y^2 = 9$$
$$x^2 + y^2 = 9$$

Yes. Thus, the graph of $x^2 + y^2 = 9$ is symmetrical with respect to the y-axis.

Next, check for origin symmetry. Does substituting $-x$ for x and $-y$ for y result in the same equation?

$$(-x)^2 + (-y)^2 = 9$$
$$x^2 + y^2 = 9$$

Yes. Thus, the graph of $x^2 + y^2 = 9$ is symmetric with respect to the origin.

Asymptotes Consider the function $y = g(x) = \frac{1}{x^2}$. The domain is $\{x \mid x \neq 0\}$. Until now, we have said that $g(0)$ is undefined and that has been the end of the discussion. Presently, we wish to examine this "undefined" phenomenon and do so by choosing values of x "close" to zero. A calculator yields the following:

x	$\frac{1}{x^2}$
1.	1.
0.1	100.
0.01	10,000.
0.001	1,000,000.
0.0001	100,000,000.
0.00008	$1.5625 \cdot 10^8$
0.00006	$2.7778 \cdot 10^8$
0.00004	$6.2500 \cdot 10^8$
0.00002	$2.500 \cdot 10^9$
0.00001	$1.000 \cdot 10^{10}$
$1 \cdot 10^{-9}$	$1.000 \cdot 10^{18}$
$1 \cdot 10^{-20}$	$1.000 \cdot 10^{40}$
$1 \cdot 10^{-40}$	$1.000 \cdot 10^{80}$
$1 \cdot 10^{-49}$	$1.000 \cdot 10^{98}$

From the table above, it should be clear that as x gets closer and closer to 0 (through positive numbers), y gets larger and larger. In symbols, we write

$$\text{as } x \longrightarrow 0^+, \ y \longrightarrow \infty$$

and this is read

"as x approaches 0 *from the right*, y *increases without bound* (or y approaches infinity)."*

In this example we can also write

$$\text{as } x \longrightarrow 0^-, y \longrightarrow \infty$$

because of the symmetry with respect to the y-axis. Actually, the graph of $g(x) = \frac{1}{x^2}$ appears as:

*$x \longrightarrow a^+$ is read "x approaches a from the right" and $x \longrightarrow a^-$ is read "x approaches a from the left."

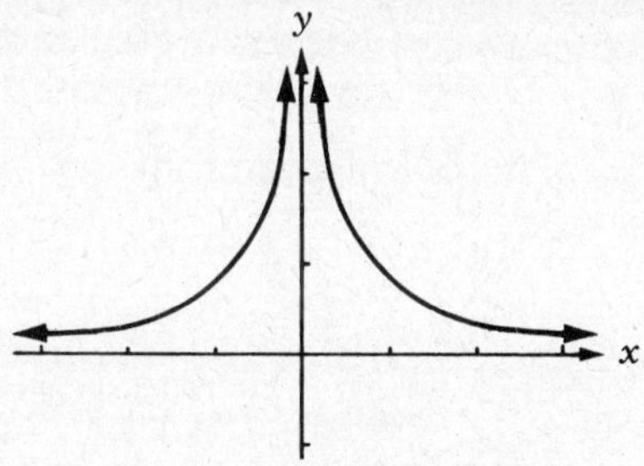

Graph of $y = g(x) = \frac{1}{x^2}$

As previously mentioned, the graph shows that as $x \longrightarrow 0^+$, $y \longrightarrow \infty$ and also as $x \longrightarrow 0^-$, $y \longrightarrow \infty$. The vertical line $x = 0$ is called a *vertical asymptote.*

In general, we say that the graph of $y = f(x)$ approaches the line $x = a$ **asymptotically** if $y \longrightarrow \infty$ as $x \longrightarrow a$. The line $x = a$ is called a **vertical asymptote**. For rational functions of the form $y = \frac{P(x)}{Q(x)}$, we can see that $y \longrightarrow \infty$ if $Q(x) \longrightarrow 0$ (and $P(x)$ doesn't approach 0). We summarize this below:

> **Vertical Asymptote**
>
> Suppose $f(x) = \frac{P(x)}{Q(x)}$ is a rational function. If $Q(a) = 0$ (and $P(a) \neq 0$), then the line $x = a$ is a vertical asymptote of the function.

EXAMPLE 6 What are the vertical asymptotes of $y = \frac{x}{x^2 - 1}$?

Solution Since $x = 1$ and $x = -1$ are zeros of $Q(x) = x^2 - 1$, and the numerator, x, is not zero, the lines $x = 1$ and $x = -1$ are vertical asymptotes.

EXAMPLE 7 Find the vertical asymptotes for $y = \frac{3x}{x^2 - 4x}$.

Solution Notice that $y = \frac{3x}{x^2 - 4x}$ is identical to $y = \frac{3}{x - 4}$, provided $x \neq 0$. The graph has the vertical asymptote $x = 4$.

Next a discussion of **horizontal asymptotes** is undertaken. As x grows large without bound (in symbols, $x \longrightarrow \infty$ or $x \longrightarrow -\infty$), the expression $\frac{1}{x^2}$ approaches 0. Thus, for the rational function

$$y = \frac{1}{x^2}$$

we see that $y \longrightarrow 0$ as $x \longrightarrow \infty$ and $y \longrightarrow 0$ as $x \longrightarrow -\infty$. Thus the line $y = 0$ is a horizontal asymptote. An example follows.

EXAMPLE 8 Consider $y = \frac{x^2 - 4}{9 - x^2}$. What are the graph's asymptotes?

Solution The vertical asymptotes are $x = 3$ and $x = -3$. To discover the horizontal asymptote we rewrite the original equation as

$$y = \frac{1 - \frac{4}{x^2}}{\frac{9}{x^2} - 1}.$$

Now, as $x \longrightarrow \infty$, the expressions $\frac{4}{x^2}$ and $\frac{9}{x^2}$ each approach 0 and

$$y \longrightarrow \frac{1 - 0}{0 - 1} = -1$$

Thus, $y = -1$ is a horizontal asymptote. The graph of $y = \frac{x^2 - 4}{9 - x^2}$ is shown below to give the reader a visual notion of how the asymptotes affect the graph.

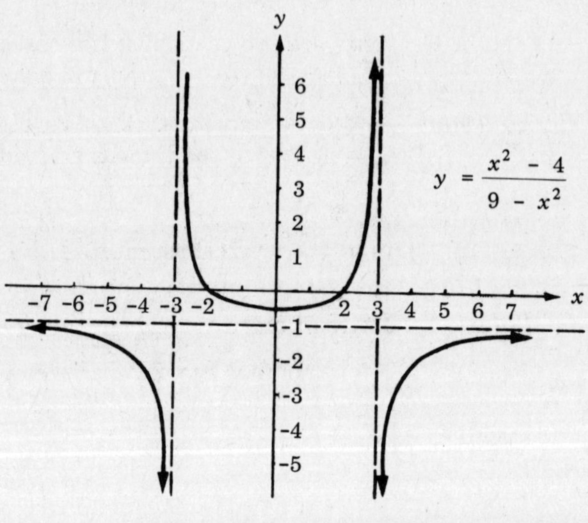

Sign Intervals for $f(x)$ A final part of the analysis of graphing that we will consider here is the examination of a function's domain to determine where the function is positive (the graph is above the x-axis) and where it is negative (the graph is below the x-axis). As the reader will shortly discover, once this type of investigation is complete, we will have certain intervals and the sign of $y = f(x)$ in each interval. Collectively, this information is called **sign intervals for $f(x)$**. Consider the equation $y = \frac{x}{x^2 - 9}$. Basically, there are three factors involving x that determine whether y is positive or negative; they are x, $(x + 3)$, and $(x - 3)$. We find the sign graphs for each factor below.

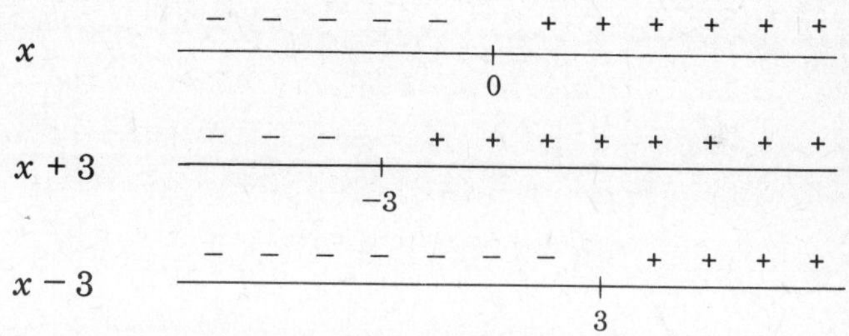

Thus, the sign intervals for y are:

$$y = \frac{x}{(x+3)(x-3)}$$

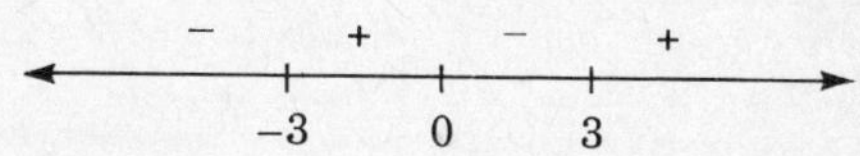

Summarizing in table form:

Intervals	*Sign of $y = f(x)$*	*Graph of $f(x)$*
$x < -3$	$-$	Below x-axis
$-3 < x < 0$	$+$	Above x-axis
$0 < x < 3$	$-$	Below x-axis
$x > 3$	$+$	Above x-axis

In Section 4.3 the information in this table will be used to graph the equation $y = \frac{x}{x^2 - 9}$.

EXAMPLE 9 Determine the sign intervals for $f(x)$.

$$f(x) = \frac{x + 3}{(x + 2)(x - 1)}$$

Solution We construct sign graphs for each factor:

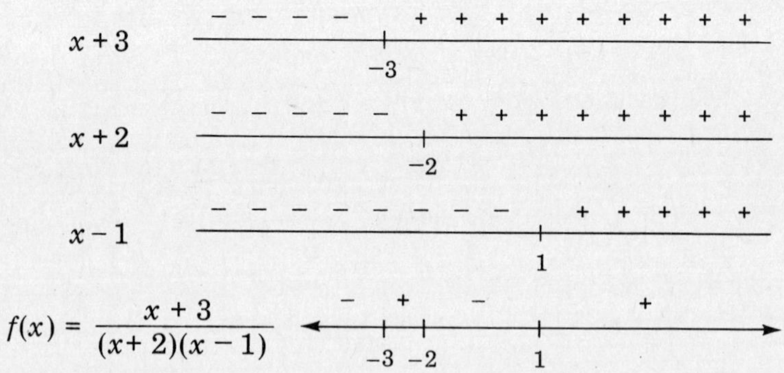

In table form:

Intervals	*Sign of* $y = f(x)$	*Graph of* $f(x)$
$x < -3$	$-$	Below x-axis
$-3 < x < -2$	$+$	Above x-axis
$-2 < x < 1$	$-$	Below x-axis
$x > 1$	$+$	Above x-axis

In Section 4.3 the information in this table will be used to graph the equation $y = \dfrac{x+3}{(x+2)(x-1)}$.

EXERCISES FOR SECTION 4.2

In Exercises 1 through 15, determine whether the graph of the given function is symmetric with respect to the y-axis, the origin, neither, or both. (See Examples 1, 2, 3, and 4.)

1. $f(x) = \dfrac{4}{x^2}$
2. $f(x) = \dfrac{x}{x^2+1}$
3. $f(x) = \dfrac{x}{x^2-1}$
4. $y = \dfrac{1}{x^2-3x-10}$
5. $y = \dfrac{1}{x^3}$
6. $y = \dfrac{x^2}{x^2-4}$
7. $y = \dfrac{x^2-9}{x^2-4}$
8. $y = \dfrac{x^2-1}{4-x^2}$
9. $y = \dfrac{1}{x}$
10. $\dfrac{3x}{x^2-4}$
11. $y = \dfrac{x^2}{9-x^2}$
12. $y = \dfrac{x^2}{(9-x)^2}$
13. $y = \dfrac{1}{(x-2)^2}$
14. $y = \dfrac{(x-1)^2}{x^2}$
15. $y = \dfrac{x^2-3x}{(x+1)(x-2)}$

For Exercises 16 through 30, use the functions of Exercises 1 through 15 to determine a) vertical asymptotes and b) horizontal asymptotes. (See Examples 6, 7, and 8.)

31. Discuss the graph of $x^2 + y^2 = 4$ for symmetry. (See Example 5.)
32. Discuss the graph of $x^2 - y^2 = 1$ for symmetry. (See Example 5.)

33. The graph of a relation is *symmetric with respect to the x-axis* if, for any point (x,y) on the graph, there is also the point $(x,-y)$ on the graph as the figure on the right illustrates.

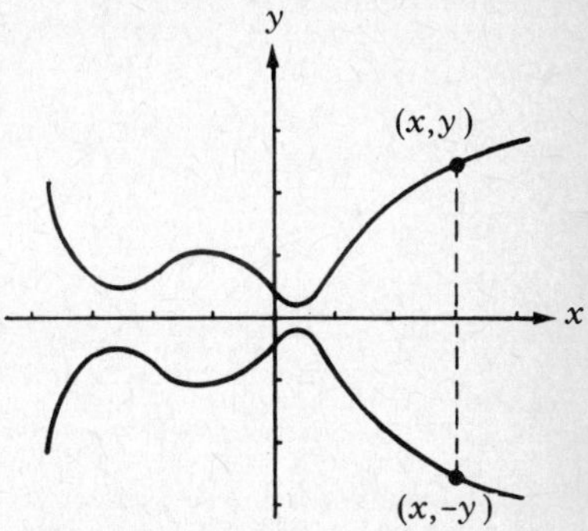

The graph of a relation symmetric with respect to the x-axis.

Determine which of the following relations are symmetric with respect to the x-axis.
a. $x^2 + y^2 = 9$
b. $x^2 - y^2 = 1$
c. $y^2 = \dfrac{1}{x}$

34. (True or False) It is possible for a relation to be symmetric to both the x- and y-axes.
35. (True or False) If a relation is symmetric with respect to both axes, it must also be symmetric with respect to the origin.
36. A function f that satisfies

$$f(-x) = f(x)$$

for all x in its domain is called an **even function**. Is every even function symmetric with respect to the y-axis?
37. A function g that satisfies

$$g(-x) = -g(x)$$

for all x in its domain is called an **odd function**. Is every odd function symmetric with respect to the origin?
38. Consider a rational function whose numerator is of degree greater than or equal to the degree of its denominator such as

$$f(x) = \frac{4x^2 + 3x + 1}{x - 1}.$$

If we perform the division $(4x^2 + 3x + 1) \div (x - 1)$ we obtain:

$$f(x) = 4x + 7 + \frac{8}{x - 1}.$$

Now as $x \longrightarrow \infty$, the term $\dfrac{8}{x-1} \longrightarrow 0$. Thus as x gets "large," y approaches the line $y = 4x + 7$. In this case $y = 4x + 7$ is called an **oblique asymptote**. Find all asymptotes for the following:

a. $y = \dfrac{3x^2 + 1}{x - 1}$ b. $y = \dfrac{x^2 + 1}{x}$ c. $y = \dfrac{x^3 + 7x + 1}{x^2 + x - 2}$

4.3 GRAPHING RATIONAL FUNCTIONS

We are now ready to graph rational functions. The graphs will be obtained by analyzing the following properties:

1. the intercepts (both x-axis and y-axis)
2. the domain (and occasionally the range)
3. symmetry
4. vertical and horizontal asymptotes
5. sign intervals

Some examples follow.

EXAMPLE 1 Graph $y = \dfrac{x}{x^2 - 9}$.

Solution Intercepts: If $x = 0$, $y = \dfrac{0}{0-9} = 0$. Thus, $(0,0)$ is the y-intercept. If $y = 0$, $\dfrac{x}{x^2 - 9} = 0$ and by a property of zero from Section 1.1 we now set the numerator equal to zero. Thus $x = 0$ and $(0,0)$ is also the x-intercept.

Domain: Since $y = \dfrac{x}{x^2 - 9}$, the domain is $\{x \mid x \neq \pm 3\}$.

Symmetry: There is origin symmetry because replacing x by $-x$ and y by $-y$ produces the same equation.

Asymptotes: Vertical asymptotes occur when $x^2 - 9 = 0$. They are $x = 3$ and $x = -3$. To find horizontal asymptotes we write $y = \dfrac{x}{x^2 - 9}$ or $y = \dfrac{\dfrac{x}{x^2}}{1 - \dfrac{9}{x^2}} = \dfrac{\dfrac{1}{x}}{1 - \dfrac{9}{x^2}}$, and we see that as $x \longrightarrow \infty, y \longrightarrow 0$. Thus $y = 0$ (the x-axis) is a horizontal asymptote.

Sign intervals: From Section 4.2,

Intervals	*Graph of f(x)*
$x < -3$	Below x-axis
$-3 < x < 0$	Above x-axis
$0 < x < 3$	Below x-axis
$x > 3$	Above x-axis

Notice that the graph is *discontinuous* at $x = 3$ and $x = -3$.

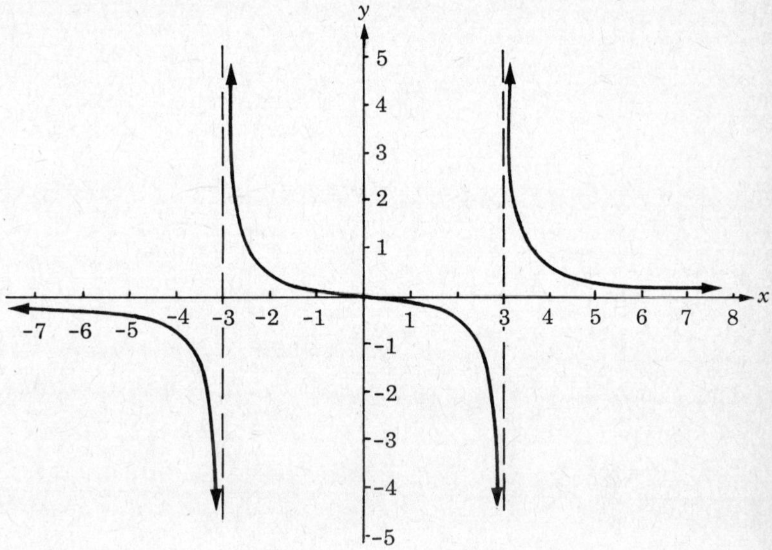

Graph of $y = \dfrac{x}{x^2 - 9}$

EXAMPLE 2 Graph $y = \dfrac{x^2}{4 - x^2}$.

Solution Intercepts: (0,0)

Domain: $\{x \mid x \neq \pm 2\}$

Symmetry: y-axis

Vertical asymptotes: $x = 2, x = -2$

Horizontal asymptote: $y = -1$

Sign intervals: The reader should verify that by examining sign graphs for the factors x^2, $2 + x$, and $2 - x$, the following result is obtained.

$y = \dfrac{x^2}{4 - x^2}$

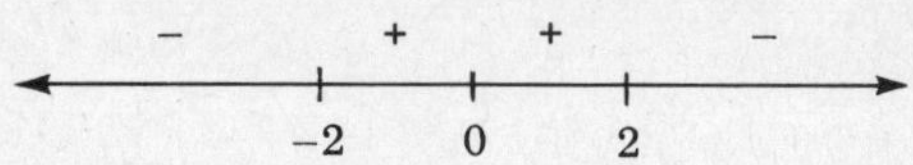

In table form:

Intervals	*Sign of f(x)*	*Graph of f(x)*
$x < -2$	–	Below x-axis
$-2 < x < 2$	+	Above x-axis (with (0,0) int.)
$x > 2$	–	Below x-axis

Notice that the graph is discontinuous at $x = 2$ and $x = -2$.

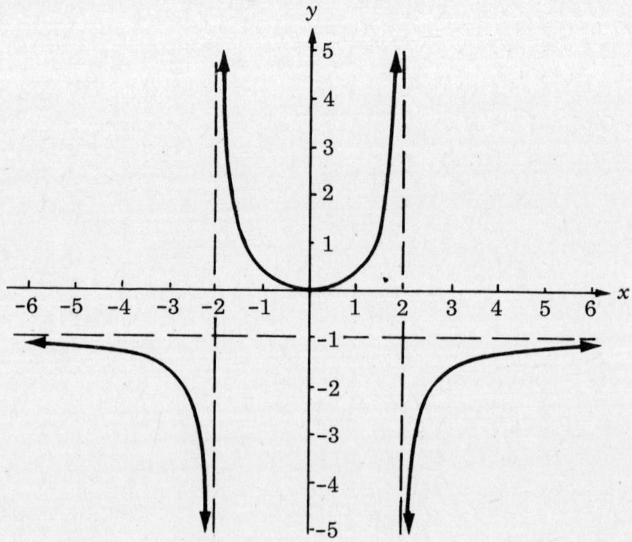

Graph of $y = \dfrac{x^2}{4 - x^2}$

EXAMPLE 3 Graph $y = \dfrac{x + 3}{(x + 2)(x - 1)}$.

Solution Intercepts: $(0, -\frac{3}{2}), (-3, 0)$

Domain: $\{x \mid x \neq -2, x \neq 1\}$

Symmetry: none

Vertical asymptotes: $x = -2, x = 1$

Horizontal asymptote: $y = 0$

Sign intervals: From Section 4.2,

Intervals	*Graph of f(x)*
$x < -3$	Below x-axis
$-3 < x < -2$	Above x-axis
$-2 < x < 1$	Below x-axis
$x > 1$	Above x-axis

Range: The technique for determining that the range is $\{y \mid y \geqslant -\frac{1}{9}$ or $y \leqslant -1\}$ is outlined in Exercise 25. The graph appears on the next page. Notice that the graph is discontinuous at $x = 1$ and $x = -2$.

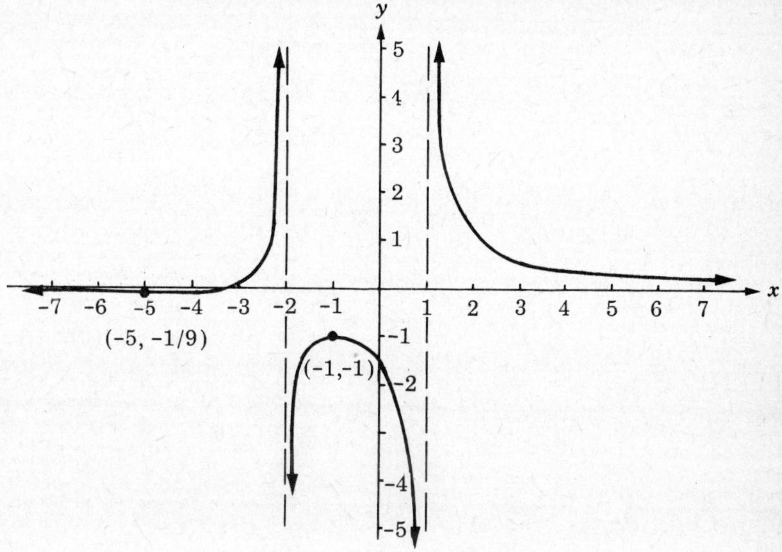

Graph of $y = \dfrac{x+3}{(x+2)(x-1)}$

EXAMPLE 4 Graph $y = \dfrac{x^2}{x^2+1}$.

Solution Intercepts: (0,0)

Domain: $\mathcal{R}$

Range: $\{y \mid 0 \leqslant y < 1\}$ (See Exercise 26.)

Symmetry: y-axis

Vertical asymptotes: none

Horizontal asymptotes: $y = 1$

Sign intervals: Graph of $y = f(x)$ is *always* above x-axis.

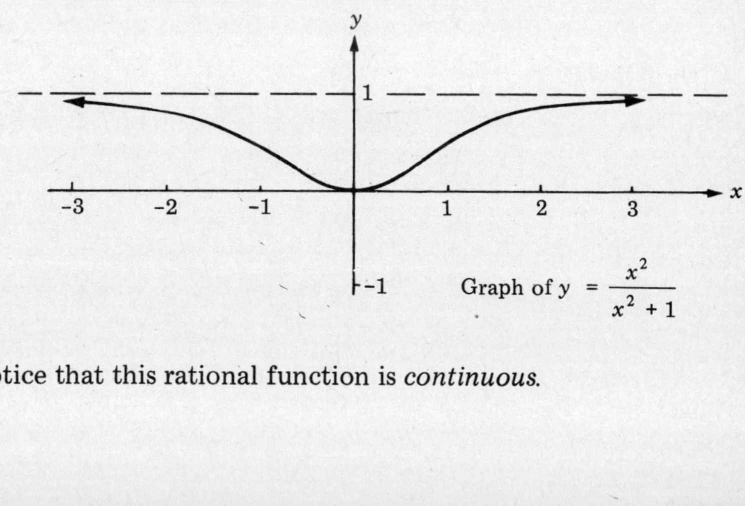

Graph of $y = \dfrac{x^2}{x^2+1}$

Notice that this rational function is *continuous.*

EXAMPLE 5 Graph $y = \dfrac{x^2 - 4}{x + 2}$.

Solution Notice that the domain is $\{x | x \neq -2\}$. Factoring the numerator yields $y = \dfrac{(x + 2)(x - 2)}{x + 2}$ which we can rewrite as

$$y = x - 2 \qquad (\text{provided } x \neq -2)$$

Thus, the graph of $y = \dfrac{x^2 - 4}{x + 2}$ is the same as the graph of $y = x - 2$, except $x \neq -2$. In other words, the point $(-2,-4)$ must *not* be included in the graph that appears below.

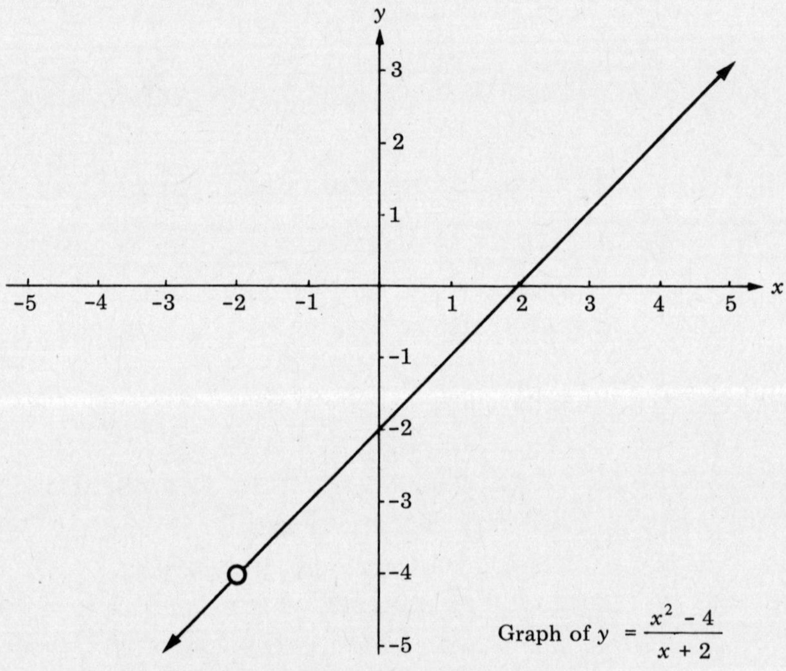

Graph of $y = \dfrac{x^2 - 4}{x + 2}$

Notice that this function is discontinuous at $x = -2$.

We conclude this section by re-emphasizing that the way in which the graphs of rational functions were examined was by means of *analysis*—not "point-plotting." The reader should be aware that being able to refine the skills of analyzing intercepts, domain and range, symmetry, asymptotes, and sign intervals will be very advantageous in the further study of mathematics.

EXERCISES FOR SECTION 4.3

In Exercises 1 through 24, graph each rational function using the methods outlined in this section. (See Examples 1 through 5.)

1. $y = \dfrac{x}{x^2 - 1}$
2. $y = \dfrac{x}{4 - x^2}$
3. $y = \dfrac{x}{x - 2}$
4. $y = \dfrac{x^2}{1 - x^2}$
5. $y = \dfrac{x^2}{x^2 - 1}$
6. $y = \dfrac{2x^2}{x^2 - 9}$
7. $y = \dfrac{4x^2}{2x^2 - 9}$
8. $y = \dfrac{1 - x^2}{x^2}$
9. $y = \dfrac{x}{(x + 1)(x + 4)}$
10. $y = \dfrac{x + 3}{(x - 2)(x + 4)}$
11. $y = \dfrac{x - 1}{(x - 2)(x + 2)}$
12. $y = \dfrac{x^2}{(x - 1)(x + 3)}$
13. $y = \dfrac{1 - x^2}{(x - 2)(x + 2)}$
14. $y = \dfrac{x^2}{x^2 + 4}$
15. $y = \dfrac{2x^2 + 9}{x^2 + 1}$
16. $y = \dfrac{4x^2}{x^2 + 1}$
17. $y = \dfrac{x^2 + 1}{x^2 + 4}$
18. $y = \dfrac{x^2 - 1}{x^2 - 9}$
19. $y = \dfrac{x}{x^2 + 1}$
20. $y = \dfrac{10}{x^2 + 1}$
21. $y = \dfrac{x^2 - 1}{x + 1}$
22. $y = \dfrac{(x - 3)^2}{x - 3}$
23. $y = \dfrac{x + 1}{x^2 - 1}$
24. $y = \dfrac{(x + 1)(x + 2)(x - 3)}{(x + 1)(x + 2)(x - 3)}$

25. The technique for finding the range of $y = \dfrac{x + 3}{(x + 2)(x - 1)}$ is to solve the equation *for* x as follows:

$$(x + 2)(x - 1)y = x + 3$$

Expand

$$yx^2 + yx - 2y = x + 3$$

Subtract $x + 3$ from both sides

$$yx^2 + yx - x - 2y - 3 = 0$$

This is a quadratic equation in x

$$yx^2 + (y - 1)x + (-2y - 3) = 0$$

Use the quadratic formula with $a = y$, $b = y - 1$, and $c = -2y - 3$ to obtain:

$$x = \frac{-y + 1 \pm \sqrt{9y^2 + 10y + 1}}{2y}$$

Finally, restrictions on y involve making sure $9y^2 + 10y + 1 \geqslant 0$:

$$9y^2 + 10y + 1 = (9y + 1)(y + 1)$$

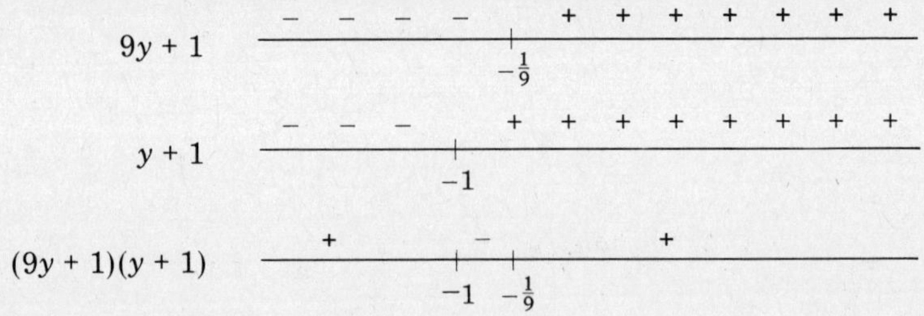

Hence we want $9y^2 + 10y + 1 \geqslant 0$, $y \leqslant -1$ or $y \geqslant -\frac{1}{9}$. Use this technique to find the range of

a. $y = \dfrac{x^2}{x^2 - 4}$ b. $y = \dfrac{(x-2)^2}{x^2}$ c. $y = \dfrac{1}{x^2 + x - 2}$

26. Find the range of $y = \dfrac{x^2}{x^2 + 1}$.

27. What is the range of $y = \dfrac{x^2 - 4}{x + 2}$? (See Example 6.)

28. Find the range of $y = \dfrac{x^2 + 2x}{x^2 - 9}$.

Use the concepts of Section 4.1 (see Examples 3 and 4 of that section) to graph the functions given in Exercises 29 and 30.

29. $y = \left|\dfrac{1}{x - 3}\right|$ 30. $y = \left|\dfrac{x}{x - 3}\right|$

4.4 THE CONIC SECTIONS

In this section we will examine the graphs of four figures: the circle, the ellipse, the parabola, and the hyperbola. Collectively, these four are called the **conic sections** because each can be obtained by examining the plane sections of a conical surface as shown below:

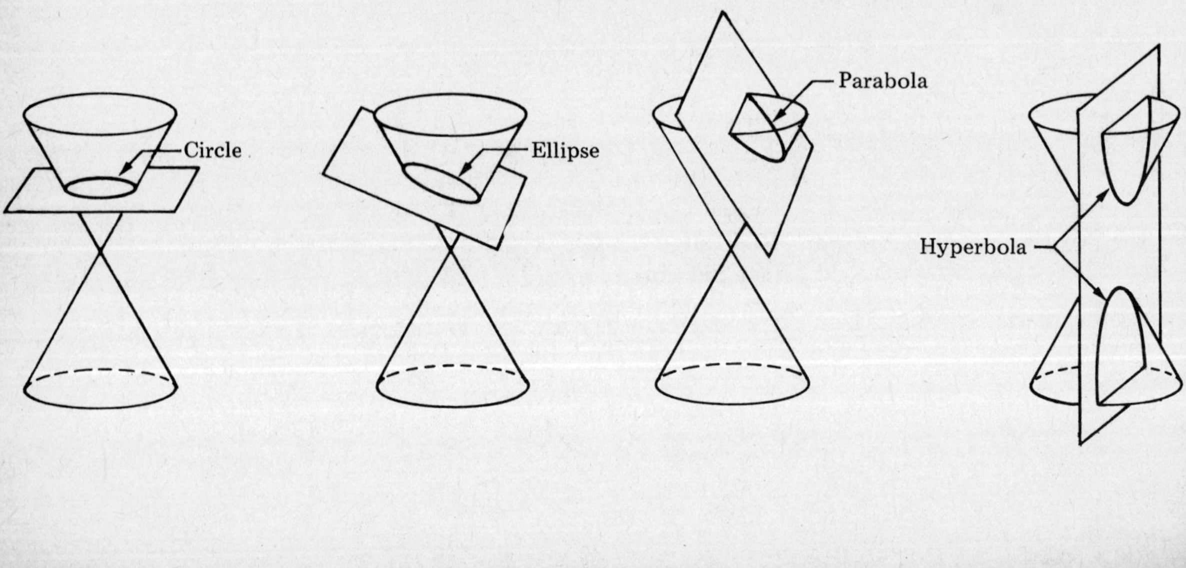

The Circle Formally, a circle is the collection of all points in a plane that are an equal distance (say, r units) from a fixed point (h, k). The point (h, k) is called the **center** of the circle and the distance r is called the **radius**. By applying the distance formula to the points (x, y) and (h, k) in the diagram on the right, we obtain an equation for *all* points on the circle:

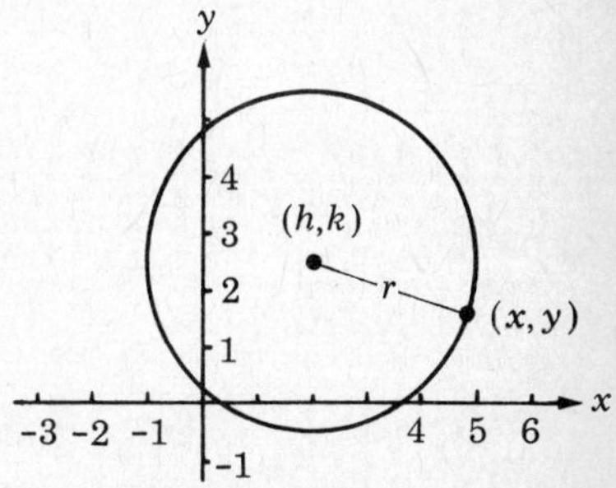

$$r = \sqrt{(x-h)^2 + (y-k)^2}$$

or

$$(x-h)^2 + (y-k)^2 = r^2$$

This leads us to the following theorem:

Equation of a Circle

THEOREM 1 If a circle is centered at (h, k) with radius r $(r > 0)$, then its equation is

$$(x-h)^2 + (y-k)^2 = r^2$$

EXAMPLE 1 Find the equation of the circle with center $(-4, 3)$ and radius 5.

Solution Since the center is $(-4, 3)$, $h = -4$ and $k = 3$. Also, $r = 5$. Thus:

$$(x-h)^2 + (y-k)^2 = r^2$$
$$(x+4)^2 + (y-3)^2 = 25.$$

EXAMPLE 2
a. Show that $x^2 + y^2 - 4x + 8y + 11 = 0$ is an equation of a circle.
b. Graph the relation.
c. What are the domain and range of this relation?

Solution
a. Rearrange the terms and use the process of completing the square (see Section 1.8).

$$x^2 - 4x + y^2 + 8y + 11 = 0$$
$$x^2 - 4x + 4 - 4 + y^2 + 8y + 16 - 16 + 11 = 0$$
$$(x-2)^2 + (y+4)^2 = 9$$

Hence, $x^2 + y^2 - 4x + 8y + 11 = 0$ is a circle of radius 3 centered at $(2,-4)$.

b.

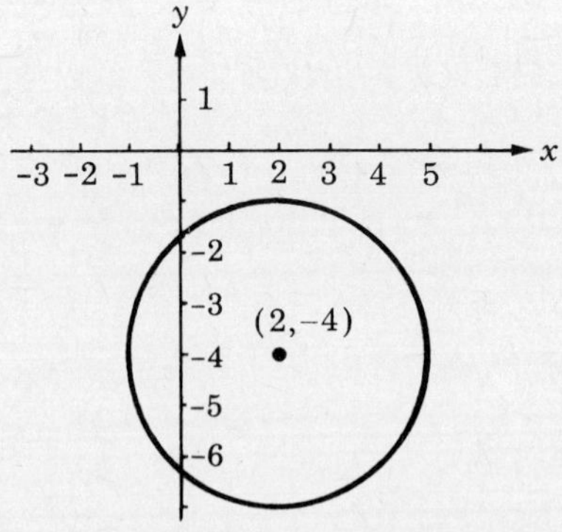

Graph of $x^2 + y^2 - 4x + 8y + 11 = 0$

c. The domain is $\{x \mid -1 \leqslant x \leqslant 5\}$.
The range is $\{y \mid -7 \leqslant y \leqslant -1\}$.

The Ellipse An ellipse is defined to be the collection of all points in a plane with the property that the sum of the distances to two fixed points (called **foci**) is constant. The proof of the following theorem is outlined in Exercise 39.

THEOREM 2

Equation of an Ellipse

If an ellipse is centered at the origin with x-intercepts $(a,0)$ and $(-a,0)$ and y-intercepts $(0,b)$ and $(0,-b)$, then its equation is

$$\frac{x^2}{a^2} + \frac{y^2}{b^2} = 1$$

It can be shown that when the foci are located on the x-axis, equally distant from the origin, the ellipse looks like the diagram on the right, $a > b$, and $c = \sqrt{a^2 - b^2}$.

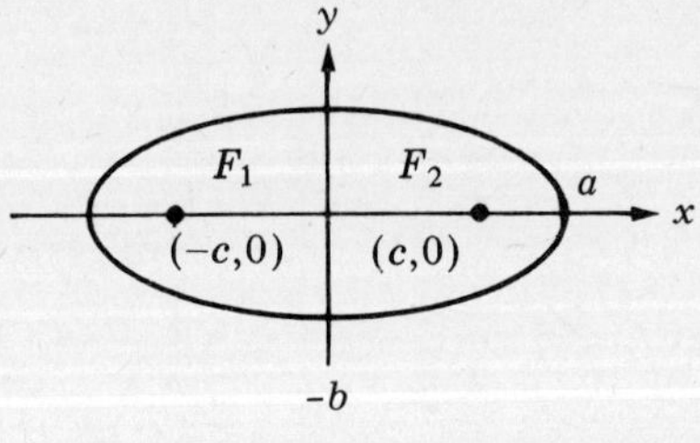

Also, when the foci are located on the y-axis, equally distant from the origin, the ellipse looks like the diagram on the right, $b > a$, and $c = \sqrt{b^2 - a^2}$.

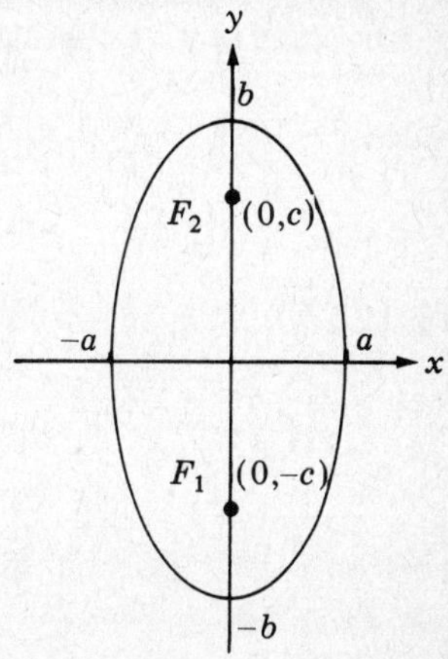

EXAMPLE 3 Graph $\frac{x^2}{25} + \frac{y^2}{16} = 1$.

Solution The x-intercepts are $(5,0)$, $(-5,0)$; the y-intercepts are $(0,4)$, $(0,-4)$. Also, notice that this (and any ellipse described in Theorem 2) is symmetric with respect to the y-axis, the x-axis, and the origin.

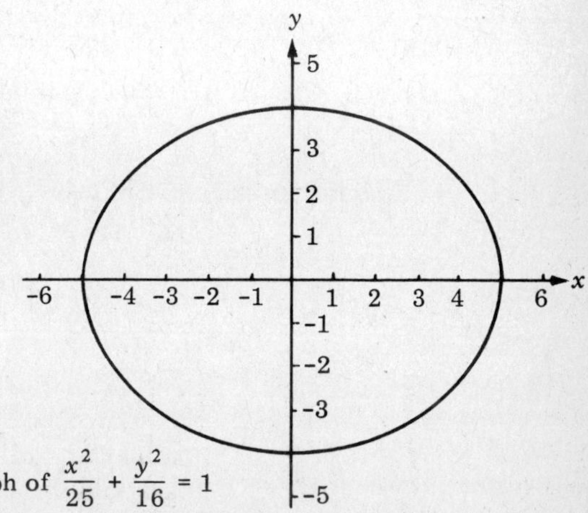

Graph of $\frac{x^2}{25} + \frac{y^2}{16} = 1$

EXAMPLE 4 a. Graph the ellipse whose equation is given by $36x^2 + 9y^2 = 324$.
b. What are the domain and range of the given relation?

Solution a. We first rewrite the equation in the form given in Theorem 2 by dividing by 324:

$$\frac{x^2}{9} + \frac{y^2}{36} = 1$$

Notice that the ellipse's intercepts are $(3,0)$, $(-3,0)$, $(0,6)$ and $(0,-6)$. The ellipse is graphed as follows.

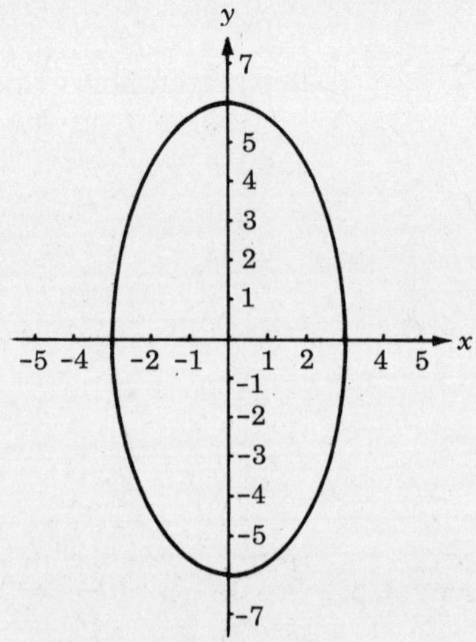

Graph of $36x^2 + 9y^2 = 324$

b. The domain is $\{x \mid -3 \leqslant x \leqslant 3\}$; the range is $\{y \mid -6 \leqslant y \leqslant 6\}$.

The Parabola In Chapter 3 we examined parabolas whose algebraic representation had one of the following two forms.

$$y = ax^2 + bx + c \qquad (a \neq 0)$$
$$x = ay^2 + by + c \qquad (a \neq 0)$$

Presently, we wish to *derive* certain other equations whose graphs are parabolas and mention some of the physical characteristics of the parabola. First, we begin with a definition.

DEFINITION

Parabola

A parabola is the collection of all points in a plane such that the distance from some fixed line L (called the **directrix**) equals the distance from some fixed point, F (called the **focus**). F does not lie on L.

To illustrate the statement above, let $y = -p$ be the given directrix L and $(0,p)$ be the focus:

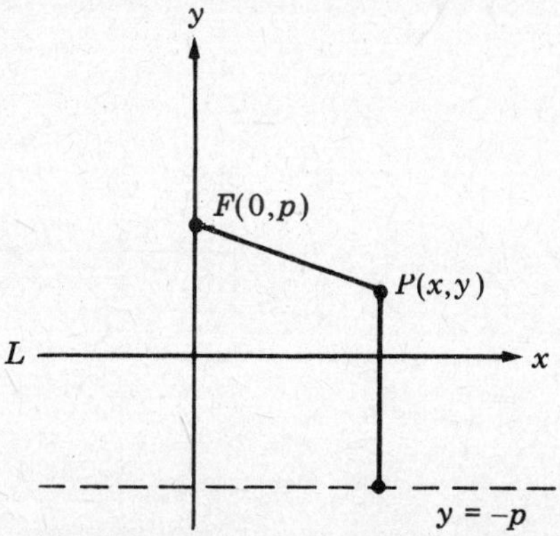

Now, according to the definition of a parabola, we have:

distance from P to L = distance $\overline{PF}$

$$y + p = \sqrt{x^2 + (y - p)^2}$$ Square both sides

$$(y + p)^2 = x^2 + (y - p)^2$$ Expand

$$y^2 + 2py + p^2 = x^2 + y^2 - 2py + p^2$$ Solve for x^2

$$x^2 = 4py$$

We leave it to the reader (in Exercise 40) to verify the other three cases of the "focus-directrix" derivations that follow:

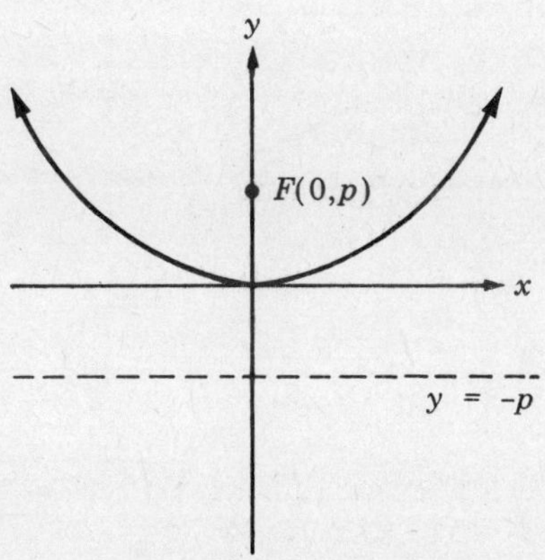

Graph of $x^2 = 4py$

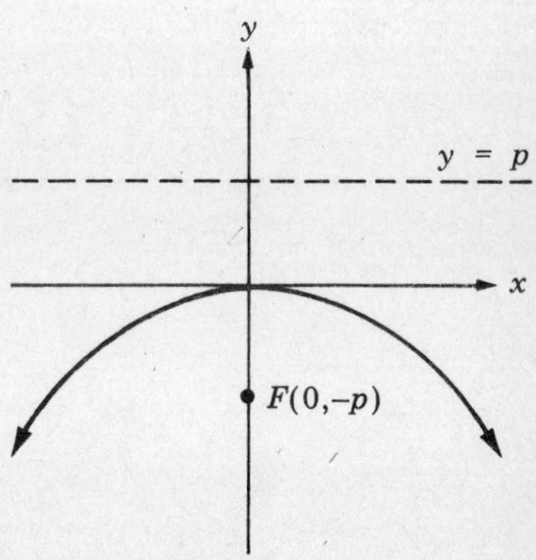

Graph of $x^2 = -4py$

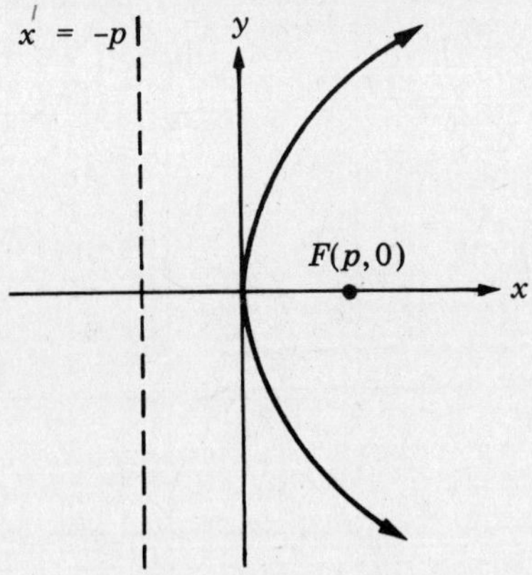

Graph of $y^2 = 4px$

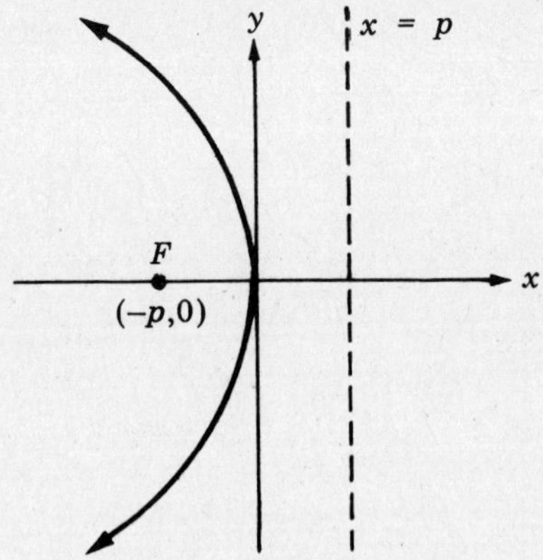

Graph of $y^2 = -4px$

EXAMPLE 5 Graph $x^2 = 8y$.

Solution In $x^2 = 8y$, we have $p = 2$.

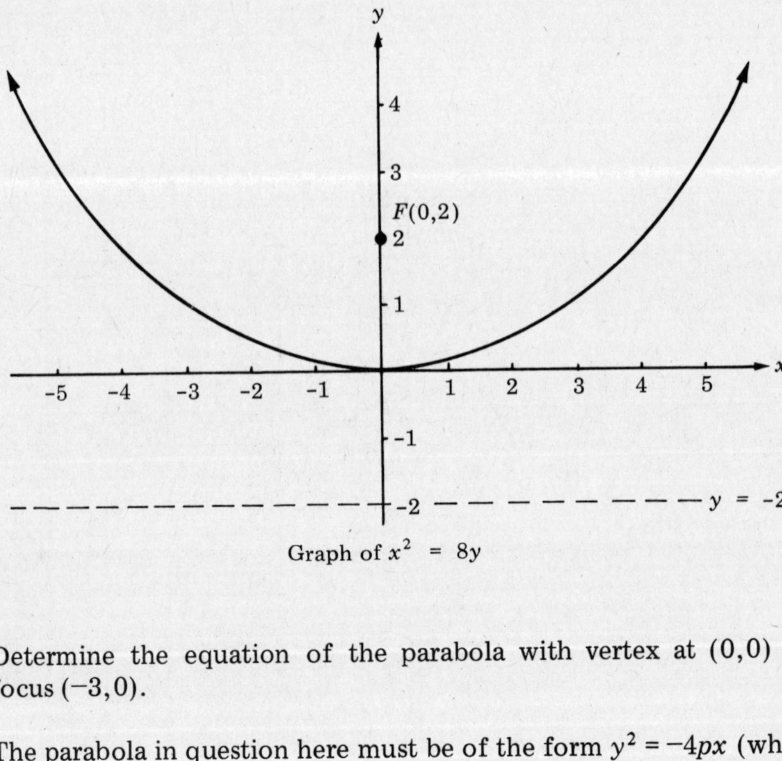

Graph of $x^2 = 8y$

EXAMPLE 6 Determine the equation of the parabola with vertex at $(0,0)$ and focus $(-3,0)$.

Solution The parabola in question here must be of the form $y^2 = -4px$ (why?). Since the focus is 3 units from the vertex, $p = 3$. Thus, the equation is $y^2 = -12x$.

Other variations of equations of parabolas are discussed in Exercises 41 and 42.

The Hyperbola Let two points (called the *foci* of the hyperbola) be fixed. A hyperbola is the collection of all points in a plane, the difference of whose distances from the foci is a constant. The hyperbolas discussed in this book will be of two varieties—one with x-intercepts and one with y-intercepts, as outlined in the following theorem (presented without proof):

THEOREM 3

Hyperbola Equations

Let a and b be positive real numbers.

1. If $\dfrac{x^2}{a^2} - \dfrac{y^2}{b^2} = 1$, then the graph is a hyperbola with x-intercepts $(a,0)$ and $(-a,0)$, and foci $(c,0)$ and $(-c,0)$ where $c = \sqrt{a^2 + b^2}$.

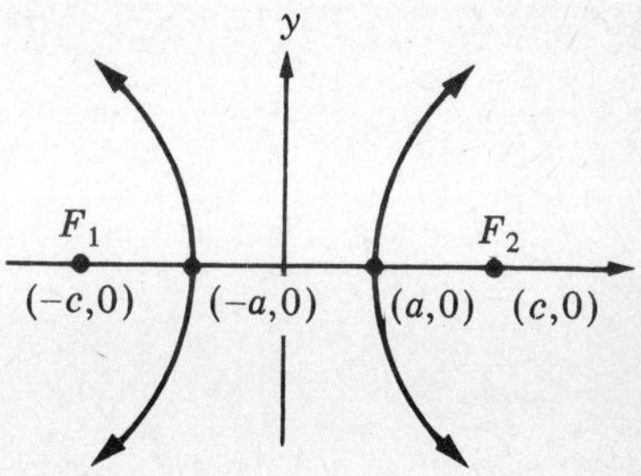

2. If $\dfrac{y^2}{b^2} - \dfrac{x^2}{a^2} = 1$, then the graph is a hyperbola with y-intercepts $(0,b)$ and $(0,-b)$, and foci $(0,c)$ and $(0,-c)$ where $c = \sqrt{a^2 + b^2}$.

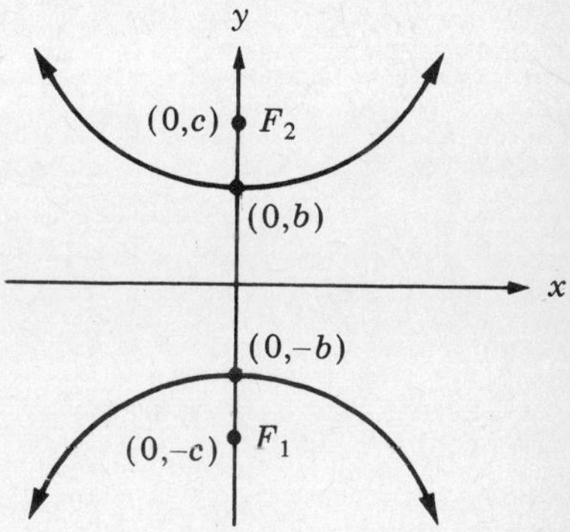

A very useful tool in graphing hyperbolas is sketching its (oblique) **asymptotes**. They are given by:

$$y = \frac{b}{a}x \quad \text{and} \quad y = -\frac{b}{a}x$$

Some examples follow.

EXAMPLE 7 Graph $\frac{x^2}{4} - \frac{y^2}{9} = 1$.

Solution We note that $a = 2$, $b = 3$ and the intercepts are $(2,0)$ and $(-2,0)$. Also, the equations of the asymptotes are $y = \frac{3}{2}x$ and $y = -\frac{3}{2}x$. The hyperbola is graphed below.

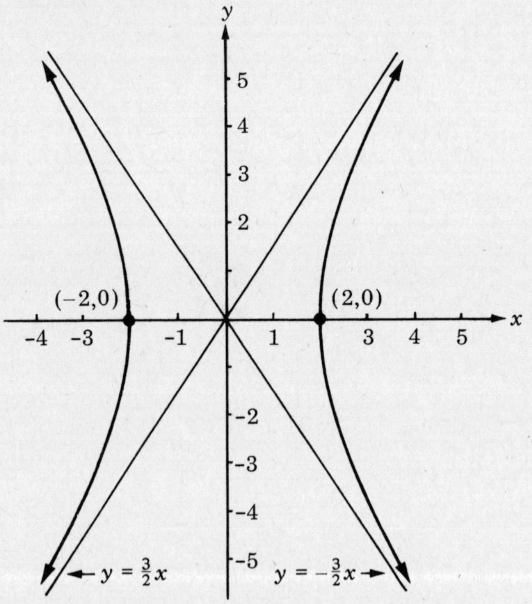

Graph of $\frac{x^2}{4} - \frac{y^2}{9} = 1$

We note here that the domain of the relation graphed above is $\{x \mid x \leqslant -2 \text{ or } x \geqslant 2\}$; the range is $\mathcal{R}$.

EXAMPLE 8 Graph $y^2 - 4x^2 = 4$.

Solution Dividing by 4 yields:

$$\frac{y^2}{4} - x^2 = 1$$

We have an equation of a hyperbola with $a = 1$, $b = 2$ and intercepts $(0,2)$ and $(0,-2)$. The asymptotes are $y = 2x$ and $y = -2x$. The hyperbola is graphed as follows:

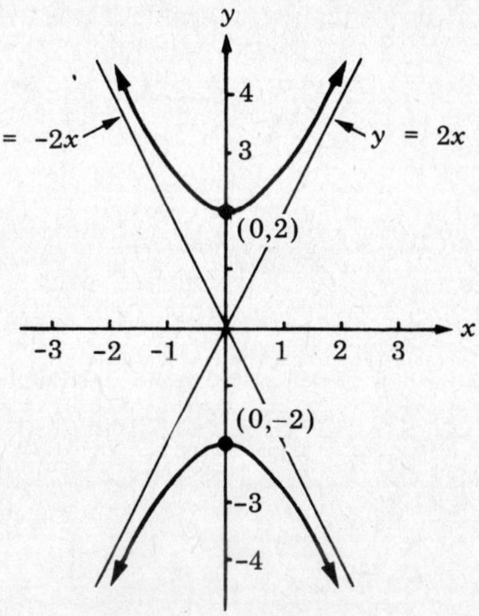

Graph of $y^2 - 4x^2 = 4$

EXERCISES FOR SECTION 4.4

In Exercises 1 through 8, a) determine the center of each circle, b) find its radius, c) graph the equation, and d) state the domain and range. (See Examples 1 and 2.)

1. $(x-3)^2+(y-2)^2=1$
2. $(x+2)^2+(y+1)^2=16$
3. $(x-1)^2+y^2=9$
4. $x^2+y^2=36$
5. $x^2+y^2-2x+4y-4=0$
6. $x^2+y^2+8x-2y+13=0$
7. $x^2+y^2=3$
8. $x^2+y^2+8x-2y+14=0$

In Exercises 9 through 16, a) find the intercepts of each ellipse, b) graph the equation, and c) state the domain and range. (See Examples 3 and 4.)

9. $\dfrac{x^2}{16}+\dfrac{y^2}{25}=1$
10. $\dfrac{x^2}{36}+\dfrac{y^2}{25}=1$
11. $\dfrac{x^2}{4}+\dfrac{y^2}{49}=1$
12. $\dfrac{x^2}{49}+\dfrac{y^2}{4}=1$
13. $\dfrac{x^2}{9}+y^2=1$
14. $9x^2+y^2=9$
15. $2x^2+4y^2=8$
16. $3x^2+2y^2=4$

In Exercises 17 through 24, a) state the coordinates of the focus of the parabola, b) determine the equation of the directrix, and c) graph the parabola. (See Examples 5 and 6.)

17. $x^2=4y$
18. $x^2=-4y$
19. $y^2=4x$
20. $y^2=-4x$
21. $x^2+y=0$
22. $y^2-x=0$
23. $x^2-2y=0$
24. $2x^2=-y$

In Exercises 25 through 32, a) find the intercepts of each hyperbola, b) find the equations of the asymptotes, c) graph the hyperbola, and d) state the domain and range. (See Examples 7 and 8.)

25. $\frac{x^2}{9} - \frac{y^2}{4} = 1$ 26. $\frac{y^2}{9} - \frac{x^2}{4} = 1$ 27. $x^2 - y^2 = 1$

28. $y^2 - x^2 = 4$ 29. $\frac{y^2}{9} - x^2 = 1$ 30. $25x^2 - y^2 = 25$

31. $9x^2 - 16y^2 = 144$ 32. $9y^2 - 16x^2 = 1$

In Exercises 33 through 38, rewrite each given equation and determine which conic section is represented by each equation.

33. $x^2 - 4x + y^2 - 6y + 8 = 0$ 34. $3x^2 + 2y^2 = 6$
35. $3x^2 - 2y = 6$ 36. $3x^2 - 2y^2 = 6$
37. $3x^2 - 3y^2 = 6$ 38. $x^2 = y^2 + 1$

39. Prove Theorem 2 for an ellipse with foci on the x-axis by providing reasons for the following argument:
    1. Let the coordinates of the foci be $F_1(c,0)$ and $F_2(-c,0)$ and the given constant be $2a$.
    2. With $P(x, y)$ being any point on the ellipse, $\overline{PF_1} + \overline{PF_2} = 2a$.
    3. $\sqrt{(x-c)^2 + y^2} + \sqrt{(x+c)^2 + y^2} = 2a$
    4. $x^2(a^2 - c^2) + a^2y^2 = a^2(a^2 - c^2)$
    5. Let $b^2 = a^2 - c^2$.
    6. $\frac{x^2}{a^2} + \frac{y^2}{b^2} = 1$

40. Use an argument similar to the one used above to derive $x^2 = 4py$ to verify each variation of the equation of a parabola.
    a. $x^2 = -4py$ b. $y^2 = 4px$ c. $y^2 = -4px$
41. The four equations of parabolas given in this section necessarily have the parabola's vertex at the origin. However, the derivation of $x^2 = 4py$ can be generalized to

$$(x-h)^2 = 4p(y-k)$$

where (h,k) are the coordinates of the vertex. Derive the new equation.
42. Using the ideas of Exercise 41, derive the three alternate forms of a parabola with vertex at (h,k).
    a. $(x-h)^2 = -4p(y-k)$ b. $(y-k)^2 = 4p(x-h)$
    c. $(y-k)^2 = -4p(x-h)$

Calculator Exercise

43. a. Consider the hyperbola $\frac{x^2}{a^2} - \frac{y^2}{b^2} = 1$. Show that equation is equivalent to $y = \pm\frac{b}{a}\sqrt{x^2 - a^2}$.

    b. Let $a = 2$ and $b = 3$. Complete the following table:

x	$\frac{3}{2}\sqrt{x^2-4}$	$\frac{3}{2}x$
10		
20		
30		
100		
1000		
1,000,000		

The table above helps to explain the fact that as $x \longrightarrow \infty$, $y \longrightarrow \frac{3}{2}x$. Note, this is precisely the behavior of an asymptote!

44. The graph of $x^2 + y^2 = 25$ is a circle centered at the origin with radius equal to 5.
   a. What is the graph of $y = \pm\sqrt{25 - x^2}$?
   b. What is the graph of $y = \sqrt{25 - x^2}$?
   c. What is the graph of $y = -\sqrt{25 - x^2}$?

45. The four figures on the first page of this section are generally referred to as the conic sections because they are the resulting configuration of a plane intersecting a double cone. There are, however, three other possibilities, called the **degenerate conics**, pictured below.

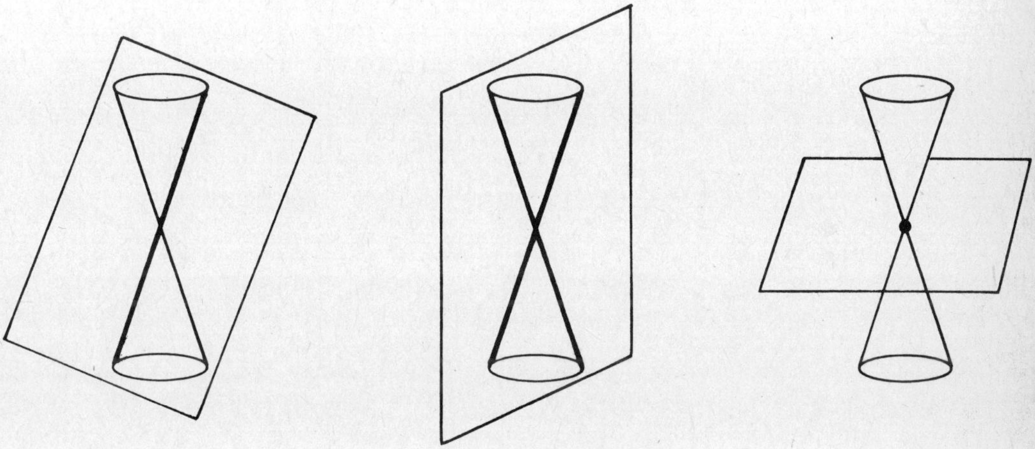

a line — intersecting lines — a point

   a. Show that the equation $y^2 = 4px$ degenerates to the equation of a line if $p = 0$.
   b. Show that the equation $\frac{x^2}{a^2} - \frac{y^2}{b^2} = d^2$ degenerates to the equations of two lines if $d = 0$.
   c. Show that the equation $x^2 + y^2 = r^2$ degenerates to a point if $r = 0$.

46. The **unit circle** is a circle of radius 1. What is the equation of the unit circle centered at the origin?

CHAPTER 4 IN REVIEW

I. Summary

The techniques of graphing three major classes of relations were examined in this chapter: piecewise functions, rational functions, and the conic sections.

In Section 4.1 we saw that piecewise functions are functions that are defined by different relations for various parts of the domain. Two important functions graphed in that section were the absolute value function, $y = |x|$, and the greatest integer function, $y = [\![x]\!]$. Also, we looked at both continuous and discontinuous piecewise functions.

In Section 4.2, the analytical tools necessary for graphing the rational functions of Section 4.3 were investigated. Basically, we used intercepts, symmetry, asymptotes, domain, range, and sign intervals to assist in the graphing.

The conic sections, circles, ellipses, parabolas, and hyperbolas, were graphed in Section 4.4.

II. Vocabulary/Symbol True-False Quiz

If the statement is true, write "TRUE." However, if the statement is false, replace the underlined word or phrase with a different word or phrase that will make the new statement true.

1. $y = |x|$ is an example of a continuous function.
2. $y = [\![x]\!]$ is an example of a discontinuous function.
3. $y = x^2 - 9$ is symmetrical with respect to the y-axis.
4. $y^2 - x^2 = 1$ is symmetrical with respect to both the y-axis and the origin.
5. The graph of $y = \dfrac{x-1}{x^2-4}$ has two vertical asymptotes.
6. The graph of $y = \dfrac{x^2}{x^2-2x}$ has a vertical asymptote $x = 0$.
7. The graph of $y = \dfrac{-1}{x^2-4}$ is below the x-axis for all values of x.
8. The equation of the circle centered at (1,2) of radius 3 is $(x+1)^2 + (y+2)^2 = 3$.
9. The graph of $x^2 - y^2 = 1$ is a hyperbola.
10. The graph of $\dfrac{x^2}{4} - y^2 = 1$ has asymptotes $y = \pm\frac{1}{4}x$.

III. Chapter Test

1. a. Graph: $f(x) = \begin{cases} 1 - x^2 & \text{if } x < 2 \\ x - 5 & \text{if } x \geqslant 2 \end{cases}$

   b. Is $f(x)$ a continuous function?
2. Graph $y = |2x - 1|$.
3. What are the domain and range of the function in Exercise 22?

For questions 4 through 9, a) find all intercepts, b) discuss symmetry, c) find all vertical and horizontal asymptotes, d) find the domain and range, e) find the sign intervals for $y = f(x)$, and f) graph the function.

4. $y = \dfrac{x}{x^2 - 16}$
5. $y = \dfrac{1}{x^2 - 16}$
6. $y = \dfrac{x^2}{x^2 - 16}$
7. $y = \dfrac{x}{x^2 + 16}$
8. $y = \dfrac{x - 4}{x^2 + x - 12}$
9. $y = \dfrac{x + 4}{x^2 + x - 12}$

In questions 10 through 15, graph the given conic section.

10. $4x^2 + 9y^2 = 36$
11. $9x^2 + 9y^2 = 36$
12. $\dfrac{x^2}{9} - \dfrac{y^2}{9} = 1$
13. $x^2 + 4y = 0$
14. $y^2 - 2x = 0$
15. $y^2 - 2x^2 = 1$

IV. Student Research Project

1. The conic sections are used in many fields. To more fully appreciate this, prepare a report on some applications of the conics. The following two sources are excellent:

   Jacobs, Harold R. *Mathematics: A Human Endeavor*, W. H. Freeman and Co., San Francisco, 1970, pp. 246–273, 300.

   Mathematics: An Introduction to Its Spirit and Use. (Readings from *Scientific American*), W. H. Freeman and Co., San Francisco, 1979, pp. 18, 25, 40–44, 90, 100–103, 136–137.

2. Among the mathematicians associated with analytic geometry are Hypatia (c. 400, usually considered to be the "first" woman mathematician) and Maria Gaetana Agnesi (1718–1799). Prepare a brief report on women in mathematics. Include in your sources the following:

   Mozans, H. J. *Woman In Science*, MIT Press, Cambridge, Mass., 1940. Consult Chapter 3, "Women in Mathematics."

   Rappaport, Karen D. "Women Mathematicians," *The MATYC Journal*, vol. 13, 1979.

   Osen, L. M. *Women in Mathematics*, MIT Press, Cambridge, Mass., 1974.

CHAPTER 5

Exponential and Logarithmic Functions

One, Two, Three, . . . , Infinity: Counting the Chemical Compounds

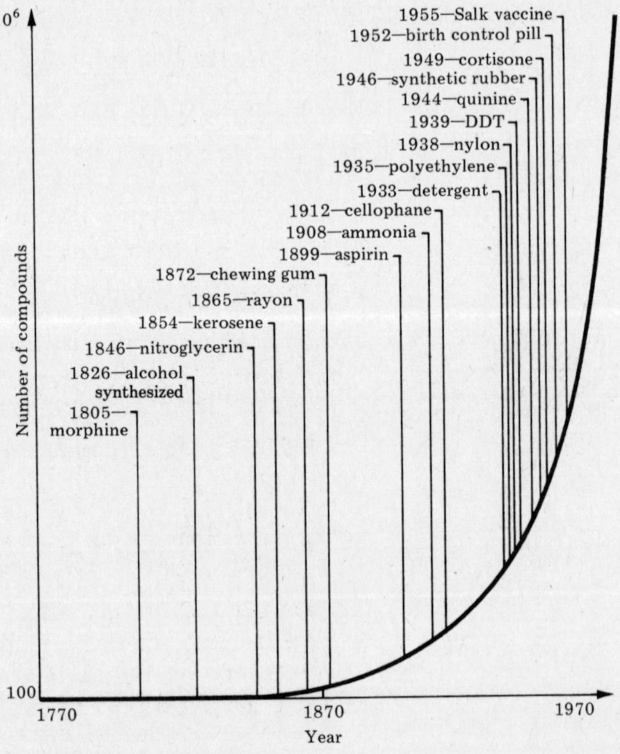

Fewer than 100 compounds had been identified at "the beginning of modern chemistry" in 1770. Several million are now known. This is an annual growth rate of almost 5%, or a doubling time of about 15 years.

From J. Arthur Campbell, Chemistry: The Unending Frontier *(Santa Monica, Calif.: Goodyear Publishing Co., Inc.) 1978, p. 128. Reprinted by permission.*

5.1 EXPONENTIAL FUNCTIONS

In Chapter 1, we defined the expression b^r, where r is any rational number. However, until now we have avoided evaluating expressions such as $2^{23/10}$, $(1.6)^{-6/5}$, or $3^{2.16}$. The laws of exponents are helpful for such evaluations:

$$2^{23/10} = \sqrt[10]{2^{23}} = (\sqrt[10]{2})^{23}$$

$$(1.6)^{-6/5} = \frac{1}{(1.6)^{6/5}} = \frac{1}{(\sqrt[5]{1.6})^6}$$

$$3^{2.16} = 3^{54/25} = (\sqrt[25]{3})^{54}.$$

Thus, to evaluate $3^{2.16}$ according to the above, the expression $\sqrt[25]{3}$ must be obtained. Fortunately, there are other ways to evaluate rational exponents: one method will be discussed later in this chapter and another method is to approximate a given expression with the use of an electronic calculator having a y^x key. (Such a device shows that $3^{2.16}$ is approximately 10.7295599.)

A concept that we will now consider is that of **irrational exponents**. For instance, consider $5^{\sqrt{2}}$. Although $\sqrt{2}$ is irrational, we can use its nonterminating decimal representation, 1.4142 . . . . Thus, as the numbers 1, 1.4, 1.41, 1.414 *approach* $\sqrt{2}$, each successive power 5^1, $5^{1.4}$, $5^{1.41}$, $5^{1.414}$ *gets closer to* $5^{\sqrt{2}}$. Hence $5^{\sqrt{2}}$ can be evaluated to any desired degree of accuracy.

The discussion thus far hints at the following theorem (which we state without proof):

THEOREM 1

> If $b > 0$ and r is any *real* number, then there exists a unique real positive number b^r.

The reader should note the restriction on the base b is required because expressions such as 0^2, $0^{3/2}$, $0^{1.2}$ are not positive, expressions such as 0^{-2}, $0^{-1/2}$ are not defined, and expressions such as $(-2)^{1/2}$, $(-\frac{1}{4})^{-1/6}$ are not real.

We are now ready to define an **exponential function.**

DEFINITION

> Let b be a positive real number, $b \neq 1$. Then $f(x) = b^x$ is an **exponential function** with base b, where x is any real number.

From the definition, it is seen that the domain of $f(x) = b^x$ is $\mathcal{R}$. Furthermore, Theorem 1 guarantees that $f(x) = b^x$ *is* a function.

In order to graph an exponential function, say $y = 2^x$, we begin by plotting several points. We choose as x values $-3, -2, -1, 0, \frac{1}{2}, 1, 2, 2.4, \frac{5}{2}$, and 3; the corresponding y values are calculated below.

x	y*
-3	$\frac{1}{8} = .125$
-2	$\frac{1}{4} = .25$
-1	$\frac{1}{2} = .5$
0	1
$\frac{1}{2}$	$\sqrt{2} \approx 1.414$
1	2
2	4
2.4	$2^{2.4} \approx 5.278$
$\frac{5}{2}$	$\sqrt{32} \approx 5.657$
3	8

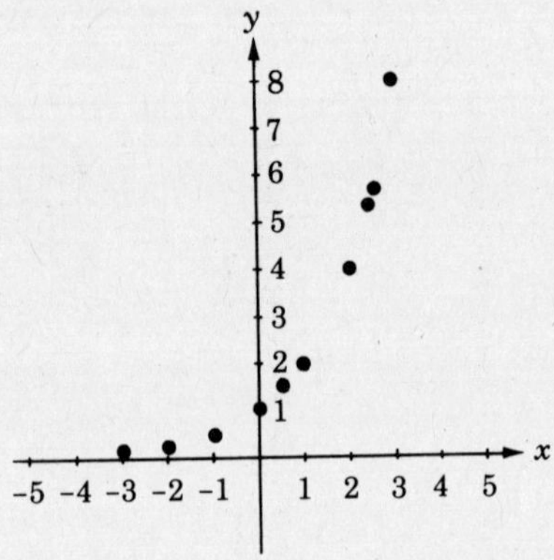

It is apparent (from the points graphed above) that the graph of $y = 2^x$ is the smooth *continuous* curve shown below.

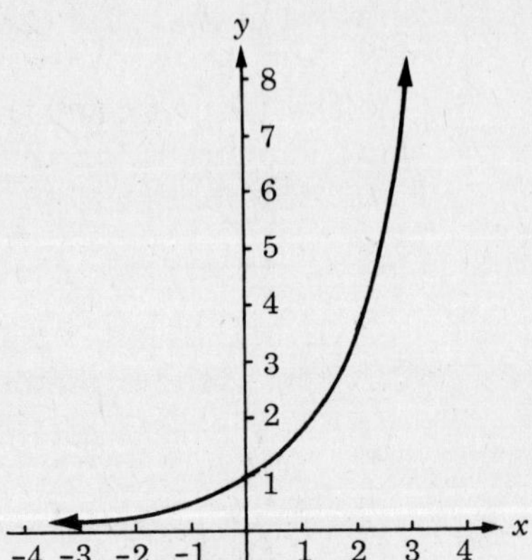

Graph of $y = 2^x$

---

*Some values of 2^x were determined by using an electronic calculator with a y^x key.

The domain $y = 2^x$ is $\mathcal{R}$ and the range is $\{y \mid y > 0\}$. We now examine the behavior of $y = 2^x$ for extreme values of x.

$$\underbrace{\text{As } x \longrightarrow -\infty,\ y \longrightarrow 0}_{\text{Thus the } x\text{-axis is a horizontal asymptote on the left.}} \quad \text{and} \quad \underbrace{\text{as } x \longrightarrow +\infty,\ y \longrightarrow +\infty}_{\text{Thus the graph rises indefinitely to the right.}}.$$

EXAMPLE 1 a. Graph the exponential function $f(x) = 4^x$.
b. State the domain and range.
c. Discuss the behavior of $f(x)$ for extreme values of x.

Solution a. We choose the x values $-2, -1, 0, \frac{1}{2}, 1$ and calculate the associated y values using $f(x) = 4^x$.

x	$f(x)$
-2	$\frac{1}{16}$
-1	$\frac{1}{4}$
0	1
$\frac{1}{2}$	2
1	4

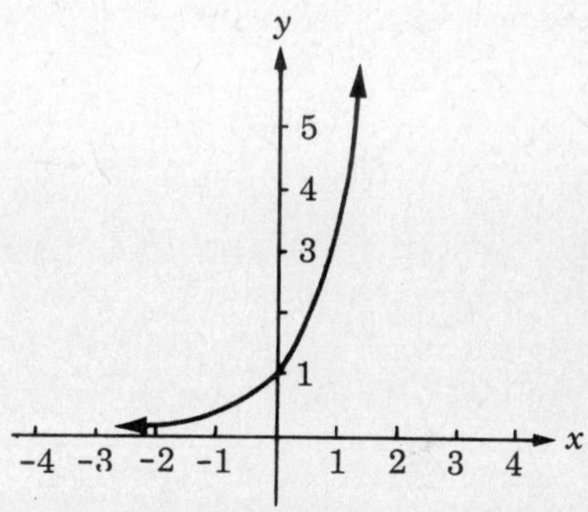

Graph of $f(x) = 4^x$

b. The domain of $f(x) = 4^x$ is $\mathcal{R}$ and the range is $\{y \mid y > 0\}$.

c. $$\underbrace{\text{As } x \longrightarrow -\infty,\ y \longrightarrow 0}_{\text{The } x\text{-axis is a horizontal asymptote on the left.}} \quad \text{and} \quad \underbrace{\text{as } x \longrightarrow +\infty,\ y \longrightarrow +\infty}_{\text{The graph rises indefinitely to the right.}}.$$

In the definition of an exponential function $f(x) = b^x$, b can be any positive real number constant, $b \neq 1$. In the next example we graph an exponential function whose base is between 0 and 1.

EXAMPLE 2 a. Graph the exponential function $y = (\frac{1}{4})^x$.
b. State the domain and range.
c. Discuss the behavior for extreme values of x.

Solution a. First we rewrite the given equation as $y = 4^{-x}$, and choose the x values $-1, -\frac{1}{2}$, 0, 1, 2.

x	y
-1	4
$-\frac{1}{2}$	2
0	1
1	$\frac{1}{4}$
2	$\frac{1}{16}$

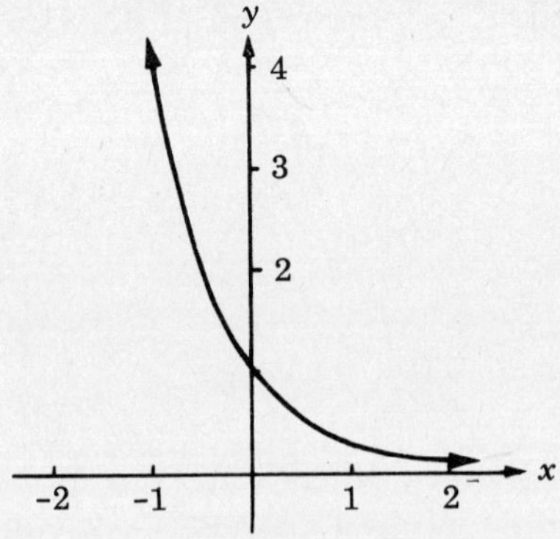

Graph of $y = (\frac{1}{4})^x$

b. The domain is $\mathcal{R}$ and the range is $\{y \mid y > 0\}$.

c. As $x \longrightarrow -\infty$, $y \longrightarrow +\infty$ and as $x \longrightarrow +\infty$, $y \longrightarrow 0$.

The graph rises indefinitely to the left. / The x-axis is a horizontal asymptote on the right.

The preceding examples serve to illustrate that two basic "shapes" of exponential functions are:

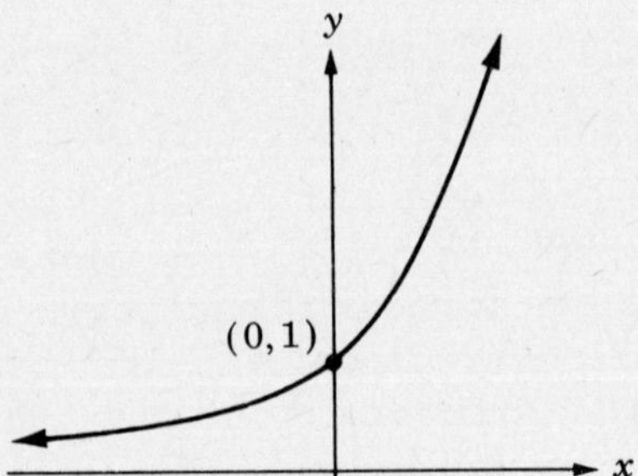

Graph of $y = b^x$, $b > 1$. Graphs of this type are referred to as exhibiting *exponential growth.*

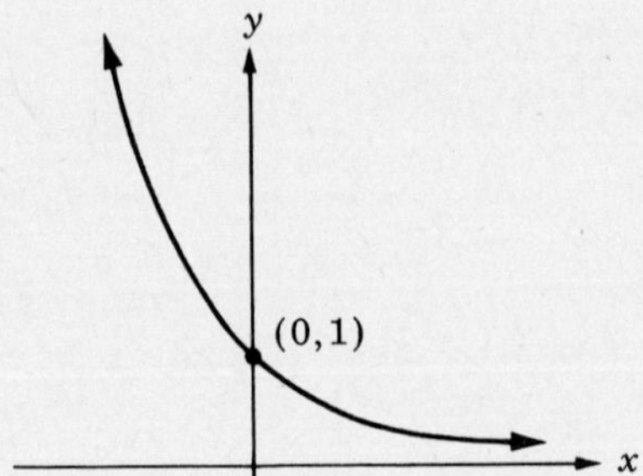

Graph of $y = b^x$, $0 < b < 1$. Graphs of this type are referred to as exhibiting *exponential decay.*

We now consider an equation of the form $y = a \cdot b^{cx}$, where b is a positive real number contant, $b \neq 1$, and a and c are nonzero constants.

EXAMPLE 3 a. Graph the exponential function $g(x) = 2 \cdot (3)^{x/2}$.
b. State the domain and range.
c. Discuss the behavior for extreme values of x.

Solution a. We choose the x values -4, $-2, 0, 1, 2$.

x	$g(x)$
-4	$\frac{2}{9}$
-2	$\frac{2}{3}$
0	2
1	3.5
2	6

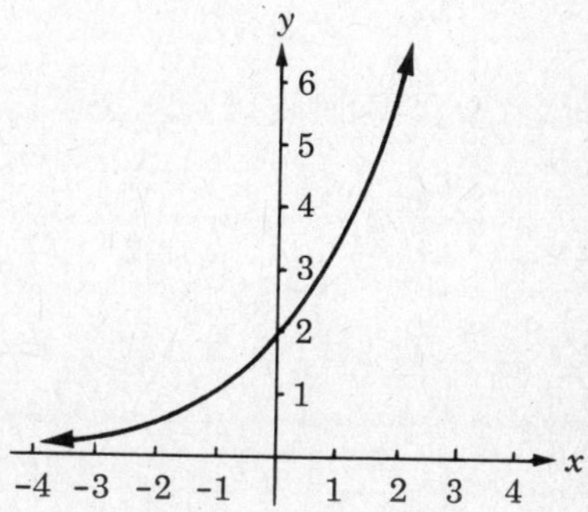

Graph of $y = 2 \cdot 3^{x/2}$

b. The domain is $\mathcal{R}$ and the range is $\{y \mid y > 0\}$.

c. As $x \longrightarrow -\infty$, $y \longrightarrow 0$ and as $x \longrightarrow +\infty$, $y \longrightarrow +\infty$.

The x-axis is a horizontal asymptote on the left. — The graph rises indefinitely to the right.

In the next example we examine an equation of the form $y = a \cdot b^{f(x)}$ where b is a real number constant, $b \neq 1$, a is a nonzero constant, and $f(x)$ is a function of x.

EXAMPLE 4 a. Graph the function $y = 3 \cdot (2)^{-x^2}$.
b. State the domain and range.
c. Discuss the behavior for extreme values of x.

Solution a. We choose the x values $-2, -1, 0, 1, 2$ and note that the graph is symmetrical with respect to the y-axis.

x	y
-2	$\frac{3}{16}$
-1	$\frac{3}{2}$
0	3
1	$\frac{3}{2}$
2	$\frac{3}{16}$

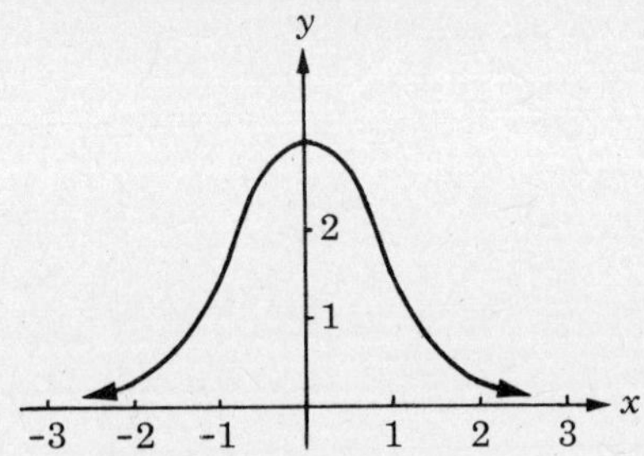

Graph of $y = 3 \cdot (2)^{-x^2}$

b. The domain is $\mathcal{R}$ and the range is $\{y \mid 0 < y \leqslant 3\}$.

c. As $x \longrightarrow -\infty$, $y \longrightarrow 0$ and as $x \longrightarrow +\infty$, $y \longrightarrow 0$.

The x-axis is a horizontal asymptote on the left. The x-axis is a horizontal asymptote on the right.

In the many applications of mathematics in business and natural and social sciences, as well as in the study of calculus, we often use a particular irrational number, denoted by e, as the base for an exponential function. To sixteen decimal places,

VALUE OF e

$$e \approx 2.7182818284590452$$

The **natural exponential function** is defined by $y = e^x$. In Exercises 29 and 46 the reader is asked to graph $y = e^x$.

For the remainder of this section we turn our attention to solving **exponential equations** in a single variable. An exponential equation is any equation that has a variable as an exponent. We will need the following theorem.

THEOREM 2 Let b be a positive real number, $b \neq 1$. If $b^r = b^s$, then $r = s$.

Proof:

Statement	*Reason*
1. $b^r = b^s$	1. Given
2. $\frac{b^r}{b^s} = 1$	2. Divide both sides by b^s.
3. $b^{r-s} = 1$	3. Law of Exponents
4. $r - s = 0$	4. Law of Exponents
5. Thus, $r = s$	5. Add s to both sides.

In Exercise 44 the reader is asked to prove the converse of Theorem 2.

In the three examples that follow, we solve exponential equations in a single variable.

EXAMPLE 5 Solve for x: $2^{1-x} = 4^{2x}$.

Solution

$$2^{1-x} = 4^{2x}$$
$$2^{1-x} = (2^2)^{2x}$$
$$\left.\begin{array}{l} 2^{1-x} = 2^{4x} \\ 1 - x = 4x \end{array}\right\} \text{Theorem 2}$$

Thus, $x = \frac{1}{5}$.

Check:

$$2^{1-x} = 4^{2x}$$
$$2^{1-(1/5)} = 4^{2(1/5)}$$
$$2^{4/5} = 4^{2/5}$$
$$\sqrt[5]{16} = \sqrt[5]{16} \checkmark$$

EXAMPLE 6 Solve for x: $27^{2x-1} = 9$.

Solution

$$27^{2x-1} = 9$$
$$(3^3)^{2x-1} = 3^2$$
$$\left.\begin{array}{l} 3^{6x-3} = 3^2 \\ 6x - 3 = 2 \end{array}\right\} \text{Theorem 2}$$
$$x = \frac{5}{6}$$

The check is left to the reader.

EXAMPLE 7 Solve for x: $2^x - 8 \cdot 2^{-x} = 7$.

Solution

$$2^x - 8 \cdot 2^{-x} = 7$$
$$(2^x)^2 - 8 = 7 \cdot 2^x \quad \text{Multiply both sides by } 2^x$$
$$(2^x)^2 - 7(2^x) - 8 = 0 \quad \leftarrow \text{This equation is in quadratic form.}$$
$$\text{Let } 2^x = p$$
$$p^2 - 7p - 8 = 0$$
$$(p - 8)(p + 1) = 0$$
$$p = 8 \mid p = -1$$

$p = -1$: This solution must be rejected because $p = 2^x > 0$.

Thus,

$$2^x = 8$$
$$2^x = 2^3$$
$$x = 3$$

Check:

$$2^x - 8 \cdot 2^{-x} = 7$$
$$2^3 - 8 \cdot 2^{-3} = 7$$
$$8 - 8 \cdot \tfrac{1}{8} = 7$$
$$7 = 7 \checkmark$$

EXERCISES FOR SECTION 5.1

In Exercises 1 through 10, use $f(x) = (\frac{1}{2})^x$ and $g(x) = 2^{3x}$ to determine each of the following.

1. $f(0)$
2. $f(3)$
3. $g(3)$
4. $f(-2)$
5. $f(-1)$
6. $g(0)$
7. $g(-1)$
8. $g(\frac{1}{3})$
9. $f(-.5)$
10. $g(\frac{1}{2})$

In Exercises 11 through 30, a) graph each exponential function, b) state the domain and range, c) discuss the behavior for extreme values of x. (See Examples 1 through 4.)

11. $f(x) = 3^x$
12. $f(x) = (\frac{1}{3})^x$
13. $f(x) = (\frac{1}{2})^x$
14. $g(x) = 5^x$
15. $y = 5^{-x}$
16. $y = 9^x$
17. $y = (\frac{1}{9})^x$
18. $f(x) = 3^{|x|}$
19. $g(x) = 3^x + 3^{-x}$
20. $y = 2^{x-1}$
21. $y = \frac{1}{3} \cdot 9^x$
22. $y = 3^{2x-1}$
23. $f(x) = \frac{1}{2} \cdot (\frac{1}{4})^{-x}$
24. $g(x) = 6 \cdot (\frac{1}{3})^{x/2}$
25. $y = 10^x$
26. $y = 10^{-x}$
27. $f(x) = -(2^x)$
28. $g(x) = 2^{x^2+1}$
29. $y = e^x$ (Use $e \approx 2.7$ unless a calculator with an e^x key is available.)
30. $y = e^{-x}$ (Use $e \approx 2.7$ unless a calculator with an e^x key is available.)

In Exercises 31 through 40, solve each exponential equation for x. (See Examples 5 through 7.)

31. $2^x = 32$
32. $10^x = 100$
33. $10^x = \frac{1}{100}$
34. $2^{3x} = 4^{x-1}$
35. $2^{2x} = 8^x$
36. $81^x = 9^{x+1}$
37. $25^x = 1$
38. $3^{x^2+2x} = \frac{1}{3}$
39. $2^x - 12 \cdot 2^{-x} = 1$
40. $3^x - 18 \cdot 3^{-x} = 7$
41. Prove: If $f(x) = b^x$, where $b > 0$, $b \neq 1$, then $f(x + y) = f(x) \cdot f(y)$.
42. Prove: If $f(x) = b^x$, where $b > 0$, $b \neq 1$, then $f(x - y) = f(x) \div f(y)$.
43. The definition of an exponential function does not allow the base, b, to be equal to 1. In order to see why, graph the function $f(x) = b^x$, where $b = 1$.
44. Prove the converse of Theorem 2, namely:

    Let b be a positive real number, $b \neq 1$. If $r = s$, then $b^r = b^s$.

Calculator Exercises

45. Graph $y = 5^x$ by letting $x = -2, -1.5, -1, 0, .2, .4, .6, .8, 1, 1.4, 1.6, 2$.
46. Graph $y = e^x$ by letting $x = -2, -1.5, -1, -.5, 0, .5, 1, 1.5, 2, 2.5$.
47. Graph $y = e^{x-1}$ by letting $x = -2, -1, -.5, 0, .5, 1, 1.5, 2, 2.5, 3$.
48. Graph $f(x) = \dfrac{e^x + e^{-x}}{2}$
49. Graph $f(x) = \dfrac{e^x - e^{-x}}{2}$
50. Graph $f(x) = e^{-x^2}$
51. Graph $y = x + e^x$
52. Graph $f(x) = xe^x$
53. Graph $g(x) = x^2e^x$
54. Graph $y = e^{-|x|}$
55. For years the following question perplexed mathematicians:

    "Which is larger e^π or π^e?"

    Using a calculator, determine which is larger.

5.2 LOGARITHMIC FUNCTIONS

Section 5.1 pertained to exponents. In the present section we will further explore exponents by defining and investigating an entity called a **logarithm**.

DEFINITION

> Let b be a positive real number, $b \neq 1$. A number u is said to be the **logarithm** of a positive real number v, if and only if $v = b^u$. In symbols,
>
> $$u = \log_b v \text{ if and only if } b^u = v.$$

There are two important observations that follow immediately from the definition. First, the equations $u = \log_b v$ and $b^u = v$ are *equivalent*.

Secondly, in the equation $b^u = v$, u is an exponent and in the equation $u = \log_b v$, u is a logarithm. Hence, a *logarithm is an exponent.*

EXAMPLE 1 Rewrite each equation in equivalent logarithmic form.

a. $3^2 = 9$ b. $(\frac{1}{2})^3 = \frac{1}{8}$ c. $4^{-2} = \frac{1}{16}$

d. $b^1 = b$ e. $b^0 = 1$

Solution

a. $3^2 = 9$ is equivalent to $\log_3 9 = 2$.

b. $(\frac{1}{2})^3 = \frac{1}{8}$ is equivalent to $\log_{1/2} \frac{1}{8} = 3$.

c. $4^{-2} = \frac{1}{16}$ is equivalent to $\log_4 \frac{1}{16} = -2$.

d. $b^1 = b$ is equivalent to $\log_b b = 1$.

e. $b^0 = 1$ is equivalent to $\log_b 1 = 0$.

EXAMPLE 2 Rewrite each equation in equivalent exponential form.

a. $\log_2 16 = 4$ b. $\log_3 \frac{1}{9} = -2$

c. $\log_2 7 = x$ d. $\log_w R = T$

Solution

a. $\log_2 16 = 4$ is equivalent to $2^4 = 16$.

b. $\log_3 \frac{1}{9} = -2$ is equivalent to $3^{-2} = \frac{1}{9}$.

c. $\log_2 7 = x$ is equivalent to $2^x = 7$.

d. $\log_w R = T$ is equivalent to $w^T = R$.

Parts d and e of Example 1 serve as a proof of the following theorem.

THEOREM 3 If b is a positive real number, $b \neq 1$, then $\log_b b = 1$ and $\log_b 1 = 0$.

Part c of Example 2 serves to illustrate that the equation $2^x = 7$ has the "solution" $x = \log_2 7$. Later in this chapter the reader will learn a method of *evaluating* $\log_2 7$.

We now further explore the relationship between exponents and logarithms by considering the *inverse of the exponential function* $y = b^x$. Interchanging x and y gives $x = b^y$ and then rewriting the latter equation in equivalent exponential form gives $y = \log_b x$.

DEFINITION Let b be a positive real number, $b \neq 1$. The **logarithmic function** with base b is defined to be $g(x) = \log_b x$.

In order to view the inverse relationship between $y = b^x$ and $y = \log_b x$ we graph $y = 3^x$ and $y = \log_3 x$ on the axes below.

$y = 3^x$

x	y
-2	$\frac{1}{9}$
-1	$\frac{1}{3}$
0	1
1	3
2	9

$y = \log_3 x$

x	y
$\frac{1}{9}$	-2
$\frac{1}{3}$	-1
1	0
3	1
9	2

These points were determined from $x = 3^y$, which is equivalent to $y = \log_3 x$, by choosing the y values and then calculating the x values.

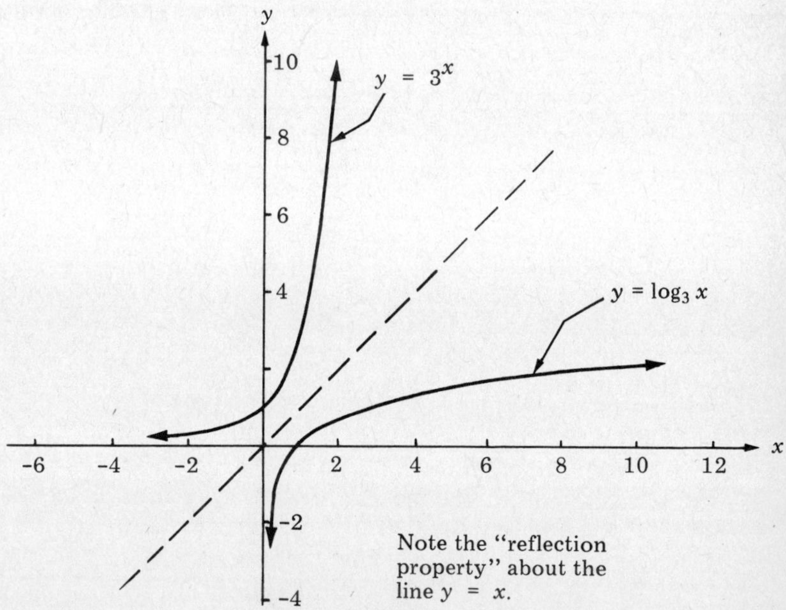

The graph of $y = 3^x$ is a smooth continuous curve as is the graph of $y = \log_3 x$.

The domain of $y = \log_3 x$ is $\{x \mid x > 0\}$.

The range of $y = \log_3 x$ is $\mathcal{R}$.

We now examine the behavior for extreme values of x. Consider the equivalent form of $y = \log_3 x$, namely $x = 3^y$.

As $x \longrightarrow 0^+$, $y \longrightarrow -\infty$ and as $x \longrightarrow +\infty$, $y \longrightarrow +\infty$.

Thus the y-axis is a vertical asymptote on the bottom.

Thus the graph rises indefinitely to the right.

EXAMPLE 3 a. Graph the logarithmic function $y = \log_{1/2} x$.
b. State the domain and range.
c. Discuss the behavior for extreme values of x.

Solution a. $y = \log_{1/2} x$ is equivalent to $(\frac{1}{2})^y = x$ or $2^{-y} = x$. We use the latter equation to obtain the table of values.

x	y
8	−3
4	−2
2	−1
1	0
$\frac{1}{2}$	1
$\frac{1}{4}$	2

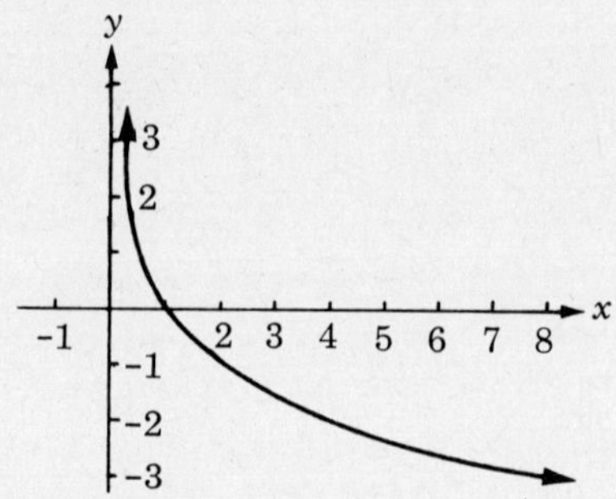

Graph of $y = \log_{1/2} x$

b. The domain of $y = \log_{1/2} x$ is $\{x \mid x > 0\}$ and the range is $\mathcal{R}$.

c. As $x \longrightarrow 0^+$, $y \longrightarrow +\infty$ and as $x \longrightarrow +\infty$, $y \longrightarrow -\infty$.

The y-axis is a vertical asymptote on the top.

The graph falls indefinitely to the right.

The preceding graphs serve to illustrate that the graph of $y = \log_b x$ is one of the two types shown on the following page. The graphs emphasize that:

1. $\log_b 1 = 0$ (This was proven in Theorem 3 also.)
2. $\log_b x$ exists only when $x > 0$. The logarithm of 0 does not exist and the logarithm of a negative number does not exist.
3. A logarithm of a positive number will, itself, be positive, zero, or negative.

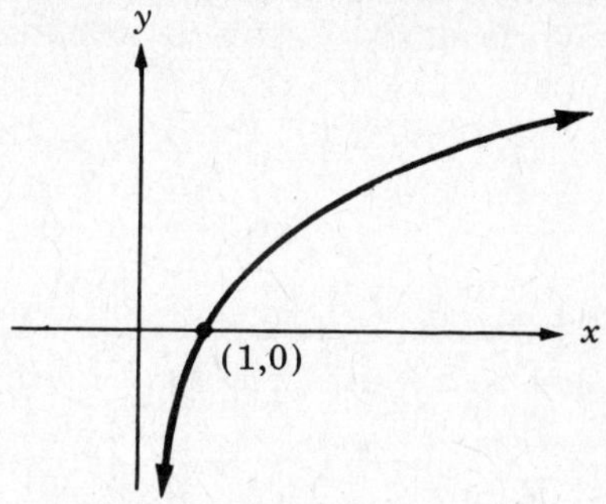

Graph of $y = \log_b x$, $b > 0$.

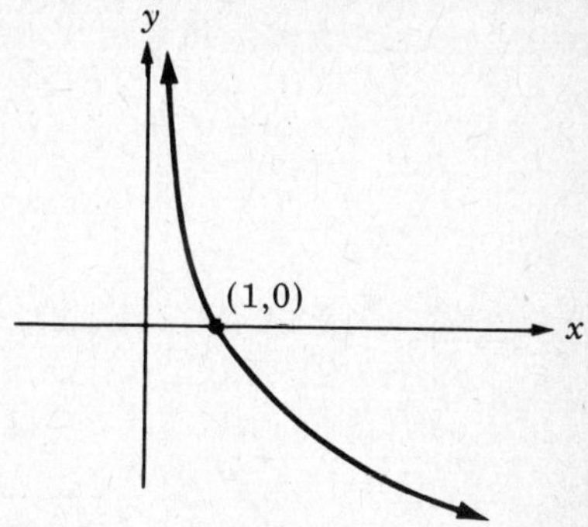

Graph of $y = \log_b x$, $0 < b < 1$.

We now consider an equation of the form $y = \log_b f(x)$ where b is a positive real number, $b \neq 1$ and $f(x)$ is a function of x. However, since the logarithm of 0 does not exist and since the logarithm of a negative number does not exist, $f(x) > 0$ in the equation $y = \log_b f(x)$.

EXAMPLE 4 Graph the logarithmic function $y = \log_4(3 - x)$.

Solution $y = \log_4(3 - x)$ is equivalent to $4^y = 3 - x$ or $x = 3 - 4^y$. We use this latter equation to obtain the table of values.

x	y
$2\frac{15}{16}$	-2
$2\frac{3}{4}$	-1
2	0
1	$\frac{1}{2}$
-1	1
-5	$\frac{3}{2}$
-13	2

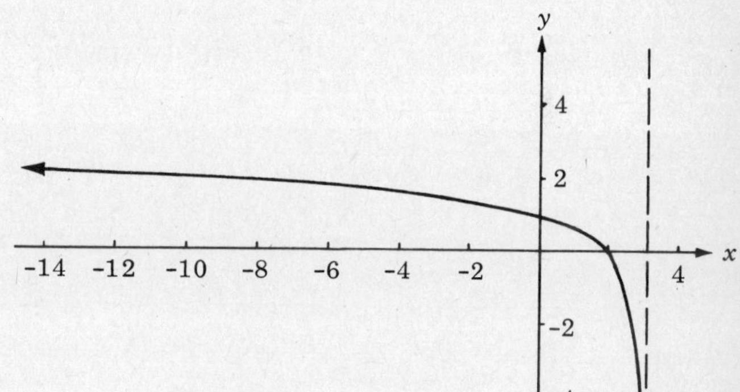

Graph of $y = \log_4(3-x)$

Note that $3-x$ *must* be positive. Thus $3-x > 0$ or $x < 3$. As $x \longrightarrow 3^-$, $y \longrightarrow -\infty$. Hence, the line $x = 3$ is a vertical asymptote on the right.

In Section 5.1 we defined the natural exponential function $y = e^x$, where $e \approx 2.718281$. Its inverse, $y = \log_e x$, more commonly written as $y = \ln x$, is referred to as the *natural logarithm function.* In order to graph this function without an electronic calculator the reader can use the approximation $e \approx 2.7$ (many calculators, however, have an $\ln x$ key). In Exercises 41 and 58 the reader is asked to graph $y = \ln x$.

For the remainder of this section we will restrict our discussion of logarithms to base 10. The equation $y = \log_{10} x$ is called the **common logarithmic function.** Its graph can be obtained, of course, by considering the equivalent exponential equation $10^y = x$. A table of values is obtained by choosing some convenient y values such as -1, $0, \frac{1}{2}, 1$ and then calculating the corresponding x values.

x	y
$\frac{1}{10}$	-1
1	0
3.16	$\frac{1}{2}$
10	1

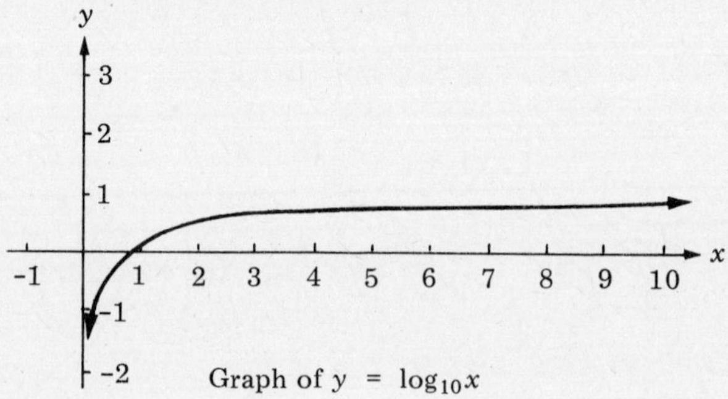

Graph of $y = \log_{10} x$

The method used thus far to graph any logarithmic function has been to graph the equivalent exponential function by choosing certain y values and then calculating the corresponding x values. Next, $y = \log_{10} x$ will be graphed by choosing the x values first, *after which* the y values will be determined. This is done by using a *table of common logarithms* (see Table I). By using this table it is possible to determine the base 10 logarithm (**common logarithm**) of any number from 1.00 through 9.99 as follows—to determine $\log_{10} 2.73$: in the "x column," find 2.7; on top of the table, find 3. The intersection of the row beginning with 2.7 and the column headed with 3, shows the number .4362.

x	0	1	2	3	4	5	6	7	8	9
⋮										⋮
2.5	.3979	.3997	.4014	.4031	.4048	.4065	.4082	.4099	.4116	.4133
2.6	.4150	.4166	.4183	.4200	.4216	.4232	.4249	.4265	.4281	.4298
2.7	.4314	.4330	.4346	.4362	.4378	.4393	.4409	.4425	.4440	.4456
2.8	.4472	.4487	.4502	.4518	.4533	.4548	.4564	.4579	.4594	.4609
2.9	.4624	.4639	.4654	.4669	.4683	.4698	.4713	.4728	.4742	.4757
⋮										⋮
x	0	1	2	3	4	5	6	7	8	9

Hence, $\log_{10} 2.73 = 0.4362$.*

We now use the table of logarithms to determine y values for $y = \log_{10} x$, when $x = 2, 4, 6, 8$:

$$\log_{10} 2 = .3010 \approx .30$$
$$\log_{10} 4 = .6021 \approx .60$$
$$\log_{10} 6 = .7782 \approx .78$$
$$\log_{10} 8 = .9031 \approx .90$$

These points along with the ones from the previous graph of $y = \log_{10} x$ are now shown below.

x	y
2	.30
4	.60
6	.78
8	.90

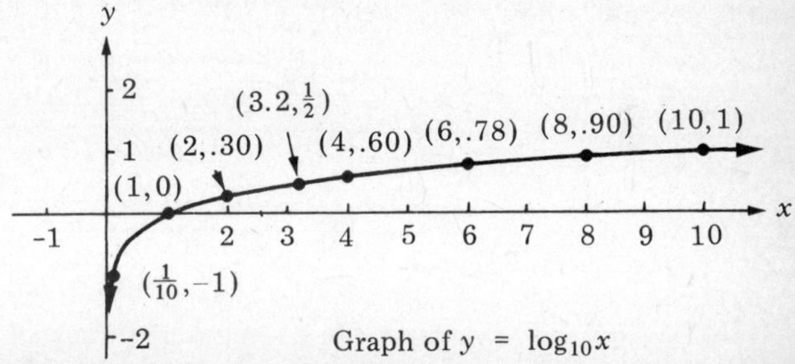

Graph of $y = \log_{10} x$

Base 10 logarithms have various uses that will be explored later in this chapter. However, for now we use Table I in order to solve certain exponential equations in a single variable.

EXAMPLE 5 Solve for x: $10^x = 6.42$.

Solution $10^x = 6.42$ is equivalent to

$$x = \log_{10} 6.42$$

Hence, $x = .8075$ — From Table I

EXAMPLE 6 What power of 10 is equal to 5?

Solution We must solve $10^x = 5$ for x.

$$10^x = 5$$
$$x = \log_{10} 5$$
$$x = .6990$$

*The four-place decimals in Table I are irrational numbers and are only expressed as approximations. Even so, when a value is obtained from Table I, we will use the "=" sign.

EXERCISES FOR SECTION 5.2

In Exercises 1 through 10, rewrite each equation in equivalent logarithmic form. (See Example 1.)

1. $2^5 = 32$
2. $P^0 = 1$
3. $9^{1/2} = 3$
4. $4^{3/2} = 8$
5. $(.1)^2 = .01$
6. $6^{-2} = \frac{1}{36}$
7. $49 = 7^2$
8. $10^2 = 100$
9. $10^3 = 1000$
10. $10^4 = 10000$

In Exercises 11 through 20, rewrite each equation in equivalent exponential form. (See Example 2.)

11. $5 = \log_2 32$
12. $\log_5 \frac{1}{5} = -1$
13. $\log_{1/5} 5 = Q$
14. $\log_7 (x^2) = y$
15. $\log_{10} 1000 = 3$
16. $P = \log_b (x + y)$
17. $P = \log_b (x - y)$
18. $S = \log_R G$
19. $\log_e 7 = M$
20. $\ln 8 = K$

In Exercises 21 through 30, a) graph each logarithmic function, b) state the domain and range, c) discuss the behavior for extreme values of x. (See Example 3.)

21. $y = \log_2 x$
22. $y = \log_5 x$
23. $y = \log_{1/4} x$
24. $y = \log_7 x$
25. $y = \log_8 x$
26. $y = \log_{12} x$
27. $y = \log_{1/3} x$
28. $y = \log_{3/2} x$
29. $y = 2 \log_9 x$
30. $y = 3 \log_4 x$

In Exercises 31 through 42, graph each logarithmic function. (See Example 4.)

31. $y = \log_2 2x$
32. $y = \log_2 (x - 1)$
33. $y = \log_2 (1 - x)$
34. $y = \log_{10} (2x + 2)$
35. $y = \log_3 \left(\dfrac{x + 1}{2}\right)$
36. $f(x) = \log_5 (4 - x)$
37. $f(x) = \log_3 (x + 4)$
38. $y = \log_6 (x + 4)$
39. $y = \log_4 (-x)$
40. $y = \log_4 |x|$
41. Graph $y = \ln x$.
42. Graph $y = \ln(x + 1)$.

In Exercises 43 through 50, solve each exponential equation for x. (See Examples 5 and 6.)

43. $10^x = 1$
44. $10^x = 9$
45. $10^x = 6.75$
46. $10^x = .09$
47. $10^{2x} = 7.65$
48. $10^{2x-1} = 9.20$
49. $10^{1-x} = 8$
50. $10^{x-1} = 8$
51. What power of 10 is equal to 7.75?
52. What power of 10 is equal to .56?
53. Prove: $a = b^{\log_b a}$ (*Hint:* use the definition of a logarithm and let $\log_b a = K$.)
54. Prove: $\log_b \left(\dfrac{1}{a}\right) = -\log_b a$ (*Hint:* let $\log_b \left(\dfrac{1}{a}\right) = x$ and let $-\log_b a = y$ and then show that x must equal y.)
55. Prove: $\log_b b^a = a$.
56. Prove: If $f(x) = b^x$ and $g(x) = \log_b x$, then $f(g(x)) = x$ and $g(f(x)) = x$.
57. The definition of the logarithmic function does not allow the base, b, to be equal to 1. In order to see why, graph the equation $y = \log_b x$, where $b = 1$.

Calculator Exercises

58. Graph $y = \ln x$ by letting $x = .2, .4, .6, .8, 1.0, 1.5, 2, 3, 4, 5, 10, 20$.
59. Graph $y = \ln(x-1)$ by letting $x = 1.2, 1.4, 1.6, 1.8, 2.0, 2.5, 3, 4, 5, 10, 20$.
60. Graph $y = \log_{10}(x^2)$ by letting $x = \pm.4, \pm.6, \pm1, \pm5, \pm10, \pm20$.
61. Graph $f(x) = x \log_{10} x$.
62. Graph $f(x) = x + \log_{10} x$.
63. Graph $f(x) = x \ln x$.
64. Graph $f(x) = x + \ln x$.
65. Graph $y = \ln(x+2)$.
66. Graph $y = \ln x + 2$.
67. Graph $y = \ln x + \log_{10} x$.
68. Graph $y = \ln|x|$.
69. This section pointed out that $f(x) = b^x$ and $g(x) = \log_b x$ are inverses of each other.
    a. On a calculator with $\ln x$ and e^x keys, perform the following sequence:
        1. "Punch in" 1.2345.
        2. Depress $\ln x$ key.
        3. Depress e^x key.
    b. On a calculator with $\log_{10} x$ and 10^x keys, perform the following squence:
        1. "Punch in" 9.8765.
        2. Depress 10^x key.
        3. Depress $\log_{10} x$ key.

    (Compare with Exercise 56.)

5.3 PROPERTIES OF LOGARITHMS

In this section properties of the logarithmic function will be investigated. Two properties that we have already proven are:

$$\left.\begin{array}{ll}\text{(i)} & \log_b b = 1 \\ \text{(ii)} & \log_b 1 = 0\end{array}\right\} \text{Theorem 3}$$

In the following theorem, four properties, usually referred to as **Laws of Logarithms**, are presented.

THEOREM 4

a. $\log_b(xy) = \log_b x + \log_b y$

b. $\log_b\left(\dfrac{x}{y}\right) = \log_b x - \log_b y$

c. $\log_b(x^n) = n \log_b x$

d. $\log_b N = \dfrac{\log_a N}{\log_a b}$

Proof of a:

Statement	*Reason*
1. Let $\log_b x = J$ Let $\log_b y = K$	1. Substitution
2. $b^J = x$ $b^K = y$	2. Definition of Logarithm
3. $b^J \cdot b^K = xy$	3. Multiply both sides of $b^J = x$ by the same quantity
4. $b^{J+K} = xy$	4. Law of Exponents
5. $\log_b(xy) = J + K$	5. Definition of Logarithm
6. Thus, $\log_b(xy) = \log_b x + \log_b y$	6. Substitution

Proof of parts b and c are left to the reader in Exercises 1 and 2.

Proof of d:

Statement	*Reason*
1. Let $\log_b N = J$ Let $\log_a N = K$ Let $\log_a b = L$	1. Substitution
2. $b^J = N = a^K$ $a^L = b$	2. Definition of Logarithm
3. $(a^L)^J = b^J$	3. Raise both sides of $a^L = b$ to the Jth power
4. $a^{LJ} = a^K$	4. Law of Exponents, substitution of a^K for b^J
5. $LJ = K$	5. Theorem 2
6. $J = \dfrac{K}{L}$	6. Divide both sides by L
7. Thus, $\log_b N = \dfrac{\log_a N}{\log_a b}$	7. Substitution

The laws a, b, and c of Theorem 4 are used in the next two examples.

EXAMPLE 1 Use the laws of logarithms to expand each of the following as a sum or difference of logarithms, if possible.

a. $\log_a 5xy$ b. $\log_{10} x^3y^2$ c. $\log_3(2 + W)$

d. $\log_a\left(\dfrac{x+h}{x}\right)$ e. $\log_2\sqrt{\dfrac{3x^2}{y^5}}$

Solution

a. $\log_a 5xy = \log_a 5x + \log_a y = \log_a 5 + \log_a x + \log_a y$

b. $\log_{10} x^3 y^2 = \log_{10} x^3 + \log_{10} y^2 = 3 \log_{10} x + 2 \log_{10} y$

c. $\log_3 (2 + W)$ is the logarithm of a *sum*. Since none of the properties of logarithms pertains to the logarithm of a sum, $\log_3 (2 + W)$ cannot be expanded.

d. $\log_a \left(\dfrac{x + h}{x}\right) = \log_a (x + h) - \log_a x$

e.
$$\log_2 \sqrt{\frac{3x^2}{y^5}} = \log_2 \left(\frac{3x^2}{y^5}\right)^{1/2} = \frac{1}{2} \log_2 \frac{3x^2}{y^5}$$
$$= \tfrac{1}{2}(\log_2 3x^2 - \log_2 y^5) = \tfrac{1}{2}(\log_2 3 + \log_2 x^2 - \log_2 y^5)$$
$$= \tfrac{1}{2}(\log_2 3 + 2 \log_2 x - 5 \log_2 y)$$
$$= \tfrac{1}{2} \log_2 3 + \log_2 x - \tfrac{5}{2} \log_2 y$$

EXAMPLE 2 Use the laws of logarithms to write each of the following as a single logarithm, if possible.

a. $\log_3 5x + 5 \log_3 y$

b. $2 \log_a W - 3 \log_a x$

c. $\log_2 x + \log_3 y$

d. $\tfrac{1}{2} \log_b (x + a) - \tfrac{1}{2} \log_b x$

Solution

a.
$$\log_3 5x + 5 \log_3 y = \log_3 5x + \log_3 y^5$$
$$= \log_3 5xy^5$$

b. $2 \log_a W - 3 \log_a x = \log_a W^2 - \log_a x^3 = \log_a \left(\dfrac{W^2}{x^3}\right)$

c. $\log_2 x + \log_3 y$ is a *sum of logarithms* with *different* bases. Since Theorem 4a pertains to a sum of logarithms with the *same* base, $\log_2 x + \log_3 y$ cannot be written as a single logarithm.

d.
$$\tfrac{1}{2} \log_b (x + a) - \tfrac{1}{2} \log_b x = \log_b (x + a)^{1/2} - \log_b x^{1/2}$$
$$= \log_b \frac{(x + a)^{1/2}}{x^{1/2}} = \log_b \sqrt{\frac{x + a}{x}}$$

Theorem 4d is a property of logarithms that can be used to determine the logarithm of a number to any base. The next example illustrates this.

EXAMPLE 3 Evaluate each logarithm below.

a. $\log_5 8$ b. $\log_5 8.23$ c. $\log_5 82.3$ d. $\ln 823$

Solution a. $\log_5 8$ can be rewritten by using Theorem 4d.

$$\log_5 8 = \frac{\log_a 8}{\log_a 5}, \text{ where } a \text{ is any positive real number, } a \neq 1.$$

Choose $a = 10$ and use Table I.

$$\log_5 8 = \frac{\log_{10} 8}{\log_{10} 5} = \frac{.9031}{.6990} \approx 1.29$$

b. $$\log_5 8.23 = \frac{\log_{10} 8.23}{\log_{10} 5} = \frac{.9154}{.6990} \approx 1.31$$

c. $$\log_5 82.3 = \frac{\log_{10} 82.3}{\log_{10} 5} = \frac{\log_{10}(8.23 \cdot 10)}{\log_{10} 5}$$

$$= \frac{\log_{10} 8.23 + \log_{10} 10}{\log_{10} 5} = \frac{.9154 + 1}{.6990}$$

$$= \frac{1.9154}{.6990} \approx 2.74$$

d. $$\ln 823 = \log_e 823 = \frac{\log_{10} 823}{\log_{10} e} \approx \frac{\log_{10}(8.23 \cdot 100)}{\log_{10} 2.72} \quad (\text{Since } e \approx 2.72)$$

$$= \frac{\log_{10}(8.23 \cdot 10^2)}{\log_{10} 2.72} = \frac{\log_{10} 8.23 + 2 \log_{10} 10}{\log_{10} 2.72}$$

$$= \frac{.9154 + 2}{.4346} = \frac{2.9154}{.4346} \approx 6.71$$

Now that the logarithm of a number to any allowable base can be determined, we can return again to solving exponential equations in a single variable.

EXAMPLE 4 Solve for x: $2^x = 7$.

Solution $2^x = 7$

$x = \log_2 7$

These are equivalent equations.

$$x = \frac{\log_{10} 7}{\log_{10} 2} = \frac{.8451}{.3010} \approx 2.81 \text{ (to the nearest hundredth)}$$

Check:

An electronic calculator with a y^x key shows $2^{2.81} \approx 7.0128458\ldots \approx 7.$ ✓

EXAMPLE 5 Solve for y: $e^y = 3.23$.

Solution $e^y = 3.23$

$y = \log_e 3.23$ — These are equivalent equations.

$$y = \frac{\log_{10} 3.23}{\log_{10} e} \approx \frac{\log_{10} 3.23}{\log_{10} 2.72} \approx \frac{.5092}{.4346}$$

$y \approx 1.17$ (to the nearest hundredth)

Check:

An electronic calculator shows $e^{1.17} \approx 3.2219926\ldots \approx 3.23$ ✓

EXERCISES FOR SECTION 5.3

1. Prove Theorem 4b. (See proof of 4a.)
2. Prove Theorem 4c.

In Exercises 3 through 20, use the properties of logarithms to expand each of the logarithms, if possible. (See Example 1.)

3. $\log_2 14$
4. $\log_2 14xy$
5. $\log_b \pi r^2$
6. $\log_{10} \frac{4}{3}\pi r^3$
7. $\log_{10}(2x + y)$
8. $\log_{10}(2x + 2y)$
9. $\log_8 \dfrac{x}{y^2}$
10. $\log_7 \dfrac{6}{xy}$
11. $\log_b \left(\dfrac{x+h}{x}\right)^{1/3}$
12. $\log_4 \sqrt{\dfrac{xy}{r}}$
13. $\log_b \dfrac{1}{a}$
14. $\log_e(x^x)$
15. $\log_e(x^{x+1})$
16. $\log_{10}(2x - y)$
17. $\log_{10}(2x - 2y)$
18. $\log_5 \sqrt{4x^2y}$
19. $\log_5 \sqrt[3]{4x^2y}$
20. $\log_2\left(\dfrac{x}{y}\right)^2$

In Exercises 21 through 40, use the properties of logarithms to write each of the following as a single logarithm, if possible. (See Example 2.)

21. $\log_2 3 - 2\log_2 t$
22. $\log_5 4 + 2\log_5 x + \log_5 y$
23. $\frac{1}{2}\log_b x - \log_b 10 - \log_b y$
24. $\log_{10}(x + a) - \log_{10} x$
25. $\log_4 2x + \log_5 y$
26. $\log_b x - \log_b y$
27. $\log_3 x - \log_3 y - \log_3 z$
28. $-\log_3 x - \log_3 y - \log_3 z$
29. $(\log_{10} R) \cdot (\log_{10} 5)$
30. $\log_{10} R + \log_{10} 5$
31. $\frac{1}{3}\log_6 x + \frac{1}{3}\log_6 3$
32. $\frac{1}{3}\log_6 x - \frac{1}{3}\log_6 3$
33. $-\log_8 x + \log_8 2x$
34. $1 - \log_3 3w$
35. $\dfrac{\log_{10} 6}{\log_8 4}$
36. $\dfrac{\log_{10} 6}{\log_{10} 2}$
37. $\ln 2x + 2\ln x - 2\ln 2x$
38. $(\ln 7x) \cdot (\ln 2x)$
39. $\dfrac{\ln 7}{\ln 2}$
40. $2\log_3 x + 3\log_4 y + 4\log_5 z$

In Exercises 41 through 56, evaluate each expression. (See Example 3.)

41. $\log_4 7$ 42. $\log_7 4$ 43. $\log_5 6.79$
44. $\log_6 6.79$ 45. $\log_{10} 6.79$ 46. $\log_5 6 + \log_4 5$
47. $\log_5 6 + \log_5 5$ 48. $\log_7 6 + \log_7 5$ 49. $\log_3 62.9$
50. $\log_3 629$ 51. $\log_3 6{,}290$ 52. $\log_3 62{,}900$
53. $\ln 1.92$ 54. $\ln 192$ 55. $\ln 192{,}000$
56. $\ln e$

In Exercises 56 through 70, solve each equation for the variable, expressing your answer to the nearest hundredth. (See Examples 4 and 5.)

57. $3^x = 2$ 58. $6^x = 9$ 59. $10^x = 6$
60. $9^x = 6$ 61. $12^r = 2$ 62. $4^t = 829$
63. $4^x = 82{,}900$ 64. $e^x = 3$ (Use $e = 2.72$)
65. $e^x = 10$ 66. $e^y = 100$ 67. $e^z = 1000$
68. $2^t = 1000$ 69. $t^2 = 1000$ 70. $6^x = 36$

71. Show that $\dfrac{\log_b(x+h) - \log_b x}{h} = \dfrac{1}{x}\log_b\left(1 + \dfrac{h}{x}\right)^{x/h}$.

Calculator Exercises

Solve each equation below. Use $e = 2.7182818$. (Express your answer to the nearest hundredth.)

72. $e^x = 3$ 73. $e^x = 10$ 74. $e^y = 100$ 75. $e^z = 1000$

Computer Exercises

In many forms of BASIC, the expressions e^x and $\ln x$ are written as EXP(X) and LOG(X), respectively.

76. Write a program to construct a table of values for x, e^x, and e^{-x} where x is an integer between 1 and 10 inclusive.
77. Write a program to construct a table of natural logarithms for $x = 1, 1.1, 1.2, \ldots, 9.9, 10$.
78. Write a program to construct a table of common logarithms for $x = 1, 1.1, 1.2, \ldots, 9.9, 10$. (*Hint:* use Theorem 4d. In particular, it may be helpful to use statement 1ØØ below.)

```
1ØØ LET Y = LOG(X)/LOG(10)
```

5.4 MORE ABOUT COMMON LOGARITHMS

Base 10 logarithms (common logarithms) are used for performing certain computations in mathematics and other disciplines. In this section common logarithms are examined more closely.

We have seen that Table I gives common logarithms of numbers between 1.00 and 9.99. Furthermore, in Example 3d of the previous section where it was necessary to evaluate $\log_{10} 823$, we proceeded as follows:

$$\begin{aligned}\log_{10} 823 &= \log_{10}(8.23 \cdot 10^2)\\ &= \log_{10} 8.23 + \log_{10} 10^2\\ &= \log_{10} 8.23 + 2\log_{10} 10\\ &= .9154 + 2\end{aligned}$$

An electronic calculator with a $\log_{10}$ key gives the slightly more precise value $\log_{10} 823 = 2.9153998$.

This number (.9154) is read from Table I. It is called the **mantissa**.

This number (2) is the power of 10 when the original number is written in scientific notation. It is called the **characteristic**.

Thus, $\log_{10} 823 = 2.9154$.

Since any positive real number can be written in scientific notation, the illustration above provides us with a method of determining common logarithms.

EXAMPLE 1 Determine each of the common logarithms below.

a. $\log_{10} 6340$ b. $\log_{10} 634$
c. $\log_{10} .634$ d. $\log_{10} .00634$

Solution a.
$$\begin{aligned}\log_{10} 6340 &= \log_{10}(6.34 \cdot 10^3)\\ &= \log_{10} 6.34 + 3\log_{10} 10\\ &= .8021 + 3\\ &= 3.8021\end{aligned}$$

Using a calculator, $\log_{10} 6340 = 3.8020893$.

b.
$$\begin{aligned}\log_{10} 634 &= \log_{10}(6.34 \cdot 10^2)\\ &= \log_{10} 6.34 + 2\log_{10} 10\\ &= .8021 + 2\\ &= 2.8021\end{aligned}$$

Using a calculator, $\log_{10} 634 = 2.8020893$.

c.
$$\begin{aligned}\log_{10} .634 &= \log_{10}(6.34 \cdot 10^{-1})\\ &= \log_{10} 6.34 - 1\log_{10} 10\\ &= .8021 - 1\end{aligned}$$

Using a calculator, $\log_{10} .634 = -.19791074$.

Note, when a characteristic is negative, we will often *not* combine it with the mantissa. Although it is correct to write $.8021 - 1 = -.1979$, this notation disguises the mantissa (which is .8021) and the characteristic (which is -1.)

Note, this result although correct, does *not* show the mantissa and characteristic (mantissas are always positive). However, $-.19791074$ can be rewritten by adding and subtracting the number 1.

$$\begin{aligned}-.19791074 &= 1 - .19791074 - 1\\ &= \underbrace{.80208926}_{\text{mantissa}} \underbrace{-\ 1}_{\text{characteristic}}\end{aligned}$$

d. $\log_{10} .00634 = \log_{10}(6.34 \cdot 10^{-3})$
$= \log_{10} 6.34 - 3 \log_{10} 10$
$= .8021 - 3$

> Using a calculator, $\log_{10} .00634 = -2.1979107$.

By adding and subtracting 3, we could write

$$-2.1979107 = 3 - 2.1979107 - 3$$
$$= \underbrace{.80208926}_{\text{mantissa}} \underbrace{- 3}_{\text{characteristic}}$$

mantissa
characteristic

At the beginning of this chapter we reached an obstacle when trying to evaluate $3^{2.16}$. The next example overcomes that obstacle.

EXAMPLE 2 Evaluate $3^{2.16}$ using logarithms.

Solution Let $x = 3^{2.16}$. Next, take the common logarithm of both sides of this equation.*

$$\log_{10} x = \log_{10} 3^{2.16}$$
$$= 2.16 \log_{10} 3$$
$$= 2.16(.4771)$$
$$= 1.030536$$
$$= 1.0305$$

characteristic — mantissa

Now we use Table I in "reverse." Find the mantissa closest in value to .0305.

x	. . .	5	6	7	8
1.0		.0212	.0253	.0294	.0334
1.1		.0607	.0645	.0682	.0719
1.2	. . .	.0969	.1004	.1038	.1072
1.3		.1303	.1335	.1367	.1399
1.4		.1614	.1644	.1673	.1703

The table shows that 1.07 has the mantissa closest to .0305. Thus, $x = 1.07 \cdot 10^1$. So, $3^{2.16} = 10.7$.

> Using a calculator with a y^x key, $3^{2.16} = 10.72956$.

*It is a fact that "If $r = s$, then $\log_b r = \log_b s$, and conversely." ($r > 0$, $s > 0$.) Thus, the equations $r = s$ and $\log_b r = \log_b s$ are *equivalent*. In Section 5.6 a proof will be presented.

EXAMPLE 3 Evaluate $(15)^{-6.9425}$.

Solution Let $x = (15)^{-6.9425}$. Take the common logarithm of both sides.

$$\begin{aligned}\log_{10} x &= \log_{10}(15)^{-6.9425}\\ &= (-6.9425)\log_{10} 15\\ &= (-6.9425)(1.1761)\\ &= -8.1651\end{aligned}$$

This is $(-8) + (-.1651)$. Be careful—the characteristic is *not* -8 and the mantissa is *not* $-.1651$. (Remember, mantissas are always positive.) Add and subtract 9 to -8.1651. Hence,

$$\log_{10} x = 9 - 8.1651 - 9$$

$$\log_{10} x = \underbrace{.8349}_{\text{mantissa}} \underbrace{-\,9}_{\text{characteristic}}$$

$$x = 6.84 \cdot 10^{-9}$$

$$(15)^{-6.9425} = 6.84 \cdot 10^{-9}$$

> Using a calculator,
> $(15)^{-6.9425} = 6.8389 \cdot 10^{-9}$

For the remainder of this section we turn our attention to a technique known as **linear interpolation**. Linear interpolation can be used to obtain more accuracy with Table I. Suppose we desire to evaluate $\log_{10} 23.76$. With the help of Table I we can determine $\log_{10} 23.70 = 1.3747$ and $\log_{10} 23.80 = 1.3766$. In picture form we have:

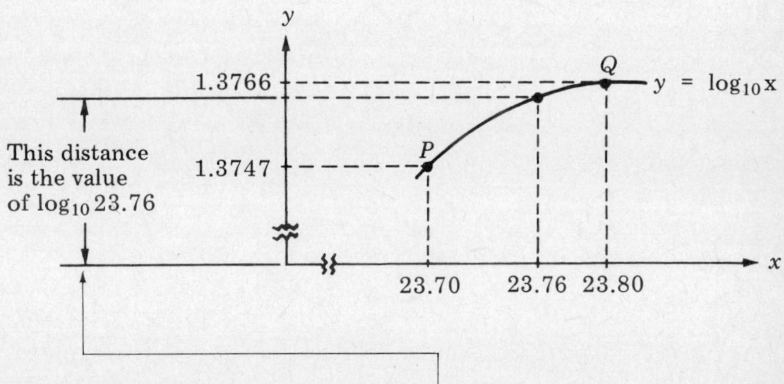

Linear interpolation approximates [this] value by assuming $y = \log_{10} x$ is *linear* between P and Q. Consider the picture on the following page.

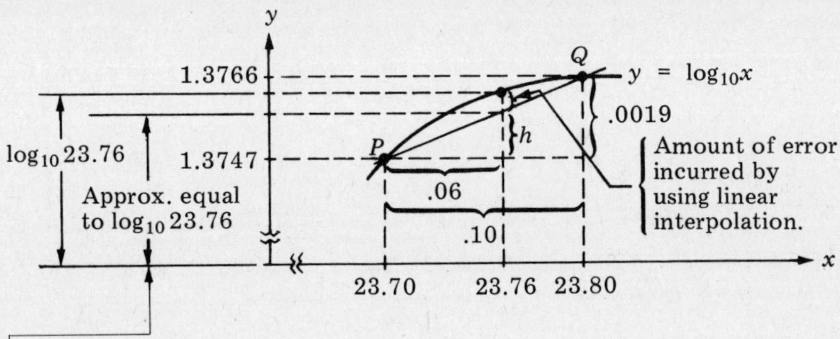

This is determined by adding $1.3747 + h$. The value of h comes from similar triangles that give the proportion $\dfrac{h}{.0019} = \dfrac{.06}{.1}$. Thus $h = .0011$ (rounded to 4 places). Hence, $\log_{10} 23.76 = 1.3747 + .0011 = 1.3758$. The diagrams above justify the procedure just shown. However in practice the reader is expected to save time by writing:

$$\begin{array}{ll} & \log_{10} 23.70 = 1.3747 \\ .10 \quad .06 & \log_{10} 23.76 = \quad ? \qquad h \qquad .0018 \\ & \log_{10} 23.80 = 1.3766 \end{array}$$

$$\frac{h}{.0019} = \frac{.06}{.10} \qquad \log_{10} 23.76 = 1.3747 + .0011$$

$$h = .0011 \qquad \log_{10} 23.76 = 1.3758$$

EXAMPLE 4 Determine the value of $\log_{10} .0006532$.

Solution

$$\begin{array}{ll} & \log_{10} .0006530 = .8149 - 4 \\ .0000010 \quad .0000002 & \log_{10} .0006532 = \quad ? \qquad h \qquad .0007 \\ & \log_{10} .0006540 = .8156 - 4 \end{array}$$

$$\frac{h}{.0007} = \frac{.0000002}{.0000010}$$

$h = .0001$ (rounded to 4 places)

$$\log_{10} .0006532 = (.8149 - 4) + (.0001)$$

Thus, $\log_{10} .0006532 = .8150 - 4$.

A calculator gives $\log_{10}$ = .0006532 = −3.1849538.

EXAMPLE 5 Determine x if $\log_{10} x = 2.8718$.

Solution With the help of Table I,

$$1\left[\begin{array}{l} h\left[\begin{array}{l}\log_{10} 744.0 = 2.8716 \\ \log_{10}\ ? \quad = 2.8718\end{array}\right] .0002 \\ \log_{10} 745.0 = 2.8722 \end{array}\right] .0006$$

$$\frac{h}{1} = \frac{.0002}{.0006}$$

$h = .3$ (rounded to 1 place)

> Since $\log_{10} x = 2.8718$, $x = 10^{2.8718}$ and a calculator gives $x = 744.38909$.

Thus, $x = 744 + h = 744.3$.

EXERCISES FOR SECTION 5.4

In Exercises 1 through 15, determine each of the common logarithms. (See Example 1.)

1. $\log_{10} 8.29$
2. $\log_{10} 829$
3. $\log_{10} 829{,}000{,}000$
4. $\log_{10} .0000829$
5. $\log_{10} .0000000829$
6. $\log_{10} 201{,}000$
7. $\log_{10} 2.01$
8. $\log_{10} .0201$
9. $\log_{10} .747$
10. $\log_{10} 747$
11. $\log_{10} 74.7$
12. $\log_{10} (\frac{7}{4})$
13. $\log_{10} (7 \cdot 4)$
14. $\log_{10} (28)^{1/2}$
15. $\log_{10} 1{,}000{,}000$

In Exercises 16 through 25, use Table I to determine x.

16. $\log_{10} x = 2.9269$
17. $\log_{10} x = 5.9269$
18. $\log_{10} x = .9269 - 4$
19. $\log_{10} x = .9269 - 1$
20. $\log_{10} x = -1.9296$
21. $\log_{10} x = 5.4855 - 6$
22. $\log_{10} x = 9.4855 - 10$
23. $\log_{10} x = -3.4192$
24. $\log_{10} x = 4.9996$
25. $\log_{10} x = 4.0043$

In Exercises 26 through 34, evaluate each expression. (See Examples 2 and 3.)

26. $2^{3.16}$
27. $5^{\sqrt{2}}$ (Use $\sqrt{2} = 1.4142$)
28. $\sqrt[3]{14.5}$ (Recall $\sqrt[3]{14.5} = (14.5)^{1/3}$)
29. $(15)^{31}$
30. $(12)^{-12}$
31. $\sqrt[5]{89{,}300}$
32. $(.00072)^{15}$
33. $\sqrt{110}$
34. $5^{\sqrt{3}}$

In Exercises 35 through 45, determine each logarithm. (See Example 4.)

35. $\log_{10} 2.345$
36. $\log_{10} 2.347$
37. $\log_{10} 63{,}240$
38. $\log_{10} 6324$
39. $\log_{10} .002914$
40. $\log_{10} 29.14$
41. $\log_{10} 1.111$
42. $\log_{10} 1111$
43. $\log_{10} 9.999$
44. $\log_{10} 9999$
45. $\log_{10} 734.6$

In Exercises 46 through 56, determine x. (See Example 5.)

46. $\log_{10} x = .8984$
47. $\log_{10} x = 2.8984$
48. $\log_{10} x = .7430$
49. $\log_{10} x = .7430 - 10$
50. $\log_{10} x = 4.3390$
51. $\log_{10} x = -4.6610$
52. $\log_{10} x = -2.4602$
53. $\log_{10} x = 1.3893$
54. $\log_{10} x = 1.3894$
55. $\log_{10} x = 2.0070$
56. $\log_{10} x = 3.0220 - 10$

Calculator Exercises

57. $\log_b x$ is defined, provided $x > 0$. Using a calculator "evaluate" $\log_{10}(-7)$.
58. Using a calculator, "evaluate" $\log_{10} 0$.

5.5 MORE ABOUT NATURAL LOGARITHMS

Base e logarithms (natural logarithms) are used for performing certain computations in mathematics and other disciplines. In this section natural logarithms are examined closely.

We have seen that natural logarithms can be determined by using the change of base formula $\log_b N = \dfrac{\log_a N}{\log_a b}$. Another way is by using Table II where natural logarithms of numbers from 1.00 through 9.99 and a few other selected values are tabulated.

EXAMPLE 1 Determine each of the natural logarithms.

a. $\ln 3.5$ b. $\ln 3.57$ c. $\ln 9.82$
d. $\ln 60$ e. $\ln 6000$ f. $\ln .39$

Solution

a. $\ln 3.5 = 1.2528$

A calculator gives $\ln 3.5 = 1.252763$.

b. $\ln 3.57 = 1.2726$

A calculator gives $\ln 3.57 = 1.2725656$.

c. $\ln 9.82 = 2.2844$

A calculator gives $\ln 9.82 = 2.2844211$.

d.
$$\begin{aligned} \ln 60 &= \ln(6 \cdot 10) \\ &= \ln 6 + \ln 10 \\ &= 1.7918 + 2.3026 \\ &= 4.0944 \end{aligned}$$

A calculator gives $\ln 60 = 4.0943446$.

e. $\ln 6000 = \ln(6 \cdot 10^3)$
$= \ln 6 + 3 \ln 10$
$= 1.7918 + 3(2.3026)$
$= 8.6996$

A calculator gives $\ln 6000 = 8.6995147$.

f. $\ln .39 = \ln(3.9 \cdot 10^{-1})$
$= \ln 3.9 - 1 \ln 10$
$= 1.3610 - 1(2.3026)$
$= -.9416$

A calculator gives $\ln .39 = -.94160854$.

If we know the numerical value of $\ln x$, it is possible to use Table II to determine the value of x, as in the next example.

EXAMPLE 2 Determine the value of x.

a. $\ln x = 1.6993$ b. $\ln x = 2.1090$
c. $\ln x = 3.4116$ d. $\ln x = -1.2344$

Solution a. $\ln x = 1.6993$
$x = 5.47$

Since $\ln x = 1.6993$, $x = e^{1.6993}$ and a calculator gives $x = 5.4701169$.

b. $\ln x = 2.1090$
$x = 8.24$

A calculator gives $x = 8.239997$.

c. $\ln x = 3.4116$

The entry 3.4116 *does not* appear in Table II. One way to proceed is to let $\ln x = .4116 + 3$.

This is closest to the table entry .4121. Thus,

$x = 1.51 \cdot e^3$
$x = 30.32$

A calculator gives $x = 30.313706$.

d. $\ln x = -1.2344$

The entry -1.2344 *does not* appear in Table II.

$\ln x = 2 - 1.2344 - 2$
$\ln x = .7656 - 2$

This is closest to the table entry .7655. Thus,

$x = 2.15 \cdot e^{-2}$
$x = .2909$

A calculator gives $x = .29100932$.

In Section 5.1 we questioned how to evaluate $2^{23/10}$. We now perform that evaluation by using natural logarithms.

EXAMPLE 3 Evaluate $2^{23/10}$.

Solution Let $x = 2^{23/10} = 2^{2.3}$. Next, take the natural logarithm of both sides of this equation.*

$$\begin{aligned} \ln x &= \ln 2^{2.3} \\ &= 2.3 \ln 2 \\ &= 2.3(.6931) \\ &= 1.59413 \end{aligned}$$

Now we use Table II "in reverse." Find the tabulated value closest to 1.5941.

x	0	1	2	3
1.0	0.0000	0.0100	0.0198	0.0296
1.1	0.0953	0.1044	0.1133	0.1222
1.2	0.1823	0.1906	0.1989	0.2070
1.3	0.2624	0.2700	0.2776	0.2852
1.4	0.3365	0.3436	0.3507	0.3577
⋮				
4.5	1.5041	1.5063	1.5085	1.5107
4.6	1.5261	1.5282	1.5304	1.5326
4.7	1.5476	1.5497	1.5518	1.5539
4.8	1.5686	1.5707	1.5728	1.5748
4.9	1.5892	1.5913	1.5933	1.5953

The table shows that ln 4.92 has the tabulated value closest to 1.5941.

$$x = 4.92$$

Thus, $2^{2.3} = 4.92$.

> Using a calculator with a y^x key, $2^{2.3} = 4.9245777$.

EXAMPLE 4 Evaluate $(1.6)^{-6/5}$.

Solution Let $x = (1.6)^{-6/5}$. Take the natural logarithm of both sides.

---

*See footnote on p. 248.

$$
\begin{aligned}
\ln x &= \ln (1.6)^{-6/5} \\
\ln x &= (-\tfrac{6}{5})(\ln 1.6) \\
&= (-\tfrac{6}{5})(.4700) \\
&= -.564 \\
&= 1 - .564 - 1 \\
\ln x &= .4360 - 1 \\
x &= 1.55 \cdot e^{-1} \\
x &= .57 \\
(1.6)^{-6/5} &= .57
\end{aligned}
$$

Using a calculator, $(1.6)^{-6/5} = .56892631.$

EXERCISES FOR SECTION 5.5

In Exercises 1 through 20, determine each of the natural logarithms. (See Example 1.)

1. $\ln 4$	2. $\ln 40$	3. $\ln 400$
4. $\ln 4000$	5. $\ln 4320$	6. $\ln 2.57$
7. $\ln 75.4$	8. $\ln 829{,}000{,}000$	9. $\ln .000000829$
10. $\ln 747$	11. $\ln .747$	12. $\ln .0203$
13. $\ln 236$	14. $\ln .23$	15. $\ln .0046$
16. $\ln (\frac{6}{5})$	17. $\ln (5.9)$	18. $\ln (9.2)^{1/2}$
19. $\ln (e)^{230}$	20. $\ln (e)^{-1}$	

In Exercises 21 through 40, determine x. (See Example 2.)

21. $\ln x = 2.2017$	22. $\ln x = 3.2017$	23. $\ln x = 1.3737$
24. $\ln x = 1.9400$	25. $\ln x = 2.9957$	26. $\ln x = 4.9957$
27. $\ln x = 2.0700$	28. $\ln x = 4.6052$	29. $\ln x = -1.2016$
30. $\ln x = 2$	31. $\ln x = -.4301$	32. $\ln x = -1.4301$
33. $\ln x = -2.4301$	34. $\ln x = 1.5974$	35. $\ln x = -8.2695$
36. $\ln x = 4$	37. $\ln x = 2.0464$	38. $\ln x = 6$
39. $\ln x = 2.3026$	40. $\ln x = 8.6052$	

In Exercises 41 through 50, evaluate each expression. (See Examples 3 and 4.)

41. $(5)^{.4}$	42. $6^{1/7}$	43. $9^{-.01}$
44. 2^{20}	45. 2^{40}	46. $(\frac{1}{2})^{1/2}$
47. $(.4)^{-.73}$	48. $(10)^{6.9}$	49. e^{7}
50. e^{-4}		

Calculator Exercises

51. $\ln x$ is defined, provided $x > 0$. Using a calculator, "evaluate" $\ln (-4)$.
52. Using a calculator, "evaluate" $\ln 0$.

5.6 EXPONENTIAL AND LOGARITHMIC EQUATIONS

In this section we continue our investigation of exponential and logarithmic equations. We begin by calling the reader's attention to the following facts:

(i) $b^u = v$ is equivalent to $\log_b v = u$ (Definition of logarithm)
(ii) $a = b^{\log_b a}$ (Exercise 53 of Section 5.2)

We now prove the following theorem.

THEOREM 5 Suppose that $r > 0$, $s > 0$, and b is a positive real number, $b \neq 1$. If $r = s$, then $\log_b r = \log_b s$, and conversely.

Proof for "If $r = s$, then $\log_b r = \log_b s$."

Statement	*Reason*
1. $r = s$	1. Given
2. $r = b^{\log_b s}$	2. Exercise 53, Section 5.2
3. Thus, $\log_b r = \log_b s$.	3. Definition of logarithm applied to step 2

Proof for "If $\log_b r = \log_b s$, then $r = s$."

Statement	*Reason*
1. $\log_b r = \log_b s$	1. Given
2. $r = b^{\log_b s}$	2. Definition of logarithm applied to step 1
3. Thus, $r = s$.	3. Exercise 53, Section 5.2

Theorem 5 guarantees that the equations $r = s$ and $\log_b r = \log_b s$ are *equivalent*, a fact that is used in the examples that follow.

EXAMPLE 1 Solve the equation $8^{2x+1} = 5$.

Solution

$$8^{2x+1} = 5$$

$$\log_{10} 8^{2x+1} = \log_{10} 5 \quad \text{Theorem 5}$$

$$(2x + 1)(\log_{10} 8) = \log_{10} 5 \quad \text{Theorem 4c}$$

$$(2x + 1)(.9031) = .6990$$

$$2x + 1 = .7740$$

$$x = -.11 \quad \text{(rounded to the nearest hundredth)}$$

Check:

$$8^{2x+1} = 5$$
$$8^{2(-.11)+1} = 5$$
$$8^{-.22+1} = 5$$
$$8^{.78} = 5$$
$$5.06 \approx 5 \checkmark$$

Here we used a calculator

The solution to Example 1 could also be obtained using natural logarithms (in Exercise 13, the reader is asked to do this). In the next example we use natural logarithms to solve the given exponential equation.

EXAMPLE 2 Solve for t: $5 \cdot 3^{2t} = 9.2$.

Solution

$$5 \cdot 3^{2t} = 9.2$$
Theorem 5
$$\ln(5 \cdot 3^{2t}) = \ln 9.2$$
Theorem 4a, c
$$\ln 5 + 2t \ln 3 = \ln 9.2$$
$$2t \ln 3 = \ln 9.2 - \ln 5$$
$$t = \frac{\ln 9.2 - \ln 5}{2 \ln 3} = \frac{2.2192 - 1.6094}{2(1.0986)}$$
$$t = .28 \quad \text{(rounded to the nearest hundredth)}$$

Check:

$$5 \cdot 3^{2t} = 9.2$$
$$5 \cdot 3^{2(.28)} = 9.2$$
$$5 \cdot 3^{.56} = 9.2$$
$$5 \cdot 1.85 = 9.2$$
$$9.25 \approx 9.2 \checkmark$$

Here we used a calculator

EXAMPLE 3 Solve for y: $\dfrac{3^y - 3^{-y}}{2} = 5$.

Solution

$$\frac{3^y - 3^{-y}}{2} = 5$$

Multiply both sides by 3^y

$$\frac{(3^y)^2 - (3^{-y})(3^y)}{2} = 5 \cdot 3^y$$

$$3^{2y} - 1 = 10 \cdot 3^y$$

$$3^{2y} - 10 \cdot 3^y - 1 = 0$$

Let $p = 3^y$

$$p^2 - 10p - 1 = 0$$

Use quadratic formula

$$p = \frac{10 \pm \sqrt{100 + 4}}{2}$$

$$\text{or } 5 \pm \sqrt{26}$$

Discard the negative solution $5 - \sqrt{26}$, since 3^y is *always* positive

$$p = 5 + \sqrt{26} \text{ or}$$

$$3^y = 5 + \sqrt{26}$$

$$\log_{10} 3^y = \log_{10}(5 + \sqrt{26})$$

$$y \log_{10} 3 = \log_{10}(5 + \sqrt{26})$$

$$y = \frac{\log_{10}(5 + \sqrt{26})}{\log_{10} 3} \approx 2.10$$

Check:

$$\frac{3^y - 3^{-y}}{2} = 5$$

$$\frac{3^{2.10} - 3^{-2.10}}{2} = 5$$

Here we used a calculator

$$\frac{10.0451 - .0995}{2} = 5$$

$$4.97 \approx 5 \checkmark$$

In the remaining examples of this section, we solve logarithmic equations.

EXAMPLE 4 Solve the logarithmic equation $\ln(1 - x) + \ln(x + 4) = \ln 6$ for x.

Solution

$$\ln(1 - x) + \ln(x + 4) = \ln 6$$

Theorem 4a

$$\ln[(1 - x)(x + 4)] = \ln 6$$

Theorem 5

$$(1 - x)(x + 4) = 6$$

$$x^2 + 3x + 2 = 0$$

$$(x + 2)(x + 1) = 0$$

$$x = -2, x = -1$$

Check for $x = -2$:

$$\begin{aligned}\ln(1-x) + \ln(x+4) &= \ln 6\\ \ln 3 + \ln 2 &= \ln 6\\ \ln 6 &= \ln 6 \checkmark\end{aligned}$$

Check for $x = -1$:

$$\begin{aligned}\ln(1-x) + \ln(x+4) &= \ln 6\\ \ln 2 + \ln 3 &= \ln 6\\ \ln 6 &= \ln 6 \checkmark\end{aligned}$$

EXAMPLE 5 Solve for w: $\log_{10}(w+1) - \log_{10}(w+2) = 2$.

Solution

$$\log_{10}(w+1) - \log_{10}(w+2) = 2$$

Theorem 4b

$$\log_{10}\frac{w+1}{w+2} = 2$$

Definition of Logarithm

$$\frac{w+1}{w+2} = 100$$

$$w + 1 = 100w + 200$$

$$w = -\frac{199}{99}$$ (Be careful, note the check below.)

Check:

$$\log_{10}(w+1) - \log_{10}(w+2) = 2$$

$$\log_{10}\left(-\frac{199}{99} + 1\right) - \log_{10}\left(-\frac{199}{99} + 2\right) = 2$$

$$\underbrace{\log_{10}\left(-\frac{100}{99}\right)} - \underbrace{\log_{10}\left(-\frac{1}{99}\right)} = 2$$

These expressions are *not* defined. Thus, the original equation *has no solution*; $w = -\frac{199}{99}$ is *extraneous*.

EXERCISES FOR SECTION 5.6

In 1 through 12, solve each exponential equation by using common logarithms. (See Example 1.)

1. $6^x = 90$
2. $8^{2y+1} = 5$
3. $10^{2y+1} = 5$
4. $4^{t-1} = 2$
5. $5 \cdot 3^{2t} = 9.2$
6. $3^{-x} = 11$
7. $(\frac{1}{2})^{3x} = 17$
8. $e^{2x} = 17$
9. $2^x = 3^{x-1}$
10. $e^{x^2} = 10$
11. $e^{x^2+1} = 10$
12. $4^{2x+1} = 6^{1-x}$

In 13 through 24, solve each exponential equation by using natural logarithms. (See Example 2.)

13. $8^{2x+5} = 5$
14. $3^{-x} = 11$
15. $e^{-x} = 12$
16. $e^{1-x} = 12$
17. $e^{1-x} = 1.2$
18. $e^{3x-1} = e^{2x+6}$
19. $2^x = 3^{x-1}$
20. $2^x = 3^x$
21. $(\frac{1}{4})^x = 8$
22. $8^{2y+1} = 5$
23. $8^{2y+1} = 8$
24. $2^x = 3^{\log_3 6}$

In 25 through 30, solve each equation for x. (See Example 3.)

25. $\dfrac{2^x - 2^{-x}}{2} = 3$ 26. $\dfrac{e^x - e^{-x}}{2} = 4$ 27. $\dfrac{3^x - 3^{-x}}{2} = 1$

28. $\dfrac{6^x + 6^{-x}}{2} = 1$ 29. $\dfrac{10^x + 10^{-x}}{2} = 10$ 30. $\dfrac{e^x + e^{-x}}{2} = e$

In 31 through 34, solve each equation for x.

31. $y = \dfrac{e^x - e^{-x}}{2}$ 32. $y = \dfrac{e^x + e^{-x}}{2}$

33. $y = \dfrac{e^x - e^{-x}}{e^x + e^{-x}}$ 34. $y = \ln\sqrt{x^2 + 1}$

In 35 through 50, solve each logarithmic equation. (See Examples 4 and 5.)

35. $\log_{10}(x + 2) = 2$ 36. $\log_5(x + 2) = 2$
37. $\ln(x + 2) = 2$ 38. $\log_{10}(x + 3) + \log_{10}(x - 1) = \log_{10} 7$
39. $\ln(1 - x) + \ln(2x + 1) = \ln 1$ 40. $\log_{10}(t - 1) + \log_{10}(t + 2) = 1$
41. $3\log_{10}x - \log_{10}2x = 2.5$ 42. $\ln\dfrac{x}{2} + \ln x = 1$
43. $\ln\dfrac{x}{2} + \ln x = \ln\dfrac{9}{2}$ 44. $\ln(x^2 - 1) - \ln(x - 1) = 3$
45. $\ln(\ln x) = 1$ 46. $\log_{10}(x^2) = \log_{10}(7x - 12)$
47. $4 + \log_{10}x = 0$ 48. $\log_4(x^2 - 9) - \log_4(x - 3) = 2$
49. $\log_5(x^2 - 9) - \log_5(x - 3) = 2$ 50. $\log_3(x - 1) + \log_3\left(\dfrac{1}{x}\right) = 1$

5.7 APPLICATIONS

There are many practical applications of exponential and logarithmic functions. In this section we investigate some of these applications.

Business Problems

An important ingredient of our economic system is the concept of compound interest. When an initial investment of P dollars earns compound interest, the equation $A = P\left(1 + \dfrac{r}{n}\right)^{nt}$ is an algebraic model for the worth or value, A, in dollars at some later time. Here,

r = annual (or nominal) interest rate
n = number of times the interest is compounded (conversion periods) per year
t = number of years the money is invested.

EXAMPLE 1 a. What is the value of a \$100 investment that earns 6% annual interest, compounded quarterly (four times a year), if it remains invested for 7 years?

b. How long will it take for a sum of money to double in value, if it earns 6% annual interest, compounded quarterly?

Solution a. $A = P\left(1 + \frac{r}{n}\right)^{nt}$

$A = \$100\left(1 + \frac{.06}{4}\right)^{28}$

$A = \$100(1.015)^{28}$

$A = \$100(1.5172221)$

By using a calculator, $(1.015)^{28} = 1.5172221$

$A = \$151.72$

b. $A = P\left(1 + \frac{r}{n}\right)^{nt}$

In order for the amount P to double, A must equal $2P$.

$2P = P\left(1 + \frac{.06}{4}\right)^{4t}$

$2 = (1.015)^{4t}$

$\log_{10} 2 = 4t \log_{10} 1.015$

Take the common logarithm of both sides

$t = \dfrac{\log_{10} 2}{4 \log_{10} 1.015} = \dfrac{.3010}{4(.0065)},$

$t = 11.58$ years (in reality, 11 years 9 months because of quarterly compounding.)

Since many banks compound interest daily, let us redo Example 1a by compounding daily.

EXAMPLE 2 What is the value of a \$100 investment that earns 6% annual interest, compounded daily (360 times a year), if it remains invested for 7 years?

Solution $A = P\left(1 + \frac{r}{n}\right)^{nt}$

$A = \$100\left(1 + \frac{.06}{360}\right)^{2520}$

$A = \$100(1.0001667)^{2520}$

$A = \$100(1.521908)$

Here we used a calculator

$A = \$152.19$

Notice that as the number of conversion periods per year increases (with everything else remaining the same), the worth of an investment increases (slightly).

What would happen if the number of conversion periods per year, n, increased even more than in Example 2? In order to investigate this question we first rewrite the compound interest formula.

$$A = P\left(1 + \frac{r}{n}\right)^{nt} \tag{1}$$

$$A = P\left[\left(1 + \frac{r}{n}\right)^{n/r}\right]^{rt} \tag{2}$$

The expression in brackets can be written as

$$\left(1 + \frac{1}{n/r}\right)^{n/r}$$

Next, let $\frac{n}{r} = w$ and obtain $\left(1 + \frac{1}{w}\right)^{w}$. Since r is a constant we can say that if $n \longrightarrow +\infty$, then $w \longrightarrow +\infty$. Consequently, we now consider the expression $\left(1 + \frac{1}{w}\right)^{w}$ as $w \longrightarrow +\infty$.

w	$\left(1 + \frac{1}{w}\right)^{w}$		
10	$(1.1)^{10}$	= 2.5937425	Determined with a calculator
100	$(1.01)^{100}$	= 2.7048138	
10,000	$(1.0001)^{10,000}$	= 2.7181459	
1,000,000	$(1.000001)^{1,000,000}$	= 2.7182818	
100,000,000	$(1.00000001)^{100,000,000}$	= 2.7182818	
⋮			

It seems as though $\left(1 + \frac{1}{w}\right)^{w} \longrightarrow e$ as $w \longrightarrow +\infty$.* In other words, when the number of conversion periods increases indefinitely, equation (2) becomes:

$$A = P e^{rt}$$

*In the study of calculus we say that the limit of $\left(1 + \frac{1}{w}\right)^{w}$ as $w \longrightarrow +\infty$ is e. In symbols, we write $\lim_{w \to \infty} \left(1 + \frac{1}{w}\right)^{w} = e$.

This equatjon is used to calculate the worth or value of an investment of P dollars when interest is *compounded continuously*. Besides that, we have just seen a situation that, in a very natural way, gives rise to the natural base e!

EXAMPLE 3 Calculate the worth of a \$100 investment that earns 6% annual interest, compounded continuously, if it remains invested for 7 years.

Solution

$A = Pe^{rt}$

$A = \$100\,e^{(.06)(7)}$

$A = \$100\,e^{.42}$ → Here we used a calculator

$A = \$100(1.5219615)$

$A = \$152.20$

Notice the result is just 1¢ more than in Example 2.

Compounding interest continuously is becoming more popular with banks in recent years.

Natural Science Problems

As a result of studying radioactive substances, scientists have determined that each radioactive isotope has a certain rate at which it changes (decays) into another element. This process is called *radioactive decay* and can be described by referring to the *half-life* of the element. For instance, the half-life of carbon-14 is 5,750 years. This means that if we started with 40 grams of carbon-14, after 5,750 years we would have 20 grams. After another 5,750 years we would have 10 grams, etc. We examine this phenomenon further in the following example.

EXAMPLE 4 Scientists can determine the age of archeological findings by a process called carbon dating. The amount of carbon-14 present (per kilogram) in a living plant or animal is the same. After death, the equation

$$N = N_o\left(\tfrac{1}{2}\right)^{\frac{x}{5,750}}$$

is an algebraic model of the decay process, where N is the number of grams of carbon-14 remaining (per kilogram) in a plant or animal x years after its death. N_o is the number of grams present (per kilogram) in the plant or animal when it was living.

A skull uncovered at an archeological digging was found to have lost 60% of its carbon-14. How old is the skull (from time of death)?

Solution 60% of the carbon-14 is gone. Thus, $N = .40N_o$.

$$N = N_o\left(\tfrac{1}{2}\right)^{\frac{x}{5{,}750}}$$

$$.40N_o = N_o\left(\tfrac{1}{2}\right)^{\frac{x}{5{,}750}}$$

$$.40 = (.5)^{\frac{x}{5{,}750}}$$

$$\log_{10} .40 = \frac{x}{5{,}750} \log_{10} .5$$

$$x = 5{,}750\left(\frac{\log .4}{\log .5}\right)$$

$$= 5{,}750\left(\frac{.6021 - 1}{.6990 - 1}\right) = 5{,}750\left(\frac{-.3979}{-.3010}\right)$$

$$\approx 7{,}600 \text{ years}$$

In the next example a natural science application of logarithms is presented.

Example 5 A city discovers that its water reservoir is polluted. In order to remedy this situation they will pump out water at the constant rate of N liters per hour and simultaneously replace it with unpolluted water until an acceptable low level of pollution is reached. An algebraic model for the time (in hours) this will take is

$$t = \frac{M}{N} \ln \frac{p_s}{p_a},$$

where M = capacity of reservoir in liters

p_s = pollution level of reservoir at the start of the pumping process (in milligrams per liter)

p_a = the acceptable pollution level of reservoir (in milligrams per liter)

How long will it take to reduce the pollution to $\frac{1}{3}$ of its initial value if the reservoir holds 20,000,000 liters of water and the pumping rate is 15,000 liters per hour?

Solution It is given that $M = 20{,}000{,}000$, $N = 15{,}000$ and $p_a = \frac{1}{3}p_s$.

$$t = \frac{M}{N} \ln \frac{p_s}{p_a}$$

$$t = \frac{20{,}000{,}000}{15{,}000} \ln \frac{p_s}{\frac{1}{3}p_s} = \frac{4{,}000}{3} \ln 3$$

$$t = 1{,}464.82 \text{ hours}$$

Social Science and Medical Problems Planet Earth surely has its share of problems. Perhaps the single most important problem to be solved is that of population control. In the next example we examine some of the mathematical concepts.

EXAMPLE 6 On our planet, the annual population growth rate is about 2%. In other words, each year, as a result of births and deaths, the population increases by 2% of what it was the previous year. An exponential function describing this growth is $N = N_o(1.02)^t$, where N is the number of people alive at any time, t, measured in years. N_o represents the number of people present initially.

a. The world population in 1979 was about 4 billion people. Assuming that the 2% growth rate remains the same, how many people will there be in 15 years?
b. Assuming the 2% growth rate remains the same, how many years will it take the world population to double?

Solution a. We must determine N, knowing that $N_o = 4 \cdot 10^9$ and $t = 15$.

$$N = N_o(1.02)^t$$
$$N = 4 \cdot 10^9(1.02)^{15}$$
$$N = 4 \cdot 10^9(1.3458683)$$
$$N = 5.38 \cdot 10^9$$

Here we used a calculator

Thus, at the present growth rate, the world population will be nearly $5.38 \cdot 10^9$ or about 5.4 billion in 15 years.

b. We must determine t, when $N = 2N_o$.

$$N = N_o(1.02)^t$$
$$2N_o = N_o(1.02)^t$$
$$2 = (1.02)^t$$
$$\ln 2 = t \ln 1.02$$
$$t = \frac{\ln 2}{\ln 1.02} = \frac{.6931}{.0198} \approx 35$$

Hence, at the present growth rate, it will take about 35 years for the world population to double.

The algebraic model in the previous example, $N = N_o(1.02)^t$ is an exponential function with base $b = 1.02$. The function is graphed at the right.

t	N
0	N_o
10	$1.2N_o$
20	$1.5N_o$
40	$2.2N_o$
60	$3.3N_o$
70	$4N_o$
100	$7.2N_o$
200	$52N_o$
300	$380N_o$

Here we used a calculator to evaluate $(1.02)^t$

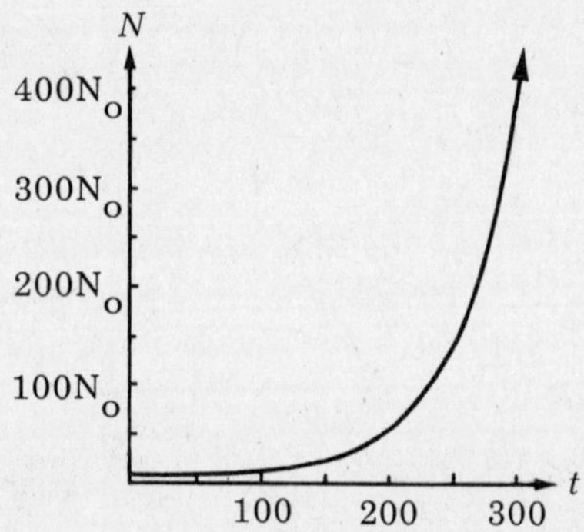

Graph of $N = N_o(1.02)^t$, $t \geqslant 0$

A little common sense tells us that although in 300 years this model predicts a population that will be 380 *times* the present population, simply put, it *is not likely*. Space, food, and energy resources would run out long before 300 years, thus causing the growth rate to slow, stop, or reverse itself. The model $N = N_o(1.02)^t$, is a model for *unbounded growth* that is derived by assuming population *can* keep growing without bound. However, if we assume that there is some upper limit, L, to the world population, it is possible to derive the equation

$$N = \frac{N_o \cdot L}{N_o + (L - N_o)e^{-LKt}}$$

where N is the number of people alive (in billions) at any time t, measured in years. N_o represents the number of people alive (in billions) initially ($t = 0$) and K is a contant that can be determined. In Exercise 16, we will investigate this model for *bounded growth*.

In the final example of this section we present an application in the field of medicine.

EXAMPLE 7 An important factor to consider when any medicine is administered to a patient is the amount of medicine present in the blood stream at a later time t. In a study with human patients a pain killing drug was administered and the drug concentration, $C(t)$, (in ng/ml*) at a later time t (in hours) was known to be given by:

$$C(t) = 54e^{-.092t},\ t \geqslant 2.$$

*The unit ng/ml means nanograms of drug per milliliter of blood. A nanogram is 10^{-9} gram.

a. Graph this function for $2 \leqslant t \leqslant 36$.
b. Determine the percent of change from $t = 2$ to $t = 3$.
c. Determine the percent of change from $t = a$ to $t = a + 1$.

Solution a.

t	C
2	44.92
6	31.09
10	21.52
20	8.58
25	5.41
30	3.42
36	1.97

Here we used a calculator

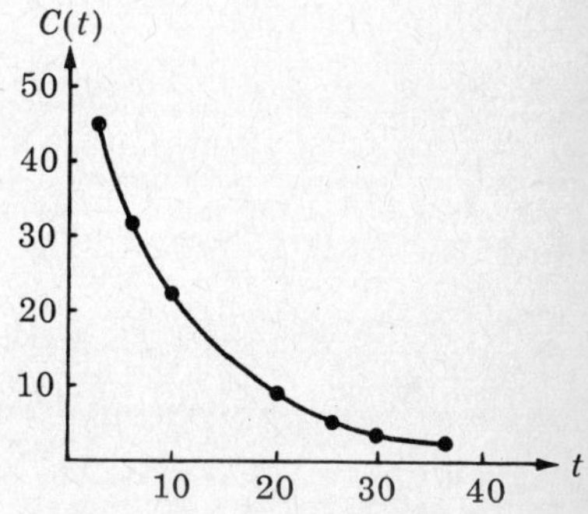

Graph of $C(t) = 54e^{-.092t}$, $t \geqslant 2$

The exponential decrease of $C(t)$ with respect to time is typical of drug concentrations in the body. Furthermore, functions like $C(t) = 54e^{-.092t}$, that represent concentration can be derived using methods of calculus.

b. The percent of change equals $\dfrac{\text{Change of concentration}}{\text{Original value of concentration}}$.

$$\text{Percent of change} = \frac{C(3) - C(2)}{C(2)} = \frac{40.98 - 44.92}{44.92}$$

$$= -.088 = -8.8\%$$

Thus the percent of change is a decrease of 8.8%.

c.
$$\text{Percent of change} = \frac{C(a+1) - C(a)}{C(a)}$$

$$= \frac{54e^{-.092(a+1)} - 54e^{-.092(a)}}{54e^{-.092(a)}}$$

$$= \frac{e^{-.092(a+1)} - e^{-.092a}}{e^{-.092a}}$$

$$= \frac{e^{-.092} - 1}{1}$$

Multiply numerator and denominator by $e^{.092a}$

$$= \frac{.91210515 - 1}{1}$$

$$= -.088$$

Thus the percent of *decrease* each hour is 8.8%.

EXERCISES FOR SECTION 5.7

1. a. What is the value of a \$200 investment that earns 8% annual interest compounded annually (once a year) if it remains invested for 10 years?
   b. How long will it take for a sum of money to double in value, if it earns 8% annual interest, compounded annually?
2. a. What is the value of a \$200 investment that earns 8% annual interest compounded semi-annually (twice a year) if it remains invested for 10 years?
   b. How long will it take for a sum of money to double in value, if it earns 8% annual interest, compounded semi-annually?
3. a. What is the value of a \$200 investment that earns 8% annual interest compounded quarterly if it remains invested for 10 years?
   b. How long will it take for a sum of money to double in value, if it earns 8% annual interest, compounded quarterly?
4. a. What is the value of a \$200 investment that earns 8% annual interest compounded daily if it remains invested for 10 years?
   b. How long will it take for a sum of money to double in value, if it earns 8% annual interest, compounded daily?
5. a. What is the value of a \$200 investment that earns 8% annual interest compounded continuously if it remains invested for 10 years?
   b. How long will it take a sum of money to double in value, if it earns 8% annual interest, compounded continuously?
6. How long will it take a sum of money to triple in value, if it earns 4% annual interest, compounded quarterly?
7. How long will it take a sum of money to triple in value, if it earns 4% annual interest, compounded continuously?
8. Workers around x-ray equipment must be properly shielded. X-rays are absorbed as they pass through any medium and the equation $x = \frac{1}{350} \ln \frac{I_o}{I}$ is an algebraic model for the thickness of lead required to cause an x-ray beam having initial intensity I_o to be reduced to intensity I.
   a. What thickness of lead (in centimeters) is required to reduce the intensity of x-rays to $\frac{1}{5}$ of their initial intensity?
   b. What thickness of lead (in centimeters) is required to reduce the intensity of x-rays to $\frac{1}{20}$ of their initial intensity?
9. Using the carbon-14 dating equation, determine the age of a bone from time of death that has lost 75% of its carbon-14.
10. Using the carbon-14 dating equation, determine the age of a bone from time of death that has lost 95% of its carbon-14.
11. The equation $M = \log_{10} \frac{A}{A_o}$ is an algebraic model that defines the Richter Scale for measuring the magnitude, M, of an earthquake. A_o is the amount of the earth's vibration for the faintest perceptible earthquake detectable by a seismograph. A is the amount of the earth's vibration of the earthquake in question as measured by a seismograph.
    a. If the amount of the earth's vibration is 1,000,000 times that of the faintest perceptible earthquake, what is the magnitude, M, on the Richter Scale?

b. An earthquake of magnitude 7.2 would cause massive damage over a wide region. Determine the amount of the earth's vibration, A, for such an earthquake in terms of A_o.

12. Using the equation that defines the magnitude, M, on the Richter Scale, determine the amount of the earth's vibration A in terms of A_o for an earthquake of magnitude 3.5.

13. The equation $N = N_o(1.01)^t$ is an unbounded growth model for the U.S. population, where N is the number of people alive at any time t (measured in years) and N_o is the number alive initially.
    a. In 1979, the U.S. population was about 200,000,000 people. Assuming the growth rate remains the same, how many people will there be in 30 years?
    b. Assuming the growth rate remains the same, how long will it take for the U.S. population to triple?

14. In the country of El Salvador, the annual population growth rate is 4%. The unbounded growth model for this is $N = N_o(1.04)^t$, where N represents the number alive at any time t (measured in years) and N_o represents the number alive initially. How long will it take the present population to triple (assuming the growth rate remains the same)?

15. The equation $N = N_o(2)^{3t}$ is an unbounded growth model for the number of bacteria, N, that exist after a time t (measured in hours), where N_o is the number of bacteria present originally.
    a. If there are 200 bacteria originally present, how many are there after 4 hours?
    b. If there are N_o bacteria originally present, how long will it take that number to quadruple?

16. The formula

$$N = \frac{N_o L}{N_o + (L - N_o)e^{-LKt}}$$

is the algebraic model for a growth rate *bounded* by the value L. Assume that the upper limit for the world population is 20 billion ($L = 20$). In 1960 ($t = 0$) the world population was 3 billion ($N_o = 3$), and in 1975 ($t = 15$) the world population was 4 billion.
    a. Find K.
    b. Using the given values of L and N_o and the computed value of K in part a, express N as a function of t.
    c. Find the world population in the year 2000.
    d. Graph the result of part b.

17. Rework Exercise 16 but this time assume the upper limit to the world population is *15* billion.

18. Assume that the upper limit to a rat population in a certain city is 50,000. Using the facts that at the end of a certain month the rat population was 10,000 and 4 months later the rat population was 20,000 and that

$$N = \frac{N_o L}{N_o + (L - N_o)e^{-LKt}}.$$

a. Determine the value of K, by measuring L, N_o, and N in thousands and by measuring t in months.
b. Using the values of L, K, and N_o rewrite the bounded growth formula, showing N a function of t.
c. Graph the result of part b.

19. An experimental drug was administered to a rat and the drug concentration in the blood stream, $C(t)$, (in μg/ml*) at a later time t (in hours) was known to be given by:

$$C(t) = .21e^{-.51t}, .2 \leqslant t \leqslant 8.$$

a. Graph this function for $.2 \leqslant t \leqslant 8$.
b. Determine the percent of change from $t = a$ to $t = a + 1$.

20. The equation $L = \log_{10}\frac{I}{I_o}$ is an algebraic model for the loudness of sound, L, measured in decibels, where I_o is the intensity of the faintest perceptible sound and I is the intensity of the sound in question. The loudness of the water at the foot of Niagara Falls is a billion times greater than the faintest perceptible sound. What is the loudness of sound, L, at the foot of Niagara Falls?

21. The equation, $p = -\log_{10}H$, is used to define the concept known as the pH of a solution, where

p represents the ph
H = the hydrogen ion concentration in the solution (measured in moles per liter)

a. Determine the pH of pure water where $H = 10^{-7}$ moles per liter.
b. Determine the pH of a shampoo, where $H = 5.5 \cdot 10^{-7}$ moles per liter.
c. If the pH of a soap is 7.7, determine H, the hydrogen ion concentration.

22. Consider the exponential function $f(x) = ab^{cx}$. Show that the percent of change in $f(x)$ is the constant $b^c - 1$ as x changes to $x + 1$.

CHAPTER 5 IN REVIEW

I. Summary

In this chapter we have seen the definition and development of the properties of two important functions—the exponential and logarithmic functions. Due to the existence of these functions we are able to give meaning to the expression b^r where $b > 0$ and r is *any real number*. We've come a long way since exponents were first discussed in Chapter 1.

In Section 5.1 we graphed various exponential functions (including the natural exponential function with base e) usually by plotting five or six carefully selected points. Furthermore, exponential equations in a single variable were solved by using the fact that the equations $r = s$ and $b^r = b^s$ are equivalent (provided $b > 0$, $b \neq 1$).

---

*The unit μg/ml means micrograms of drug per milliliter of blood. A microgram is 10^{-6} gram.

In Section 5.2 we graphed various logarithmic functions (including the natural logarithmic function with base e). It was observed that the functions $f(x) = b^x$ and $g(x) = \log_b x$ are inverses of each other. Since a logarithm is an exponent, it became possible to use logarithms to solve exponential equations in a single variable.

Many properties of logarithms were studied and proven in Section 5.3. As a result, we broadened our capability of expanding and simplifying logarithmic expressions. Calculations of logarithms with various bases were performed and then in Sections 5.4 and 5.5 we centered our attention on common logarithms and natural logarithms. Tables of common and natural logarithms are provided in the Appendix; however, the use of a calculator was seen to be a great aid.

In Section 5.6 logarithmic equations were solved by using the fact that the equations $r = s$ and $\log_b r = \log_b s$ are equivalent (provided $r > 0, s > 0, b > 0, b \neq 1$).

At the close of the chapter, applications of exponential and logarithmic functions were viewed in the business, natural science, social science, and medicine fields.

II. Vocabulary/Symbol True–False Quiz

If the statement is true, write "TRUE." However, if the statement is false, replace the $\underline{\text{underlined}}$ word or phrase with a different word or phrase that will make the new statement true.

1. $f(x) = b^x$ is an exponential function provided $\underline{b > 0}$.
2. The graph of $y = 10^x$ approaches the x-axis as x approaches $\underline{-\infty}$.
3. Suppose $b > 0$ and $b \neq 1$. Then $p = b^q$ is equivalent to $\underline{\log_b p = q}$.
4. The inverse of $f(x) = e^x$ is $g(x) = \underline{\ln x}$.
5. The domain of $y = \ln(x - 3)$ is $\underline{\{x \mid x \geq 3\}}$.
6. $\log_{10}\sqrt{200} = \underline{1 + \frac{1}{2}\log_{10} 2}$.
7. $\log_{10} N = \underline{\dfrac{\ln N}{\ln 10}}$.
8. Using Table I, $\log_{10} 87.5 = \underline{1.9420}$.
9. Using Table II, $\ln 12.7 = \underline{2.5416}$.
10. If $a = \ln 27$, then $e^a = \underline{27}$.

III. Chapter Test

1. Graph $y = e^x$ and $y = \ln x$ on the same set of axes.
2. a. Graph $y = 5^{-x}$. b. State the domain and range.
3. Find the value of each of the following:
   a. $\ln 5.88$ b. $\log_{10}\sqrt{3.4}$ c. t if $10^t = .19$

4. Rewrite $\log_{10}(x^2 - 1) = y$ in an equivalent exponential form.
5. Solve for x: $e^{2x} = 150$.
6. Write $\log_3 x + 3\log_3 y - \log_3 z$ as a single logarithm.
7. Solve for x: $\log_{10} x = -1.9101$.
8. Solve for x: $8^{1-x} = 2$.
9. Which is larger: $5^{\sqrt{3}}$ or $(\sqrt{3})^5$?
10. Solve for t: $5 \cdot 3^{2t} = 11.8$.
11. Solve for x: $2^x = 3^{x^2}$.
12. Solve for y: $\log_{10} y^2 = \log_{10}(8y - 15)$.
13. Solve for z: $y = \dfrac{e^z + e^{-z}}{2}$.
14. Solve for w: $\log_{10}(w + 3) + \log_{10}(w + 1) = 3$.
15. a. What is the value of a $1000.00 investment that earns 6% interest compounded daily (360 days per year), on deposit for 5 years?
    b. How long will it take for that $1000.00 investment to become $5000.00 if left in that same account?

IV. Student Research Project

1. John Napier (1550–1617) is usually credited for the first use of logarithms. Consult texts on the history of mathematics and prepare a brief report on Napier and the early use of logarithms. One interesting fact that could be included in such a report is that logarithms were used *before* exponents!
2. The slide rule is none other than a logarithmic scale. Read pp. 171–177 in *Mathematics: A Human Endeavor* by Harold Jacobs (W. H. Freeman, San Francisco, 1970) and pp. 178–182 in *Mathematics in the Making* by Lancelot Hogben (Galahad Books, London, 1972). What other logarithmic scales can you name?

CHAPTER **6**

Trigonometric Functions

Sun Power

Global, Direct and Diffuse Radiation

The availability of solar energy in any location in the world can be studied by two methods. The first involves measurements from a radiation monitoring network and the second is based on the use of physical formulae and constants. Direct solar radiation, I, is the solar radiation flux associated with the direct solar beam from the direction of the sun's disc, which may be assumed to be a point source, and is measured normal to the beam (that is on a plane which is perpendicular to the direction of the sun). Diffuse radiation, D, reaches the ground from the rest of the whole sky hemisphere from which it has been scattered in passing through the atmosphere. Global solar radiation, G, includes all the radiation, direct and diffuse, incident on a horizontal plane. The distribution of diffuse radiation is not uniform over the whole sky hemisphere and is more intense from a zone of about 5 degrees radius surrounding the sun. This is known as circumsolar radiation. Radiation may also be reflected from the ground onto any inclined surface, though this is very difficult to assess. The relationship between direct radiation, I, the diffuse radiation, D, and the global radiation, G, is given by:

$$G = D + I \sin \gamma$$

where γ is the solar altitude above the horizon.

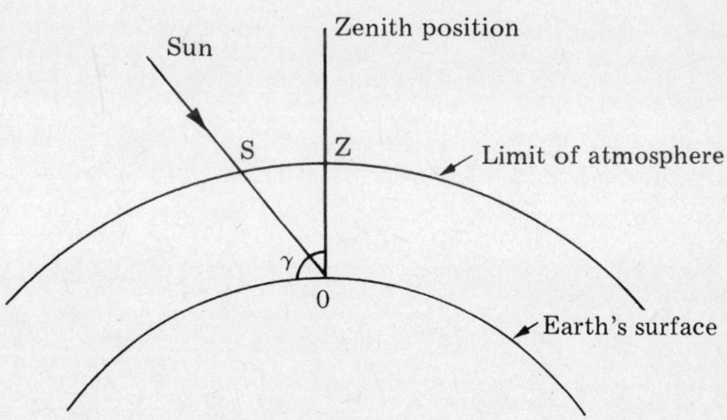

From J. C. McVeigh, Sun Power: An Introduction to the Applications of Solar Energy *(Oxford: Pergamon Press, Ltd.), 1977, p. 12.*

6.1 ANGLES AND THEIR MEASUREMENT

The word "trigonometry" literally means "triangle measure." As a result, this section begins with an examination of angles and triangles. When two half lines meet at a point, an **angle** is formed.

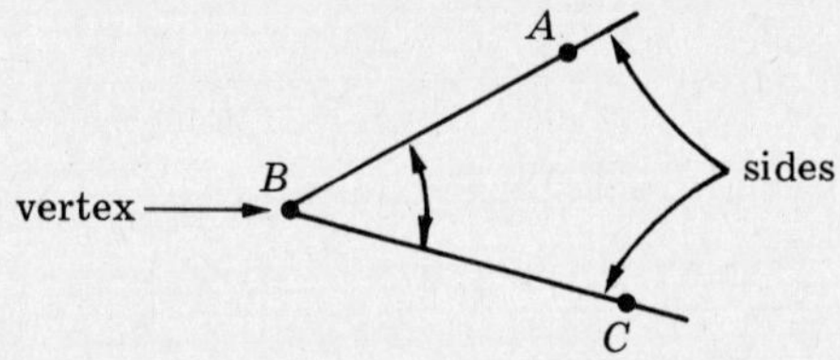

The "angle ABC" is written as $\measuredangle ABC$ or just $\measuredangle B$ if there is only one angle under consideration whose vertex is at B. The size of an angle is measured by considering the amount of rotation about the vertex that is required to cause one of its sides to coincide with the other. One side (either one) is called the **initial side** and the other is called the **terminal side**. There are two systems for measuring angles that are commonly in use.

DEFINITION

Degree Measurement of Angles

1 **degree**, denoted 1°, is the measurement of an angle that is $\frac{1}{360}$th of a revolution about any circle.

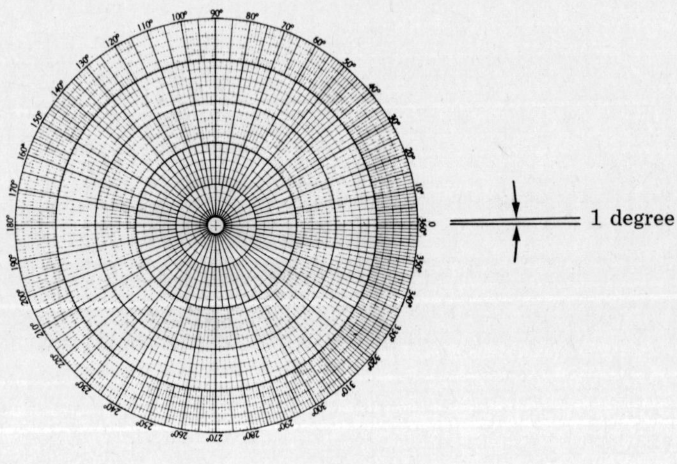

Since 1° is the measurement of an angle that is $\frac{1}{360}$ of a circle, a whole circle has a measure of 360°, a semi-circle has a measure of 180°, and a quarter circle has a measure of 90°. The following diagrams depict various angles.

Straight Angle

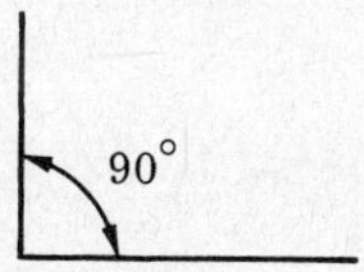

Right Angle

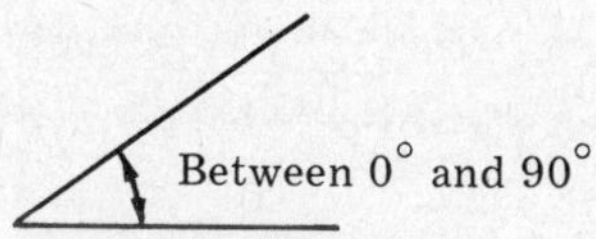

Acute Angle

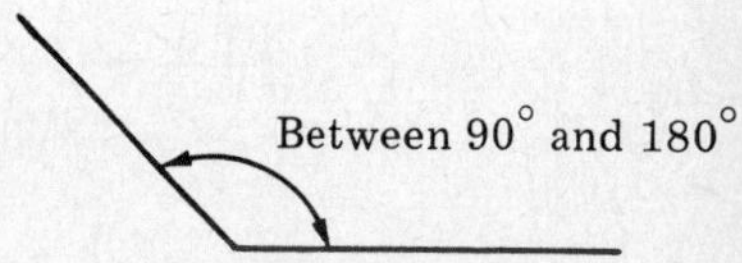

Obtuse Angle

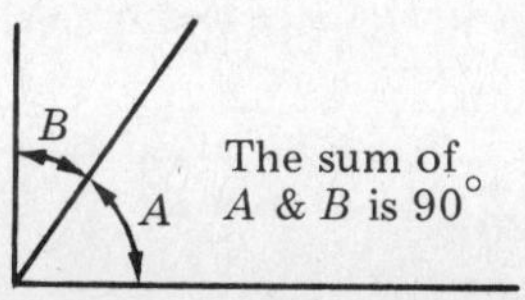

Complementary Angles A & B

(A is the *complement* of B and vice versa.)

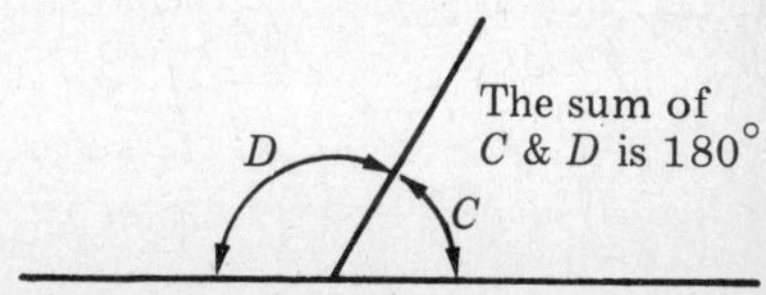

Supplementary Angles C & D

(C is the *supplement* of D and vice versa.)

DEFINITION

Radian Measurement of Angles

1 **radian,** is the measurement of an angle that is $\dfrac{1}{2\pi}$ th of a complete circumference of any circle.

Since circumference, C, is given by $C = 2\pi r$,

$$1 \text{ radian} = \frac{1}{2\pi}(2\pi r) = r$$

(*on the circumference*).

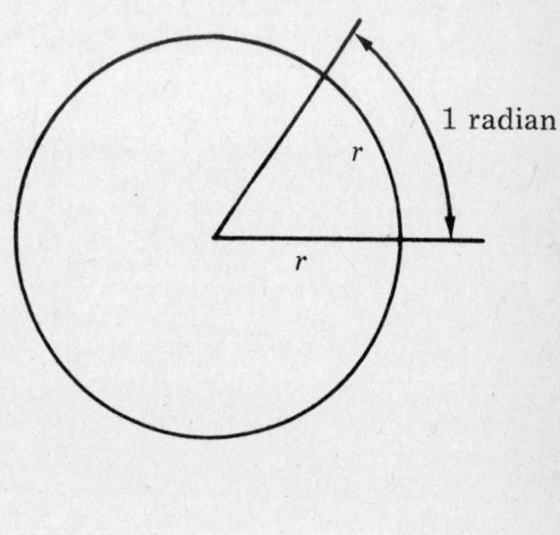

In other words, one radian is the size of an angle whose vertex is at the center of any circle and whose sides intersect (or *subtend*)

an arc on the circle where the length of the arc equals the radius of the circle. Since 1 radian is the measurement of an angle that is $\frac{1}{2\pi}$ th of a whole circle, a whole circle has a measure of 2π radians. That is, if the radius of a circle is repeatedly "laid down" end to end on the (curved) circumference there will be $2\pi \approx 6.28$ radii present.

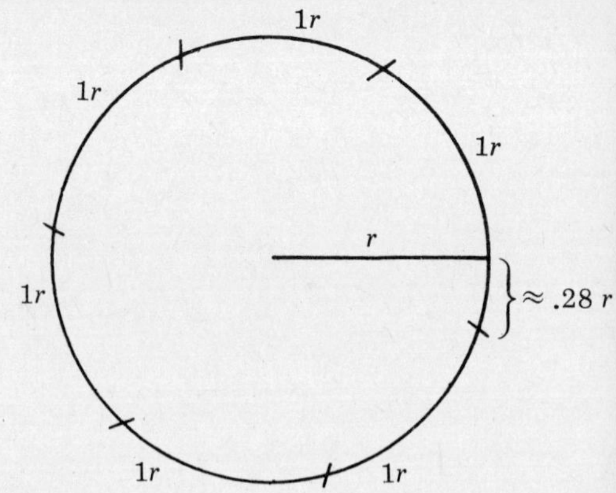

Thus, there are two ways to visualize the amount of rotation between the initial and terminal sides of an angle.

1. In degree measure an angle is visualied as chopped into a certain number of little "wedge shaped" regions called degrees.

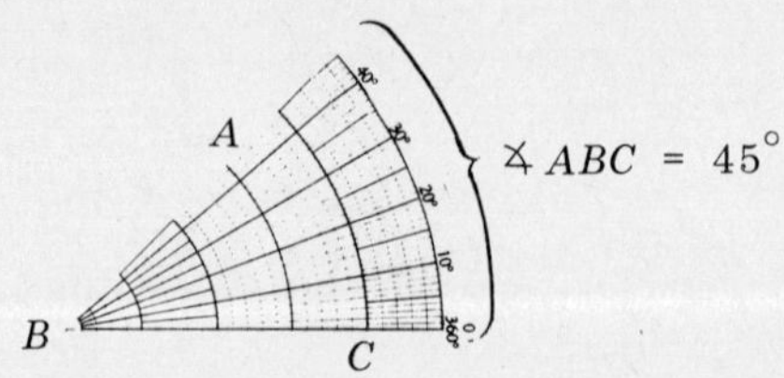

2. In radian measure an angle is visualized as a *ratio* of arc length to radius.

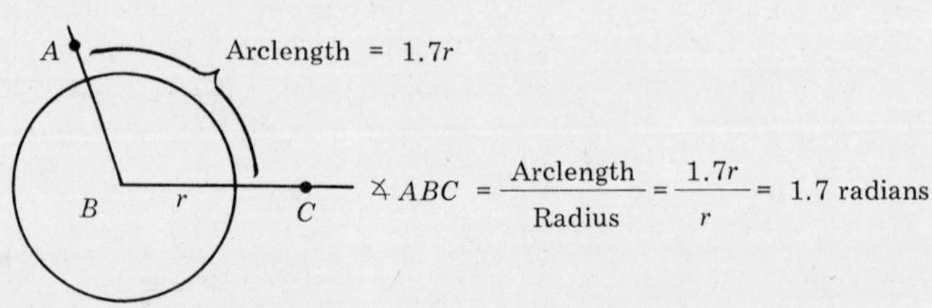

Since one complete trip around any circle is 360° or 2π radians we surely can say that $360^\circ = 2\pi$ radians. Multiplying both sides of this statement by an appropriate quantity produces other degree-radian equations:

> **Relationship Between Radians and Degrees**
>
> $$1^\circ = \frac{\pi}{180} \text{ radians} \approx 0.01745 \text{ radians}$$
>
> $$1 \text{ radian} = \frac{360^\circ}{2\pi} \approx 57.2958^\circ$$

The two equations above are now used to convert between the two systems of angle measurement.

EXAMPLE 1 a. Convert 90° to radians. b. Convert 135° to radians.

c. Convert $\frac{7\pi}{6}$ radians to degrees.

d. Convert $\frac{\pi}{4}$ radians to degrees.

Solution a. To change degrees to radians, multiply by $\frac{\pi}{180}$:

$$90^\circ = 90\left(\frac{\pi}{180}\right) = \frac{\pi}{2} \text{ radians}$$

b. $135^\circ = 135\left(\frac{\pi}{180}\right) = \frac{3\pi}{4}$ radians

c. To change radians to degrees, multiply by $\frac{180}{\pi}$:

$$\frac{7\pi}{6} \text{ radians} = \frac{7\pi}{6}\left(\frac{180}{\pi}\right) = 210^\circ$$

d. $\frac{\pi}{4}$ radians $= \frac{\pi}{4}\left(\frac{180}{\pi}\right) = 45^\circ$

Both measuring systems mentioned above place the vertex of the angle in question at the center of a circle. Such an angle is called a **central angle**. Suppose a central angle, θ, (theta) subtends an arclength s in a circle of radius r. Since the radian measure of an angle is a ratio of arclength to radius, then $\theta = \frac{s}{r}$, where θ is *expressed in radians.* We have proven the following theorem.

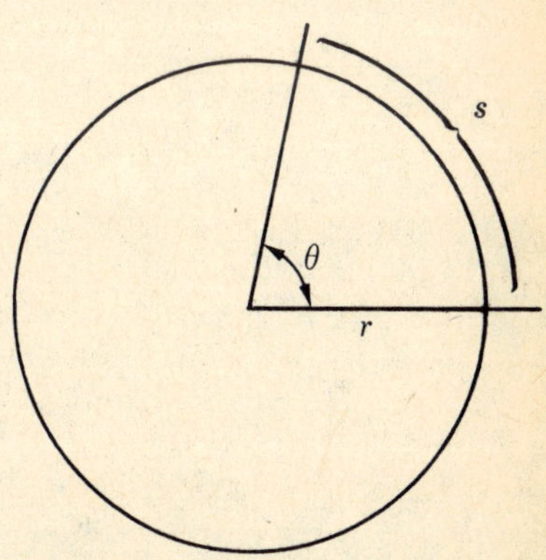

THEOREM 1

Arclength Formula

If a central angle θ in a circle of radius r subtends an arclength s, then

$$\theta = \frac{s}{r}, \text{ or equivalently, } s = \theta r,$$

where θ is *expressed in radians.*

EXAMPLE 2 Suppose a central angle θ subtends an arc that is 12 centimeters long in a circle of radius 5 centimeters.

a. Determine the radian measure of θ.
b. Determine the degree measure of θ.

Solution a. $\theta = \dfrac{s}{r} = \dfrac{12 \text{ cm}}{5 \text{ cm}} = 2.4$ radians

b. 2.4 radians $= 2.4\left(\dfrac{180°}{\pi}\right) \approx 137.5°$

In the formula, $\theta = \dfrac{s}{r}$, the ratio of s to r is a ratio of two distances. As a result, θ is a unitless or dimensionless quantity. Hence we can write $\theta = 2.4$ in place of $\theta = 2.4$ radians. A word of caution: the angles α (alpha) and β (beta) where $\alpha = 4$ and $\beta = 4°$ are *not* the same: $\alpha = 4$ means 4 radians and $\beta = 4°$ means 4 degrees.

EXAMPLE 3 a. Determine the arclength subtended by a central angle of 2 (radians) in a circle with radius 4 inches.
b. Determine the arclength subtended by a central angle of 150° in a circle with radius 4 inches.

Solution a.

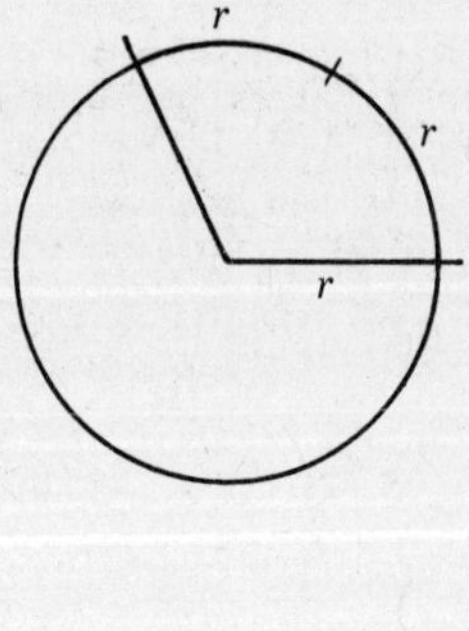

$s = \theta r = (2)(4)$

$s = 8$ inches

b.

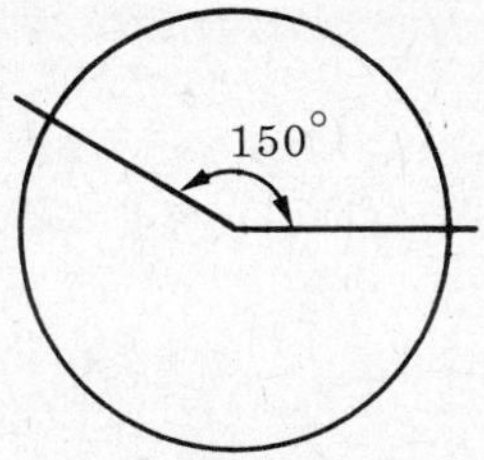

$$150° = 150\left(\frac{\pi}{180}\right) = \left(\frac{5\pi}{6}\right)\text{(radians)}$$

$$s = \theta r = \left(\frac{5\pi}{6}\right)(4)$$

$$s \approx 10.47 \text{ inches}$$

Sometimes it is desireable to know the area, A, of a **circular sector** having central angle θ in a circle of radius r.

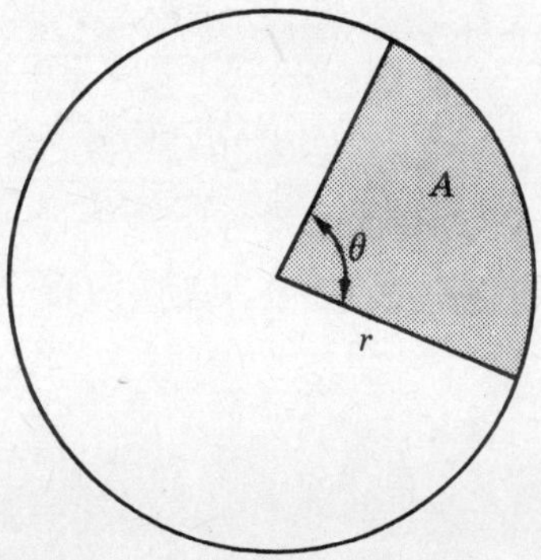

The following theorem has its proof outlined in Exercise 82.

THEOREM 2 The area, A, of a circular sector with central angle θ in a circle of radius r is $A = \frac{1}{2}r^2\theta$, where θ *is measured in radians.*

EXAMPLE 4 Determine the area of a circular sector having a central angle of 120° and a radius of 2 meters.

Solution $120° = 120\left(\frac{\pi}{180}\right) = \frac{2}{3}\pi$

$$A = \frac{1}{2}r^2\theta = \frac{1}{2}(2)^2\left(\frac{2}{3}\pi\right) = \frac{4\pi}{3} \text{ square meters}$$

From time to time fractional parts of angles will be encountered. One way to deal with this is to use decimals, as in 23.6°, 245.44°, 2.4 radians, etc. In the degree measurement system **minutes** and **seconds** are also used; 1 degree = 60 minutes and 1 minute = 60 seconds. Minutes are denoted by $'$ and seconds by $''$.

EXAMPLE 5
a. Convert $87^\circ 45'$ to decimal form.
b. Convert $15^\circ 15' 15''$ to decimal form.
c. Convert 67.6° to degree-minute-second form.
d. Convert 23.47° to degree-minute-second form.

Solution

a. $45' = \left(\frac{45}{60}\right)^\circ = 0.75^\circ$ Thus, $87^\circ 45' = 87.75^\circ$

b. $15' = \left(\frac{15}{60}\right)^\circ = 0.25^\circ$
$15'' = \left(\frac{15}{3600}\right)^\circ = 0.0042^\circ$ Thus, $15^\circ 15' 15'' \approx 15.2542^\circ$

c. $0.6^\circ = 0.6(60)' = 36'$ Thus, $67.6^\circ = 67^\circ 36'$

d. $0.47^\circ = 0.47(60)' = 28.2'$
$0.2' = 0.2(60)'' = 12''$ Thus, $23.47^\circ = 23^\circ 28' 12''$

EXERCISES FOR SECTION 6.1

In Exercises 1 through 18, convert each degree measurement into radian measurement. (See Example 1.)

1. 45°	2. 90°	3. 135°	4. 180°
5. 225°	6. 270°	7. 315°	8. 360°
9. 30°	10. 60°	11. 120°	12. 150°
13. 240°	14. 330°	15. 100°	16. 40°
17. 140°	18. 190°		

In Exercises 19 through 35, convert each radian measurement into degree measurement. (See Example 1.)

19. $\frac{3\pi}{4}$	20. $\frac{\pi}{4}$	21. $\frac{5\pi}{4}$	22. $\frac{7\pi}{4}$
23. $\frac{5\pi}{6}$	24. $\frac{\pi}{9}$	25. $\frac{\pi}{12}$	26. $\frac{\pi}{18}$
27. $\frac{5\pi}{12}$	28. $\frac{\pi}{6}$	29. $\frac{2\pi}{3}$	30. $\frac{\pi}{3}$
31. $\frac{7\pi}{6}$	32. $\frac{13\pi}{18}$	33. π	34. $\frac{4\pi}{9}$
35. 2π			

In Exercises 36 through 45, determine the central angle θ by using the given values of arclength s and radius r. (See Example 2.)

36. $s = 20$ in, $r = 10$ in, find θ in radians
37. $s = 20$ in, $r = 10$ in, find θ in degrees
38. $s = 6$ cm, $r = 1$ cm, find θ in radians
39. $s = 6$ cm, $r = 1$ cm, find θ in degrees
40. $s = 2$ ft, $r = 6$ in, find θ in radians

41. $s = 2$ ft, $r = 6$ in, find θ in degrees
42. $s = 8$ in, $r = 8$ in, find θ in radians
43. $s = 8$ in, $r = 8$ in, find θ in degrees
44. $s = 20$ cm, $r = 8$ cm, find θ in radians
45. $s = 20$ cm, $r = 8$ cm, find θ in degrees

In Exercises 46 through 55, determine the arclength, s, subtended in the circle with given values of central angle θ and radius r. (See Example 3.)

46. $\theta = 3, r = 6$ in
47. $\theta = 30^\circ, r = 6$ in
48. $\theta = \frac{\pi}{3}, r = 2$ in
49. $\theta = \pi, r = 5$ cm
50. $\theta = 45^\circ, r = 4$ cm
51. $\theta = 270^\circ, r = 3$ cm
52. $\theta = 2, r = 2$ in
53. $\theta = 4, r = 4$ in
54. $\theta = \frac{3\pi}{4}, r = 1$ in
55. $\theta = 2.5, r = 2$ in

In Exercises 56 through 61, determine the area, A, of a circular sector having the given values of central angle θ and radius r. (See Example 4.)

56. $\theta = \frac{\pi}{4}, r = 2$ cm
57. $\theta = \frac{\pi}{6}, r = 2$ in
58. $\theta = 2, r = 6$ in
59. $\theta = 50^\circ, r = 1$ in
60. $\theta = 200^\circ, r = 3$ in
61. $\theta = 100^\circ, r = 3$ in

In Exercises 62 through 69, convert each angle as specified. (See Example 5.)

62. $45^\circ 30'$ into decimal form
63. 76.7° into degree-minute-second form
64. $144^\circ 30' 30''$ into decimal form
65. 76.72° into degree-minute-second form
66. $54^\circ 34'$ into decimal form
67. $30^\circ 14' 22''$ into decimal form
68. $120^\circ 20' 20''$ into decimal form
69. 10.125° into degree-minute-second form

In Exercises 70 through 75, determine the complement of each angle.

70. 30°
71. 70°
72. 30.5°
73. $70^\circ 30'$
74. 56°
75. 84°

In Exercises 76 through 81, determine the supplement of each angle.

76. 60°
77. 90°
78. 45°
79. 135°
80. $50^\circ 30'$
81. $40^\circ 20' 10''$

82. Theorem 2 can be proven by setting up the proportion

$$\frac{\text{Area of sector}}{\text{Area of circle}} = \frac{\theta}{2\pi}.$$

Complete the proof.

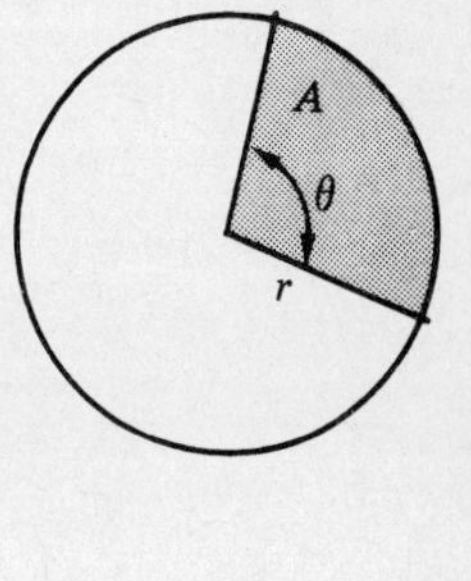

6.2 TRIANGLES AND TRIGONOMETRIC RATIOS

This section begins with a brief view of some basic facts about **triangles**. Recall, of course, that a triangle is the geometric figure formed when any three noncollinear points are connected with straight line segments. The sides of a triangle are often labeled with lower-case letters while the three interior angles are often labeled with upper-case letters, such that side a is opposite $\measuredangle A$, side b is opposite $\measuredangle B$, and side c is opposite $\measuredangle C$.

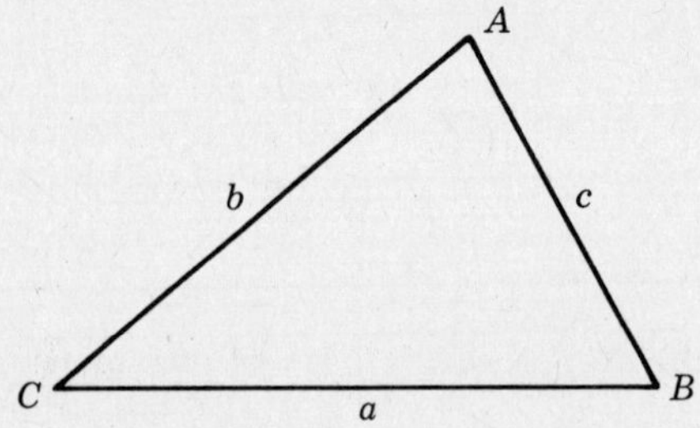

Triangle ABC or ΔABC

Some special triangles are depicted below.

Isoceles Triangle	**Equilateral Triangle**	**Right Triangle**
Two sides are equal.	All three sides are equal.	Has one right angle.

Congruent Triangles: Have the same size and shape.
Corresponding angles are *equal.*
Corresponding sides are *equal.*

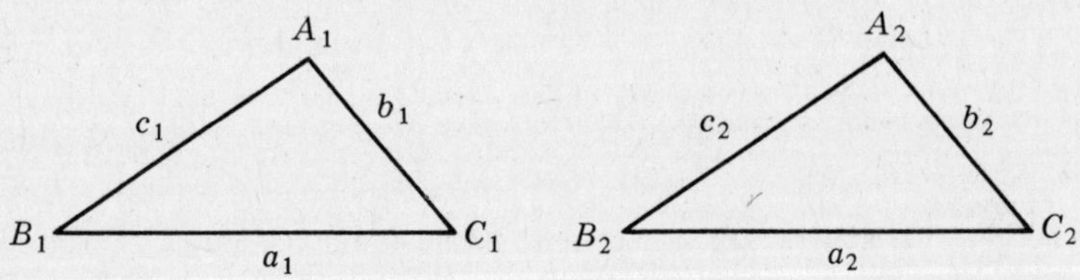

$$\measuredangle A_1 = \measuredangle A_2 \qquad a_1 = a_2$$

$$\measuredangle B_1 = \measuredangle B_2 \qquad b_1 = b_2$$

$$\measuredangle C_1 = \measuredangle C_2 \qquad c_1 = c_2$$

Similar Triangles: Have the same shape.
Corresponding angles are *equal.*
Corresponding sides are *in proportion.*

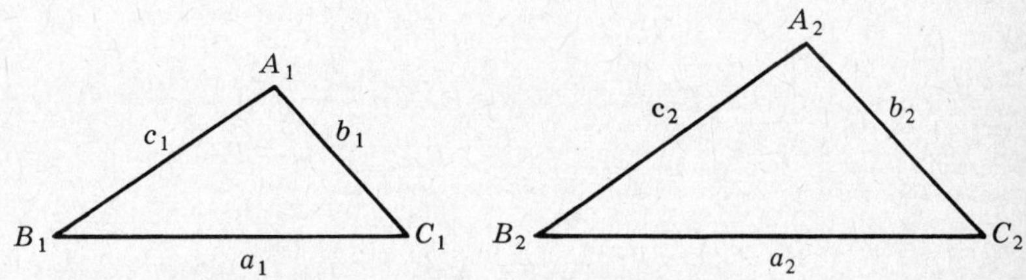

$$\angle A_1 = \angle A_2$$
$$\angle B_1 = \angle B_2$$
$$\angle C_1 = \angle C_2$$

$$\frac{a_1}{a_2} = \frac{b_1}{b_2} \quad \text{or} \quad \frac{a_1}{b_1} = \frac{a_2}{b_2}$$

$$\frac{b_1}{b_2} = \frac{c_1}{c_2} \quad \text{or} \quad \frac{b_1}{c_1} = \frac{b_2}{c_2}$$

$$\frac{a_1}{a_2} = \frac{c_1}{c_2} \quad \text{or} \quad \frac{a_1}{c_1} = \frac{a_2}{c_2}$$

Right Triangles:

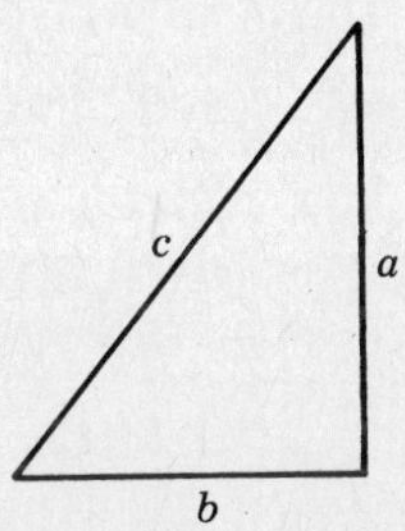

The **Pythagorean Theorem** states that in any right triangle with hypotenuse c, and sides (legs) a and b:

$$c^2 = a^2 + b^2$$

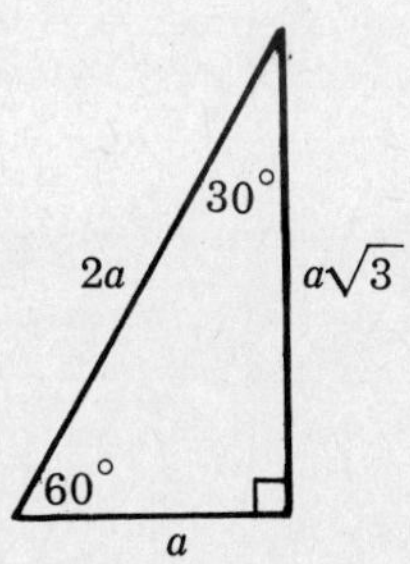

30-60-90 Triangle

In any 30-60-90 triangle, the hypotenuse is *always* twice as long as the shorter leg. Note the relationship among the sides in the picture.

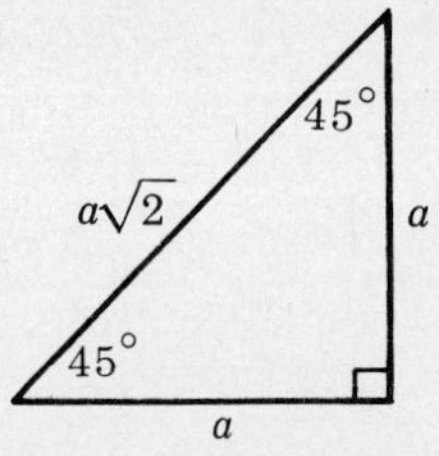

45-45-90 Triangle

In any 45-45-90 triangle, the legs are equal. Note the relationship among the sides in the picture.

Interior Angles: In *any* triangle the sum of all three interior angles is 180°.

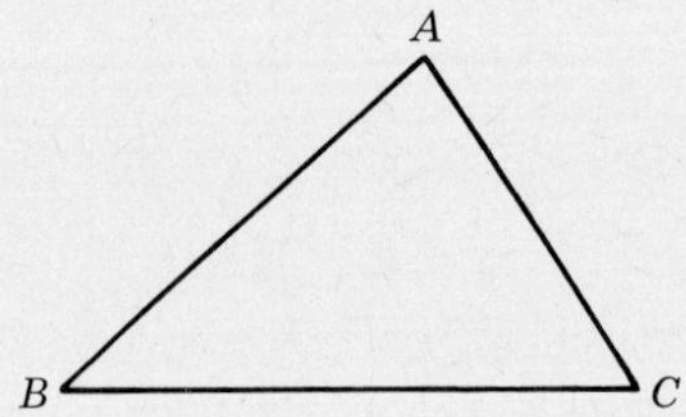

$\measuredangle A + \measuredangle B + \measuredangle C = 180°$

For the remainder of this section we turn our attention to right triangles and the ratio of one side to another. In the right triangle ABC, there are six possible ratios of one side to another, namely, $\frac{a}{c}, \frac{b}{c}, \frac{a}{b}, \frac{c}{a}, \frac{c}{b}, \frac{b}{a}$.

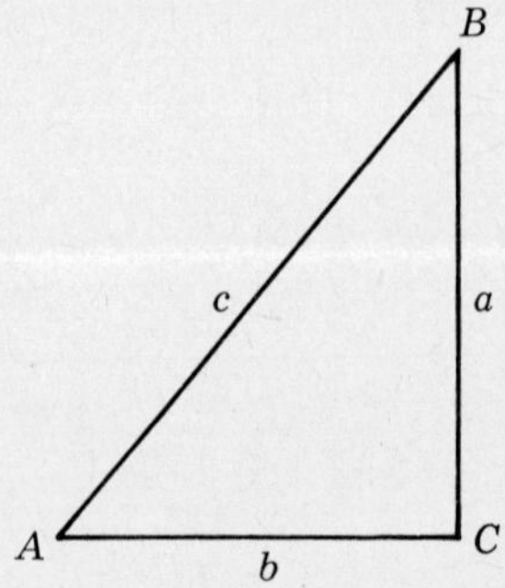

In the definition that follows, each of these ratios is given a name. The Greek letter θ is used to represent one of the acute angles.

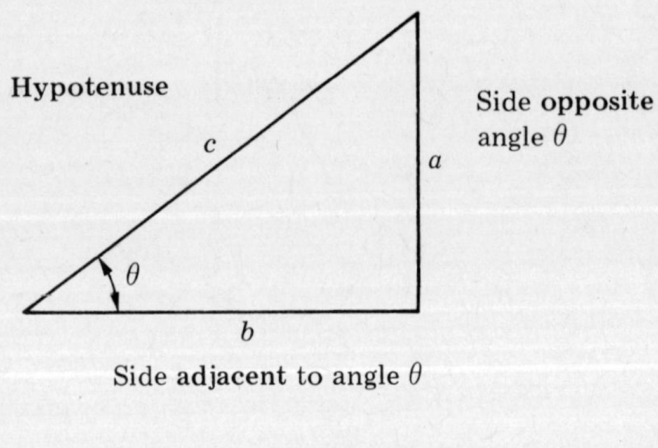

DEFINITION

sine $\theta = \dfrac{a}{c}$	or	$\sin\theta = \dfrac{\text{opposite}}{\text{hypotenuse}}$
cosine $\theta = \dfrac{b}{c}$	or	$\cos\theta = \dfrac{\text{adjacent}}{\text{hypotenuse}}$
tangent $\theta = \dfrac{a}{b}$	or	$\tan\theta = \dfrac{\text{opposite}}{\text{adjacent}}$
cosecant $\theta = \dfrac{c}{a}$	or	$\csc\theta = \dfrac{\text{hypotenuse}}{\text{opposite}}$
secant $\theta = \dfrac{c}{b}$	or	$\sec\theta = \dfrac{\text{hypotenuse}}{\text{adjacent}}$
cotangent $\theta = \dfrac{b}{a}$	or	$\cot\theta = \dfrac{\text{adjacent}}{\text{opposite}}$

This definition will now be used to determine the values of the trigonometric ratios for some particular angles.

EXAMPLE 1 Determine the values of each of the six trigonometric ratios of the angle whose measure is 45°.

Solution We use the 45-45-90 triangle at the right.

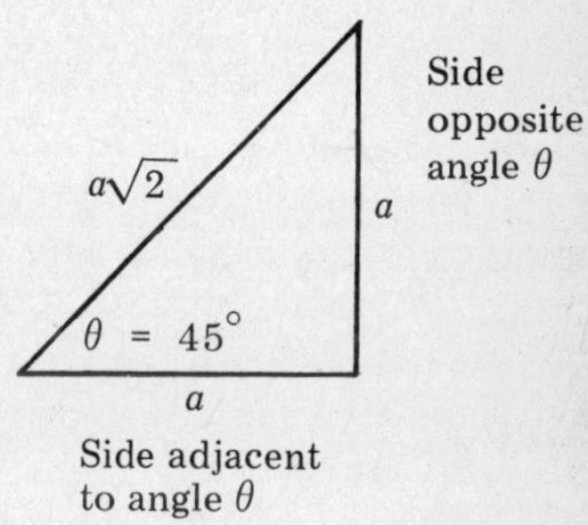

$$\sin 45^\circ = \frac{\text{opp}}{\text{hyp}} = \frac{a}{a\sqrt{2}} = \frac{1}{\sqrt{2}} = \frac{\sqrt{2}}{2}$$

$$\cos 45^\circ = \frac{\text{adj}}{\text{hyp}} = \frac{a}{a\sqrt{2}} = \frac{1}{\sqrt{2}} = \frac{\sqrt{2}}{2}$$

$$\tan 45^\circ = \frac{\text{opp}}{\text{adj}} = \frac{a}{a} = 1$$

$$\csc 45^\circ = \frac{\text{hyp}}{\text{opp}} = \frac{a\sqrt{2}}{a} = \sqrt{2}$$

$$\sec 45^\circ = \frac{\text{hyp}}{\text{adj}} = \frac{a\sqrt{2}}{a} = \sqrt{2}$$

$$\cot 45^\circ = \frac{\text{adj}}{\text{opp}} = \frac{a}{a} = 1$$

EXAMPLE 2 Determine the values of each of the six trigonometric ratios of a 30° angle.

Solution We use the 30-60-90 triangle at the right.

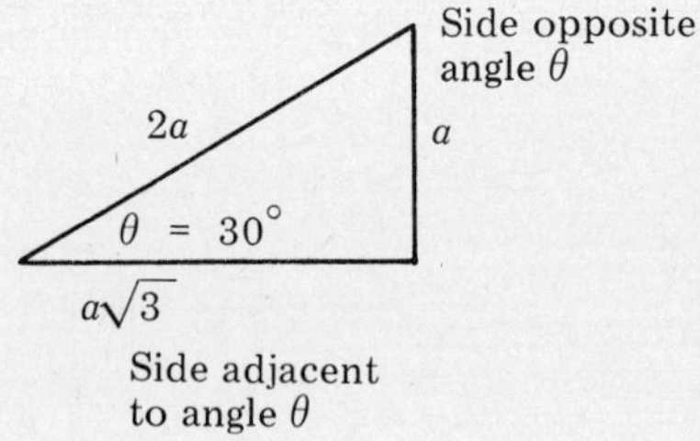

$$\sin 30° = \frac{a}{2a} = \frac{1}{2}$$

$$\cos 30° = \frac{a\sqrt{3}}{2a} = \frac{\sqrt{3}}{2}$$

$$\tan 30° = \frac{a}{a\sqrt{3}} = \frac{1}{\sqrt{3}} = \frac{\sqrt{3}}{3}$$

$$\csc 30° = \frac{2a}{a} = 2$$

$$\sec 30° = \frac{2a}{a\sqrt{3}} = \frac{2}{\sqrt{3}} = \frac{2\sqrt{3}}{3}$$

$$\cot 30° = \frac{a\sqrt{3}}{a} = \sqrt{3}$$

The results in Examples 1 and 2 could be written in decimal form, but for the time being we leave answers in radical form (denominators rationalized). Also, any of the results could be written by replacing the degree measure by the radian measure. That is, $\cos 30° = \frac{\sqrt{3}}{2}$ or $\cos\frac{\pi}{6} = \frac{\sqrt{3}}{2}$, etc. The trigonometric ratios of 0° will be examined closer later in this chapter but for the present, they will be determined by imagining the side a in the right triangle collapsed to 0 in which case $\theta = 0°$ and $b = c$. The ratios would then be:

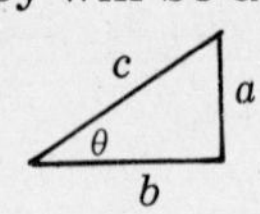

$$\sin 0° = \frac{a}{c} = \frac{0}{c} = 0 \qquad \csc 0° = \frac{c}{a} = \frac{c}{0} = \text{undefined}$$

$$\cos 0° = \frac{b}{c} = \frac{c}{c} = 1 \qquad \sec 0° = \frac{c}{b} = \frac{c}{c} = 1$$

$$\tan 0° = \frac{a}{b} = \frac{0}{b} = 0 \qquad \cot 0° = \frac{b}{a} = \frac{b}{0} = \text{undefined}$$

In a similar way, the trigonometric ratios of 90° can be determined by imagining the base, b, in the right triangle collapsed to 0 in which case $\theta = 90°$ and $a = c$. The values of the trigonometric ratios calculated so far, along with those for 60° and 90°, are summarized on the following page. The reader is asked to verify those for 60° and 90° in Exercises 28 and 29.

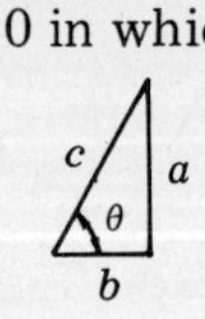

θ (radians)	θ (degrees)	$\sin\theta$	$\csc\theta$	$\cos\theta$	$\sec\theta$	$\tan\theta$	$\cot\theta$
0	$0°$	0	undefined	1	1	0	undefined
$\frac{\pi}{6}$	$30°$	$\frac{1}{2}$	2	$\frac{\sqrt{3}}{2}$	$\frac{2\sqrt{3}}{3}$	$\frac{\sqrt{3}}{3}$	$\sqrt{3}$
$\frac{\pi}{4}$	$45°$	$\frac{\sqrt{2}}{2}$	$\sqrt{2}$	$\frac{\sqrt{2}}{2}$	$\sqrt{2}$	1	1
$\frac{\pi}{3}$	$60°$	$\frac{\sqrt{3}}{2}$	$\frac{2\sqrt{3}}{3}$	$\frac{1}{2}$	2	$\sqrt{3}$	$\frac{\sqrt{3}}{3}$
$\frac{\pi}{2}$	$90°$	1	1	0	undefined	undefined	0

The values above occur commonly in the study of trigonometry and should be memorized. Values of the trigonometric ratios of *any* angle between $0°$ and $90°$ can be calculated by using other methods and these values appear in decimal form in Table III. By using this table it is possible to find any of the trigonometric ratios of an angle from $0.0°$ to $90.0°$ as follows.

$$0.0° \leqslant \theta \leqslant 45.0°$$

To determine $\sin 37.6°$: In the "θ column" *on the left*, find $37.6°$. *On the top* of the table find sin. The intersection gives:

$$\sin 37.6° = .6101*$$

θ (degrees)	s	$\sin\theta$ or $\sin s$	$\csc\theta$ or $\csc s$	$\cos\theta$ or $\cos\theta$
⋮	⋮	⋮	⋮	⋮
37.0	0.6458	0.6018	1.6616	0.7986
37.1	0.6475	0.6032	1.6578	0.7976
37.2	0.6493	0.6046	1.6540	0.7965
37.3	0.6510	0.6060	1.6502	0.7955
37.4	0.6528	0.6074	1.6464	0.7944
37.5	0.6545	0.6088	1.6427	0.7934
37.6	0.6562	0.6101	1.6390	0.7923
37.7	0.6580	0.6115	1.6353	0.7912
37.8	0.6597	0.6129	1.6316	0.7902
37.9	0.6615	0.6143	1.6279	0.7891

An electronic calculator with a sin key gives the slightly more precise value

$$\sin 37.6° = .61014516$$

*The tabulated decimals in Table III are mostly irrational numbers and are only expressed as approximations. Even so, when a value comes from Table III, we will use the "=" sign.

$$45.0° \leqslant \theta \leqslant 90.0°$$

To determine tan 77.3°: In the "θ column" *on the right*, find 77.3°. *On the bottom* of the table find tan. The intersection gives:

$$\tan 77.3° = 4.4374$$

1.0227	0.2144	4.6646	1.3596	77.9
1.0231	0.2162	4 6252	1.3579	77.8
1.0235	0.2180	4.5864	1.3561	77.7
1.0239	0.2199	4.5483	1.3544	77.6
1.0243	0.2217	4.5107	1.3526	77.5
1.0247	0.2235	4.4737	1.3509	77.4
1.0251	0.2254	4.4374	1.3491	77.3
1.0255	0.2272	4.4015	1.3474	77.2
1.0259	0.2290	4.3662	1.3456	77.1
1.0263	0.2309	4.3315	1.3439	77.0
⋮	⋮	⋮	⋮	⋮
csc θ or csc s	cot θ or cot s	tan θ or tan s	s	θ (degrees)

An electronic calculator with a tan key gives the slightly more precise value

tan 77.3° = 4.43734995

EXAMPLE 3 Determine each of the trigonometric ratios.

a. sin 89.2° b. sec 57.3° c. cot 12°

Solution a. sin 89.2° = .9999 b. sec 57.3° = 1.8510
c. cot 12° = 4.7046

One important use of trigonometric ratios is in determining the measure of the unknown sides or angles of right triangles as shown in the next two examples.

EXAMPLE 4 Determine the lengths of the unknown sides (to the nearest hundredth) and unknown angle in the right triangle ABC.

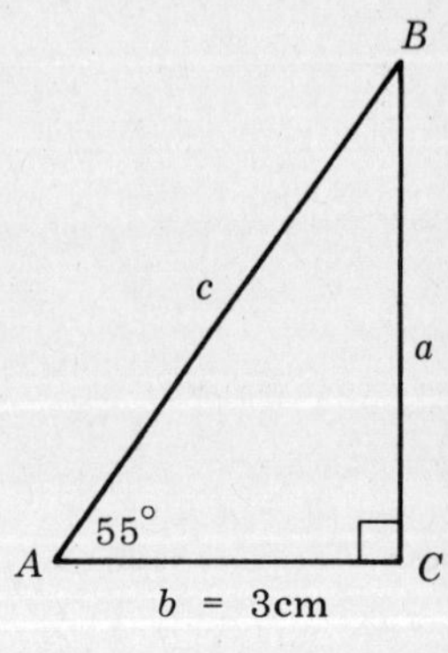

Solution Since $\measuredangle A + \measuredangle B + \measuredangle C = 180°$,

$$B = 180° - \measuredangle A - \measuredangle C = 35°$$

Next, let us determine side c. Since we know $b = 3$ cm, choose a trigonometric ratio involving b and c. There are various choices possible. One is:

$$\cos 55^\circ = \frac{b}{c}$$

$$.5736 = \frac{3\text{ cm}}{c}$$

$$c = \frac{3\text{ cm}}{.5736} = 5.23\text{ cm} \quad \text{(to the nearest hundredth)}$$

Side a can be found from the Pythagorean Theorem.

$$c^2 = a^2 + b^2$$

$$a = \sqrt{c^2 - b^2} = \sqrt{(5.23)^2 - 3^2} = 4.28\text{ cm} \quad \text{(to the nearest hundredth)}$$

In "solving a triangle" as in Example 4, there are often several choices possible. For instance, side a could have been found by utilizing $\tan B = \frac{b}{a}$. The reader is asked to explore such alternatives in Exercises 42 through 44.

EXAMPLE 5 Determine the length of the unknown side and each of the unknown angles in the right triangle DFG.

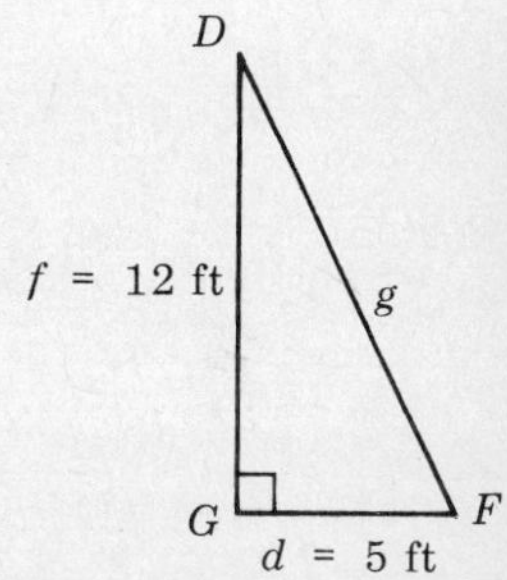

Solution Side g is found using the Pythagorean Theorem.

$$g^2 = d^2 + f^2 = 25 + 144 = 169$$
$$g = 13\text{ ft}$$

Continuing, $\sin D = \frac{5}{13} = .3846$. Now use Table III "in reverse." Find the four-place tabulated value in the sin column nearest to .3846. The associated angle is 22.6°.

$$\measuredangle D = 22.6^\circ$$
$$\measuredangle F = 180^\circ - \measuredangle G - \measuredangle D, \ \measuredangle F = 67.4^\circ$$

> Using a calculator, enter $\frac{5}{13} = .38461538$. Find $\measuredangle D$ by then pushing the keys [INV] [SIN] *
>
> $\measuredangle D = 22.619865^\circ$

If just one trigonometric ratio of an angle in a right triangle is known, all five others can be determined easily as shown in the next example.

*Variations exist from calculator to calculator; check your owner's manual.

EXAMPLE 6 Given that $\cos \alpha = \frac{2}{3}$, where α is an acute angle in a right triangle, find the other five trigonometric ratios of α.

Solution Sketch a right triangle with an acute angle α such that the ratio of adjacent to hypotenuse (relative to α) is 2:3.

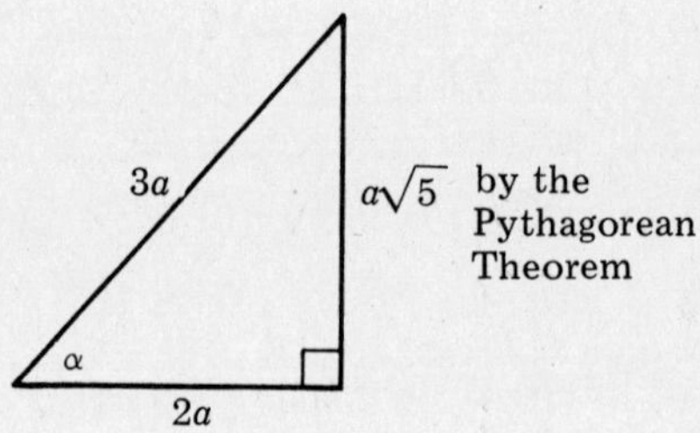

$$\sin \alpha = \frac{a\sqrt{5}}{3a} = \frac{\sqrt{5}}{3}$$

$$\tan \alpha = \frac{a\sqrt{5}}{2a} = \frac{\sqrt{5}}{2}$$

$$\csc \alpha = \frac{3a}{a\sqrt{5}} = \frac{3}{\sqrt{5}} = \frac{3\sqrt{5}}{5}$$

$$\sec \alpha = \frac{3a}{2a} = \frac{3}{2}$$

$$\cot \alpha = \frac{2a}{a\sqrt{5}} = \frac{2}{\sqrt{5}} = \frac{2\sqrt{5}}{5}$$

Notice also that we could now determine the trigonometric ratios of the *other* acute angle if desired.

EXERCISES FOR SECTION 6.2

In 1 through 9, determine the lengths of the two unknown sides in each triangle. (Use the fact that legs are equal in any 45-45-90 triangle and also the Pythagorean Theorem.)

1.

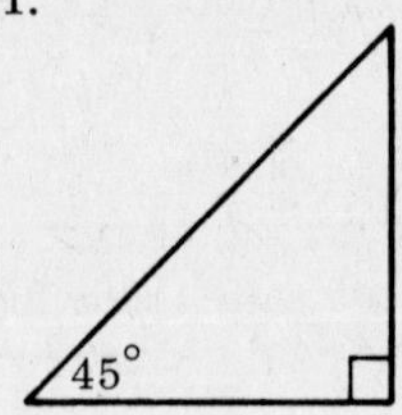

2.

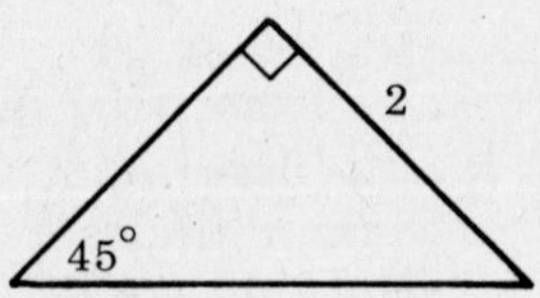

3.

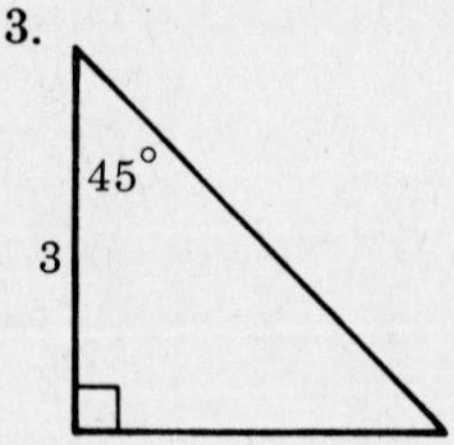

4.

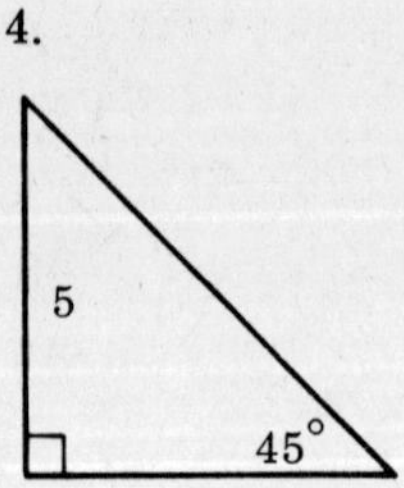

5.

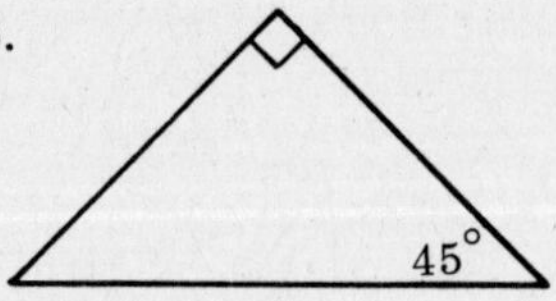

6.

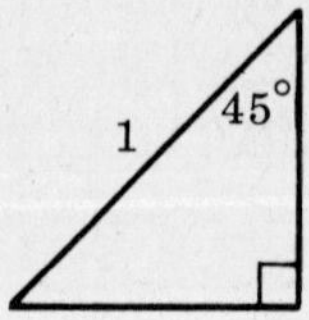

7. 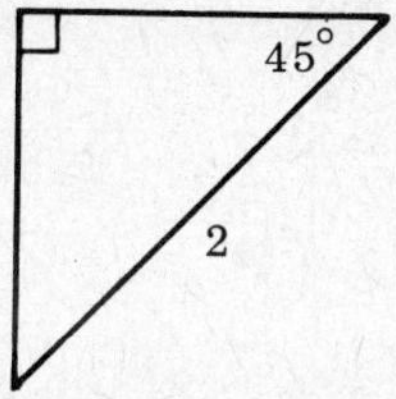

8. 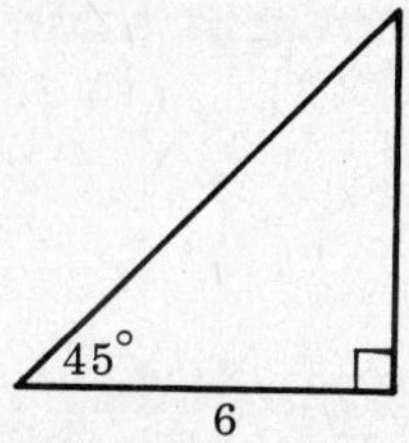

9. 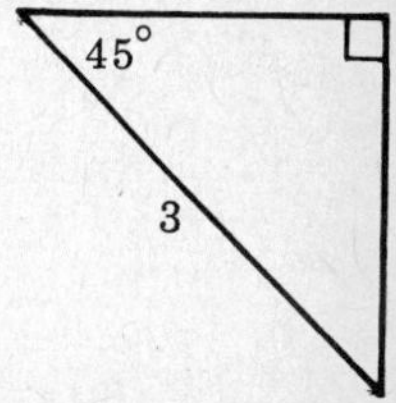

In 10 through 18, determine the lengths of the two unknown sides in each triangle. (Use the fact that the hypotenuse is twice the length of the shorter leg in any 30-60-90 triangle and also the Pythagorean Theorem.)

10. 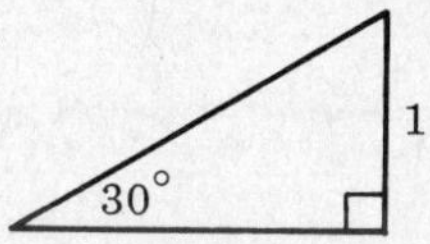

11. 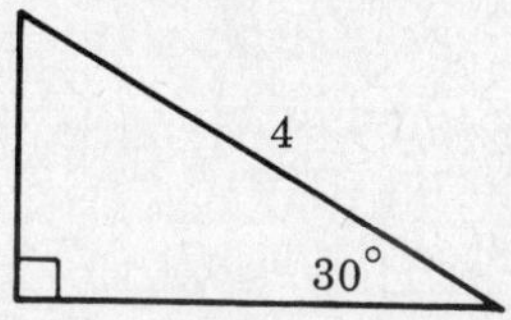

12. 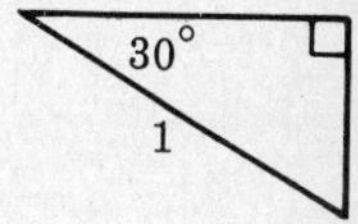

13. 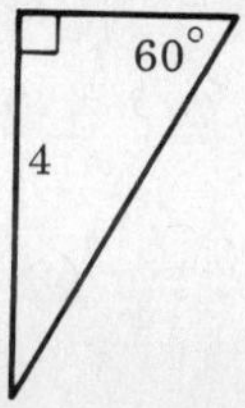

14. 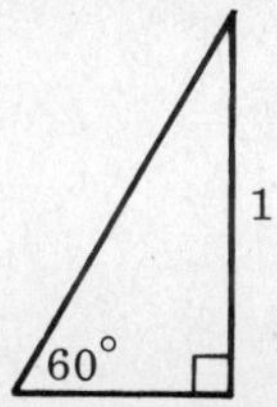

15. 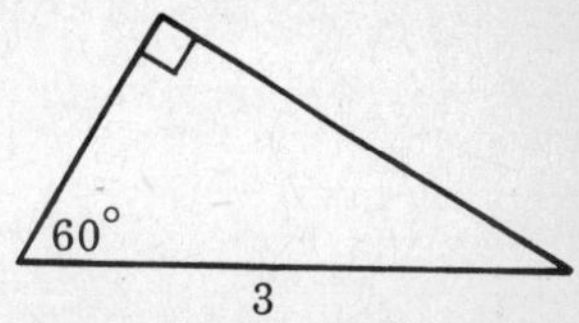

16. 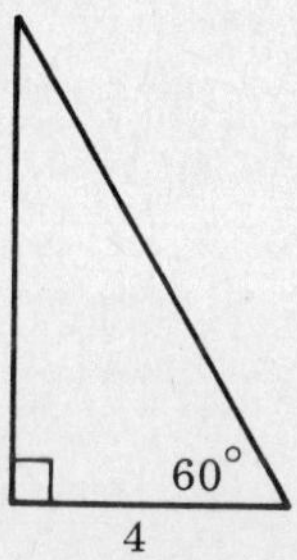

17. 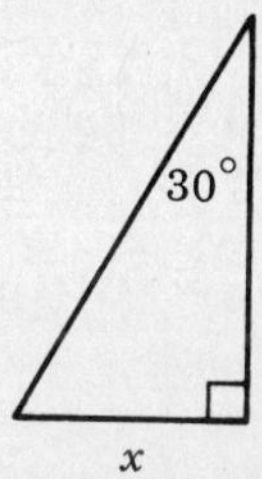

18. 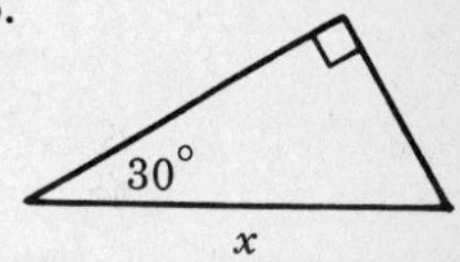

In 19 through 27, express the given trigonometric ratio in terms of the sides of the given triangle. (See definition of trigonometric ratios.)

19. 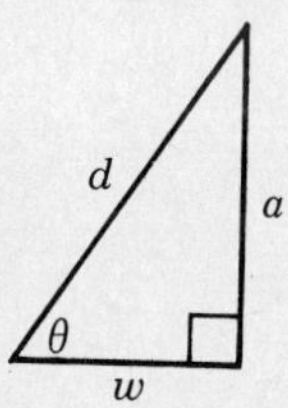

$\sin \theta = ?$

20. 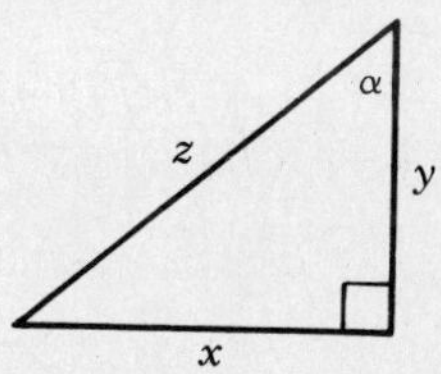

$\cos \alpha = ?$

21. 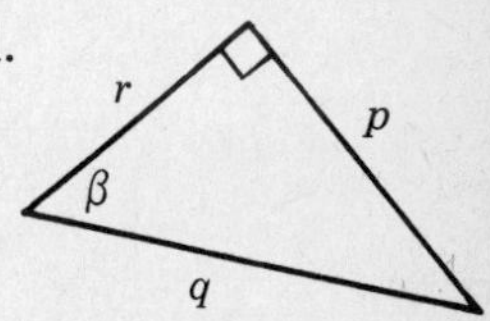

$\tan \beta = ?$

22.

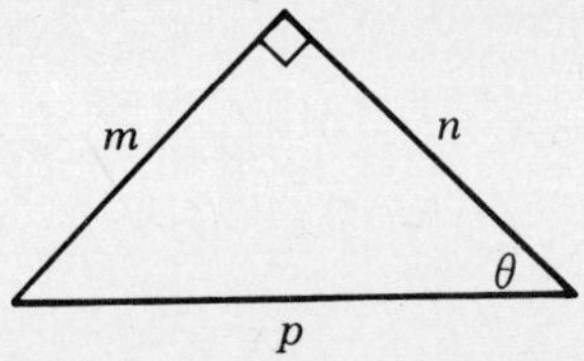

csc θ = ?

23.

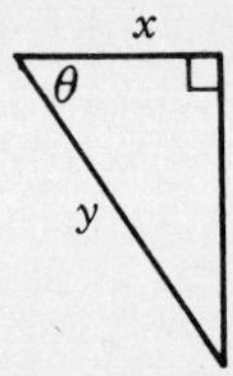

sec θ = ?

24.

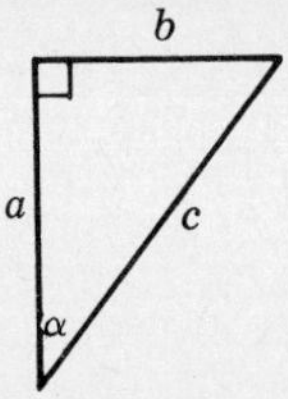

cot α = ?

25.

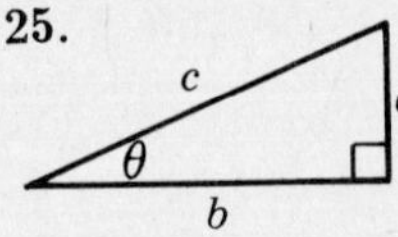

sin θ = ?

26.

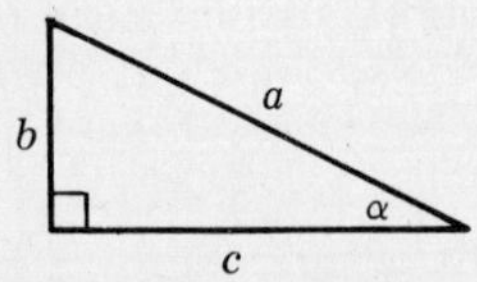

cos α = ?

27.

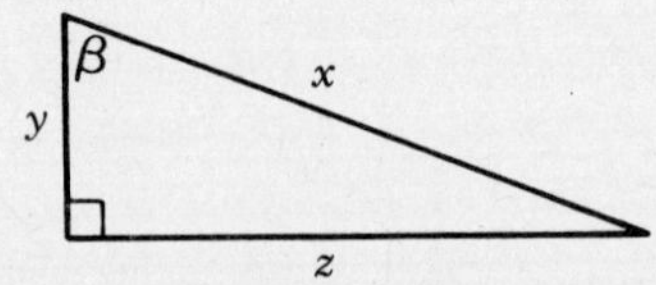

tan β = ?

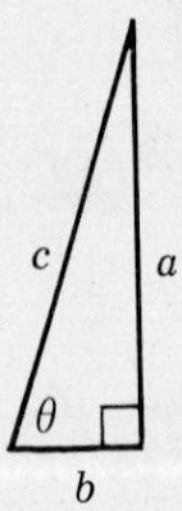

28. Determine each of the trigonometric ratios of a 60° angle. (See Examples 1 and 2.)

29. Determine each of the trigonometric ratios of a 90° angle by imagining the base, b, in the triangle at the left collapsed to 0.

In 30 through 41, determine each trigonometric ratio. (See Example 3.)

30. sin 29.8°	31. cos 82.3°	32. tan 82.3°	33. csc 25°
34. cot 75.5°	35. cos 10°	36. sec 90°	37. sin 67°
38. sec 40.1°	39. tan 10°	40. cot 15.6°	41. csc 90°

42. In Example 4, determine side a by using $\tan B = \frac{b}{a}$.

43. In Example 4, determine side c by using $\sec A = \frac{c}{b}$.

44. In Example 4, determine side a by using $\tan A = \frac{a}{b}$.

In 45 through 52, "solve each triangle"; that is, determine the unknown sides and angles in each given right triangle. Round off answers to the nearest tenth. (See Examples 4 and 5.)

45.

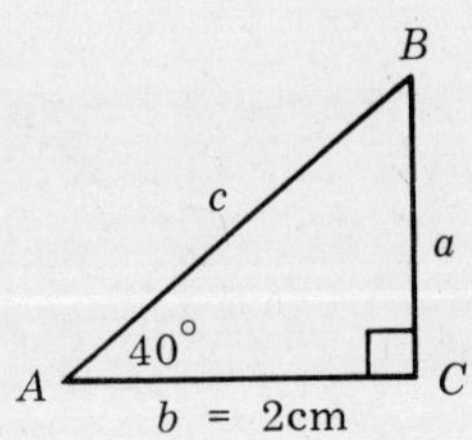

46.

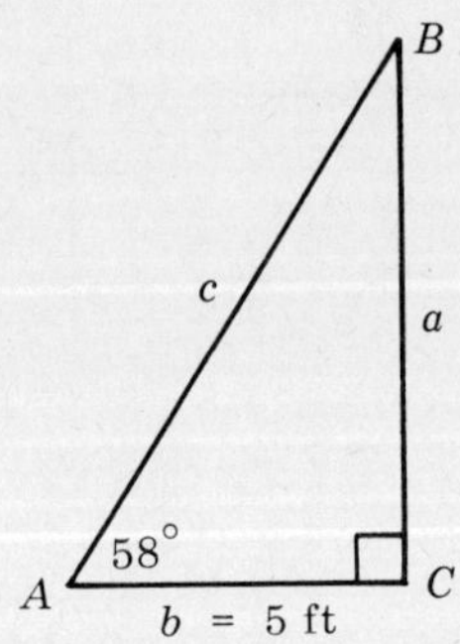

47.

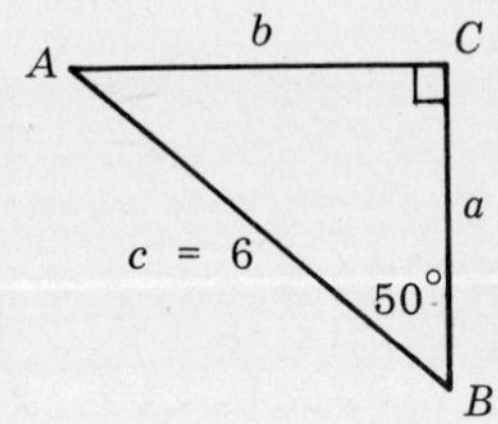

48.

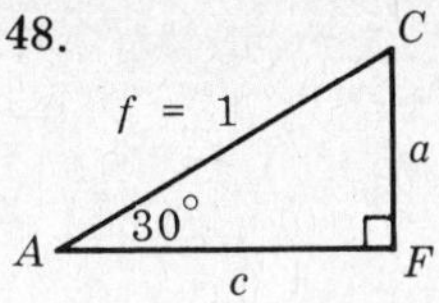

49.

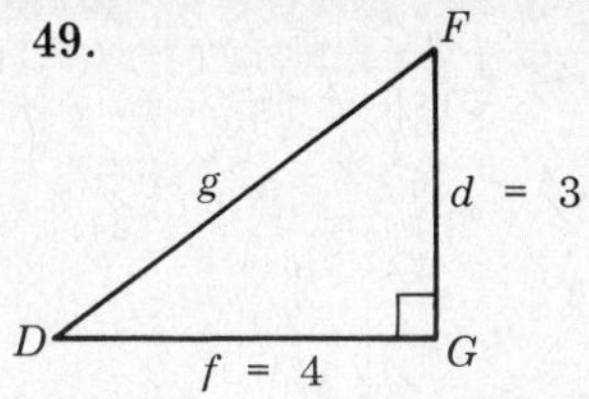

50.

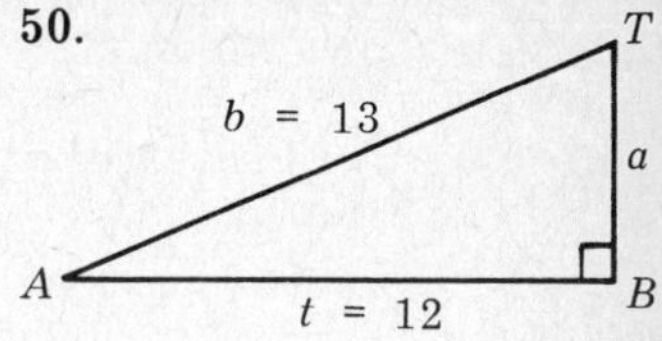

51.

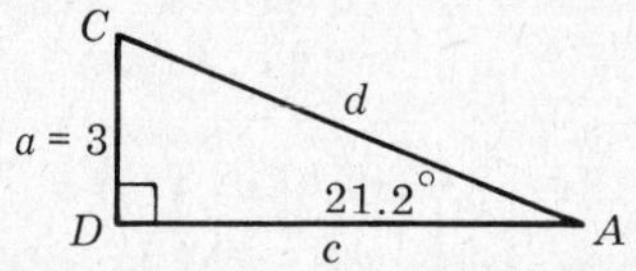

52.

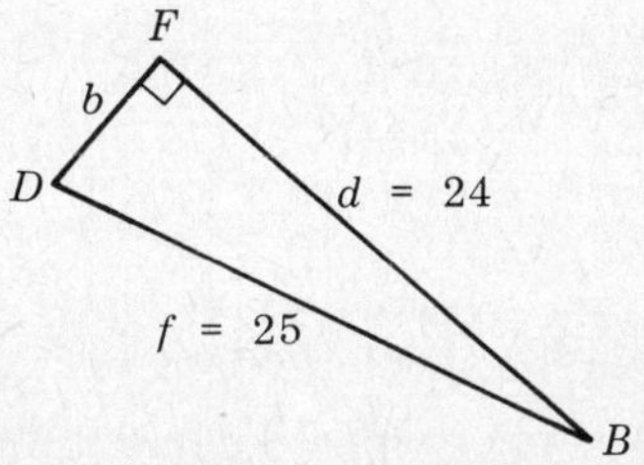

In 53 through 59, use the given trigonometric ratio of θ to determine each of the other trigonometric ratios of θ. (See Example 6.)

53. $\sin\theta = \frac{3}{5}$ **54.** $\cos\theta = \frac{3}{5}$ **55.** $\tan\theta = \frac{3}{5}$ **56.** $\csc\theta = \frac{25}{24}$

57. $\sin\theta = y$ 58. $\cos\theta = x$ 59. $\cot\theta = a$

60. Using the definitions of the trigonometric ratios, show that:

a. $\sin\theta = \dfrac{1}{\csc\theta}$ or equivalently $\csc\theta = \dfrac{1}{\sin\theta}$

b. $\cos\theta = \dfrac{1}{\sec\theta}$ or equivalently $\sec\theta = \dfrac{1}{\cos\theta}$

c. $\tan\theta = \dfrac{1}{\cot\theta}$ or equivalently $\cot\theta = \dfrac{1}{\tan\theta}$

d. $\tan\theta = \dfrac{\sin\theta}{\cos\theta}$ e. $\cot\theta = \dfrac{\cos\theta}{\sin\theta}$

Interpolation. Interpolation with trigonometric tables is performed in the same manner as with logarithm tables. For example to find sin 47.86°, write:

$$.10\left[\;.06\left[\begin{array}{l}\sin 47.8^\circ = .7408\\ \sin 47.86^\circ = \;?\end{array}\right]h\right.\\ \left.\sin 47.9^\circ = .7420\;\right] .0012$$

$$\frac{.06}{.10} = \frac{h}{.0012}$$

Thus, $h = .0007$ and $\sin 47.86^\circ = .7408 + .0007$

$$\sin 47.86^\circ = .7415$$

In Exercises 61 through 70, determine the value of each trigonometric ratio.

61. sin 42.63° 62. tan 42.63° 63. cos 42.63° 64. sec 59.48°

65. tan 59.48° 66. sin 59.48° 67. csc 59.48° 68. cot 89.25°

69. cos 89.25° 70. csc 89.25°

Calculator Exercises 71-80. Determine the value of each trigonometric ratio from Exercises 61-70.

6.3 TRIGONOMETRIC FUNCTIONS OF A REAL NUMBER

We begin this section by recalling the graph of the equation $x^2 + y^2 = 1$ (see Section 4.4, Exercise 46). The graph is a circle of radius 1 unit having center at the origin and, thus, is commonly referred to as the **unit circle**. Let the coordinates of a moveable point P on the unit circle be (x, y). We want to determine how far point P has moved *on the unit circle* having started from the point $(1,0)$. The distance traveled will be the arclength s on the unit circle, measured from $(1,0)$ and will be considered *positive* for counterclockwise rotation and *negative* for clockwise rotation.

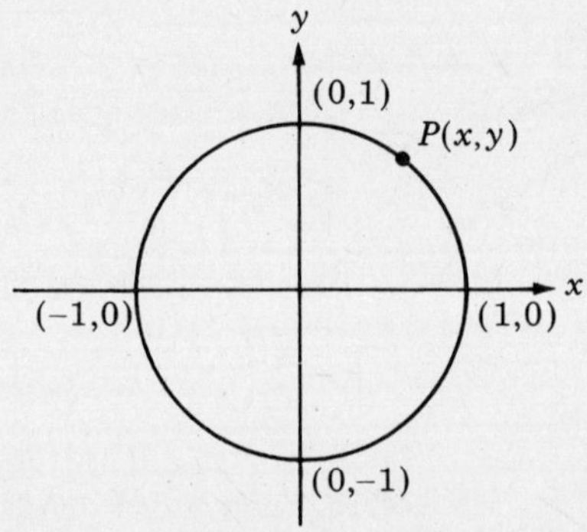

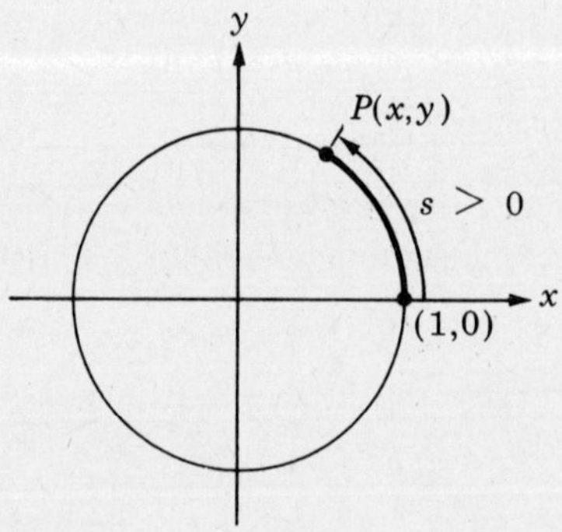

P moves counterclockwise (ccw)

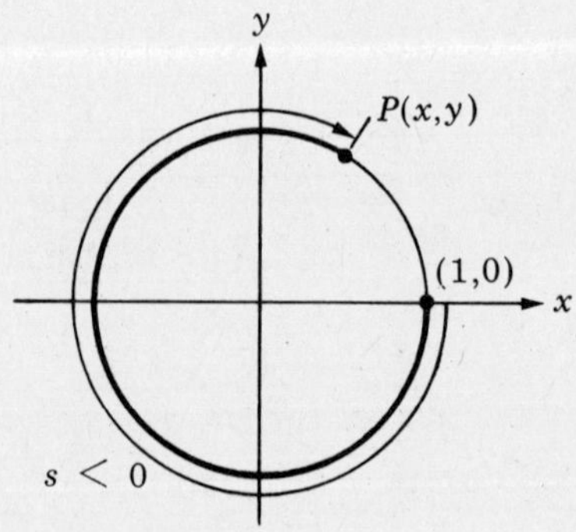

P moves clockwise (cw)

First, we will construct a table of values showing how x varies with s and also a table of values showing how y varies with s. Thus we need a way of measuring s and also a way of determining the x value and the y value for a known s. Recall that the circumference of the unit circle is $C = 2\pi r = 2\pi(1) = 2\pi$ units.

1/4 of a whole circumference (ccw)
$s = 1/4\ (2\pi) = \pi/2$

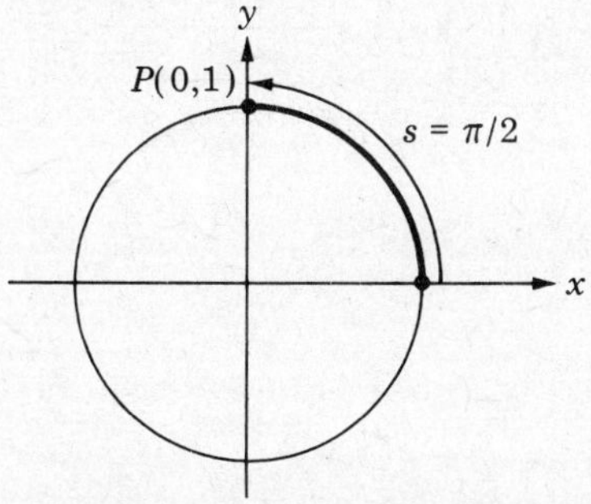

$s = \frac{\pi}{2},\ x = 0,\ y = 1$

1/2 of a whole circumference (ccw)
$s = 1/2\ (2\pi) = \pi$

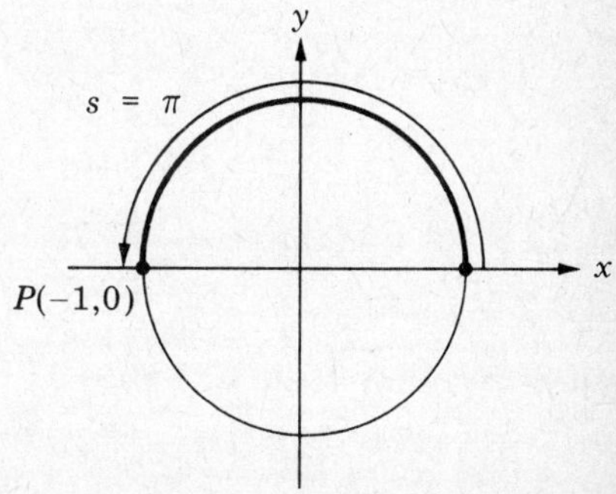

$s = \pi,\ x = -1,\ y = 0$

EXAMPLE 1 If P has moved $\frac{1}{8}$ of a whole circumference (ccw) what are the values of s, x, and y?

Solution For $\frac{1}{8}$ of a whole circumference (ccw)

$$s = \frac{1}{8}(2\pi) = \frac{\pi}{4}.$$

Also, $\frac{1}{8}$ of a whole circumference puts P on the 45° line bisecting the first quadrant, because $\frac{1}{8}(360°) = 45°$. Thus $x = y$. Substituting into the unit circle equation $x^2 + y^2 = 1$ gives

$$x^2 + x^2 = 1$$
$$2x^2 = 1$$
$$x = \frac{\sqrt{2}}{2} \quad \text{(Negative solution rejected—why?)}$$

Thus $$y = \frac{\sqrt{2}}{2}$$

Hence for $\frac{1}{8}$ of a whole circumference, $s = \frac{\pi}{4}, x = \frac{\sqrt{2}}{2}, y = \frac{\sqrt{2}}{2}$.

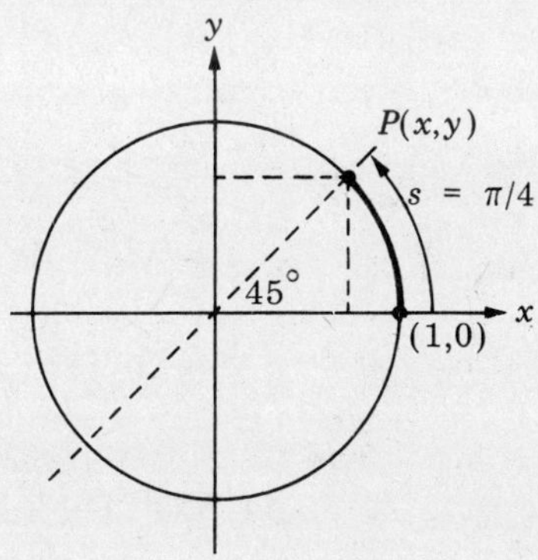

The results of Example 1 can be used to obtain certain additional corresponding values of s, x, and y.

$\frac{3}{8}$ of a whole circumference (ccw)

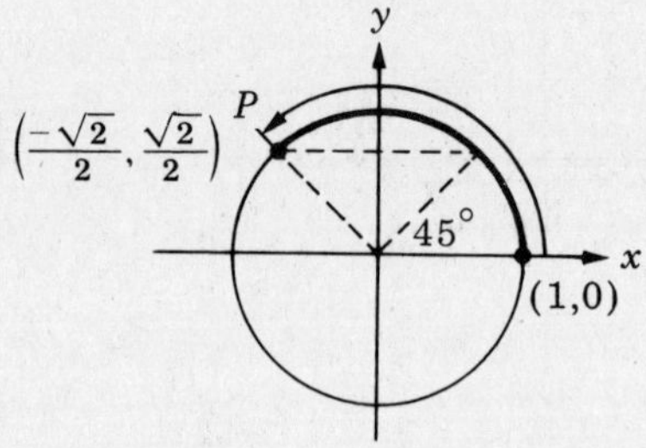

$s = \frac{3}{4}\pi,\ x = \frac{-\sqrt{2}}{2},\ y = \frac{\sqrt{2}}{2}$

$\frac{5}{8}$ of a whole circumference (ccw)

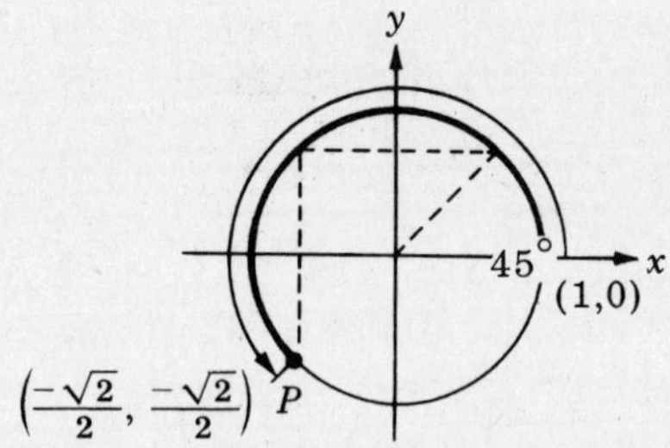

$s = \frac{5}{4}\pi,\ x = \frac{-\sqrt{2}}{2},\ y = \frac{-\sqrt{2}}{2}$

$\frac{7}{8}$ of a whole circumference (ccw)

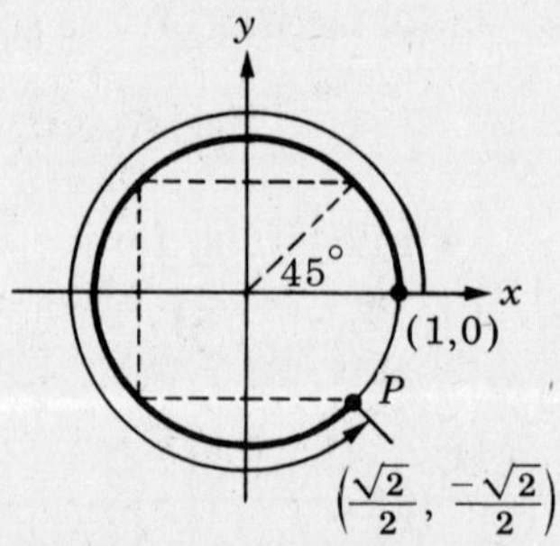

$s = \frac{7}{4}\pi,\ x = \frac{\sqrt{2}}{2},\ y = \frac{-\sqrt{2}}{2}$

$\frac{1}{8}$ of a whole circumference (ccw)

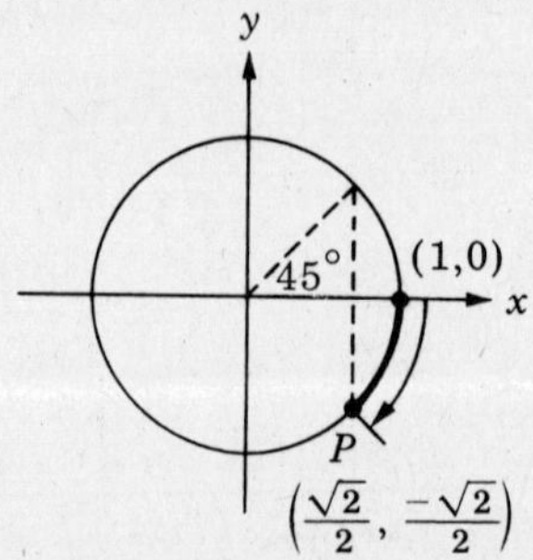

$s = -\frac{\pi}{4},\ x = \frac{\sqrt{2}}{2},\ y = \frac{-\sqrt{2}}{2}$

EXAMPLE 2 If P has moved $\frac{1}{12}$ of a complete circumference (ccw), what are the values of s, x, and y?

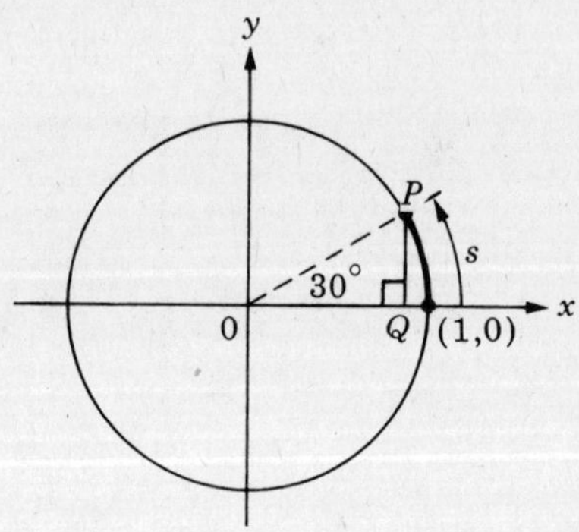

Solution For $\frac{1}{12}$ of a complete circumference (ccw)

$$s = \frac{1}{12}(2\pi) = \frac{\pi}{6}$$

Also $\frac{1}{12}$ of a whole circumference puts P on a 30° line, because $\frac{1}{12}(360°) = 30°$. A perpendicular from P to the x-axis gives a right triangle POQ. The height, y, must be $\frac{1}{2}$ since in any 30-60-90 triangle the hypotenuse is twice the length of the shorter side. Using $y = \frac{1}{2}$ in the unit circle equation gives:

$$x^2 + y^2 = 1$$
$$x^2 + (\tfrac{1}{2})^2 = 1$$
$$x^2 = \tfrac{3}{4}$$
$$x = \frac{\sqrt{3}}{2} \quad \text{(Negative solution rejected—why?)}$$

Hence, for $\frac{1}{12}$ of a whole circumference (ccw) $s = \frac{\pi}{6}$, $x = \frac{\sqrt{3}}{2}$, $y = \frac{1}{2}$.

As with Example 1 we can now find additional corresponding values of s, x, and y.

$\frac{5}{12}$ of a whole circumference (ccw)

$s = \frac{5}{12}(2\pi) = \frac{5}{6}\pi$

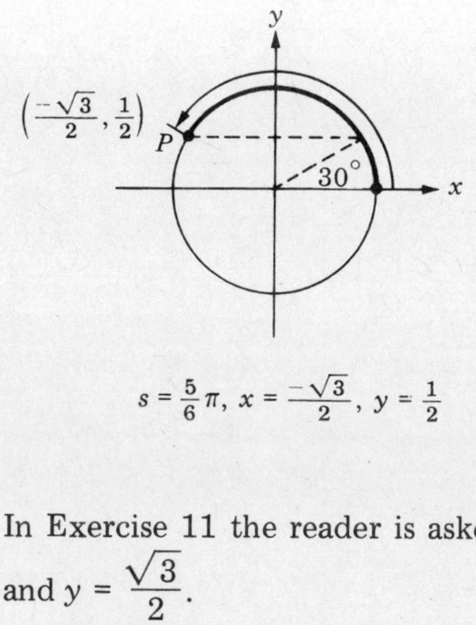

$s = \frac{5}{6}\pi$, $x = \frac{-\sqrt{3}}{2}$, $y = \frac{1}{2}$

$\frac{7}{12}$ of a whole circumference (ccw)

$s = \frac{7}{12}(2\pi) = \frac{7}{6}\pi$

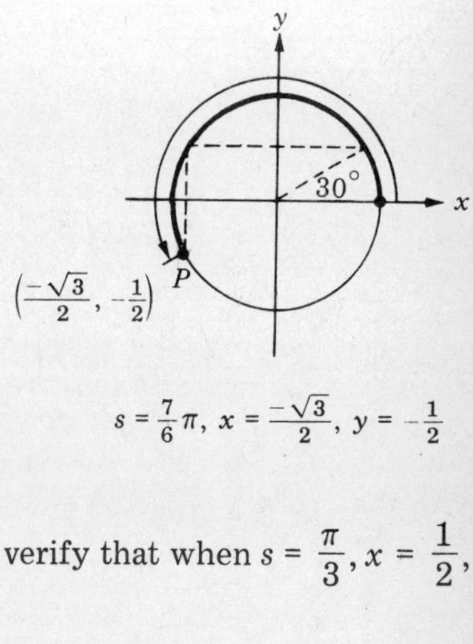

$s = \frac{7}{6}\pi$, $x = \frac{-\sqrt{3}}{2}$, $y = -\frac{1}{2}$

In Exercise 11 the reader is asked to verify that when $s = \frac{\pi}{3}$, $x = \frac{1}{2}$, and $y = \frac{\sqrt{3}}{2}$.

The values of s, x, and y calculated thus far are now summarized along with a few others that the reader should check. The summary is presented in two tables; one for x versus s and the other for y versus s. Also, decimal equivalents (to the nearest hundredth) have been supplied for graphing that follows.

s	x	s	x
$-\frac{\pi}{4}(-.79)$	$\frac{\sqrt{2}}{2}(.71)$	$\pi(3.14)$	-1
0	1	$\frac{7\pi}{6}(3.67)$	$-\frac{\sqrt{3}}{2}(-.87)$
$\frac{\pi}{6}(.52)$	$\frac{\sqrt{3}}{2}(.87)$	$\frac{5\pi}{4}(3.93)$	$-\frac{\sqrt{2}}{2}(-.71)$
$\frac{\pi}{4}(.79)$	$\frac{\sqrt{2}}{2}(.71)$	$\frac{4\pi}{3}(4.19)$	$-\frac{1}{2}(-.50)$
$\frac{\pi}{3}(1.05)$	$\frac{1}{2}(.50)$	$\frac{3\pi}{2}(4.71)$	0
$\frac{\pi}{2}(1.57)$	0	$\frac{5\pi}{3}(5.24)$	$\frac{1}{2}(.50)$
$\frac{2\pi}{3}(2.09)$	$-\frac{1}{2}(-.50)$	$\frac{7\pi}{4}(5.50)$	$\frac{\sqrt{2}}{2}(.71)$
$\frac{3\pi}{4}(2.36)$	$-\frac{\sqrt{2}}{2}(-.71)$	$\frac{11\pi}{6}(5.76)$	$\frac{\sqrt{3}}{2}(.87)$
$\frac{5\pi}{6}(2.62)$	$-\frac{\sqrt{3}}{2}(-.87)$	$2\pi(6.28)$	1

The graph of x as a function of s appears on the following page.

s	y	s	y
$-\frac{\pi}{4}(-.79)$	$-\frac{\sqrt{2}}{2}(-.71)$	$\pi(3.14)$	0
0	0	$\frac{7\pi}{6}(3.67)$	$-\frac{1}{2}(-.50)$
$\frac{\pi}{6}(.52)$	$\frac{1}{2}(.50)$	$\frac{5\pi}{4}(3.93)$	$-\frac{\sqrt{2}}{2}(-.71)$
$\frac{\pi}{4}(.79)$	$\frac{\sqrt{2}}{2}(.71)$	$\frac{4\pi}{3}(4.19)$	$-\frac{\sqrt{3}}{2}(-.87)$
$\frac{\pi}{3}(1.05)$	$\frac{\sqrt{3}}{2}(.87)$	$\frac{3\pi}{2}(4.71)$	-1
$\frac{\pi}{2}(1.57)$	1	$\frac{5\pi}{3}(5.24)$	$-\frac{\sqrt{3}}{2}(-.87)$
$\frac{2\pi}{3}(2.09)$	$\frac{\sqrt{3}}{2}(.87)$	$\frac{7\pi}{4}(5.50)$	$-\frac{\sqrt{2}}{2}(-.71)$
$\frac{3\pi}{4}(2.36)$	$\frac{\sqrt{2}}{2}(.71)$	$\frac{11\pi}{6}(5.76)$	$-\frac{1}{2}(-.50)$
$\frac{5\pi}{6}(2.62)$	$\frac{1}{2}(.50)$	$2\pi(6.28)$	0

The graph of y as a function of s appears on the following page.

(1)

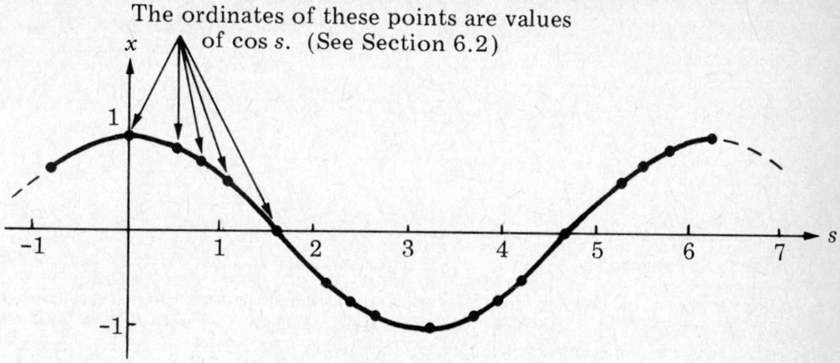

(2)

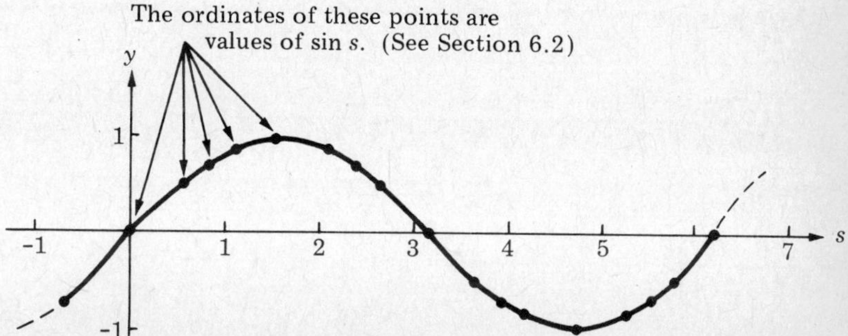

Both graphs continue in the same *oscillating* manner indefinitely to the left and right and the Vertical Line Test shows that *each is a function*. Since graph (1) is associated with the cosine ratio and graph (2) is associated with the sine ratio (from Section 6.2), the following definitions are motivated.

DEFINITION

Given the point P with coordinates (x, y) on the unit circle $x^2 + y^2 = 1$, and s the distance of P from $(1,0)$, then

$y = \sin s$ is the **standard sine function**

and

$x = \cos s$ is the **standard cosine function,**

where $s > 0$ if measured ccw and $s < 0$ if measured cw.

In this defintion, s is sometimes referred to as the **argument** of the functions. Note that the argument, s, is a *real number*. Thus, $y = \sin s$ and $x = \cos s$ are **trigonometric functions of a real number.** (In Section 6.2, a more limited view is taken where the trigonometric ratios of acute angles are presented.) Another name for the trigonometric functions is the **circular functions** since they are generated by the point P moving on a circle.

When displaying sine or cosine graphs, it is easier to scale the horizontal axis in units of π. Furthermore, in order to conform with past graphing practices, the horizontal axis is labeled with an x and the vertical axis with a y. This x represents, then, arclength on the unit circle from $(1,0)$ to P. In each graph the maximum vertical dis-

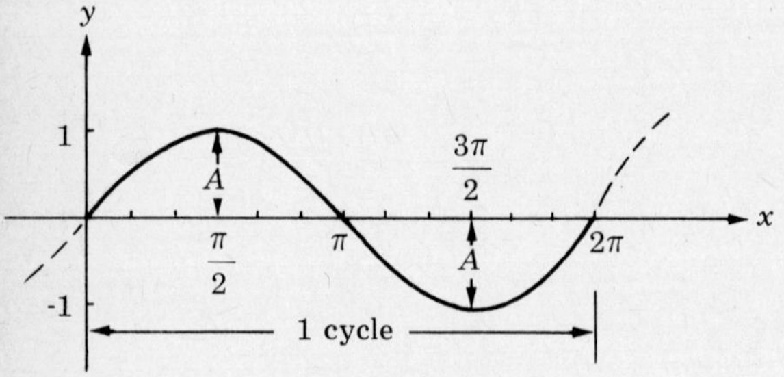

Graph of $y = \sin x$ or equivalently $f(x) = \sin x$

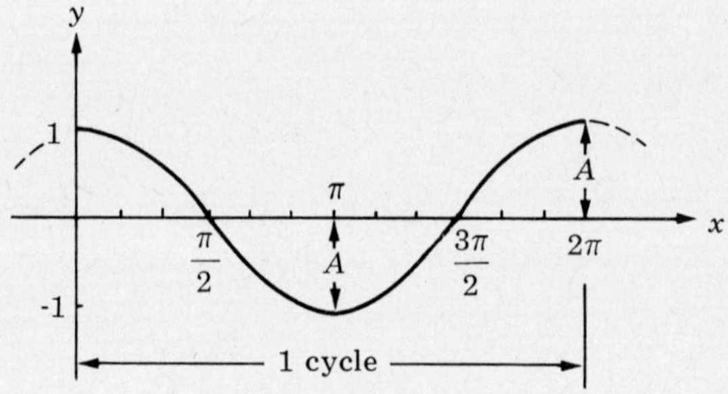

Graph of $y = \cos x$ or equivalently $g(x) = \cos x$

placement in either direction from the x-axis is labeled A. The absolute value of this distance is called the **amplitude**. The amplitude of $f(x) = \sin x$ and $g(x) = \cos x$ is 1. Also, in the interval $[0, 2\pi]$ there is one complete **cycle** of both $f(x) = \sin x$ and $g(x) = \cos x$. We say that the **period** is length 2π, or simply, the period is 2π. Knowing the shape of the graph in one cycle gives sufficient information to graph as many cycles as desired by repeating the same shape again and again to the left and right. Compared to $[0, 2\pi]$, the functional values of $f(x) = \sin x$ and $g(x) = \cos x$ repeat themselves on $[-2\pi, 0]$, $[-4\pi, -2\pi]$, . . . , and also on $[2\pi, 4\pi]$, $[4\pi, 6\pi]$, ... Thus we can write:

$f(x + 2\pi n) = f(x)$ and $g(x + 2\pi n) = g(x)$, where n is *any integer.**

Another quality of the graphs of $y = \sin x$ and $y = \cos x$ worthy of examining is the intercepts.

*In general, a function $h(x)$ is **periodic** if there exists a number $p > 0$, such that $h(x + p) = h(x)$ for all x in the domain of $h(x)$. The *smallest* such p is called the period.

EXAMPLE 3 Determine the intercepts of $y = \sin x$.

Solution From the graph, it is seen that when $x = 0$, $y = 0$. Thus the y-intercept is 0. The x-intercepts (or zeros) are seen to be $x = 0, \pm\pi, \pm 2\pi, \pm 3\pi, \ldots$ Thus $\sin x = 0$ when its argument $x = k\pi$, where k is any integer.

EXAMPLE 4 Determine the intercepts of $y = \cos x$.

Solution From the graph, it is seen that when $x = 0$, $y = 1$. Thus the y-intercept is 1. The x-intercepts (or zeros) are seen to be $x = \pm\frac{\pi}{2}, \pm\frac{3\pi}{2}, \pm\frac{5\pi}{2}, \ldots$. Thus $\cos x = 0$, when its argument $x = \frac{(2k+1)\pi}{2}$, where k is any integer.

We can now state the following theorem.

THEOREM 3

a. If $\sin x = 0$, then $x = k\pi$ where k is any integer.

b. If $\cos x = 0$, then $x = \frac{(2k+1)\pi}{2}$ where k is any integer.

Theorem 3 will now be used to solve some trigonometric equations in a single variable.

EXAMPLE 5 Solve for x: $\sin 3x = 0$.

Solution By Theorem 3a, $\sin 3x = 0$ when the *argument* $3x = k\pi$, k is any integer. Thus, $x = \frac{k\pi}{3}$; k is any integer.

EXAMPLE 6 Solve for w: $\cos\left(4w + \frac{\pi}{2}\right)$.

Solution By Theorem 3b, $\cos\left(4w + \frac{\pi}{2}\right) = 0$ when the argument $4w + \frac{\pi}{2} = \frac{(2k+1)\pi}{2}$, k is any integer. Thus,

$$4w + \frac{\pi}{2} = \frac{(2k+1)\pi}{2}$$

$$4w = \frac{(2k+1)\pi}{2} - \frac{\pi}{2}$$

$$w = \frac{k\pi}{4};\ k \text{ is any integer.}$$

This section began with an investigation of the variation of y with s and x with s where the point (x, y) moves around the unit circle a distance s from $(1,0)$. The functional relationship of y with s is $y = \sin s$ and of x with s is $x = \cos s$. That is, the coordinates of the point (x, y) on the unit circle are $(\cos s, \sin s)$. Next, the definition for the tangent (tan) function is presented.

DEFINITION

> Given the point P with coordinates (x, y) on the unit circle $x^2 + y^2 = 1$, and s the distance of P from $(1,0)$, then $r = \frac{y}{x} = \tan s$ is the **standard tangent function**, where $s > 0$ if measured ccw and $s < 0$ if measured cw.

In effect, this definition says that the tangent function is the functional relationship of the quantity $\frac{y}{x}$ with s. Using a portion of the tables earlier in this section we obtain:

s	x	y	$\frac{y}{x} = r$
0	1	0	0
$\frac{\pi}{4}$	.71	.71	1
$\frac{\pi}{3}$	.50	.87	1.73
$\frac{\pi}{2}$	0	1	undefined
$\frac{3\pi}{4}$	−.71	.71	−1
$\frac{5\pi}{6}$	−.87	.50	−.58
π	−1	0	0
$\frac{5\pi}{4}$	−.71	−.71	1
$\frac{4\pi}{3}$	−.50	−.87	1.73
$\frac{3\pi}{2}$	0	1	undefined
$\frac{5\pi}{3}$	.50	−.87	−1.73
$\frac{11\pi}{6}$	.87	−.50	−.58
2π	1	0	0

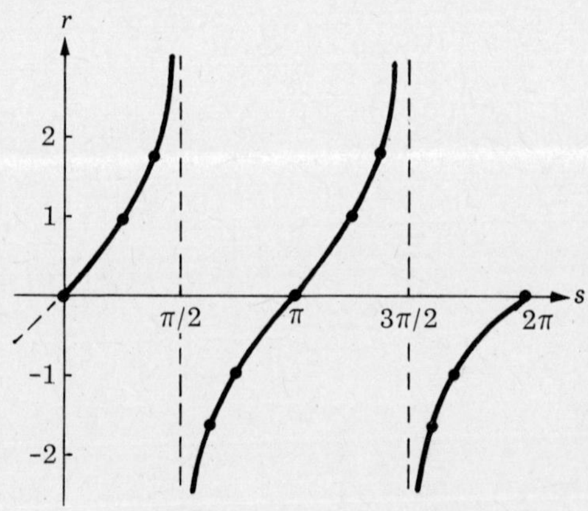

Graph of $r = \tan s$ on $[0, 2\pi]$ (Vertical Line Test shows this *is* a function.)

The concept of amplitude is not defined for the tangent function since there is no maximum vertical displacement. Notice that the period is π because knowing the shape of the graph on $[0,\pi]$ gives sufficient information to reproduce the graph on $[\pi,2\pi]$, $[2\pi,3\pi]$, $[-\pi,0]$, etc. Thus, for $f(x) = \tan x$, which is graphed below:

$$f(x + n\pi) = f(x), \text{ where } n \text{ is any integer.}$$

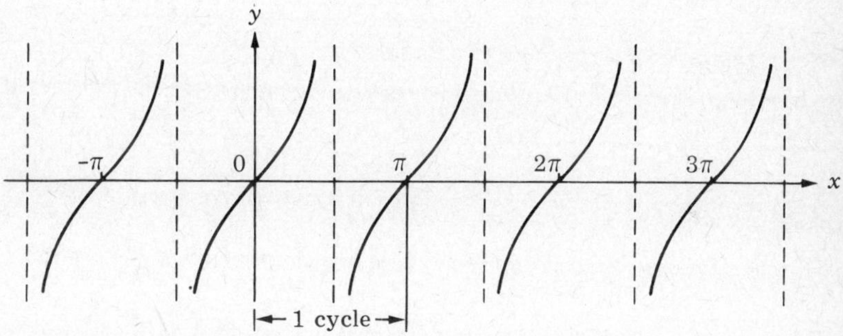

Graph of $y = \tan x$ or equivalently $f(x) = \tan x$

The graph shows $\tan x = 0$, when the argument $x = k\pi$ where k is any integer.

EXAMPLE 7 Solve for x: $\tan\left(x + \frac{\pi}{4}\right) = 0$.

Solution $\tan\left(x + \frac{\pi}{4}\right) = 0$ when the argument $x + \frac{\pi}{4} = k\pi$, where k is any integer.

$$x + \frac{\pi}{4} = k\pi$$

$$x = k\pi - \frac{\pi}{4}$$

$$x = (k - \tfrac{1}{4})\pi, \ k \text{ is any integer}$$

In Exercises 41, 42, and 43 the reader is guided to drawing the graphs of the cosecant (csc), secant (sec), and cotangent (cot) functions that appear on the following page.

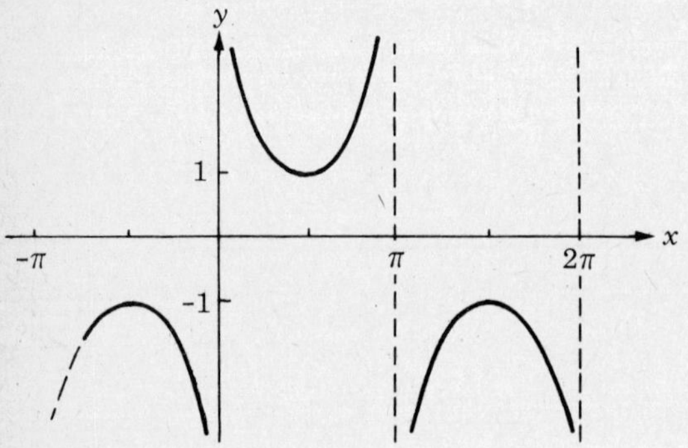

Graph of $y = \csc x$
(1 period $= 2\pi$)

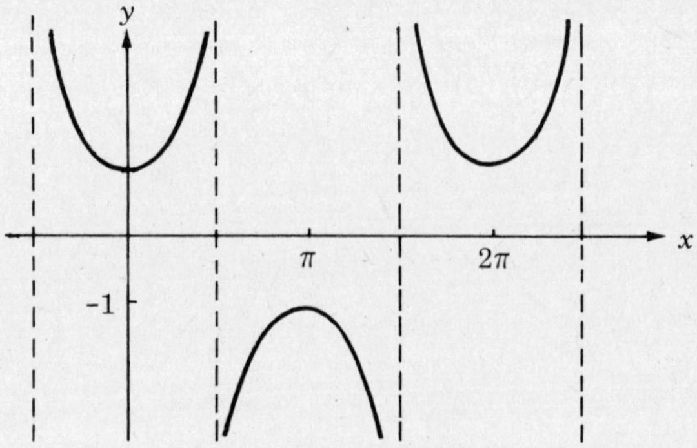

Graph of $y = \sec x$
(1 period $= 2\pi$)

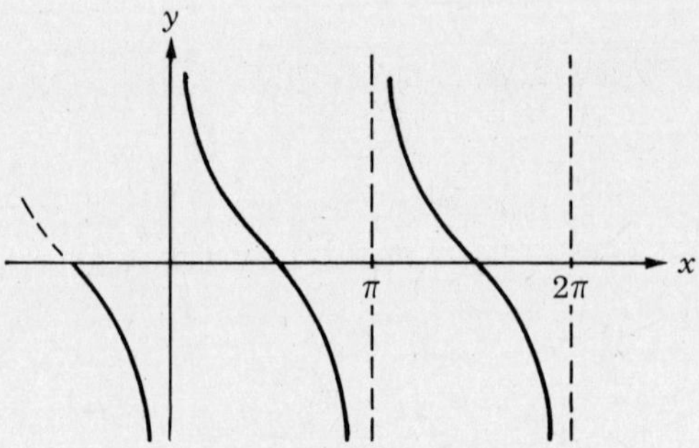

Graph of $y = \cot x$
(1 period $= \pi$)

The reader is advised to memorize the graphs of the six standard trigonometric functions presented in this section.

EXERCISES FOR SECTION 6.3

In Exercises 1 through 8, P is the point (x,y) on the unit circle $x^2 + y^2 = 1$. Find the values of s, x, and y for the specified positions of P. (See Examples 1 and 2.)

1. P has moved $\frac{1}{8}$ of a whole circumference (cw).
2. P has moved $\frac{1}{4}$ of a whole circumference (cw).
3. P has moved $\frac{3}{8}$ of a whole circumference (cw).
4. P has moved $\frac{1}{2}$ of a whole circumference (cw).
5. P has moved $\frac{5}{8}$ of a whole circumference (cw).
6. P has moved $\frac{3}{4}$ of a whole circumference (cw).
7. P has moved $\frac{7}{8}$ of a whole circumference (cw).
8. P has moved a whole circumference (cw).
9. Using the results of Exercises 1 through 8,
   a. Create a graph of $y = \sin s$ on the interval $[-2\pi, 0]$.
   b. Create a graph of $x = \cos s$ on the interval $[-2\pi, 0]$.
10. a. Using your answer to 9a, replace s by the letter x and graph $y = \sin x$ on $[-2\pi, 0]$.

b. Using your answer to 9b, replace x and s by the letters y and x respectively and graph $y = \cos x$ on $[-2\pi, 0]$.

11. Show that when $s = \frac{\pi}{3}$, it follows that $x = \frac{1}{2}, y = \frac{\sqrt{3}}{2}$. (See Example 2.)

In Exercises 12 through 20, a) draw the graph of each function on the specified interval and b) determine how many cycles are present. (See graphs of $y = \sin x$, $y = \cos x$, $y = \tan x$.)

12. $y = \sin x$ on $[2\pi, 4\pi]$
13. $y = \cos x$ on $[-4\pi, -2\pi]$
14. $y = \tan x$ on $[2\pi, 4\pi]$
15. $y = \sin x$ on $[-4\pi, 4\pi]$
16. $y = \cos x$ on $[-2\pi, 6\pi]$
17. $y = \tan x$ on $[-2\pi, \pi]$
18. $y = \sin x$ on $[0, 3\pi]$
19. $y = \cos x$ $[0, \pi]$
20. $y = \tan x$ on $\left[-\frac{\pi}{2}, \frac{\pi}{2}\right]$

In Exercises 21 through 40, solve for the variable present in the argument. (See Examples 5, 6, and 7.)

21. $\sin 2x = 0$
22. $\cos 4t = 0$
23. $\cos 3x = 0$
24. $\tan 2\theta = 0$
25. $\sin 3x = 0$
26. $\sin 6x = 0$
27. $\tan 6w = 0$
28. $\cos 6x = 0$
29. $\cos(2y + \pi) = 0$
30. $\sin(2x + \pi) = 0$
31. $\sin(2x - \pi) = 0$
32. $\tan(2x + \pi) = 0$
33. $\sin\left(2x + \frac{\pi}{2}\right) = 0$
34. $\sin\left(2t - \frac{\pi}{2}\right) = 0$
35. $\sin\left(2x + \frac{3\pi}{2}\right) = 0$
36. $\tan(x - \pi) = 0$
37. $\cos\left(3x - \frac{\pi}{2}\right) = 0$
38. $\cos\left(3x + \frac{\pi}{2}\right) = 0$
39. $\sin\left(4\theta + \frac{3\pi}{2}\right) = 0$
40. $\tan\left(\alpha + \frac{\pi}{2}\right) = 0$

For Exercises 41 through 43, P is the point (x, y) on the unit circle $x^2 + y^2 = 1$ and s is the distance of P from $(1,0)$.

41. The definition of the cosecant function is $r = \frac{1}{y} = \csc s$. Use the numerical tables in this section to obtain a table of values showing the functional relationship of $\frac{1}{y}$ with s, for $0 \leqslant s \leqslant 2\pi$. Bearing in mind that $r = \frac{1}{y}$, graph r as a function of s to obtain the graph of $r = \csc s$.

42. The definition of the secant function is $r = \frac{1}{x} = \sec s$. Use the numerical tables of this section to obtain a table of values showing the functional relationship of $\frac{1}{x}$ with s, for $0 \leqslant s \leqslant 2\pi$. Bearing in mind that $r = \frac{1}{x}$, graph r as a function of s to obtain the graph of $r = \sec s$.

43. The definition of the cotangent function is $r = \frac{x}{y} = \cot s$. Use the numerical tables of this section to obtain a table of values showing the functional relationship of $\frac{x}{y}$ with s, for $0 \leqslant s \leqslant 2\pi$. Bearing in mind that $r = \frac{x}{y}$, graph r as a function of s to obtain the graph of $r = \cot s$.

6.4 EVALUATING TRIGONOMETRIC FUNCTIONS

In Section 6.2 the trigonometric ratios of an acute angle θ (domain of $\theta: 0° \leqslant \theta \leqslant 90°$) were defined and in Section 6.3 the trigonometric functions of any real number s were presented. We now begin to unify these two concepts. In the diagram below, the right triangle

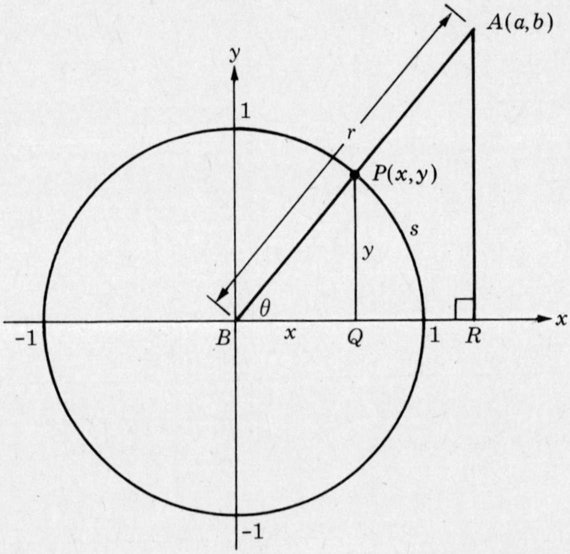

ABR has been superimposed (in the first quadrant) on the unit circle. The angle θ has its vertex at the origin and its initial side on the x-axis. We say that θ is in **standard position**. The terminal side passes through the points (x, y) and (a, b), and the distance of a point from the origin is called the **radius vector** of that point. The radius vector of (x, y) is 1 and the radius vector of (a, b) is r. Using the definitions of the trigonometric ratios, the fact that $\triangle ABR$ is similar to $\triangle PBQ$, and the definitions of the trigonometric functions we obtain:

$$\sin\theta = \underbrace{\frac{b}{r} = \frac{y}{1}}_{\dfrac{\text{ordinate}}{\text{radius vector}}} = \sin s \qquad \csc\theta = \underbrace{\frac{r}{b} = \frac{1}{y}}_{\dfrac{\text{radius vector}}{\text{ordinate}}} = \csc s$$

$$\cos\theta = \underbrace{\frac{a}{r} = \frac{x}{1}}_{\dfrac{\text{abscissa}}{\text{radius vector}}} = \cos s \qquad \sec\theta = \underbrace{\frac{r}{a} = \frac{1}{x}}_{\dfrac{\text{radius vector}}{\text{abscissa}}} = \sec s$$

$$\tan\theta = \underbrace{\frac{b}{a} = \frac{y}{x}}_{\dfrac{\text{ordinate}}{\text{abscissa}}} = \tan s \qquad \cot\theta = \underbrace{\frac{a}{b} = \frac{x}{y}}_{\dfrac{\text{abscissa}}{\text{ordinate}}} = \cot s$$

A look at the previous six equalities reveals that the sine and cosecant functions are *reciprocals* of each other, the cosine and secant functions are *reciprocals* of each other, and the tangent and cotangent functions are *reciprocals* of each other. Stated in terms of θ, we can write the following **trigonometric identities** that are true for all values of θ (except when denominators are zero).

Reciprocal Identities

$$\sin\theta = \frac{1}{\csc\theta} \quad \text{or equivalently} \quad \csc\theta = \frac{1}{\sin\theta}$$

$$\cos\theta = \frac{1}{\sec\theta} \quad \text{or equivalently} \quad \sec\theta = \frac{1}{\cos\theta}$$

$$\tan\theta = \frac{1}{\cot\theta} \quad \text{or equivalently} \quad \cot\theta = \frac{1}{\tan\theta}$$

Two other identities that follow are:

$$\tan\theta = \frac{\sin\theta}{\cos\theta} \quad \text{and} \quad \cot\theta = \frac{\cos\theta}{\sin\theta}$$

Before beginning the next few examples, we examine two special types of angles in the following diagrams.

Quadrantal Angles: the terminal side lies on a coordinate axis.

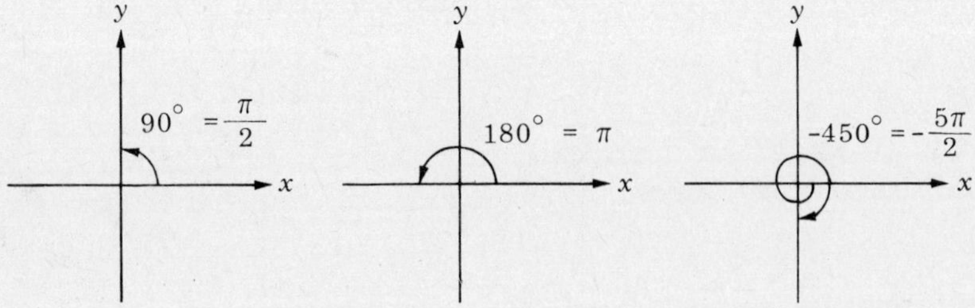

Coterminal Angles: angles with the same initial and terminal sides.

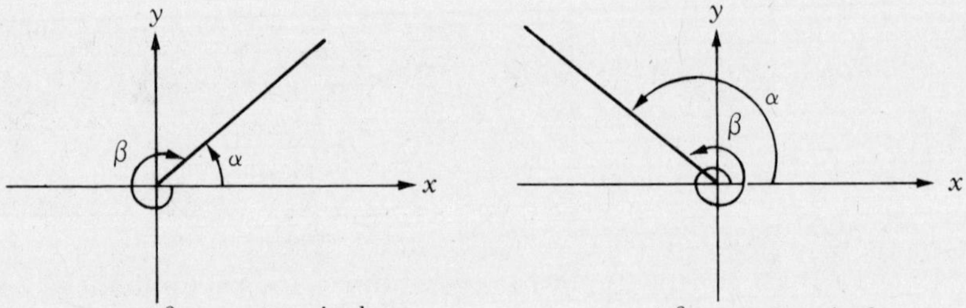

α and β are coterminal α and β are coterminal

EXAMPLE 1 Given the angle θ in standard position in a rectangular coordinate system, and the point A with coordinates $(-6,-2)$ on the terminal side of θ, determine each of the trigonometric functions of θ.

Solution The radius vector has length

$$r = \sqrt{(-6-0)^2 + (-2-0)^2}$$
$$= \sqrt{40} = 2\sqrt{10}.$$

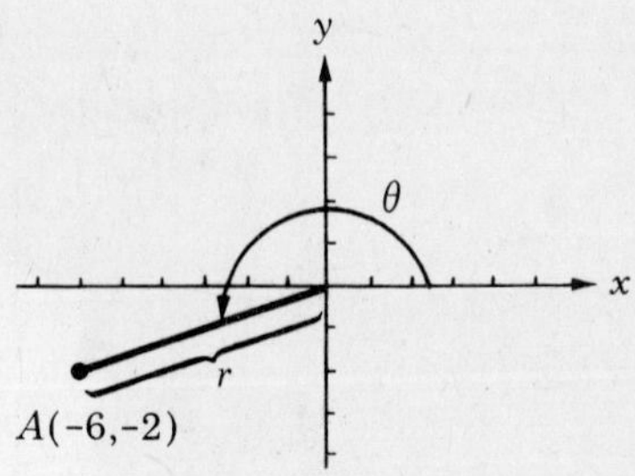

The abscissa of A is $a = -6$.
The ordinate of A is $b = -2$.

$$\sin\theta = \frac{\text{ordinate}}{\text{rad. vec.}} = \frac{-2}{2\sqrt{10}} = \frac{-\sqrt{10}}{10} \qquad \csc\theta = \frac{1}{\sin\theta} = -\sqrt{10}$$

$$\cos\theta = \frac{\text{abscissa}}{\text{rad. vec.}} = \frac{-6}{2\sqrt{10}} = \frac{-3\sqrt{10}}{10} \qquad \sec\theta = \frac{1}{\cos\theta} = -\frac{\sqrt{10}}{3}$$

$$\tan\theta = \frac{\text{ordinate}}{\text{abscissa}} = \frac{-2}{-6} = \frac{1}{3} \qquad \cot\theta = \frac{1}{\tan\theta} = 3$$

In the next example, the trigonometric ratios of quadrantal angles are examined.

EXAMPLE 2 Determine the trigonometric functions of the quadrantal angle, 90°.

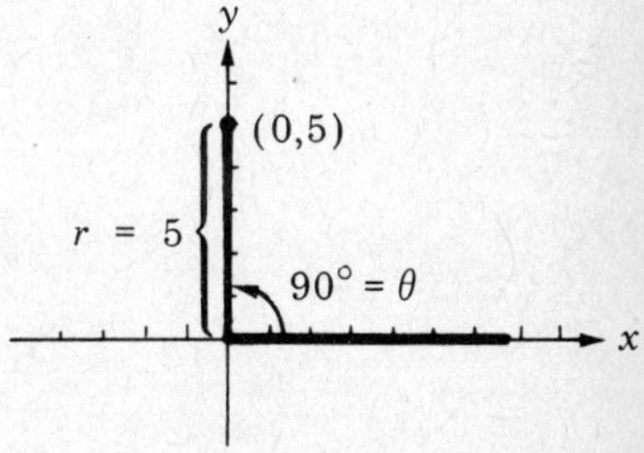

Solution Pick any point on the terminal side, say (0,5).

$$\sin\theta = \frac{\text{ordinate}}{\text{rad. vec.}} = \frac{5}{5} = 1$$

$$\cos\theta = \frac{\text{abscissa}}{\text{rad. vec.}} = \frac{0}{5} = 0$$

$$\tan\theta = \frac{\text{ordinate}}{\text{abscissa}} = \frac{5}{0} = \text{undefined} \qquad \csc\theta = \frac{\text{rad. vec.}}{\text{ordinate}} = \frac{5}{5} = 1$$

$$\sec\theta = \frac{\text{rad. vec.}}{\text{abscissa}} = \frac{5}{0} = \text{undefined} \qquad \cot\theta = \frac{\text{abscissa}}{\text{ordinate}} = \frac{0}{5} = 0$$

EXAMPLE 3 Determine each of the trigonometric functions of the quadrantal angle, 180°.

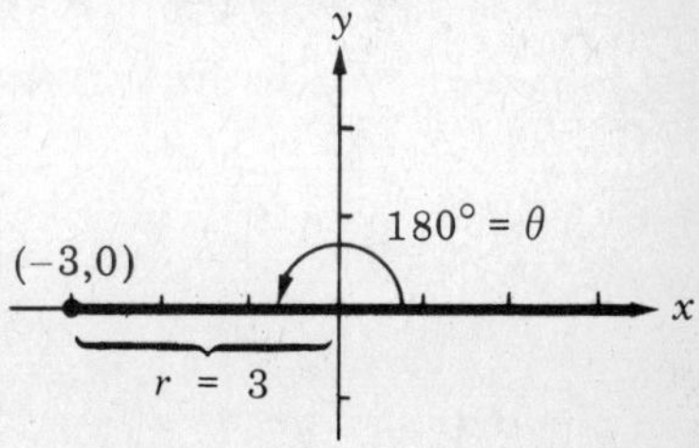

Solution Pick any point on the terminal side, say (−3,0).

$$\sin\theta = \frac{0}{3} = 0 \qquad \csc\theta = \frac{3}{0} = \text{undefined}$$

$$\cos\theta = \frac{-3}{3} = -1 \qquad \sec\theta = \frac{3}{-3} = -1$$

$$\tan\theta = \frac{0}{-3} = 0 \qquad \cot\theta = \frac{-3}{0} = \text{undefined}$$

In Exercises 21, 22, and 23, the reader is asked to obtain the trigonometric functions of the quadrantal angles 0°, 270°, and 360°. These quadrantal values are now summarized.

θ (degrees)	s	$\sin\theta$ or $\sin s$	$\csc\theta$ or $\csc s$	$\cos\theta$ or $\cos s$	$\sec\theta$ or $\sec s$	$\tan\theta$ or $\tan s$	$\cot\theta$ or $\cot s$
0	0	0	undefined	1	1	0	undefined
90°	$\frac{\pi}{2}$	1	1	0	undefined	undefined	0
180°	π	0	undefined	−1	−1	0	undefined
270°	$\frac{3\pi}{2}$	−1	−1	0	undefined	undefined	0
360°	2π	0	undefined	1	1	0	undefined

By examining the graphs of Section 6.3, one can determine the *signs* of each of the trigonometric functions in each quadrant. For instance, consider the graph of $r = \sin s$.

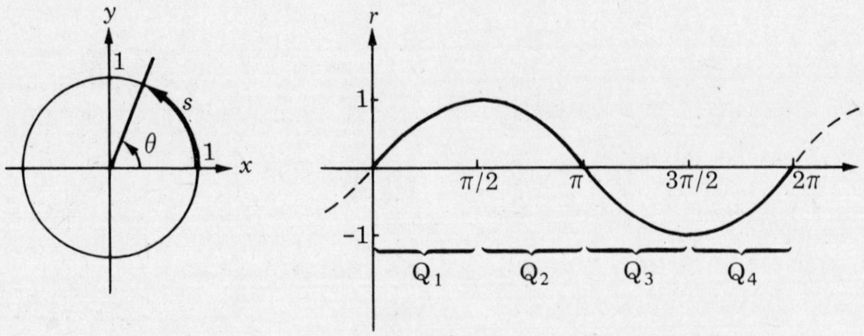

In Q_1: $0 < s < \frac{\pi}{2}$, $0° < \theta < 90°$ and $r = \sin s$ is *positive.*

In Q_2: $\frac{\pi}{2} < s < \pi$, $90° < \theta < 180°$ and $r = \sin s$ is *positive.*

In Q_3: $\pi < s < \frac{3\pi}{2}$, $180° < \theta < 270°$ and $r = \sin s$ is *negative.*

In Q_4: $\frac{3\pi}{2} < s < 2\pi$, $270° < \theta < 360°$ and $r = \sin s$ is *negative.*

Doing the same analysis for each of the other graphs of Section 6.3 we find:

Quadrant	$\sin\theta$ or $\sin s$	$\csc\theta$ or $\csc s$	$\cos\theta$ or $\cos s$	$\sec\theta$ or $\sec s$	$\tan\theta$ or $\tan s$	$\cot\theta$ or $\cot s$
1	+	+	+	+	+	+
2	+	+	–	–	–	–
3	–	–	–	–	+	+
4	–	–	+	+	–	–

We now turn our attention to the evaluation of a trigonometric function of *any* angle (or *any* real number). The reader will see how this can be done by using Table III which lists the trigonometric functions for an angle θ, $0° \leqslant \theta \leqslant 90°$, and for a real number s, $0 \leqslant s \leqslant 1.5708$. First we need the following definition.

DEFINITION

The **reference angle** θ_r associated with an angle θ is the acute angle formed by the terminal side of θ and the x-axis. (See illustrations on the following page, where θ_r is shaded.)

Each picture below shows the angle being measured by θ (degrees) as well as by the real number s (arclength on the unit circle). The symbol s_r is the **reference number** and is analogous to θ_r.

In Q_1: In Q_2:

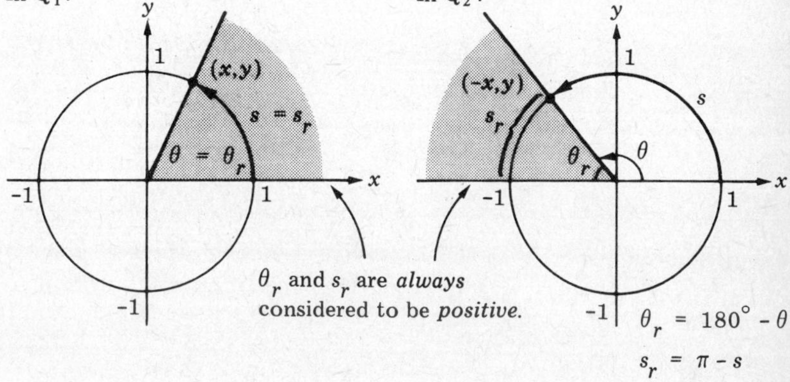

In Q_3: In Q_4:

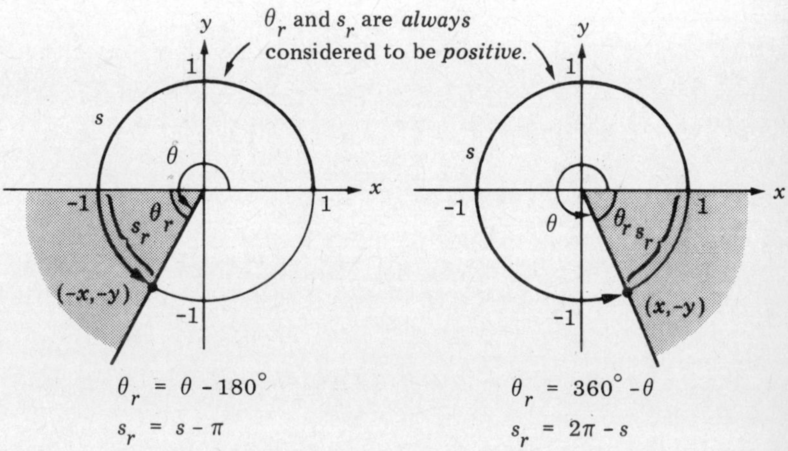

As will now be shown, a trigonometric function of an angle θ (or real number s) in *any* quadrant has the same value as the trigonometric function of the associated reference angle θ_r (or reference number s_r), *except possibly for its sign*. We will verify this statement for the cases when the terminal side lies in the first or the second quadrants. In Exercises 56 and 57 the reader can perform the verification for quadrants three and four.

In the first quadrant, with $\theta = \theta_r$ and $s = s_r$, we have $\sin\theta = \sin\theta_r$, $\sin s = \sin s_r$, $\cos\theta = \cos\theta_r$, $\cos s = \cos s_r$, etc. In the second quadrant,

$$\sin\theta = \frac{y}{1} = y = \sin\theta_r \qquad \text{or} \qquad \sin s = \sin s_r$$

$$\cos\theta = \frac{-x}{1} = -(x) = -\cos\theta_r \qquad \text{or} \qquad \cos s = -\cos s_r$$

$$\tan\theta = \frac{y}{-x} = -\left(\frac{y}{x}\right) = -\tan\theta_r \qquad \text{or} \qquad \tan s = -\tan s_r$$

$$\csc\theta = \frac{1}{y} = \csc\theta_r \qquad \text{or} \qquad \csc s = \csc s_r$$

$$\sec\theta = \frac{1}{-x} = -\left(\frac{1}{x}\right) = -\sec\theta_r \qquad \text{or} \qquad \sec s = -\sec s_r$$

$$\cot\theta = \frac{-x}{y} = -\left(\frac{x}{y}\right) = -\cot\theta_r \qquad \text{or} \qquad \cot s = -\cot s_r$$

The sign differences in the equalities above are contained in the table on page 310. In summary, to determine the value of a trigonometric function of an angle θ (or real number s):

1. Determine the quadrant in which the terminal side lies.
2. Determine the reference angle θ_r (reference number s_r).
3. Use Table III to obtain the trigonometric function of θ_r (s_r).
4. Attach the appropriate algebraic sign.

EXAMPLE 4 Determine each of the following.

a. $\sin 141^\circ$ b. $\cos 141^\circ$ c. $\tan 250^\circ$ d. $\csc 660^\circ$

Solution a. The sine function is positive in the second quadrant.

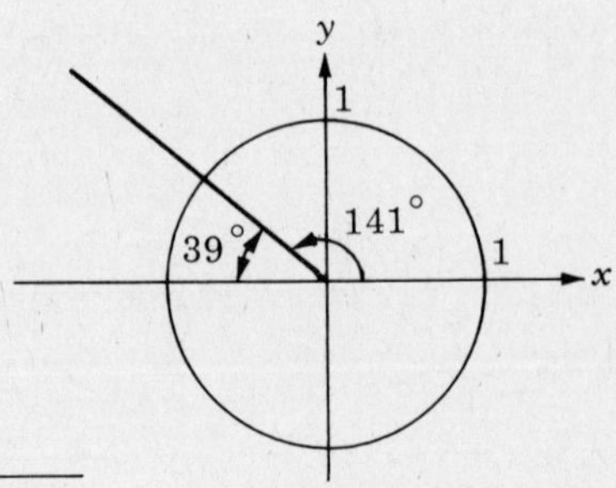

θ (degrees)	s	$\sin\theta$ or $\sin s$	$\csc\theta$ or $\csc s$	$\cos\theta$ or $\cos s$
39.0	0.6807	0.6293	1.5890	0.7771
39.1	0.6824	0.6307	1.5856	0.7760
39.2	0.6842	0.6320	1.5822	0.7749
39.3	0.6859	0.6334	1.5788	0.7738
39.4	0.6877	0.6347	1.5755	0.7727

$\sin 141^\circ = \sin 39^\circ = .6293$

A calculator with a sin key gives the slightly more precise value $\sin 141^\circ = .62932039$

b. Since the cosine function is negative in the second quadrant,

$$\cos 141^\circ = -\cos 39^\circ = -.7771$$

> Using a calculator,
> $\cos 141^\circ = -.77714596$

c. The tangent function is positive in the third quadrant.

$$\tan 250^\circ = \tan 70^\circ = 2.7475$$

> Using a calculator
> $\tan 250^\circ = 2.7474774$

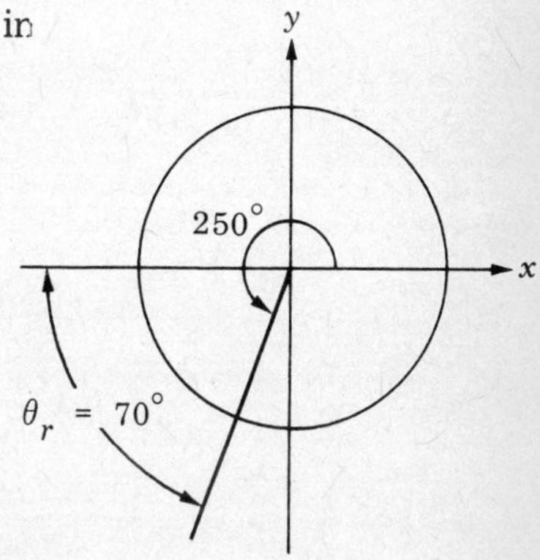

d. 660° consists of one complete circumference (360°) plus 300° more. Thus 660° is a fourth quadrant angle. Since the cosecant function is negative in the fourth quadrant,

$$\csc 660^\circ = -\csc 60^\circ = -1.1547$$

> Using a calculator, $\sin 660^\circ = -.86602541$, and thus
> $$\csc 660^\circ = \frac{1}{\sin 660^\circ} = -1.1547005$$

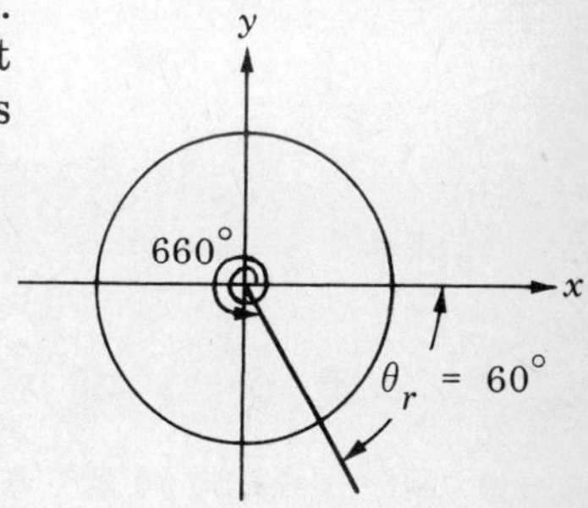

EXAMPLE 5 Determine each of the following.

a. cos 4 b. sec 4
c. sin (−4.56)

Solution a. The cosine function is negative in the third quadrant.

$$\cos 4 = -\cos .8584$$

In Table III, the s value closest to .8584 is .8587. We use that.

$$\cos 4 = -.6534$$

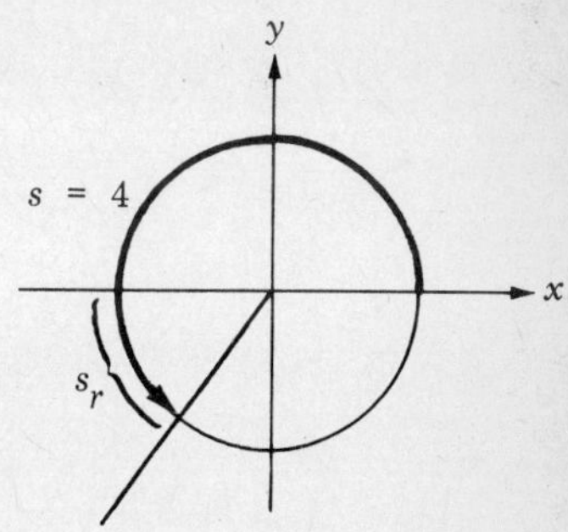

$s_r = 4 - 3.1416 = .8584$

> Using a calculator
> $\cos 4 = -.65364362$

b. $\sec 4 = -\sec .8584$

$\sec 4 = -1.5304$

> Using a calculator,
> $\sec 4 = -1.5298857$

c. Since the sine function is positive in the second quadrant,

$$\sin(-4.56) = \sin 1.4184$$

In Table III, the s value closest to 1.4184 is 1.4190. We use that.

$$\sin(-4.56) = .9885$$

> Using a calculator,
> $\sin(-4.56) = .98841125$

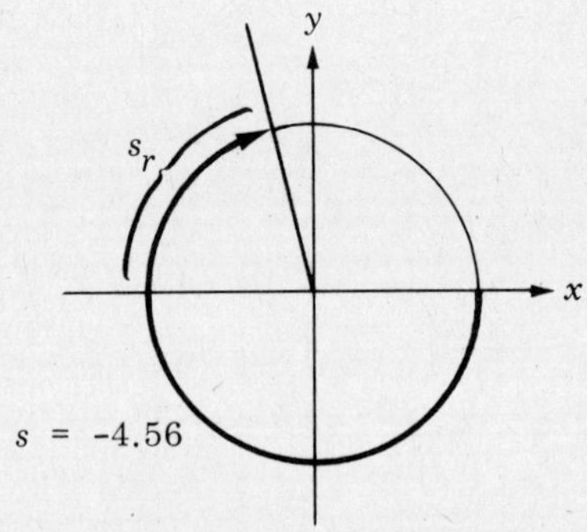

$s_r = 4.56 - 3.1416 = 1.4184$

EXERCISES FOR SECTION 6.4

In Exercises 1 through 20, determine each of the trigonometric functions of the angle θ, where θ is in standard position in a rectangular coordinate system and its terminal side passes through the given point. (See Example 1.)

1. (1,3) 2. (2,6) 3. (3,9) 4. (−4,2)
5. (1,9) 6. (−1,−1) 7. (0,−6) 8. (3,−2)
9. (3,1) 10. (2,1) 11. (−1,3) 12. (1,4)
13. (−1,2) 14. (1,0) 15. (−2,6) 16. (−1,0)
17. (−1,5) 18. (−1,−3) 19. (−4,−3) 20. (5,−1)

In Exercises 21 through 35, determine each of the trigonometric functions of the given quadrantal angle. (See Examples 2 and 3.)

21. $0°$ 22. $270°$ 23. $360°$ 24. $-90°$
25. $180°$ 26. $450°$ 27. $540°$ 28. $630°$
29. $720°$ 30. $900°$ 31. $-270°$ 32. $-180°$
33. $810°$ 34. $-630°$ 35. $-360°$

In Exercises 36 through 45, determine the reference angle θ_r associated with each given angle θ.

36. $\theta = 45°$ 37. $\theta = 160°$ 38. $\theta = 200°$ 39. $\theta = 330°$
40. $\theta = 470°$ 41. $\theta = -100°$ 42. $\theta = 400°$ 43. $\theta = 1000°$
44. $\theta = -60°$ 45. $\theta = -280°$

In Exercises 46 through 55, determine the reference number s_r associated with each given real number s. (Use $\pi = 3.1416$.)

46. $s = 1.26$ 47. $s = 1.89$ 48. $s = 1.9016$ 49. $s = 4$
50. $s = 5$ 51. $s = -4$ 52. $s = -5$ 53. $s = 6$
54. $s = 20$ 55. $s = -2.98$

56. Show that the trigonometric functions of an angle θ in standard position whose terminal side lies in the third quadrant have the same value as the trigonometric functions of the associated reference angle θ_r, *except possibly for the sign.*
57. Show that the trigonometric functions of an angle θ in standard position whose terminal side lies in the fourth quadrant have the same values as the trigonometric functions of the associated reference angle θ_r, *except possibly for the sign.*

In Exercises 58 through 72, use Table III to determine each of the trigonometric functions of the given angle θ (or real number s). (See Examples 4 and 5.)

58. $\theta = 29^\circ$ 59. $\theta = 68^\circ$ 60. $s = 2.6529$ 61. $s = 3.8397$
62. $\theta = 300^\circ$ 63. $\theta = 144.6^\circ$ 64. $s = 1.5691$ 65. $s = 1.5708$
66. $\theta = 500^\circ$ 67. $\theta = -105^\circ$ 68. $s = .0436$ 69. $s = -3$
70. $\theta = -45^\circ$ 71. $\theta = 212^\circ$ 72. $s = -2$

Calculator Exercises

73. Graph $y = \sin x$ on the interval $[0, 2\pi]$ by choosing $x = 0, .5, 1.5, 2, 2.5, \ldots$
74. Graph $y = \cos x$ on the interval $[0, 2\pi]$ by choosing $x = 0, .5, 1.5, 2, 2.5, \ldots$
75. Graph $y = \tan x$ on the interval $[0, \pi]$ by choosing $x = 0, .2, .4, .6, \ldots$
76. Graph $y = \csc x$ on the interval $[0, 2\pi]$ by choosing $x = 0, .5, 1.5, 2, 2.5, \ldots$
77. Graph $y = \sec x$ on the interval $[0, 2\pi]$ by choosing $x = 0..5, 1.5, 2, 2.5, \ldots$
78. Graph $y = \cot x$ on the interval $[0, \pi]$ by choosing $x = 0, .2, .4, .6, \ldots$

Computer Exercises

In most versions of BASIC the three trigonometric functions sine, cosine, and tangent are expressed SIN(X), COS(X), TAN(X) where X *is in radians.* A student wrote the following program to generate a partial table of trigonometric values:

```
10 REM TRIG TABLE
20 FOR X = 0 TO 3.14 STEP .1
30 PRINT X; SIN(X); COS(X); TAN(X)
40 NEXT X
50 STOP
```

79. Refine the above program to:
    a. include column headings
    b. include columns for $\csc x$, $\sec x$, and $\cot x$. (*Hint:* write $1/\sin x$ for $\csc x$.)
    c. provide checks for division by zero
    d. skip a line after 5 lines of table values
80. Rewrite the program to print the six trigonometric functions for every angle (in *degree* measure) from 0° to 90°.

6.5 MORE GRAPHS OF THE TRIGONOMETRIC FUNCTIONS

In Section 6.3 each of the standard trigonometric functions was graphed. We repeat each below for handy reference because in this section, more trigonometric functions having similar shapes will be graphed.

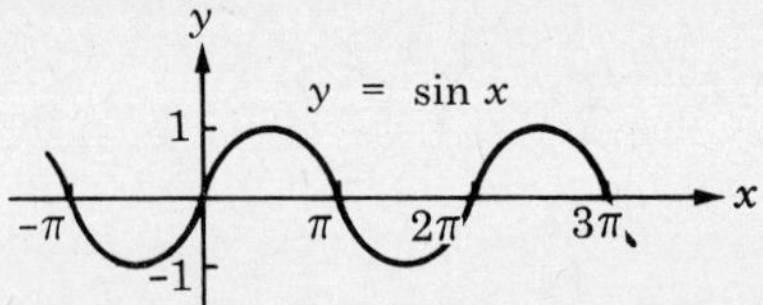

Domain: $\mathcal{R}$
Range: $-1 \leqslant y \leqslant 1$

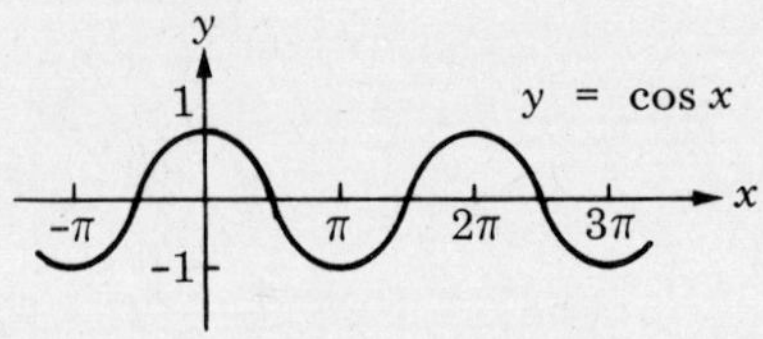

Domain: $\mathcal{R}$
Range: $-1 \leqslant y \leqslant 1$

Amplitude = 1
Period = 2π
(that is, as the argument x varies from 0 to 2π, one complete cycle is obtained).

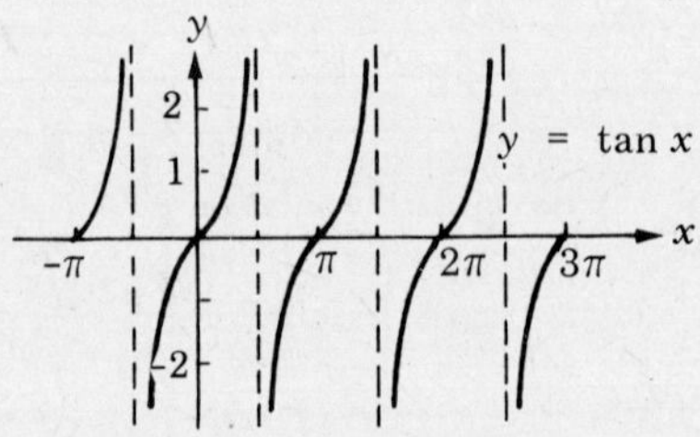

Domain:
$x \neq \pm\frac{\pi}{2}, \pm\frac{3\pi}{2}, \ldots$
Range: $\mathcal{R}$

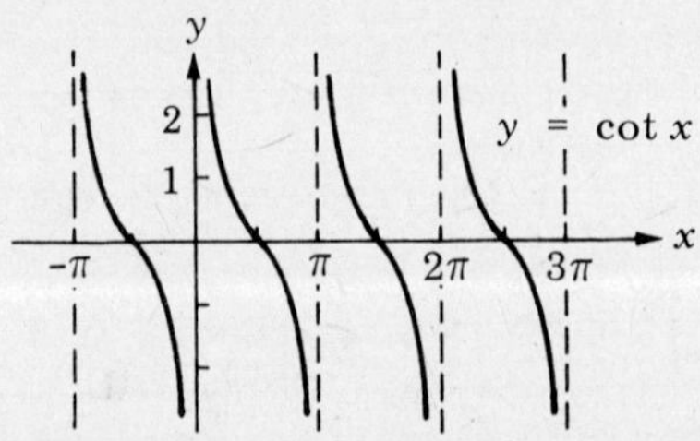

Domain:
$x \neq 0, \pm\pi, \pm 2\pi, \ldots$
Range: $\mathcal{R}$

No amplitude
Period = π
(that is, as the argument x varies from 0 to π, one complete cycle is obtained).

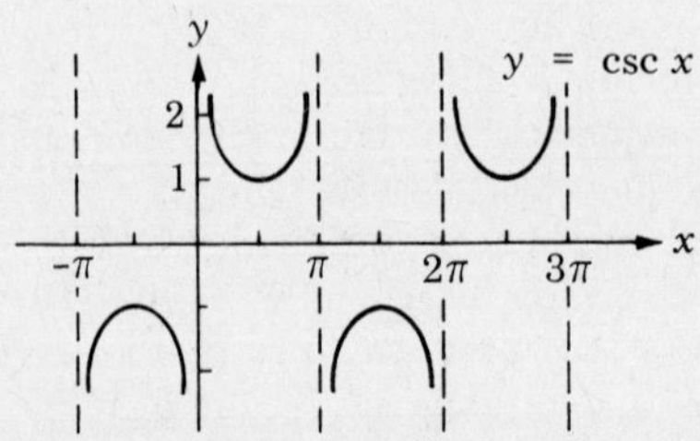

Domain:
$x \neq 0, \pm\pi, \pm 2\pi, \ldots$
Range:
$y \leqslant -1$ or $y \geqslant 1$

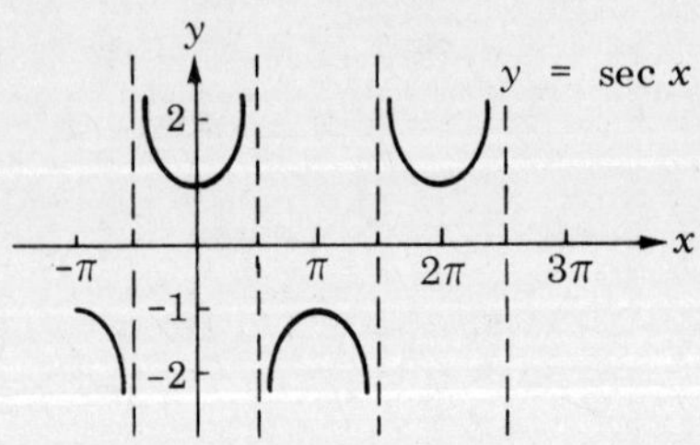

Domain:
$x \neq \pm\frac{\pi}{2}, \pm\frac{3\pi}{2}, \ldots$

Range: $y \leqslant -1$ or $y \geqslant 1$

No amplitude
Period = 2π
(that is, as the argument x varies from 0 to 2π, one complete cycle is obtained).

The notions that will be explored in the first three examples will be used to motivate Theorem 4.

EXAMPLE 1 Graph $y = 2 \sin x$.

Solution The x-intercepts are found by letting $y = 0$.

$$2 \sin x = 0$$
$$\sin x = 0$$
$$x = k\pi, \text{ where } k \text{ is any integer}$$

Theorem 3a

The y-intercept is found by letting $x = 0$.

$$y = 2 \sin 0$$
$$y = 0$$

The intercepts of $y = 2 \sin x$ are identical to those of the standard sine function. However, every ordinate on $y = 2 \sin x$ is *double* that of the corresponding ordinate on $y = \sin x$. The graph of $y = 2 \sin x$ is obtained from the graph of $y = \sin x$ (dashed line below) by **multiplying ordinates** by 2.

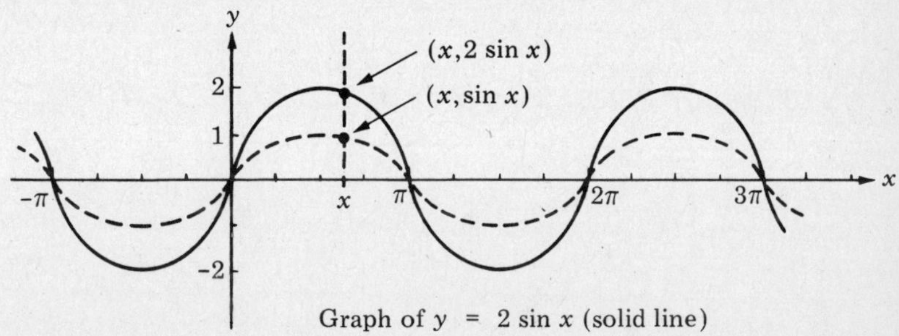

Graph of $y = 2 \sin x$ (solid line)

Notice that a real number factor multiplying $\sin x$ affects only the amplitude. In particular, the real number 2 multiplying $\sin x$ *doubles* the amplitude of $y = \sin x$.

Next we examine the effect of a real number factor in the argument.

EXAMPLE 2 Graph $y = \sin 2x$.

Solution The x-intercepts are found by letting $y = 0$.

$$\sin 2x = 0$$
$$2x = k\pi, \text{ where } k \text{ is any integer}$$
$$x = \frac{k\pi}{2}, \text{ where } k \text{ is any integer}$$

Theorem 3a

The y-intercept is $y = \sin 2(0) = \sin 0 = 0$. To complete the graph, a few carefully selected points are chosen.

x	y
$\frac{\pi}{4}$	1
$\frac{3\pi}{4}$	−1
$\frac{5\pi}{4}$	1

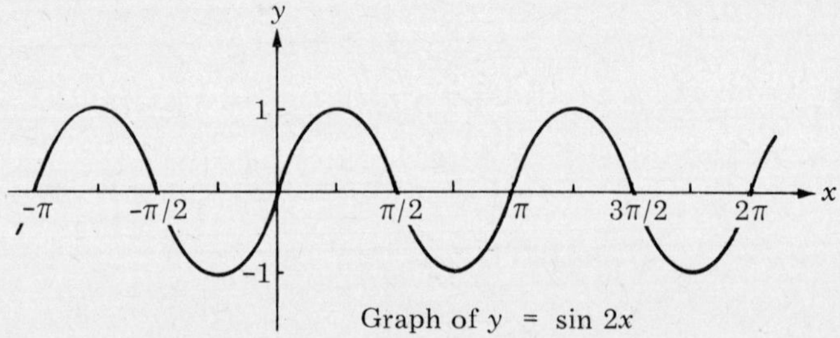

Graph of $y = \sin 2x$

Notice that a real number factor within the argument affects only the period. In particular, the factor 2 within the argument ($y = \sin 2x$) results in a period of *half* that of $y = \sin x$.

In the next example the effect of addition within the argument is examined.

EXAMPLE 3 Graph $y = \sin\left(2x + \frac{\pi}{2}\right)$.

Solution First determine the x-intercepts.

$$\sin\left(2x + \frac{\pi}{2}\right) = 0$$
$$2x + \frac{\pi}{2} = k\pi, \text{ where } k \text{ is any integer}$$

Theorem 3a

$$2x = k\pi - \frac{\pi}{2} = \left(k - \frac{1}{2}\right)\pi$$
$$x = \left(k - \frac{1}{2}\right)\frac{\pi}{2}$$

k	0	1	−1	2	−2	3	−3 . . .
x-intercepts	$-\frac{\pi}{4}$	$\frac{\pi}{4}$	$-\frac{3\pi}{4}$	$\frac{3\pi}{4}$	$-\frac{5\pi}{4}$	$\frac{5\pi}{4}$	$-\frac{7\pi}{4}$

The y-intercept is $y = \sin\left(2 \cdot 0 + \frac{\pi}{2}\right) = \sin\frac{\pi}{2} = 1$.

To complete the graph, a few carefully selected points are chosen.

x	y
$\frac{\pi}{2}$	-1
$-\frac{\pi}{2}$	-1
π	1

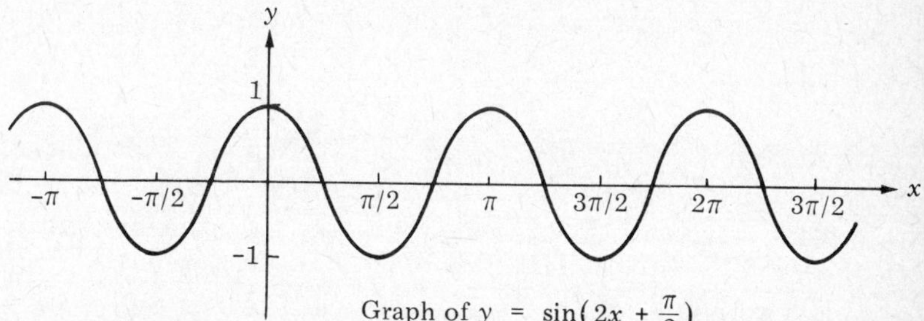

Graph of $y = \sin(2x + \frac{\pi}{2})$

Notice that the graph of $y = \sin 2x$ (from Example 2), if shifted along the x-axis $\frac{\pi}{4}$ units to the *left*, would become the graph of $y = \sin\left(2x + \frac{\pi}{2}\right)$. Thus, addition of a positive real number within the argument of the sine function causes a shift along the x-axis to the left.

The amount of the shift is called the **phase shift** or **phase angle**. The concepts of amplitude, period, and phase shift are summarized in the following theorem for sine and cosine functions.

THEOREM 4 Given the real numbers a, b, c ($a > 0$, $b > 0$), then

a. $f(x) = a \sin(bx + c)$ has amplitude $= a$, period $= \frac{2\pi}{b}$, and phase shift $= -\frac{c}{b}$.

b. $f(x) = a \cos(bx + c)$ has amplitude $= a$, period $= \frac{2\pi}{b}$ and phase shift $= -\frac{c}{b}$.

A negative phase shift indicates shift to the left and a positive phase shift indicates shift to the right.

Theorem 4 makes no mention of what happens when either $a < 0$ or $b < 0$. These situations are examined in the exercises. Also in the exercises, the reader is asked to prove parts of Theorem 4. Here we show how the theorem can be used to graph sine and cosine functions.

EXAMPLE 4 Graph $y = 3 \sin\left(\frac{1}{2}x - \frac{\pi}{4}\right)$.

Solution The amplitude is 3. The period is $\frac{2\pi}{b} = \frac{2\pi}{\frac{1}{2}} = 4\pi$. The phase shift is $-\frac{c}{b} = -\frac{-\pi/4}{\frac{1}{2}} = \frac{\pi}{2}$ (to the right).

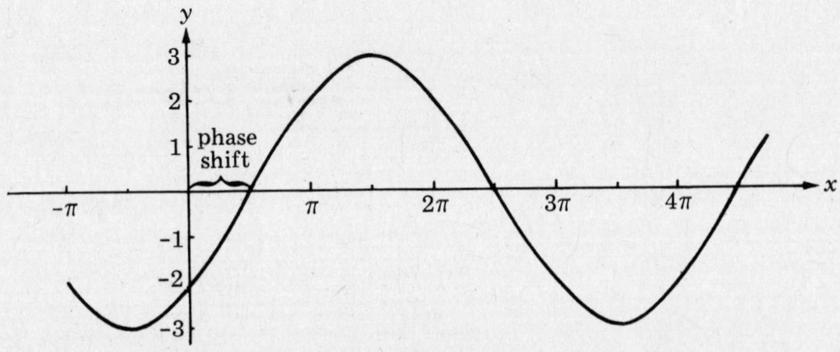

Graph of $y = 3\sin(\frac{1}{2}x - \frac{\pi}{4})$

EXAMPLE 5 Graph $y = 2\cos(3x + \pi)$.

Solution The amplitude is 2. The period is $\frac{2\pi}{b} = \frac{2\pi}{3}$. The phase shift is $-\frac{c}{b} = -\frac{\pi}{3}$ (to the left).

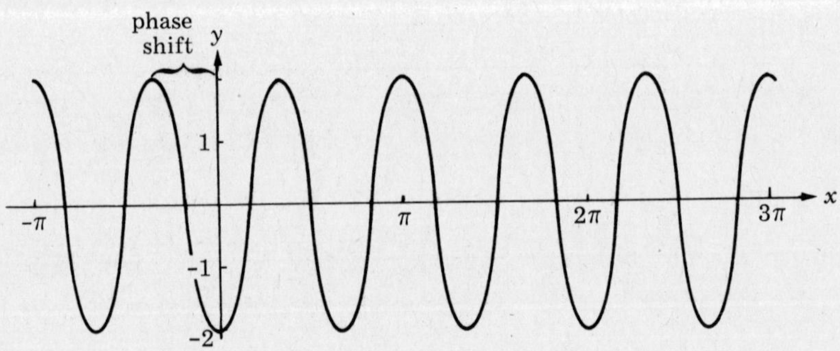

Graph of $y = 2\cos(3x + \pi)$

Next the graph of $y = \tan(bx + c)$ will be examined. We know that the standard tangent function $y = \tan x$ has a period of π. That is, as the argument x varies from 0 to π, one complete cycle of $y = \tan x$ is obtained. Reasoning in a similar manner, as the argument $bx + c$ varies from 0 to π, one complete cycle of $y = \tan(bx + c)$ is obtained. Assuming $b > 0$, we have:

$0 \leqslant bx + c \leqslant \pi$ — Subtract c from each member of the inequality

$-c \leqslant bx \leqslant \pi - c$ — Divide each member of the inequality by the number $b > 0$

$-\dfrac{c}{b} \leqslant x \leqslant \dfrac{\pi}{b} - \dfrac{c}{b}$

This interval starts at $-\dfrac{c}{b}$ and has a length of $\dfrac{\pi}{b}$. We are thus able to state the following theorem.

THEOREM 5 Given the real numbers b, c $(b > 0)$, then $f(x) = \tan(bx + c)$ has period $\dfrac{\pi}{b}$ and phase shift $-\dfrac{c}{b}$.

The case of $b < 0$ in $f(x) = \tan(bx + c)$ is treated in the exercises.

EXAMPLE 6 Graph $f(x) = \tan\left(\dfrac{1}{2}x - \dfrac{\pi}{4}\right)$.

Solution The period is $\dfrac{\pi}{b} = \dfrac{\pi}{\frac{1}{2}} = 2\pi$. The phase shift is $-\dfrac{c}{b} = -\dfrac{-\pi/4}{\frac{1}{2}} = \dfrac{\pi}{2}$ (to the right). We also graph the point $(\pi, 1)$.

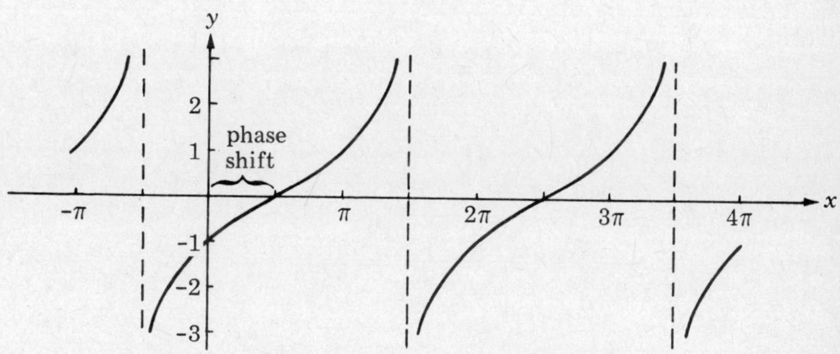

Graph of $f(x) = \tan\left(\frac{1}{2}x - \frac{\pi}{4}\right)$

The graph of $f(x) = a\tan(bx + c)$ can be obtained by using Theorem 5 to first graph $g(x) = \tan(bx + c)$. Then each ordinate is multiplied by a.

EXAMPLE 7 Graph $y = \frac{1}{2}\tan\frac{1}{3}x$.

Solution First consider $y = \tan \frac{1}{3}x$. The period is $\frac{\pi}{b} = \frac{\pi}{\frac{1}{3}} = 3\pi$. The phase shift is $-\frac{c}{b} = 0$. On the axes below, $y = \tan \frac{1}{3}x$ is the curved dashed line. Multiplying ordinates on the graph by $\frac{1}{2}$ gives the final result (solid line).

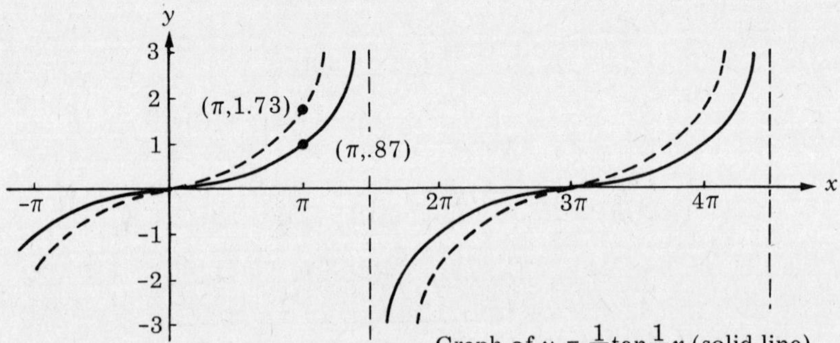

Graph of $y = \frac{1}{2}\tan\frac{1}{3}x$ (solid line)

In Section 6.4 the reciprocal identities were presented. They are: $\csc s = \dfrac{1}{\sin s}$, $\sec s = \dfrac{1}{\cos s}$, $\cot s = \dfrac{1}{\tan s}$. These can be conveniently used as a graphing tool to graph $y = a \csc(bx + c)$, $y = a \sec(bx + c)$, or $y = a \cot(bx + c)$. An example follows.

EXAMPLE 8 Graph $g(x) = 2 \csc 2x$.

Solution First, graph $y = \csc 2x$. Since $\csc 2x = \dfrac{1}{\sin 2x}$, the graph of $y = \csc 2x$ can be obtained from $y = \sin 2x$ (Example 2) by taking reciprocals of

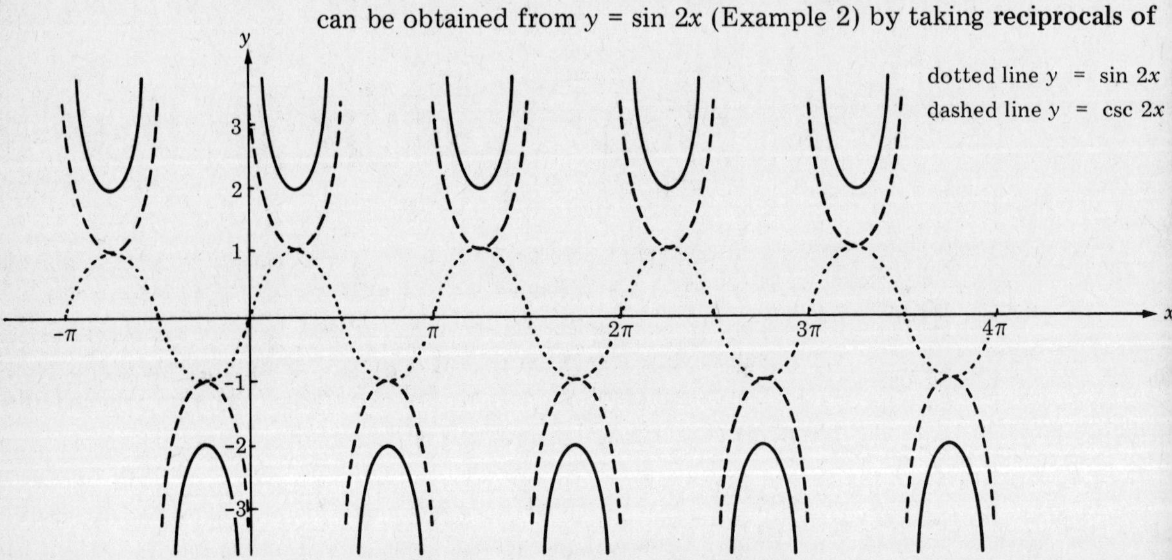

Graph of $g(x) = 2 \csc 2x$ (solid line)

ordinates. Then the graph of $g(x) = 2 \csc 2x$ is obtained from $y = \csc 2x$ by multiplication of ordinates.

In the following example a technique known as **addition of ordinates** is used. This method can be applied to sums or differences of any functions.

EXAMPLE 9 Graph $f(x) = \sin x + 2 \cos x$.

Solution First, graph $f_1(x) = \sin x$ and $f_2(x) = 2 \cos x$ on the same set of axes. Next, proceeding from left to right select several x values and add the two ordinates from $f_1(x)$ and $f_2(x)$ in order to obtain the ordinate of $f(x)$.

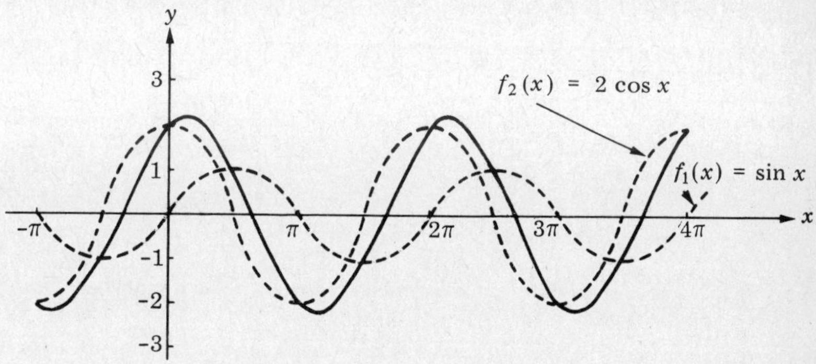

Graph of $y = \sin x + 2 \cos x$ (solid line)

EXERCISES FOR SECTION 6.5

In Exercises 1 through 24, graph each function. (See Theorem 4 and Examples 4 and 5.)

1. $y = 3 \sin x$
2. $y = \sin 3x$
3. $y = 3 \cos x$
4. $y = \cos 3x$
5. $y = 3 \sin 3x$
6. $y = 3 \cos 3x$
7. $y = 2 \sin 4x$
8. $y = 4 \sin 2x$
9. $y = 4 \cos 4x$
10. $f(x) = 5 \sin \frac{1}{2}x$
11. $g(x) = 2 \cos \dfrac{\pi}{3} x$
12. $f(x) = \sin(2x + \pi)$
13. $g(x) = \sin(2x - \pi)$
14. $h(x) = \dfrac{1}{2} \cos\left(\dfrac{1}{2}x + \dfrac{\pi}{4}\right)$
15. $r(x) = \dfrac{1}{2} \cos\left(\dfrac{1}{2}x - \dfrac{\pi}{4}\right)$
16. $y = \sin\left(x + \dfrac{\pi}{4}\right)$
17. $y = \cos\left(x - \dfrac{\pi}{4}\right)$
18. $y = \sin\left(x - \dfrac{\pi}{2}\right)$

19. $y = 4\cos\left(x - \frac{\pi}{8}\right)$
20. $f(x) = 2\sin(2x + 2\pi)$
21. $y = \sin(x - \pi)$
22. $g(x) = 8\cos 4\pi x$
23. $h(x) = \frac{1}{4}\cos 2\pi x$
24. $f(x) = \frac{1}{2}\cos\left(\frac{1}{2}x - \frac{\pi}{2}\right)$

In Exercises 25 through 36, graph each function. (See Theorem 5 and Examples 6 and 7.)

25. $y = \tan x$
26. $y = 2\tan \pi x$
27. $y = \tan 2x$
28. $y = 2\tan 2x$
29. $y = 2\tan(2x + \pi)$
30. $y = 2\tan(2x - \pi)$
31. $f(x) = \tan(4x - 2\pi)$
32. $g(x) = \tan(4x + \pi)$
33. $h(x) = 3\tan\left(x + \frac{\pi}{4}\right)$
34. $y = \frac{1}{4}\tan 4x$
35. $r(x) = \frac{1}{2}\tan\left(x - \frac{\pi}{8}\right)$
36. $y = \tan\left(2x - \frac{\pi}{4}\right)$

In Exercises 37 through 48, graph each function. (See Example 8.)

37. $y = \sec 2x$
38. $y = \csc 4x$
39. $y = \cot \frac{1}{2}x$
40. $y = 3\sec 2x$
41. $y = 2\csc 4x$
42. $y = \frac{1}{2}\cot \frac{1}{2}x$
43. $f(x) = 3\sec(2x - \pi)$
44. $g(x) = 2\csc(4x + 2\pi)$
45. $h(x) = \frac{1}{2}\cot(\frac{1}{2}x + \pi)$
46. $f(x) = 3\sec\left(2x - \frac{\pi}{2}\right)$
47. $g(x) = 2\csc(4x + \pi)$
48. $h(x) = \frac{1}{2}\cot\left(\frac{1}{2}x - \frac{\pi}{2}\right)$

In Exercises 49 through 58, graph each function by using addition of ordinates (See Example 9.)

49. $y = \sin x + \cos x$
50. $y = 2\sin x + \cos x$
51. $y = \sin x - \cos x$ (*Hint:* graph $f_1(x) = \sin x$ and $f_2(x) = -\cos x$.)
52. $y = \sin x + \cos 2x$
53. $y = \sin x - \cos 2x$
54. $y = 2 + \sin x$
55. $y = 3 - \cos x$
56. $y = x + \sin x$
57. $y = \sin 2x + \cos x$
58. $y = |x| + \sin x$
59. Show that $f(x) = a\sin(bx + c)$ has period $\frac{2\pi}{b}$ and phase shift $-\frac{c}{b}$ (assume $b > 0$). *Hint:* let the argument $bx + c$ vary between 0 and 2π.
60. Show that $f(x) = a\cos(bx + c)$ has period $\frac{2\pi}{b}$ and phase shift $-\frac{c}{b}$ (assume $b > 0$). *Hint:* let the argument $bx + c$ vary between 0 and 2π.

Theorem 4 does not cover the situation $a < 0$ in $y = a\sin(bx + c)$. If $a < 0$, first graph $y = |a|\sin(bx + c)$ by using Theorem 4 and then obtain $y = a\sin(bx + c)$ by multiplying ordinates by -1. This *reflects the graph* of $y = |a|\sin(bx + c)$ *across the x-axis.* A similar technique can be used with the other trigonometric functions. Using this method, graph each function in Exercises 61 through 69.

61. $y = -3\sin x$
62. $y = -\cos 3x$
63. $y = -2\sin 4x$
64. $y = -5\sin \frac{1}{2}x$
65. $y = -\tan x$
66. $y = -2\tan 2x$
67. $y = -3\sec(2x - \pi)$
68. $y = -\cot \frac{1}{2}x$
69. $y = -\csc 2x$

Theorems 4 and 5 do not cover the situation $b < 0$ in $y = a \sin(bx + c)$, $y = a \cos(bx + c)$ or $y = \tan(bx + c)$. In Chapter 7 it is shown that:

$$\sin(-s) = -\sin s, \ \cos(-s) = \cos s, \ \tan(-s) = -\tan s.$$

By using these identities, graph each function in Exercises 70 through 75.

70. $y = \sin(-2x)$
71. $y = -\sin(-2x)$
72. $y = \cos(-4x)$
73. $y = \tan(-\frac{1}{2}x)$
74. $y = 2 \sin(-2x)$
75. $y = \cos(-3x)$

In Exercises 76 through 81, determine $f \circ g(x)$.

76. $f(u) = \sin u$, $u = g(x) = x^2$
77. $f(u) = u^2$, $u = g(x) = \sin x$
78. $f(u) = \cos u$, $u = g(x) = 3x + 1$
79. $f(u) = \tan u$, $u = g(x) = \sqrt{x}$
80. $f(u) = (\sin u)^2 + 1$, $u = g(x) = 2x$
81. $f(u) = (2u + 1)^{1/2}$, $u = g(x) = \cos x$

Calculator Exercises

82. Graph $y = \sin(x + 1)$, by letting $x = 0, \pm.5, \pm1.0, \pm1.5, \ldots, \pm6.5$.
83. Graph $y = (\sin x)^2$, by letting $x = 0, \pm.5, \pm1.0, \pm1.5, \ldots, \pm6.5$.
84. Graph $y = \sin x + \cos x$, by letting $x = 0, \pm.5, \pm1.0, \pm1.5, \ldots, \pm6.5$.

6.6 APPLICATIONS

In this section some of the many practical applications of trigonometry and the trigonometric functions are investigated.

Natural and Forensic Science Problems

When any object is projected upward from the ground at an angle θ with respect to the horizontal position, and with initial velocity, v_o, the equation $R = \frac{v_o^2}{32} \sin 2\theta$ is an algebraic model for the horizontal distance or range, R, traveled (in feet) (air resistance is neglected).

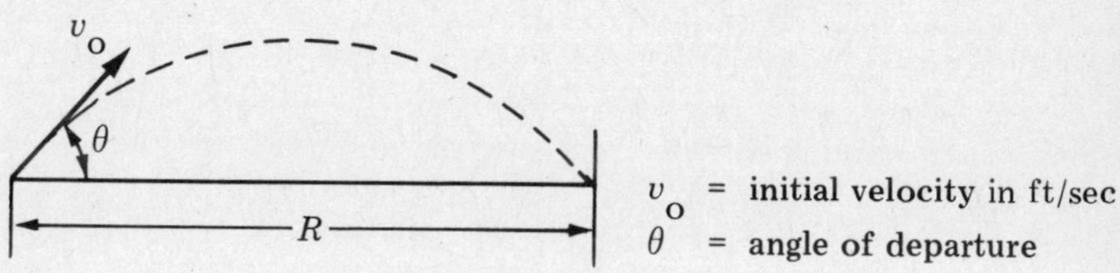

Thus, once v_o is known, R is a function of θ.

EXAMPLE 1 a. A football is kicked upward from the ground with an initial velocity of 40 ft/sec. If $\theta = 30°$, what is the range R?

b. Graph the function $R = \frac{v_o^2}{32} \sin 2\theta$ with R on the vertical axis. $\left(0 \leqslant \theta \leqslant \frac{\pi}{2}, \text{ and treat } v_o \text{ as a constant.}\right)$

c. For what angle θ will the range R be a maximum?

Solution a. $R = \frac{v_o^2}{32} \sin 2\theta = \frac{(40)^2}{32} \sin 60°$

$R = 43.3$ ft.

b. $R = \frac{v_o^2}{32} \sin 2\theta$ has amplitude $\frac{v_o^2}{32}$, period π and phase shift 0.

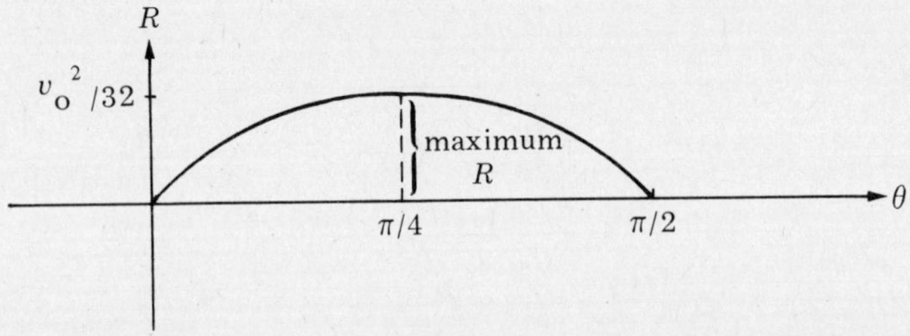

c. From the graph above, maximum R occurs when $\theta = \frac{\pi}{4}$ or 45°.

EXAMPLE 2 A bullet is discovered in the wall of a room, with an entry point 2 meters above the floor. The path of the bullet in the wall is inclined upward at an angle of 25°. How far from the wall was the gun when fired, if the gun was fired from 1 meter above the floor? (Assume the bullet traveled in a straight line.)

Solution The distance x must be determined.

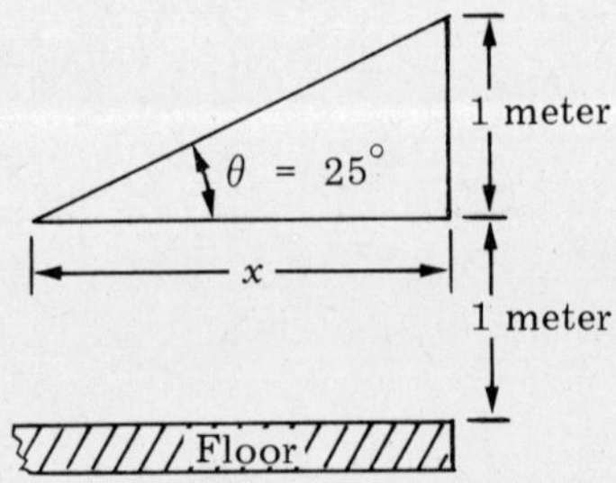

$$\tan 25° = \frac{1}{x}$$

$$x = \frac{1}{\tan 25°} = 2.14 \text{ meters}$$

Navigation and Surveying Problems Knowledge of trigonometry is essential in navigation and surveying. Direction is specified by stating either the **true bearing** or the **relative bearing**. The former expresses the angle between a ship's direction and a north-south line. The latter expresses the angle between a ship's direction and the direction of some other object. In the examples that follow both types of bearings are employed.

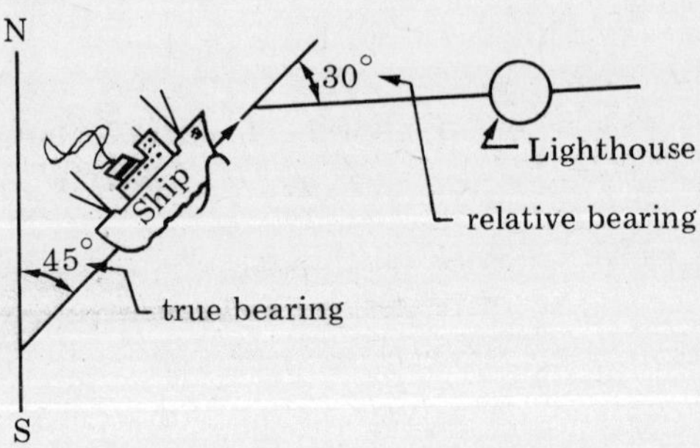

EXAMPLE 3 A ship is sailing on the course shown below.

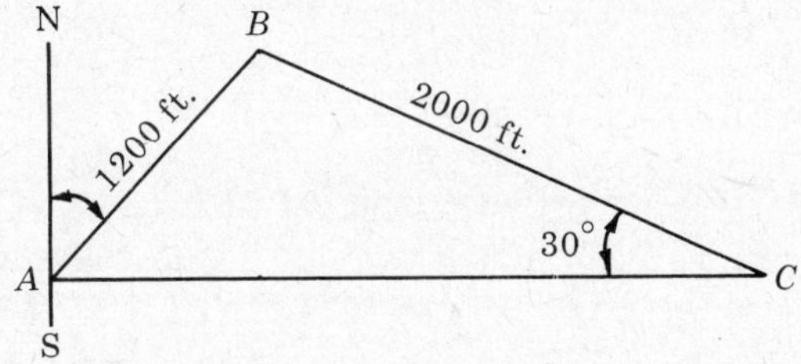

a. Determine the distance $\overline{AC}$.
b. Determine the true bearing $\measuredangle NAB$.

Solution a. First, draw a perpendicular line from point B to $\overline{AC}$. Since $\overline{BD}$ is the shorter leg of the right triangle BDC, $\overline{BD}$ must be one-half the hypotenuse.

$$\overline{BD} = 1000 \text{ ft}$$

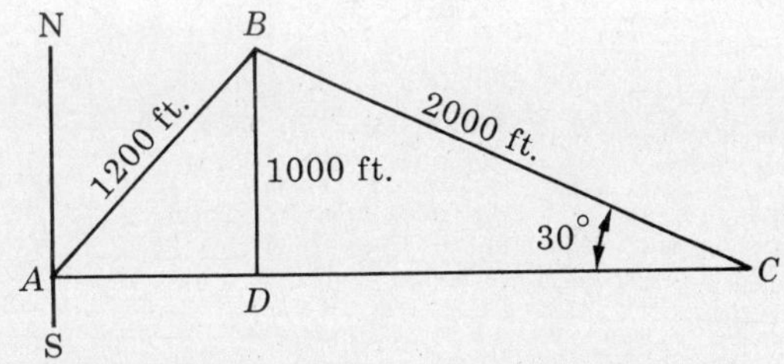

$\overline{AC}$ will be determined by adding $\overline{AD}$ and $\overline{DC}$.

$$\overline{AD} = \sqrt{(1200)^2 - (1000)^2} = \sqrt{440{,}000} \approx 663.32 \text{ ft}$$
$$\overline{DC} = \sqrt{(2000)^2 - (1000)^2} = \sqrt{3{,}000{,}000} \approx 1732.05 \text{ ft}$$
$$\overline{AC} \approx 2395.4 \text{ ft}$$

b. $\sin \measuredangle BAD = \dfrac{1000}{1200} = .8333$

Using Table III,

$$\measuredangle BAD = 56.4^\circ$$

Hence the true bearing

$$\measuredangle NAB = 90^\circ - 56.4^\circ = 33.6^\circ$$

The distance of a ship from a fixed object can always be determined from two consecutive bearings on that object if there is a straight line of known distance between the two observations (see Exercise 13). This same idea is used by a surveyor to determine an unknown height by sighting to the top of the object and measuring the acute angle between the line of sight and the horizontal (called the **angle of elevation**). Then by moving a known distance toward or away and again measuring the angle of elevation, the height can be determined. The next example illustrates this.

EXAMPLE 4 A surveyor measures the angle of elevation to the top of a building to be 32°. She then moves 100 ft closer to the building and measures the angle of elevation to be 48.2°. How high is the building?

Solution We desire to determine h. Using the right triangles ADC and BDC,

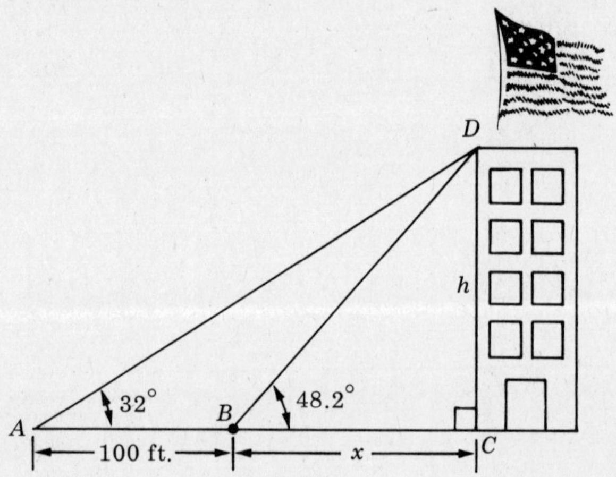

$$\tan 32° = \frac{h}{100 + x}$$

$$\tan 48.2° = \frac{h}{x}$$

Solve each equation for x.

$$x = \frac{h}{\tan 32°} - 100$$

$$x = \frac{h}{\tan 48.2°}$$

Since the two right-hand sides of the equations above equal the same quantity, x, they can be set equal to each other.

$$\frac{h}{\tan 32°} - 100 = \frac{h}{\tan 48.2°}$$

$$\frac{h}{\tan 32°} - \frac{h}{\tan 48.2°} = 100$$

$$h\left(\frac{1}{\tan 32°} - \frac{1}{\tan 48.2°}\right) = 100$$

$$h(\cot 32° - \cot 48.2°) = 100$$

$$h = \frac{100}{\cot 32° - \cot 48.2°} = \frac{100}{1.6003 - .8941} = \frac{100}{.7062}$$

$$h = 141.6 \text{ ft.}$$

Industrial and Engineering Problems There are numerous uses of trigonometry in industry. In many settings (automobile production, plumbing, etc.) one must determine the length of pipe needed to fit a certain curve. The next example is typical.

EXAMPLE 5 Determine the total length of pipe along the center line in the following diagram.

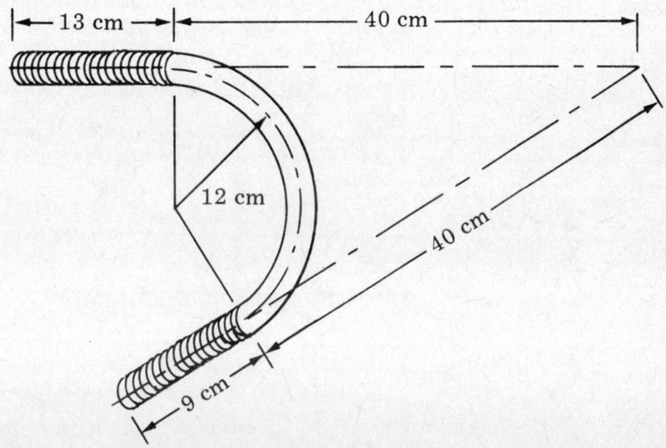

Solution First let us redraw the essential parts of the diagram. We need to know the length of circular arc from B to D and once known, it will be added to $\overline{FB}$ and $\overline{ED}$.

$$\tan \measuredangle BCA = \frac{40}{12} = 3.3333$$
$$\measuredangle BCA = 73.3^\circ$$

Hence $\measuredangle BCD = 2(73.3^\circ) = 146.6^\circ = 146.6^\circ\left(\frac{\pi}{180^\circ}\right) = 2.56$ (radians).

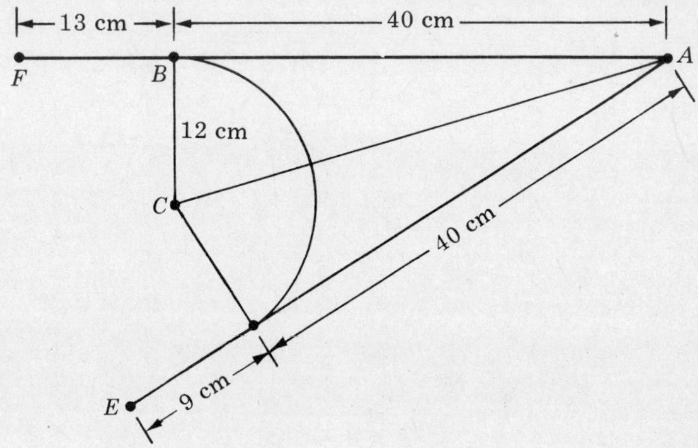

Using the arclength formula $s = \theta r$, arclength $\overline{BD} = (2.56)(12 \text{ cm}) = 30.72$ cm.

$$\text{Length of pipe} = 30.72 \text{ cm} + 13 \text{ cm} + 9 \text{ cm}$$
$$\text{Length of pipe} = 52.72 \text{ cm}$$

Engineers frequently work with systems in which some quantity is changing with time. A simple system is that of a **frictionless harmonic oscillator** represented by a mass oscillating up and down on the end of a spring. Being frictionless, the up and down motion continues forever and it can be shown that the displacement of the mass from the "at rest" position varies *sinusoidally* with time.* More realistic than this is the **damped harmonic oscillator** that is discussed next.

EXAMPLE 6 The equation $x = 2^{-t} \cos \pi t$ is an algebraic model for a certain damped haromonic oscillator where x is displacement and t is time. Graph this equation and comment on the nature of the motion. $(t \geqslant 0)$.

Solution In order to graph $x = 2^{-t} \cos \pi t$ $(t \geqslant 0)$ we will graph $f(t) = 2^{-t}$ and $g(t) = \cos \pi t$ and then multiply ordinates to obtain $x = 2^{-t} \cos \pi t$. Since 2^{-t} is *never* zero, the zeros of $x = 2^{-t} \cos \pi t$ will occur when $\cos \pi t = 0$. From Theorem 3b, $\pi t = \dfrac{(2k+1)\pi}{2}$ or $t = \dfrac{2k+1}{2}$. Thus the zeros of $x = 2^{-t} \cos \pi t$ $(t \geqslant 0)$ are $t = \frac{1}{2}, \frac{3}{2}, \frac{5}{2}, \frac{7}{2}, \ldots$.

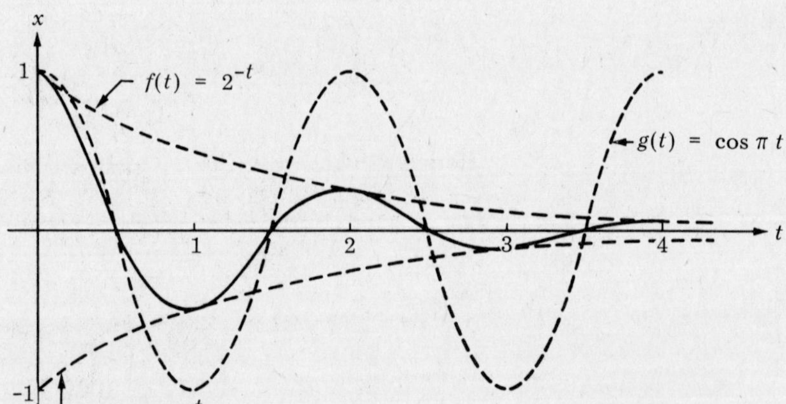

Graph of $x = 2^{-t} \cos \pi t$ (solid line)

The variable x oscillates with time, but its maximum and minimum values are decreasing with time.

*The equation relating displacement and time is a sine or cosine function.

EXERCISES FOR SECTION 6.6

1. A football is kicked upward from the ground with an initial velocity of 60 ft/sec. If the angle of departure is 60°, what is the range R?
2. A golfball is hit upward from the ground with an initial velocity of 150 ft/sec. If the angle of departure is 42°, what is the range R?
3. If the initial velocity of a projectile is 50 ft/sec, graph the function

$$R = \frac{v_o^2}{32} \sin 2\theta \left(0 \leqslant \theta \leqslant \frac{\pi}{2}\right).$$

4. If the angle of departure of a projectile is 15°, graph the function

$$R = \frac{v_o^2}{32} \sin 2\theta \ (v_o \geqslant 0).$$

The equation for the range of a projectile, namely $R = \frac{v_o}{32} \sin 2\theta$, gives the range R when v_o is expressed in ft/sec. However, if v_o is expressed in m/sec, one must use the equation $R = \frac{v_o^2}{9.8} \sin 2\theta$ to obtain R in meters.

5. A soccerball is kicked upward from the ground with an initial velocity of 15 m/sec at a 20° angle of departure. What is the range R?
6. Graph the equation $R = \frac{v_o^2}{9.8} \sin 2\theta$. $\left(0 \leqslant \theta \leqslant \frac{\pi}{2}\right.$ and treat v_o as a constant.$\left.\right)$
7. A bullet is discovered in the wall of a room, with an entry point 3 meters above the floor. The path of the bullet in the wall is inclined upward with a 28° angle of elevation. Determine how far from the wall the gun was when fired, if:
   a. the gun was fired from the floor.
   b. the gun was fired from 1.2 meters above the floor.
8. Visual identification is extremely important in criminal work and testifying. For a person with normal vision, experience has shown that two points on an object must subtend an angle of at least one minute in the observer's eye in order that they appear separate.
   a. At a distance of 100 feet, what angle is subtended by two points 1 in. apart?
   b. At a distance of 1000 ft, what angle is subtended by two points 1 in. apart?
   c. For a person with normal vision, would the points in part a appear separate?
   d. For a person with normal vision, would the points in part b appear separate?
9. In an arson investigation it was necessary to construct a model of a building where a suspicious fire had occurred in a stairway. The height and base of the stairs are shown in the diagram.
   a. Determine $\overline{AD}$.
   b. Determine $\measuredangle ADC$.
   c. Determine $\measuredangle BDC$.

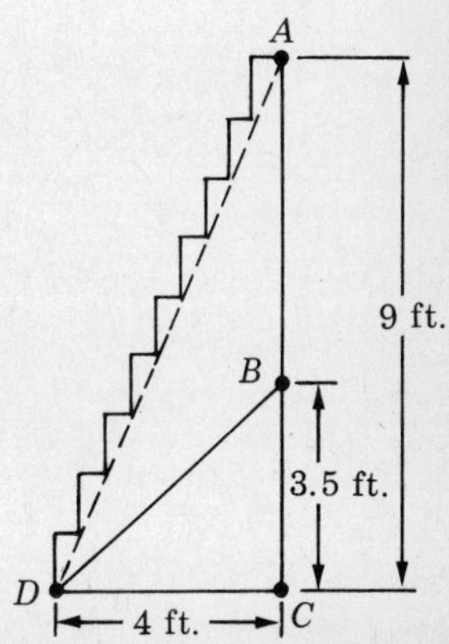

10. A ship is sailing on the course shown at the right.
   a. Determine $\overline{AC}$.
   b. Determine the true bearing $\measuredangle NAB$.
   c. Determine $\measuredangle B$.

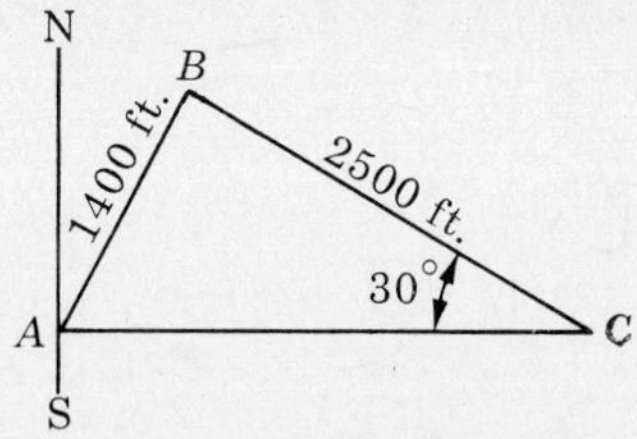

11. A ship is sailing on the course shown at the right.
   a. Determine $\overline{AC}$.
   b. Determine the true bearing $\measuredangle NAB$.

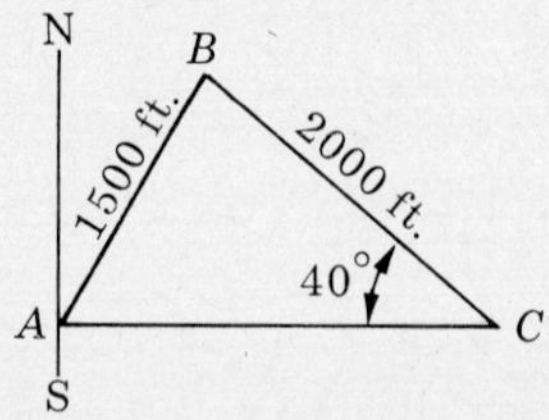

12. Suppose a sailboat has a sail area A. As she heels (tips to one side) the wind "sees" less sail. The amount of sail seen by the wind while heeling is called **effective sail area**, E. The equation $E = A \cos \theta$ is an algebraic model for the effective sail area E for an angle of heel θ.
   a. Determine the effective sail area if $\theta = 10°$.
   b. Determine the effective sail area if $\theta = 20°$.
   c. Determine the effective sail area if $\theta = 30°$.
   d. Between 0° and 10°, what is the percent of loss of sail area?
   e. Between 10° and 20°, what is the percent of loss of sail area?
   f. Between 20° and 30°, what is the percent of loss of sail area?
13. The distance of a ship from a fixed object can always be determined from two consecutive bearings on that object if there is a straight line of known distance between the two observations. Show that

$$x = \frac{d}{\cot \theta - \cot \alpha}$$

and

$$y = \frac{d}{\dfrac{\tan \alpha}{\tan \theta} - 1}$$

and hence $BC = \sqrt{x^2 + y^2}$ is able to be determined.

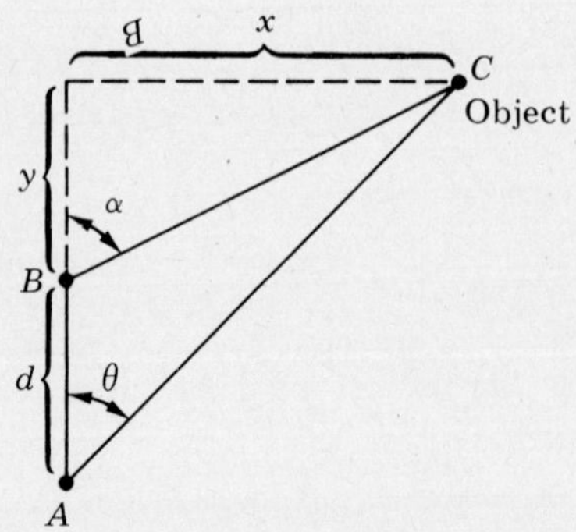

14. An airplane is flying at a height of 20,000 feet. Sighting ahead to a town gives an **angle of depression** of 40°. (The angle of depression is the acute angle between the line of sight and the horizon-

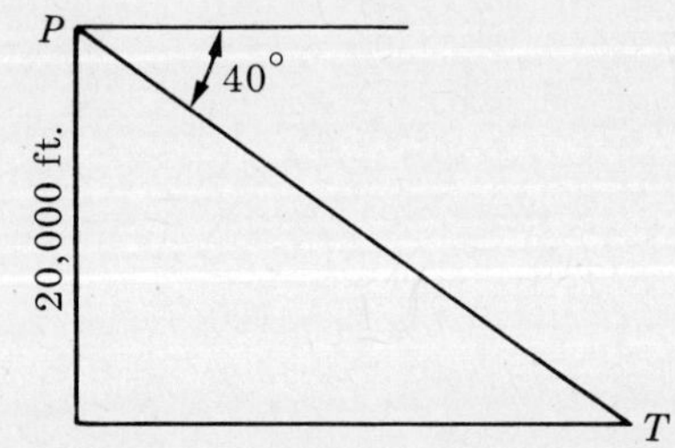

tal.) Determine how far away the airplane is from the town.

15. An airplane is flying at a height of 25,000 feet. Sighting ahead to a town gives a 40° angle of depression. Sighting ahead to another town gives a 20° angle of depression. Determine the distance between the towns, d.

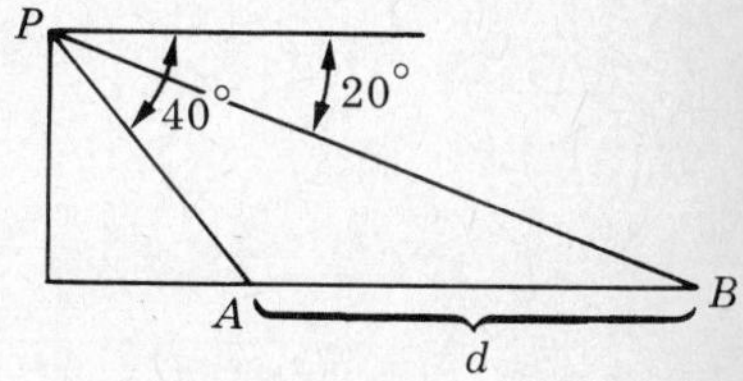

16. In order to minimize undesirable reflections of light the Federal Aviation Administration specifies that the windows in air traffic control cabs at airports be tilted outward from the sill to the ceiling at an angle of 15° from the vertical. Show that $t = w + 2h \tan 15^\circ$.

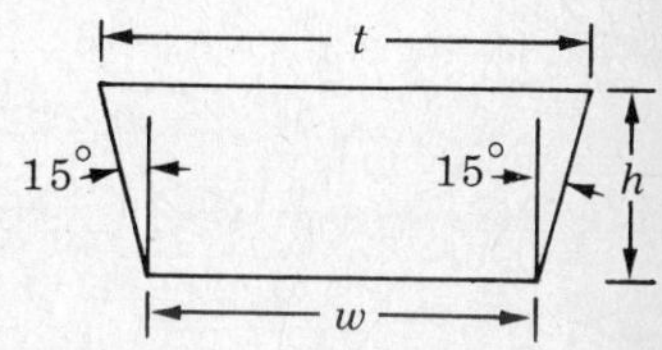

17. The Federal Aviation Administration requires that during landing the angle of descent (angle of depression) be 3° for the last 4 miles of descent. How high will an airplane be at the moment of beginning this last stage of descent?

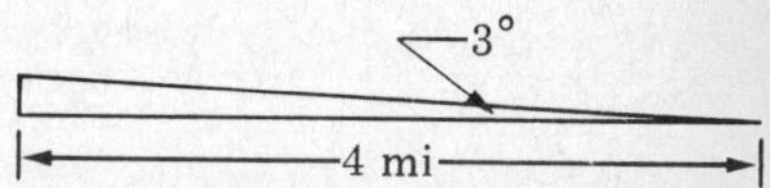

18. A radio antenna ($\overline{AC}$) 100 ft high is to be installed and will be supported by two guy wires $\overline{BA}$ and $\overline{DA}$.
    a. Determine the length of wire $\overline{AD}$.
    b. Determine the angle of elevation of wire $\overline{BA}$.

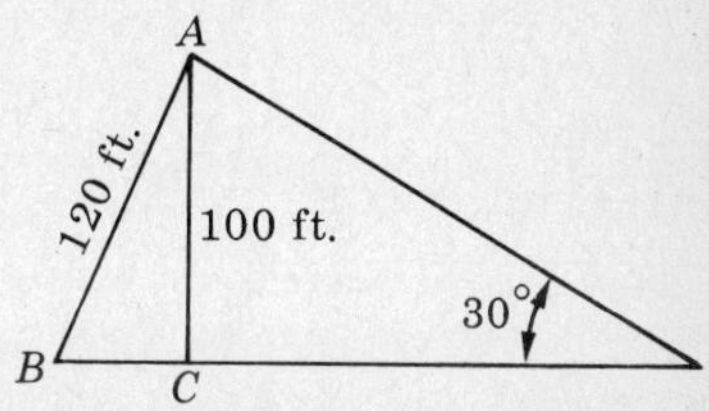

19. A triangular piece of property has a perimeter of 1000 ft. (That is, $x + y + z = 1000$.)
    a. Show that

$$y = \frac{1000}{1 + \frac{1}{\tan 25^\circ} + \frac{1}{\sin 25^\circ}}$$

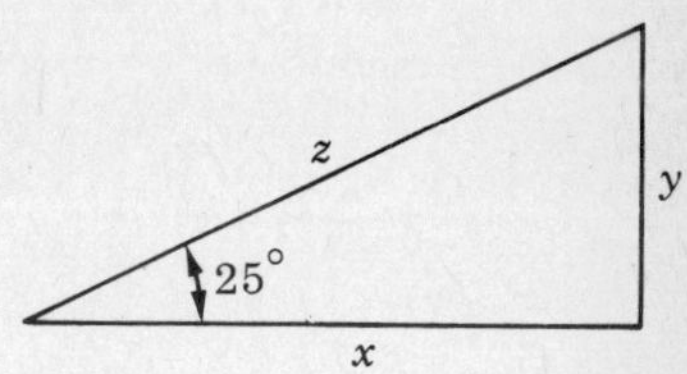

    b. Show that the area

$$A = \frac{1}{2}\left(\frac{1000}{1 + \frac{1}{\tan 25^\circ} + \frac{1}{\sin 25^\circ}}\right)^2\left(\frac{1}{\tan 25^\circ}\right)$$

    c. Evaluate the expression in part b. (A calculator will save time.)

20. Determine the length of road along the center line in the following diagram.

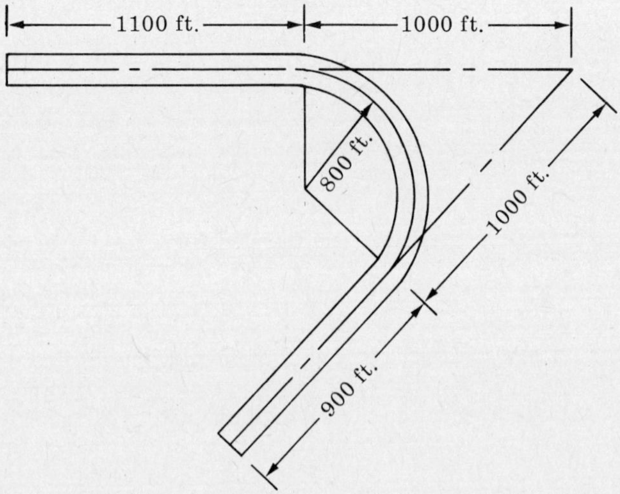

21. Determine the total length of pipe along the center line in the adjacent diagram.

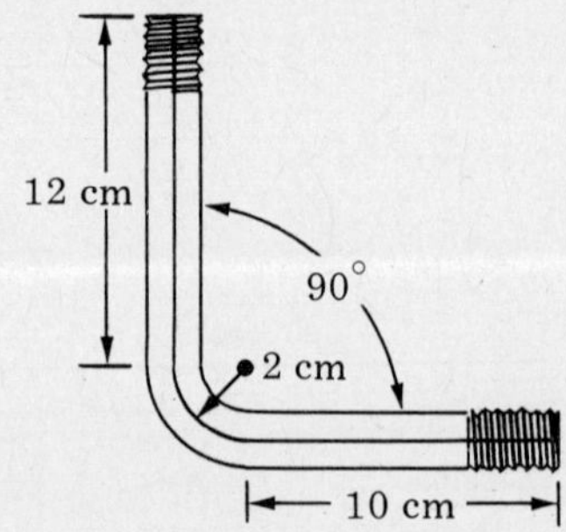

22. Determine the total length of pipe along the center line in the adjacent diagram.

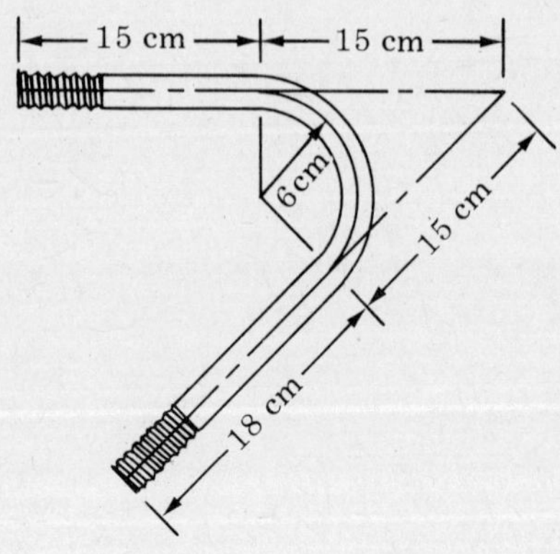

23. Determine the angles A and B in the funnel shaped adapter used to connect two different sizes of pipe.

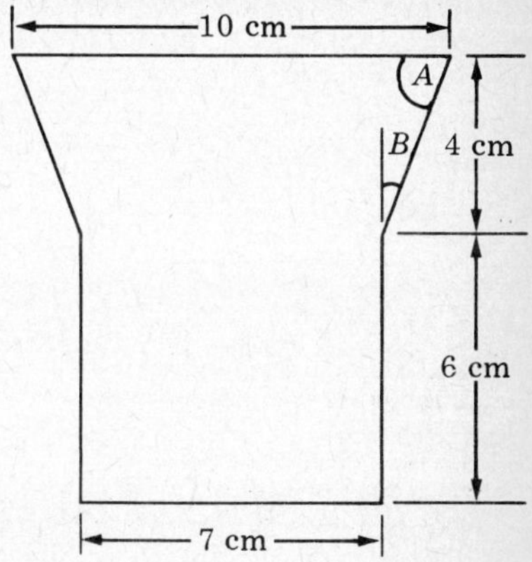

24. Graph the equation $x = 2^{-t} \sin \pi t$, $t \geqslant 0$.
25. Graph the equation $x = e^{-t} \cos \pi t$, $t \geqslant 0$.
26. Graph the equation $x = 3^{-t} \cos \frac{\pi}{2} t$, $t \geqslant 0$.
27. Graph the equation $x = e^{-t} \sin 2\pi t$, $t \geqslant 0$.

CHAPTER 6 IN REVIEW

I. Summary

In this chapter many concepts of trigonometry were developed. In Section 6.1, measurement of angles with degrees and radians was defined. Conversion between the two systems was seen to be an easy matter.

In Section 6.2, the six trigonometric ratios of an acute angle in a right triangle were investigated. Values of these six ratios were calculated for some commonly occurring angles; and by using Table III, we learned how to determine any trigonometric ratio of any acute angle.

The unit circle was used in Section 6.3 to define the trigonometric functions of a real number. Each trigonometric function represents a relationship between certain variables on that circle.

In Section 6.4, the trigonometric ratios of a standard position angle with terminal side in any quadrant were determined by using Table III. In Section 6.5 we learned how to graph equations of the form $y = a \sin(bx + c)$, $y = a \cos(bx + c)$, $y = a \tan(bx + c)$, $y = a \csc(bx + c)$, $y = a \sec(bx + c)$, $y = a \cot(bx + c)$ by using the amplitude, period, and phase shift.

Section 6.6 contained various applications of trigonometry to real life situations.

II. Vocabulary/Symbol True-False Quiz

If the statement is true, write "TRUE." However, if the statement is false, replace the underlined word or phrase with a different word or phrase that will make the new statement true.

1. An angle that measures one radian is larger than an angle that measures one degree.
2. If the radius of a circular sector is one foot and its central angle is 18°, then its area is 9 square feet.
3. The longest side of a right triangle is called its hypotenuse.
4. A 30-60-90 triangle whose hypotenuse is 90 inches has legs of length 30 inches and 60 inches.
5. The sum of the interior angles of any triangle is 180°.
6. $\sin \pi + \cos 90^\circ + \tan 30^\circ + \cot 300^\circ + \sec 2\pi + \csc 270^\circ = 0$.
7. Corresponding angles of similar triangles are equal.
8. If $\tan \theta = 2/5$ then $\cot \theta = 5/2$.
9. The period of $y = 3 \sin \pi x$ is 2.
10. The phase shift of $y = 2 \sin(3x + 4)$ is -2.

III. Chapter Test

1. Convert each degree measurement into radian measurement.
   a. 45° b. 750° c. -90°
2. Convert each radian measurement into degree measurement.
   a. $5\pi/4$ b. $\pi/4$ c. -2
3. a. Convert $100^\circ\, 35'20''$ into decimal form.
   b. Convert 12.76° into degree-minute-second form.
4. Find the six trigonometric ratios of a 60° angle.
5. Determine all the unknown sides and angles of right triangle *PQR*:

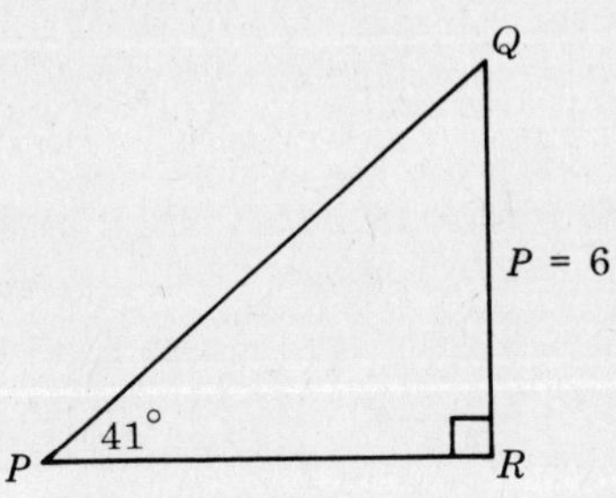

6. Determine the value of each trigonometric ratio:
   a. $\sin 42.8°$ b. $\cos 71.2°$ c. $\tan 89.4°$

In questions 7 through 9, solve for x.

7. $\sin 5x = 0$ 8. $\cos(2x + \pi) = 0$ 9. $\tan(2x - \pi) = 0$
10. Determine the value of each of the following:
    a. $\sin 151°$ b. $\cos 151°$ c. $\tan 285°$
    d. $\cot 191°$ e. $\sec -100°$ f. $\csc -180°$
11. Determine the value of each of the following:
    a. $\sin 2$ b. $\cos 3$ c. $\tan 4$
    d. $\cot 5$ e. $\sec 6$ f. $\csc 6$
12. If θ is an angle in standard position and its terminal side passes through the point $(5,-12)$, determine each of the trigonometric functions of θ.

In questions 13 through 18, graph each function. Also, state the amplitude (if any), period, and phase shift (if any) for each.

13. $y = \sin\left(2x + \dfrac{\pi}{2}\right)$ 14. $y = 2\cos \pi x$ 15. $y = 3\sin\left(2x - \dfrac{\pi}{4}\right)$
16. $y = \tan\left(\dfrac{1}{2}x + \dfrac{\pi}{4}\right)$ 17. $y = \csc\dfrac{\pi}{2}x$ 18. $y = \cot\left(x - \dfrac{\pi}{4}\right)$
19. Graph $y = |\sin x|$.
20. A ship is sailing on the course shown at the right.
    a. Determine the distance $\overline{AC}$.
    b. Determine the true bearing $\measuredangle NAB$.

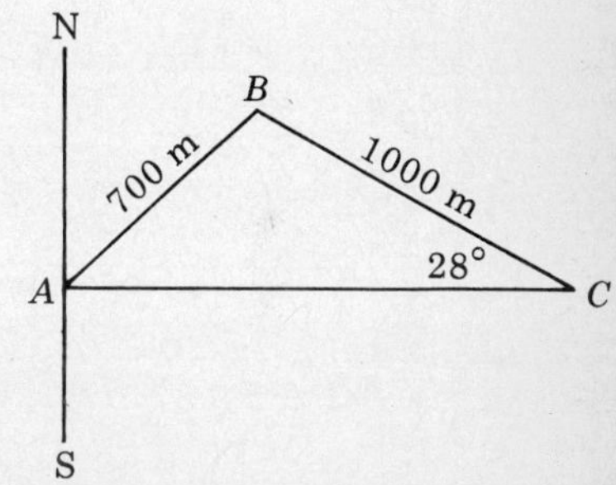

IV. Student Research Project

Trigonometric functions of the form $f(t) = a + b\sin \omega t$ are useful in neurophysiology when studying the activity of the human brain.

a. Graph $f(t) = a + b\sin \omega t$ when $a = 2$, $b = 6$, and $\omega = 10\pi$.
b. Graph $f(t) = a + b\sin \omega t$ when $a = 1$, $b = 4$, and $\omega = 20\pi$.
c. Prepare a brief report (less than 100 words) on this topic by reading the following source:

   Daniel Saltz, *A Short Calculus: An Applied Approach*, 2nd ed. (Santa Monica, Calif.: Goodyear Publishing Co., Inc.) pp. 339, 340.

CHAPTER **7**

More About Trigonometry

As an example in biomechanics, consider the forearm in the function of supporting an object in the hand, as shown below. Imagine that the object weights 10 pounds, and that the elbow is flexed 30°. Assume that the biceps is attached 2 inches from the nominal center of rotation of the elbow joint, and that the load is applied 14 inches from the center of rotation. A free body diagram of this problem is shown below. To determine the magnitude of the force T, it is necessary to find first the perpendicular distances of the two force action lines from the center of rotation, depicted as r_1 and r_2.

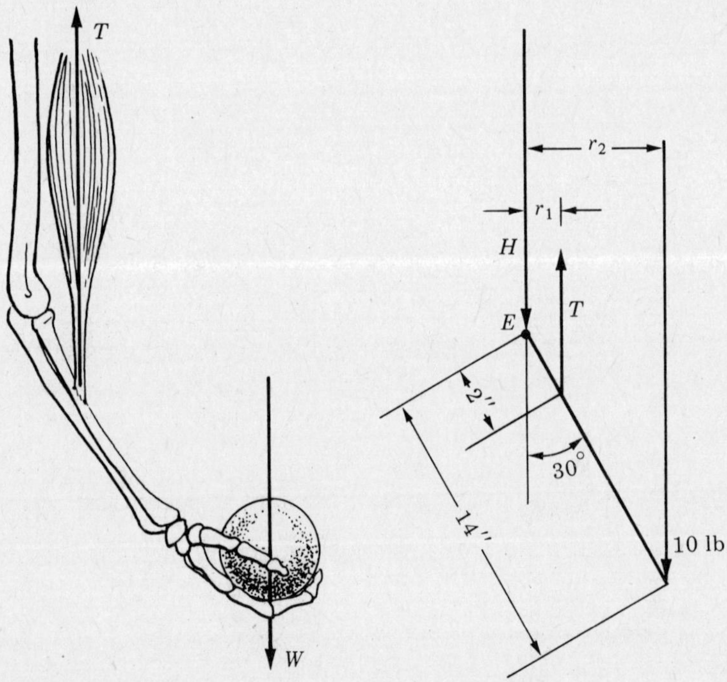

Diagram illustrating the forearm acting as a lever to support a weight, W, by biceps tension, T. The freebody diagram gives pertinent information to solve for biceps tension and the force of the humerus on the ulna, H.

From Frank H. Krusan, Frederick J. Kottke, and Paul M. Ellwood (eds.), Handbook of Physical Medicine and Rehabilitation, *2nd ed. (Philadelphia: W. B. Saunders Co.) 1971, pp. 825–826.*

7.1 LAW OF SINES AND LAW OF COSINES

The trigonometric ratios have been used in right triangles for the purpose of determining either an angle or a side. In this section we will see an extension of this use made to a **general** triangle ABC (that is, *any* triangle). For notational uniformity side a is opposite angle A, side b is opposite angle B, and side c is opposite angle C. By drawing a perpendicular to a side opposite an angle in any triangle, right triangles are formed. Consider the general triangle ABC with the perpenducular drawn to $\overline{BC}$.

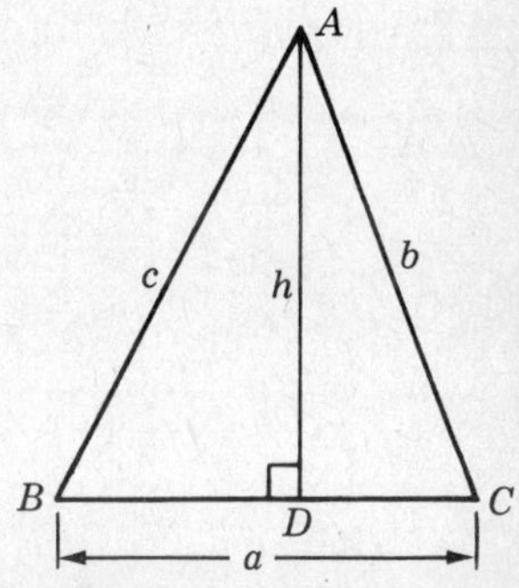

$$\sin B = \frac{h}{c} \quad \text{and} \quad \sin C = \frac{h}{b}$$

$$h = c \sin B \quad \text{and} \quad h = b \sin C$$

Thus $c \sin B = b \sin C$ and dividing both sides by bc gives: $\dfrac{\sin B}{b} = \dfrac{\sin C}{c}$.

A perpendicular from B to $\overline{AC}$ along with a similar sequence of steps produces $\dfrac{\sin A}{a} = \dfrac{\sin C}{c}$ (see Exercise 49).

The two proportions just obtained can be derived from a triangle of any shape—although it might be necessary to extend a side in order that a perpendicular meets it (see Exercise 50). In any case, by combining results the theorem below is obtained.

THEOREM 1

Law of Sines

In any triangle ABC,

$$\frac{\sin A}{a} = \frac{\sin B}{b} = \frac{\sin C}{c}.$$

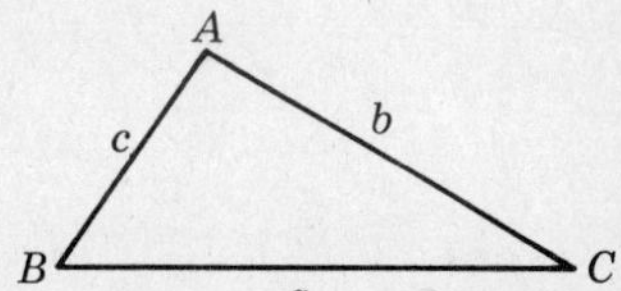

The Law of Sines can be used to solve a triangle when we know:

(i) any two angles and a side or
(ii) any two sides and an angle opposite one of them.

Case (i) is illustrated in the example that follows. An electronic calculator is a convenience (but not a necessity) in this section.

EXAMPLE 1 Solve the triangle in the diagram on the left.

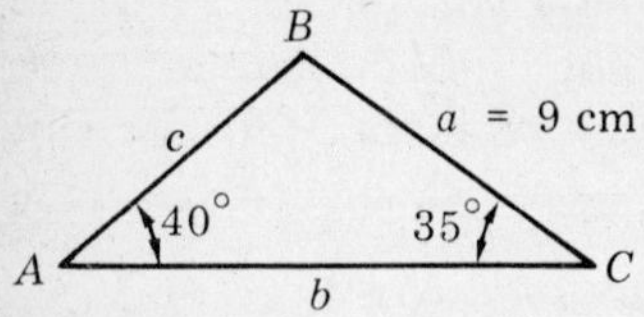

Solution Since $\measuredangle A + \measuredangle B + \measuredangle C = 180^\circ$,

$$\measuredangle B = 180^\circ - \measuredangle A - \measuredangle C = 180^\circ - 40^\circ - 35^\circ = 105^\circ.$$

Next, since we know A, B, and a, this suggests using $\dfrac{\sin A}{a} = \dfrac{\sin B}{b}$. We solve for b:

$$b = \frac{a \sin B}{\sin A} = \frac{9(\sin 105^\circ)}{\sin 40^\circ}$$

$$b = \frac{9(.9659)}{(.6428)} = 13.5 \text{ cm}$$

Finally, since we know A, C, and a, this suggests using $\dfrac{\sin A}{a} = \dfrac{\sin C}{c}$. We solve for c:

$$c = \frac{a \sin C}{\sin A} = \frac{9(\sin 35^\circ)}{\sin 40^\circ} = 8.0 \text{ cm}$$

Case (ii) is now illustrated. It is called the **ambiguous case** because there might be no triangle, one triangle, or two triangles.

EXAMPLE 2 Given $b = 9$, $c = 5$, and $C = 40^\circ$, find the remaining parts of the triangle ABC.

Solution Knowing b, c, and C suggests using $\dfrac{\sin B}{b} = \dfrac{\sin C}{c}$.

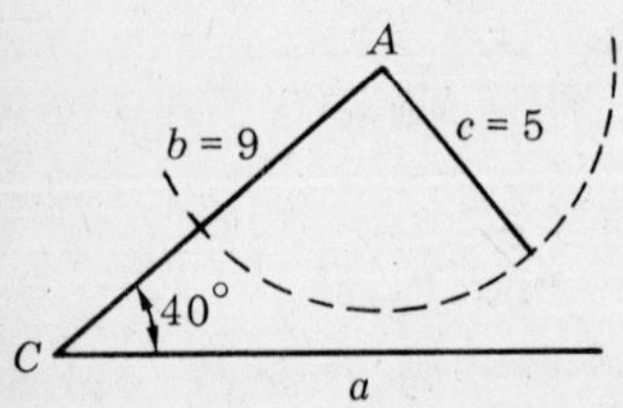

$$\sin B = \frac{b \sin C}{c} = \frac{9 \sin 40^\circ}{5} = 1.1570$$

Since $\sin B > 1$ (impossible), there is no solution. In picture form, we have the diagram at the left. Side c is too short to reach side a. Thus no triangle is formed.

EXAMPLE 3 Given $a = 10$, $b = 12$, and $A = 20^\circ$, determine angle B.

Solution $\dfrac{\sin A}{a} = \dfrac{\sin B}{b}$, so $\sin B = \dfrac{b \sin A}{a} = \dfrac{12(\sin 20^\circ)}{10} = .4104$.

Now there are *two* possibilities for angle B, since the sine function is positive in the first and second quadrants.

$$B_1 = 24.2°$$
$$B_2 = 180° - 24.2° = 155.8°$$

In picture form we have:

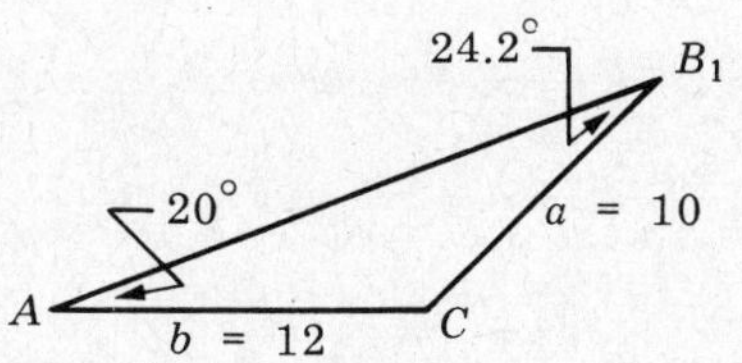

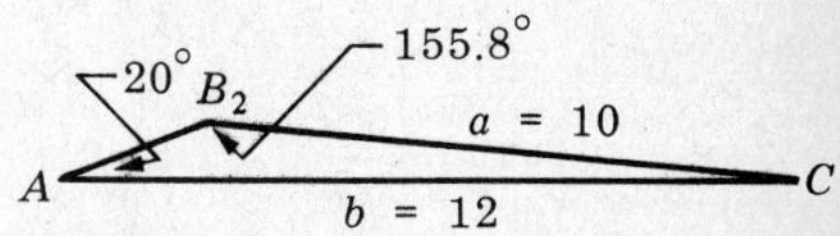

If necessary, each of these triangles can be completely solved as in Example 1.

Next, another theorem useful for solving triangles is stated.

THEOREM 2

Law of Cosines

In any triangle ABC,

$$a^2 = b^2 + c^2 - 2bc \cos A$$
$$b^2 = a^2 + c^2 - 2ac \cos B$$
$$c^2 = a^2 + b^2 - 2ab \cos C$$

Proof (of $a^2 = b^2 + c^2 - 2bc \cos A$)

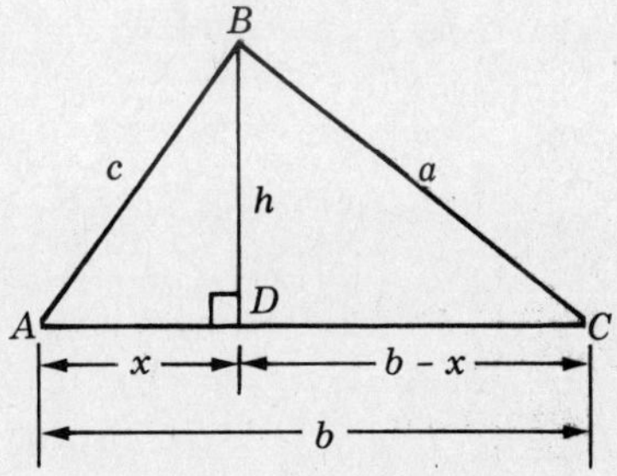

Statement	*Reason*
1. From $\triangle ADB$, $h^2 = c^2 - x^2$ From $\triangle BDC$, $h^2 = a^2 - (b-x)^2$	1. Pythagorean Theorem
2. $c^2 - x^2 = a^2 - (b-x)^2$	2. Equate the right-hand sides from step 1
3. $c^2 - x^2 = a^2 - b^2 + 2bx - x^2$	3. Simplify
4. $a^2 = b^2 + c^2 - 2bx$	4. Solve for a^2
5. $\cos A = \dfrac{x}{c}$	5. $\text{Cosine} = \dfrac{\text{adjacent}}{\text{hypotenuse}}$
6. $x = c \cos A$	6. Solve for x
7. $a^2 = b^2 + c^2 - 2bc \cos A$	7. Substitute step 6 into step 4

The other two equations in Theorem 2 can be proven similarly and all three equations hold true for any triangle (even though the one in the proof above is special in the sense that it has three acute angles). Furthermore, it is not necessary to memorize the three equations of Theorem 2. Instead, memorize the verbal interpretation:

> In any triangle, the square of any side equals the sum of squares of the other two sides minus twice their product times the cosine of the angle between them.

The Law of Cosines can be used to solve a triangle when we know:

(i) any two sides and the included angle or
(ii) all three sides.

Case (i) is illustrated in the next example.

EXAMPLE 4 Solve the triangle in the diagram on the left.

Solution $b^2 = a^2 + c^2 - 2ac \cos B = 64 + 25 - 2(8)(5) \cos 100°$

$\cos 100° = -\cos 80°$, since the cosine is negative in the second quadrant

$b^2 = 64 + 25 - 2(8)(5)(-.1736) = 102.888$

$b = 10.1$

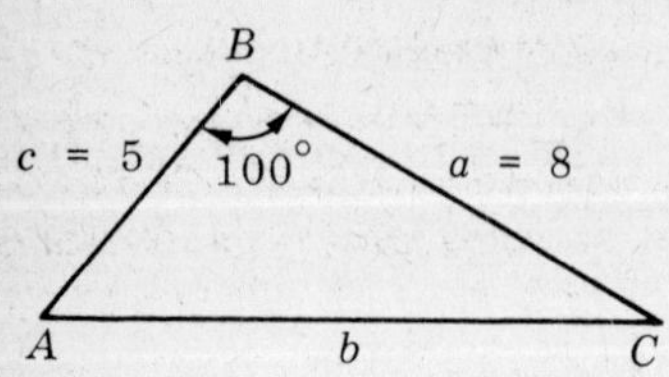

Angle A is most easily found using the Law of Sines.

$$\frac{\sin A}{a} = \frac{\sin B}{b}, \text{ so } \sin A = \frac{a \sin B}{b}$$

$$\sin A = \frac{8(\sin 100^\circ)}{10.1} = .7800$$

$$\measuredangle A = 51.3^\circ$$

$$\measuredangle C = 180^\circ - (\measuredangle A + \measuredangle B)$$
$$= 180^\circ - (151.3^\circ) = 28.7^\circ$$

Although the sine is positive in the first and second quadrants, we use only the first quadrant solution (why?).

EXAMPLE 5 Solve the triangle in the diagram on the left.

Solution

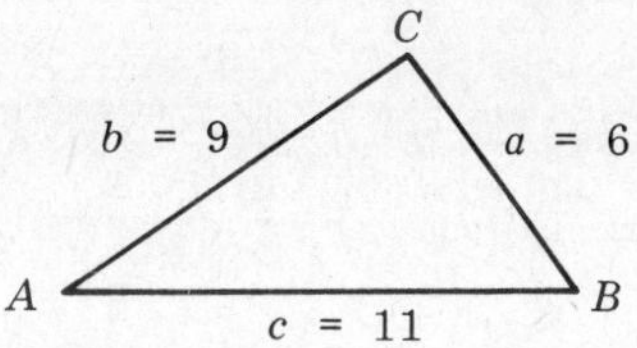

$$a^2 = b^2 + c^2 - 2bc \cos A$$
$$36 = 81 + 121 - 2(9)(11) \cos A$$
$$\cos A = \frac{202 - 36}{198} = .8383$$
$$\measuredangle A = 33.0^\circ$$

Although the cosine is positive in the first and fourth quadrants we use only the first quadrant solution (why?).

$$c^2 = a^2 + b^2 - 2ab \cos C$$
$$121 = 36 + 81 - 2(6)(9) \cos C$$
$$\cos C = \frac{117 - 121}{108} = -.0370$$
$$\measuredangle C = 92.1^\circ$$
$$\measuredangle B = 180^\circ - (\measuredangle A + \measuredangle C)$$
$$= 54.9^\circ$$

Although the cosine is negative in the second and third quadrants we use only the second quadrant solution (why?).

EXAMPLE 6 A roof of a house addition is being built to accept solar energy panels.*

a. Determine $\overline{AC}$.
b. Determine $\measuredangle ACB$.

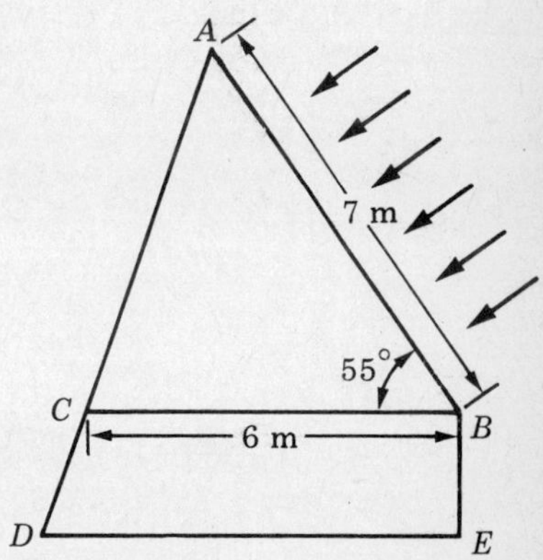

Solution a. The Law of Cosines gives

$$(\overline{AC})^2 = (\overline{AB})^2 + (\overline{BC})^2 - 2(\overline{AB})(\overline{BC}) \cos \measuredangle ABC$$
$$(\overline{AC})^2 = 49 + 36 - 2(7)(6) \cos 55^\circ$$
$$(\overline{AC})^2 = 36.82$$
$$\overline{AC} = 6.1 \text{ m}$$

*With $\measuredangle ABC = 55^\circ$, this roof inclination would be appropriate for cities such as New York, NY, Philadelphia, PA, Indianapolis, IN, Decatur, IL, St. Joseph, MO, Topeka, KS, Denver, CO, Provo, UT, Reno, NV, and Eureka, CA. See the Student Research Project at the end of this chapter.

b. Using the Law of Sines,

$$\frac{\sin \measuredangle ABC}{\overline{AC}} = \frac{\sin \measuredangle ACB}{\overline{AB}}$$

$$\sin \measuredangle ACB = \frac{(\overline{AB})(\sin \measuredangle ABC)}{\overline{AC}} = \frac{(7)(\sin 55^\circ)}{6.1} = .9400$$

$$\measuredangle ACB = 70.1^\circ$$

EXERCISES FOR SECTION 7.1

In Exercises 1 through 20, solve each triangle. (See Examples 1, 2, and 3.)

1. $A = 40^\circ$, $B = 35^\circ$, $a = 10$ cm
2. $B = 60^\circ$, $C = 70^\circ$, $c = 12$ ft
3. $B = 72^\circ$, $C = 60^\circ$, $b = 4$
4. $B = 72^\circ$, $C = 60^\circ$, $a = 5$ m
5. $A = 20^\circ$, $a = 20$, $b = 24$
6. $C = 55^\circ$, $b = 22$ m, $c = 20$ m
7. $C = 40^\circ$, $b = 18$, $c = 10$
8. $A = 120^\circ$, $B = 45^\circ$, $a = 6$
9. $B = 110^\circ$, $b = 10$, $c = 9$
10. $B = 50^\circ$, $b = 3$, $c = 6$
11. $B = 50^\circ$, $b = 5$, $c = 6$
12. $B = 50^\circ$, $b = 6$, $c = 6$
13. $B = 150^\circ$, $b = 10$, $c = 6$
14. $A = 30^\circ$, $a = 5$ cm, $b = 10$ cm
15. $A = 30^\circ$, $a = 6$, $b = 10$
16. $B = 100^\circ$, $C = 10^\circ$, $c = 2.5$ in
17. $A = 90^\circ$, $b = 4$, $a = 8$
18. $A = 120^\circ$, $B = 45.5^\circ$, $a = 5$
19. $A = 30.5^\circ$, $b = 10.5$, $a = 4$
20. $A = 22.4^\circ$, $B = 112.6^\circ$, $c = 5$

In Exercises 21 through 40, solve each triangle. (See Examples 4 and 5.)

21. $A = 38^\circ$, $b = 5$ cm, $c = 4$ cm
22. $B = 38^\circ$, $a = 5$ cm, $c = 5$ cm
23. $C = 60^\circ$, $a = 5$ in, $b = 5$ in
24. $C = 120^\circ$, $a = 5$ in, $b = 5$ in
25. $a = 2$, $b = 4$, $c = 5$
26. $a = 4$, $b = 8$, $c = 10$
27. $a = 10$, $b = 1$, $c = 2$
28. $a = 10$, $b = 1$, $c = 9$
29. $a = 10$, $b = 1$, $c = 10$
30. $A = 44.5^\circ$, $b = 2$ ft, $c = 3$ ft
31. $a = 2c$, $b = 2c$, $c = c$
32. $C = 69^\circ$, $a = 5$ cm, $b = 10$ cm
33. $B = 130^\circ$, $a = 4.4$ m, $c = 5.4$ m
34. $A = 40^\circ$, $b = 22.5$, $c = 10.5$
35. $a = 5$, $b = 13$, $c = 12$
36. $a = 3$, $b = 4$, $c = 5$
37. $B = 85^\circ$, $a = 10$, $b = 20$
38. $C = 10.6^\circ$, $a = 100$, $b = 50$
39. $C = 100^\circ$, $a = 6$ cm, $b = 6$ cm
40. $C = 145^\circ$, $a = 6$ cm, $b = 5$ cm

In Exercises 41 through 50, solve each problem. (See Examples 1 through 6.)

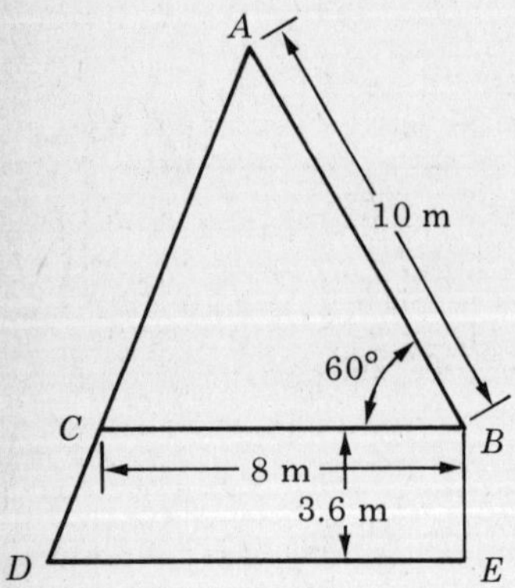

41. A roof of a house addition is being built to accept solar energy panels as shown in the diagram on the left.
    a. Determine $\overline{AC}$.
    b. Determine $\measuredangle ACB$.
    c. Determine $\overline{CD}$.
    d. Determine $\overline{AD}$.
42. Solve Exercise 41 if just the dimension $\overline{BC}$ is changed to 9 m.
43. Solve Exercise 41 if just the $\measuredangle ABC$ is changed to 65°.
44. A piece of land has the shape of a right triangle as shown in the diagram on the right.
    a. Show that $a = b \sin A$.
    b. Show that the Area $= \frac{1}{2} bc \sin A$.
    c. Show that the Area $= \frac{1}{2} ab \sin C$.

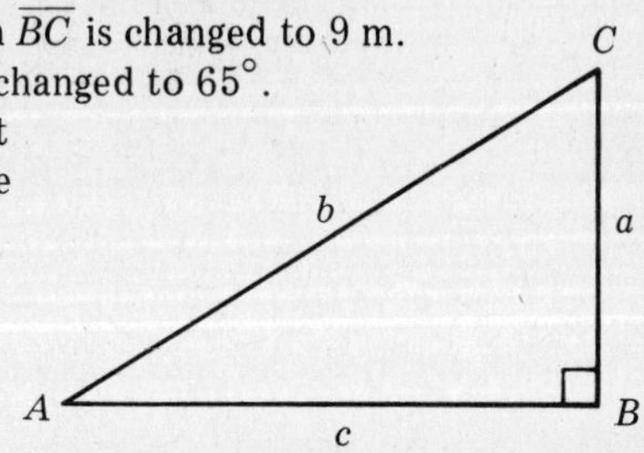

45. Show that the area of *any* triangle is equal to one-half the product of any two sides times the sine of the angle between them.

46. A surveyor sites to the top and bottom of an antenna from point E. Determine the height of the antenna $\overline{AB}$ (see diagram on the right).

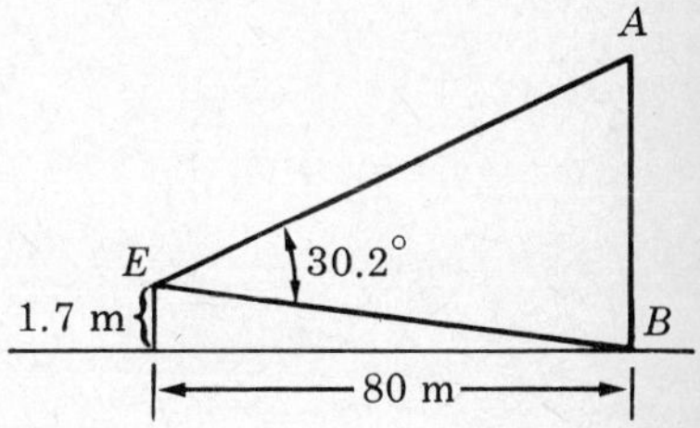

47. At two different points A and B, a ship's navigator sites to the buoy at point C.
    a. Determine the distance $\overline{BC}$.
    b. If the ship continues on its present course, determine how much further it must travel until it reaches the point of closest approach to the buoy.

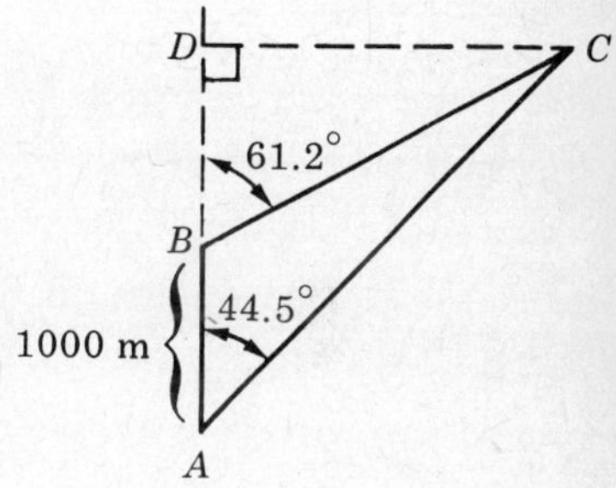

48. Show that the area of the shaded portion of the circle in the adjacent diagram is

$$\frac{r^2(\theta - \sin\theta)}{2},$$

where θ is measured in radians.

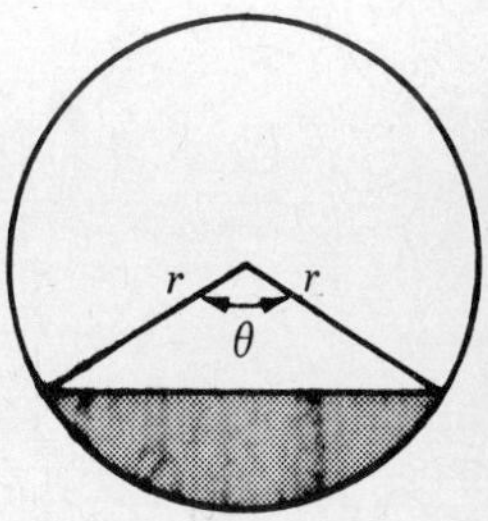

49. In the triangle at the right, draw a perpendicular from B to $\overline{AC}$. Using right triangle trigonometry, show that $\dfrac{\sin A}{a} = \dfrac{\sin C}{c}$.

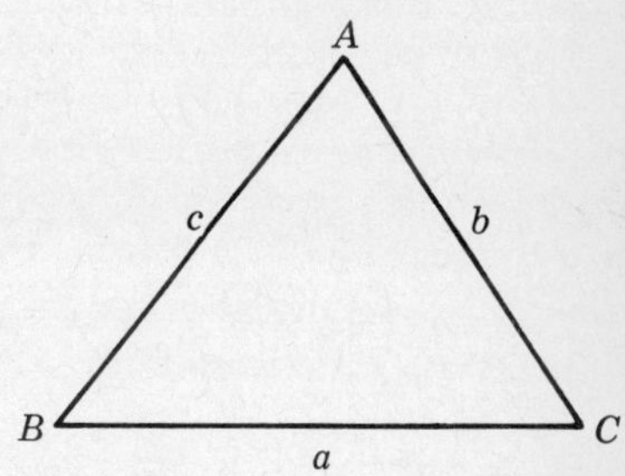

50. In the triangle at the right, draw a perpendicular from B to $\overline{AC}$. Using right triangle trigonometry, show that $\dfrac{\sin\angle BAC}{a} = \dfrac{\sin\angle ACB}{c}$.

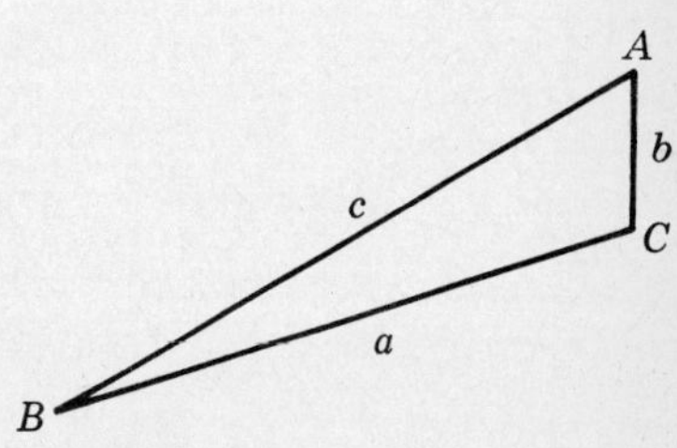

51. In Exercises 27 and 28 of this section, there is no triangle formed from the given three sides. State what conditions must be met by the three sides a, b, and c in order to form a triangle.

Calculator Exercises

In Exercises 52 through 55, solve each triangle.

52. $A = 56.758°$, $B = 41.2216°$, $a = 6.2$
53. $C = 40.29°$, $b = 18.26$, $c = 9.564$
54. $B = 15.77°$, $b = 1.1$, $c = 4.27$
55. $A = 19.941°$, $a = 20.02$, $b = 24.61$

7.2 TRIGONOMETRIC IDENTITIES

A **trigonometric identity** is a statement of equality or a formula involving various trigonometric functions that is true for all possible values of the arguments for which the functions are defined. For example, we have already seen and used (in Section 6.4) the following **Fundamental Identities**. Remember that the argument s can be thought of as representing real numbers or degrees.

Fundamental Identities (1)

$$\sin s = \frac{1}{\csc s} \qquad \cos s = \frac{1}{\sec s}$$

$$\tan s = \frac{1}{\cot s} \qquad \tan s = \frac{\sin s}{\cos s}$$

$$\cot s = \frac{\cos s}{\sin s}$$

There are three other well-known formulas that are usually included in this same category.

Fundamental Identities (2)

$$\sin^2 s + \cos^2 s = 1^*$$

$$\tan^2 s + 1 = \sec^2 s$$

$$\cot^2 s + 1 = \csc^2 s$$

EXAMPLE 1 Derive the identity $\sin^2 s + \cos^2 s = 1$.

*Usually when a trigonometric function is raised to a power as with $(\sin s)^2$, custom dictates that we write $\sin^2 s$ ("sine square s").

Solution Consider the unit circle at the right, $x^2 + y^2 = 1$. By definition, $x = \cos s$ and $y = \sin s$. Thus, $(\cos s)^2 + (\sin s)^2 = 1$ or equivalently $\sin^2 s + \cos^2 s = 1$.

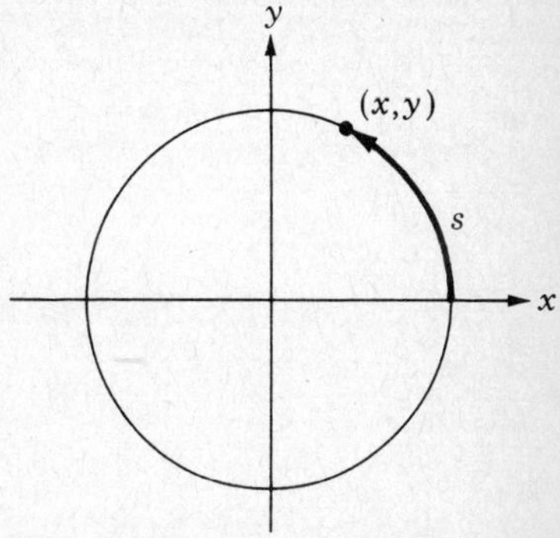

In Exercises 1 and 2, the reader is asked to derive $\tan^2 s + 1 = \sec^2 s$ and $\cot^2 s + 1 = \csc^2 s$.

There are a variety of uses for trigonometric identities. They can be used in various numerically oriented problems. They can be used to help *derive* other identities. Also they can be used to help *verify* known identities. The reader will see each of these uses presented.

EXAMPLE 2 a. If $\tan s = -\frac{2}{5}$, and s is in the fourth quadrant, determine the value of $\sec s$.

b. If $\cos s = Q$, and s is in the third quadrant, derive an expression for $\sin s$ in terms of Q.

Solution a. Since $\tan^2 s + 1 = \sec^2 s$, $\sec^2 s = \tan^2 s + 1$ and $\sec s = \pm\sqrt{\tan^2 s + 1}$. Since s is in the fourth quadrant, $\sec s$ must be positive. Hence,

$$\sec s = +\sqrt{\tan^2 s + 1} = +\sqrt{(-\tfrac{2}{5})^2 + 1} = \sqrt{\tfrac{29}{25}} = \tfrac{1}{5}\sqrt{29}$$

b. Since $\sin^2 s + \cos^2 s = 1$, $\sin^2 s = 1 - \cos^2 s$ or $\sin s = \pm\sqrt{1 - \cos^2 s}$. Since s is in the third quadrant, $\sin s$ must be negative. Hence,

$$\sin s = -\sqrt{1 - Q^2}$$

When working with trigonometric functions it is often convenient to have the following collection of trigonometric identities.

(3)

$\sin(-s) = -\sin s$	$\csc(-s) = -\csc s$
$\cos(-s) = \cos s$	$\sec(-s) = \sec s$
$\tan(-s) = -\tan s$	$\cot(-s) = -\cot s$

Each of these can be proven quite easily (see Exercises 17 through 22).

Next we present a collection of trigonometric identities known as the **addition formulas**.

Addition Formulas (4)

$$\sin(\alpha + \beta) = \sin\alpha\cos\beta + \cos\alpha\sin\beta$$
$$\sin(\alpha - \beta) = \sin\alpha\cos\beta - \cos\alpha\sin\beta$$
$$\cos(\alpha + \beta) = \cos\alpha\cos\beta - \sin\alpha\sin\beta$$
$$\cos(\alpha - \beta) = \cos\alpha\cos\beta + \sin\alpha\sin\beta$$

EXAMPLE 3 Derive the identity $\sin(\alpha + \beta) = \sin\alpha\cos\beta + \cos\alpha\sin\beta$.

Solution Consider the triangle ACE at the right. Applying the Law of Sines gives

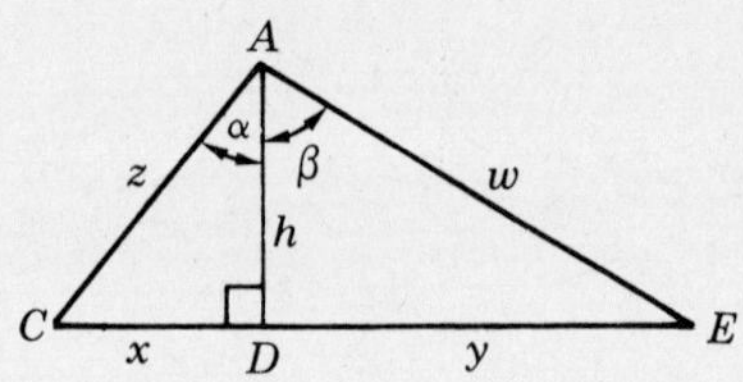

$$\frac{\sin \measuredangle CAE}{\overline{CE}} = \frac{\sin \measuredangle ACE}{\overline{AE}}$$

or

$$\frac{\sin(\alpha + \beta)}{x + y} = \frac{h/z}{w}$$

$$\sin(\alpha + \beta) = \frac{h}{zw}(x + y) = \frac{hx}{zw} + \frac{hy}{zw} = \frac{x}{z} \cdot \frac{h}{w} + \frac{h}{z} \cdot \frac{y}{w}$$

$$\sin(\alpha + \beta) = \sin\alpha\cos\beta + \cos\alpha\sin\beta$$

The derivation above is really only valid for the case where α and β are acute angles (why?) but it is possible to derive the same result for any two angles (or real numbers) α and β. In Exercises 23, 24, and 25, the reader is guided in deriving the other three addition formulas in (4).

EXAMPLE 4 a. Determine the exact value of $\sin 75^\circ$ by using the fact that $75^\circ = 45^\circ + 30^\circ$.

b. Determine the exact value of $\cos\frac{7\pi}{12}$ by using the fact that $\frac{7\pi}{12} = \frac{\pi}{3} + \frac{\pi}{4}$.

Solution a. $\sin 75° = \sin(45° + 30°) = \sin 45° \cos 30° + \cos 45° \sin 30°$

$$= \left(\frac{\sqrt{2}}{2}\right)\left(\frac{\sqrt{3}}{2}\right) + \left(\frac{\sqrt{2}}{2}\right)\left(\frac{1}{2}\right)$$

$$= \frac{\sqrt{6} + \sqrt{2}}{4}$$

b. $\cos \frac{7\pi}{12} = \cos\left(\frac{\pi}{3} + \frac{\pi}{4}\right) = \cos \frac{\pi}{3} \cos \frac{\pi}{4} - \sin \frac{\pi}{3} \sin \frac{\pi}{4}$

$$= \left(\frac{1}{2}\right)\left(\frac{\sqrt{2}}{2}\right) - \left(\frac{\sqrt{3}}{2}\right)\left(\frac{\sqrt{2}}{2}\right)$$

$$= \frac{\sqrt{2} - \sqrt{6}}{4}$$

By using the addition formulas in (4) it is easy to obtain the following identities (see Exercises 36 through 41).

(5)
$$\sin\left(\frac{\pi}{2} - \beta\right) = \cos \beta \quad \tan\left(\frac{\pi}{2} - \beta\right) = \cot \beta \quad \sec\left(\frac{\pi}{2} - \beta\right) = \csc \beta$$
$$\cos\left(\frac{\pi}{2} - \beta\right) = \sin \beta \quad \cot\left(\frac{\pi}{2} - \beta\right) = \tan \beta \quad \csc\left(\frac{\pi}{2} - \beta\right) = \sec \beta$$

The angles $\frac{\pi}{2} - \beta$ and β are defined as complementary angles. It is for this reason, along with the identities in (5), that we say sine and *co*sine are *co*functions of each other, tangent and *co*tangent are *co*functions of each other, secant and *co*secant are *co*functions of each other.

The next collection of identities that will be discussed is known as the **double-angle formulas.**

Double-Angle Formulas

(6)
$$\sin 2u = 2 \sin u \cos u \qquad \cos 2u = \cos^2 u - \sin^2 u$$
$$\cos 2u = 1 - 2 \sin^2 u \qquad \cos 2u = 2 \cos^2 u - 1$$

EXAMPLE 5 Derive the identity $\sin 2u = 2 \sin u \cos u$.

Solution From (4) we have $\sin(\alpha + \beta) = \sin \alpha \cos \beta + \cos \alpha \sin \beta$. Let $\alpha = \beta = u$.

$$\sin(u + u) = \sin u \cos u + \sin u \cos u$$
$$\sin 2u = 2 \sin u \cos u$$

In Exercises 42, 43, and 44, the reader is guided in deriving the other double-angle formulas in (6).

EXAMPLE 6 a. Determine the value of $\sin 2u$ if $\sin u = \frac{2}{3}$ and u is in the second quadrant.

b. Determine the value of $\cos 2u$ if $\sin u = \frac{2}{3}$ and u is in the second quadrant.

Solution a. $\sin 2u = 2 \sin u \cos u$ — $\cos^2 u = 1 - \sin^2 u$

$\sin 2u = 2 \sin u \, (-\sqrt{1 - \sin^2 u}),$

where the negative sign must be used since cosine is negative in the second quadrant. Thus,

$$\sin 2u = 2(\tfrac{2}{3})(-\sqrt{1 - (\tfrac{2}{3})^2}) = \tfrac{4}{3}(-\sqrt{\tfrac{5}{9}})$$
$$\sin 2u = -\tfrac{4}{9}\sqrt{5}$$

b. $\cos 2u = 1 - 2 \sin^2 u$

$\cos 2u = 1 - 2(\frac{2}{3})^2 = \frac{1}{9}$

EXAMPLE 7 Derive a formula for $\sin 3u$ in terms of $\sin u$.

Solution

$$\begin{aligned}\sin 3u &= \sin(2u + u)\\ &= \sin 2u \cos u + \cos 2u \sin u\\ &= (2 \sin u \cos u) \cos u + (1 - 2 \sin^2 u) \sin u\\ &= 2 \sin u(\cos^2 u) + \sin u - 2 \sin^3 u\\ &= 2 \sin u(1 - \sin^2 u) + \sin u - 2 \sin^3 u\\ &= 3 \sin u - 4 \sin^3 u\end{aligned}$$

By using two of the double-angle formulas in (6) it is possible to derive the **half-angle formulas**. Consider.

$$\cos 2u = 1 - 2 \sin^2 u \quad \text{and} \quad \cos 2u = 2 \cos^2 u - 1$$
$$2 \sin^2 u = 1 - \cos 2u \quad \text{and} \quad 2 \cos^2 u = 1 + \cos 2u$$
$$\sin^2 u = \frac{1 - \cos 2u}{2} \quad \text{and} \quad \cos^2 u = \frac{1 + \cos 2u}{2}$$

Taking the square root of both sides of the latter two identities gives:

Half-Angle Formulas (7)

$$\sin u = \pm\sqrt{\frac{1 - \cos 2u}{2}} \qquad \cos u = \pm\sqrt{\frac{1 + \cos 2u}{2}}$$

If u is replaced by $\frac{t}{2}$, alternate forms of the half-angle formulas are obtained, namely,

$$\sin\frac{t}{2} = \pm\sqrt{\frac{1-\cos t}{2}} \qquad \cos\frac{t}{2} = \pm\sqrt{\frac{1+\cos t}{2}}$$

When using these half-angle formulas, the choice of sign depends upon what quadrant u or $\frac{t}{2}$ is located in and whether the particular trigonometric function is positive or negative there.

EXAMPLE 8 If $\cos s = -\frac{1}{8}$ and s is in the third quadrant, determine:

a. $\sin\frac{s}{2}$ b. $\cos\frac{s}{2}$ c. $\tan\frac{s}{2}$

Solution a. $\sin\frac{s}{2} = \pm\sqrt{\frac{1-\cos s}{2}}$. Since $\pi < s < \frac{3\pi}{2}$, $\frac{\pi}{2} < \frac{s}{2} < \frac{3\pi}{4}$. Thus $\frac{s}{2}$ is in the *second* quadrant.

$$\sin\frac{s}{2} = +\sqrt{\frac{1-\cos s}{2}} = +\sqrt{\frac{1-(-\frac{1}{8})}{2}} = \sqrt{\frac{9}{16}}$$

$$\sin\frac{s}{2} = \frac{3}{4}$$

b. $\cos\frac{s}{2} = \pm\sqrt{\frac{1+\cos s}{2}}$. Since $\frac{s}{2}$ is in the *second* quadrant,

$$\cos\frac{s}{2} = -\sqrt{\frac{1+(-\frac{1}{8})}{2}} = -\sqrt{\frac{7}{16}}$$

$$\cos\frac{s}{2} = -\frac{1}{4}\sqrt{7}$$

c. $$\tan\frac{s}{2} = \frac{\sin\frac{s}{2}}{\cos\frac{s}{2}} = \frac{\frac{3}{4}}{-\frac{1}{4}\sqrt{7}} = -\frac{3}{\sqrt{7}}$$

$$\tan\frac{s}{2} = -\frac{3}{7}\sqrt{7}$$

EXERCISES FOR SECTION 7.2

1. Derive the identity $\tan^2 s + 1 = \sec^2 s$ by starting with the known identity $\sin^2 s + \cos^2 s = 1$ and dividing both sides by $\cos^2 s$.
2. Derive the identity $\cot^2 s + 1 = \csc^2 s$ by starting with the known identity $\sin^2 s + \cos^2 s = 1$ and dividing both sides by $\sin^2 s$.

In Exercises 3 through 16, determine the specified trigonometric functions. (See Example 2 and identities (1).)

3. $\tan s = \frac{1}{2}$, s is in the first quadrant, determine $\sec s$.
4. $\tan s = \frac{1}{2}$, s is in the third quadrant, determine $\sec s$.
5. $\sin s = \frac{2}{3}$, s is in the second quadrant, determine $\cos s$.
6. $\sin s = \frac{2}{3}$, s is in the second quadrant, determine $\tan s$.
7. $\sin s = \frac{2}{3}$, s is in the second quadrant, determine $\sec s$.
8. $\csc s = -5$, s is in the fourth quadrant, determine $\cot s$.
9. $\csc s = -5$, s is in the fourth quadrant, determine $\tan s$.
10. $\cos s = \frac{3}{4}$, s is in the fourth quadrant, determine $\sin s$.
11. $\cos s = -\frac{1}{3}$, s is in the second quadrant, determine $\sin s$.
12. $\cos s = -\frac{1}{3}$, s is in the third quadrant, determine $\sin s$.
13. If $\cos s = Q$, and s is in the second quadrant, derive an expression for $\sin s$ in terms of Q.
14. If $\tan s = Q$ and s is in the second quadrant, derive an expression for $\sec s$ in terms of Q.
15. If $\csc s = \dfrac{K}{2}$, and s is in the first quadrant, derive an expression for $\cot s$ in terms of K.
16. If $\sin s = \dfrac{t}{4}$, and s is in the second quadrant, derive an expression for $\cos s$ in terms of t.
17. a. Prove that $\sin(-s) = -\sin s$, where s is in the first quadrant (see diagram).
    b. Prove that $\sin(-s) = -\sin s$, where s is in the second quadrant.
    c. Prove that $\sin(-s) = -\sin s$, where s is in the third quadrant.
    d. Prove that $\sin(-s) = -\sin s$, where s is in the fourth quadrant.

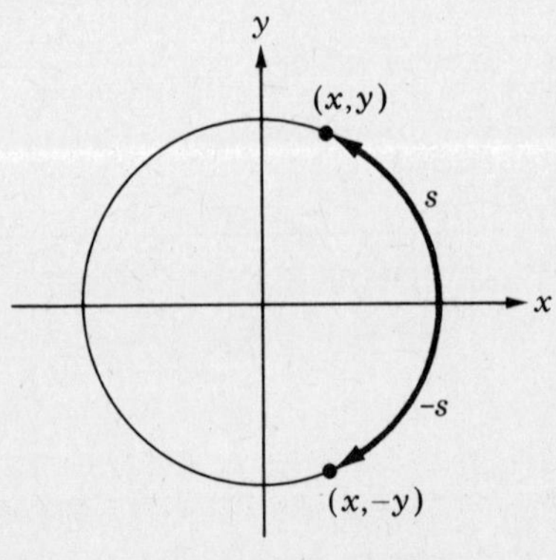

18. Prove that $\cos(-s) = \cos s$. (*Hint:* separate your proof into 4 cases, as in Exercise 17.)
19. Prove that $\tan(-s) = -\tan s$. $\left(\textit{Hint: this proof is short, use the identity } \tan\alpha = \dfrac{\sin\alpha}{\cos\alpha}.\right)$
20. Prove that $\csc(-s) = -\csc s$. $\left(\textit{Hint: use the identity } \csc\alpha = \dfrac{1}{\sin\alpha}.\right)$
21. Prove that $\sec(-s) = \sec s$. $\left(\textit{Hint: use the identity } \sec\alpha = \dfrac{1}{\cos\alpha}.\right)$
22. Prove that $\cot(-s) = -\cot s$. $\left(\textit{Hint: use the identity } \cot\alpha = \dfrac{1}{\tan\alpha}.\right)$

23. Derive the identity $\cos(\alpha + \beta) = \cos\alpha\cos\beta - \sin\alpha\sin\beta$. (*Hint:* follow the outline below. Also see Example 3.)
   a. Apply the Law of Cosines to $\triangle ACE$, namely:
$$(\overline{CE})^2 = (\overline{CA})^2 + (\overline{AE})^2 - 2(\overline{CA})(\overline{AE})\cos\measuredangle CAE$$
   b. Substitute the various letters in the appropriate places of step a and solve for $\cos(\alpha + \beta)$.
   c. In the numerator, eliminate z^2 and w^2 by using the Pythagorean Theorem.
   d. Write the resulting expression as a difference of fractions and then go to the *right* triangles ACD and AED.

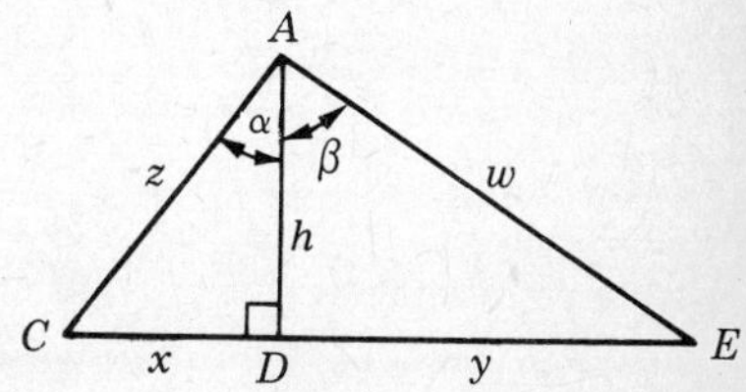

24. Derive the identity $\sin(u - v) = \sin u\cos v - \cos u\sin v$ by using the known identity for $\sin(\alpha + \beta)$ and letting $\alpha = u$ and $\beta = -v$.
25. Derive the identity $\cos(u - v) = \cos u\cos v + \sin u\sin v$ by using the known identity for $\cos(\alpha + \beta)$ and letting $\alpha = u$ and $\beta = -v$.
26. Derive the identity $\tan(u + v) = \dfrac{\tan u + \tan v}{1 - \tan u\tan v}$.
$\left(\textit{Hint:}\text{ start with } \tan(u + v) = \dfrac{\sin(u + v)}{\cos(u + v)}.\right)$
27. Derive the identity $\tan(u - v) = \dfrac{\tan u - \tan v}{1 + \tan u\tan v}$.

In Exercises 28 through 35, determine the exact value of each trigonometric function by using an appropriate addition formula. (See Example 4.)

28. $\cos 15°$, $15° = 45° - 30°$
29. $\sin 15°$, $15° = 45° - 30°$
30. $\cos 75°$, $75° = 45° + 30°$
31. $\sin 135°$, $135° = 90° + 45°$
32. $\cos 15°$, $15° = 60° - 45°$
33. $\cos 105°$, $105° = 60° + 45°$
34. $\sin\dfrac{5\pi}{6}$, $\dfrac{5\pi}{6} = \dfrac{\pi}{2} + \dfrac{\pi}{3}$
35. $\cos\dfrac{\pi}{3}$, $\dfrac{\pi}{3} = \dfrac{\pi}{2} - \dfrac{\pi}{6}$
36. Derive the identity $\sin\left(\dfrac{\pi}{2} - \beta\right) = \cos\beta$. (*Hint:* start with an appropriate addition formula.)
37. Derive the identity $\cos\left(\dfrac{\pi}{2} - \beta\right) = \sin\beta$. (*Hint:* start with an appropriate addition formula.)
38. Derive the identity $\tan\left(\dfrac{\pi}{2} - \beta\right) = \cot\beta$. (*Hint:* use Exercises 36 and 37.)
39. Derive the identity $\cot\left(\dfrac{\pi}{2} - \beta\right) = \tan\beta$. $\left(\textit{Hint:}\text{ use the identity } \cot u = \dfrac{1}{\tan u}\right)$
40. Derive the identity $\sec\left(\dfrac{\pi}{2} - \beta\right) = \csc\beta$.
41. Derive the identity $\csc\left(\dfrac{\pi}{2} - \beta\right) = \sec\beta$.
42. Derive the identity $\cos 2u = \cos^2 u - \sin^2 u$. (*Hint:* see Example 5.)

43. Derive the identity $\cos 2u = 1 - 2\sin^2 u$. (*Hint:* start with Exercise 42 and also use $\sin^2 u + \cos^2 u = 1$.)
44. Derive the identity $\cos 2u = 2\cos^2 u - 1$.

In Exercises 45 through 52, determine the specified trigonometric function. (See Example 6.)

45. Determine the value of $\sin 2u$ if $\sin u = \frac{2}{3}$ and u is in the first quadrant.
46. Determine the value of $\sin 2u$ if $\sin u = -\frac{2}{3}$ and u is in the fourth quadrant.
47. Determine the value of $\sin 2u$ if $\sin u = -\frac{2}{3}$ and u is in the third quadrant.
48. Determine the value of $\cos 2u$ if $\sin u = \frac{1}{5}$ and u is in the first quadrant.
49. Determine the value of $\cos 2u$ if $\cos u = \frac{1}{4}$ and u is in the first quadrant.
50. Determine the value of $\cos 2u$ if $\sin u = -\frac{2}{5}$ and u is in the third quadrant.
51. Determine the value of $\sin 2u$ if $\cos u = -\frac{1}{4}$ and u is in the second quadrant.
52. Determine the value of $\sin 2u$ if $\cos u = -\frac{2}{3}$ and u is in the third quadrant.

In Exercises 53 through 57, derive the specified formula. (See Example 7.)

53. Derive a formula for $\cos 3u$ in terms of $\cos u$.
54. Derive a formula for $\cos 4u$ in terms of $\cos u$.
55. Derive a formula for $\cos 5u$ in terms of $\cos u$.
56. Derive a formula for $\sin 4u$ in terms of $\sin u$ and $\cos u$.
57. Derive a formula for $\sin 4u$ in terms of $\sin u$.

In Exercises 58 through 69, determine the specified trigonometric function. (See Example 8.)

58. If $\cos s = -\dfrac{1}{4}$ and s is in the third quadrant, determine $\sin \dfrac{s}{2}$.

59. If $\cos s = -\dfrac{1}{4}$ and s is in the third quadrant, determine $\cos \dfrac{s}{2}$.

60. If $\cos s = -\dfrac{1}{4}$ and s is in the third quadrant, determine $\tan \dfrac{s}{2}$.

61. If $\cos s = \dfrac{1}{3}$ and s is in the fourth quadrant, determine $\sin \dfrac{s}{2}$.

62. If $\cos s = \dfrac{1}{3}$ and s is in the fourth quadrant, determine $\cos \dfrac{s}{2}$.

63. If $\cos s = \dfrac{1}{3}$ and s is in the fourth quadrant, determine $\tan \dfrac{s}{2}$.

64. If $\cos s = -\dfrac{2}{5}$ and s is in the second quadrant, determine $\sin \dfrac{s}{2}$.

65. If $\cos s = -\dfrac{2}{5}$ and s is in the second quadrant, determine $\cos \dfrac{s}{2}$.

66. If $\cos s = -\dfrac{2}{5}$ and s is in the second quadrant, determine $\tan \dfrac{s}{2}$.

67. If $\sin s = \dfrac{3}{5}$ and s is in the first quadrant, determine $\sin \dfrac{s}{2}$.

68. If $\sin s = \dfrac{4}{5}$ and s is in the second quadrant, determine $\cos \dfrac{s}{2}$.

69. If $\sin s = -\dfrac{4}{5}$ and s is in the third quadrant, determine $\sin \dfrac{s}{2}$.

70. Show that by multiplying numerator and denominator of $\dfrac{\cos h - 1}{h}$ by $\cos h + 1$, the expression $\dfrac{-\sin^2 h}{h(\cos h + 1)}$ is obtained.

71. Using $f(x) = \sin x$ and $h \neq 0$, show that

$$\frac{f(x+h) - f(x)}{h} = \sin x\left(\frac{\cos h - 1}{h}\right) + \cos x\left(\frac{\sin h}{h}\right).$$

72. If $f(u) = \sin u$, and $u = g(x) = 4x^2$, determine $f \circ g(x)$.
73. If $f(u) = \cos u$, and $u = g(x) = \sqrt{1 - 2x}$, determine $f \circ g(x)$.

74. By using the half-angle formulas (7), show that $\tan\dfrac{t}{2} = \pm\sqrt{\dfrac{1 - \cos t}{1 + \cos t}}$.

In Exercises 75 through 78, derive each identity by using appropriate addition formulas (4).

75. $\sin(u + v) + \sin(u - v) = 2 \sin u \cos v$
76. $\sin(u + v) - \sin(u - v) = 2 \cos u \sin v$
77. $\cos(u + v) + \cos(u - v) = 2 \cos u \cos v$
78. $\cos(u - v) - \cos(u + v) = 2 \sin u \sin v$

In Exercises 79 through 82, express each product as a sum or difference. (Use Exercises 75 through 78.)

79. $\cos 5x \cos 4x$
80. $\sin 2t \cos 4t$
81. $4 \sin 5s \cos 2s$
82. $\sin 3\theta \sin 5\theta$

In Exercises 83 through 86, derive each identity. (Use Exercises 75 through 78 with $u + v = a$ and $u - v = b$.)

83. $\sin a + \sin b = 2 \sin\dfrac{a+b}{2}\cos\dfrac{a-b}{2}$

84. $\sin a - \sin b = 2 \cos\dfrac{a+b}{2}\sin\dfrac{a-b}{2}$

85. $\cos a + \cos b = 2 \cos\dfrac{a+b}{2}\cos\dfrac{a-b}{2}$

86. $\cos b - \cos a = 2 \sin\dfrac{a+b}{2}\sin\dfrac{a-b}{2}$

In Exercises 87 through 90, write each expression as a product. (See Exercises 83 through 86.)

87. $\sin 2x + \sin 3x$
88. $\sin 4x - \sin x$
89. $\cos 6x + \cos 2x$
90. $\cos x - \cos 5x$

7.3 TRIGONOMETRIC IDENTITIES AND EQUATIONS

In this section a little more work with **trigonometric identities** will be presented and then we will be concerned with **trigonometric equations**. Let us make clear the distinction between these two. A trigonometric identity is true for all values of the arguments for which the functions involved are defined. For example, $\sin^2\theta + \cos^2\theta = 1$ is true for all values of θ. However, consider the trigonometric equation $3 \sin \theta = 1 + \sin \theta$. This is not an identity because it is not true for all values of θ $\left(\text{try } \theta = 0 \text{ or } \frac{\pi}{4}\right)$. Any values of θ that *do* make the equation true are *solutions.* This particular equation is solved in Example 4.

In the previous section many identities were presented and proved. Now we will use some of those to gain additional practice in the manipulation of trigonometric expressions. In particular certain trigonometric identities will be *verified.* Verifying a trigonometric identity is usually accomplished by transforming the left-hand side into the right-hand side or vice versa.

EXAMPLE 1 Verify the identity $\dfrac{\cot \alpha - \sin \alpha}{\cos \alpha} = \csc \alpha - \tan \alpha$.

Solution The left side will be transformed into the right side.

$$\frac{\cot \alpha - \sin \alpha}{\cos \alpha} = \frac{\cot \alpha}{\cos \alpha} - \frac{\sin \alpha}{\cos \alpha} = \frac{\frac{\cos \alpha}{\sin \alpha}}{\cos \alpha} - \tan \alpha$$

$$= \frac{\cos \alpha}{\sin \alpha} \cdot \frac{1}{\cos \alpha} - \tan \alpha = \frac{1}{\sin \alpha} - \tan \alpha$$

$$= \csc \alpha - \tan \alpha$$

EXAMPLE 2 Verify the identity $\cos u \sin u = \dfrac{1}{\cot u + \tan u}$.

Solution The right side will be transformed into the left side.

$$\frac{1}{\cot u + \tan u} = \frac{1}{\frac{\cos u}{\sin u} + \frac{\sin u}{\cos u}} = \frac{1}{\frac{\cos u \cos u}{\sin u \cos u} + \frac{\sin u \sin u}{\cos u \sin u}}$$

$$= \frac{1}{\frac{\cos^2 u}{\cos u \sin u} + \frac{\sin^2 u}{\cos u \sin u}} = \frac{1}{\frac{\cos^2 u + \sin^2 u}{\cos u \sin u}}$$

$$= \frac{\cos u \sin u}{\cos^2 u + \sin^2 u} = \cos u \sin u$$

In the next example we verify the identity by individually transforming each side into the same common expression and thus reasoning that the left-hand and right-hand sides must be equal.

EXAMPLE 3 Verify the identity:

$$\frac{\cos x}{1-\sin x} + \tan x = \frac{\tan^2 x + 2\sec x \tan x + 1}{\sec x}$$

Solution Denote the left-hand side by L and the right-hand side by R.

$$L = \frac{\cos x}{1-\sin x} + \tan x = \frac{(\cos x)(1+\sin x)}{(1-\sin x)(1+\sin x)} + \frac{\sin x}{\cos x}$$

$$= \frac{(\cos x)(1+\sin x)}{1-\sin^2 x} + \frac{\sin x}{\cos x} = \frac{(\cos x)(1+\sin x)}{\cos^2 x} + \frac{\sin x}{\cos x}$$

$$= \frac{1+\sin x}{\cos x} + \frac{\sin x}{\cos x} = \frac{1+2\sin x}{\cos x}$$

$$R = \frac{\tan^2 x + 2\sec x \tan x + 1}{\sec x} = \frac{\tan^2 x + 1 + 2\sec x \tan x}{\sec x}$$

$$= \frac{\sec^2 x + 2\sec x \tan x}{\sec x} = \frac{\sec x(\sec x + 2\tan x)}{\sec x}$$

$$= \sec x + 2\tan x = \frac{1}{\cos x} + \frac{2\sin x}{\cos x} = \frac{1+2\sin x}{\cos x} = L$$

It is important when working with identities to note that the operations used in transforming a side are performed on *only one side at a time* (see Exercise 46 for what can happen if this procedure is violated).

In contrast to this procedure, when working with equations it is necessary to perform the same operation on *both* sides simultaneously.

EXAMPLE 4 Solve the trigonometric equation $3\sin x = 1 + \sin x$.

Solution $3\sin x = 1 + \sin x$
$\sin x = \frac{1}{2}$ — Add $-\sin x$ to both sides

Now we want all values of x such that $\sin x = \frac{1}{2}$. Immediately, $\frac{\pi}{6}$ is one such value. There is also a second quadrant value whose sine is $\frac{1}{2}$ (because sine is positive in the first and second quadrants). That value is $\frac{5\pi}{6}$. Finally, any integral multiple of 2π added to $\frac{\pi}{6}$ or $\frac{5\pi}{6}$ will solve the equation. Thus,

$$x = \begin{cases} \dfrac{\pi}{6} + 2\pi k \\ \dfrac{5\pi}{6} + 2\pi k \end{cases} \quad \text{where } k \text{ is any integer.}$$

(We sometimes call $\frac{\pi}{6}$ and $\frac{5\pi}{6}$ the **basic solutions** to the equation.)

In degree form the answers are

$$x = \begin{cases} 30° + 360°k \\ 150° + 360°k \end{cases} \quad \text{where } k \text{ is any integer.}$$

EXAMPLE 5 Solve $\cos x \tan x = \cos x$.

Solution

$$\cos x \tan x = \cos x$$
$$\cos x \tan x - \cos x = 0$$
$$\cos x(\tan x - 1) = 0$$

$\cos x = 0$

The basic solutions are

$$x = \frac{\pi}{2} \text{ and } x = \frac{3\pi}{2}$$

Thus,

$$x = \begin{cases} \frac{\pi}{2} + 2\pi k \\ \frac{3\pi}{2} + 2\pi k \end{cases}$$

are solutions to $\cos x = 0$.

$\tan x - 1 = 0$

$\tan x = 1$

The basic solutions are $x = \frac{\pi}{4}$ and $x = \frac{5\pi}{4}$. (Recall the tangent is positive in the first and third quadrants.) Thus $\frac{\pi}{4} + 2\pi k$ and $\frac{5\pi}{4} + 2\pi k$ are solutions to $\tan x = 1$.

Finally, the solutions are

$$x = \begin{cases} \frac{\pi}{4} + 2\pi k \\ \frac{\pi}{2} + 2\pi k \\ \frac{5\pi}{4} + 2\pi k \\ \frac{3\pi}{2} + 2\pi k \end{cases} \quad \text{where } k \text{ is any integer.}$$

In degree form we have

$$x = \begin{cases} 45° + 360°k \\ 90° + 360°k \\ 225° + 360°k \\ 270° + 360°k \end{cases} \quad \text{where } k \text{ is any integer.}$$

The previous two examples illustrate that the method of solving trigonometric equations involves three steps. They are:

1. Reduce the given equation into simpler equations of the form $\sin x = r$, $\cos x = r$, $\tan x = r$, etc., where r is a real number.
2. Find the *basic solutions* to the reduced equations of step 1—they are in the interval $[0, 2\pi)$.
3. Add all integral multiples of 2π to the basic solutions.

EXAMPLE 6 Solve $\sin x + \sin x \sin 3x = 0$.

Solution

$$\sin x + \sin x \sin 3x = 0$$
$$\sin x(1 + \sin 3x) = 0$$

$\sin x = 0$	$1 + \sin 3x = 0$
$x = 0, x = \pi$	$\sin 3x = -1$
Thus $x = \begin{cases} 0 + 2\pi k \\ \pi + 2\pi k \end{cases}$	Thus $3x = \frac{3\pi}{2} + 2\pi k$
or simply	or
$x = \pi k$ (why?)	$x = \frac{\pi}{2} + \frac{2}{3}\pi k$

Finally,

$$x = \begin{cases} \pi k \\ \frac{\pi}{2} + \frac{2}{3}\pi k \end{cases} \quad \text{where } k \text{ is any integer.}$$

EXAMPLE 7 a. Solve $\cos 2t = \sin t$ on the interval $[0, 2\pi)$.
b. Solve $\cos 2t = \sin t$.

Solution a.

$$\cos 2t = \sin t \qquad \cos 2t = 1 - 2\sin^2 t$$
$$1 - 2\sin^2 t = \sin t$$
$$2\sin^2 t + \sin t - 1 = 0 \quad \leftarrow \text{This is a quadratic equation in } \sin t.$$
$$(2\sin t - 1)(\sin t + 1) = 0$$

$2\sin t - 1 = 0$	$\sin t + 1 = 0$
$\sin t = \frac{1}{2}$	$\sin t = -1$
$t = \frac{\pi}{6}, \frac{5\pi}{6}$	$t = \frac{3\pi}{2}$

The three values of t are $\frac{\pi}{6}, \frac{5\pi}{6}, \frac{3\pi}{2}$. (These are the basic solutions.)

b.
$$t = \begin{cases} \dfrac{\pi}{6} + 2\pi k \\ \dfrac{5\pi}{6} + 2\pi k \\ \dfrac{3\pi}{2} + 2\pi k \end{cases} \quad \text{where } k \text{ is any integer.}$$

EXAMPLE 8 a. Solve $\sin^4\theta = 1$ on the interval $[0, 2\pi)$.
b. Solve $\sin^4\theta = 1$.

Solution a.

$$\sin^4\theta = 1$$
$$\sin^4\theta - 1 = 0 \leftarrow \text{This is the difference of two squares.}$$
$$(\sin^2\theta + 1)(\sin^2\theta - 1) = 0$$

$\sin^2\theta + 1 = 0$	$\sin^2\theta - 1 = 0$
$\sin^2\theta = -1$	$(\sin\theta + 1)(\sin\theta - 1) = 0$
No real solutions, since the square of $\sin\theta$ must be nonnegative.	$\sin\theta = -1$, $\theta = \frac{3\pi}{2}$; $\sin\theta = 1$, $\theta = \frac{\pi}{2}$

b.
$$\theta = \begin{cases} \dfrac{3\pi}{2} + 2\pi k \\ \dfrac{\pi}{2} + 2\pi k \end{cases} \quad \text{where } k \text{ is any integer.}$$

EXAMPLE 9 Solve $2\sin^2 x - 3\sin x - 2 = 0$ on the interval $[0, 4\pi)$.

Solution

$$2\sin^2 x - 3\sin x - 2 = 0$$
$$(2\sin x + 1)(\sin x - 2) = 0$$

$2\sin x + 1 = 0$	$\sin x - 2 = 0$
$\sin x = -\frac{1}{2}$	$\sin x = 2$
$x = \frac{7\pi}{6}, \frac{7\pi}{6} + 2\pi$	No real solutions, since $\lvert\sin x\rvert$ must be less than or equal to 1.
$x = \frac{11\pi}{6}, \frac{11\pi}{6} + 2\pi$	

$$x = \frac{7\pi}{6}, \frac{11\pi}{6}, \frac{19\pi}{6}, \frac{23\pi}{6}$$

In degree measure, $x = 210°, 330°, 570°, 690°$.

EXERCISES FOR SECTION 7.3

In Exercises 1 through 40, verify each identity. (See Examples 1, 2, and 3.)

1. $\sin\theta\csc\theta = 1$
2. $\tan u \cot u = 1$
3. $\sec\theta\cos\theta + 1 = 2$
4. $\tan x \cos x = \sin x$
5. $\sin^2 t \cot^2 t = \cos^2 t$
6. $\sin^2\theta\cos\theta = \dfrac{1-\cos^2\theta}{\sec\theta}$
7. $\dfrac{\tan x}{\sin x \sec x} = 1$
8. $\dfrac{1+\tan^2 y}{\tan^2 y} = \csc^2 y$
9. $\dfrac{\sin x}{1-\cos x} = \dfrac{1+\cos x}{\sin x}$
10. $\tan^2\alpha\cos^2\alpha = 1-\cos^2\alpha$
11. $(\sin^2\theta + \cos^2\theta)^2 = 1$
12. $(\sin\theta + \cos\theta)^2 = 1 + 2\sin\theta\cos\theta$
13. $\sin 2x = 2\tan x\cos^2 x$
14. $\cos 2x = (\cos x - \sin x)(\cos x + \sin x)$
15. $\sin 4x = 4\sin x\cos x\cos 2x$
16. $(\sin\theta + \cos\theta)^2 = 1 + \sin 2\theta$
17. $\sin t + \sin t\cot^2 t = \cos t\csc t\sec t$
18. $\sin 4x = 4(\sin x\cos^3 x - \sin^3 x\cos x)$
19. $\dfrac{1}{\tan\beta + \sin\beta} = \dfrac{\cos\beta}{\sin\beta(1+\cos\beta)}$
20. $\sin 4x = \tan 4x\cos 4x$
21. $(\sin\theta + \cos\theta)^2 = 1 + 2\tan\theta - 2\sin^2\theta\tan\theta$
22. $1 - \sin^4 x = \cos^2 x + (\sin x\cos x)^2$
23. $2\csc 2s = \csc s\sec s$
24. $\sin\theta = \dfrac{\sec\theta}{\tan\theta + \cot\theta}$
25. $\dfrac{\sin\alpha}{1-\cos\alpha} - \cot\alpha = \dfrac{1}{\sin\alpha}$
26. $\csc\theta + \sec\theta = \dfrac{2\sin\theta + 2\cos\theta}{\sin 2\theta}$
27. $\dfrac{1}{\cos x} - \dfrac{\cos x}{1+\sin x} = \tan x$
28. $\left(\sin\dfrac{x}{2} + \cos\dfrac{x}{2}\right)^2 - \sin x = 1$
29. $\sec^2 t\csc^2 t = \csc^2 t + \sec^2 t$
30. $\dfrac{\cos(u-v)}{\sin u\cos v} = \cot u + \tan v$
31. $\dfrac{1}{1-\sin x} - \dfrac{1}{1+\sin x} = 2\tan x\sec x$
32. $\dfrac{\sin(u+v)}{\sin u} = \cos v + \cot u\sin v$
33. $\sin 2\alpha = \dfrac{2}{\tan\alpha + \cot\alpha}$
34. $\cos 2A + 2\sin^2 A = 1$
35. $\sin(\pi + A) = -\sin A$
36. $\cos(\pi + A) = -\cos A$
37. $\sin\left(\dfrac{\pi}{2} + A\right) = \cos A$
38. $\cos\left(\dfrac{\pi}{2} + A\right) = -\sin A$
39. $\tan\dfrac{\theta}{2} = \dfrac{\sin\theta}{2\cos^2\dfrac{\theta}{2}}$
40. $1 + \dfrac{1}{\cos B} = \dfrac{\tan^2 B}{\sec B - 1}$

In Section 7.2 the expressions $\sin^2 x = \frac{1}{2}(1-\cos 2x)$ and $\cos^2 x = \frac{1}{2}(1+\cos 2x)$ were encountered (see discussion preceding the half-angle formulas (7)). These can be used to rewrite expressions involving even powers of sine and cosine. For example, $\sin^2 2u = \frac{1}{2}(1-\cos 4u)$, $\cos^2 3t = \frac{1}{2}(1+\cos 6t)$, etc.

In Exercises 41 through 45, verify each identity by using the previous formulas.

41. $1 + \sin^2 x = \dfrac{3 - \cos 2x}{2}$
42. $1 + \cos^2 x = \dfrac{3 + \cos 2x}{2}$
43. $\sin^2 x \cos^2 x = \frac{1}{8}(1 - \cos 4x)$
44. $\cos^4 t = \frac{3}{8} + \frac{1}{2}\cos 2t + \frac{1}{8}\cos 4t$
45. $\sin^4 x = \frac{3}{8} - \frac{1}{2}\cos 2x + \frac{1}{8}\cos 4x$
46. It is *incorrect procedure* to attempt to verify an identity by performing the same operation on both sides. Below is such an attempt.

$$\sqrt{1 - \sin^2 x} = \frac{\sin x}{\tan x}$$

$$\sqrt{1 - \sin^2 x} = \frac{\sin x}{\dfrac{\sin x}{\cos x}} \qquad \tan x = \frac{\sin x}{\cos x}$$

$$\sqrt{1 - \sin^2 x} = \cos x$$

$$1 - \sin^2 x = \cos^2 x \qquad \text{Square both sides}$$

$$\cos^2 x = \cos^2 x$$

Even though it appears as though the original expression is an identity, *it is not*.

a. What step violated proper procedure for working with identities?
b. Find (by trial and error) three values of x for which the original statement $\sqrt{1 - \sin^2 x} = \dfrac{\sin x}{\tan x}$ is not true.

In Exercises 47 through 70, solve each trigonometric equation. (See Examples 4 through 9.)

47. $\sin \theta + 1 = 0$
48. $\tan \theta + 1 = 0$
49. $\cos \theta - \frac{1}{2} = 0$
50. $\cos 2\theta - \frac{1}{2} = 0$
51. $2 \sin \theta - 1 = 0$
52. $4 \sin x = \frac{1}{2} + 3 \sin x$
53. $\tan x = 2 \tan x - 1$
54. $\sin A \tan A = \sin A$
55. $\sec B \sin B = \sec B$
56. $2 \sin C = \sqrt{3}$
57. $2 \sin C = -\sqrt{3}$
58. $(\cos \theta - 1)(\tan \theta - 1) = 0$
59. $(\sin 2x + 1)(\cos x + 1) = 0$
60. $\sin^2 \theta - 1 = 0$
61. $\tan^2 \theta - 1 = 0$
62. $\tan^2 \theta - 3 = 0$
63. $\sin 2x = 2 \sin x$
64. $\cos 2x + \sin^2 x = 0$
65. $2 \sin^2 x + \sin x - 1 = 0$
66. $\cos^4 A = 1$
67. $\cos^4 2A = 1$
68. $2 \sin^2 B + 3 \sin B - 2 = 0$
69. $\cos^2 \theta - 7 \cos \theta + 12 = 0$
70. $\sin x \cos x + \dfrac{\sqrt{3}}{2} \sin x - \dfrac{1}{2} \cos x - \dfrac{\sqrt{3}}{4} = 0$ (Factor by grouping.)

In Exercises 71 through 82, solve each trigonometric equation on the specified interval and give the solution in degree form. (See Examples 4 through 9.)

71. $\cos x - \dfrac{\sqrt{3}}{2} = 0$ on $[0, 2\pi)$
72. $2 \cos 4\theta = 1$ on $[0, 2\pi)$
73. $2 \cos 6\theta = 1$ on $[0, 2\pi)$
74. $2 \cos 6\theta = 1$ on $[0, 4\pi)$
75. $\tan 2\alpha = 1$ on $[0, \pi)$
76. $\sin 2A = \sqrt{2} \cos A$ on $[0, 2\pi)$

77. $2\sin^2 t - 3\sin t - 2 = 0$ on $[0, 2\pi)$
78. $(\csc\theta + 1)(\sin\theta + 1) = 0$ on $[0, 2\pi)$
79. $\tan^2\theta + \sec\theta + 1 = 0$ on $[0, 2\pi)$
80. $\sin 2x + \sin x = 0$ on $[0, 3\pi)$
81. $\cot^2\theta + 3\csc\theta + 3 = 0$ on $[0, 2\pi)$
82. $\cos\theta\tan\theta + \cos\theta + \tan\theta + 1 = 0$ on $[0, 4\pi)$ (Factor by grouping.)

7.4 INVERSE TRIGONOMETRIC FUNCTIONS

In this present section the **inverse trigonometric functions** will be defined and illustrated. First recall some things that are true for any relation g and the inverse relation g^{-1}:

1. the equation of g^{-1} can be obtained from the equation of g by interchanging the variables,
2. the domain of g is the range of g^{-1},
3. the range of g is the domain of g^{-1},
4. the graph of g^{-1} is the "mirror image" of g about the line $y = x$,
5. the inverse of a relation might or might not, itself, be a function.

Consider the graph of $y = \sin x$. Interchanging the variables x and y and reflecting the graph of $y = \sin x$ across the line $y = x$ produces the inverse relation of $y = \sin x$, namely $x = \sin y$. The vertical line test shows that this relation $x = \sin y$ is *not* a function. However by a suitable restriction on the domain of $y = \sin x$ or equivalently, a restriction on the range of $x = \sin y$, we can obtain a function. Such is accomplished in the following definition.

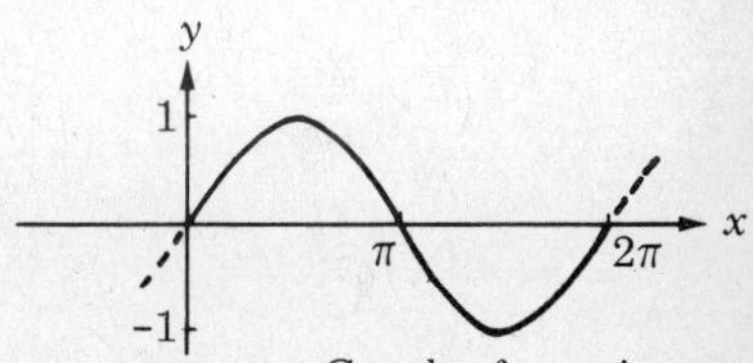

Graph of $y = \sin x$

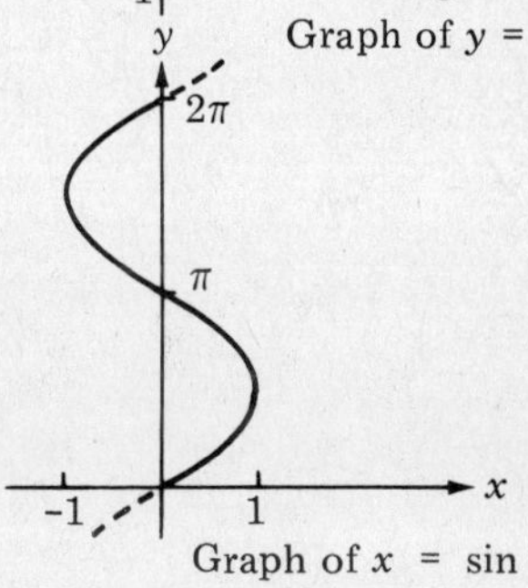

Graph of $x = \sin y$

DEFINITION

The **inverse sine function** (or **arcsine function**) symbolized by $\sin^{-1}$ (or arcsin) is defined by

$$\sin^{-1} u = v \qquad \text{or} \qquad \arcsin u = v$$

if and only if $u = \sin v$, where $-1 \leqslant u \leqslant 1$ and $-\frac{\pi}{2} \leqslant v \leqslant \frac{\pi}{2}$.*

*As always with inverse functions, the "−1" on $\sin^{-1}$ does *not* represent an exponent.

EXAMPLE 1 Graph $y = \arcsin x$.

Solution

$$y = \arcsin x \text{ means}$$

$$x = \sin y, \text{ where } -\frac{\pi}{2} \leqslant y \leqslant \frac{\pi}{2}$$

Thus we use the graph of $x = \sin y$ from the previous page restricted to the interval $-\frac{\pi}{2} \leqslant y \leqslant \frac{\pi}{2}$.

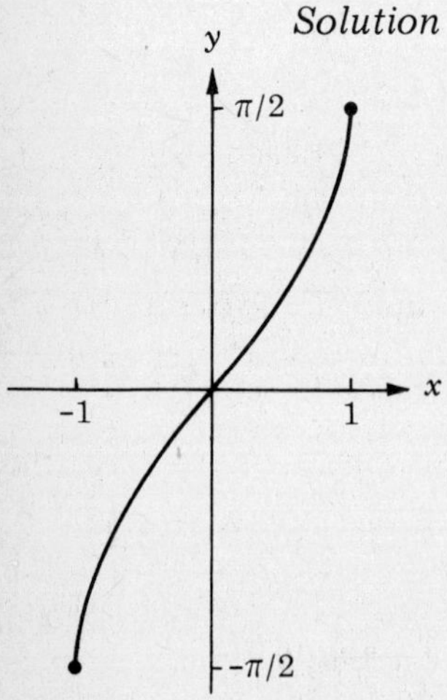

Graph of $y = \arcsin x$ or $y = \sin^{-1} x$

In the following definition we define the five other inverse trigonometric functions.

DEFINITION

$v = \arccos u$ if and only if $u = \cos v,\ 0 \leqslant v \leqslant \pi$

$v = \arctan u$ if and only if $u = \tan v,\ -\frac{\pi}{2} < v < \frac{\pi}{2}$

$v = \text{arccsc } u$ if and only if $u = \csc v,\ -\frac{\pi}{2} \leqslant v \leqslant \frac{\pi}{2}, v \neq 0$

$v = \text{arcsec } u$ if and only if $u = \sec v,\ 0 \leqslant v \leqslant \pi,\ v \neq \frac{\pi}{2}$

$v = \text{arccot } u$ if and only if $u = \cot v,\ 0 < v < \pi$

With $u = x$ and $v = y$, the graphs of each of these five functions appear as follows.

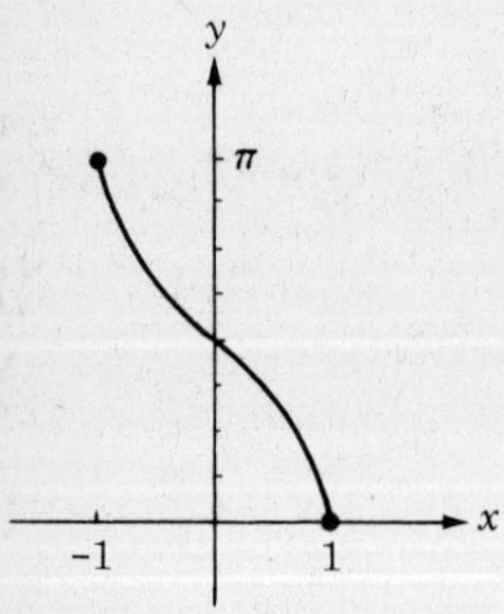

Graph of $y = \arccos x$ or $y = \cos^{-1} x$

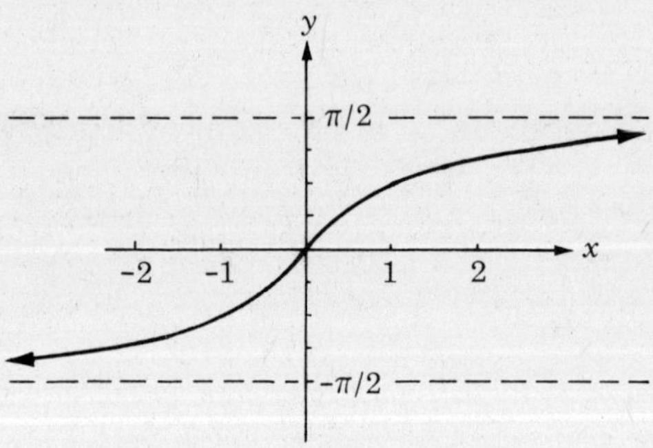

Graph of $y = \arctan x$ or $y = \tan^{-1} x$

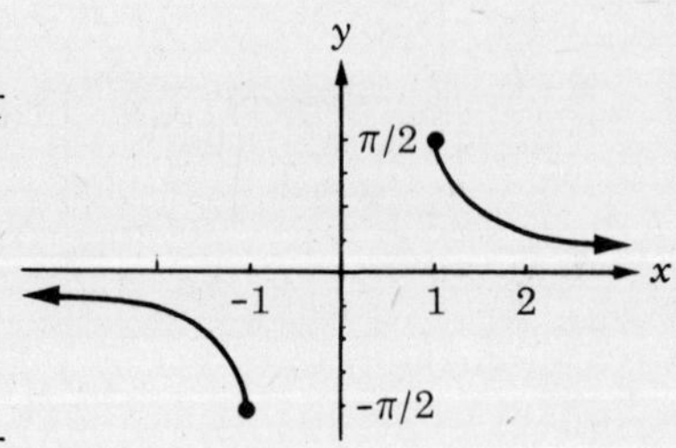

Graph of $y = \text{arccsc } x$ or $y = \csc^{-1} x$

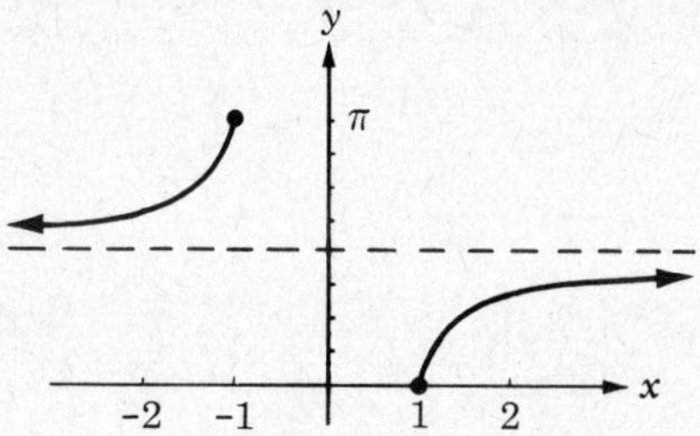

Graph of y = arcsec x or $y = \sec^{-1} x$

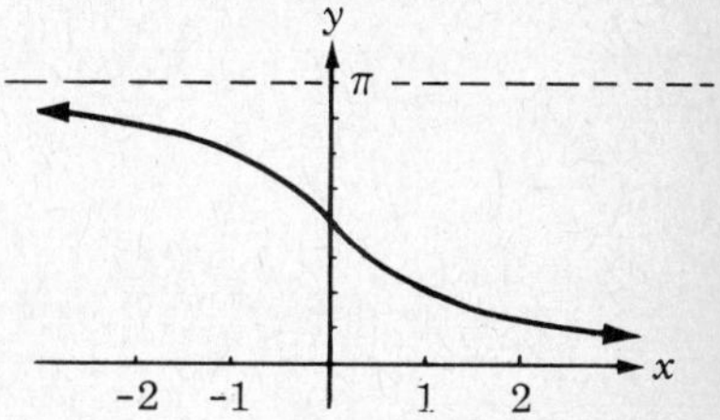

Graph of y = arccot x or $y = \cot^{-1} x$

Graphs of other inverse trigonometric functions can be obtained by using the definitions above and known properties of trigonometric functions as shown in the next two examples.

EXAMPLE 2 Graph $y = 4 \arcsin 2x$.

Solution $y = 4 \arcsin 2x$ can be rewritten as $\frac{1}{4}y = \arcsin 2x$. Comparing this to the definition, $v = \frac{1}{4}y$ and $u = 2x$. Thus, $\frac{1}{4}y = \arcsin 2x$ if and only if $\sin \frac{1}{4}y = 2x$, where $-\dfrac{\pi}{2} \leqslant \dfrac{1}{4}y \leqslant \dfrac{\pi}{2}$. Hence, $x = \frac{1}{2} \sin \frac{1}{4}y$, where $-2\pi \leqslant y \leqslant 2\pi$. Thus we graph that portion of $x = \frac{1}{2} \sin \frac{1}{4}y$ in the interval $-2\pi \leqslant y \leqslant 2\pi$.

Amplitude = $\frac{1}{2}$.

Period = $\dfrac{2\pi}{b} = \dfrac{2\pi}{\frac{1}{4}} = 8\pi$.

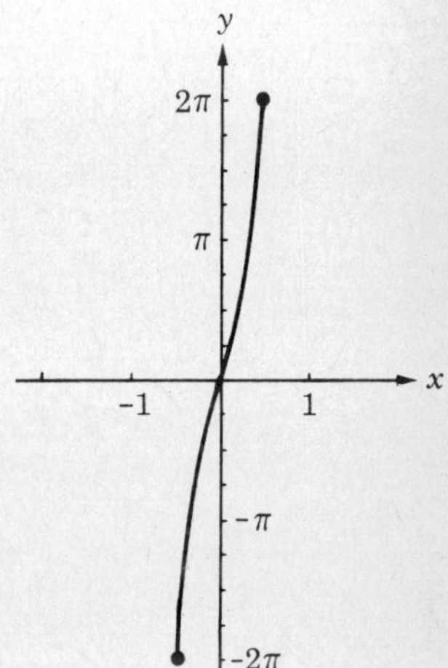

Graph of $y = 4 \arcsin 2x$

EXAMPLE 3 Graph $y = 2 \tan^{-1} 3x$.

Solution $y = 2 \tan^{-1} 3x$ can be rewritten as $\frac{1}{2}y = \tan^{-1} 3x$. Comparing this to the definition, $v = \frac{1}{2}y$ and $u = 3x$. Thus $\frac{1}{2}y = \tan^{-1} 3x$ if and only if $\tan \frac{1}{2}y = 3x$, where $-\dfrac{\pi}{2} < \dfrac{1}{2}y < \dfrac{\pi}{2}$. Hence, $x = \frac{1}{3} \tan \frac{1}{2}y$, where $-\pi < y < \pi$.

Period = $\dfrac{\pi}{b} = 2\pi$.

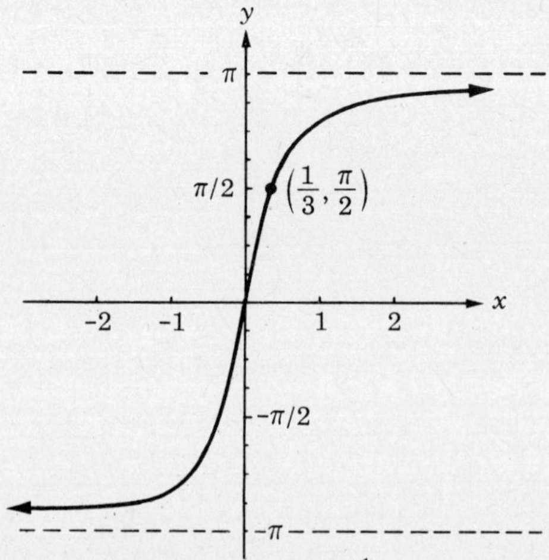

Graph of $y = 2\tan^{-1} 3x$

Next, we consider the evaluation of inverse trigonometric functions. First it is important to point out the distinction between the symbols $\sin t$ and $\arcsin t$. The former is read "the sine of t," whereas the latter is read as "the angle (or real number) whose sine is t."

EXAMPLE 4 Determine the value of each of the following.

a. $\arcsin \dfrac{\sqrt{2}}{2}$ b. $\cos^{-1}(-\frac{1}{2})$ c. $\arctan(-.7002)$

Solution a. Let $\arcsin \dfrac{\sqrt{2}}{2} = x$. This is equivalent to $\sin x = \dfrac{\sqrt{2}}{2}$, where $-\dfrac{\pi}{2} \leqslant x \leqslant \dfrac{\pi}{2}$. Thus, $x = \dfrac{\pi}{4}$.

b. Let $\cos^{-1}(-\frac{1}{2}) = x$. This is equivalent to $\cos x = -\frac{1}{2}$, where $0 \leqslant x \leqslant \pi$. Thus, $x = \frac{2}{3}\pi$.

c. Let $\arctan(-.7002) = x$. This is equivalent to $\tan x = -.7002$, where $-\dfrac{\pi}{2} < x < \dfrac{\pi}{2}$. From Table III, the reference number is .6109. Thus, $x = -.6109$.

> Using a calculator with an inverse tangent key gives the slightly more precise value $\tan^{-1}(-.7002) = -.61086018$.

EXAMPLE 5 Determine the value of each of the following.

a. $\sin[\tan^{-1}(-1)]$ b. $\cos\left[\cos^{-1}\dfrac{\sqrt{3}}{2}\right]$ c. $\sin[\arcsin \frac{4}{5} + \arctan \frac{2}{3}]$

Solution a. First, let $\tan^{-1}(-1) = x$. Thus $\tan x = -1$, where $-\frac{\pi}{2} < x < \frac{\pi}{2}$. Hence $x = -\frac{\pi}{4}$.

$$\sin[\tan^{-1}(-1)] = \sin\left(-\frac{\pi}{4}\right) = -\sin\frac{\pi}{4} = -\frac{\sqrt{2}}{2}$$

b. First, let $\cos^{-1}\frac{\sqrt{3}}{2} = x$. Thus $\cos x = \frac{\sqrt{3}}{2}$, where $0 \leqslant x \leqslant \pi$. Hence $x = \frac{\pi}{6}$.

$$\cos\left[\cos^{-1}\frac{\sqrt{3}}{2}\right] = \cos\frac{\pi}{6} = \frac{\sqrt{3}}{2}$$

c. Let $\arcsin\frac{4}{5} = A$ and $\arctan\frac{2}{3} = B$. Then $\sin A = \frac{4}{5}$ and $\tan B = \frac{2}{3}$. We need the value of

$$\sin(A + B) = \sin A \cos B + \cos A \sin B$$

$$\sin(A + B) = \frac{4}{5} \cdot \frac{3}{\sqrt{13}} + \frac{3}{5} \cdot \frac{2}{\sqrt{13}},$$

from the right triangles at the right.

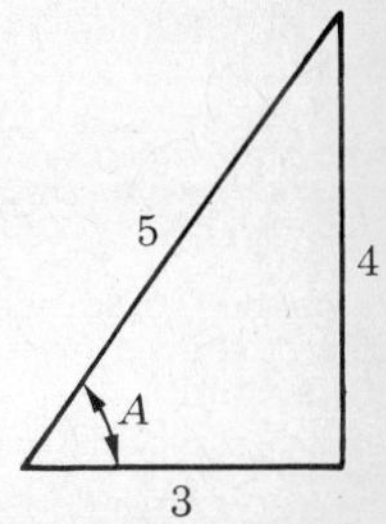

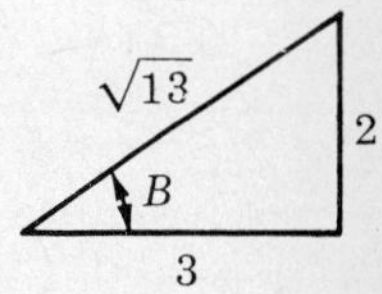

$$\sin(A + B) = \frac{12}{5\sqrt{13}} + \frac{6}{5\sqrt{13}}$$

$$= \frac{18}{5\sqrt{13}} = \frac{18\sqrt{13}}{65}$$

EXAMPLE 6 a. Obtain a formula for $\sin(\cos^{-1}x)$ in terms of x.
b. Obtain a formula for $\sin(\tan^{-1}x)$ in terms of x.

Solution a. Let $\cos^{-1}x = Q$. Thus, $\cos Q = x$, $0 \leqslant Q \leqslant \pi$. We need $\sin Q$ which will be positive in the interval $0 \leqslant Q \leqslant \pi$. Thus

$$\sin Q = \sqrt{1 - \cos^2 Q} = \sqrt{1 - x^2}$$

b. Let $\tan^{-1}x = Q$. Thus, $\tan Q = x$, $-\frac{\pi}{2} < Q < \frac{\pi}{2}$. We need $\sin Q$ which will be negative if $-\frac{\pi}{2} < Q < 0$ and positive if $0 < Q < \frac{\pi}{2}$.

$$\sin Q = \begin{cases} \dfrac{x}{\sqrt{x^2 + 1}}, & \text{if } 0 < Q < \dfrac{\pi}{2} \\ -\dfrac{x}{\sqrt{x^2 + 1}}, & \text{if } -\dfrac{\pi}{2} < Q < 0 \end{cases}$$

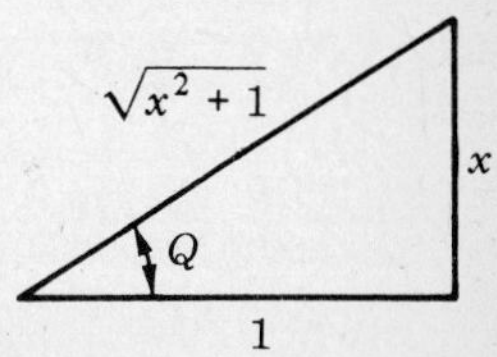

EXAMPLE 7 Solve the trigonometric equation $\cos^2\theta + 2\cos\theta - 1 = 0, 0 \leqslant \theta \leqslant \pi$.

Solution The given equation is quadratic in $\cos\theta$, but it does not factor. We use the quadratic formula to solve for $\cos\theta$, with $a = 1, b = 2, c = -1$.

$$\cos\theta = \frac{-2 \pm \sqrt{(2)^2 - 4(1)(-1)}}{2(1)} = \frac{-2 \pm \sqrt{8}}{2} = -1 \pm \sqrt{2}$$

$$\cos\theta = -1 + \sqrt{2}$$
$$\theta = \cos^{-1}(-1 + \sqrt{2})$$

Using a calculator,
$\theta = \cos^{-1}(.41421356)$
$\theta = 1.1437177$

$$\cos\theta = -1 - \sqrt{2}$$

No real solutions, since $|\cos x|$ must be less than or equal to 1.

EXERCISES FOR SECTION 7.4

In Exercises 1 through 12, solve each equation for r. (See Example 1 and Definitions of the inverse trigonometric functions.)

1. $t = \arcsin r$
2. $2t = \cos^{-1} r$
3. $2t = 3 \arctan r$
4. $4t = 3 \operatorname{arccot}\left(\frac{r}{2}\right)$
5. $\frac{t}{2} = 1 + 4 \operatorname{arccsc}\left(\frac{r}{3}\right)$
6. $t + 2 = \frac{1}{2} \operatorname{arcsec}(r + 3)$
7. $1 - t = 4 \arcsin\left(\frac{r+2}{3}\right)$
8. $|t| = 2 + 2\sec^{-1}(\sqrt{r})$
9. $1 - t = 4 \tan\left(\frac{1}{r}\right)$
10. $\csc^{-1}\left(\frac{1}{r}\right) = \log_{10} t$
11. $\cot^{-1}(r) = \cot(t + 1)$
12. $\arccos\left(\frac{1}{r+1}\right) = t$

In Exercises 13 through 24, graph each inverse function. (See Examples 2 and 3.)

13. $y = \arcsin 2x$
14. $y = 2 \arcsin x$
15. $y = \cos^{-1} 2x$
16. $y = 2\cos^{-1} x$
17. $y = \tan^{-1} 2x$
18. $y = 2\tan^{-1} x$
19. $y = \frac{1}{2}\sin^{-1} 2x$
20. $y = 2 \arccos 4x$
21. $y = \sec^{-1} 2x$
22. $y = 2\sec^{-1} x$
23. $y = \csc^{-1} 4x$
24. $y = \cot^{-1} 4x$

In Exercises 25 through 44, determine the value of each of the following. (See Examples 4 and 5.)

25. $\arccos \frac{\sqrt{2}}{2}$
26. $\arcsin \frac{\sqrt{3}}{2}$
27. $\arcsin\left(-\frac{\sqrt{3}}{2}\right)$
28. $\cot^{-1}(\frac{1}{2})$
29. $\tan^{-1} 1$
30. $\tan^{-1}\sqrt{3}$
31. $\tan^{-1}(-\sqrt{3})$
32. $\arccos\left(-\frac{\sqrt{3}}{2}\right)$
33. $\sec^{-1} 2$
34. $\csc^{-1} 2$
35. $\tan^{-1}(2.565)$
36. $\sin^{-1}(.3923)$
37. $\tan^{-1}(-2.565)$
38. $\sin^{-1}(-.3923)$

39. $\sin(\sin^{-1}\frac{1}{2} + \cos^{-1}\frac{1}{2})$
40. $\sin(\cos^{-1}\frac{1}{2} + \tan^{-1} 1)$
41. $\cos(\cos^{-1}\frac{1}{2} + \tan^{-1} 1)$
42. $\cos(\arcsin\frac{1}{2} - \arctan\frac{1}{3})$
43. $\sin(\arcsin\frac{3}{5} - \arccos\frac{1}{2})$
44. $\cos(\arcsin\frac{1}{2} + \arctan\frac{1}{3})$

In Exercises 45 through 52, obtain a formula for each expression in terms of x. (See Example 6.)

45. $\sin(\cos^{-1} 2x)$
46. $\sin(\tan^{-1} 2x)$
47. $\cos(\sin^{-1} x)$
48. $\cos[\sin^{-1}(x + 1)]$
49. $\tan(\sin^{-1} x)$
50. $\tan(\cot^{-1} x)$
51. $\cos(\sin^{-1} 2x)$
52. $\cot(\tan^{-1} x)$

In Exercises 53 through 58, solve each equation. (See Example 7.)

53. $\cos^2\theta + 3\cos\theta - 1 = 0$ on $[0, 2\pi)$
54. $\sin^2\theta - .9 = 0$ on $[0, 2\pi)$
55. $2\sin^2 t - 3\sin t - 1 = 0$
56. $\cos^2\theta + 2\cos\theta - 5 = 0$
57. $\sin^2 u + 4\sin u + 1 = 0$
58. $2\tan^2 y - \tan y - 1 = 0$ on $[0, \pi)$

Calculator Exercises

59. Graph $y = \arcsin x$ for $x = 0, \pm.2, \pm.4, \pm.6, \pm.8, \pm1.0$.
60. Graph $y = \arccos x$ for $x = 0, \pm.2, \pm.4, \pm.6, \pm.8, \pm1.0$.
61. Graph $y = \arctan x$ for $x = -10, -8, -6, -4, -2, 0, 2, 4, 6, 8, 10$.
62. Determine the value of $\frac{\sin x}{x}$ where $x = 2, 1, .1, .01, .001, .0001$.
63. Determine the value of $\frac{\cos x}{x}$ where $x = 2, 1, .1, .01, .001, .0001$.
64. Determine the value of $\frac{\tan x}{x}$ where $x = 2, 1, .1, .01, .001, .0001$.
65. Solve for x: $\sin(\tan^{-1} x) = .4$.
66. Solve for x: $\log_{10}(\tan^{-1} x) = .4$.
67. Solve for x: $\ln(\tan^{-1} x) = .4$.
68. Solve for x: $e^{\cos^{-1} x} = .4$.
69. Solve for x: $\sin^{-1}(e^x) = .4$.
70. Solve for x: $\sin(\ln x) = 0$.

7.5 THE POLAR COORDINATE SYSTEM

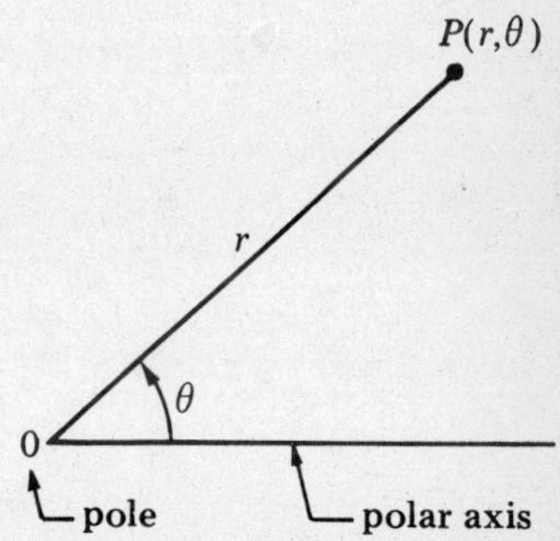

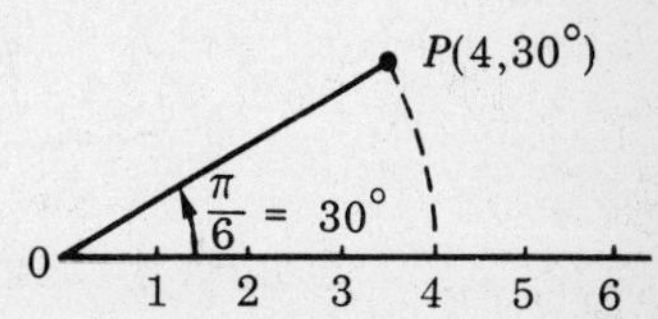

When a point P lies in a plane, its exact location can be specified by coordinates in a Cartesian coordinate system. Another system commonly used is the **polar coordinate system** that specifies the location of a point P by means of the **polar coordinates** r and θ shown at the right. The polar coordinates of a point are written as (r, θ). When r and θ are given, a unique point is determined, but the reverse is not true. The point P at the right can be specified by the coordinates $(4, 30°)$, $(4, 390°)$, $(4, 750°)$, $(4, -330°)$, etc. Each of these pairs was obtained by starting at the pole and moving $r = 4$ units in the direction of the polar axis (to the right) and then rotating through the specified angle to "arrive at" P. The sign convention on θ is positive for CCW rotation and negative for CW rotation. The coordinate r can also be specified as negative by starting at the pole and moving on the polar axis (to the left) 4 units and then from there rotating through an angle to "arrive at" P. Doing so in the diagram on the previous page gives the following as possible coordinates of P: $(-4, 210°)$, $(-4, -150°)$, $(-4, 570°)$, etc.

EXAMPLE 1 Specify three pairs of polar coordinates for each point on the polar graph paper below.

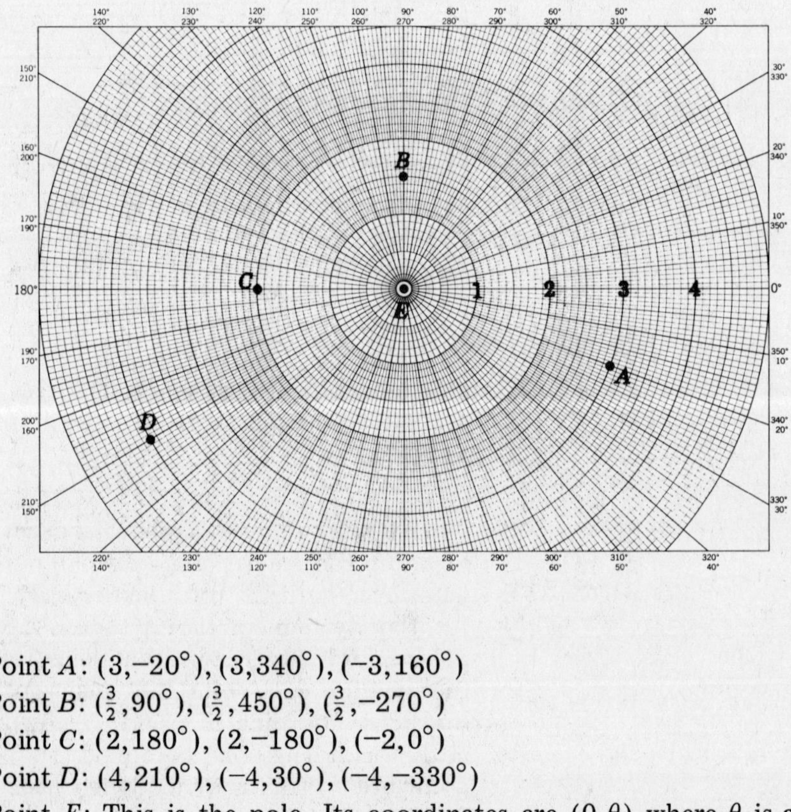

Solution Point A: $(3, -20°)$, $(3, 340°)$, $(-3, 160°)$
Point B: $(\frac{3}{2}, 90°)$, $(\frac{3}{2}, 450°)$, $(\frac{3}{2}, -270°)$
Point C: $(2, 180°)$, $(2, -180°)$, $(-2, 0°)$
Point D: $(4, 210°)$, $(-4, 30°)$, $(-4, -330°)$
Point E: This is the pole. Its coordinates are $(0, \theta)$ where θ is any angle.

Next, graphs of polar equations are considered.

EXAMPLE 2 Graph $r = 3$.

Solution θ does not appear in the given equation. Thus, θ can assume any value as long as $r = 3$. Hence the graph is a circle of radius 3 centered at the pole.

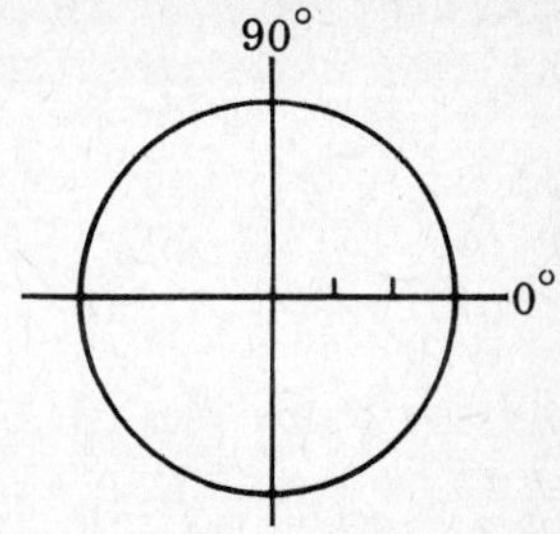

Graph of $r = 3$

EXAMPLE 3 Graph $\theta = \frac{\pi}{6}$ (or equivalently $\theta = 30°$).

Solution Since r does not appear in the equation, it can assume any value provided $\theta = \frac{\pi}{6}$. So, the graph is the straight line through the pole shown at the right.

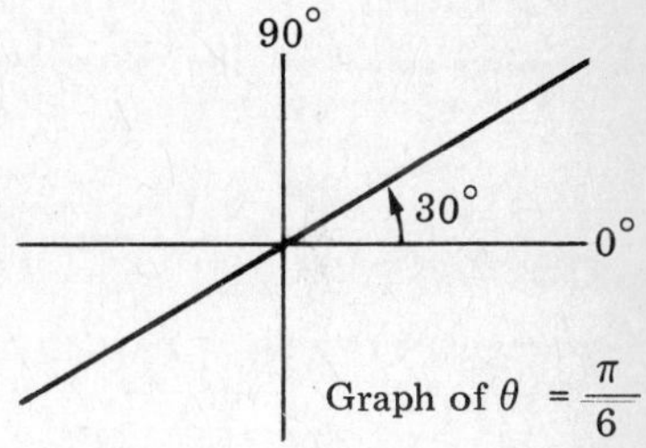

Graph of $\theta = \frac{\pi}{6}$

EXAMPLE 4 Graph $r = 2 \cos \theta$.

Solution By selecting various values of θ, the corresponding values of r can be calculated.

r	θ
2	0°
$\sqrt{3}$	30°
1	60°
0	90°
−1	120°
$-\sqrt{3}$	150°
−2	180°

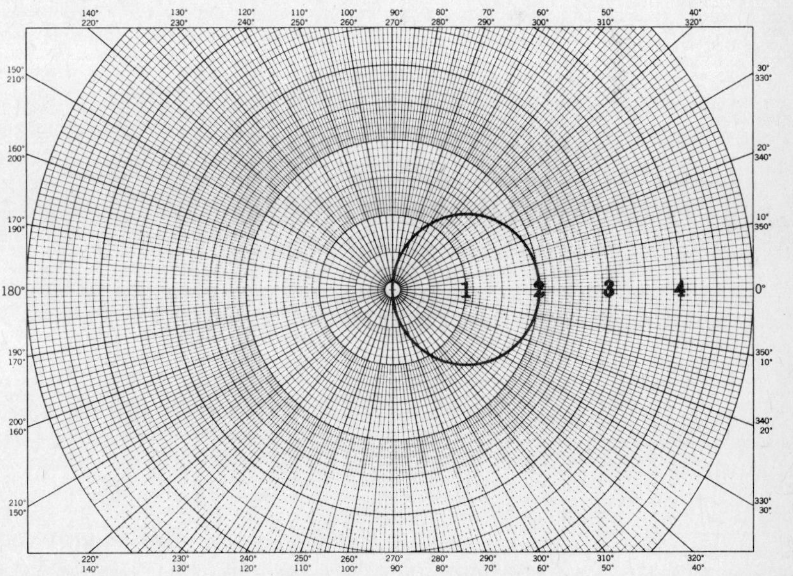

Graph of $r = 2 \cos \theta$

Taking values of θ less than 0° or greater than 180° produces more points that lie on the same curve. In Exercise 40 the reader is asked to show that the curve in this example is a circle.

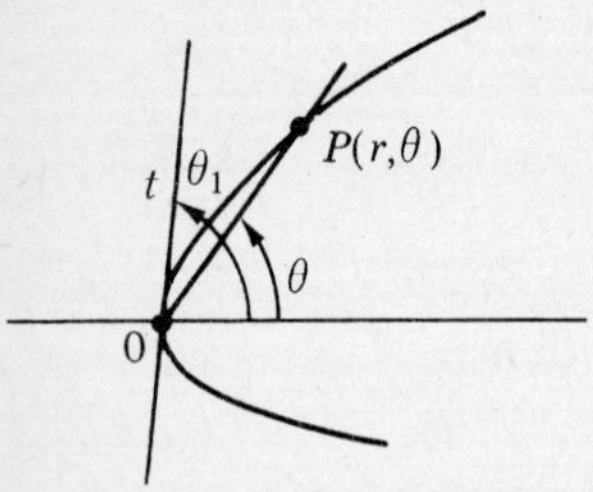

Rather than totally relying on point plotting we will mention one helpful technique for graphing polar curves.

Many polar curves pass through the pole. When such is the case there is generally a tangent line to the curve at the pole. Suppose that $P(r, \theta)$ is a point on a polar curve, and further suppose that r equals zero when $\theta = \theta_1$. Thus, it seems reasonable to say that if point P approaches point 0 along the curve, r approaches zero and θ approaches θ_1 as a limiting position. Thus, t is a tangent to the polar curve at the pole and we have:

Tangents at the Pole

To obtain the tangents to a polar curve at the pole, let $r = 0$ and solve for θ.

For example, consider $r = 2 \sin 2\theta$. If $r = 0$, $2 \sin 2\theta = 0$, and thus

$$\sin 2\theta = 0$$

$$2\theta = 0^\circ, 180^\circ \quad \text{(Integral multiples of } 180^\circ \text{ will give the same tangent lines)}$$

$$\theta = 0^\circ, 90^\circ \quad \text{(Tangent lines)}$$

EXAMPLE 5 Graph $r = 2 \sin 2\theta$.

Solution From the discussion above, $\theta = 0^\circ, 90^\circ$ are tangent lines at the pole. Next plot points.

r	θ
0	0°
1	15°
1.4	22.5°
1.7	30°
2	45°
1.7	60°
1	75°
0	90°
−1.7	120°
−2	135°

r	θ
−1.7	150°
0	180°
1.7	210°
2	225°
1.7	240°
0	270°
−1.7	300°
−2	315°
−1.7	330°
0	360°

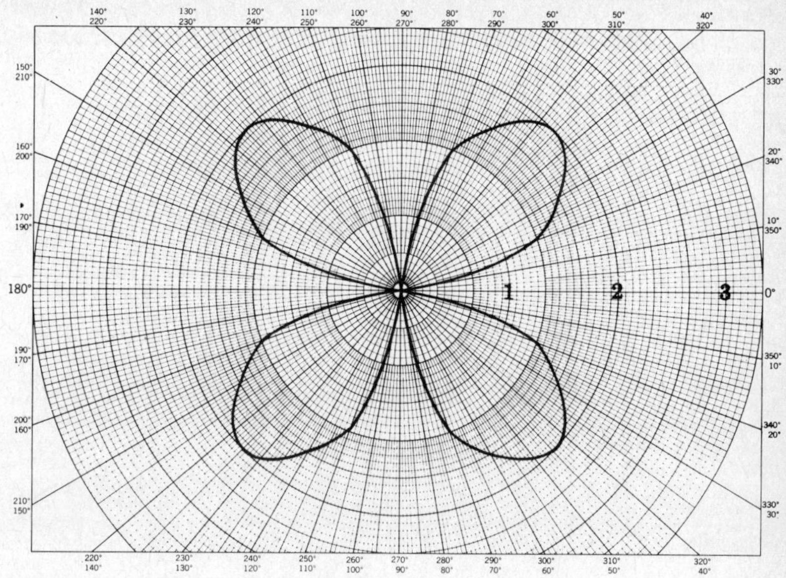

Graph of $r = 2 \sin 2\theta$

The curve is a so-called 4-leaf rose.

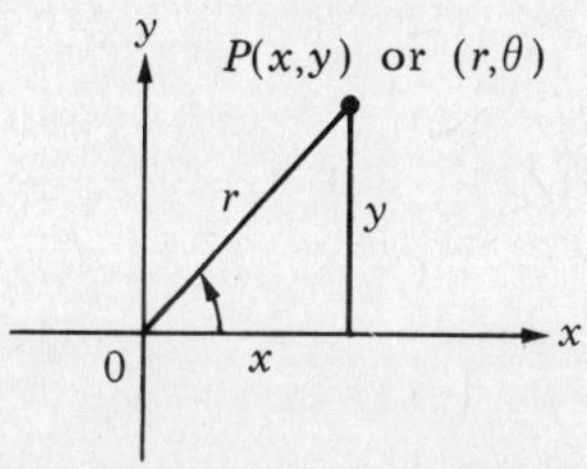

The Cartesian coordinate system and the polar coordinate system can be related to each other by viewing the two systems together. If r is positive,

$$\cos \theta = \frac{x}{r} \text{ and } \sin \theta = \frac{y}{r}.$$

Solving for x and y, we obtain:

$$\boxed{x = r \cos \theta \quad \text{and} \quad y = r \sin \theta}$$

In Exercise 50 the reader is asked to show that the same two equations can be derived when r is negative. Also true are:

$$\boxed{\tan \theta = \frac{y}{x} \quad \text{and} \quad r^2 = x^2 + y^2}$$

These equations can be used to transform polar coordinate equations into Cartesian coordinate equations and vice versa.

EXAMPLE 6 Determine the Cartesian coordinate equation corresponding to

$$r^2 = 9 \sin 2\theta.$$

Solution

$$r^2 = 9 \sin 2\theta$$
$$r^2 = 9(2 \sin \theta \cos \theta)$$
$$x^2 + y^2 = 18 \cdot \frac{y}{r} \cdot \frac{x}{r} = \frac{18xy}{r^2}$$
$$x^2 + y^2 = \frac{18xy}{x^2 + y^2} \quad \text{or} \quad (x^2 + y^2)^2 = 18xy$$

EXAMPLE 7 Determine the polar coordinate equation corresponding to

$$x^2 + y^2 - 2y = 0.$$

Solution

$$x^2 + y^2 - 2y = 0$$
$$(r \cos \theta)^2 + (r \sin \theta)^2 - 2(r \sin \theta) = 0$$
$$r^2(\cos^2\theta + \sin^2\theta) - 2r \sin \theta = 0$$
$$r^2(1) - 2r \sin \theta = 0$$
$$r(r - 2 \sin \theta) = 0$$
$$r = 0 \quad \text{or} \quad r = 2 \sin \theta$$

The equation $r = 0$ can be deleted because $r = 0$ represents only the pole and this is included in the equation $r = 2 \sin \theta$ (when $\theta = 0$).

Thus, $r = 2 \sin \theta$.

EXERCISES FOR SECTION 7.5

In Exercises 1 through 9, specify two other pairs of polar coordinates for each given point below. (See Example 1.)

1. $(2, 20°)$
2. $(2, -20°)$
3. $(-2, 20°)$
4. $(-2, -20°)$
5. $\left(3, \frac{\pi}{2}\right)$
6. $\left(-2, \frac{\pi}{4}\right)$
7. $\left(4, \frac{3\pi}{4}\right)$
8. $\left(5, \frac{7\pi}{6}\right)$
9. $\left(5, \frac{5\pi}{3}\right)$

In Exercises 10 through 30, graph each polar equation. (See Examples 2 through 5.)

10. $r = 3$
11. $\theta = \frac{\pi}{3}$
12. $r = -3$
13. $\theta = -\frac{\pi}{3}$
14. $r = 4 \cos \theta$ (circle)
15. $r = \sin \theta$ (circle)
16. $r = 2 \sin \theta$ (circle)
17. $r = \sin 2\theta$ (rose)
18. $r = \sin 3\theta$ (rose)
19. $r = \theta$—express θ in radians (spiral)
20. $r = 2\theta$—express θ in radians (spiral)
21. $r = \cos 3\theta$ (rose)
22. $r = 2 - 2 \cos \theta$ (cardioid)
23. $r = 2 + 2 \cos \theta$ (cardioid)
24. $r = 2 + 2 \sin \theta$ (cardioid)

25. $r = 1 - 2\cos\theta$ (limaçon with loop)

26. $r = 2 - \cos\theta$ (limaçon without loop)

27. $r^2 = 4\cos 2\theta$ (lemniscate)

28. $r^2 = 4\sin 2\theta$ (lemniscate)

29. $r = \dfrac{5}{\theta}$—express θ in radians (spiral)

30. $r\cos\theta = 1$

In Exercises 31 through 39, determine the corresponding Cartesian coordinate equation. (See Example 6.)

31. $r = \sin\theta$
32. $r = \cos\theta$
33. $r = \cos 2\theta$
34. $r = \sin 2\theta$
35. $r = 3$
36. $r^2 = \sin^2\theta$
37. $r = 2 - \cos\theta$
38. $\theta = \dfrac{\pi}{6}$
39. $r = \theta$

40. a. Find the Cartesian coordinate equation corresponding to $r = 2\cos\theta$ and show by putting the equation in the form $(x-h)^2 + (y-k)^2 = a^2$ that a circle is represented.
    b. What are the center and radius of the circle?

In Exercises 41 through 49, determine the corresponding polar coordinate equation. (See Example 7.)

41. $x^2 + y^2 = 9$
42. $x^2 - y^2 = 9$
43. $2xy = 9$
44. $(x^2 + y^2)^2 = x^2 - y^2$
45. $3x + 2y = 1$
46. $x^2 + y^2 - 2x = 0$
47. $x^2 + y^2 - 2x - 4y = 0$
48. $x = 2$
49. $y = 4$

50. Consider the Cartesian coordinate and polar coordinate systems at the right, where r is negative. The angle in standard position with OP as its terminal side is $180° + \theta$.
    a. Write expressions for $\sin(180° + \theta)$ and $\cos(180° + \theta)$.
    b. Using your answers to part a, show that $x = r\cos\theta$ and $y = r\sin\theta$.

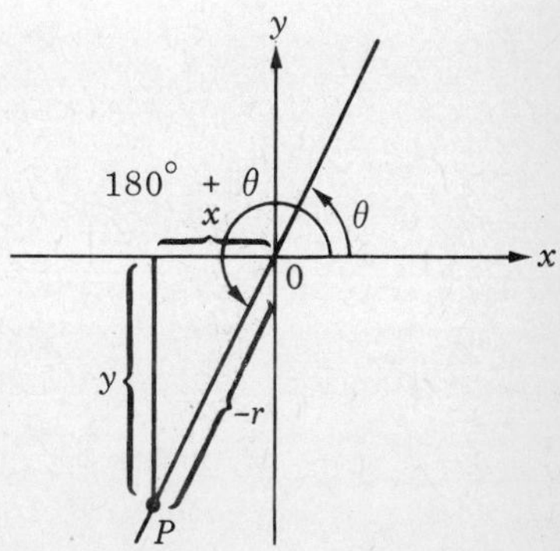

Computer Exercises

51. Write a BASIC program to convert any given Cartesian coordinate point (x, y) to polar coordinates (r, θ).

52. Write a BASIC program to convert a given polar coordinate point (r, θ) to its Cartesian coordinate equivalent.

7.6 VECTORS

In many areas where mathematics is applied there are quantities such as temperature, pressure, and resistance, whose description requires one number. Such a quantity is called a **scalar quantity** and the real number that represents it is called a **scalar**. There are other quantities, called **vector quantities**, whose description requires two numbers referred to as **magnitude** and **direction**. Examples of vector quantities are velocity, force, and momentum. A vector quantity is often represented pictorially by a directed line segment called a **vector**. The vector at the right from point O to point P can be symbolized as **OP** or **A**. The magnitude (length) of **A** is denoted by $|\mathbf{A}|$ and the direction can be indicated by specifying the angle between **A** and some reference direction (this is done later). In order to develop more ideas the following definitions are needed.

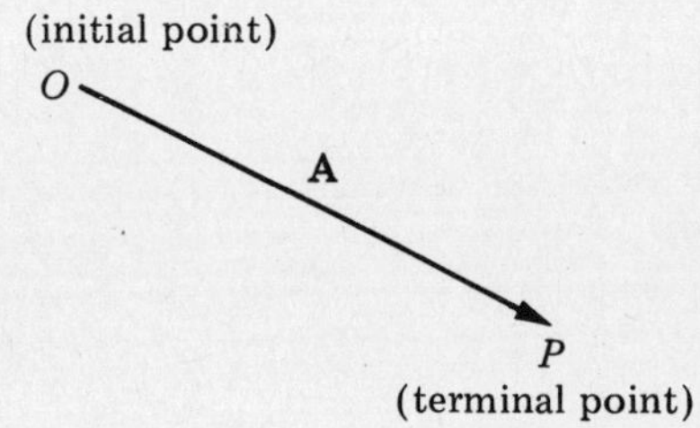

DEFINITION

Equal Vectors

Two nonzero* vectors **A** and **B** are equal provided they have the same magnitude *and* direction.

DEFINITION

Addition of Two Vectors (Parallelogram Method)

In order to add two vectors **A** and **B**, place their initial points together without altering the magnitude or direction of either one and then complete the parallelogram having **A** and **B** as a pair of adjacent sides. The **sum** or **resultant**, **A** + **B**, is that vector whose length equals the diagonal of the parallelogram with the same initial point as **A** and **B**.

EXAMPLE 1 a. Which of the vectors below are equal?

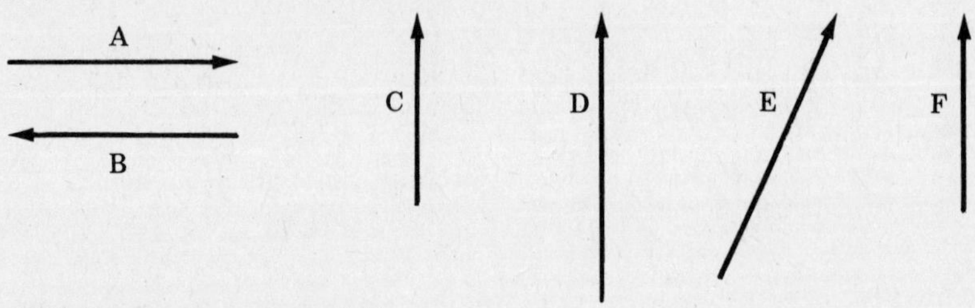

b. Add the vectors **A** and **E**.

Solution a. The only two vectors that have the same magnitude and direction are **C** and **F**. Thus, **C** = **F**.

*A zero vector, denoted by the symbol O, has magnitude 0 and arbitrary direction.

b. We place the initial points of **A** and **E** together.

The parallelogram is completed (dotted lines) and the resultant **A** + **E** is indicated.

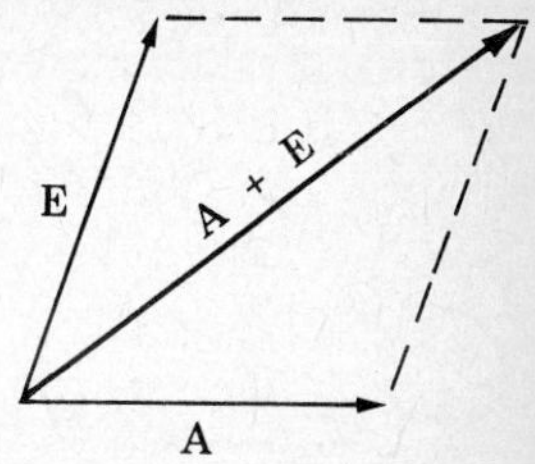

Note in the above example that **A** and **B** have the same magnitudes, but *opposite* directions. When such is the case, for any two vectors, **A** and **B**, we write

$$\mathbf{A} = -\mathbf{B} \quad \text{or} \quad \mathbf{B} = -\mathbf{A}.$$

In order to use the definition above to add three vectors, the sum of any two can be added to the third. However, in the case of three or more vectors it is easier to use the following (equivalent) definition.

DEFINITION

Addition of Vectors (Polygon Method)

In order to add any number of vectors, place them in any order with the terminal point of one touching the initial point of the next. The **sum** or **resultant**, is that vector drawn from the initial point of the first to the terminal point of the last.

EXAMPLE 2 Add the vectors below.

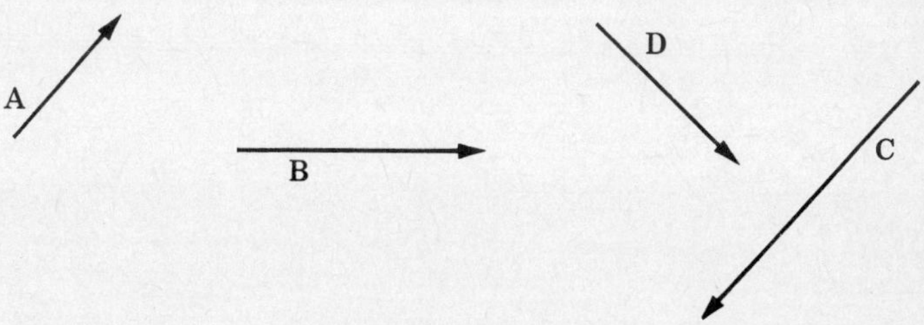

Solution

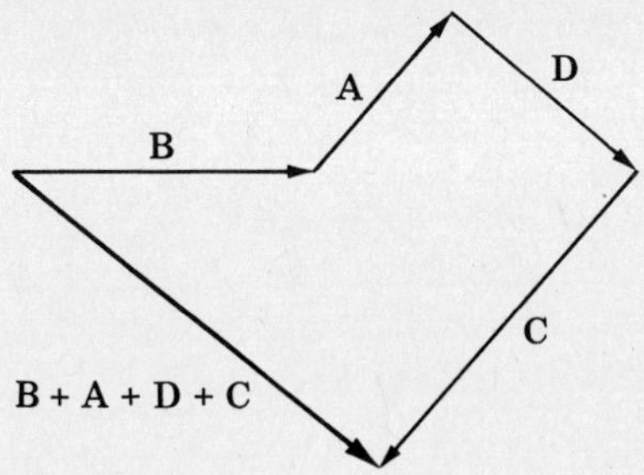

To *subtract* vector **B** from vector **A** (denoted **A** − **B**), we add −**B** to **A**. That is, **A** − **B** = **A** + (−**B**).

EXAMPLE 3 Determine **A** − **B**.

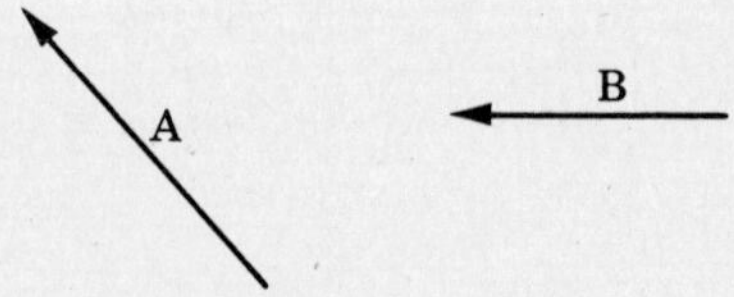

Solution We add **A** and −**B**.

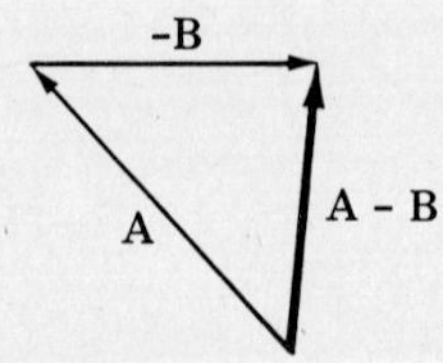

Notice in the example above that if **A** and **B** have their initial points placed in common then **A** − **B** is seen to be a vector from the terminal point of **B** to the terminal point of **A** (see the diagram at the right).

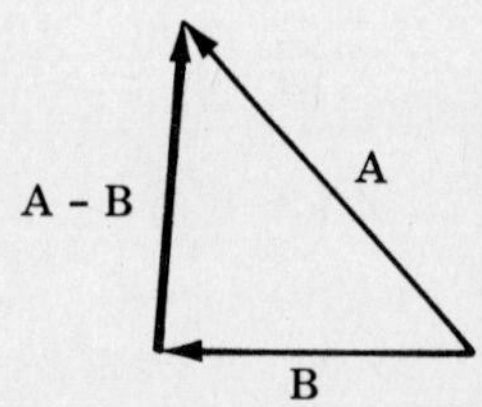

DEFINITION

Multiplication of a Vector by a Scalar

In order to multiply the vector **A** by the scalar c, the result, denoted by c**A**, is a vector $|c|$ times as long as **A** and having the same direction as **A** if $c > 0$, or the opposite direction of **A** if $c < 0$.

EXAMPLE 4 Given the vectors **C** and **D** below, determine

a. 3**C** b. $-\frac{1}{2}$**D** c. 3**C** $-\frac{1}{2}$**D**

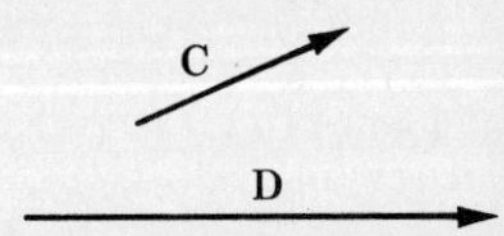

Solution

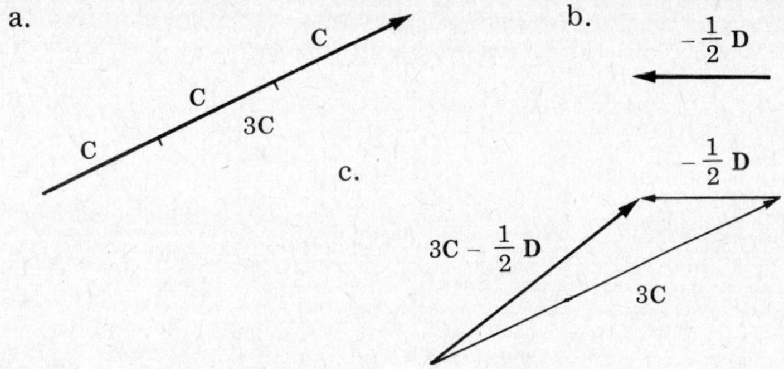

The definitions cited thus far give rise to a number of **laws of vector algebra** that are now listed without proof; c and d represent scalars.

Laws of Vector Algebra

1. $\mathbf{A} + \mathbf{B} = \mathbf{B} + \mathbf{A}$	6. $\mathbf{O} + \mathbf{A} = \mathbf{A}$
2. $(\mathbf{A} + \mathbf{B}) + \mathbf{C} = \mathbf{A} + (\mathbf{B} + \mathbf{C})$	7. $\mathbf{A} + (-\mathbf{A}) = \mathbf{O}$
3. $c(d\mathbf{A}) = d(c\mathbf{A}) = (cd)\mathbf{A}$	8. $1\mathbf{A} = \mathbf{A}$
4. $c(\mathbf{A} + \mathbf{B}) = c\mathbf{A} + c\mathbf{B}$	9. $0\mathbf{A} = \mathbf{O}$
5. $(c + d)\mathbf{A} = c\mathbf{A} + d\mathbf{A}$	10. $c\mathbf{O} = \mathbf{O}$

In Exercises 19 through 22 and 73, the reader is asked to investigate these laws.

Consider part b of Example 1. Vectors **A** and **E** were added to produce the resultant **A** + **E**. Individually, **A** and **E** are called **components** of the resultant. It is possible to separate or decompose any vector into various pairs of components. This is easily done in a coordinate system with the aid of **unit vectors** (vectors of length or magnitude 1 unit).

DEFINITION

Unit Vectors i and j

The unit vector **i** is a vector with magnitude of 1 unit and the direction of the positive x-axis. The unit vector **j** is a vector with magnitude of 1 unit and the direction of the positive y-axis.

Any vector in the xy-plane can be expressed in terms of **i** and **j**.

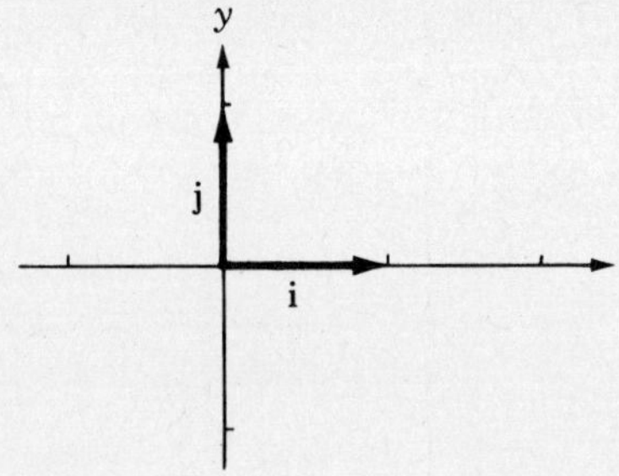

EXAMPLE 5 A vector is drawn from the initial point $P_1(-1,2)$ to the terminal point $P_2(4,6)$. Write an expression for the vector $\mathbf{P_1P_2}$ in terms of **i** and **j**.

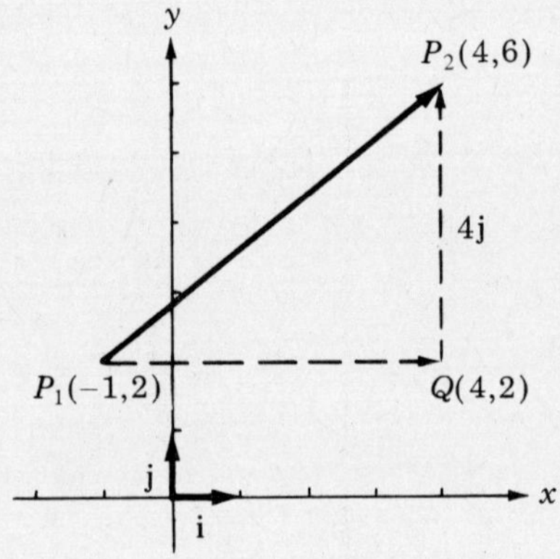

Solution Refer to the diagram at the right.

$$\begin{aligned}\mathbf{P_1P_2} &= \mathbf{P_1Q} + \mathbf{QP_2}\\ &= [4-(-1)]\mathbf{i} + [6-2]\mathbf{j}\\ &= 5\mathbf{i} + 4\mathbf{j}\end{aligned}$$

In the answer above, $5\mathbf{i}$ is called the **x-component** and $4\mathbf{j}$ the **y-component** of $\mathbf{P_1P_2}$.

The example above illustrates Theorem 3, which the reader is asked to prove in Exercise 71.

THEOREM 3

Let P_1 and P_2 be any two points in the xy-plane with coordinates (x_1, y_1) and (x_2, y_2) respectively. Then

$$\mathbf{P_1P_2} = (x_2 - x_1)\mathbf{i} + (y_2 - y_1)\mathbf{j}.$$

The next theorem follows immediately from the distance formula developed in Chapter 3.

THEOREM 4

Let P_1 and P_2 be any two points in the xy-plane with coordinates (x_1, y_1) and (x_2, y_2) respectively. Then

$$|\mathbf{P_1P_2}| = \sqrt{(x_2 - x_1)^2 + (y_2 - y_1)^2}.$$

EXAMPLE 6 Given $P_1(-2,3)$ and $P_2(-4,-5)$, determine:

a. $\mathbf{P_1P_2}$ b. $\mathbf{P_2P_1}$ c. $|\mathbf{P_1P_2}|$

Solution a. $\mathbf{P_1P_2} = [-4-(-2)]\mathbf{i} + [-5-3]\mathbf{j}$

$\mathbf{P_1P_2} = -2\mathbf{i} - 8\mathbf{j}$

b. $\mathbf{P_2P_1} = -\mathbf{P_1P_2} = 2\mathbf{i} + 8\mathbf{j}$

c. $|\mathbf{P_1P_2}| = \sqrt{(-2)^2 + (-8)^2} = \sqrt{68}$

The next theorem, which follows from the Laws of Vector Algebra, states that two vectors can be added by individually adding their x- and y-components.

THEOREM 5

If $\mathbf{A} = a_1\mathbf{i} + a_2\mathbf{j}$ and $\mathbf{B} = b_1\mathbf{i} + b_2\mathbf{j}$, then
$\mathbf{A} + \mathbf{B} = (a_1 + b_1)\mathbf{i} + (a_2 + b_2)\mathbf{j}$.

EXAMPLE 7 Given $\mathbf{A} = 2\mathbf{i} + 3\mathbf{j}$, $\mathbf{B} = -\mathbf{i} + 4\mathbf{j}$, determine

a. $\mathbf{A} + \mathbf{B}$ b. $\mathbf{A} - \mathbf{B}$ c. $2\mathbf{A} - \mathbf{B}$

Solution a. $\mathbf{A} + \mathbf{B} = (2\mathbf{i} + 3\mathbf{j}) + (-\mathbf{i} + 4\mathbf{j})$

$\mathbf{A} + \mathbf{B} = \mathbf{i} + 7\mathbf{j}$

b. $\mathbf{A} - \mathbf{B} = (2\mathbf{i} + 3\mathbf{j}) - (-\mathbf{i} + 4\mathbf{j})$

$= 2\mathbf{i} + 3\mathbf{j} + \mathbf{i} - 4\mathbf{j}$

$= 3\mathbf{i} - \mathbf{j}$

c. $2\mathbf{A} - \mathbf{B} = 2(2\mathbf{i} + 3\mathbf{j}) - (-\mathbf{i} + 4\mathbf{j})$

$= 4\mathbf{i} + 6\mathbf{j} + \mathbf{i} - 4\mathbf{j}$

$= 5\mathbf{i} + 2\mathbf{j}$

In the definition that follows a form of vector multiplication called **scalar product** is presented.

DEFINITION

Scalar Product (Dot Product)

The scalar (or dot) product of two vectors **A** and **B**, symbolized as $\mathbf{A} \bullet \mathbf{B}$, is equal to the product of their magnitudes times the cosine of the angle, θ, between them. That is,

$$\mathbf{A} \bullet \mathbf{B} = |\mathbf{A}|\,|\mathbf{B}| \cos\theta.^*$$

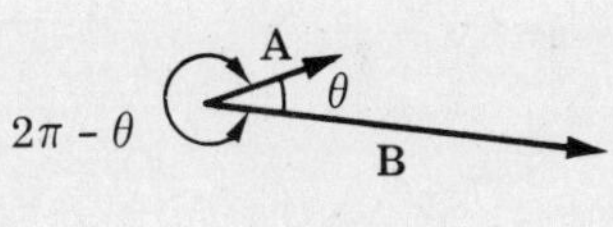

*Actually, between any two vectors, there are *two angles*, θ and $2\pi - \theta$. However, since $\cos(2\pi - \theta) = \cos\theta$, either angle can be used.

In the next example, some special dot products are computed.

EXAMPLE 8 Determine

a. $\mathbf{i} \bullet \mathbf{i}$ b. $\mathbf{j} \bullet \mathbf{j}$ c. $\mathbf{i} \bullet \mathbf{j}$ d. $\mathbf{j} \bullet \mathbf{i}$

Solution

a. $\mathbf{i} \bullet \mathbf{i} = |\mathbf{i}|\,|\mathbf{i}| \cos 0^\circ = 1$ b. $\mathbf{j} \bullet \mathbf{j} = |\mathbf{j}|\,|\mathbf{j}| \cos 0^\circ = 1$
c. $\mathbf{i} \bullet \mathbf{j} = |\mathbf{i}|\,|\mathbf{j}| \cos 90^\circ = 0$ d. $\mathbf{j} \bullet \mathbf{i} = |\mathbf{j}|\,|\mathbf{i}| \cos 90^\circ = 0$

The previous example somewhat oversimplifies matters because generally with two given vectors we do not know the angle θ between them. As a result, the scalar product cannot be easily computed. However, the following theorem alleviates the difficulty.

THEOREM 6

Given two vectors $\mathbf{A} = a_1\mathbf{i} + a_2\mathbf{j}$ and $\mathbf{B} = b_1\mathbf{i} + b_2\mathbf{j}$, then

$$\mathbf{A} \bullet \mathbf{B} = a_1 b_1 + a_2 b_2.$$

The proof is outlined in Exercise 72.

EXAMPLE 9 Given $\mathbf{A} = \mathbf{i} - 3\mathbf{j}$ and $\mathbf{B} = 4\mathbf{i} + 2\mathbf{j}$,

a. determine $\mathbf{A} \bullet \mathbf{B}$
b. determine the angle between $\mathbf{A}$ and $\mathbf{B}$.

Solution

a. $\mathbf{A} \bullet \mathbf{B} = (1)(4) + (-3)(2) = -2$
b. By definition, $\mathbf{A} \bullet \mathbf{B} = |\mathbf{A}|\,|\mathbf{B}| \cos \theta$. Thus

$$|\mathbf{A}|\,|\mathbf{B}| \cos \theta = -2$$

$$\cos \theta = \frac{-2}{|\mathbf{A}|\,|\mathbf{B}|} = \frac{-2}{\sqrt{10}\sqrt{20}} = \frac{-2}{\sqrt{200}} = \frac{-\sqrt{2}}{10}$$

$$\theta = \cos^{-1} \frac{-\sqrt{2}}{10}$$

Using a calculator, $\theta = 98.130102^\circ$

Thus by combining Theorem 6 with the definition of scalar product, it is a fairly easy matter to determine the angle between any two vectors in the xy-plane.

EXERCISES FOR SECTION 7.6

In Exercises 1 through 4, determine which of the vectors are equal. (See Example 1.)

1. 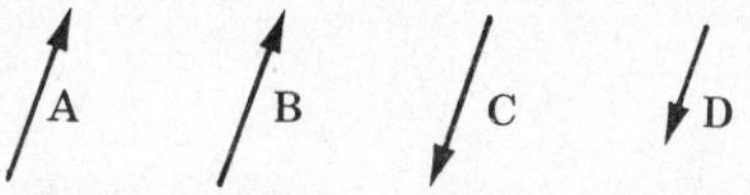

2. 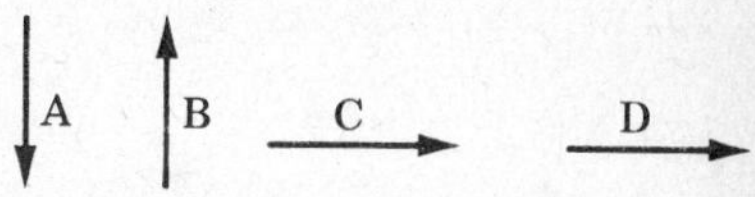

3. 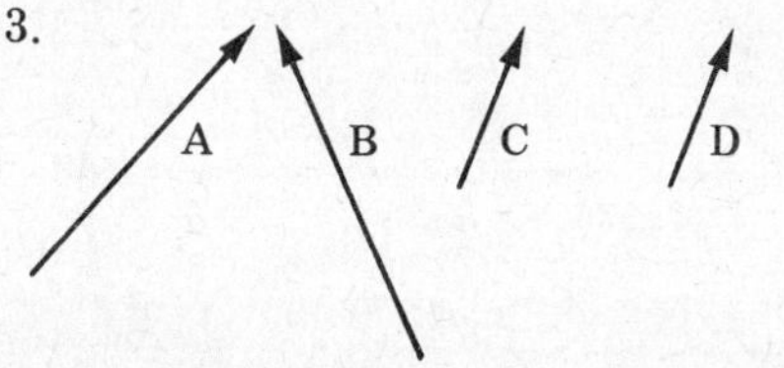

4. 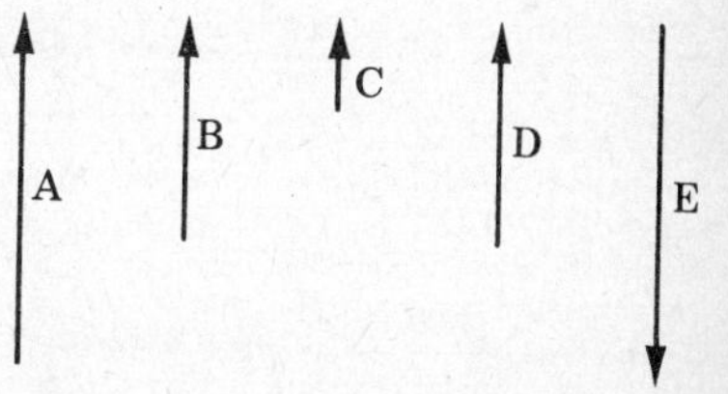

In Exercises 5 through 22, perform the vector operations with the given vectors **u**, **v**, and **w**. (See Examples 2, 3, and 4.)

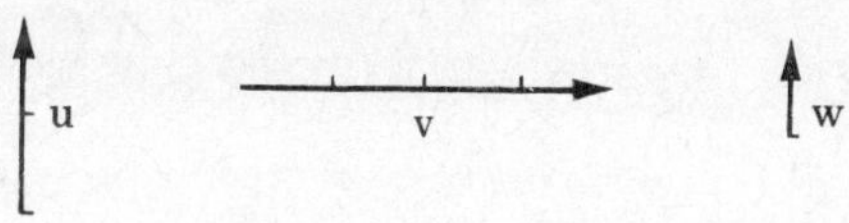

5. u + v
6. u − v
7. u + 2w
8. u + $\frac{1}{2}$v
9. u − 2w
10. u + 2v
11. u + v + w
12. u + v − 2w
13. 3u − w
14. w + v + u
15. −2w + u + v
16. −w + 3u
17. $\frac{1}{2}$u + w
18. $\frac{1}{2}$u − w
19. 2(u + v)
20. 2u + 2v
21. (2 + 3)u
22. 2u + 3u

In Exercises 23 through 34, a) determine $\mathbf{P_1P_2}$, b) determine $\mathbf{P_2P_1}$, c) determine $|\mathbf{P_1P_2}|$. (See Examples 5 and 6.)

23. $P_1(1,2)$ $P_2(-1,4)$
24. $P_1(2,1)$ $P_2(1,2)$
25. $P_1(3,-1)$ $P_2(1,3)$
26. $P_1(0,0)$ $P_2(1,2)$
27. $P_1(0,0)$ $P_2(-1,-2)$
28. $P_1(-1,4)$ $P_2(2,1)$
29. $P_1(7,1)$ $P_2(6,2)$
30. $P_1(-1,2)$ $P_2(3,0)$
31. $P_1(4,2)$ $P_2(-1,-3)$
32. $P_1(1,2)$ $P_2(3,4)$
33. $P_1(4,3)$ $P_2(2,1)$
34. $P_1(3,2)$ $P_2(3,2)$

In Exercises 35 through 50, a) determine **A** + **B**, b) determine **A** − **B**, c) determine **A** − 2**B**. (See Example 7.)

35. A = i + 2j, B = i + 2j
36. A = i + j, B = i − j
37. A = 0i + 0j, B = −i + 4j
38. A = 3i, B = −2j
39. A = i, B = 2i − 3j
40. A = 6i − 7j, B = 2i + 10j
41. A = i − j, B = 2i + 5j
42. A = −3i − j, B = 3i − j
43. A = 2i, B = 5i − j
44. A = 4i + 2j, B = −4i − 2j
45. A = −i + j, B = 4i + 2j
46. A = −5i + j, B = j + 5i
47. A = 3i − 2j, B = 3i − 3j
48. A = −i + j, B = j − 6i
49. A = −2i + 9j, B = 2i − 2j
50. A = 2i + j, B = j + i

In Exercises 51 through 60, determine whether the given vectors are perpendicular or not. (See Example 9 and Theorem 6.)

51. V_1 = 2i + j, V_2 = i − 2j
52. V_1 = −2i + j, V_2 = −2i − 4j
53. V_1 = 2i + j, V_2 = 3i − 6j
54. V_1 = 2i + j, V_2 = 4i + 2j
55. V_1 = −3i + j, V_2 = −2i + 6j
56. V_1 = 3i + j, V_2 = 2i − 5j

57. $\mathbf{V}_1 = 3\mathbf{i} + \mathbf{j}$, $\mathbf{V}_2 = 2\mathbf{i} - 6\mathbf{j}$
58. $\mathbf{V}_1 = \frac{3}{2}\mathbf{i} + \frac{5}{2}\mathbf{j}$, $\mathbf{V}_2 = 5\mathbf{i} - 3\mathbf{j}$
59. $\mathbf{V}_1 = 2\mathbf{i}$, $\mathbf{V}_2 = -3\mathbf{j}$
60. $\mathbf{V}_1 = 2\mathbf{i}$, $\mathbf{V}_2 = 2\mathbf{i} - 7\mathbf{j}$

In Exercises 61 through 70, determine the angle between the given vectors. (See Example 9.)

61. $\mathbf{A} = \mathbf{i} + 6\mathbf{j}$, $\mathbf{B} = 2\mathbf{i} + 3\mathbf{j}$
62. $\mathbf{A} = \mathbf{i} + 6\mathbf{j}$, $\mathbf{B} = -2\mathbf{i} + 3\mathbf{j}$
63. $\mathbf{A} = 2\mathbf{i} + \mathbf{j}$, $\mathbf{B} = -3\mathbf{i} + 4\mathbf{j}$
64. $\mathbf{A} = -2\mathbf{i} + \mathbf{j}$, $\mathbf{B} = -3\mathbf{i}$
65. $\mathbf{A} = 3\mathbf{i} + 2\mathbf{j}$, $\mathbf{B} = 2\mathbf{i} - 3\mathbf{j}$
66. $\mathbf{A} = -3\mathbf{i} - 2\mathbf{j}$, $\mathbf{B} = -3\mathbf{i}$
67. $\mathbf{A} = 3\mathbf{i} - \mathbf{j}$, $\mathbf{B} = -3\mathbf{j}$
68. $\mathbf{A} = -3\mathbf{i} - \mathbf{j}$, $\mathbf{B} = -\mathbf{i} + 5\mathbf{j}$
69. $\mathbf{A} = -\mathbf{i} - 5\mathbf{j}$, $\mathbf{B} = 2\mathbf{i} - 2\mathbf{j}$
70. $\mathbf{A} = 4\mathbf{i} - \mathbf{j}$, $\mathbf{B} = \frac{1}{2}\mathbf{i} + 2\mathbf{j}$

71. Using the picture at the right, prove that if P_1 and P_2 are any two points in the xy-plane with coordinates (x_1, y_1) and (x_2, y_2) respectively, then

$$\mathbf{P_1P_2} = (x_2 - x_1)\mathbf{i} + (y_2 - y_1)\mathbf{j}.$$

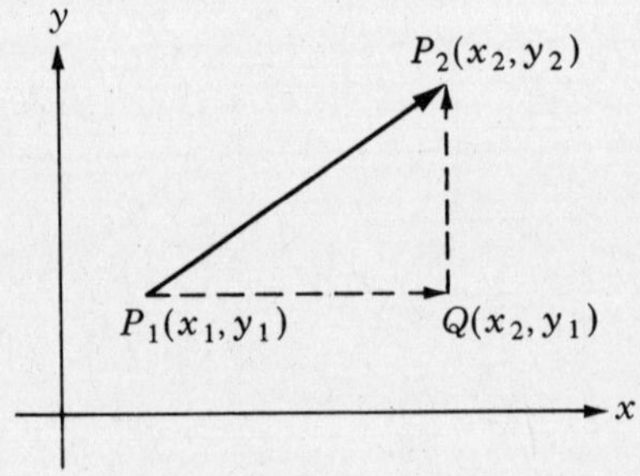

72. Following the suggested outline, prove that if $\mathbf{A} = a_1\mathbf{i} + a_2\mathbf{j}$ and $\mathbf{B} = b_1\mathbf{i} + b_2\mathbf{j}$, then $\mathbf{A} \bullet \mathbf{B} = a_1b_1 + a_2b_2$.

Proof: From the Law of Cosines,

$$|\mathbf{A} - \mathbf{B}|^2 = |\mathbf{A}|^2 + |\mathbf{B}|^2 - 2\underbrace{|\mathbf{A}|\,|\mathbf{B}|\cos\theta}_{\mathbf{A}\bullet\mathbf{B}}$$

Substitute expressions for the magnitudes $|\mathbf{A} - \mathbf{B}|$, $|\mathbf{A}|$, $|\mathbf{B}|$ into the equation above. Then square, simplify and solve for $\mathbf{A} \bullet \mathbf{B}$.

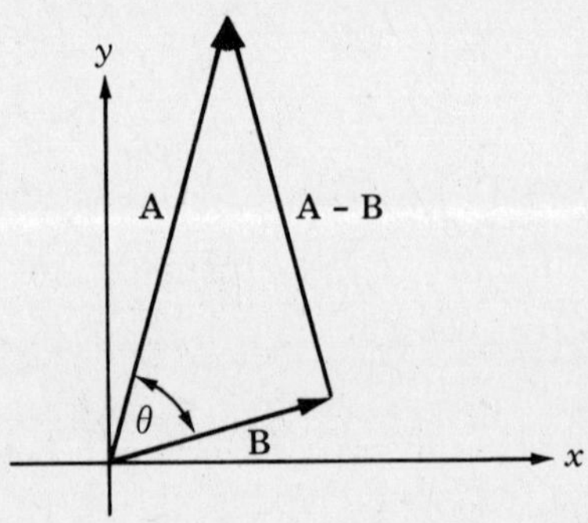

73. Given $\mathbf{A} = 2\mathbf{i} + 3\mathbf{j}$, $\mathbf{B} = 3\mathbf{i} - 5\mathbf{j}$, $\mathbf{C} = -\mathbf{i} + \mathbf{j}$, $c = 2$, $d = -5$, verify each of the 10 Laws of Vector Algebra.

7.7 DeMOIVRE'S THEOREM

When complex numbers were studied in Section 1.9, we learned how to add, subtract, multiply, and divide them. Moreover, complex numbers were expressed in the form $a + bi$, where a and b are real numbers and $i = \sqrt{-1}$. In this section we will again multiply and divide complex numbers, but in a more streamlined fashion and with the eventual goal of learning and understanding DeMoivre's Theorem.

Complex numbers can be displayed graphically in an **Argand diagram**. Such a diagram consists of two perpendicular axes: the **real**

axis (labeled x) and the **imaginary axis** (labeled y) forming what is called the **complex plane**. Also pictured below are the complex numbers $2 + i$, $2 - i$, $-2 + 3i$, and $-4 - 2i$. Notice that the method of graphing the complex number $x + yi$ in the complex plane is the same as the method of graphing the point (x, y) in the ordinary (real) xy-plane. In fact, the complex number $x + yi$ is often represented as the ordered pair (x, y). The context of the problem should make it

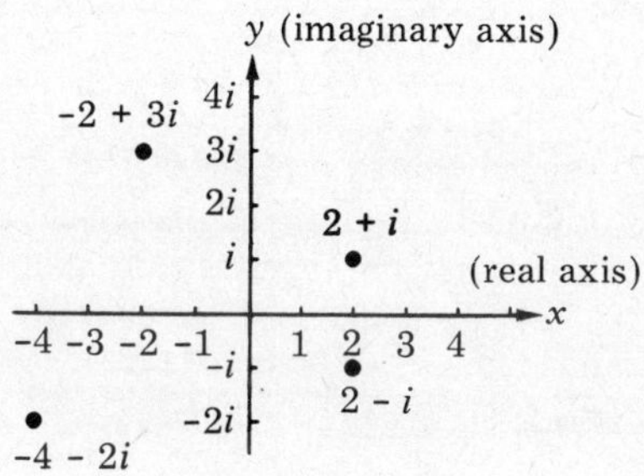

Argand diagram

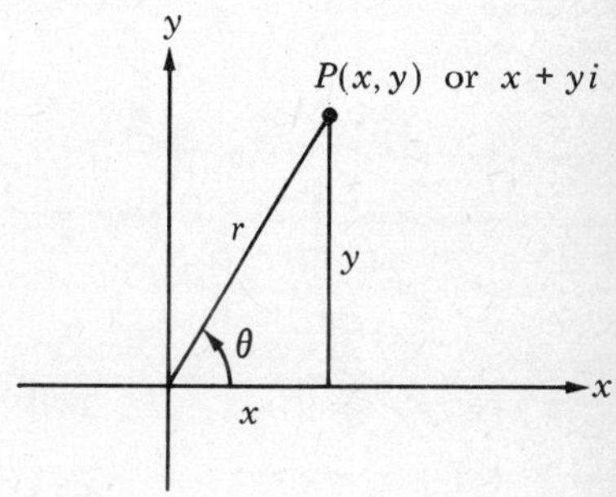

clear as to whether (x, y) denotes a point in the ordinary xy-plane or the complex plane. By introducing the polar coordinates r and θ we can write $x = r \cos \theta$ and $y = r \sin \theta$. We have the following:

$$x + yi = r \cos \theta + ir \sin \theta = r(\cos \theta + i \sin \theta)$$

DEFINITION

The **rectangular form** of a complex number is $x + yi$. The **polar form** of a complex number is $r(\cos \theta + i \sin \theta)$. The polar form is often abbreviated $r \operatorname{cis} \theta$ and θ is called the **argument** of the complex number.

DEFINITION

The **absolute value** (or **modulus**) of the complex number $x + yi$ is the distance r and is symbolized as $|x + yi|$. Hence $r = |x + yi| = \sqrt{x^2 + y^2}$. Note that r is real and nonnegative.

EXAMPLE 1 Rewrite each complex number in polar form.

a. $1 + i\sqrt{3}$ b. $3 - 3i$

Solution a. We need r and θ.

$$r = \sqrt{1+3} = 2$$

$$\text{Since } \tan\theta = \frac{y}{x} = \frac{\sqrt{3}}{1} = \sqrt{3}$$

$$\theta = 60^\circ = \frac{\pi}{3}$$

(we only use the first quadrant value—why?)

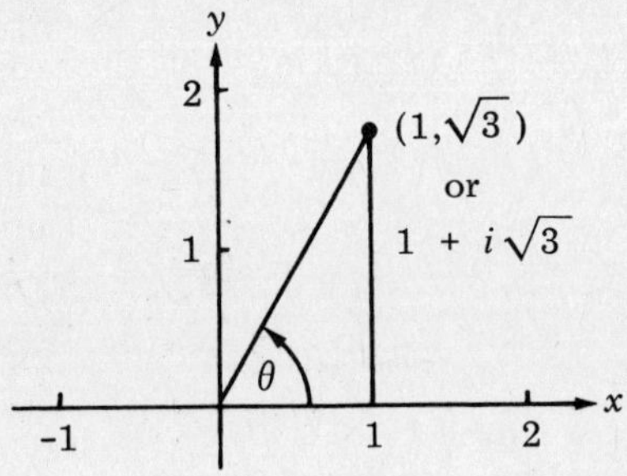

Thus,

$$1 + i\sqrt{3} = 2\left(\cos\frac{\pi}{3} + i\sin\frac{\pi}{3}\right)$$
$$= 2 \text{ cis } \frac{\pi}{3}$$

b. $r = \sqrt{9+9} = \sqrt{18} = 3\sqrt{2}$

$$\tan\theta = \frac{y}{x} = \frac{-3}{3} = -1$$

$\theta_R = 45^\circ$ and thus

$$\theta = 315^\circ \text{ or } \frac{7\pi}{4}$$

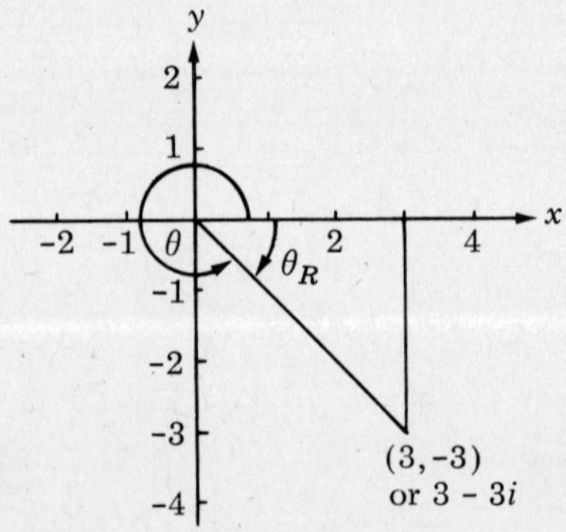

So,

$$3 - 3i = 3\sqrt{2}\left(\cos\frac{7\pi}{4} + i\sin\frac{7\pi}{4}\right)$$
$$= 3\sqrt{2} \text{ cis } \frac{7\pi}{4}$$

EXAMPLE 2 Rewrite each complex number in rectangular form.

a. $2 \text{ cis } \frac{3\pi}{4}$ b. $3 \text{ cis } \frac{3\pi}{2}$

Solution a.

$$2 \text{ cis } \frac{3\pi}{4} = 2\left(\cos\frac{3\pi}{4} + i\sin\frac{3\pi}{4}\right)$$
$$= 2\left(-\cos\frac{\pi}{4} + i\sin\frac{\pi}{4}\right)$$

Since $\theta = \frac{3\pi}{4}$, $\theta_R = \frac{\pi}{4}$

$$= 2\left(-\frac{\sqrt{2}}{2} + i\frac{\sqrt{2}}{2}\right)$$
$$= -\sqrt{2} + i\sqrt{2}$$

b. $$3 \text{ cis } \frac{3\pi}{2} = 3\left(\cos\frac{3\pi}{2} + i\sin\frac{3\pi}{2}\right) = 3[0 + i(-1)] = 0 - 3i$$

Now we turn our attention to multiplication and division of complex numbers in polar form.

THEOREM 7

Given the complex numbers $r_1 \text{cis } \theta_1$ and $r_2 \text{cis } \theta_2$, then

a. $(r_1 \text{cis } \theta_1)(r_2 \text{cis } \theta_2) = r_1 r_2 \text{ cis}(\theta_1 + \theta_2)$

b. $\dfrac{r_1 \text{cis } \theta_1}{r_2 \text{cis } \theta_2} = \dfrac{r_1}{r_2} \text{ cis}(\theta_1 - \theta_2)$

The proofs are outlined in Exercises 75 and 76. In words the theorem says:

a. When multiplying two complex numbers, the modulus of the product equals the *product* of their moduli and the argument of the product equals the *sum* of their arguments.
b. When dividing two complex numbers, the modulus of the quotient equals the *quotient* of the moduli (note order) and the argument of the quotient equals the *difference* of their arguments (note order).

EXAMPLE 3 Evaluate:

a. $\left(2 \text{ cis } \frac{\pi}{3}\right)\left(3 \text{ cis } \frac{\pi}{2}\right)$ b. $\dfrac{8 \text{ cis } \frac{7\pi}{4}}{2 \text{ cis } \frac{3\pi}{2}}$

Solution a. $$\left(2 \text{ cis } \frac{\pi}{3}\right)\left(3 \text{ cis } \frac{\pi}{2}\right) = 6 \text{ cis}\left(\frac{\pi}{3} + \frac{\pi}{2}\right) = 6 \text{ cis } \frac{5\pi}{6}$$

b. $$\frac{8 \text{ cis } \frac{7\pi}{4}}{2 \text{ cis } \frac{3\pi}{2}} = 4 \text{ cis}\left(\frac{7\pi}{4} - \frac{3\pi}{2}\right) = 4 \text{ cis } \frac{\pi}{4}$$

Recall that the *complex conjugate* of the complex number $a + bi$ is $a - bi$.

EXAMPLE 4 Multiply the complex number $-\sqrt{3} + i$ by its complex conjugate:

a. by converting each to polar form and using Theorem 7a,
b. by using the definition in Section 1.9.

Solution a. For $-\sqrt{3} + i$, $r = 2$ and $\theta = 150°$.
For $-\sqrt{3} - i$, $r = 2$ and $\theta = 210°$.

$$\begin{aligned}(-\sqrt{3} + i)(-\sqrt{3} - i) &= (2 \text{ cis } 150°)(2 \text{ cis } 210°)\\ &= 4 \text{ cis}(150° + 210°)\\ &= 4 \text{ cis } 360°\\ &= 4(\cos 360° + i \sin 360°)\\ &= 4\end{aligned}$$

b. $$\begin{aligned}(-\sqrt{3} + i)(-\sqrt{3} - i) &= 3 + i\sqrt{3} - i\sqrt{3} - i^2\\ &= 4\end{aligned}$$

Next consider the operation of raising a complex number in polar form to a power.

$$\begin{aligned}(r \text{ cis } \theta)^2 &= (r \text{ cis } \theta)(r \text{ cis } \theta) = r^2 \text{ cis } 2\theta && \text{by Theorem 7a.}\\ (r \text{ cis } \theta)^3 &= (r \text{ cis } \theta)^2(r \text{ cis } \theta)\\ &= (r^2 \text{ cis } 2\theta)(r \text{ cis } \theta) = r^3 \text{ cis } 3\theta && \text{by Theorem 7a.}\\ (r \text{ cis } \theta)^4 &= (r \text{ cis } \theta)^3(r \text{ cis } \theta)\\ &= (r^3 \text{ cis } 3\theta)(r \text{ cis } \theta) = r^4 \text{ cis } 4\theta && \text{by Theorem 7a.}\end{aligned}$$

By repeated application of Theorem 7a we have DeMoivre's Theorem.

THEOREM 8

DeMoivre's Theorem

$(r \text{ cis } \theta)^n = r^n \text{ cis } n\theta$, where n is a positive integer.

EXAMPLE 5 Use DeMoivre's Theorem to evaluate each of the following and express answers in rectangular form.

a. $[2 \text{ cis } 30°]^5$ b. $(1 + i)^{10}$

Solution a. $$\begin{aligned}[2 \text{ cis } 30°]^5 &= 2^5 \text{ cis } 150°\\ &= 32(\cos 150° + i \sin 150°)\\ &= 32\left(-\frac{\sqrt{3}}{2} + \frac{1}{2}i\right)\\ &= -16\sqrt{3} + 16i\end{aligned}$$

b. First note $1 + i = \sqrt{2}\left(\cos\frac{\pi}{4} + i\sin\frac{\pi}{4}\right) = \sqrt{2}\text{ cis }\frac{\pi}{4}$.

$$(1 + i)^{10} = \left(\sqrt{2}\text{ cis }\frac{\pi}{4}\right)^{10} = 2^5\text{ cis }\frac{5\pi}{2} = 32\left(\cos\frac{\pi}{2} + i\sin\frac{\pi}{2}\right)$$
$$= 32(0 + i) = 32i$$

Example 5 demonstrates that raising a complex number to a positive integral power is most easily accomplished by using DeMoivre's Theorem. Another use of DeMoivre's Theorem is in determining nth roots of a complex number. As we shall see, *any nonzero complex number has exactly n distinct nth roots.*

Suppose $R\text{ cis }\beta$ is an nth root of $r\text{ cis }\theta$. Then it must be that $(R\text{ cis }\beta)^n = r\text{ cis }\theta$. This can be rewritten as

$$R^n\text{ cis }n\beta = r\text{ cis }\theta.$$

Equating moduli gives $R^n = r$ or $R = r^{1/n}$. Equating arguments (bearing in mind the periodicity of $\cos\theta$ and $\sin\theta$) gives:

$$n\beta = \theta + 2\pi k, \text{ where } k \text{ is any integer}$$

or

$$\beta = \frac{\theta}{n} + \frac{2\pi k}{n}.$$

Thus, $R\text{ cis }\beta$ (an nth root of $r\text{ cis }\theta$) can be written as:

$$r^{1/n}\text{ cis}\left(\frac{\theta}{n} + \frac{2\pi k}{n}\right), \text{ where } k \text{ is any integer}$$

In the examples that follow, the reader will see that by letting $k = 0, 1, 2, \ldots, n - 1$, the n distinct nth roots of $r\text{ cis }\theta$ are obtained and that other values of k do not produce other roots. In summary then,

$$(r\text{ cis }\theta)^{1/n} = r^{1/n}\text{ cis}\left(\frac{\theta}{n} + \frac{2\pi k}{n}\right), \text{ where } k = 0, 1, 2, \ldots, n - 1.$$

In solving problems, it is often easier to replace 2π with $360°$.

EXAMPLE 6 Determine the three cube roots of 8.

Solution $8 = 8 + 0i = r\text{ cis }\theta = 8\text{ cis }0°$.

$$8^{1/3} = (8\text{ cis }0°)^{1/3} = 8^{1/3}\text{ cis}\left(\frac{0°}{3} + \frac{360°k}{3}\right), k = 0, 1, 2.$$

for $k = 0$, $8^{1/3}$ cis $0° = 2(\cos 0° + i \sin 0°) = 2$
for $k = 1$, $8^{1/3}$ cis $120° = 2(\cos 120° + i \sin 120°) = -1 + i\sqrt{3}$
for $k = 2$, $8^{1/3}$ cis $240° = 2(\cos 240° + i \sin 240°) = -1 - i\sqrt{3}$

Note that other values of k produce no other roots: for example, if $k = 3$, $8^{1/3} = 2$, or if $k = 4$, $8^{1/3} = -1 + i\sqrt{3}$, or if $k = -1$, $8^{1/3} = -1 - i\sqrt{3}$, etc. In Exercise 77, the reader is asked to show that each answer above, when cubed, produces 8.

EXAMPLE 7 Determine the four fourth roots of $-1 + i\sqrt{3}$.

Solution $-1 + i\sqrt{3} = r \text{ cis } \theta = 2 \text{ cis } 120°$

$$(-1 + i\sqrt{3})^{1/4} = (2 \text{ cis } 120°)^{1/4} = 2^{1/4} \text{ cis } \frac{120°}{4} + \frac{360°k}{4}, k = 0, 1, 2, 3.$$

for $k = 0$, $2^{1/4}$ cis $30° = 2^{1/4}(\cos 30° + i \sin 30°)$
for $k = 1$, $2^{1/4}$ cis $120° = 2^{1/4}(\cos 120° + i \sin 120°)$
for $k = 2$, $2^{1/4}$ cis $210° = 2^{1/4}(\cos 210° + i \sin 210°)$
for $k = 3$, $2^{1/4}$ cis $300° = 2^{1/4}(\cos 300° + i \sin 300°)$

EXAMPLE 8 Solve the equation $x^6 = 1$.

Solution Since $x^6 = 1$, $x = 1^{1/6}$. Thus we need the six sixth roots of 1.

$$1 = 1 + 0i = r \text{ cis } \theta = 1 \text{ cis } 0°$$
$$1^{1/6} = (1 \text{ cis } 0°)^{1/6} = 1^{1/6} \text{ cis}\left(\frac{0°}{6} + \frac{360°k}{6}\right), k = 0, 1, 2, 3, 4, 5.$$

The roots are:

$$1, \text{ cis } 60°, \text{ cis } 120°, -1, \text{ cis } 240°, \text{ cis } 300°$$

The roots of Example 8 are referred to as the 6th roots of unity. In general, the solutions to the equation $x^n - 1 = 0$ are called the **nth roots of unity.**

EXERCISES FOR SECTION 7.7

In Exercises 1 through 15, rewrite each complex number in polar form. (See Example 1.)

1. $1 + i$
2. $1 - i$
3. $-1 + i$
4. $-1 - i$
5. $\frac{1}{2} + i\frac{\sqrt{3}}{2}$
6. $\frac{1}{2} - i\frac{\sqrt{3}}{2}$
7. $-3 + 3\sqrt{3}\,i$
8. $-3 - 3\sqrt{3}\,i$
9. $-4i$
10. 6
11. $6 + 6i$
12. $2 + i$
13. $3 + 4i$
14. $3 - 4i$
15. $4 + 2i$

In Exercises 16 through 30, rewrite each complex number in rectangular form. (See Example 2.)

16. $\text{cis}\,\dfrac{\pi}{4}$ 17. $3\,\text{cis}\,\dfrac{\pi}{4}$ 18. $4\,\text{cis}\,\dfrac{5\pi}{6}$

19. $2\,\text{cis}\left(-\dfrac{\pi}{4}\right)$ 20. $\text{cis}\,\dfrac{\pi}{6}$ 21. $\text{cis}\,\dfrac{5\pi}{4}$

22. $\text{cis}\,\dfrac{7\pi}{4}$ 23. $3\,\text{cis}(-60^\circ)$ 24. $3\,\text{cis}\,300^\circ$

25. $2\,\text{cis}\,20^\circ$ 26. $\text{cis}(-10^\circ)$ 27. $4\,\text{cis}\,\dfrac{\pi}{2}$

28. $10\,\text{cis}\,30^\circ$ 29. $2\,\text{cis}\,\pi$ 30. $2\,\text{cis}\,2\pi$

In Exercises 31 through 44, perform the indicated operation and write the answer in polar form. (See Example 3.)

31. $\left(\text{cis}\,\dfrac{\pi}{2}\right)\left(\text{cis}\,\dfrac{3\pi}{2}\right)$ 32. $\dfrac{\text{cis}\,5\pi}{2\,\text{cis}\,4\pi}$ 33. $\left(\text{cis}\,\dfrac{\pi}{4}\right)\left(\text{cis}\,\dfrac{\pi}{6}\right)$

34. $\dfrac{\text{cis}\,20^\circ}{\text{cis}\,5^\circ}$ 35. $\dfrac{\text{cis}\,\dfrac{\pi}{4}}{\text{cis}\,\dfrac{\pi}{2}}$ 36. $\dfrac{4\,\text{cis}\,100^\circ}{2\,\text{cis}\,40^\circ}$

37. $\left(\text{cis}\,\dfrac{\pi}{4}\right)\left(6\,\text{cis}\,\dfrac{\pi}{4}\right)$ 38. $\left(\text{cis}\,\dfrac{\pi}{6}\right)\left(\text{cis}\,\dfrac{2\pi}{3}\right)$ 39. $\dfrac{\left(\text{cis}\,\dfrac{\pi}{3}\right)\left(\text{cis}\,\dfrac{\pi}{3}\right)}{\text{cis}\,\dfrac{\pi}{2}}$

40. $(\text{cis}\,10^\circ)(2\,\text{cis}\,20^\circ)(3\,\text{cis}\,5^\circ)$ 41. $\dfrac{(2\,\text{cis}\,10^\circ)(3\,\text{cis}\,30^\circ)}{\text{cis}\,40^\circ}$

42. $\left(2\,\text{cis}\,\dfrac{3\pi}{2}\right)\left[3\,\text{cis}\left(-\dfrac{\pi}{2}\right)\right]$ 43. $\left(2\,\text{cis}\,\dfrac{\pi}{6}\right)\left(5\,\text{cis}\,\dfrac{4\pi}{3}\right)$ 44. $\dfrac{8\,\text{cis}\,\dfrac{5\pi}{3}}{3\,\text{cis}\,\dfrac{\pi}{2}}$

In Exercises 45 through 58, perform each indicated operation by converting to polar form if necessary, and using Theorem 7 or 8. (See Examples 4 and 5.)

45. $(-2 + 2\sqrt{3}\,i)(-2 - 2\sqrt{3}\,i)$ 46. $(\frac{1}{2} + \frac{1}{2}i)(\frac{1}{2} - \frac{1}{2}i)$

47. $(1 + i)^{20}$ 48. $(1 - i)^{10}$

49. $(4 + 4i)\left(\dfrac{1}{2} + \dfrac{\sqrt{3}}{2}i\right)$ 50. $[3\,\text{cis}\,10^\circ]^3$

51. $(2 + 2i)^5$ 52. $(4\,\text{cis}\,15^\circ)^2$

53. $(-1 - i)^{100}$ 54. $(\sqrt{3} + i)(\sqrt{3} - i)$

55. $\dfrac{[3\,\text{cis}\,10^\circ]^4}{\text{cis}\,25^\circ}$ 56. $(\sqrt{3} + i)^{10}$

57. $\left(-\dfrac{1}{2} - \dfrac{\sqrt{3}}{2}i\right)\left(-\dfrac{1}{2} + \dfrac{\sqrt{3}}{2}i\right)$ 58. $(-2i)(-3i)$

In Exercises 59 through 68, determine the specified roots. (See Examples 6 and 7.)

59. The three cube roots of 27 60. The three cube roots of -27

61. The two square roots of i 62. The four fourth roots of i

63. The four fourth roots of -1 64. The eight eighth roots of unity

65. The three cube roots of $\sqrt{3} - i$
66. The two square roots of $1 + i$
67. The three cube roots of $1 + i$
68. The three cube roots of $8 + 8i$

In Exercises 69 through 74, solve each equation for x. (See Example 8.)

69. $x^3 + 1 = 0$
70. $x^4 + 1 = 0$
71. $x^5 = 1$
72. $x^5 = i$
73. $x^4 + 1 + i\sqrt{3} = 0$
74. $x^3 = 8i$
75. Prove that $(r_1 \text{ cis } \theta_1)(r_2 \text{ cis } \theta_2) = r_1 r_2 \text{ cis}(\theta_1 + \theta_2)$ by following the outline below.
    a. Rewrite each $r \text{ cis } \theta$ factor as $r(\cos\theta + i\sin\theta)$, keeping track of subscripts.
    b. Perform the multiplication and collect real and imaginary parts.
    c. Use the identities for $\cos(\theta_1 + \theta_2)$ and $\sin(\theta_1 + \theta_2)$.
76. Prove that $\dfrac{r_1 \text{ cis } \theta_1}{r_2 \text{ cis } \theta_2} = \dfrac{r_1}{r_2} \text{ cis}(\theta_1 - \theta_2)$ by following the outline below.
    a. Rewrite each $r \text{ cis } \theta$ factor as $r(\cos\theta + i\sin\theta)$.
    b. Perform the division.
    c. Use the identities for $\cos(\theta_1 - \theta_2)$ and $\sin(\theta_1 - \theta_2)$.
77. Show that the cube of each answer in Example 6 is equal to 8.
78. Consider the equality $(\cos\theta + i\sin\theta)^2 = \cos 2\theta + i\sin 2\theta$.
    a. Expand and simplify the left-hand side.
    b. Comparing left- and right-hand sides, equate the real parts and equate the imaginary parts in order to obtain formulas for $\cos 2\theta$ and $\sin 2\theta$.
79. Consider the equality $(\cos\theta + i\sin\theta)^3 = \cos 3\theta + i\sin 3\theta$. Follow steps a and b of Exercise 78 in order to obtain formulas for $\cos 3\theta$ and $\sin 3\theta$.
80. Show that the three cube roots of 8 (from Example 6) lie equally spaced on a circle of radius 2 whose center is at the origin.
81. Show that the four fourth roots of $-1 + i\sqrt{3}$ (from Example 7) lie equally spaced on a circle of radius $\sqrt[4]{2}$ whose center is at the origin.
82. Show that the six sixth roots of 1 (from Example 8) lie equally spaced on a circle of radius 1 whose center is at the origin.

CHAPTER 7 IN REVIEW

I. Summary

In this chapter our study of trigonometry proceeded in many directions. In Section 7.1, the Law of Sines and Law of Cosines were used to solve a general triangle. This adds greatly to our capability of handling applied problems. In Section 7.2, many trigonometric identities were presented and many exercises of a manipulatory nature were encountered. Success in handling such exercises is important when trigonometry is used as a basic tool in calculus. Furthermore, experience in handling trigonometric identities was found to be useful in solving the trigonometric equations of Section 7.3.

In Section 7.4, we saw a return to graphing of inverse functions; namely, the inverse trigonometric functions. Recall that it was necessary to restrict the range (in all six cases) in order that the inverse trigonometric relation be a function.

In Section 7.5, some basic notions of trigonometry were used to develop the polar coordinate system and we had an introduction to graphing polar curves. In Section 7.6, geometric and trigonometric ideas were utilized to discuss and develop many notions about vector quantities.

Finally, in Section 7.7, we revisited complex numbers, this time by combining the basic ideas from Section 1.9 with the polar coordinate system in order to multiply and divide complex numbers in polar form. DeMoivre's Theorem was used to obtain powers and roots of complex numbers.

II. Vocabulary/Symbol True–False Quiz

If the statement is always true, write "TRUE." However, if the statement can be false, replace the underlined word or phrase with a difference word or phrase that will make the new statement true.

1. If three sides of a triangle are known, we can solve the triangle by using the Law of Cosines.
2. If $-\pi < t < 0$, then $\sin \frac{t}{2} = -\sqrt{\frac{1-\cos t}{2}}$.
3. If $\cos \theta = \frac{3}{5}$, then $\sin \theta = \frac{4}{5}$.
4. If $\sin \theta = x$ and $\frac{\pi}{2} < \theta < \pi$ then $\tan \theta = \frac{-x}{\sqrt{1-x^2}}$.
5. $\sin(\sin^{-1}x) = x$.
6. $\sin^{-1}(\sin 400°) = 40°$.
7. If $\sin^{-1}P = Q$, then $-1 \leqslant P \leqslant 1$.
8. Suppose $(-1, \sqrt{3})$ are the Cartesian coordinates of a point. The polar coordinates of that point are $(2, 120°)$.
9. If **A** is the vector from (0,0) to (1,5) and **B** is the vector from (0,0) to (2,−6), then **A** + **B** is the vector from (1,5) to (2,−6).
10. A cube root of 27 is $-1.5(1 - i\sqrt{3})$.

III. Chapter Test

In Exercises 1 through 4, solve each triangle.

1. $A = 40°$, $B = 30°$, $c = 4$
2. $b = 10, c = 12, C = 24°$
3. $a = 1$, $b = 2$, $c = 1.5$
4. $b = 10$, $c = 12$, $A = 24°$
5. If $\cos u = -\frac{2}{5}$ and $\pi < u < \frac{3\pi}{2}$ determine

   a. $\sin u$ b. $\sin \frac{u}{2}$ c. $\tan u$ d. $\cos \frac{u}{2}$

In questions 6 through 10, verify each identity.

6. $\dfrac{1 + \cos x}{\sin x} = \dfrac{\sin x}{1 - \cos x}$

7. $\cos 2\theta + 2 \sin^2\theta = 1$

8. $\sin 2A = \dfrac{2}{\tan A + \cot A}$

9. $\dfrac{\tan^2 B}{\sec B - 1} = \dfrac{1 + \cos B}{\cos B}$

10. $\cos\left(\dfrac{\pi}{2} + 2A\right) = -2 \sin A \cos A$

11. Graph $y = \arcsin x$

In questions 12 through 15, solve each trigonometric equation.

12. $2 \sin \theta = \sqrt{3}$

13. $\cos \theta - \frac{1}{2} = 0$

14. $\sin^2 2x = 1$

15. $6 \sin^2 x - 5 \sin x + 1 = 0$

16. Determine the value of each of the following.
    a. $\arcsin \dfrac{\sqrt{2}}{2}$
    b. $\arccos\left(-\dfrac{\sqrt{2}}{2}\right)$
    c. $\arcsin(-1)$
    d. $\tan^{-1}(-\sqrt{3})$
17. Express $\tan(\sin^{-1}x)$ in terms of x.
18. Graph the polar equation $r = \cos 3\theta$.
19. Graph the polar equation $r = 3 - \cos \theta$.
20. Suppose $\mathbf{A} = 2\mathbf{i} + 3\mathbf{j}$ and $\mathbf{B} = 3\mathbf{i} - 4\mathbf{j}$. Find
    a. $|\mathbf{A}|$
    b. $|\mathbf{B}|$
    c. $\mathbf{A} + \mathbf{B}$
    d. $|\mathbf{A} + \mathbf{B}|$
    e. $\mathbf{A} - \mathbf{B}$
    f. $3\mathbf{A} - 2\mathbf{B}$
    g. $\mathbf{A} \bullet \mathbf{B}$
21. Find the angle between $\mathbf{A} = \mathbf{i} - 2\mathbf{j}$ and $\mathbf{B} = 4\mathbf{i} + 7\mathbf{j}$.
22. Find $(1 + i)^8$.
23. Find the three cube roots of 27.
24. Find the four fourth roots of -1.
25. Solve for x: $x^3 + 1 = 0$. (Find all complex numbers, x)

IV. Student Research Project

In the northern hemisphere it is recommended that the angle between the plane of a solar collector and the horizontal be equal to the local latitude plus 15°. Also, it is recommended that the collector face due south. When these two criteria are not met, the effectiveness of a solar collector is somewhat reduced. Prepare a report of 150-200 words discussing this reduction of effectiveness. In your report you should cite examples. See: *Solar Heating and Cooling of Residential Buildings: Sizing, Installation and Operation of Systems*, prepared by Solar Energy Applications Laboratory, Colorado State University, 1977, pp. 14-7, 14-8.

CHAPTER 8

Systems of Equations and Inequalities

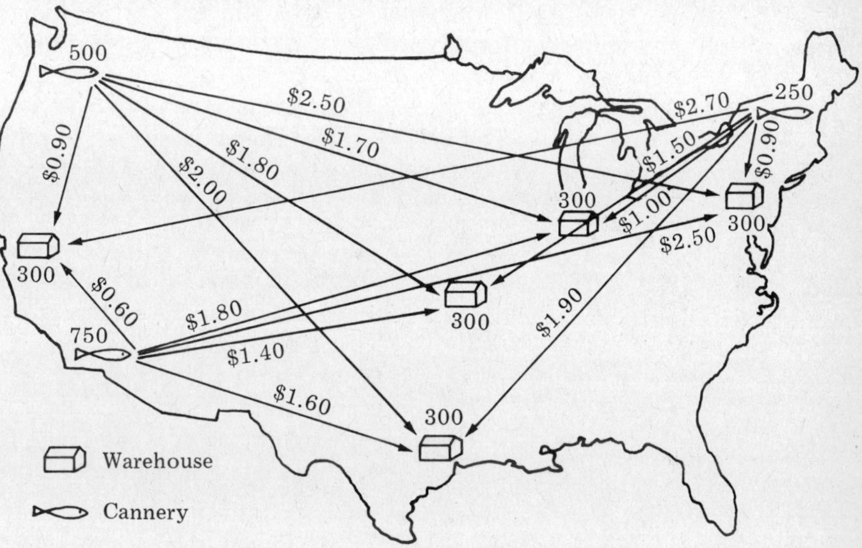

The Problem: Find a least cost plan of shipping from canneries to warehouse (the costs per case, availabilities and requirements are as indicated).

Although historically linear programming had its initial application in optimizing warfare strategies, minimizing the costs in the transportation of goods is another application.

G. B. Dantzig, Linear Programming and Its Extension *(Princeton, N.J.: Princeton University Press) 1963, p. 3.*

8.1 SYSTEMS OF LINEAR EQUATIONS: 2-BY-2 SYSTEMS

In Chapter 3, we graphed straight lines and examined the general linear equation in two variables. In this section, we will examine *systems* of two linear equations in two unknowns (sometimes called **"2-by-2" systems** or **simultaneous equations**). We begin with a definition:

DEFINITION

System of Two Linear Equations in Two Variables

If a_1, a_2, b_1, b_2, c_1, and c_2 are any real numbers, then we call

$$a_1x + b_1y = c_1$$
$$a_2x + b_2y = c_2$$

a system of two (linear) equations in the two variables, x and y.

Since we know that a linear equation is graphically represented by a straight line, when we consider two such equations *simultaneously*, the following cases can occur:

Case I: The two straight lines might intersect at exactly *one* point. This one point represents the **unique simultaneous solution** to the system.

Case II: The two straight lines might be parallel. There is *no* point common to both lines and we say the system is **inconsistent**.

Case III: The two straight lines might coincide. In this case, we say the system is **dependent**.

An example of each case follows.

EXAMPLE 1 Graph (on the same set of axes) the system

$$2x - y = 4$$
$$x + y = 5$$

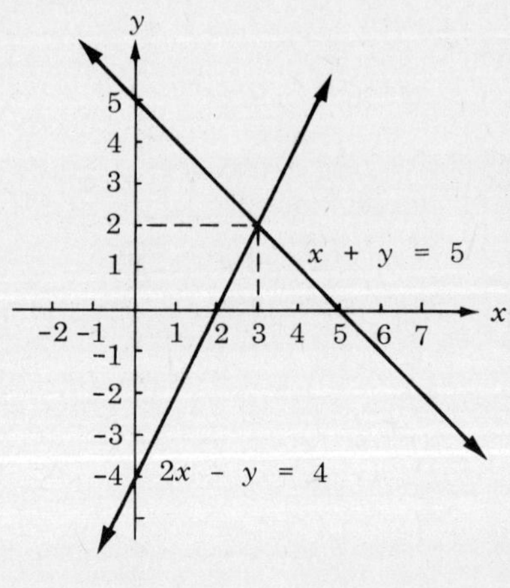

Solution We can read from the graph that the point of intersection is (3,2). Thus, $x = 3$, $y = 2$ is the solution to the system.

EXAMPLE 2 Graphically represent the system

$$x - \ \ y = 2$$
$$3x - 3y = 0$$

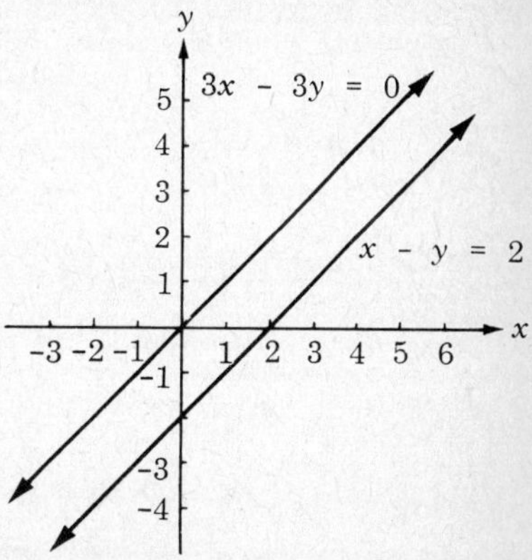

Solution This system is inconsistent; there is no solution.

EXAMPLE 3 Graphically represent the system

$$x + \ \ y = 3$$
$$2x + 2y = 6$$

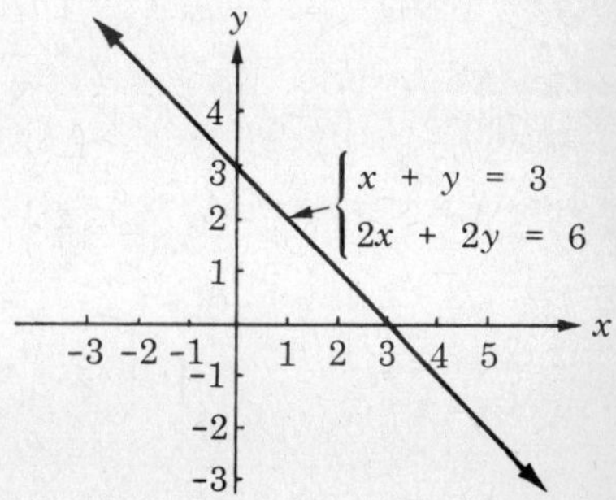

Solution The system is dependent; there are infinitely many solutions, each of which is graphically represented as a point on the line $x + y = 3$.

The graphical technique for solving systems of equations is somewhat limited. For example, when the coordinates of an intersection are fractions or possibly irrational numbers, we can only approximate the solution by "reading" the graph. Also, this graphical technique becomes difficult or impossible in systems involving three or more variables. What follows, then, are two alternative techniques for solving systems of equations *algebraically*.

The Substitution Method The substitution method for solving a system basically involves solving for a certain variable in one equation and using that result in the other equation.

EXAMPLE 4 Solve the system

$$y = 2x - 1 \quad (1)$$
$$2x + y = 9 \quad (2)$$

Solution Equation (1) is already solved for y (in terms of x). We now *substitute* $2x - 1$ for y in equation (2):

$$\begin{aligned} 2x + y &= 9 \\ 2x + (2x - 1) &= 9 \end{aligned} \quad \text{Substitute } 2x - 1 \text{ for } y$$

$$\begin{aligned} 4x - 1 &= 9 \\ x &= \tfrac{5}{2} \end{aligned} \quad \text{Solve for } x$$

If $x = \frac{5}{2}$, then from equation (1), $y = 2(\frac{5}{2}) - 1 = 4$. The solution is $x = \frac{5}{2}$, $y = 4$.

Check in equation (1):

$$\begin{aligned} y &= 2x - 1 \\ 4 &= 2(\tfrac{5}{2}) - 1 \\ 4 &= 4 \ \checkmark \end{aligned}$$

Check in equation (2):

$$\begin{aligned} 2x + y &= 9 \\ 2(\tfrac{5}{2}) + 4 &= 9 \\ 9 &= 9 \ \checkmark \end{aligned}$$

The Elimination Method

By certain "manipulations" of the equations in a given system, we may be able to eliminate certain variables. For instance, in the system

$$x + y = 14 \tag{3}$$
$$2x - y = -26 \tag{4}$$

we can add the two equations to "eliminate" y:

$$\begin{aligned} x + y &= 14 \\ 2x - y &= -26 \end{aligned} \quad \text{Add}$$

$$\begin{aligned} 3x &= -12 \\ x &= -4 \end{aligned}$$

That is, by reducing the system to an equation in just one variable, we can solve and find the value of that variable. To complete the solution, we can use the known value of x in either original equation to find the value of y. The reader should verify that $y = 18$.

In the illustration above, we chose to add the two equations to eliminate y. It is important to point out that had we first multiplied equation (3) by -2 and then added, the x's would have been eliminated. In short, there are three types of operations (or **transformations**) that can be used in the elimination method. They are:

Transformations used in Elimination Method

1. Any equation can be multiplied by any nonzero number.
2. Any multiple of one equation can be added to (or subtracted from) another equation.
3. Any two equations can be interchanged.

EXAMPLE 5 Solve the system

$$2x + 3y = 4 \tag{5}$$
$$4x + 5y = 6 \tag{6}$$

Solution We will eliminate y by multiplying equation (5) by -5 and equation (6) by 3 and adding:

$$\begin{array}{rcl} -10x - 15y &=& -20 \\ 12x + 15y &=& 18 \\ \hline 2x \quad\quad &=& -2 \\ x \quad\quad &=& -1 \end{array} \quad \text{Add}$$

Thus, $x = -1$. To find y, we can substitute -1 for x in an original equation:

$$\begin{aligned} 2x + 3y &= 4 \\ 2(-1) + 3y &= 4 \\ 3y &= 6 \\ y &= 2 \end{aligned}$$

Thus, $x = -1$ and $y = 2$. The checks are left to the reader.

The reader should note that instead of substituting -1 for x in order to find y, we could have repeated the elimination method, this time eliminating the variable x. See Exercise 31.

Earlier in this section we saw that a system of two linear equations in two variables might not have a unique solution. The next two examples show the algebraic interpretations of inconsistency and dependence.

EXAMPLE 6 Solve the system

$$\begin{aligned} 2x - 6y &= 4 \\ -x + 3y &= 5 \end{aligned}$$

Solution Using the elimination method and multiplying the second equation by 2, we have:

$$\begin{aligned} 2x - 6y &= 4 \\ -2x + 6y &= 10 \end{aligned}$$

Adding, we obtain:

$$0 = 14$$

which is impossible, *regardless* of values of x and y. Thus, *no pair of values* of x and y can satisfy the two equations simultaneously and the system is inconsistent. Graphically, the given equations represent parallel lines.

EXAMPLE 7 Solve the system

$$\begin{aligned} 3x - 6y &= 12 \\ 2x - 4y &= 8 \end{aligned}$$

Solution Multiply the first equation by 4 and the second equation by −6 and add:

$$\begin{array}{r} 12x - 24y = 48 \\ -12x + 24y = -48 \\ \hline 0 = 0 \end{array} \quad \text{Add}$$

Since $0 = 0$ is always true, we have infinitely many solutions to the system; it is dependent. The infinitely many solutions are graphically represented as the points on the line $3x - 6y = 12$ or the coincident line $2x - 4y = 8$.

We conclude this section with an applied problem.

EXAMPLE 8 A company has two cost concerns: material (measured in tons) and labor (measured in hours). January's total cost was $2,500 involving 3 tons of material and 100 hours of labor. February's total cost was $3,500 involving 4 tons of material and 150 hours of labor. Find the unit costs for material and labor.

Solution

STEP 1 Problem translation

We need to find two things—we use two variables:

Let x = the cost of one ton of material

Let y = the cost of one hour of labor

From January's cost statement we obtain $3x + 100y = 2500$.
From February's cost statement we obtain $4x + 150y = 3500$.

STEP 2 Solution of system

$$\begin{aligned} 3x + 100y &= 2500 \leftarrow \text{Multiply by } 3 \\ 4x + 150y &= 3500 \leftarrow \text{Multiply by } -2 \end{aligned}$$

$$\begin{array}{r} 9x + 300y = 7500 \\ -8x - 300y = -7000 \\ \hline x \quad = \$500 \end{array} \quad \text{Add}$$

Substitute 500 for x in January's equation:

$$3(500) + 100y = 2500$$
$$y = \$10$$

Thus, the cost of one ton of material is \$500 and the cost of one hour of labor is \$10.

STEP 3 Check

January:	3 tons of material at \$500 per ton	= \$1500
	100 hours of labor at \$10 per hour	= \$1000
	January total	= \$2500 ✓
February:	4 tons of material at \$500 per ton	= \$2000
	150 hours of labor at \$10 per hour	= \$1500
	February total	= \$3500 ✓

EXERCISES FOR SECTION 8.1

In Exercises 1 through 12, graph each system and determine the unique simultaneous solution, if it exists. If not, state whether the system is inconsistent or dependent. (See Examples 1, 2, and 3.)

1. $y = x + 1$, $y = 2x - 2$
2. $x + y = 5$, $x - y = 1$
3. $x + y = 5$, $x + y = 1$
4. $x + y = -3$, $2x - y = 0$
5. $y = x - 1$, $y = x + 3$
6. $x + y = 1$, $2x + 2y = 2$
7. $y = 2x - 3$, $y = 3x - 6$
8. $x + y = -8$, $2x - y = 2$
9. $y = -x + 3$, $2x + 2y = 6$
10. $x + 3y = 7$, $2x + 6y = 1$
11. $2x + 3y = 4$, $-2x + 3y = -4$
12. $x = 2$, $y = 3$

In Exercises 13 through 18, solve each system by using the substitution method. (See Example 4.)

13. $y = x + 1$, $2x - y = 2$
14. $y = x - 1$, $x + y = 5$
15. $y = 5 - x$, $x + y = 1$
16. $y = x - 2$, $x - y = 2$
17. $x + 2y = 5$, $3x - 2y = -1$
18. $x = 3y$, $x - 6y = 1$

In Exercises 19 through 30, solve each system by using the elimination method. (See Examples 5, 6, and 7.)

19. $2x - y = 4$, $x + y = 5$
20. $x + 2y = 1$, $-x - 3y = 0$
21. $3x + 3y = 3$, $2x + 3y = -2$
22. $x + 3y = 5$, $2x + 5y = 8$
23. $-2x + 3y = 11$, $-x - 2y = 2$
24. $-2x + 3y = 11$, $-5x - 10y = 10$
25. $4x + 3y = 9$, $6x + 6y = 17$
26. $3x + 2y = -2$, $3x + 4y = -3$
27. $3x + 4y = 9$, $-6x - 8y = 3$
28. $x - y = 3$, $3x + 3y = -3$
29. $4x + 7y = 10$, $8x + 14y = 20$
30. $8x - 3y = 9$, $7x + 11y = -2$

31. In Example 5, we solved the system

$$2x + 3y = 4$$
$$4x + 5y = 6$$

by eliminating y. Redo the problem by eliminating the variable x.

32. a. Use the elimination method to solve for y in the general system

$$a_1x + b_1y = c_1$$
$$a_2x + b_2y = c_2$$

b. Use the elimination method to solve for x.

c. What condition (involving a_1, b_1, a_2, b_2) must be true in order for a unique solution to exist?

In Exercises 33 through 38, solve each applied problem. (See Example 8.)

33. The length of a rectangle is 5 meters more than its width. The perimeter is 24 meters. What are the dimensions of the rectangle?
34. The sum of two numbers is 119. Their difference is 45. Find the numbers.
35. Thirty coins in a piggy bank, composed only of quarters and half-dollars, total \$14. How many quarters are there? How many half-dollars are there?
36. A factory has two main costs—labor and material. One week the factory uses 2 tons of material and 100 hours of labor for a total weekly cost of \$2800. The next week it uses 3 tons of material and 250 hours of labor for a total weekly cost of \$5000. What is the cost of 1 ton of material? What does one hour of labor cost?
37. Two adults and four children pay admission for the circus and the cost of tickets is \$16. When one adult and six children enter the same circus, the cost is \$18. Determine each admission price.
38. Janet invested a total of \$1000 in two business ventures—one that yielded a 5% profit after a year and one that yielded a 6% profit. If she made \$54 in profit after a year, how much did she invest in each business?
39. Suppose a company's costs are as follows:

	labor	*materials*	*total*
January	150 hours	6 tons	\$ 5100
February	300 hours	12 tons	\$10,200

Why *can't* the cost of labor and materials be determined?

40. Comment on the information that a company has stated below:

	labor	*materials*	*total*
January	150 hours	6 tons	\$ 5100
February	300 hours	12 tons	\$10,000

41. Consider the system

$$\frac{2}{s} + \frac{3}{t} = -5$$
$$\frac{1}{s} - \frac{1}{t} = 5$$

Although this system is *not* linear in the variables s and t, it is linear in x and y if $x = \frac{1}{s}$ and $y = \frac{1}{t}$. Make that substitution and find s and t.

In Exercises 42 through 47 use the method of Exercise 41 to solve each system.

42. $\dfrac{1}{s}+\dfrac{2}{t}=2$
$\dfrac{3}{s}-\dfrac{2}{t}=2$

43. $\dfrac{4}{s}+\dfrac{1}{t}=2$
$\dfrac{8}{s}+\dfrac{3}{t}=5$

44. $\dfrac{2}{r}+\dfrac{3}{s}=7$
$\dfrac{1}{r}-\dfrac{1}{s}=-\dfrac{3}{2}$

45. $\dfrac{1}{r}-\dfrac{1}{s}=\dfrac{1}{6}$
$-\dfrac{1}{r}+\dfrac{1}{s}=-\dfrac{1}{6}$

46. $\dfrac{16}{s^2}+\dfrac{1}{t^2}=2$
$\dfrac{64}{s^2}+\dfrac{1}{t^2}=5$

47. $\dfrac{4}{m^2}-\dfrac{4}{n^2}=-1$
$\dfrac{8}{m^2}-\dfrac{2}{n^2}=1$

Calculator Exercises

In Exercise 32, the following formulas for x and y were found:

$$x=\frac{b_2c_1-b_1c_2}{a_1b_2-a_2b_1} \quad \text{and} \quad y=\frac{a_1c_2-a_2c_1}{a_1b_2-a_2b_1}$$

Use them to find x and y *directly* from a calculator in each system in Exercises 48 through 53.

48. $3x+4y=5$
$5x-2y=17$

49. $x+7y=18$
$3x-10y=7$

50. $1.8x-3.7y=9.1$
$-.5x+11.4y=10.4$

51. $2.1x-3.2y=-1.6$
$x+1.7y=0$

52. $\sqrt{2}\,x-\sqrt{3}\,y=3$
$\sqrt{5}\,x+y=4$

53. $x+\dfrac{1}{y}=2$
$3.2x-\dfrac{\sqrt{3}}{y}=1$

8.2 SYSTEMS OF THREE LINEAR EQUATIONS IN THREE VARIABLES

In this section we will use the elimination method that was discussed in the previous section to solve systems of three linear equations in three variables. The general system is

$$\begin{aligned} a_1x+b_1y+c_1z&=d_1\\ a_2x+b_2y+c_2z&=d_2\\ a_3x+b_3y+c_3z&=d_3 \end{aligned}$$

and the basic plan of "attack" is to eliminate the same variable in two equations and then proceed as in the previous section. Some examples follow.

EXAMPLE 1 Solve the system

$$\begin{aligned} x+y+z&=0 &\quad(1)\\ -x+3y+z&=2 &\quad(2)\\ -3x+y-5z&=14 &\quad(3) \end{aligned}$$

Solution We eliminate the variable x by adding equations (1) and (2) and also by adding equation (3) to 3 times equation (1):

$$\begin{array}{lrcl} (1) & x + y + z &=& 0 \\ (2) & -x + 3y + z &=& 2 \\ \hline & 4y + 2z &=& 2 \end{array} \quad \text{Add}$$

$4y + 2z = 2$ ← We now have a system of two linear equations → $4y - 2z = 14$ (4)

3 times equation (1)

$$\begin{array}{rcll} 3x + 3y + 3z &=& 0 & \\ -3x + y - 5z &=& 14 & (3) \\ \hline 4y - 2z &=& 14 & (4) \end{array}$$

$$\begin{array}{rcl} 4y + 2z &=& 2 \\ 4y - 2z &=& 14 \\ \hline 8y &=& 16 \\ y &=& 2 \end{array} \quad \text{Add}$$

Now, we substitute 2 for y in equation (4) to find z:

$$\begin{aligned} 4y - 2z &= 14 \\ 4(2) - 2z &= 14 \\ -2z &= 6 \\ z &= -3 \end{aligned}$$

With $y = 2$ and $z = -3$, we return to any original equation:

$$\begin{aligned} x + y + z &= 0 \qquad (1) \\ x + 2 - 3 &= 0 \\ x &= 1 \end{aligned}$$

Thus, $x = 1$, $y = 2$, and $z = -3$. We can write the solution as the *ordered triple* $(1,2,-3)$. We leave the checks to the reader.

EXAMPLE 2 Solve the system

$$\begin{aligned} y + z &= 1 \\ 2x - y - 2z &= 1 \\ 4x - 2y - z &= -7 \end{aligned}$$

Solution Since there is no x term in the first equation, we choose to eliminate x from the other equations by adding -2 times the second equation to the third:

$$\begin{array}{rcl} -4x + 2y + 4z &=& -2 \\ 4x - 2y - z &=& -7 \\ \hline 3z &=& -9 \\ z &=& -3 \end{array} \quad \text{Add}$$

With $z = -3$, we can find y by using the first equation

$$\begin{aligned} y + z &= 1 \\ y - 3 &= 1 \\ y &= 4 \end{aligned}$$

Finally, to find x, we return to the second equation:

$$\begin{aligned} 2x - y - 2z &= 1 \\ 2x - 4 - 2(-3) &= 1 \\ 2x &= -1 \\ x &= -\tfrac{1}{2} \end{aligned}$$

Thus, $x = -\frac{1}{2}$, $y = 4$, and $z = -3$. The checks are left to the reader.

We point out here, that, as in the case with 2-by-2 systems of the previous section, there are three possibilities for solutions to a 3-by-3 system: 1) a unique solution might exist; 2) there might be infinitely many solutions (a dependent system); 3) there might be no solution (inconsistent system). The latter two cases are discussed more in the next chapter (see also Exercises 26 through 29). We continue this section with an applied problem that involves solving a 3-by-3 system.

EXAMPLE 3 Joan has \$35,000 that she split into three investments. Part of the money is invested in a savings account that yields 6% annual interest, part is in 7% annual-yield bonds, and the remainder is invested in a business. In 1979, when the business investment *lost* at an annual rate of 6%, her net income from all three investments was \$660. In 1980, when the business gained at an annual rate of 10%, her net income rose to \$2740. How much money did Joan put in each investment?

Solution

STEP 1 Problem translation

Let x = amount invested in savings account
y = amount invested in bonds
z = amount invested in business

Since the total is \$35,000, the first equation is

$$x + y + z = 35{,}000 \qquad (1)$$

The 1979 information yields the following:

$$0.06x + 0.07y - 0.06z = 660 \qquad (2)$$

The 1980 information yields:

$$0.06x + 0.07y + 0.10z = 2740 \qquad (3)$$

Thus, the system to be solved is

$$x + y + z = 35{,}000 \qquad (1)$$
$$0.06x + 0.07y - 0.06z = 660 \qquad (2)$$
$$0.06x + 0.07y + 0.10z = 2740 \qquad (3)$$

STEP 2 Solution of system

Subtracting equation (2) from equation (3) will eliminate both x and y and produce an equation in z:

$$.16z = 2080$$
$$z = \$13{,}000$$

Now, we substitute 13,000 for z in both equations (1) and (2):

$$x + y + 13{,}000 = 35{,}000$$
$$0.06x + 0.07y - 0.06(13{,}000) = 660$$

or

$$x + y = 22{,}000 \qquad (4)$$
$$0.06x + 0.07y = 1440 \qquad (5)$$

Next, to eliminate y, we add equation (5) to -0.07 times equation (4).

$$\begin{array}{r} -0.07x - 0.07y = -1540 \\ 0.06x + 0.07y = 1440 \\ \hline -.01x = -100 \end{array} \quad \text{Add}$$
$$x = \$10{,}000$$

Finally, we find y by using equation (4):

$$x + y = 22{,}000$$
$$10{,}000 + y = 22{,}000$$
$$y = 12{,}000$$

Hence, Joan invested \$10,000 in savings, \$12,000 in bonds, and \$13,000 in the business.

STEP 3 Check

The sum of investments = \$10,000 + \$12,000 + \$13,000 = \$35,000 ✓
The 1979 income = 0.06(\$10,000) + 0.07(\$12,000) − 0.06(\$13,000) = \$660 ✓
The 1980 income = 0.06(\$10,000) + 0.07(\$12,000) + 0.10(\$13,000) = \$2740 ✓

We conclude this section by mentioning that systems of n equations in n unknowns can be solved using the methods presented in this section. Such systems are examined in the next chapter.

EXERCISES FOR SECTION 8.2

In Exercises 1 through 25, solve each system. (See Examples 1 and 2.)

1. $$\begin{aligned} x + y + z &= 6 \\ -x + 2y + 3z &= 12 \\ -2x - y + 5z &= 11 \end{aligned}$$

2. $$\begin{aligned} x + y + z &= 0 \\ 3x - y + z &= 2 \\ x - 3y - 5z &= 14 \end{aligned}$$

3. $$\begin{aligned} x + y + z &= 2 \\ 3x - 4y + z &= 3 \\ x - y - z &= 4 \end{aligned}$$

4. $$\begin{aligned} x + y + z &= 9 \\ 2x - y - z &= -3 \\ x - 5y + 7z &= 15 \end{aligned}$$

5. $$\begin{aligned} x + y + z &= 9 \\ 2x - y - z &= -3 \\ x + 7y - 5z &= 15 \end{aligned}$$

6. $$\begin{aligned} x + y + z &= 9 \\ -x + 2y - z &= -3 \\ -5x + y + 7z &= 15 \end{aligned}$$

7. $$\begin{aligned} x - 5y + z &= 7 \\ 3x + y - z &= -7 \\ x + 3y &= -4 \end{aligned}$$

8. $$\begin{aligned} x + y + 2z &= 2 \\ 3x - y - z &= 21 \\ y + z &= -3 \end{aligned}$$

9. $$\begin{aligned} 2x + y + z &= 2 \\ -x - y + 3z &= 21 \\ x + y &= -3 \end{aligned}$$

10. $$\begin{aligned} 2x + y + z &= 40 \\ -x - y + 3z &= 10 \\ x + y &= 20 \end{aligned}$$

11. $$\begin{aligned} x + y + 2z &= 0 \\ 3x - 4z &= -2 \\ 5y - 6z &= -8 \end{aligned}$$

12. $$\begin{aligned} x + 4y + z &= 3 \\ 3x + 2z &= 0 \\ x - 2y + 4z &= 9 \end{aligned}$$

13. $$\begin{aligned} x + 3y - z &= 11 \\ 3x + y - z &= 13 \\ x + y - 3z &= 11 \end{aligned}$$

14. $$\begin{aligned} 4x + 2y - 5z &= 1 \\ 2x + y - 3z &= 1 \\ y - 3z &= 2 \end{aligned}$$

15. $$\begin{aligned} x + y + z &= 0 \\ x + 2y + 3z &= 0 \\ -2x + 5y - z &= 0 \end{aligned}$$

16. $$\begin{aligned} x - y + z &= 0 \\ x - y - z &= 6 \\ 2x + y + z &= -3 \end{aligned}$$

17. $$\begin{aligned} 2x + 3y + 4z &= 3 \\ 4x - 3y + 4z &= 2 \\ 4x + 6y - 4z &= 3 \end{aligned}$$

18. $$\begin{aligned} 2x + 3y + 4z &= 3 \\ 4x - 3y + 4z &= -3 \\ 4x + 6y - 4z &= 6 \end{aligned}$$

19. $$\begin{aligned} x + y + z &= 6 \\ y - z &= 1 \\ x - y &= -2 \end{aligned}$$

20. $$\begin{aligned} x + 2y + 12z &= 7 \\ -x - 7y + 6z &= 1 \\ \tfrac{1}{2}x + y &= 0 \end{aligned}$$

21. $$\begin{aligned} 2x + y + 12z &= 7 \\ -7x - y + 6z &= 1 \\ x + \tfrac{1}{2}y &= 0 \end{aligned}$$

22. $$\begin{aligned} a + b + 3c &= 3 \\ 4b - c &= -4 \\ a &= 4 \end{aligned}$$

23. $$\begin{aligned} p + q - 3r &= 3 \\ 2p - q - r &= 2 \\ p + q + r &= 1 \end{aligned}$$

24. $$\begin{aligned} x_1 + x_2 + x_3 &= 13 \\ x_1 - x_2 - 3x_3 &= 3 \\ x_2 + 2x_3 &= 5 \end{aligned}$$

25. $$\begin{aligned} x + y + 4 &= z \\ y - 3 &= x + z \\ 4x - y + 2z + 5 &= 0 \end{aligned}$$

26. It was pointed out in this section that there might be *no solution* to a 3-by-3 system (the system is inconsistent). Attempt to use the elimination method on the system below and show that the system is inconsistent. Compare with Example 6 of the previous section.

$$\begin{aligned} x + y + z &= 6 \\ x - 2y + 3z &= 6 \\ 2x - y + 4z &= 10 \end{aligned}$$

27. Show that the following system is inconsistent:

$$\begin{aligned} 2x + y - 4z &= 7 \\ x - y + 4z &= 3 \\ x &= 3 \end{aligned}$$

28. It was pointed out in this section that there might be *infinitely many* solutions to a 3-by-3 system (the system is dependent). Attempt to use the elimination method on the system below and show that the system is dependent. Compare with Example 7 of the previous section.

$$\begin{aligned} x + y + z &= 6 \\ x - 2y + 3z &= 6 \\ 2x - y + 4z &= 12 \end{aligned}$$

29. Show that the following system is dependent:

$$\begin{aligned} 2x + y - 4z &= 6 \\ x - y + 4z &= 3 \\ x &= 3 \end{aligned}$$

In Exercises 30 through 35, solve each applied problem. (See Example 3.)

30. Jack invested a total of \$10,000 in three savings accounts: one is a 6% account, one is a 7% account, and one is an 8% account. If the simple annual interest for the three accounts totalled \$740 and three times as much was invested in the 8% account as the 6% account, find out how much Jack invested in each savings account.
31. Ricardo has one-dollar, two-dollar, and five-dollar bills in his cash register at closing time. In all there are 30 bills adding up to \$95. If there are twice as many one-dollar bills as two-dollar bills, how many of each denomination does Ricardo have?
32. A toy-making company produces 3 items—dolls, train sets, and tricycles. In the months of June, July, and August, 1980, they sold the following quantities:

	June	*July*	*August*
Dolls	10	4	15
Train sets	6	5	20
Tricycles	40	100	200
Monthly income from dolls, train sets, and tricycles	\$2030	\$4657	\$9620

Determine the selling price for each of the three toys.

33. A souvenir barker at a carnival sells three items—balloons, pennants, and hats. His revenue from selling these items is given below:

	Balloons	*Pennants*	*Hats*	*Income*
Monday	10	4	1	\$10
Tuesday	11	2	6	\$19
Wednesday	0	8	30	\$66

Determine the selling price of each of these items.

34. Nancy is a nutritionist at the Crossroads Restaurant. A luncheon specialty is ground steak, which Nancy has the cooks make by mixing top round, sirloin, and chuck steak. There must be five times as much top round as sirloin and two times as much sirloin as chuck. In total, there must be 32 pounds of meat. Let x = pounds of top round, y = pounds of sirloin, and z = pounds of chuck.
   a. Set up a system of three equations using the variables, x, y, z that represents this problem.
   b. Solve the system in part a in order to determine how much meat of each type Nancy should buy.

35. A parabola that can be described by the equation $y = ax^2 + bx + c$, is known to pass through the points $(2,3)$, $(-2,7)$, and $(3,12)$. Find the equation of the parabola.

8.3 NONLINEAR SYSTEMS

Thus far in this chapter, we have solved systems that involve only *linear* equations. In this section, we will be concerned with solving systems that involve at least one nonlinear equation. For example, the system

$$x^2 + y^2 = 25$$
$$y = x^2 - 5$$

is a system of two second degree equations in two variables. Graphically, this system is represented by a circle and a parabola; its solution

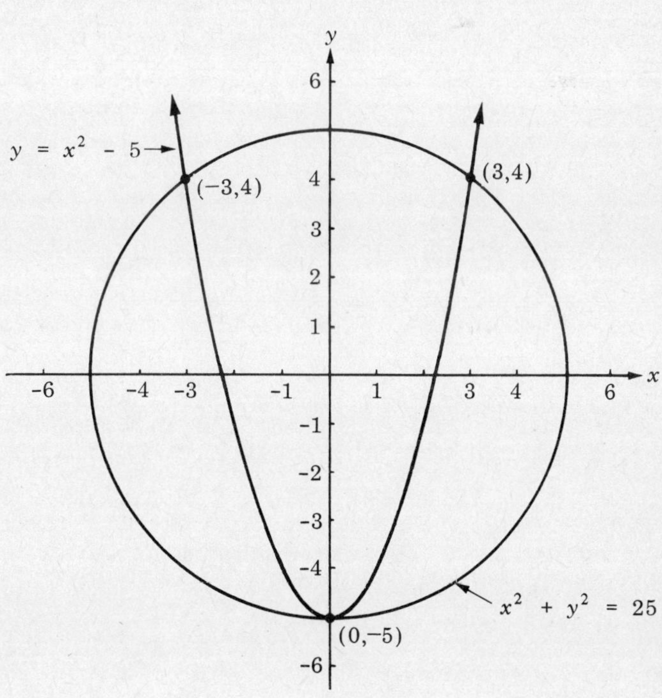

consists of the points of intersection of the circle with the parabola. Although we will be concerned primarily with *algebraic* methods of solution, the graphical solution to the above system is presented here as review. Since the graphs of the parabola and the circle intersect in exactly three points, the system

$$x^2 + y^2 = 25$$
$$y = x^2 - 5$$

has exactly three solutions. They are $x = -3$, $y = 4$; $x = 0$, $y = -5$; $x = 3$, $y = 4$. In ordered pair notation, we write: $(-3,4)$, $(0,-5)$, and $(3,4)$.

To solve this type of system algebraically, we will use either the substitution method or the elimination method. Several examples follow.

EXAMPLE 1 Solve the system algebraically:

$$x^2 + y^2 = 25 \quad (1)$$
$$y = x^2 - 5 \quad (2)$$

Solution Since one equation is already solved for the variable y, we *substitute* $x^2 - 5$ for y in equation (1):

$$x^2 + y^2 = 25$$
$$x^2 + (x^2 - 5)^2 = 25$$
$$x^4 - 9x^2 = 0$$
$$x^2(x + 3)(x - 3) = 0$$

$x^2 = 0$	$x + 3 = 0$	$x - 3 = 0$
$x = 0$	$x = -3$	$x = 3$

There are three x values for which we must find corresponding y values. By returning to one of the original equations, we find y:

$x = 0$:	$x = 3$:	$x = -3$:
$y = x^2 - 5$	$y = x^2 - 5$	$y = x^2 - 5$
$y = 0 - 5 = -5$	$y = 9 - 5 = 4$	$y = (-3)^2 - 5 = 4$

There are three ordered pairs: $(0,-5)$, $(3,4)$ and $(-3,4)$. The reader should check *each* of these in *each* original equation to verify that all three are solutions.

EXAMPLE 2 Solve the system

$$x^2 + y^2 = 9$$
$$x^2 - y^2 = 1$$

Solution Since neither equation is solved for a variable, it is simpler to *eliminate* the y^2 terms by adding the equations:

$$\begin{aligned} x^2 + y^2 &= 9 \\ x^2 - y^2 &= 1 \\ \hline 2x^2 \quad &= 10 \\ x^2 &= 5 \\ x &= \pm\sqrt{5} \end{aligned}$$

For these two x values, we find corresponding y values by returning to either original equation:

$$\begin{array}{cc} x = \sqrt{5}: & x = -\sqrt{5}: \\ x^2 + y^2 = 9 & x^2 + y^2 = 9 \\ 5 + y^2 = 9 & 5 + y^2 = 9 \\ y = \pm 2 & y = \pm 2 \end{array}$$

Thus, there are *four* solutions; they are $(\sqrt{5},2)$, $(\sqrt{5},-2)$, $(-\sqrt{5},2)$, $(-\sqrt{5},-2)$. The checks are left to the reader.

In the next example a system involving logarithmic equations is solved.

EXAMPLE 3 Solve the system

$$y = \log_{10}x \tag{3}$$
$$y + \log_{10}(x - 9) = 1 \tag{4}$$

Solution We substitute $\log_{10}x$ for y in the second equation:

$$\log_{10}x + \log_{10}(x - 9) = 1$$

We solve this new equation using the techniques of Section 5.6:

$$\begin{aligned} \log_{10}x(x - 9) &= 1 \\ x(x - 9) &= 10 \\ x^2 - 9x - 10 &= 0 \\ (x - 10)(x + 1) &= 0 \\ x = 10,\ x &= -1 \end{aligned}$$

We must reject $x = -1$ because it does not yield a y value. If we substitute 10 for x in equation (3), we have $y = \log_{10} 10 = 1$. The reader should check the ordered pair (10,1) in both original equations in order to verify that it is a solution.

EXAMPLE 4 Solve the system

$$y = 3^x + 4$$
$$y = 3^{2x} - 2$$

Solution We substitute $3^{2x} - 2$ for y in the first equation:

$$\begin{aligned} 3^{2x} - 2 &= 3^x + 4 \\ 3^{2x} - 3^x - 6 &= 0 \\ t^2 - t - 6 &= 0 \\ (t-3)(t+2) &= 0 \\ t = 3,\ t &= -2 \end{aligned}$$

Let $t = 3^x$ (*Note:* then $x = \log_3 t$.)

Thus, $x = \log_3 3 = 1$. (The -2 value is extraneous.) To find y, $y = 3^x + 4 = 3 + 4 = 7$. The solution is $x = 1$, $y = 7$. The checks are left to the reader.

We conclude this section with an example to show how a nonlinear system can be the representation for a verbal problem.

EXAMPLE 5 The perimeter of a rectangular room is 13 meters and the area of the floor is 10 square meters. Find the length and width.

Solution Let x = the width
y = the length

We have the following system

$$2x + 2y = 13 \leftarrow \text{perimeter}$$
$$xy = 10 \leftarrow \text{area}$$

To solve the system, rewrite the perimeter equation as $y = \frac{1}{2}(13 - 2x)$ and substitute in the area equation:

$$\begin{aligned} x[\tfrac{1}{2}(13 - 2x)] &= 10 \\ 2x^2 - 13x + 20 &= 0 \\ (2x - 5)(x - 4) &= 0 \\ x = 2.5,\ x &= 4 \end{aligned}$$

If $x = 2.5$, $y = 4$; if $x = 4$, $y = 2.5$. The dimensions of the room are 2.5 meters by 4 meters.

EXERCISES FOR SECTION 8.3

In Exercises 1 through 30, solve the given system. (See Examples 1, 2, 3, and 4.)

1. $y = x^2 - x - 12$
   $y = x - 4$
2. $y = x^2 - 4x + 4$
   $y = 2x - 4$
3. $x^2 + y^2 = 36$
   $x^2 - y^2 = 36$
4. $y = x^2$
   $x^2 + (y-1)^2 = 1$
5. $\frac{x^2}{9} + \frac{y^2}{16} = 1$
   $y = x$
6. $\frac{x^2}{9} - \frac{y^2}{4} = 1$
   $x^2 + y^2 = 9$
7. $x^2 + y^2 = 10$
   $9x^2 + y^2 = 18$
8. $y = x^2 - x - 6$
   $y = 2x - 8$
9. $x^2 + y^2 = 1$
   $x - y = 0$
10. $x^2 + y^2 = 4$
    $x^2 - y^2 = 4$
11. $y = x^2 - 2x + 1$
    $x + y = 3$
12. $x^2 + y^2 = 1$
    $x - y = 1$
13. $x^2 - y^2 = 1$
    $x + y = 0$
14. $y = \frac{4}{x}$
    $x + y = 5$
15. $y = \frac{4}{x^2}$
    $x^2 + y = 5$
16. $x^2 + y^2 = 20$
    $x^2 - y^2 = 12$
17. $xy + 6 = 0$
    $3x + y = 7$
18. $x = y^2$
    $y = x^2$
19. $y = \log_{10}(3x + 4) - 5$
    $y = \log_{10}(x - 1) - 4$
20. $y = \log_{10}(x + 1)$
    $y + \log_{10}(x - 8) = 1$
21. $y = \ln x$
    $y + \ln x = 2$
22. $y - \ln(2x - 1) = 0$
    $3y + \ln(2x - 1) = 8$
23. $\log_{10}(x + 16) = y - 1$
    $\log_{10}(\frac{1}{4}x - 11) = y - 2$
24. $\log_{10}(3 - x) + y = 3$
    $y = \log_{10}(x + 8) + 2$
25. $y = 2^x + 3$
    $y = 2^{2x} + 1$
26. $y = 2(2^x) + 3$
    $y = 2^{2x}$
27. $y = 2(3^x) + 3$
    $y = 3^{2x}$
28. $y = 2^x + 3$
    $y = 4^x + 1$
29. $x = 3^y$
    $3^{2y} - 3x + 2 = 0$
30. $x^2 - 3^{y+1} + 2 = 0$
    $x = 3^y$
31. Graph $x^2 + y^2 = 9$ and $x^2 - y^2 = 1$ on the same axes to obtain a geometric interpretation of Example 2.
32. Graph $y = x^2 - x - 12$ and $y = x - 4$ on the same axes. Compare with Exercise 1, above.
33. Graph $x^2 + y^2 = 4$ and $x^2 - y^2 = 4$ on the same axes. Compare with Exercise 10, above.
34. Graph $x^2 - y^2 = 1$ and $x + y = 0$ on the same axes. What does this graphical solution say about the system? See Exercise 13, above.
35. Graph $x^2 + y^2 = 9$ and $x^2 - y^2 = 25$ on the same axes. How many solutions are there to the system?

In Exercises 36 through 43, solve each applied problem. (See Example 5.)

36. The perimeter of a rectangle is 18 feet; its area is 20 square feet. Find its dimensions.
37. The sum of two numbers is 6; their product is −72. What are the numbers?
38. Find two positive real numbers whose sum is 6 and whose product is 8.
39. Find two positive real numbers whose sum is 4 and whose product is 2.
40. Find two negative real numbers whose sum is −11 and whose product is 18.
41. Find two real numbers whose sum is $\frac{7}{6}$ and whose product is $\frac{1}{3}$.
42. The difference of two positive real numbers is 5 and the sum of their squares is 73. Find the numbers.
43. The hypotenuse of a right triangle is 8.2 meters long and the area of the triangle is 7.2 square meters. Find the lengths of the two legs.
44. Sometimes more imaginative types of substitutions must be made when trying to solve a given system. For example, consider

$$x^2 + 4xy + y^2 = 88 \qquad (5)$$
$$x^2 - xy - 6y^2 = 0 \qquad (6)$$

Solve the system by factoring equation (6) and substituting expressions found equivalent to x (involving y) in equation (5).

45. Follow the suggestion in Exercise 44 to solve:

$$x^2 + 2xy - 4y^2 = 1$$
$$x^2 - 5xy + 6y^2 = 0$$

8.4 SYSTEMS OF INEQUALITIES AND LINEAR PROGRAMMING

In Section 3.5 we graphed linear inequalities in two variables. We now wish to examine the graph of a *system* of such inequalities. The method is straight-forward—the system's solution is the simultaneous intersection of the graphs of all of the inequalities. Two examples follow.

EXAMPLE 1 Graph the following system:

$$y \leqslant 2x + 3 \qquad (1)$$
$$x + y \leqslant 6 \qquad (2)$$

Solution From Section 3.5, the graph of (1) and (2) appear separately below.

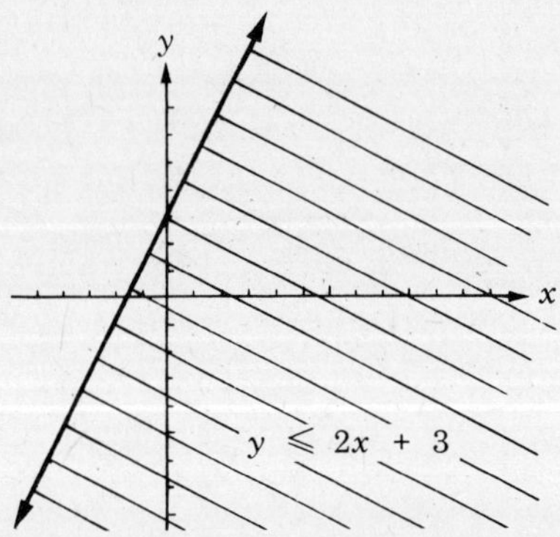

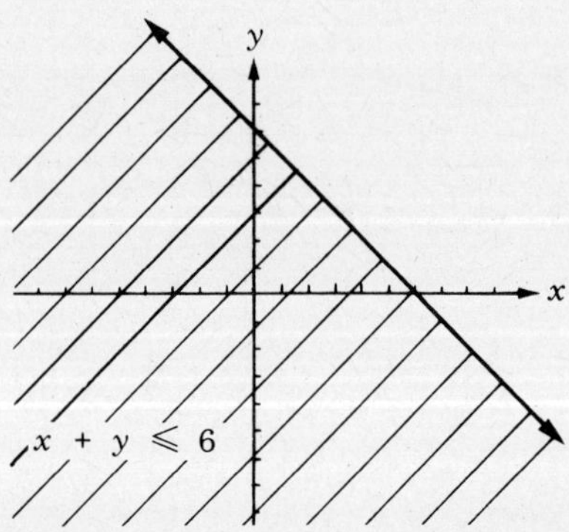

To graph the *system* (that is, (1) and (2) together), we plot the points common to *both* graphs. The solution consists of the solid line boundary and the shaded region. Notice that the point (1,5) is the point common to both lines. In other words, $x = 1$ and $y = 5$ is the solution to the system of *equations*:

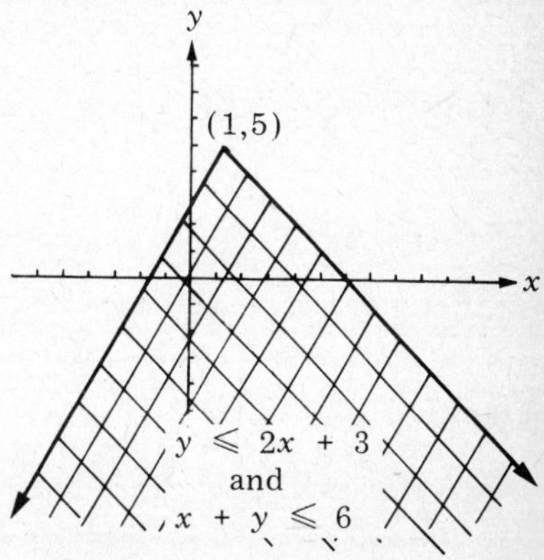

$$y = 2x + 3$$
$$x + y = 6$$

This point of intersection will be important in our work later in this section; it should be labeled.

A system of linear inequalities can involve any number of inequalities as the next example illustrates.

EXAMPLE 2 Graph the following system:

$$y \leqslant 2x + 3 \quad (1)$$
$$x + y \leqslant 6 \quad (2)$$
$$4x + 3y \geqslant 12 \quad (3)$$
$$x \geqslant 0 \quad (4)$$
$$y \geqslant 0 \quad (5)$$

Solution Notice that inequalities (4) and (5) confine the solution set to the first quadrant. Inequalities (1) and (2) were graphed in Example 1 and inequality (3) appears below:

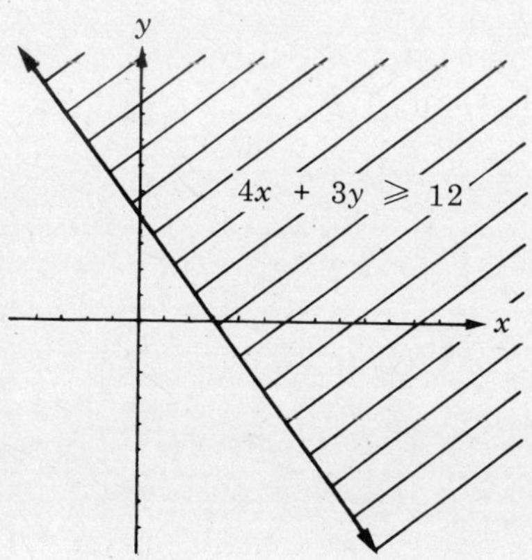

The intersection of all five inequalities appears below:

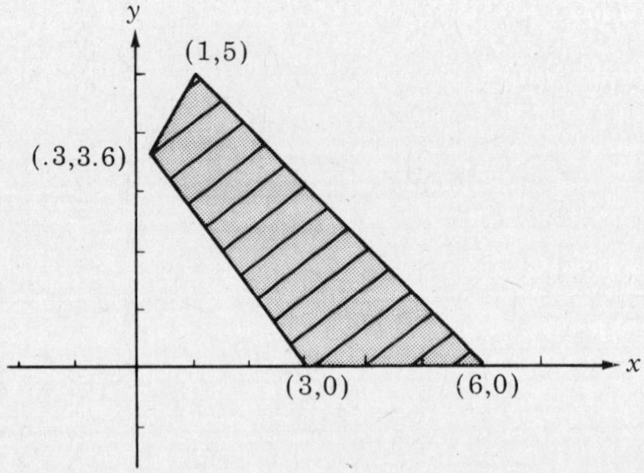

$$\begin{cases} y \leqslant 2x + 3 \\ x + y \leqslant 6 \\ 4x + 3y \geqslant 12 \\ x \geqslant 0 \\ y \geqslant 0 \end{cases}$$

The point (.3,3.6) is the solution to $\begin{cases} y = 2x + 3 \\ 4x + 3y = 12 \end{cases}$

The point (3,0) is the solution to $\begin{cases} y = 0 \\ 4x + 3y = 12 \end{cases}$

The point (6,0) is the solution to $\begin{cases} y = 0 \\ x + y = 6 \end{cases}$

The figure resulting from the previous example is called a **convex bounded set**. Intuitively, a convex set of points has the property that a line segment connecting any two points on its perimeter falls within the set.

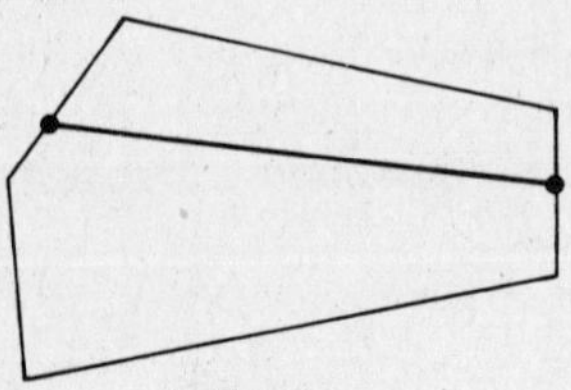

Convex
(Line segment falls within set.)

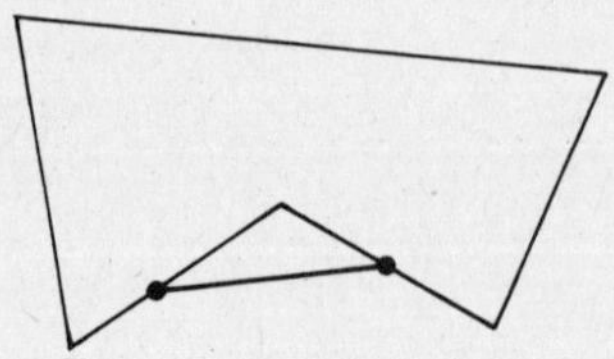

Not Convex.
(Line segment is outside set.)

If the area is finite, the convex set is said to be **bounded**. If the area is infinite, the convex set is **unbounded**.

Systems of linear inequalities and their corresponding convex sets are an important basis in the study of a branch of mathematics called **linear programming**. Linear programming, although having its roots in solving complex military strategies, has been a very useful tool in solving problems related to business and the social sciences.

Basically, a linear programming problem involves the *objective* of either maximizing something (profit, for instance) or minimizing something (like cost), subject to certain restrictions or constraints. For example, suppose the monthly profit P for the Lealo Company is given by

$$P = 20x + 30y$$

where x denotes the number of small desks sold and y denotes the number of executive desks sold per month. The objective is to maximize P; P is called the **objective function.** Suppose further that the company knows, because of consumers' demands, that they should produce at least twice as many small desks as executive desks and, because of the factory's physical limitations they can produce no more than a total of 45 desks. Also, there must be at least 5 small desks produced (because of backorders). This problem is summarized below:

$$\text{Maximize } P = 20x + 30y \text{ subject to:}$$

$$\left.\begin{array}{rl} x \geqslant 2y & \text{(c1)} \\ x + y \leqslant 45 & \text{(c2)} \\ x \geqslant 5 & \text{(c3)} \\ y \geqslant 0 & \text{(c4)} \end{array}\right\} \text{constraints}$$

Notice that constraint (c4) is not stated in the original problem but exists for obvious reasons.

We focus our attention on the constraints—each must be satisfied in order to find the appropriate x and y that maximize P. We graph the constraints as a system of linear inequalities on the following page; notice that a bounded convex set of points is formed. The **vertices** (or corners) of the convex set are called **feasible solutions** to the problem. One of these feasible solutions, called the **optimal feasible solution** is the one at which P is maximum.*

Now, we find the value of P at each point:

Point	$20x + 30y = P$
(5,0)	$20 \cdot 5 + 30 \cdot 0 = 100$
(5,2.5)	$20 \cdot 5 + 30 \cdot 2.5 = 175$
(30,15)	$20 \cdot 30 + 30 \cdot 15 = 1050$
(45,0)	$20 \cdot 45 + 30 \cdot 0 = 900$

The maximum profit, \$1050, occurs at point (30,15).

*This statement is one of the fundamental theorems in the theory of linear programming; its proof is beyond the scope of this text. Our goal in this section is to present the reader with an introduction to the topic.

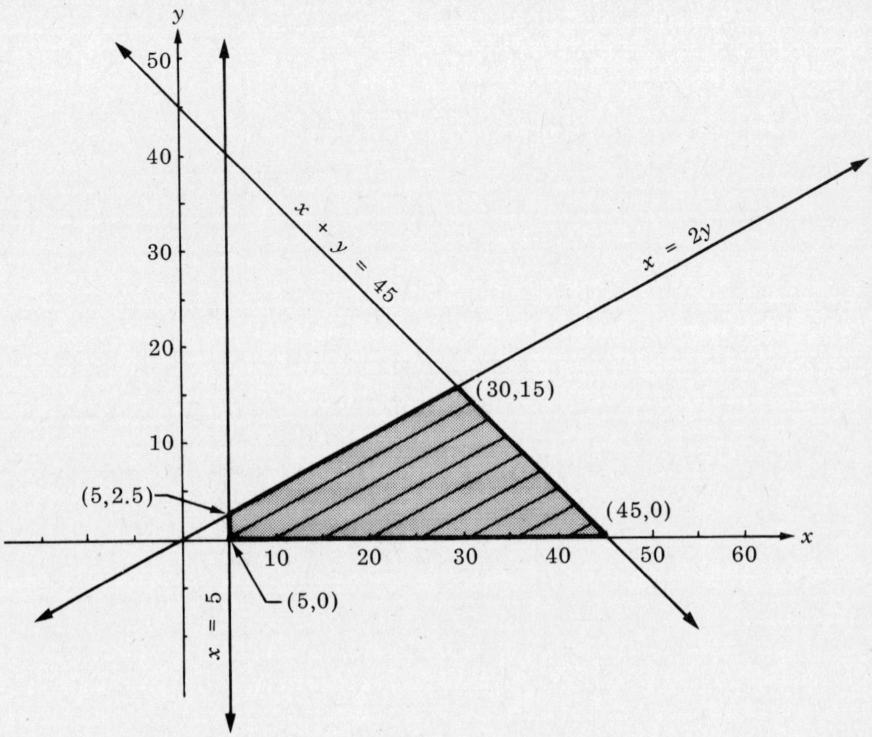

Thus, when 30 small desks and 15 executive desks are produced, the Lealo Company maximizes its profit subject to certain given conditions (the constraints). Some examples follow.

EXAMPLE 3 Maximize $P = 7x + 10y$ subject to:

$$\begin{aligned} x + 2y &\geqslant 10 \\ 7x - 8y &\geqslant -40 \\ x + y &\leqslant 20 \\ 4x - y &\leqslant 40 \end{aligned}$$

Solution The convex set is depicted on the following page. (The reader is asked to verify this convex set, including vertices, in Exercise 24.) Now, we examine $P = 7x + 10y$ at each vertex:

Point	$7x + 10y = P$
(0, 5)	$7 \cdot 0 + 10 \cdot 5 = 50$
(8, 12)	$7 \cdot 8 + 10 \cdot 12 = 176$
(12, 8)	$7 \cdot 12 + 10 \cdot 8 = 164$
(10, 0)	$7 \cdot 10 + 10 \cdot 0 = 70$

Hence, P achieves its maximum value when $x = 8$ and $y = 12$. That value is 176.

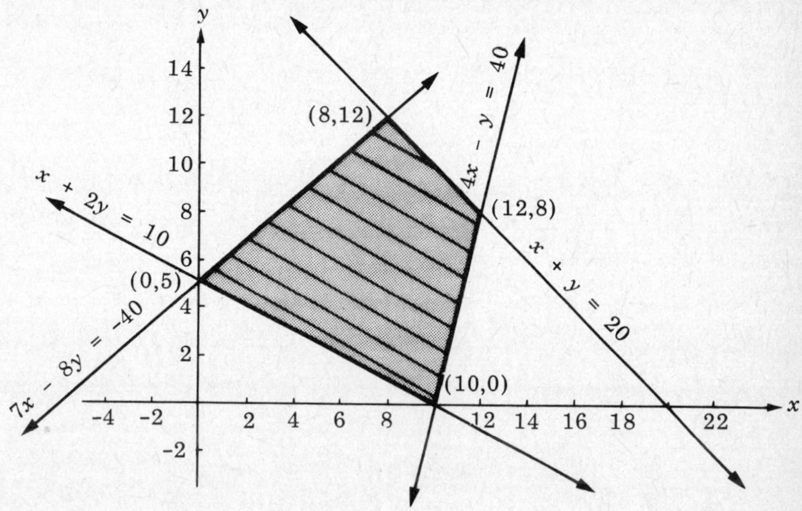

EXAMPLE 4 Minimize $C = 2x - 2y$ subject to:

$$\begin{aligned} -5x + 6y &\leqslant 80 \\ 7x - 5y &\leqslant 80 \\ 5x + 8y &\geqslant 115 \\ 10x - y &\leqslant 60 \\ x &\leqslant 20 \end{aligned}$$

Solution The convex set is pictured below.

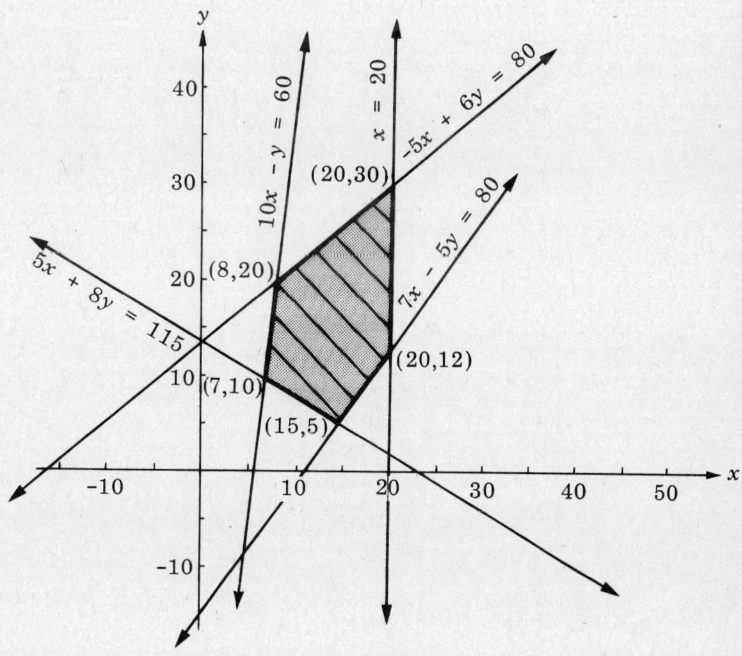

We search for the optimal feasible solution:

Point	$2x - 2y = C$
(7,10)	$2 \cdot 7 - 2 \cdot 10 = -6$
(8,20)	$2 \cdot 8 - 2 \cdot 20 = -24$
(20,30)	$2 \cdot 20 - 2 \cdot 30 = -20$
(20,12)	$2 \cdot 20 - 2 \cdot 12 = 16$
(15,5)	$2 \cdot 15 - 2 \cdot 5 = 20$

The minimum value of C is -24 when $x = 8$ and $y = 20$.

EXERCISES FOR SECTION 8.4

In Exercises 1 through 23, graph each system of linear inequalities. Be sure to label all points of intersection. (See Examples 1 and 2.)

1. $y \leqslant x$, $y \leqslant -\frac{1}{2}x + 4$
2. $y \leqslant -x + 5$, $2x - y \geqslant 4$
3. $x - y \leqslant 2$, $3x - 3y \geqslant 0$
4. $y \leqslant 2x - 1$, $2x + y \leqslant 9$
5. $y \geqslant 2x - 1$, $2x + y \geqslant 9$
6. $y \geqslant 2x - 1$, $2x + y \leqslant 9$
7. $y \leqslant 2x - 1$, $2x + y \geqslant 9$
8. $y \leqslant 2x - 3$, $y \geqslant 3x - 6$
9. $x + y \leqslant 8$, $2x - y \leqslant 2$
10. $y \leqslant -x + 3$, $y \leqslant 2x - 3$
11. $y \leqslant x$, $y \leqslant -\frac{1}{2}x + 4$, $y \geqslant 0$
12. $y \leqslant x$, $y \leqslant -\frac{1}{2}x + 4$, $y \geqslant 0$, $x \geqslant 1$
13. $x - 2y \leqslant 3$, $x + y \leqslant 6$
14. $x - 2y \leqslant 3$, $x + y \leqslant 6$, $x \geqslant 0$
15. $x - 2y \leqslant 3$, $x + y \leqslant 6$, $x \geqslant 0$, $y \geqslant 0$
16. $y \leqslant x + 5$, $7x + 2y \leqslant 28$
17. $y \leqslant x + 5$, $7x + 2y \leqslant 28$, $x \geqslant 0$
18. $y \leqslant x + 5$, $7x + 2y \leqslant 28$, $x \geqslant 0$, $x \leqslant 3$
19. $y \leqslant x + 5$, $7x + 2y \leqslant 28$, $x \geqslant 0$, $x \leqslant 3$, $y \geqslant 0$
20. $y \leqslant x + 5$, $7x + 2y \leqslant 28$, $x \geqslant 0$, $x \leqslant 3$, $y \geqslant 0$, $y \leqslant 6$
21. $x + y \geqslant 50$, $2x + 5y \geqslant 150$, $2x + y \geqslant 80$, $x + y \leqslant 100$, $x \geqslant 0$, $y \geqslant 0$
22. $x \leqslant y$, $x + y \leqslant 5$, $x \geqslant 0$, $x \leqslant 2$
23. $y \leqslant x$, $x + y \leqslant 5$, $y \geqslant 0$, $x \leqslant 4$
24. Verify the convex set (and vertices) obtained in Example 3.

In Exercises 25 through 33, solve each linear programming problem. (See Examples 3 and 4.)

25. Maximize $F = 10x + 10y$ subject to:

$$\begin{aligned} 3x + 2y &\leqslant 22 \\ y &\leqslant 5 \\ x &\leqslant 6 \\ x &\geqslant 0 \\ y &\geqslant 0 \end{aligned}$$

26. Minimize $F = 10x + 10y$ subject to the constraints of Exercise 25.
27. Minimize $G = -6x - 2y$ subject to the constraints of Exercise 25.
28. Using the constraints of Exercise 25 maximize $10x + y$.
29. Using the constraints of Exercise 25, maximize $10x - y$.
30. Using the constraints of Exercise 25, maximize $-x + 2y$.
31. Maximize $x + y$ subject to:

$$\begin{aligned} -x + 6y &\leqslant 30 \\ 3x + y &\leqslant 24 \\ x &\geqslant 0 \\ y &\geqslant 0 \end{aligned}$$

32. Minimize $x - y$ subject to the constraints of Exercise 31.
33. Maximize $5x + y$ subject to the constraints of Exercise 31.

In Exercises 34 through 43, solve each applied linear programming problem. (See the illustration preceding Example 3.)

34. The Luxuro Company operates two factories, each producing three accessories—cigarette lighters, seat harnesses, and door handles. The following table depicts their daily output:

	Cigarette lighters	*Seat harnesses*	*Door handles*
Factory 1	100	200	400
Factory 2	100	500	200

It is known that the daily cost of operating Factory 1 is \$10,000; the daily cost of operating Factory 2 is \$20,000. To meet demands, Luxuro must produce at least 5,000 cigarette lighters, 15,000 seat harnesses, and 16,000 door handles next quarter. How many days should each factory be in production in order to minimize cost if, in addition (because of union rules), the sum of the days both factories can be open per quarter cannot exceed 100?

35. a. The Embryo Toy Company manufactures two toys—wooden block sets and wooden trains. Due to limited facilities, the company cannot produce more than 40 block sets per week nor more than 60 trains per week, and the total number of toys produced cannot exceed 80. Represent the constraints of this problem as inequalities. Let x = number of block sets produced and y = number of trains produced each week.

    b. Graph the constraints of part a to obtain the convex set and feasible solutions.

    c. The Embryo Toy Company has determined that the profit for each block set is \$2.00, the profit for each train set is \$3.00. How many of each toy should the company produce to maximize profit?

36. a. Art is having a difficult time maintaining his car. His crankcase must contain at least 2 quarts of oil, but no more than 5. His 20-quart capacity radiator must contain at least 10 quarts of antifreeze. The sum of 5 times the number of quarts of oil and twice the number of quarts of antifreeze cannot exceed 55 because of storage problems. Represent the constraints of this problem as inequalities. Let x = number of quarts of oil and y = number of quarts of antifreeze.
b. Graph the constraints of part a to obtain the convex set and feasible solutions.
c. If, in addition, each quart of oil costs \$.75 and each quart of antifreeze costs \$.25, minimize Art's cost.

37. The Octavius Music Company manufactures pianos and organs. Because of customers' demands, they do not make more than 30 pianos or more than 15 organs in any given month, but must produce at least 30 instruments each month. Into the production of each piano and organ goes raw materials and labor according to the table below.

	Units of raw materials	*Units of labor*
Each piano	1	1
Each organ	1	2

Due to costs, there are at most 35 units of raw materials and 40 units of labor available. How many units of each instrument must be produced to maximize profit, if the profit for each piano is \$200 and for each organ \$100? What is that profit?

38. A moonlighting carpenter, Joseph, produces two types of furniture—bookcases and end tables—in his boss's shop after hours. He can produce no more than a total of nine items. In addition, he uses two tools—a lathe and a table saw—with units as described below:

	Hours on lathe	*Hours on saw*
Each book case	3	1
Each end table	5	3

The boss limits Joseph's hours on the lathe to 29 per week and limits the hours Joseph can spend on the saw to 15 per week. Joseph can sell each bookcase for a \$20 profit and each end table for a \$40 profit. How many items of each must he make to maximize his profit? What is that profit?

39. Use the constraints of Exercise 38 to minimize Joseph's cost when Joseph figures each bookcase costs \$5 in materials and each end table costs \$6. What is that cost?

40. All linear programming problems discussed thus far have employed the testing of the vertices of a bounded convex set. It could happen that the intersection of constraints yields an unbounded convex set of points. For example, consider the following two problems and show that each convex set is unbounded:

a. Maximize $P = x + y$ subject to:
$$\begin{aligned} y - x &\leqslant 6 \\ y &\leqslant 8 \\ y &\geqslant 0 \\ x &\geqslant 0 \end{aligned}$$

b. Minimize $C = x - y$ subject to:
$$\begin{aligned} y - x &\leqslant 6 \\ y &\leqslant 8 \\ y &\geqslant 0 \\ x &\geqslant 0 \end{aligned}$$

41. a. Examine the points on $y = 8$ in problem 40a, and note that the optimal solution is also unbounded.
    b. Examine the points between (0,6) and (2,8) in problem 40b, and note that the optimal solution need not be unique.
42. a. Maximize $P = x + y$ subject to:
$$\begin{aligned} y - x &\geqslant 6 \\ y &\leqslant 8 \\ y &\geqslant 0 \\ x &\geqslant 0 \end{aligned}$$
    b. Compare your results above to the results in Exercise 40a.
43. Robert is about to go on a self-induced "milk and sardines" diet. He has obtained the following information:

	Protein (in grams)	*Riboflavin (in mg)*	*Calories*
Sardines (per serving)	20	.18	175
Milk (per serving)	12	.54	210

A person in Robert's age-weight-activity classification should have at least 60 grams of protein daily and at least .90 mg of riboflavin. How many servings of sardines and how many servings of milk should Robert have daily to minimize his caloric intake?

CHAPTER 8 IN REVIEW

I. Summary

This chapter began with a discussion of "2-by-2" systems (two linear equations, each involving two variables). Such systems were solved in three ways: by graphing, the elimination method, and the substitution method. The latter two ways were also used to solve 3-by-3 systems (Section 8.2) and systems that involved nonlinear equations (Section 8.3). In each section, applied problems that involved solving such systems were also studied.

In Section 8.4, systems of linear inequalities involving two variables were solved and the branch of mathematics called linear programming was introduced.

II. Vocabulary/Symbol True–False Quiz

If the statement is true, write "TRUE." However, if the statement is false, replace the underlined word or phrase with a different word or phrase that will make the statement true.

1. $\left.\begin{aligned} 2x + 3y &= 7 \\ x - y &= 10 \end{aligned}\right\}$ is an example of a system of two linear equation in two variables.
2. The elimination method and substitution method are algebraic methods for solving linear systems.
3. There are infinitely many solutions to a dependent system.
4. There are no solutions to an inconsistent system.
5. $\left.\begin{aligned} x^2 + y^2 &= 25 \\ x + y &= 8 \end{aligned}\right\}$ is an example of a nonlinear system.
6. The graphical solution to a system of linear inequalities is a convex set.
7. The inequalities in a linear programming problem are called objectives.
8. The expression to be maximized (or minimized) in a linear programming problem is called a constraint.

III. Chapter Test

In questions 1 through 6, solve each system graphically:

1. $\begin{aligned} 4x - y &= 4 \\ y &= -4x - 2 \end{aligned}$
2. $\begin{aligned} 4x - y &= 4 \\ y &= 4x - 2 \end{aligned}$
3. $\begin{aligned} x^2 + y^2 &= 25 \\ 2x - y &= 0 \end{aligned}$
4. $\begin{aligned} -x + 2y &\leqslant 3 \\ 7x + 2y &\leqslant 13 \end{aligned}$
5. $\begin{aligned} -x + 2y &\leqslant 3 \\ 7x + 2y &\leqslant 13 \\ x &\geqslant 0 \\ y &\geqslant 0 \end{aligned}$
6. $\begin{aligned} -x + 2y &\leqslant 3 \\ 7x + 2y &\leqslant 13 \\ x + 3y &\geqslant -3 \end{aligned}$

In questions 7 through 12, solve each system algebraically.

7. $\begin{aligned} 2x + 3y &= 10 \\ 5x - 4y &= -25 \end{aligned}$
8. $\begin{aligned} 2x - y &= 7 \\ y &= 2x - 7 \end{aligned}$
9. $\begin{aligned} x + y - z &= 4 \\ -x + 2y + 11z &= 12 \\ 2x - 3y + 5z &= 5 \end{aligned}$
10. $\begin{aligned} x + z &= -2 \\ 2y - z &= 2 \\ x + y &= 0 \end{aligned}$
11. $\begin{aligned} x^2 + y^2 &= 25 \\ x + y &= 1 \end{aligned}$
12. $\begin{aligned} y &= \log_{10}(3x + 7) - 5 \\ y &= \log_{10}x - 4 \end{aligned}$
13. The sum of two numbers is 5; their difference is 6. What are the numbers?
14. A piggy bank contains 30 coins made up only of pennies, nickels, and dimes that total \$2.30. If there are the same number of pennies as nickels, how many of each coin are there?

15. a. Maximize $P = 2x + y$ subject to
$$\begin{aligned} -3x + 5y &\leqslant 5 \\ x + y &\leqslant 9 \\ x &\geqslant 0 \\ y &\geqslant 0 \end{aligned}$$
b. Minimize $C = x + \frac{1}{2}y$ subject to the constraints in part a.

IV. Student Research Project

We have only solved linear programming problems using a graphical method. A more elaborate method, called the **Simplex Method**, can be used for problems involving any number of variables.

1. Consult Chapter Six of *Finite Mathematics with Applications to Life*, 2nd ed., by L. Gilligan and R. Nenno, Goodyear Publishing Co., Santa Monica, California, 1979. Prepare a summary of the Simplex Method.
2. To appreciate how linear programming can be applied to many fields, prepare a bibliography of linear programming articles in a field that interests you (business, medicine, biology, etc.). Your librarian can probably suggest many helpful sources; some obvious ones include: The Card Catalog, Readers' Guide to Periodical Literature, Business Periodicals Index, The New York Times Index, Public Affairs Information Bulletin, Applied Science and Technology Index, International Encyclopedia of Social Science. Some examples of items you might find under the title "Linear Programming" are:

Ledley, Robert Steven. *Use of Computers in Biology and Medicine*, McGraw-Hill, New York, 1965, Chapter 12.

"Mathematical Methods of Organizing and Planning Production," L. V. Kantorvich, *Management Science*, v. 6, 1960, pp. 342-366.

"Applications of Linear Programming to Plywood Manufacture," D. B. Kotak, *Interfaces*, v. 7, 1976, p. 56-68.

"Linear Programming: A New Rate Tool," A. Schoor, *Electrical World*, v. 190, 1978, pp. 56-58.

"The Cost of Subsistence," George Stigler, *American Journal of Agricultural Economics*, v. 27, 1945, pp. 303-314.

Daetz, D. and Pantell, R., eds. *Environmental Modeling: Analysis and Management*, Dowden Hutchinson and Ross, Stroudsberg, Pennsylvania, 1974, pp. 224ff.

CHAPTER 9

Matrices and Systems of Linear Equations

For over two decades, Wassily W. Leontief, Henry Lee Professor of Economics at Harvard University, has been recognized as having made major advances in economic analysis and technique through his development of the input-output method. It was almost predictable then that his name would be added to the distinguished ranks of the Nobel laureats in economic science.

Using vector and matrix notation

$$(I - a)X = Y$$

where X and Y are $(n \times 1)$ column vectors, I is an $(n \times n)$ identity matrix, and a is the $(n \times n)$ constant production coefficients matrix

$$a = \begin{bmatrix} a_{11}, \ldots, a_{1n} \\ \vdots \qquad \vdots \\ a_{n1}, \ldots, a_{nn} \end{bmatrix} \tag{3}$$

With the use of the inverse matrix $A \equiv (I - a)^{-1}$, and given the conditions under which a meaningful economy normally operates, we obtain the solution

$$X = AY \tag{4}$$

Equations 4 thus project (forecast) the output of any sector, whether it be agriculture, textiles, chemicals, or housing, on the basis of the given final demand vector Y. If any component of this vector changes, say the final demand for steel, Y_j, increases, the consequences for the output of every economic sector, including steel, are given by Eqs. 4.

As you can see, one application of matrices is to economics. The excerpts above are from Science *(Washington, D.C.: American Association for the Advancement of Science) 9 November 1973, pp. 568–569.*

9.1 MATRIX SOLUTIONS TO SYSTEMS OF LINEAR EQUATIONS

In the previous chapter we examined three ways of solving linear systems (the graphical, elimination, and substitution methods). In this section we extend the elimination method to solve larger systems of equations.

Recall the three types of operations "allowed" in transforming the equations of a linear system into an equivalent system:

Transformations used in the Elimination Method

1. Any equation can be multiplied by any nonzero number.
2. Any multiple of one equation can be added to (or subtracted from) another equation.
3. Any two equations can be interchanged.

Consider the following system

$$\begin{aligned} 6x + 10y &= 60 \\ 10x + 5y &= 65 \end{aligned} \qquad (1)$$

We solve the system with the elimination method, keeping in mind that we want to transform system (1) into

$$\begin{aligned} 1x + 0y &= \square \\ 0x + 1y &= \triangle \end{aligned}$$

or simply $x = \square$ and $y = \triangle$. Consider the following sequence:

$$\boxed{\begin{aligned} 6x + 10y &= 60 \\ 10x + 5y &= 65 \end{aligned}}$$ Multiply top equation by $\frac{1}{6}$. → $$\boxed{\begin{aligned} 1x + \tfrac{5}{3}y &= 10 \\ 10x + 5y &= 65 \end{aligned}}$$ Multiply top equation by −10 and add to bottom equation →

$$\boxed{\begin{aligned} 1x + \tfrac{5}{3}y &= 10 \\ 0x - \tfrac{35}{3}y &= -35 \end{aligned}}$$ Multiply bottom equation by $-\frac{3}{35}$. → $$\boxed{\begin{aligned} 1x + \tfrac{5}{3}y &= 10 \\ 0x + 1y &= 3 \end{aligned}}$$ Multiply bottom equation by $-\frac{5}{3}$ and add to top equation. →

$$\boxed{\begin{aligned} 1x + 0y &= 5 \\ 0x + 1y &= 3 \end{aligned}}$$ (2) Thus $x = 5$ and $y = 3$.

Before proceeding to examples, it should be observed that, as the original system (1) was transformed into system (2), the variables x and y served merely as "place-holders." Actually, we could have performed the same sequence of manipulations USING ONLY THE COEFFICIENTS! That is, the system

$$\begin{aligned} 6x + 10y &= 60 \\ 10x + 5y &= 65 \end{aligned}$$

can be represented by the rectangular array

$$\left[\begin{array}{cc|c} 6 & 10 & 60 \\ 10 & 5 & 65 \end{array}\right]$$

This rectangular array of numbers is called the **augmented matrix** for the system. For completeness, we mention here that *any* rectangular array of numbers written between brackets is called a **matrix** and that the numbers in a matrix are called **entries, elements,** or **components**. More on the properties of matrices appears in the next section.

The three transformations in the elimination method are now restated in terms of operations on the **rows** of the augmented matrix for the system:

Elementary Row Transformations

1. Any row can be multiplied by any nonzero number.
2. Any multiple of one row can be added to (or subtracted from) another row.
3. Any two rows can be interchanged.

The next example illustrates how the augmented matrix can be transformed into an *equivalent* matrix by using the elementary row transformations.

EXAMPLE 1 Solve the system

$$\begin{aligned} x - y + 6z &= 8 \\ 3x - y - 2z &= 6 \\ 2x - 3y + 4z &= 16 \end{aligned}$$

Solution The augmented matrix is

$$\left[\begin{array}{ccc|c} 1 & -1 & 6 & 8 \\ 3 & -1 & -2 & 6 \\ 2 & -3 & 4 & 16 \end{array}\right]$$

We show the step-by-step transformation of the matrix below, keeping in mind that our goal is to end up with a matrix resembling

$$\left[\begin{array}{ccc|c} 1 & 0 & 0 & \square \\ 0 & 1 & 0 & \triangle \\ 0 & 0 & 1 & \star \end{array}\right]$$

equivalently, $x = \square$, $y = \triangle$, and $z = \star$.

$$\left[\begin{array}{ccc|c} 1 & -1 & 6 & 8 \\ 3 & -1 & -2 & 6 \\ 2 & -3 & 4 & 16 \end{array}\right]$$

Add −3 times top row to middle row. Add −2 times top row to bottom row. →

$$\left[\begin{array}{ccc|c} 1 & -1 & 6 & 8 \\ 0 & 2 & -20 & -18 \\ 0 & -1 & -8 & 0 \end{array}\right]$$

Divide middle row by 2. →

$$\left[\begin{array}{ccc|c} 1 & -1 & 6 & 8 \\ 0 & 1 & -10 & -9 \\ 0 & -1 & -8 & 0 \end{array}\right]$$

Add middle row to top row. Add middle row to bottom row. →

$$\left[\begin{array}{ccc|c} 1 & 0 & -4 & -1 \\ 0 & 1 & -10 & -9 \\ 0 & 0 & -18 & -9 \end{array}\right]$$

Divide bottom row by −18. →

$$\left[\begin{array}{ccc|c} 1 & 0 & -4 & -1 \\ 0 & 1 & -10 & -9 \\ 0 & 0 & 1 & \frac{1}{2} \end{array}\right]$$

Add 4 times bottom row to top row. Add 10 times bottom row to middle row. →

$$\left[\begin{array}{ccc|c} 1 & 0 & 0 & 1 \\ 0 & 1 & 0 & -4 \\ 0 & 0 & 1 & \frac{1}{2} \end{array}\right]$$

Thus, $x = 1$, $y = -4$, and $z = \frac{1}{2}$. The checks are left to the reader.

The final matrix in the previous example was arrived at after a step-by-step procedure was followed. The matrix is a **reduced matrix** and satisfies the following definition.

DEFINITION

Reduced Matrix

A matrix is called a reduced matrix if all three of the following conditions are satisfied:

1. The first nonzero entry in each row is 1 and every other entry in that column is 0.
2. If any row contains all zeros, it is below a row that contains a nonzero entry.
3. The first nonzero entry in each row is to the right of the first nonzero entry of the preceding row.

Our goal then is to transform the augmented matrix into another matrix with the "strategy" of obtaining a $\begin{matrix} 1 & 0 & \\ 0 & 1 & \dots \\ \vdots & & \ddots \end{matrix}$ configuration to the left of the vertical line, *if possible*. One advantage of this procedure is that it is an *iterative procedure* (it can be easily adapted for computer solution) and it can be used to solve *larger* systems.

We have the following definition:

DEFINITION **The generalized system of n linear equations in n variables is given by**

$$\begin{aligned} a_{11}x_1 + a_{12}x_2 + \dots + a_{1n}x_n &= b_1 \\ a_{21}x_1 + a_{22}x_2 + \dots + a_{2n}x_n &= b_2 \\ &\vdots \\ a_{n1}x_1 + a_{n2}x_2 + \dots + a_{nn}x_n &= b_n \end{aligned}$$

where a_{ij} represents a real number coefficient of the jth variable in the ith equation and each b_i is a real number.

EXAMPLE 2 Solve the 4-by-4 system:

$$\begin{aligned} x_1 + x_2 + x_3 + x_4 &= 3 \\ x_1 - x_2 &= -3 \\ 3x_1 + x_3 + 2x_4 &= 2 \\ x_2 + x_3 + 6x_4 &= 31 \end{aligned}$$

Solution The augmented matrix is

$$\left[\begin{array}{cccc|c} 1 & 1 & 1 & 1 & 3 \\ 1 & -1 & 0 & 0 & -3 \\ 3 & 0 & 1 & 2 & 2 \\ 0 & 1 & 1 & 6 & 31 \end{array}\right]$$

and the reader should verify using elementary row transformations that this matrix can be transformed into the reduced matrix.

$$\left[\begin{array}{cccc|c} 1 & 0 & 0 & 0 & -3 \\ 0 & 1 & 0 & 0 & 0 \\ 0 & 0 & 1 & 0 & 1 \\ 0 & 0 & 0 & 1 & 5 \end{array}\right]$$

Hence, $x_1 = -3, x_2 = 0, x_3 = 1$, and $x_4 = 5$ is the solution to the given system. The checks are left to the reader.

We already know that there may not always be a *unique* solution to a system—there may be infinitely many solutions (a dependent system) or no solution (an inconsistent system). These cases are examined in the next two examples.

EXAMPLE 3 Solve the system

$$\begin{aligned} x_1 + x_2 + x_3 &= 4 \\ 2x_1 - x_2 + x_3 &= 0 \\ 4x_1 + x_2 + 3x_3 &= 9 \end{aligned}$$

Solution The reader should verify the following sequence of row transformations:

$$\left[\begin{array}{rrr|r} 1 & 1 & 1 & 4 \\ 2 & -1 & 1 & 0 \\ 4 & 1 & 3 & 9 \end{array}\right] \rightarrow \left[\begin{array}{rrr|r} 1 & 1 & 1 & 4 \\ 0 & -3 & -1 & -8 \\ 0 & -3 & -1 & -7 \end{array}\right] \rightarrow \underbrace{\left[\begin{array}{rrr|r} 1 & 0 & \frac{2}{3} & \frac{4}{3} \\ 0 & 1 & \frac{1}{3} & \frac{8}{3} \\ 0 & 0 & 0 & 1 \end{array}\right]}_{\text{reduced matrix}}$$

The bottom row of the matrix on the right is interpreted as

$$0x_1 + 0x_2 + 0x_3 = 1$$

or

$$0 = 1$$

which is NEVER true. The system NEVER has a solution; it is inconsistent.

EXAMPLE 4 Solve the system

$$\begin{aligned} x_1 + x_2 + x_3 &= 4 \\ 2x_1 - x_2 + x_3 &= 0 \\ 4x_1 + x_2 + 3x_3 &= 8 \end{aligned}$$

Solution Again, the reader should verify the sequence of row transformations:

$$\left[\begin{array}{rrr|r} 1 & 1 & 1 & 4 \\ 2 & -1 & 1 & 0 \\ 4 & 1 & 3 & 8 \end{array}\right] \rightarrow \left[\begin{array}{rrr|r} 1 & 1 & 1 & 4 \\ 0 & -3 & -1 & -8 \\ 0 & -3 & -1 & -8 \end{array}\right] \rightarrow \underbrace{\left[\begin{array}{rrr|r} 1 & 0 & \frac{2}{3} & \frac{4}{3} \\ 0 & 1 & \frac{1}{3} & \frac{8}{3} \\ 0 & 0 & 0 & 0 \end{array}\right]}_{\text{reduced matrix}}$$

The matrix on the right is interpreted as

$$\begin{aligned} x_1 + \tfrac{2}{3}x_3 &= \tfrac{4}{3} & & & x_1 &= \tfrac{4}{3} - \tfrac{2}{3}x_3 & (3) \\ x_2 + \tfrac{1}{3}x_3 &= \tfrac{8}{3} & &\text{or equivalently} & x_2 &= \tfrac{8}{3} - \tfrac{1}{3}x_3 & (4) \\ 0 &= 0 & & & x_3 &= x_3 & (5) \end{aligned}$$

We can see that from equation (5), x_3 can be *any* real number and from equations (3) and (4) x_1 and x_2 *depend* on the value of x_3. Thus, there are infinitely many* solutions; the system is dependent.

Later in this chapter and occasionally in applications, it is necessary to solve different systems having identical coefficients. Such is the case in the next example.

EXAMPLE 5 Solve each system.

$$\begin{aligned} x_1 - x_2 + x_3 &= 7 \\ 2x_1 + 3x_2 - x_3 &= 4 \\ 3x_1 + 6x_2 - 4x_3 &= 3 \end{aligned} \qquad \begin{aligned} x_1 - x_2 + x_3 &= 3 \\ 2x_1 + 3x_2 - x_3 &= 13 \\ 3x_1 + 6x_2 - 4x_3 &= 20 \end{aligned}$$

Solution Since the coefficients are identical, we combine the systems by only writing the coefficients once and then reducing. The augmented matrix is:

$$\left[\begin{array}{rrr|rr} 1 & -1 & 1 & 7 & 3 \\ 2 & 3 & -1 & 4 & 13 \\ 3 & 6 & -4 & 3 & 20 \end{array}\right] \quad \text{which can be transformed into the reduced matrix} \quad \left[\begin{array}{rrr|rr} 1 & 0 & 0 & 5 & 4 \\ 0 & 1 & 0 & -2 & 2 \\ 0 & 0 & 1 & 0 & 1 \end{array}\right]$$

Thus, the solution to the first system is $x_1 = 5, x_2 = -2, x_3 = 0$. And, the solution to the second system is $x_1 = 4, x_2 = 2, x_3 = 1$.

The method described in this section is often called the **Gaussian elimination method** (after the nineteenth century mathematician Carl F. Gauss) and is summarized below.

Summary of Gaussian Elimination for n Linear Equations in n Variables

Step 1. If necessary, interchange two rows so that the component in the first row first column, a_{11}, is nonzero.
Step 2. Divide each component in row one by a_{11}, thereby changing the new a_{11} component to 1.
Step 3. Convert all other components in the first column to zero by adding an appropriate multiple of the first row to all other rows.

*Be careful. Not just any triplet of numbers will solve this system—the triplet must satisfy (3), (4), and (5). One such solution is (0, 2, 2); another is $(\frac{4}{3}, \frac{8}{3}, 0)$. But (1, 2, 3) is not.

Step 4. Repeat steps 1 through 3 for each diagonal component. That is, convert a_{kk} to 1 and all other components in column k to 0 for $k = 1, 2, \ldots, n$ if possible.

Step 5. Interpreting the reduced augmented matrix:

a. Whenever a row of zeros exists to the left of the vertical line and a nonzero number appears to the right, the system is inconsistent.

b. Whenever a complete row of zeros exists (that is, zeros on both sides of the vertical line), the n-by-n system is dependent.

c. Otherwise, the unique solution for the system is found in the column to the right of the vertical line.

EXERCISES FOR SECTION 9.1

1. Use the Gaussian elimination method to solve the system

$$\begin{aligned} 6x + 10y &= 60 \\ 10x + 5y &= 65 \end{aligned}$$

(Compare this method with the solution that appears preceding Example 1.)

In Exercises 2 through 25, solve each system, if possible. (See Examples 1, 2, 3, and 4.)

2. $\begin{aligned} 3x - y &= 2 \\ 11x - 4y &= 6 \end{aligned}$

3. $\begin{aligned} 5x + 2y &= 19 \\ 2x + y &= 8 \end{aligned}$

4. $\begin{aligned} 3x_1 + 4x_2 &= 5 \\ 5x_1 - 2x_2 &= 17 \end{aligned}$

5. $\begin{aligned} x_1 + x_2 + x_3 &= 7 \\ 2x_1 - x_2 + x_3 &= 4 \\ 3x_1 - 5x_2 + 2x_3 &= 1 \end{aligned}$

6. $\begin{aligned} 2x + 3y + 5z &= 0 \\ 4x - 6y + z &= 0 \\ 2x - 3y &= 0 \end{aligned}$

7. $\begin{aligned} 2x + 3y + 5z &= 2 \\ 4x - 6y + z &= 0 \\ 2x - 3y &= 0 \end{aligned}$

8. $\begin{aligned} x_1 + x_2 + x_3 &= 0 \\ -x_1 + 3x_2 + x_3 &= 2 \\ -3x_1 + x_2 - 5x_3 &= 14 \end{aligned}$

9. $\begin{aligned} x + y + 2z &= 0 \\ 3x - 4z &= -2 \\ 5y - 6z &= -8 \end{aligned}$

10. $\begin{aligned} x_1 + 3x_2 - x_3 &= 11 \\ 3x_1 + x_2 - x_3 &= 13 \\ x_1 + x_2 - 3x_3 &= 11 \end{aligned}$

11. $\begin{aligned} x_1 + x_2 + x_3 &= 6 \\ x_1 - 2x_2 + 3x_3 &= 6 \\ 2x_1 - x_2 + 4x_3 &= 10 \end{aligned}$

12. $\begin{aligned} x_1 + x_3 &= 7 \\ x_1 + x_2 &= 4 \\ 2x_1 + x_2 + x_3 &= 10 \end{aligned}$

13. $\begin{aligned} 2x_1 + x_2 - 4x_3 &= 6 \\ x_1 - x_2 + 4x_3 &= 3 \\ x_1 &= 3 \end{aligned}$

14. $\begin{aligned} x_2 + x_3 &= 1 \\ 2x_1 - x_2 - 2x_3 &= 1 \\ 4x_1 - 2x_2 - x_3 &= -7 \end{aligned}$

15. $\begin{aligned} x + y + z &= 8 \\ x - y - 2z &= 10 \\ x + 3y + 4z &= 6 \end{aligned}$

16. $$\begin{aligned} 3x_1 + x_2 &= 10 \\ x_1 - x_3 &= 2 \\ 5x_2 - x_3 &= 4 \end{aligned}$$

17. $$\begin{aligned} x_1 - x_2 + 5x_3 &= 11 \\ 2x_1 + x_2 &= 0 \\ 2x_1 - 5x_2 - 10x_3 &= 32 \end{aligned}$$

18. $$\begin{aligned} 2x_1 - 3x_2 + 7x_3 &= 0 \\ x_1 &= x_2 \\ 4x_1 + x_2 - 5x_3 &= 0 \end{aligned}$$

19. $$\begin{aligned} x_1 + x_2 + x_3 + x_4 &= 10 \\ x_1 + x_2 - x_3 + 2x_4 &= 8 \\ 2x_1 - x_2 + 4x_3 - 3x_4 &= 0 \\ x_3 + 5x_4 &= 23 \end{aligned}$$

20. $$\begin{aligned} x_1 + x_2 + x_3 &= 5 \\ 2x_1 + 3x_3 - x_4 &= 5 \\ -x_1 + x_2 - 5x_4 &= 26 \\ x_2 + x_3 + x_4 &= -3 \end{aligned}$$

21. $$\begin{aligned} x_1 + x_2 + x_3 &= 5 \\ 2x_2 + 3x_3 - x_4 &= 5 \\ x_1 - x_2 - 5x_4 &= 26 \\ x_1 + x_3 + x_4 &= -3 \end{aligned}$$

22. $$\begin{aligned} x_2 + x_3 + x_4 &= 5 \\ -x_1 + 3x_3 + 2x_4 &= 5 \\ -5x_1 + x_2 - x_4 &= 26 \\ x_1 + x_2 + x_3 &= -3 \end{aligned}$$

23. $$\begin{aligned} 2x_1 - 2x_3 + x_4 &= 1 \\ 3x_2 + 4x_3 - 2x_4 &= -1 \\ -x_1 + x_2 + x_3 &= 0 \\ -x_1 + x_2 - 2x_3 + 3x_4 &= 3 \end{aligned}$$

24. $$\begin{aligned} x_1 + x_2 + x_3 + x_4 &= 7 \\ x_1 - x_2 - x_3 + x_4 &= 11 \\ x_1 + x_4 &= 5 \\ 3x_1 - x_2 - x_3 &= 6 \end{aligned}$$

25. $$\begin{aligned} x_1 + x_2 + x_3 &= 9 \\ 2x_1 - x_2 - x_3 &= -3 \\ x_1 - 5x_2 + 7x_3 &= 15 \\ x_4 &= 6 \end{aligned}$$

26. The general system of n equations in n variables is called **homogeneous** if each b_i is zero:

$$\begin{aligned} a_{11}x_1 + a_{12}x_2 + \ldots + a_{1n}x_n &= 0 \\ a_{21}x_1 + a_{22}x_2 + \ldots + a_{2n}x_n &= 0 \\ &\vdots \\ a_{n1}x_1 + a_{n2}x_2 + \ldots + a_{nn}x_n &= 0 \end{aligned}$$

Obviously, $x_i = 0$ for all i is *always* a solution to the homogeneous system —but there may be others! Solve each homogeneous system below:

a. $$\begin{aligned} x_1 + x_2 + x_3 &= 0 \\ x_1 - x_2 - x_3 &= 0 \\ 5x_1 - 6x_2 + 7x_3 &= 0 \end{aligned}$$

b. $$\begin{aligned} x_1 + x_2 + x_3 &= 0 \\ x_1 - x_2 + x_3 &= 0 \\ x_1 + x_3 &= 0 \end{aligned}$$

c. $$\begin{aligned} x_1 + 2x_2 + 3x_3 &= 0 \\ x_1 - x_2 - x_3 &= 0 \\ 2x_1 + x_2 + 2x_3 &= 0 \end{aligned}$$

27. a. Use the methods of this section to solve the *general* 2-by-2 system:

$$\begin{aligned} a_{11}x_1 + a_{12}x_2 &= b_1 \\ a_{21}x_1 + a_{22}x_2 &= b_2 \end{aligned}$$

b. What restrictions must there be on the coefficients in order for a solution to exist?

In Exercises 28 and 29, solve each system. (See Example 5.)

28. a. $$\begin{aligned} 2x_2 + x_3 &= 1 \\ 2x_1 + x_2 - x_3 &= 0 \\ -x_1 + x_3 &= 0 \end{aligned}$$

b. $$\begin{aligned} 2x_2 + x_3 &= 0 \\ 2x_1 + x_2 - x_3 &= 1 \\ -x_1 + x_3 &= 0 \end{aligned}$$

c. $$\begin{aligned} 2x_2 + x_3 &= 0 \\ 2x_1 + x_2 - x_3 &= 0 \\ -x_1 + x_3 &= 1 \end{aligned}$$

29. a.
$$\begin{aligned} x_1 + 3x_2 &= 1 \\ 2x_1 + 4x_3 &= 0 \\ x_1 - 3x_2 + 3x_3 &= 0 \end{aligned}$$
b.
$$\begin{aligned} x_1 + 3x_2 &= 0 \\ 2x_1 + 4x_3 &= 1 \\ x_1 - 3x_2 + 3x_3 &= 0 \end{aligned}$$
c.
$$\begin{aligned} x_1 + 3x_2 &= 0 \\ 2x_1 + 4x_3 &= 0 \\ x_1 - 3x_2 + 3x_3 &= 1 \end{aligned}$$

In Exercises 30 through 32, solve each applied problem.

30. Katie has \$35,000 that has been divided into three investments. Part of the money is invested in a savings account with an annual rate of 6% interest, part in 7% annual-yield bonds, and the remainder is invested in a business. In 1979, when she lost 6% of the money that she invested in the business, her net income from all three investments was \$660. If she invested \$3000 more in the business than in the savings account, how much was invested in each?

31. A toy company produces three main items—dolls, train sets, and tricycles. In the months of June, July, and August, 1980 they sold the following quantities:

	Dolls	*Train sets*	*Tricycles*	*Total monthly income*
June	10	6	40	\$2030
July	4	5	100	4657
August	15	20	200	9620

Determine the selling price for each of the three toys.

32. A certain recipe includes top round, sirloin, and chuck steak. There must be five times as much top round as sirloin and twice as much sirloin as chuck. In total, there must be 19.5 pounds of meat. How much of each type of meat is to be used?

Calculator Exercises

In Exercise 27a we found

$$x_1 = \frac{a_{22}b_1 - a_{12}b_2}{a_{11}a_{22} - a_{12}a_{21}} \quad \text{and} \quad x_2 = \frac{a_{11}b_2 - a_{21}b_1}{a_{11}a_{22} - a_{12}a_{21}}.$$

Use these formulas directly on a calculator to solve the systems in Exercises 33 through 40.

33. $\begin{aligned} 3x_1 + 4x_2 &= 5 \\ 5x_1 - 2x_2 &= 17 \end{aligned}$

34. $\begin{aligned} x_1 + 7x_2 &= 18 \\ 3x_1 - 10x_2 &= 7 \end{aligned}$

35. $\begin{aligned} 1.8x_1 - 3.7x_2 &= 9.1 \\ -.5x_1 + 11.4x_2 &= 10.4 \end{aligned}$

36. $\begin{aligned} 2x_1 - 1.5x_2 &= 1.5 \\ x_1 + x_2 &= 0 \end{aligned}$

37. $\begin{aligned} x_1 - 2.7x_2 &= 1 \\ 3.1x_1 - 1.2x_2 &= -4 \end{aligned}$

38. $\begin{aligned} 1.1x_1 + 2.2x_2 &= 3.3 \\ 4.4x_1 + 5.5x_2 &= 1.0 \end{aligned}$

39. $\begin{aligned} \sqrt{2}x_1 - \sqrt{3}x_2 &= 0 \\ 3x_1 + x_2 &= \sqrt{5} \end{aligned}$

40. $\begin{aligned} \sqrt{3}x_1 + \sqrt{3}x_2 &= 4 \\ \sqrt{5}x_1 - 1.2x_2 &= -2 \end{aligned}$

9.2 MATRIX ALGEBRA

In the previous section we saw that a matrix is a rectangular array of numbers written between brackets. In this section we will examine operations and uses of matrices. Consider the matrix

$$A = \begin{bmatrix} 1 & 4 & -6 & \pi \\ \sqrt{2} & 4 & 11 & 3.7 \\ 1 & 2 & 0 & -\frac{1}{2} \end{bmatrix}$$

Notice that A contains 12 entries (or components) arranged in three rows and four columns. We say A is a 3 x 4 (read "3 by 4") matrix and "3 x 4" are called the **dimensions** of A. Thus, the matrix

$$\begin{bmatrix} a_{11} & a_{12} & \cdots & a_{1n} \\ a_{21} & a_{22} & \cdots & a_{2n} \\ \vdots & & & \\ a_{m1} & a_{m2} & \cdots & a_{mn} \end{bmatrix}$$

is an $m \times n$ matrix. Notice that a_{21}, for example, is the component in the second row, first column. If B is the letter used to represent a particular matrix, b_{ij} is used to symbolize the general component of B, and $[b_{ij}]$ denotes the matrix.

We now wish to examine some *algebraic* properties of matrices. To do so, we shall need the following definitions.

DEFINITION

Matrix Equality

If A and B are two $m \times n$ matrices, then

$$A = B \text{ if and only if } a_{ij} = b_{ij} \text{ for every } i \text{ and } j.$$

DEFINITION

Matrix Addition

If A and B are two $m \times n$ matrices then the sum of A and B is defined by

$$A + B = [a_{ij} + b_{ij}]$$

(That is, matrices are added in a component-by-component fashion.)

DEFINITION

Matrix Multiplication by a Scalar*

If **A** is an $m \times n$ matrix and k is any real number then

$$k \cdot \mathbf{A} = [ka_{ij}]$$

(That is, *each* component of **A** is multiplied by k.)

DEFINITION

Matrix Subtraction

If **A** and **B** are two $m \times n$ matrices, then their subtraction is defined to be

$$\mathbf{A} - \mathbf{B} = \mathbf{A} + (-1 \cdot \mathbf{B})$$

EXAMPLE 1 Given $\mathbf{A} = \begin{bmatrix} -2 & 3 & 4 \\ 6 & 1 & 2 \end{bmatrix}$

$$\mathbf{B} = \begin{bmatrix} 5 & 1.6 & 3 \\ 4 & 0 & 0 \end{bmatrix} \quad \text{and} \quad \mathbf{C} = \begin{bmatrix} 1 & 3 \\ 2 & 4 \end{bmatrix}$$

Find a. **A + B** b. **A − B** c. **A + C** d. **2A + 3B**

Solution a. $\mathbf{A} + \mathbf{B} = \begin{bmatrix} -2+5 & 3+1.6 & 4+3 \\ 6+4 & 1+0 & 2+0 \end{bmatrix} = \begin{bmatrix} 3 & 4.6 & 7 \\ 10 & 1 & 2 \end{bmatrix}$

b. $\mathbf{A} - \mathbf{B} = \begin{bmatrix} -7 & 1.4 & 1 \\ 2 & 1 & 2 \end{bmatrix}$

c. **A** is a 2×3 matrix and **C** is a 2×2 matrix. Since they do not have the same dimensions, their addition is *undefined.*

d. $2\mathbf{A} = \begin{bmatrix} -4 & 6 & 8 \\ 12 & 2 & 4 \end{bmatrix} \qquad 3\mathbf{B} = \begin{bmatrix} 15 & 4.8 & 9 \\ 12 & 0 & 0 \end{bmatrix}$

Thus,

$$2\mathbf{A} + 3\mathbf{B} = \begin{bmatrix} 11 & 10.8 & 17 \\ 24 & 2 & 4 \end{bmatrix}$$

*In matrix theory the word *scalar* is used to represent a real number.

A useful theorem from matrix theory is now stated.

THEOREM 1

> Suppose **A** and **B** are $m \times n$ matrices. Then the following are true:
>
> a. $\mathbf{A} + \mathbf{B} = \mathbf{B} + \mathbf{A}$ (That is, matrix addition is *commutative.*)
> b. $\mathbf{A} + (\mathbf{B} + \mathbf{C}) = (\mathbf{A} + \mathbf{B}) + \mathbf{C}$ (That is, matrix addition is *associative.*)
> c. $k(\mathbf{A} + \mathbf{B}) = k\mathbf{A} + k\mathbf{B}$

Proof:

We prove part a here; the proofs of parts b and c are left to the reader (in Exercises 26 and 27).

Let $\mathbf{A} = [a_{ij}]$, $\mathbf{B} = [b_{ij}]$. Then

$$\mathbf{A} + \mathbf{B} = [a_{ij} + b_{ij}]$$

but since a_{ij} and b_{ij} are real numbers,

$$a_{ij} + b_{ij} = b_{ij} + a_{ij}.$$

Hence,

$$[a_{ij} + b_{ij}] = [b_{ij} + a_{ij}] = \mathbf{B} + \mathbf{A}.$$

Now we wish to examine a more complex matrix operation, multiplication. Consider

$$\mathbf{X} = \begin{bmatrix} 7 & 4 \\ 3 & 2 \\ 0 & 1 \end{bmatrix} \quad \text{and} \quad \mathbf{Y} = \begin{bmatrix} 5 & 9 & 20 & 40 \\ 6 & 10 & 30 & 50 \end{bmatrix}$$

To find the product of **X** and **Y**, denoted $\mathbf{X} \bullet \mathbf{Y}$, we *must* have the number of columns of **X** equal to the number of rows of **Y**. The product $\mathbf{X} \bullet \mathbf{Y}$ will be a 3×4 matrix having the same number of rows as **X** and the same number of columns as **Y**. Each component, c_{ij}, of $\mathbf{X} \bullet \mathbf{Y}$, is the sum of the component-by-component products of the *ith* row of **X** by the *jth* column of **Y**. That is, c_{24}, for example, is found as follows:

$$\begin{bmatrix} 7 & 4 \\ 3 & 2 \\ 0 & 1 \end{bmatrix} \bullet \begin{bmatrix} 5 & 9 & 20 & 40 \\ 6 & 10 & 30 & 50 \end{bmatrix} = \begin{matrix} \_\_ & \_\_ & \_\_ & \_\_ \\ \_\_ & \_\_ & \_\_ & 220 \\ \_\_ & \_\_ & \_\_ & \_\_ \end{matrix}$$

$$3 \cdot 40 + 2 \cdot 50 = 220$$

The process continues until all 12 multiplications have been calculated. The reader should verify the final matrix below:

$$\begin{bmatrix} 7 & 4 \\ 3 & 2 \\ 0 & 1 \end{bmatrix} \bullet \begin{bmatrix} 5 & 9 & 20 & 40 \\ 6 & 10 & 30 & 50 \end{bmatrix} = \begin{bmatrix} 59 & 103 & 260 & 480 \\ 27 & 47 & 120 & 220 \\ 6 & 10 & 30 & 50 \end{bmatrix}$$

We formally present the definition of matrix multiplication below:

DEFINITION

Matrix Multiplication

If $\mathbf{A} = [a_{ij}]$ is an $m \times p$ matrix and $\mathbf{B} = [b_{ij}]$ is a $p \times n$ matrix, then the matrix product $\mathbf{A} \bullet \mathbf{B} = [c_{ij}]$ is an $m \times n$ matrix where each c_{ij} is given by

$$c_{ij} = a_{i1} b_{1j} + a_{i2} b_{2j} + \ldots + a_{ip} b_{pj}$$

EXAMPLE 2 Let $\mathbf{A} = \begin{bmatrix} 0 & -1 & -2 \\ 5 & 6 & 2 \end{bmatrix}$ and $\mathbf{B} = \begin{bmatrix} 1 & 2 \\ -3 & -2 \\ -2 & 0 \end{bmatrix}$

Find: a. $\mathbf{A} \bullet \mathbf{B}$ b. $\mathbf{B} \bullet \mathbf{A}$

Solution a. Since **A** is a 2×3 matrix and **B** is a 3×2 matrix, $\mathbf{A} \bullet \mathbf{B}$ is a 2×2 matrix:

$$\mathbf{A} \bullet \mathbf{B} = \begin{bmatrix} 0 & -1 & -2 \\ 5 & 6 & 2 \end{bmatrix} \bullet \begin{bmatrix} 1 & 2 \\ -3 & -2 \\ -2 & 0 \end{bmatrix} = \begin{bmatrix} 7 & 2 \\ -17 & -2 \end{bmatrix}$$

b. $\mathbf{B} \bullet \mathbf{A}$ is a 3×3 matrix:

$$\mathbf{B} \bullet \mathbf{A} = \begin{bmatrix} 1 & 2 \\ -3 & -2 \\ -2 & 0 \end{bmatrix} \bullet \begin{bmatrix} 0 & -1 & -2 \\ 5 & 6 & 2 \end{bmatrix} = \begin{bmatrix} 10 & 11 & 2 \\ -10 & -9 & 2 \\ 0 & 2 & 4 \end{bmatrix}$$

The previous example illustrates that matrix multiplication is *not* commutative. That is, in general $\mathbf{A} \bullet \mathbf{B} \neq \mathbf{B} \bullet \mathbf{A}$.

The next example illustrates a use of matrix multiplication.

EXAMPLE 3 The Lightning Bolt Company produces three sizes of bolts (3 cm, 7 cm, and 12 cm) in each of its two factories. The Los Angeles factory produces 400, 500, 100, respectively, of each type and the San Francisco factory produces 100, 200, 700, respectively, of each type. In addition, it is known that each 3 cm bolt costs \$.12 to produce and \$.03 to ship; each 7 cm bolt costs \$.15 to produce and \$.03 to ship; each 12 cm bolt costs \$.20 to produce and \$.05 to ship. Let **A** be the 2 x 3 matrix representing factory output and let **B** represent the 3 x 2 matrix showing costs for each bolt. Find the costs for each factory, **A • B**.

Solution

$$\mathbf{A} = \begin{array}{l} \\ \text{L.A. Factory} \\ \text{S.F. Factory} \end{array} \begin{array}{c} \begin{array}{ccc} \text{3 cm} & \text{7 cm} & \text{12 cm} \end{array} \\ \begin{bmatrix} 400 & 500 & 100 \\ 100 & 200 & 700 \end{bmatrix} \end{array}$$

$$\mathbf{B} = \begin{array}{l} \\ \text{3 cm} \\ \text{7 cm} \\ \text{12 cm} \end{array} \begin{array}{c} \begin{array}{cc} \text{production costs per bolt} & \text{transportation costs per bolt} \end{array} \\ \begin{bmatrix} .12 & .03 \\ .15 & .03 \\ .20 & .05 \end{bmatrix} \end{array}$$

$$\mathbf{A} \bullet \mathbf{B} = \begin{array}{l} \\ \text{L.A.} \\ \text{S.F.} \end{array} \begin{array}{c} \begin{array}{cc} \text{total production costs} & \text{total transportation costs} \end{array} \\ \begin{bmatrix} 143 & 32 \\ 182 & 44 \end{bmatrix} \end{array}$$

The matrix **A • B** can be interpreted as follows:

total production costs for the Los Angeles factory is \$143,
total production costs for the San Francisco factory is \$182,
total transportation cost for the Los Angeles factory is \$32, and
total transportation cost for the San Francisco factory is \$44.

Notice, in the previous example, that the columns of **A** and the rows of **B** refer to the same thing—bolts. Thus, the product **A • B** has meaning in terms of the aspects of this problem; the product **B • A** does not.

EXERCISES FOR SECTION 9.2

In Exercises 1 through 12, perform the indicated matrix operations. (See Example 1.) Use:

$$\mathbf{A} = \begin{bmatrix} 1 & 5 & 10 \\ -1 & 0 & 2 \end{bmatrix} \qquad \mathbf{B} = \begin{bmatrix} 0 & 1 & 2 \\ 3 & 4 & 5 \end{bmatrix} \qquad \mathbf{C} = \begin{bmatrix} 0 & 0 \\ 0 & 1 \\ 0 & 2 \end{bmatrix}$$

1. $\mathbf{A} + \mathbf{B}$
2. $\mathbf{A} - \mathbf{B}$
3. $2\mathbf{C}$
4. $3\mathbf{A}$
5. $5\mathbf{B}$
6. $3\mathbf{A} + 5\mathbf{B}$
7. $\frac{1}{2}\mathbf{A} + \frac{1}{2}\mathbf{B}$
8. $\frac{1}{2}(\mathbf{A} + \mathbf{B})$
9. $\frac{1}{2}(4\mathbf{A} + 2\mathbf{B})$
10. $2\mathbf{A} + \mathbf{B}$
11. $4\mathbf{A} + \mathbf{C}$
12. $\mathbf{B} - \mathbf{C}$

In Exercises 13 through 25 find **A • B** and **B • A**, if possible. (See Example 2.)

13. $\mathbf{A} = \begin{bmatrix} 1 & 2 \\ -3 & 5 \end{bmatrix}$ $\mathbf{B} = \begin{bmatrix} 0 & 2 \\ 1 & -2 \end{bmatrix}$

14. $\mathbf{A} = \begin{bmatrix} 1 & -3 \\ 2 & 5 \end{bmatrix}$ $\mathbf{B} = \begin{bmatrix} 0 & 1 \\ 2 & -2 \end{bmatrix}$

15. $\mathbf{A} = \begin{bmatrix} 0 & 0 \\ 0 & 5 \end{bmatrix}$ $\mathbf{B} = \begin{bmatrix} 3 & -7 \\ 6 & 4 \end{bmatrix}$

16. $\mathbf{A} = \begin{bmatrix} 1 & 5 \\ 8 & -10 \end{bmatrix}$ $\mathbf{B} = \begin{bmatrix} 0 & -1 \\ -3 & 1 \end{bmatrix}$

17. $\mathbf{A} = \begin{bmatrix} 1 & 8 \\ 5 & -10 \end{bmatrix}$ $\mathbf{B} = \begin{bmatrix} 1 & 0 \\ 0 & 1 \end{bmatrix}$

18. $\mathbf{A} = \begin{bmatrix} 1 & 1 & -2 \\ 3 & 2 & 5 \\ 6 & 3 & 0 \end{bmatrix}$ $\mathbf{B} = \begin{bmatrix} 4 & -6 & 0 \\ 0 & 1 & 2 \\ 1 & -1 & 5 \end{bmatrix}$

19. $\mathbf{A} = \begin{bmatrix} 1 & 1 & 1 \\ 0 & 1 & 1 \\ 0 & 0 & 1 \end{bmatrix}$ $\mathbf{B} = \begin{bmatrix} 1 & 0 & 0 \\ 1 & 1 & 0 \\ 1 & 1 & 1 \end{bmatrix}$

20. $\mathbf{A} = \begin{bmatrix} 1 & 2 & 3 \\ 4 & 5 & 6 \\ 7 & 8 & 9 \end{bmatrix}$ $\mathbf{B} = \begin{bmatrix} 1 & 0 & 0 \\ 0 & 1 & 0 \\ 0 & 0 & 1 \end{bmatrix}$

21. $\mathbf{A} = \begin{bmatrix} 1 & -2 \\ 0 & 4 \\ 8 & 5 \end{bmatrix}$ $\mathbf{B} = \begin{bmatrix} 1 & 2 & -1 \\ 0 & 4 & -3 \end{bmatrix}$

22. $\mathbf{A} = \begin{bmatrix} 4 & 5 \\ -3 & 6 \end{bmatrix}$ $\mathbf{B} = \begin{bmatrix} 1 & 5 & 6 \\ -2 & 0 & 3 \end{bmatrix}$

23. $\mathbf{A} = \begin{bmatrix} 4 & -1.5 \\ 3.6 & 4.1 \end{bmatrix}$ $\mathbf{B} = \begin{bmatrix} 0.1 & 0.2 & -1.1 \\ 0.3 & 0.4 & -4.8 \end{bmatrix}$

24. $\mathbf{A} = \begin{bmatrix} 1 & 4 \\ -3 & 10 \end{bmatrix}$ $\mathbf{B} = \begin{bmatrix} 1 & 4 & 6 & -5 \\ 2 & 3 & 1 & 0 \end{bmatrix}$

25. $\mathbf{A} = \begin{bmatrix} 1 & 2 & 3 & 4 \\ 5 & 6 & 7 & 8 \\ 9 & 10 & 11 & 12 \\ 13 & 14 & 15 & 16 \end{bmatrix}$ $\mathbf{B} = \begin{bmatrix} 1 & 0 & 0 & 0 \\ 0 & 1 & 0 & 0 \\ 0 & 0 & 1 & 0 \\ 0 & 0 & 0 & 1 \end{bmatrix}$

26. Prove Theorem 1 part b.

27. Prove Theorem 1 part c.

28. We often use **0** to denote an $m \times n$ matrix whose components are all zero; **0** is called the **additive identity** (or sometimes the zero matrix). Verify that **A** + **0** = **A** for each **A** in Exercises 13 through 25.

29. Thę **multiplicative identity, I**, has the property that **A • I** = **I • A** = **A**. Exhibit such a matrix. (*Hint:* see Exercises 17, 20, and 25.)

30. Certain properties that hold for real numbers do not hold in the algebra of matrices. We have already seen that **A • B** ≠ **B • A**, for example. Show that the property of real numbers:

$$\text{if } xy = 0, \text{ then } x = 0 \text{ or } y = 0$$

cannot be extended to matrices by calculating **X • Y**:

$$\mathbf{X} = \begin{bmatrix} 1 & -5 & 4 \\ 6 & 2 & 0 \end{bmatrix} \qquad \mathbf{Y} = \begin{bmatrix} -2 & -1 \\ 6 & 3 \\ 8 & 4 \end{bmatrix}$$

31. A row vector is a $1 \times n$ matrix (that is, a matrix with only one row) and a column vector is an $m \times 1$ matrix (or a matrix with only one column). Perform the indicated operation using

$$\mathbf{A} = [1 \quad 5 \quad 10] \qquad \mathbf{B} = \begin{bmatrix} 6 \\ -1 \\ 2 \end{bmatrix}$$

a. $3\mathbf{A}$ b. $4\mathbf{B}$ c. $\mathbf{A} \bullet \mathbf{B}$ d. $\mathbf{B} \bullet \mathbf{A}$

In Exercises 32 through 34, use matrices to solve each applied problem. (See Example 3.)

32. A textbook publisher categorizes costs into: production costs, advertising expense, payroll costs, and miscellaneous expenses. For a certain biology text they are 40%, 10%, 5%, 5% of the wholesale price, respectively. For a literature text, they are 20%, 15%, 5%, 5%, respectively, and for a mathematics text they are 40%, 15%, 10%, 5%, respectively.
    a. Depict the above cost percents in a 4×3 matrix **C**.
    b. The wholesale price of the biology text is \$8; the wholesale price of the literature text is \$6; the wholesale price of the mathematics text is \$10. Using suitable dimensions, depict a wholesale price matrix **P**, and find the publisher's categorized costs by multiplying **C** and **P** in the correct order.
    c. Multiply the 4×1 matrix solution of b, above, by the matrix $\mathbf{S} = [1 \; 1 \; 1 \; 1]$. That is, find $\mathbf{S} \bullet (\mathbf{C} \bullet \mathbf{P})$. What does the result represent?
33. Jennifer is about to panel her house. She is interested in four different types, A, B, C, D. There are three lumber yards that carry the panelling, Yard 1, Yard 2, Yard 3. Yard 1 regularly offers the panelling at \$4, \$3, \$2, \$1, respectively. Yard 2 regularly offers the panelling at \$4.50, \$3.50, \$2.50, \$1.50, respectively. Yard 3 regularly offers the panelling at \$5, \$3, \$1, \$.75, respectively. Jennifer has decided she needs either five A panels, three B panels, four C panels, and one D panel, or four A panels, four B panels, four C panels, and one D panel. Use matrix multiplication to find the cheapest choice and lumber yard for Jennifer.
34. Rework Exercise 33 if lumber yard 2 offers a "25% off" sale.
35. Suppose matrix **A** is an $m \times n$ matrix. By interchanging rows for columns, we form a new matrix $\mathbf{A}^T$, called the **transpose** of **A**. ($\mathbf{A}^T$ is, of course, an $n \times m$ matrix.) For example, if

$$\mathbf{A} = \begin{bmatrix} 1 & -1 & 6 \\ -3 & 2 & 4 \end{bmatrix} \quad \text{then} \quad \mathbf{A}^T = \begin{bmatrix} 1 & -3 \\ -1 & 2 \\ 6 & 4 \end{bmatrix}$$

Find the transpose of each matrix below.

a. $\begin{bmatrix} 1 & 2 & 3 \\ 4 & 5 & 6 \end{bmatrix}$ b. $[5 \quad 7 \quad -6]$ c. $\begin{bmatrix} 1 \\ 2 \\ \frac{1}{2} \end{bmatrix}$

d. $\begin{bmatrix} 1 & 0 & 0 & 0 \\ 0 & 1 & 0 & 0 \\ 0 & 0 & 1 & 0 \\ 0 & 0 & 0 & 1 \end{bmatrix}$ e. $\begin{bmatrix} \frac{1}{2} & \frac{2}{5} & \frac{1}{10} \\ 1 & 0 & 0 \\ \frac{1}{2} & \frac{1}{3} & \frac{1}{6} \end{bmatrix}$

Computer Exercises

In many versions of BASIC, matrix operations can easily be performed. First, however, the matrices must be dimensioned with a DIM statement. For example, the statement

```
1Ø DIM A(2,3), B(4,4)
```

allows a variable **A** to have 2 rows and 3 columns; the variable **B** is a 4 x 4 matrix. Also, if we write

```
15 LET A(1,2) = 14.Ø6
```

this means that the component in the first row, second column of **A** takes on the value 14.06.

The following program was written by a student to perform certain matrix operations

```
  5 REM MATRIX ARITHMETIC
 1Ø DIM A(2,3), B(2,3), C(3,3), D(3,3), E(2,3), F(2,3)
 15 DIM G(2,3), H(3,3)
 2Ø MAT READ A, B, C, D
 3Ø MAT E = A + B
 4Ø MAT PRINT E
 5Ø MAT G = 2 * A
 6Ø MAT E = G - B
 7Ø MAT PRINT E
 8Ø MAT F = A * C
 9Ø MAT H = C * D
1ØØ MAT PRINT F
11Ø MAT PRINT H
9ØØ DATA
91Ø DATA
92Ø DATA
93Ø DATA
999 STOP
```

36. What matrix is calculated in statement 3Ø?
37. What matrix is calculated in statement 5Ø?
38. What matrix is calculated in statement 6Ø?
39. What matrix is calculated in statement 8Ø?
40. Supply the appropriate DATA statements for the matrices below:

$$\mathbf{A} = \begin{bmatrix} 1 & 2 & 7 \\ 1 & 5 & -6 \end{bmatrix} \qquad \mathbf{B} = \begin{bmatrix} -1 & 7 & 8 \\ 10 & 3 & 0 \end{bmatrix}$$

$$\mathbf{C} = \begin{bmatrix} 1 & 4 & 0 \\ 2 & 8 & 1 \\ 6 & -5 & 0 \end{bmatrix} \qquad \mathbf{D} = \begin{bmatrix} 3 & 10 & 4 \\ -5 & 0 & 8 \\ 0 & 1 & -2 \end{bmatrix}$$

41. Run the program to compute **A** + **B**, 2**A** − **B**, **A** • **C**, and **C** • **D**.
42. Refine the program to compute **A** + **B**, k**A** − **B**, **A** • **C**, and **C** • **D** for *any* (appropriately dimensioned) matrices **A**, **B**, **C**, **D**.

9.3 THE MATRIX INVERSE*

Just as the real number 1 is called the **multiplicative identity** in algebra because multiplying by 1 does not "change" a given number, there also exists a multiplicative identity for matrix multiplication:

DEFINITION

> **Identity Matrix**
>
> The $n \times n$ matrix whose entries are defined as $a_{ij} = 0$ if $i \neq j$ and $a_{ij} = 1$ if $i = j$ is called the **identity matrix** and is denoted by $\mathbf{I}_n$ (or simply, $\mathbf{I}$ if the dimensions are known).

Thus,
$$\mathbf{I}_2 = \begin{bmatrix} 1 & 0 \\ 0 & 1 \end{bmatrix} \qquad \mathbf{I}_3 = \begin{bmatrix} 1 & 0 & 0 \\ 0 & 1 & 0 \\ 0 & 0 & 1 \end{bmatrix}$$

The next example illustrates the main property of $\mathbf{I}$: $\mathbf{A} \bullet \mathbf{I} = \mathbf{I} \bullet \mathbf{A} = \mathbf{A}$. (The reader should also see Exercise 29 of the previous section.)

EXAMPLE 1 Let $\mathbf{A} = \begin{bmatrix} 1 & 4 & -3 \\ 0 & 2 & 11 \\ 14 & -6 & 8 \end{bmatrix}$.

Show that $\mathbf{A} \bullet \mathbf{I} = \mathbf{I} \bullet \mathbf{A} = \mathbf{A}$.

Solution With $\mathbf{I} = \begin{bmatrix} 1 & 0 & 0 \\ 0 & 1 & 0 \\ 0 & 0 & 1 \end{bmatrix}$ we have:

$$\mathbf{A} \bullet \mathbf{I} = \begin{bmatrix} 1 \cdot 1 + 0 + 0 & 0 + 4 \cdot 1 + 0 & 0 + 0 + -3 \cdot 1 \\ 0 \cdot 1 + 0 + 0 & 0 + 2 \cdot 1 + 0 & 0 + 0 + 11 \cdot 1 \\ 14 \cdot 1 + 0 + 0 & 0 + -6 \cdot 1 + 0 & 0 + 0 + 8 \cdot 1 \end{bmatrix}$$

$$= \begin{bmatrix} 1 & 4 & -3 \\ 0 & 2 & 11 \\ 14 & -6 & 8 \end{bmatrix} = \mathbf{A}$$

$$\mathbf{I} \bullet \mathbf{A} = \begin{bmatrix} 1 \cdot 1 + 0 + 0 & 1 \cdot 4 + 0 + 0 & 1 \cdot -3 + 0 + 0 \\ 0 + 1 \cdot 0 + 0 & 0 + 1 \cdot 2 + 0 & 0 + 1 \cdot 11 + 0 \\ 0 + 0 + 1 \cdot 14 & 0 + 0 + 1 \cdot -6 & 0 + 0 + 1 \cdot 8 \end{bmatrix}$$

$$= \begin{bmatrix} 1 & 4 & -3 \\ 0 & 2 & 11 \\ 14 & -6 & 8 \end{bmatrix} = \mathbf{A}$$

*All matrices for the remainder of this chapter will be square matrices, that is, $n \times n$ matrices.

When a matrix **A** is multiplied by a matrix **B** to yield the identity matrix, **I**, then **B** is called the **multiplicative inverse** of **A** (or simply, the **inverse** of **A**).

DEFINITION

The Matrix Inverse

A matrix **B** is called the **inverse** of **A** if and only if

$$\mathbf{A} \bullet \mathbf{B} = \mathbf{I}$$

We usually denote the inverse of **A** by $\mathbf{A}^{-1}$.

EXAMPLE 2 Show that if $\mathbf{A} = \begin{bmatrix} 1 & 2 \\ 3 & 4 \end{bmatrix}$, then $\mathbf{A}^{-1} = \begin{bmatrix} -2 & 1 \\ \frac{3}{2} & -\frac{1}{2} \end{bmatrix}$

Solution We must show $\mathbf{A} \bullet \mathbf{A}^{-1} = \mathbf{I}$.*

$$\mathbf{A} \bullet \mathbf{A}^{-1} = \begin{bmatrix} 1 & 2 \\ 3 & 4 \end{bmatrix} \bullet \begin{bmatrix} -2 & 1 \\ \frac{3}{2} & -\frac{1}{2} \end{bmatrix} = \begin{bmatrix} 1 & 0 \\ 0 & 1 \end{bmatrix} = \mathbf{I}$$

Given two matrices then, we can determine whether they are inverses of one another by multiplying them to see if their product is the identity matrix, **I**. We now examine the procedure for *finding* the inverse of a given matrix, **A**.

Suppose $\mathbf{A} = \begin{bmatrix} 2 & 5 \\ 1 & 3 \end{bmatrix}$. We wish to find a matrix $\mathbf{A}^{-1}$ such that $\mathbf{A} \bullet \mathbf{A}^{-1} = \mathbf{I}$. Let's label the unknown components of $\mathbf{A}^{-1}$ as

$$\mathbf{A}^{-1} = \begin{bmatrix} a_{11} & a_{12} \\ a_{21} & a_{22} \end{bmatrix}$$

Thus,

$$\mathbf{A} \bullet \mathbf{A}^{-1} = \begin{bmatrix} 2 & 5 \\ 1 & 3 \end{bmatrix} \bullet \begin{bmatrix} a_{11} & a_{12} \\ a_{21} & a_{22} \end{bmatrix} = \begin{bmatrix} 1 & 0 \\ 0 & 1 \end{bmatrix}$$

and, if we perform the matrix multiplication we have

$$\begin{bmatrix} 2a_{11} + 5a_{21} & 2a_{12} + 5a_{22} \\ a_{11} + 3a_{21} & a_{12} + 3a_{22} \end{bmatrix} = \begin{bmatrix} 1 & 0 \\ 0 & 1 \end{bmatrix}$$

*It can be shown that if $\mathbf{A} \cdot \mathbf{A}^{-1} = \mathbf{I}$ then $\mathbf{A}^{-1} \cdot \mathbf{A} = \mathbf{I}$, also.

and by the definition of matrix equality, we get the following two systems of equations:

$$\begin{aligned} 2a_{11} + 5a_{21} &= 1 \\ a_{11} + 3a_{21} &= 0 \end{aligned} \qquad\qquad \begin{aligned} 2a_{12} + 5a_{22} &= 0 \\ a_{12} + 3a_{22} &= 1 \end{aligned}$$

Solving these systems the same way we solved systems with identical coefficients in Section 9.1 (see Example 5 and Exercises 28 and 29 of that section), we have:

$$\left[\begin{array}{cc|cc} 2 & 5 & 1 & 0 \\ 1 & 3 & 0 & 1 \end{array}\right] \begin{array}{l}\text{Interchange}\\ \text{the two rows.}\end{array} \rightarrow \left[\begin{array}{cc|cc} 1 & 3 & 0 & 1 \\ 2 & 5 & 1 & 0 \end{array}\right] \begin{array}{l}\text{Add } -2 \text{ times the}\\ \text{top row to the}\\ \text{bottom row.}\end{array} \rightarrow$$

$$\left[\begin{array}{cc|cc} 1 & 3 & 0 & 1 \\ 0 & -1 & 1 & -2 \end{array}\right] \begin{array}{l}\text{Multiply the}\\ \text{bottom row}\\ \text{by } -1.\end{array} \rightarrow \left[\begin{array}{cc|cc} 1 & 3 & 0 & 1 \\ 0 & 1 & -1 & 2 \end{array}\right] \begin{array}{l}\text{Add } -3 \text{ times the}\\ \text{bottom row to}\\ \text{the top row.}\end{array} \rightarrow$$

$$\left[\begin{array}{cc|cc} 1 & 0 & 3 & -5 \\ 0 & 1 & -1 & 2 \end{array}\right]$$

This final augmented matrix is interpreted as follows:

$$\begin{aligned} a_{11} &= 3 \\ a_{21} &= -1 \end{aligned} \qquad\qquad \begin{aligned} a_{12} &= -5 \\ a_{22} &= 2 \end{aligned}$$

Thus, $\mathbf{A}^{-1} = \begin{bmatrix} 3 & -5 \\ -1 & 2 \end{bmatrix}$. A check will verify that $\mathbf{A} \bullet \mathbf{A}^{-1} = \mathbf{A}^{-1} \bullet \mathbf{A} = \mathbf{I}$.

The method above is called the **elimination method** for finding the inverse.

EXAMPLE 3 Find $\mathbf{C}^{-1}$ if $\mathbf{C} = \begin{bmatrix} 1 & 3 & 0 \\ 2 & 0 & 4 \\ 1 & -3 & 3 \end{bmatrix}$.

Solution First, we write

$$\left[\begin{array}{ccc|ccc} 1 & 3 & 0 & 1 & 0 & 0 \\ 2 & 0 & 4 & 0 & 1 & 0 \\ 1 & -3 & 3 & 0 & 0 & 1 \end{array}\right]$$

and then perform the row transformations to "convert" the left-hand side of the augmented matrix "into" the identity matrix. The reader should supply some of the missing steps and reasons in the elimination method that follows:

$$\left[\begin{array}{rrr|rrr} 1 & 3 & 0 & 1 & 0 & 0 \\ 2 & 0 & 4 & 0 & 1 & 0 \\ 1 & -3 & 3 & 0 & 0 & 1 \end{array}\right] \to \left[\begin{array}{rrr|rrr} 1 & 3 & 0 & 1 & 0 & 0 \\ 0 & -6 & 4 & -2 & 1 & 0 \\ 0 & -6 & 3 & -1 & 0 & 1 \end{array}\right] \to$$

$$\left[\begin{array}{rrr|rrr} 1 & 0 & 2 & 0 & \frac{1}{2} & 0 \\ 0 & 1 & -\frac{2}{3} & \frac{1}{3} & -\frac{1}{6} & 0 \\ 0 & 0 & -1 & 1 & -1 & 1 \end{array}\right] \to \left[\begin{array}{rrr|rrr} 1 & 0 & 0 & 2 & -\frac{3}{2} & 2 \\ 0 & 1 & 0 & -\frac{1}{3} & \frac{1}{2} & -\frac{2}{3} \\ 0 & 0 & 1 & -1 & 1 & -1 \end{array}\right]$$

Hence,

$$\mathbf{C}^{-1} = \begin{bmatrix} 2 & -\frac{3}{2} & 2 \\ -\frac{1}{3} & \frac{1}{2} & -\frac{2}{3} \\ -1 & 1 & -1 \end{bmatrix}$$

The reader should check that the inverse of **C** is correct by calculating to verify $\mathbf{C} \bullet \mathbf{C}^{-1} = \mathbf{C}^{-1} \bullet \mathbf{C} = \mathbf{I}$.

The reader is cautioned that the matrix inverse *may not always exist*. For instance there is no matrix that can multiply $\begin{bmatrix} 9 & 12 \\ 3 & 4 \end{bmatrix}$ to yield $\begin{bmatrix} 1 & 0 \\ 0 & 1 \end{bmatrix}$ (see Exercise 21, also).

Although there are many applications that involve the matrix inverse, one use is to solve systems of linear equations. Matrix notation is used to represent a given linear system as follows. The system

$$\begin{aligned} x + 2y &= 1 \\ 3x + 4y &= -1 \end{aligned} \tag{1}$$

is rewritten as

$$\begin{bmatrix} 1 & 2 \\ 3 & 4 \end{bmatrix} \bullet \begin{bmatrix} x \\ y \end{bmatrix} = \begin{bmatrix} 1 \\ -1 \end{bmatrix}$$

or, in general terms as

$$\mathbf{A} \bullet \mathbf{X} = \mathbf{B} \tag{2}$$

where **A** (called the **coefficient matrix** for the system), **X**, and **B** are:

$$\mathbf{A} = \begin{bmatrix} 1 & 2 \\ 3 & 4 \end{bmatrix}, \qquad \mathbf{X} = \begin{bmatrix} x \\ y \end{bmatrix}, \qquad \mathbf{B} = \begin{bmatrix} 1 \\ -1 \end{bmatrix}.$$

To solve for x and y, we isolate $\mathbf{X}$. This can be achieved by "left-multiplying" both sides of (2) by $\mathbf{A}^{-1}$ (if it exists):

$$\begin{aligned} \mathbf{A} \bullet \mathbf{X} &= \mathbf{B} \\ \mathbf{A}^{-1} \bullet \mathbf{A} \bullet \mathbf{X} &= \mathbf{A}^{-1} \bullet \mathbf{B} \\ \mathbf{I} \bullet \mathbf{X} &= \mathbf{A}^{-1} \bullet \mathbf{B} \\ \mathbf{X} &= \mathbf{A}^{-1} \bullet \mathbf{B} \end{aligned} \tag{3}$$

In particular, $\mathbf{A}^{-1} = \begin{bmatrix} -2 & 1 \\ \frac{3}{2} & -\frac{1}{2} \end{bmatrix}$ (from Example 2) and system (1) is solved below:

$$\mathbf{X} = \mathbf{A}^{-1} \bullet \mathbf{B}$$

$$\begin{bmatrix} x \\ y \end{bmatrix} = \mathbf{X} = \begin{bmatrix} -2 & 1 \\ \frac{3}{2} & -\frac{1}{2} \end{bmatrix} \bullet \begin{bmatrix} 1 \\ -1 \end{bmatrix} = \begin{bmatrix} -3 \\ 2 \end{bmatrix}$$

That is, $x = -3$ and $y = 2$. An example follows.

EXAMPLE 4 Solve the system

$$\begin{aligned} x_1 + 3x_2 &= 6 \\ 2x_1 + 4x_3 &= 4 \\ x_1 - 3x_2 + 3x_3 &= 0 \end{aligned}$$

Solution We want to solve $\mathbf{A} \bullet \mathbf{X} = \mathbf{B}$ where

$$\mathbf{A} = \begin{bmatrix} 1 & 3 & 0 \\ 2 & 0 & 4 \\ 1 & -3 & 3 \end{bmatrix}, \qquad \mathbf{X} = \begin{bmatrix} x_1 \\ x_2 \\ x_3 \end{bmatrix}, \qquad \mathbf{B} = \begin{bmatrix} 6 \\ 4 \\ 0 \end{bmatrix}$$

We found $\mathbf{A}^{-1}$ in Example 3; it is

$$\mathbf{A}^{-1} = \begin{bmatrix} 2 & -\frac{3}{2} & 2 \\ -\frac{1}{3} & \frac{1}{2} & -\frac{2}{3} \\ -1 & 1 & -1 \end{bmatrix}$$

Thus,

$$\mathbf{X} = \mathbf{A}^{-1} \bullet \mathbf{B} = \begin{bmatrix} 2 & -\frac{3}{2} & 2 \\ -\frac{1}{3} & \frac{1}{2} & -\frac{2}{3} \\ -1 & 1 & -1 \end{bmatrix} \bullet \begin{bmatrix} 6 \\ 4 \\ 0 \end{bmatrix} = \begin{bmatrix} 6 \\ 0 \\ -2 \end{bmatrix}$$

and $x_1 = 6, x_2 = 0, x_3 = -2$.

Since several methods have been given for solving systems of equations (in this chapter and the previous one), it is natural for the reader to be curious as to when the inverse method should be used. Actually it is most convenient to use either when the inverse of the coefficient matrix is already known (as in Example 4) or when there are *several* systems to solve, each having the same coefficient matrix.

EXERCISES FOR SECTION 9.3

1. Verify that $\mathbf{A} \bullet \mathbf{I} = \mathbf{I} \bullet \mathbf{A} = \mathbf{A}$ for each of the following matrices. (See Example 1.)

a. $\mathbf{A} = \begin{bmatrix} 4 & 9 \\ -1 & 3 \end{bmatrix}$ b. $\mathbf{A} = \begin{bmatrix} 1 & 2 & 3 \\ 4 & 5 & 6 \\ 7 & 8 & 9 \end{bmatrix}$ c. $\mathbf{A} = \begin{bmatrix} 0 & 0 & 1 \\ 1 & 0 & 1 \\ -2 & 5 & 17 \end{bmatrix}$

In Exercises 2 through 10, determine whether or not the two given matrices are inverses of one another. (See Example 2.)

2. $\begin{bmatrix} 2 & 0 \\ 0 & 7 \end{bmatrix}, \begin{bmatrix} \frac{1}{2} & 0 \\ 0 & \frac{1}{7} \end{bmatrix}$

3. $\begin{bmatrix} 2 & -4 \\ 1 & 3 \end{bmatrix}, \begin{bmatrix} \frac{3}{10} & \frac{2}{5} \\ -\frac{1}{10} & 1 \end{bmatrix}$

4. $\begin{bmatrix} 2 & -4 \\ 1 & 3 \end{bmatrix}, \begin{bmatrix} \frac{3}{10} & \frac{2}{5} \\ -\frac{1}{10} & \frac{1}{5} \end{bmatrix}$

5. $\begin{bmatrix} 2 & 1 \\ 7 & 4 \end{bmatrix}, \begin{bmatrix} 4 & -1 \\ -7 & 1 \end{bmatrix}$

6. $\begin{bmatrix} 4 & -1 \\ -7 & 2 \end{bmatrix}, \begin{bmatrix} 2 & 1 \\ 7 & 4 \end{bmatrix}$

7. $\begin{bmatrix} 4 & 3 \\ 5 & 4 \end{bmatrix}, \begin{bmatrix} 4 & -3 \\ -5 & 4 \end{bmatrix}$

8. $\begin{bmatrix} 1 & 0 & 8 \\ 5 & 10 & -20 \\ 0 & 0 & 2 \end{bmatrix}, \begin{bmatrix} 1 & 0 & 4 \\ -.5 & .1 & 3 \\ 0 & 0 & .5 \end{bmatrix}$

9. $\begin{bmatrix} 1 & -2.5 & 1.25 \\ 0 & .25 & -.125 \\ -3 & 7.25 & -3.125 \end{bmatrix}, \begin{bmatrix} 1 & 10 & 0 \\ 3 & 5 & 1 \\ 6 & 2 & 2 \end{bmatrix}$

10. $\begin{bmatrix} 0 & 0 & 0 & 5 \\ 5 & 0 & 0 & 0 \\ 0 & 5 & 0 & 0 \\ 0 & 0 & 5 & 0 \end{bmatrix}, \begin{bmatrix} 0 & .2 & 0 & 0 \\ 0 & 0 & .2 & 0 \\ 0 & 0 & 0 & .2 \\ .2 & 0 & 0 & 0 \end{bmatrix}$

In Exercises 11 through 20, find the inverse of the given matrix, if it exists. Check each answer by showing $\mathbf{A} \bullet \mathbf{A}^{-1} = \mathbf{A}^{-1} \bullet \mathbf{A} = \mathbf{I}$. (See Example 3.)

11. $\begin{bmatrix} 1 & 0 \\ 2 & 1 \end{bmatrix}$

12. $\begin{bmatrix} -4 & 2 \\ 3 & -1 \end{bmatrix}$

13. $\begin{bmatrix} -.5 & .5 \\ -.5 & -.5 \end{bmatrix}$

14. $\begin{bmatrix} 6 & 12 \\ 0 & 4 \end{bmatrix}$

15. $\begin{bmatrix} 5 & 7 \\ 3 & 4 \end{bmatrix}$

16. $\begin{bmatrix} 8 & 2 \\ 16 & 4 \end{bmatrix}$

17. $\begin{bmatrix} 8 & -20 & 10 \\ 0 & 2 & -1 \\ -24 & 58 & 25 \end{bmatrix}$

18. $\begin{bmatrix} -5 & -2 & 0 \\ 4 & 1 & 2 \\ -2 & 0 & -2 \end{bmatrix}$

19. $\begin{bmatrix} 2 & 3 & -4 \\ 1 & 1 & 1 \\ -1 & 7 & -10 \end{bmatrix}$

20. $\begin{bmatrix} 2 & 4 & 2 & 2 \\ 0 & 4 & 0 & 2 \\ 0 & 0 & -1 & 2 \\ 0 & 0 & 0 & -2 \end{bmatrix}$

21. A matrix **S** that has no inverse is called **singular** (if a matrix has an inverse, it is called **nonsingular**). An indication that the matrix is singular is the appearance of a row of zeros during the elimination process. Determine which of the matrices below are singular.

a. $\begin{bmatrix} 9 & 12 \\ 3 & 4 \end{bmatrix}$ b. $\begin{bmatrix} 9 & 4 \\ 3 & 12 \end{bmatrix}$ c. $\begin{bmatrix} 1 & -2 \\ -1 & -13 \end{bmatrix}$

In Exercises 22 through 33, use the method of this section to solve each system. (See Example 4.)

22. $\begin{aligned} x + y &= 11 \\ x - y &= 3 \end{aligned}$

23. $\begin{aligned} x_1 + x_2 + x_3 &= 6 \\ x_1 - x_2 &= -1 \\ 10x_1 + 6x_2 - x_3 &= 19 \end{aligned}$

24. $\begin{aligned} x_1 + x_2 + x_3 &= 5 \\ x_1 - x_2 &= 0 \\ 10x_1 + 6x_2 - x_3 &= -5 \end{aligned}$

25. $\begin{aligned} x_1 + x_2 + x_3 &= 6 \\ x_1 - x_2 &= 6 \\ 10x_1 + 6x_2 - x_3 &= 65 \end{aligned}$

26. $\begin{aligned} x_1 + x_2 + x_3 &= -18 \\ x_1 - x_2 &= 1 \\ 10x_1 + 6x_2 - x_3 &= -79 \end{aligned}$

27. $\begin{aligned} x_1 + x_2 - x_3 &= -17 \\ x_1 + 2x_2 + 2x_3 &= 3 \\ 7x_2 + x_3 &= -40 \end{aligned}$

28. $\begin{aligned} x_1 + x_2 - x_3 &= 5 \\ x_1 + 2x_2 + 2x_3 &= 5 \\ 7x_2 + x_3 &= 0 \end{aligned}$

29. $\begin{aligned} x_1 + x_2 - x_3 &= -1 \\ x_1 + 2x_2 + 2x_3 &= 1 \\ 7x_2 + x_3 &= 4 \end{aligned}$

30. $\begin{aligned} x_1 + x_2 - x_3 &= 5 \\ x_1 + 2x_2 + 2x_3 &= 6 \\ 7x_2 + x_3 &= -3 \end{aligned}$

31. $\begin{aligned} x_1 + x_2 + x_3 + x_4 &= 11 \\ x_2 - 2x_3 + x_4 &= -4 \\ 10x_1 + x_3 + 2x_4 &= 17 \\ x_1 + 2x_2 + 3x_3 + 4x_4 &= 39 \end{aligned}$

32. $\begin{aligned} x_1 + x_2 + x_3 + x_4 &= 0 \\ x_2 - 2x_3 + x_4 &= 0 \\ 10x_1 + x_3 + 2x_4 &= 0 \\ x_1 + 2x_2 + 3x_3 + 4x_4 &= 0 \end{aligned}$

33. $\begin{aligned} x_1 + x_2 + x_3 + x_4 &= 0 \\ x_2 - 2x_3 + x_4 &= 10 \\ 10x_1 + x_3 + 2x_4 &= -1 \\ x_1 + 2x_2 + 3x_3 + 4x_4 &= 10 \end{aligned}$

Computer Exercises

The inverse, **B**, of a matrix **A** is computed in most versions of BASIC as

```
1Ø MAT B = INV(A)
```

34. Write a program that reads a 3 x 3 matrix **A** and prints its inverse.
35. Generalize the program to handle *any* square matrix **A**.
36. a. Write a program that reads data for two 3 x 3 matrices (say **C** and **D**). Also, compute **C • D**, $(\mathbf{C} \bullet \mathbf{D})^{-1}$, $\mathbf{C}^{-1}$, $\mathbf{D}^{-1}$, and $\mathbf{D}^{-1} \bullet \mathbf{C}^{-1}$ in your program.
    b. Verify that $(\mathbf{C} \bullet \mathbf{D})^{-1} = \mathbf{D}^{-1} \bullet \mathbf{C}^{-1}$ for the 3 x 3 matrices of your choice.
37. The following program was written to solve an $n \times n$ system of linear equations:

```
1Ø REM SYSTEM OF N EQUATIONS IN N VARIABLES
2Ø READ N
3Ø DIM A(N,N), B(N,1), X(N,1), C(N,N)
35 REM C IS A INVERSE
```

```
 40 MAT READ A
 45 MAT READ B
 50 MAT C = INV(A)
 60 MAT X = C * B
 70 MAT PRINT X
 80 DATA
 90 DATA
100 DATA
999 STOP
```

a. Use

```
 80 DATA 4
 90 DATA 1,1,1,1,0,1,-2,1,10,0,1,2,1,2,3,4
100 DATA 11,-4,17,39
```

to solve the system of Exercise 31.

b. Solve the system of Exercise 32.

c. Solve the system of Exercise 33.

9.4 DETERMINANTS AND THEIR PROPERTIES

In this section we will examine yet another operation on square matrices—one that yields a *real number* (not another matrix). It is called the **determinant** and will first be defined for 2 x 2 matrices.

DEFINITION

Determinant of a 2 x 2 Matrix

If $\mathbf{A} = \begin{bmatrix} a_{11} & a_{12} \\ a_{21} & a_{22} \end{bmatrix}$, then the **determinant of A** is

$$\det A = \begin{vmatrix} a_{11} & a_{12} \\ a_{21} & a_{22} \end{vmatrix} = a_{11}a_{22} - a_{12}a_{21}.^*$$

Often we will write $\det A = |\mathbf{A}|$.

EXAMPLE 1 Find the determinant of each matrix below.

$$\mathbf{A} = \begin{bmatrix} 5 & 6 \\ -2 & 7 \end{bmatrix} \qquad \mathbf{B} = \begin{bmatrix} 0 & 3 \\ 2 & 5 \end{bmatrix}$$

Solution $\det A = |\mathbf{A}| = 5 \cdot 7 - 6(-2) = 35 + 12 = 47$

$\det B = |\mathbf{B}| = 0 \cdot 5 - 3 \cdot 2 = -6$

*Actually, a determinant can be thought of as a function whose domain is the collection of all square matrices and whose range is $\mathcal{R}$.

Before we can proceed with the definition of determinants of larger matrices, we need the following two definitions:

DEFINITION

Minor of a_{ij}

Let **A** be an $n \times n$ matrix $(n > 2)$. Let M_{ij} be the determinant of the $(n-1) \times (n-1)$ matrix obtained from **A** by deleting the ith row and jth column of **A**. Then M_{ij} is called the minor of the element a_{ij}.

DEFINITION

Cofactor of a_{ij}

Let **A** and M_{ij} be defined as above. Then

$$A_{ij} = (-1)^{i+j} M_{ij}$$

is called the **cofactor** of the element a_{ij}.

EXAMPLE 2 Find M_{11}, M_{12}, and M_{13} for

$$\mathbf{A} = \begin{bmatrix} 3 & 1 & -7 \\ 0 & 2 & 4 \\ 8 & -1 & 6 \end{bmatrix}$$

Solution

$$M_{11} = \begin{vmatrix} 3 & 1 & -7 \\ 0 & 2 & 4 \\ 8 & -1 & 6 \end{vmatrix} = \begin{vmatrix} 2 & 4 \\ -1 & 6 \end{vmatrix} = 16$$

$$M_{12} = \begin{vmatrix} 3 & 1 & -7 \\ 0 & 2 & 4 \\ 8 & -1 & 6 \end{vmatrix} = \begin{vmatrix} 0 & 4 \\ 8 & 6 \end{vmatrix} = -32$$

$$M_{13} = \begin{vmatrix} 3 & 1 & -7 \\ 0 & 2 & 4 \\ 8 & -1 & 6 \end{vmatrix} = \begin{vmatrix} 0 & 2 \\ 8 & -1 \end{vmatrix} = -16$$

EXAMPLE 3 Find A_{11}, A_{12}, A_{13} for

$$\mathbf{A} = \begin{bmatrix} 3 & 1 & -7 \\ 0 & 2 & 4 \\ 8 & -1 & 6 \end{bmatrix}$$

Solution Notice that M_{11}, M_{12}, M_{13} were found in Example 2.

$$\begin{aligned} A_{11} &= (-1)^{1+1}M_{11} = 16 \\ A_{12} &= (-1)^{1+2}M_{12} = 32 \\ A_{13} &= (-1)^{1+3}M_{13} = -16 \end{aligned}$$

We are now ready to define the determinant of *any* $n \times n$ matrix.

DEFINITION

Determinant of $n \times n$ Matrix

Let **A** be any $n \times n$ matrix ($n > 2$). Then the determinant of **A** is given by

$$|\mathbf{A}| = a_{11}A_{11} + a_{12}A_{12} + \ldots + a_{1n}A_{1n} \tag{1}$$

More thoroughly, the value for $|\mathbf{A}|$ given in (1) is called the determinant of **A** by **expansion along the** ***first*** **row.** It can be shown (see Exercise 21) that *any* row or column can be used. Thus,

Expansion along the ith row

$$|\mathbf{A}| = a_{i1}A_{i1} + a_{i2}A_{i2} + \ldots + a_{in}A_{in} \tag{2}$$

Expansion along the jth column

$$|\mathbf{A}| = a_{1j}A_{1j} + a_{2j}A_{2j} + \ldots + a_{nj}A_{nj} \tag{3}$$

Some examples follow.

EXAMPLE 4 If $\mathbf{A} = \begin{bmatrix} 3 & 1 & -7 \\ 0 & 2 & 4 \\ 8 & -1 & 6 \end{bmatrix}$, find $|\mathbf{A}|$.

Solution We have already found (in Example 3) A_{11}, A_{12}, A_{13}. So, by expansion along the first row:

$$\begin{aligned} |\mathbf{A}| &= a_{11}A_{11} + a_{12}A_{12} + a_{13}A_{13} \\ |\mathbf{A}| &= 3(16) + 1(32) - 7(-16) = 192 \end{aligned}$$

EXAMPLE 5 If $\mathbf{B} = \begin{bmatrix} 5 & 3 & 14 \\ 0 & 8 & -6 \\ 4 & 0 & 0 \end{bmatrix}$, find $|\mathbf{B}|$.

Solution Because of the appearance of two zeros in the third row, we expand along it using (2) with $i = 3$ and $n = 3$. Note that

$$B_{31} = (-1)^{3+1}M_{31} = \begin{vmatrix} 3 & 14 \\ 8 & -6 \end{vmatrix} = -130$$

Thus,

$$|\mathbf{B}| = 4(-130) + 0 \cdot B_{32} + 0 \cdot B_{33} = -520$$

EXAMPLE 6 If $\mathbf{C} = \begin{bmatrix} 0 & 2 & 0 & 1 \\ 3 & 1 & 1 & -7 \\ 0 & 0 & 2 & 4 \\ 8 & 2 & -1 & 6 \end{bmatrix}$, find $|\mathbf{C}|$.

Solution Expanding along the first row gives

$$\begin{aligned} |\mathbf{C}| &= c_{11}C_{11} + c_{12}C_{12} + c_{13}C_{13} + c_{14}C_{14} \\ &= 0 \cdot C_{11} + c_{12}(-1)^{1+2}M_{12} + 0 \cdot C_{13} + c_{14}(-1)^{1+4}M_{14} \\ &= -2\underbrace{\begin{vmatrix} 3 & 1 & -7 \\ 0 & 2 & 4 \\ 8 & -1 & 6 \end{vmatrix}}_{\substack{\text{This determinant} \\ \text{equals 192, from} \\ \text{Example 4.}}} - 1\underbrace{\begin{vmatrix} 3 & 1 & 1 \\ 0 & 0 & 2 \\ 8 & 2 & -1 \end{vmatrix}}_{\substack{\text{This determinant} \\ \text{equals 4, by expand-} \\ \text{ing along the} \\ \text{second row.}}} \end{aligned}$$

So, $|\mathbf{C}| = -2(192) - 4 = -388$.

We conclude this section with a partial listing of some of the properties of determinants; proofs are given for the $n = 2$ case only. In each of the following theorems, $\mathbf{A}$ represents an $n \times n$ matrix.

THEOREM 2 If $\mathbf{A}'$ is the matrix obtained from $\mathbf{A}$ by interchanging two rows, then

$$|\mathbf{A}'| = -|\mathbf{A}|$$

Proof (for $n = 2$):

Let $\mathbf{A} = \begin{bmatrix} a_{11} & a_{12} \\ a_{21} & a_{22} \end{bmatrix}$. Then $\mathbf{A}' = \begin{bmatrix} a_{21} & a_{22} \\ a_{11} & a_{12} \end{bmatrix}$.

$$|\mathbf{A}| = a_{11}a_{22} - a_{12}a_{21}$$

and

$$|\mathbf{A}'| = a_{21}a_{12} - a_{22}a_{11} = a_{12}a_{21} - a_{11}a_{22}$$

Thus, $|\mathbf{A}'| = -|\mathbf{A}|$.

THEOREM 3 If $\mathbf{A}'$ is the matrix obtained from $\mathbf{A}$ by multiplying each entry in a row of $\mathbf{A}$ by a constant real number k, then

$$|\mathbf{A}'| = k|\mathbf{A}|$$

Proof (for $n = 2$):

Let $\mathbf{A} = \begin{bmatrix} a_{11} & a_{12} \\ a_{21} & a_{22} \end{bmatrix}$ and $\mathbf{A}' = \begin{bmatrix} ka_{11} & ka_{12} \\ a_{21} & a_{22} \end{bmatrix}$.

$$|\mathbf{A}| = a_{11}a_{22} - a_{12}a_{21}$$

and

$$|\mathbf{A}'| = ka_{11}a_{22} - ka_{12}a_{21} = k(a_{11}a_{22} - a_{12}a_{21})$$

Thus, $|\mathbf{A}'| = k|\mathbf{A}|$. The same results apply if row two is multiplied by k.

THEOREM 4 If $\mathbf{A}'$ is the matrix obtained from $\mathbf{A}$ by adding a multiple of one row to another, then $|\mathbf{A}'| = |\mathbf{A}|$.

Proof (for $n = 2$):

Let $\mathbf{A} = \begin{bmatrix} a_{11} & a_{12} \\ a_{21} & a_{22} \end{bmatrix}$ then $|\mathbf{A}| = a_{11}a_{22} - a_{12}a_{21}$.

Now add a multiple (say k) of row 1 to row 2:

$$\mathbf{A}' = \begin{bmatrix} a_{11} & a_{12} \\ ka_{11} + a_{21} & ka_{12} + a_{22} \end{bmatrix}$$

Hence,

$$\begin{aligned} |\mathbf{A}'| &= a_{11}(ka_{12} + a_{22}) - a_{12}(ka_{11} + a_{21}) \\ &= ka_{11}a_{12} + a_{11}a_{22} - ka_{11}a_{12} - a_{12}a_{21} \\ &= a_{11}a_{22} - a_{12}a_{21} = |\mathbf{A}| \end{aligned}$$

THEOREM 5

> If two rows of matrix **A** are identical, then
>
> $$|\mathbf{A}| = 0$$

Proof (the general case):

Without loss of generality, suppose the two identical rows are rows 1 and 2. We write:

$$\mathbf{A} = \begin{bmatrix} a_{11} & a_{12} & \ldots & a_{1n} \\ a_{11} & a_{12} & \ldots & a_{1n} \\ a_{31} & a_{32} & \ldots & a_{3n} \\ \vdots & \vdots & & \vdots \\ a_{n1} & a_{n2} & \ldots & a_{nn} \end{bmatrix}$$

Now, by Theorem 4, we can add any multiple k of row 2 to row 1. Let $k = -1$:

$$\mathbf{A}' = \begin{bmatrix} 0 & 0 & \ldots & 0 \\ a_{11} & a_{12} & \ldots & a_{1n} \\ a_{31} & a_{32} & \ldots & a_{3n} \\ \vdots & \vdots & & \vdots \\ a_{n1} & a_{n2} & \ldots & a_{nn} \end{bmatrix}$$

and we know

$$|\mathbf{A}| = |\mathbf{A}'| = 0 \cdot A_{11} + 0 \cdot A_{12} + \ldots + 0 \cdot A_{1n} = 0$$

We mention here that Theorems 2 through 5 remain true if the word "row(s)" is replaced with "column(s)" (see Exercises 22 through 25.)

EXERCISES FOR SECTION 9.4

In Exercises 1 through 9, find the determinant of each matrix. (See Example 1.)

1. $\begin{bmatrix} 5 & 6 \\ 8 & 9 \end{bmatrix}$
2. $\begin{bmatrix} 4 & 6 \\ 7 & 9 \end{bmatrix}$
3. $\begin{bmatrix} 4 & 5 \\ 7 & 8 \end{bmatrix}$
4. $\begin{bmatrix} 2 & 3 \\ 8 & 9 \end{bmatrix}$
5. $\begin{bmatrix} 1 & -3 \\ 7 & 9 \end{bmatrix}$
6. $\begin{bmatrix} -1 & 2 \\ 7 & 8 \end{bmatrix}$
7. $\begin{bmatrix} 1 & -4 \\ -8 & -3 \end{bmatrix}$
8. $\begin{bmatrix} 0 & \pi \\ 0 & \sqrt{7} \end{bmatrix}$
9. $\begin{bmatrix} 0 & 0 \\ 0 & -\frac{\sqrt{7}}{18} \end{bmatrix}$

In Exercises 10 through 15, find the indicated minor and cofactor. (See Examples 2 and 3.) Use

$$\mathbf{A} = \begin{bmatrix} 1 & 2 & 3 \\ 4 & 5 & 6 \\ 7 & 8 & 9 \end{bmatrix}$$

10. M_{11} and A_{11} **11.** M_{12} and A_{12} **12.** M_{13} and A_{13}

13. M_{21} and A_{21} **14.** M_{22} and A_{22} **15.** M_{23} and A_{23}

In Exercises 16 through 20, find the determinant of the given matrix. (See Examples 4 and 5.)

16. $\begin{bmatrix} 1 & 2 & 3 \\ 4 & 5 & 6 \\ 7 & 8 & 9 \end{bmatrix}$ **17.** $\begin{bmatrix} 0 & 5 & 8 \\ 6 & -2 & 3 \\ -1 & 0 & 6 \end{bmatrix}$ **18.** $\begin{bmatrix} -3 & 5 & 8 \\ 4 & -2 & 3 \\ 9 & 0 & 6 \end{bmatrix}$

19. $\begin{bmatrix} -3 & 0 & 8 \\ 4 & 6 & 3 \\ 9 & -1 & 6 \end{bmatrix}$ **20.** $\begin{bmatrix} -3 & 0 & 5 \\ 4 & 6 & -2 \\ 9 & -1 & 0 \end{bmatrix}$

21. In Example 4, the determinant of

$$\mathbf{A} = \begin{bmatrix} 3 & 1 & -7 \\ 0 & 2 & 4 \\ 8 & -1 & 6 \end{bmatrix}$$

was evaluated by expansion along row 1.

a. Find $|\mathbf{A}|$ by expansion along row 2.

b. Find $|\mathbf{A}|$ by expansion along row 3.

c. Find $|\mathbf{A}|$ by expansion along column 3.

The statements in Exercises 22 through 25 are the "column" equivalents of Theorems 2 through 5. Prove each statement for the 2 x 2 case of

$$\mathbf{A} = \begin{bmatrix} a_{11} & a_{12} \\ a_{21} & a_{22} \end{bmatrix}.$$

22. If $\mathbf{A}'$ is the matrix obtained from $\mathbf{A}$ by interchanging two columns, then $|\mathbf{A}'| = -|\mathbf{A}|$.

23. If $\mathbf{A}'$ is the matrix obtained from $\mathbf{A}$ by multiplying each entry in a column of $\mathbf{A}$ by a constant real number k, then $|\mathbf{A}'| = k|\mathbf{A}|$.

24. If $\mathbf{A}'$ is the matrix obtained from $\mathbf{A}$ by adding a multiple of one column to another, then $|\mathbf{A}'| = |\mathbf{A}|$.

25. If two columns of $\mathbf{A}$ are identical, then $|\mathbf{A}| = 0$.

26. Prove: $|\mathbf{A}| = |\mathbf{A}^T|$. (See Exercise 35 of Section 9.2.) Use

$$\mathbf{A} = \begin{bmatrix} a_{11} & a_{12} \\ a_{21} & a_{22} \end{bmatrix}.$$

27. Let

$$\mathbf{A} = \begin{bmatrix} a_{11} & a_{12} & a_{13} \\ a_{21} & a_{22} & a_{23} \\ a_{31} & a_{32} & a_{33} \end{bmatrix}$$

Show

$$|\mathbf{A}| = a_{11}a_{22}a_{33} + a_{12}a_{23}a_{31} + a_{13}a_{21}a_{32} - a_{11}a_{23}a_{32} - a_{12}a_{21}a_{33} - a_{13}a_{22}a_{31}$$

28. Let

$$\mathbf{A} = \begin{bmatrix} a_{11} & a_{12} \\ a_{21} & a_{22} \end{bmatrix} \quad \text{and} \quad \mathbf{B} = \begin{bmatrix} b_{11} & b_{12} \\ b_{21} & b_{22} \end{bmatrix}$$

Prove or disprove:

$$|\mathbf{A} \cdot \mathbf{B}| = |\mathbf{A}| \cdot |\mathbf{B}|$$

(The result is also true for $n \times n$ matrices, **A** and **B**.)

In Exercises 29 through 31, find the determinant of each matrix. (See Example 6.)

29. $\begin{bmatrix} 1 & 5 & 6 & 7 \\ -3 & 0 & 5 & 8 \\ 4 & 6 & -2 & 3 \\ 9 & -1 & 0 & 6 \end{bmatrix}$ (*Hint:* Expand along row 1 and use Exercises 17 through 20.)

30. $\begin{bmatrix} 1 & 5 & 0 & 6 & 7 \\ -3 & 0 & 0 & 5 & 8 \\ 4 & 6 & 0 & -2 & 3 \\ 18 & 4.6 & 4 & 1 & 7 \\ 9 & -1 & 0 & 0 & 6 \end{bmatrix}$ (*Hint:* Expand along column 3 and use Exercise 29.)

31. $\begin{bmatrix} 1 & 2 & 3 & 4 & 5 \\ 14 & -8 & 7 & 6 & 11 \\ 6 & 9 & 8 & \pi & 6 \\ 4 & -1 & 0 & 8 & 1.75 \\ 10 & 20 & 30 & 40 & 50 \end{bmatrix}$

Computer Exercise

32. a. Write a program that will evaluate the determinant of any 2×2 matrix.
    b. Write a program that will evaluate the determinant of any 3×3 matrix. (*Hint:* Use Exercise 27, above.)
    c. Many versions of BASIC have a "built-in" determinant function. The statement

```
50 D = DET(A)
```

will find $|\mathbf{A}|$. Write a program to incorporate this statement to find the determinant of the matrix of Exercise 31 directly.

9.5 CRAMER'S RULE

This section applies the use of determinants to solving systems of n linear equations in n unknowns. Such systems are called $n \times n$ or **square** systems. It can be shown that the solution of

$$\begin{aligned} a_{11}x + a_{12}y &= b_1 \\ a_{21}x + a_{22}y &= b_2 \end{aligned} \tag{1}$$

is given by

$$x = \frac{a_{22}b_1 - a_{12}b_2}{a_{11}a_{22} - a_{12}a_{21}} \qquad y = \frac{a_{11}b_2 - a_{21}b_1}{a_{11}a_{22} - a_{12}a_{21}} \text{ provided} \tag{2}$$
$$a_{11}a_{22} - a_{12}a_{21} \neq 0$$

(See Exercise 27, Section 9.1.)

The next example shows how (2) can be rewritten in terms of determinants.

EXAMPLE 1 Evaluate each of the following determinants

$$|\mathbf{D}| = \begin{vmatrix} a_{11} & a_{12} \\ a_{21} & a_{22} \end{vmatrix} \qquad |\mathbf{D}_x| = \begin{vmatrix} b_1 & a_{12} \\ b_2 & a_{22} \end{vmatrix} \qquad |\mathbf{D}_y| = \begin{vmatrix} a_{11} & b_1 \\ a_{21} & b_2 \end{vmatrix}$$

Solution $|\mathbf{D}| = a_{11}a_{22} - a_{12}a_{21}$, $|\mathbf{D}_x| = a_{22}b_1 - a_{12}b_2$, $|\mathbf{D}_y| = a_{11}b_2 - a_{21}b_1$

Notice that $|\mathbf{D}|$ is the determinant of the coefficient matrix of (1) and that $|\mathbf{D}_x|$ is the determinant of the matrix $\begin{bmatrix} b_1 & a_{12} \\ b_2 & a_{22} \end{bmatrix}$ which is obtained from the coefficient matrix by replacing the column of x's coefficients with the constants b_1, b_2. Also, $|\mathbf{D}_y|$ is obtained from the determinant of the coefficient matrix by replacing the column of y's coefficients with the constants b_1, b_2.

The example above lets us rewrite (2) in determinant form as

$$x = \frac{|\mathbf{D}_x|}{|\mathbf{D}|} \qquad y = \frac{|\mathbf{D}_y|}{|\mathbf{D}|} \quad \text{provided } |\mathbf{D}| \neq 0$$

The above is referred to as **Cramer's Rule** and will be extended to larger square systems. First, however, we present some examples.

EXAMPLE 2 Use Cramer's Rule to solve the system

$$\begin{aligned} 2x + y &= 1 \\ 8x - 2y &= 22 \end{aligned}$$

Solution $|\mathbf{D}| = \begin{vmatrix} 2 & 1 \\ 8 & -2 \end{vmatrix}, \quad |\mathbf{D}_x| = \begin{vmatrix} 1 & 1 \\ 22 & -2 \end{vmatrix}, \quad |\mathbf{D}_y| = \begin{vmatrix} 2 & 1 \\ 8 & 22 \end{vmatrix}$

Thus, $|\mathbf{D}| = -12$, $|\mathbf{D}_x| = -24$, $|\mathbf{D}_y| = 36$. Hence,

$$x = \frac{|\mathbf{D}_x|}{|\mathbf{D}|} = \frac{-24}{-12} = 2 \quad \text{and} \quad y = \frac{|\mathbf{D}_y|}{|\mathbf{D}|} = \frac{36}{-12} = -3.$$

The reader should check these values.

EXAMPLE 3 Use Cramer's Rule to solve the system

$$\begin{aligned} x - 4y &= 5 \\ 3x + 6y &= -5 \end{aligned}$$

Solution

$$x = \frac{|\mathbf{D}_x|}{|\mathbf{D}|} = \frac{\begin{vmatrix} 5 & -4 \\ -5 & 6 \end{vmatrix}}{\begin{vmatrix} 1 & -4 \\ 3 & 6 \end{vmatrix}} = \frac{10}{18} = \frac{5}{9}$$

$$y = \frac{|\mathbf{D}_y|}{|\mathbf{D}|} = \frac{\begin{vmatrix} 1 & 5 \\ 3 & -5 \end{vmatrix}}{\begin{vmatrix} 1 & -4 \\ 3 & 6 \end{vmatrix}} = \frac{-20}{18} = -\frac{10}{9}$$

Before we proceed to generalize Cramer's Rule to larger systems, keep in mind that $|\mathbf{D}|$, the determinant of the coefficient matrix, cannot be zero. If $|\mathbf{D}| = 0$, then a unique solution to the system (1) does not exist (see Exercises 21 and 22).

Cramer's Rule

The system

$$\begin{aligned} a_{11}x_1 + a_{12}x_2 + \ldots + a_{1n}x_n &= b_1 \\ a_{21}x_1 + a_{22}x_2 + \ldots + a_{2n}x_n &= b_2 \\ \vdots \qquad \vdots \qquad\qquad \vdots \quad &\quad \vdots \\ a_{n1}x_1 + a_{n2}x_2 + \ldots + a_{nn}x_n &= b_n \end{aligned} \tag{3}$$

has the unique solution

$$x_1 = \frac{|\mathbf{D}_{x_1}|}{|\mathbf{D}|}, \quad x_2 = \frac{|\mathbf{D}_{x_2}|}{|\mathbf{D}|}, \ldots, x_n = \frac{|\mathbf{D}_{x_n}|}{|\mathbf{D}|}$$

provided $|\mathbf{D}| \neq 0$, and where $|\mathbf{D}|$ is the determinant of the coefficient matrix of (3). Also, each $|\mathbf{D}_{x_i}|$ is the determinant of the matrix obtained from the coefficient matrix by replacing the column of x_i's coefficients with the constants $b_1, b_2, \ldots, b_n$.

EXAMPLE 4 Use Cramer's Rule to solve the system

$$\begin{aligned} x - y + 6z &= 8 \\ 3x - y - 2z &= 6 \\ 2x - 3y + 4z &= 16 \end{aligned}$$

Solution The reader should verify the values of the various determinants below:

$$x = \frac{|\mathbf{D}_x|}{|\mathbf{D}|} = \frac{\begin{vmatrix} 8 & -1 & 6 \\ 6 & -1 & -2 \\ 16 & -3 & 4 \end{vmatrix}}{\begin{vmatrix} 1 & -1 & 6 \\ 3 & -1 & -2 \\ 2 & -3 & 4 \end{vmatrix}} = \frac{-36}{-36} = 1$$

$$y = \frac{|\mathbf{D}_y|}{|\mathbf{D}|} = \frac{\begin{vmatrix} 1 & 8 & 6 \\ 3 & 6 & -2 \\ 2 & 16 & 4 \end{vmatrix}}{-36} = \frac{144}{-36} = -4$$

$$z = \frac{|\mathbf{D}_z|}{|\mathbf{D}|} = \frac{\begin{vmatrix} 1 & -1 & 8 \\ 3 & -1 & 6 \\ 2 & -3 & 16 \end{vmatrix}}{-36} = \frac{-18}{-36} = \frac{1}{2}$$

To use Cramer's Rule to solve a 4 x 4 system is extremely time consuming. As the size of the system increases, Cramer's Rule is inefficient and the Gaussian elimination method of Section 9.1 is generally a better method to use.

EXERCISES FOR SECTION 9.5

In Exercises 1 through 20, use Cramer's Rule to solve each system. (See Examples 2, 3, and 4.)

1. $2x + 3y = 5$
   $6x - 2y = 4$
2. $x + 2y = 5$
   $3x + 2y = 3$
3. $2x - 4y = -3$
   $3x + y = -1$
4. $x + 3y = 9$
   $2x - 6y = 14$
5. $x + y = \frac{1}{2}$
   $4x - 8y = -7$
6. $x_1 + x_2 = 4$
   $5x_1 - 5x_2 = -7$
7. $6x_1 - x_2 = 0$
   $3x_1 + 2x_2 = 0$
8. $x_1 + x_2 = \sqrt{5}$
   $x_1 + \sqrt{5}x_2 = 5$
9. $.5x_1 + .1x_2 = 3.0$
   $.2x_1 - .5x_2 = -4.2$
10. $x + y + z = 11$
    $x + y - z = -1$
    $3x + 5y - 2z = 11$
11. $2x + y + z = 8$
    $3x - y + 10z = 93$
    $-x + y + z = 20$
12. $4y - z = 3$
    $8x + y - z = 5$
    $7x - 2y = 8$

13. $$\begin{aligned} 2x - y &= 0 \\ x - z &= -7 \\ 3y + 8z &= 56 \end{aligned}$$

14. $$\begin{aligned} x + y + z &= 6 \\ x + y - z &= 0 \\ 8x - y + 3z &= 15 \end{aligned}$$

15. $$\begin{aligned} x - 2y - 2z &= 12 \\ -x + y - 6z &= 16 \\ 3x + y - 10z &= 20 \end{aligned}$$

16. $$\begin{aligned} x_1 - 2x_2 - 2x_3 &= -15 \\ -x_1 + x_2 - 6x_3 &= -1 \\ 3x_1 + x_2 - 10x_3 &= 7 \end{aligned}$$

17. $$\begin{aligned} x_1 - 2x_2 - 2x_3 &= -16 \\ -x_1 + x_2 - 6x_3 &= -6 \\ 3x_1 + x_2 - 10x_3 &= -14 \end{aligned}$$

18. $$\begin{aligned} x_1 - 2x_2 - 2x_3 &= 7 \\ -x_1 + x_2 - 6x_3 &= -7 \\ 3x_1 + x_2 - 10x_3 &= 21 \end{aligned}$$

19. $$\begin{aligned} x_1 + x_2 + x_3 &= 0 \\ 2x_1 - x_2 &= 0 \\ 10x_1 + 5x_2 - 5x_3 &= .5 \end{aligned}$$

20. $$\begin{aligned} x_1 + 7x_2 - 8x_3 &= 4 \\ 2x_1 - x_2 - 10x_3 &= 7 \\ 3x_1 + 5x_2 + 7x_3 &= 0 \end{aligned}$$

21. Consider the system

$$\begin{aligned} x_1 + x_2 + x_3 &= 4 \\ 2x_1 - x_2 + x_3 &= 0 \\ 4x_1 + x_2 + 3x_3 &= 9 \end{aligned}$$

   a. Find $|\mathbf{D}|$, $|\mathbf{D}_{x_1}|$, $|\mathbf{D}_{x_2}|$, $|\mathbf{D}_{x_3}|$.
   b. This system has the following properties:
      i) $|\mathbf{D}| = 0$
      ii) at least one $|\mathbf{D}_{x_i}| \neq 0$

      Show that this system is inconsistent. (See Example 3, Section 9.1.)

22. Consider the system

$$\begin{aligned} x_1 + x_2 + x_3 &= 4 \\ 2x_1 - x_2 + x_3 &= 0 \\ 4x_1 + x_2 + 3x_3 &= 8 \end{aligned}$$

   a. Find $|\mathbf{D}|$, $|\mathbf{D}_{x_1}|$, $|\mathbf{D}_{x_2}|$, $|\mathbf{D}_{x_3}|$.
   b. This system has the following properties:
      i) $|\mathbf{D}| = 0$
      ii) each $|\mathbf{D}_{x_i}| = 0$

      Show that this system is dependent. (See Example 4, Section 9.1.)

Computer Exercise

23. Using the DET(X) function (see Section 9.4), write a program that will solve a square system using Cramer's Rule. Be sure to build in a "test" to check if $|\mathbf{D}| = 0$.

CHAPTER 9 IN REVIEW

I. Summary

In this chapter we saw that the system of n linear equations in n variables

$$\begin{aligned} a_{11}x_1 + a_{12}x_2 + \ldots + a_{1n}x_n &= b_1 \\ a_{21}x_1 + a_{22}x_2 + \ldots + a_{2n}x_n &= b_2 \\ \vdots \qquad\quad \vdots \qquad\qquad\quad \vdots \quad &\ \ \vdots \\ a_{n1}x_1 + a_{n2}x_2 + \ldots + a_{nn}x_n &= b_n \end{aligned}$$

can be represented by the augmented matrix

$$\left[\begin{array}{cccc|c} a_{11} & a_{12} & \ldots & a_{1n} & b_1 \\ a_{21} & a_{22} & \ldots & a_{2n} & b_2 \\ \vdots & \vdots & & \vdots & \vdots \\ a_{n1} & a_{n2} & \ldots & a_{nn} & b_n \end{array}\right]$$

Elementary row transformations are performed on this augmented matrix in order to obtain a solution (if one exists). This method of solution is called the Gaussian elimination method. In Section 9.2 we examined the algebra of matrices. Matrix addition and multiplication were studied, as well as applied problems involving matrices.

The inverse of a matrix **A**, denoted $\mathbf{A}^{-1}$, has the property that $\mathbf{A} \bullet \mathbf{A}^{-1} = \mathbf{A}^{-1} \bullet \mathbf{A} = \mathbf{I}$, the identity matrix. We learned how to find $\mathbf{A}^{-1}$ using the elimination procedure and saw how the matrix inverse can be used to solve $n \times n$ systems of equations.

Finally, in Sections 9.4 and 9.5, determinants and Cramer's Rule were studied as an alternative method of solving systems of linear equations.

II. Vocabulary/Symbol True–False Quiz

If the statement is always true, write "TRUE." However, if the statement can be false, replace the underlined word or phrase with a different word or phrase that will make the new statement true.

1. $\left[\begin{array}{cc|c} 1 & 5 & 7 \\ 3 & 8 & 6 \end{array}\right]$ is an example of an <u>augmented matrix.</u>
2. If $\left[\begin{array}{ccc|c} 1 & 5 & 0 & 1 \\ 0 & 0 & 1 & 2 \\ 0 & 0 & 0 & 3 \end{array}\right]$ is the result obtained by using Gaussian elimination, then the system is <u>inconsistent</u>.
3. A matrix with 3 rows and 5 columns has <u>8</u> components.
4. Two matrices are equal if <u>they have the same dimension.</u>
5. In order for **A** + **B** to be defined, **A** and **B** must be matrices that <u>have the same dimension.</u>
6. In order for **A** • **B** to be defined, <u>the number of rows of **A** must equal the number of columns of **B**.</u>
7. If **A** is a 3 x 4 matrix and **B** is a 4 x 5 matrix, then **A** • **B** is <u>a 3 x 5 matrix.</u>
8. If $\mathbf{A}^{-1}$ is the inverse of matrix **A**, then $\underline{\mathbf{A} \bullet \mathbf{A}^{-1} = \mathbf{I}}$.
9. If $|\mathbf{A}| = x$ and $|\mathbf{B}| = y$ then $|\mathbf{A} \bullet \mathbf{B}| = \underline{xy}$.
10. If **I** is the identity matrix, then $|\mathbf{I}| = \underline{1}$.

III. Chapter Test

In questions 1 through 6, use the Gaussian elimination method to solve each system.

1. $\begin{aligned} x_1 - 4x_2 &= 11 \\ 5x_1 + 3x_2 &= 9 \end{aligned}$

2. $\begin{aligned} 2x_1 + 5x_2 &= 6 \\ 3x_1 - 4x_2 &= 10 \end{aligned}$

3. $\begin{aligned} x_1 + x_2 + x_3 &= 5 \\ 2x_1 - x_2 + 6x_3 &= 0 \\ 4x_1 + x_2 + 8x_3 &= 10 \end{aligned}$

4. $\begin{aligned} x_1 + 2x_2 + x_3 &= -1 \\ 3x_1 - 4x_2 - x_3 &= 16 \\ 4x_1 - 10x_2 + 2x_3 &= -1 \end{aligned}$

5. $\begin{aligned} x_1 + x_2 - x_3 &= 7 \\ 2x_1 - 3x_2 + 2x_3 &= 8 \\ 3x_1 - 2x_2 + x_3 &= 15 \end{aligned}$

6. $\begin{aligned} x_1 + x_2 + x_3 + x_4 &= 4 \\ 2x_1 + 3x_2 - 2x_3 - x_4 &= 4 \\ x_2 - 5x_3 + 4x_4 &= -21 \\ x_1 - 10x_4 &= 21 \end{aligned}$

In questions 7 through 11, perform the indicated operation. Use

$$\mathbf{W} = \begin{bmatrix} 4 & 1 \\ 0 & 2 \\ 3 & -1 \end{bmatrix} \quad \mathbf{X} = \begin{bmatrix} 0 & 1 \\ 2 & 3 \\ 4 & 5 \end{bmatrix} \quad \mathbf{Y} = \begin{bmatrix} 1 & -1 & 6 \\ 0 & 0 & 5 \\ 1 & 4 & -2 \end{bmatrix}$$

$$\mathbf{Z} = \begin{bmatrix} 1 & -10 \\ 5 & 2 \end{bmatrix} \quad \mathbf{P} = \begin{bmatrix} \frac{1}{2} & \frac{1}{2} \\ \frac{1}{4} & \frac{3}{4} \end{bmatrix}$$

7. $\mathbf{X} \bullet \mathbf{P}$
8. $\mathbf{Y} \bullet (\mathbf{W} + 2\mathbf{X})$
9. $\mathbf{W} \bullet (4\mathbf{P} - \mathbf{Z})$
10. $\mathbf{Y} \bullet \mathbf{Y}$
11. $\mathbf{Z} \bullet \mathbf{P} - \mathbf{P} \bullet \mathbf{Z}$

In questions 12 through 15, find the inverse of the given matrix.

12. $\begin{bmatrix} 5 & 11 \\ 4 & 9 \end{bmatrix}$

13. $\begin{bmatrix} 1 & 3 \\ -8 & 6 \end{bmatrix}$

14. $\begin{bmatrix} 1 & -1 & 6 \\ 0 & 0 & 1 \\ 1 & 4 & -2 \end{bmatrix}$

15. $\begin{bmatrix} 4 & 0 & 0 & 0 \\ 0 & 4 & 0 & 0 \\ 0 & 0 & 4 & 0 \\ 0 & 0 & 0 & 4 \end{bmatrix}$

16. Consider the system

$$\begin{aligned} x - 3y &= 20 \\ -8x + 6y &= -34 \end{aligned}$$

a. What is the coefficient matrix of the system?
b. Use the matrix inverse of your answer to (a) to solve the system.

17. Find $|\mathbf{A}|$ if $\mathbf{A} = \begin{bmatrix} 6 & 4 \\ 3 & 2 \end{bmatrix}$.

18. Find $|\mathbf{B}|$ if $\mathbf{B} = \begin{bmatrix} 1 & -1 & 6 \\ 0 & 0 & 1 \\ 1 & 4 & -2 \end{bmatrix}$.

19. Find $|\mathbf{C}|$ if $\mathbf{C} = \begin{bmatrix} 3 & 0 & 0 & 0 \\ 4 & 1 & -1 & 6 \\ 19 & 0 & 0 & 1 \\ \pi & 1 & 4 & -2 \end{bmatrix}$.

20. Use Cramer's Rule to solve the system:

$$\begin{aligned} 4x_1 - x_2 + x_3 &= 5 \\ x_1 + x_2 + x_3 &= 2 \\ 3x_1 - 5x_2 + 7x_3 &= 10 \end{aligned}$$

IV. Student Research Project

1. *Uses of Matrices.* The algebra of matrices has been applied to many fields—often as a solution device to solve systems of linear equations. Construct a bibliography of matrix applications in a field that interests you. One source is Chapter Five of *Finite Mathematics with Applications to Life* by L. Gilligan and R. Nenno, Goodyear Publishing Company, Santa Monica, California, 1979.

2. *History of Matrices.* Prepare a report on the history of matrices and matrix notation. Three useful sources are:

Bell, E. T. *Men of Mathematics.* New York: Simon and Schuster, 1937. (Chapter 21 gives an account of Arthur Cayley, often cited as the inventor of the matrix notation.)

Eves, Howard. *An Introduction to the History of Mathematics,* 4th ed. New York: Holt, Rinehart, and Winston, 1976.

Hogben, Lancelot. *Mathematics in the Making.* Garden City, N.J.: Doubleday & Co., 1960. See Chapter 7.

CHAPTER **10**

Polynomials

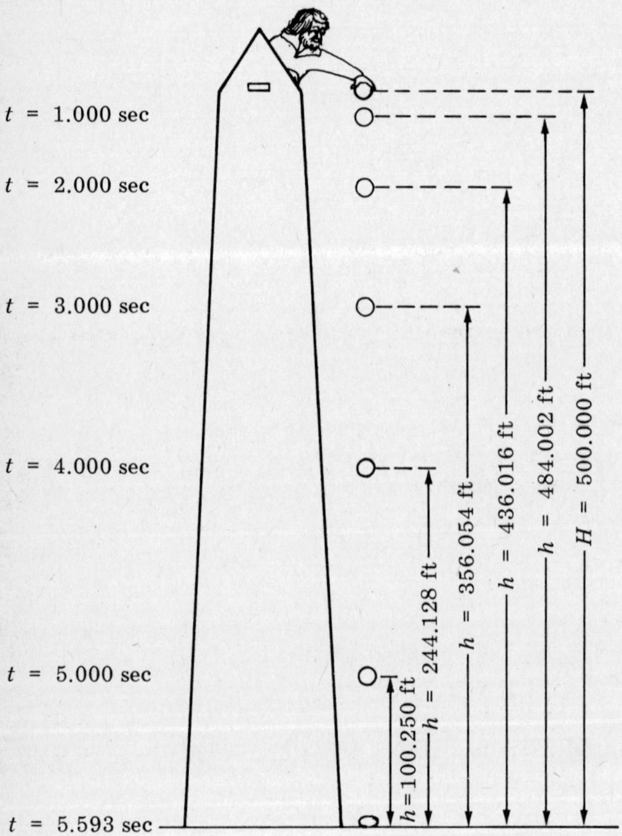

Dropping a ball from the Washington Monument.

THE PROBLEM. Consider a modern-day Galileo dropping a ball from the Washington Monument. The problem is to compute the height of the ball each ½ sec after it is dropped. If we take into account air resistance as well as gravity, one approximate formula that will give us the successive heights is

$$h = H - \frac{g}{2}t^2 + \frac{a}{w}(0.001)t^3$$

where

- h = height of ball above ground, ft
- t = time after ball has been dropped, sec
- a = radius of ball, ft
- w = weight of ball, lb
- g = acceleration of gravity (a constant)
- H = height of Washington Monument (a constant)

The information above is from pages 56 and 57 of: Use of Computers in Biology and Medicine, *by Robert Steven Ledley with the assistance of James Bruce Wilson, McGraw-Hill Book Company, New York, 1965.*

10.1 THE REMAINDER AND FACTOR THEOREMS

Our study of polynomials began in Section 1.4. There, we used polynomial notation and synthetic division. In the present chapter, we will continue to examine polynomials; in particular, we shall solve polynomial equations in one variable and graph polynomial equations in two variables.

We begin here by recalling a definition:

DEFINITION

$$a_n x^n + a_{n-1}x^{n-1} + \ldots + a_1 x + a_0$$

is called a **polynomial** (in x) **of degree n with real coefficients** provided $a_n \neq 0$, each a_i is a real number, and n is a nonnegative integer.

Recall also, that when a polynomial $P(x)$ is divided by another polynomial $D(x)$, a *quotient*, $Q(x)$ and *remainder*, $R(x)$ are formed. We write

$$\frac{P(x)}{D(x)} = Q(x) + \frac{R(x)}{D(x)} \qquad \text{Form I}$$

or

$$P(x) = D(x) \cdot Q(x) + R(x) \qquad \text{Form II}$$

With this information recalled from Chapter 1, we are ready to examine the **zeros** of the polynomial function $y = P(x)$. The zeros of $y = P(x)$ are the solutions (or **roots**) of the polynomial equation $P(x) = 0$. Of course, if $P(x) = 0$ is a linear equation (i.e., $P(x)$ has degree one) or a quadratic equation (i.e., $P(x)$ has degree two), we already know how to solve it. Primarily, we will be concerned with polynomials of degree three or more.

The importance of the following three theorems will be shown in the examples that follow them.

THEOREM 1

Remainder Theorem

If a polynomial $P(x)$ is divided by $x - c$, where c is a real number, then the remainder is equal to $P(c)$.

Proof: With divisor $x - c$, we use form II to write

$$P(x) = (x - c)Q(x) + R \tag{1}$$

where $Q(x)$ is a polynomial of degree $n - 1$ and R is a constant. Since (1) is true for all values of x, we set $x = c$:

$$P(c) = (c - c)Q(c) + R$$

Thus, $P(c) = R$.

EXAMPLE 1 Suppose $P(x) = x^3 + x^2 - 4x + 5$. Use the Remainder Theorem to find $P(2)$.

Solution We use synthetic division (see Section 1.4) to find the remainder when $P(x)$ is divided by $x - 2$:

$$\begin{array}{r|rrrr} 2 & 1 & 1 & -4 & 5 \\ & & 2 & 6 & 4 \\ \hline & 1 & 3 & 2 & \boxed{9} \end{array} \leftarrow \text{Remainder}$$

Thus, according to the Remainder Theorem, $P(2) = 9$.

An important consequence of the Remainder Theorem is the Factor Theorem.

THEOREM 2

Factor Theorem

If $P(x)$ is a polynomial, $x - c$ is a factor of $P(x)$ if and only if $P(c) = 0$, where c is a real number.

Proof: If $x - c$ is a factor of $P(x)$, the remainder must be zero when $P(x)$ is divided by $x - c$. By the Remainder Theorem, that zero remainder is $P(c)$. Thus $P(c) = 0$.

Conversely, if $P(c) = 0$, again by the Remainder Theorem,

$$P(x) = (x - c)Q(x) + 0 = (x - c)Q(x)$$

Thus, $x - c$ is a factor of $P(x)$.

EXAMPLE 2 Show that $x + 1$ is a factor of $P(x) = x^4 + 7x^3 + x^2 - 7x - 2$.

Solution With $c = -1$, we must show the remainder of $P(x) \div (x + 1)$ is zero—or equivalently $P(-1) = 0$:

$$P(-1) = (-1)^4 + 7(-1)^3 + (-1)^2 - 7(-1) - 2$$
$$= 1 - 7 + 1 + 7 - 2 = 0$$

EXAMPLE 3 Suppose $P(x) = x^2 + 5ax + a$. Find a so that $x - 2$ is a factor of $P(x)$.

Solution In order for $x - 2$ to be a factor of $P(x)$, by the Factor Theorem, $P(2) = 0$

$$P(2) = 0 = 2^2 + 5a(2) + a = 11a + 4$$

Hence,

$$11a + 4 = 0$$
$$a = -\tfrac{4}{11}$$

EXAMPLE 4 a. Is $x = 3$ a root of $x^3 - 3x^2 - 5x + 15 = 0$?
b. Find all zeros of $P(x) = x^3 - 3x^2 - 5x + 15$.

Solution a. We use synthetic division to find $P(3)$:

$$\begin{array}{r|rrrr} 3 & 1 & -3 & -5 & 15 \\ & & 3 & 0 & -15 \\ \hline & 1 & 0 & -5 & \boxed{0} \end{array}$$

Since the remainder is 0, 3 *is* a root of $x^3 - 3x^2 - 5x + 15 = 0$.

b. According to the division in part a,

$$x^3 - 3x^2 - 5x + 15 = (x - 3)(x^2 - 5)$$

The other zeros of $P(x)$ occur when $x^2 - 5 = 0$. Thus the zeros of $P(x)$ are 3, $\sqrt{5}, -\sqrt{5}$.

The next theorem is stated without proof:

THEOREM 3

Fundamental Theorem of Algebra

Every polynomial of degree $n \geqslant 1$ has at least one zero (perhaps complex).

In Exercise 30, the reader is asked to prove Theorem 4, which is a consequence of the Fundamental Theorem and the Factor Theorem.

THEOREM 4 If $P(x) = a_n x^n + a_{n-1}x^{n-1} + \ldots + a_1 x + a_0$ is a polynomial of degree $\geqslant 1$, then there exist numbers $c_1, c_2, \ldots, c_n$ (perhaps complex) such that

$$P(x) = a_n(x - c_1)(x - c_2)\ldots(x - c_n)$$

EXAMPLE 5 a. Find a polynomial $P(x)$ of degree 3 with zeros 1, 2, and -4.

b. Find a polynomial $P(x)$ of degree 3 with zeros 1, 2, and -4 such that $P(-1) = 9$.

Solution a. By Theorem 4,

$$P(x) = a(x - 1)(x - 2)(x + 4)$$

Since this polynomial has degree three, it satisfies the conditions; a is an arbitrary nonzero number.

b. With $P(x) = a(x - 1)(x - 2)(x + 4)$, we must further satisfy $P(-1) = 9$.

$$\begin{aligned} P(-1) = a(-1 - 1)(-1 - 2)(-1 + 4) &= 9 \\ 18a &= 9 \\ a &= \tfrac{1}{2} \end{aligned}$$

Thus,

$$P(x) = \tfrac{1}{2}(x - 1)(x - 2)(x + 4) = \tfrac{1}{2}x^3 + \tfrac{1}{2}x^2 - 5x + 4.$$

EXAMPLE 6 Express $P(x) = x^4 - 8x^3 + 41x^2$ as a product of 4 linear factors.

Solution By Theorem 4, we need only find the zeros of $P(x)$. Clearly,

$$P(x) = x^2(x^2 - 8x + 41)$$

which we can write as

$$P(x) = (x - 0)(x - 0)(x^2 - 8x + 41).$$

Notice that 0 is a root *twice*—we say it is a **root of multiplicity two** or a **double root**. The other zeros of $P(x)$ are the zeros of $x^2 - 8x + 41$, found by the quadratic formula:

$$\frac{8 \pm \sqrt{64 - 164}}{2} = \frac{8 \pm 10i}{2} = 4 \pm 5i$$

The linear factors $(x - 4 - 5i)$ and $(x - 4 + 5i)$ complete the result:

$$P(x) = (x)(x)(x - 4 - 5i)(x - 4 + 5i)$$

EXERCISES FOR SECTION 10.1

In Exercises 1 through 15, find the value of the stated expression by using the Remainder Theorem. (See Example 1.)

1. $P(x) = x^3 - 4x^2 - 3x + 12$
   Find $P(2)$
2. $P(x) = x^3 - 4x^2 - 3x + 12$
   Find $P(1)$
3. $P(x) = x^3 - 4x^2 - 3x + 12$
   Find $P(-2)$
4. $P(x) = x^3 - 4x^2 - 3x + 12$
   Find $P(4)$
5. $P(x) = x^4 - 3x^3 + 3x - 1$
   Find $P(2)$
6. $P(x) = x^4 - 3x^3 + 3x - 1$
   Find $P(-2)$
7. $P(x) = x^4 - 3x^3 + 3x - 1$
   Find $P(-4)$
8. $P(x) = x^4 - 3x^3 + 3x - 1$
   Find $P(-1)$
9. $P(x) = 2x^4 - x^2 - 1$
   Find $P(1)$
10. $P(x) = 2x^4 - x^2 - 1$
    Find $P(4)$
11. $P(x) = 2x^4 - x^2 - 1$
    Find $P(-3)$
12. $P(x) = 2x^4 - x^2 - 1$
    Find $P(-1)$
13. $P(x) = 3x^5 - x^4 + 2x^2 - 1$
    Find $P(1)$
14. $P(x) = 3x^5 - x^4 + 2x^2 - 1$
    Find $P(-2)$
15. $P(x) = 3x^5 - x^4 + 2x^2 - 1$
    Find $P(\frac{1}{3})$
16. If $x + k$ is a factor of $P(x) = x^3 - 3x + 14$, what is the value of $-k^3 + 3k + 14$?
17. (True or False) If $P(-7) = 0$, then $x + 7$ is a factor of $P(x)$, where $P(x)$ is a polynomial.
18. (True or False) If $P(-7) = 0$, then -7 is a zero of $P(x)$, where $P(x)$ is a polynomial.
19. (True or False) If the remainder is zero when the polynomial $P(x)$ is divided by $x + 4$, the point $(-4, 0)$ is an x-intercept of the graph $y = P(x)$.

In Exercises 20 through 29, determine whether or not the given linear expression is a factor of the given polynomial, $P(x)$. (See Example 2.)

20. $x - 1$; $P(x) = x^3 - x^2 + x - 1$
21. $x + 1$; $P(x) = x^3 - x^2 + x - 1$
22. $x + 3$; $P(x) = 3x^3 - x^2 + 21$
23. $x - 4$; $P(x) = x^3 + 3x^2 - 16x - 48$
24. $x + 4$; $P(x) = x^3 + 3x^2 - 16x - 48$
25. $x - \frac{1}{3}$; $P(x) = 3x^2 - 14x - 5$
26. $x - 3$; $P(x) = x^4 - 81$
27. $x + 2$; $P(x) = x^6 + 3x^5 + 2x^4 - x^2 - 3x - 2$
28. $x + \frac{1}{2}$; $P(x) = 2x^3 - 15x^2 + 31x - 12$
29. $2x + 1$; $P(x) = 2x^3 - 15x^2 + 31x - 12$
30. Prove Theorem 4.

In Exercises 31 through 35, find all zeros of the given polynomial. (See Examples 4 and 6.)

31. Find all zeros of $P(x) = x^3 + 5x^2 - 2x - 24$ given that $x - 2$ is a factor of $P(x)$.

32. Find all zeros of $P(x) = 2x^3 + 6x^2 - x - 3$ given that -3 is a zero.
33. Find all zeros of $P(x) = x^4 - 7x^3 + 14x^2 - 14x + 24$ given that $x - 3$ and $x - 4$ are linear factors of $P(x)$.
34. Find all zeros of $P(x) = 2x^4 + 3x^3 - 3x^2 - 7x - 3$ given that -1 is a root of multiplicity three of the equation $P(x) = 0$.
35. Find all zeros of $P(x) = 2x^4 - x^2 - 1$. (See Exercises 9 and 12.)
36. For what value(s) of c is $x^2 + 7x + c$ divisible by $x - c$?
37. What restrictions exist on the positive integer n if $x + a$ is a factor of $x^n - a^n$?
38. Show that $x - 2$ is a factor of $x^{99} - 2x^{98}$.
39. Show that $x + 2$ is not a factor of $x^{99} - 2x^{98}$.
40. Show that $x + 1$ is a factor of $93x^{97} + 100x^{44} - 7$.

Calculator Exercise

41. Find all zeros of the polynomial $P(x) = .5x^3 - .2x^2 + .5x - .2$ if one zero is .4.

10.2 THE RATIONAL ROOT THEOREM

In the previous section we examined ways of determining whether or not a *given* number c is a zero of a certain polynomial $P(x)$. In this section, we want to examine ways of *finding* a real number c such that $P(c) = 0$. The key to finding a **rational root** is the following theorem. Its proof is outlined in Exercise 40.

THEOREM 5

Rational Root Theorem

If $P(x) = a_nx^n + a_{n-1}x^{n-1} + \ldots + a_1x + a_0$ is a polynomial with integer coefficients ($a_n \neq 0$), $P\left(\frac{p}{q}\right) = 0$, and $\frac{p}{q}$ is a rational number in lowest terms, then p is a factor of a_0 and q is a factor of a_n.

To illustrate the usefulness of the theorem, consider the polynomial $P(x) = 2x^3 + 3x^2 - 6x - 9$. *If* a rational root of $P(x) = 0$ exists*, it is of the form $\frac{p}{q}$ where p must be a factor of -9 and q is a factor of 2. The possibilities for p are $\pm 1, \pm 3, \pm 9$; the possibilities for q are $\pm 1, \pm 2$. The collection of all possible $\frac{p}{q}$'s is

$$\pm 1, \pm \tfrac{1}{2}, \pm 3, \pm \tfrac{3}{2}, \pm 9, \pm \tfrac{9}{2}$$

*Note: Theorem 5 does not guarantee the existence of any rational roots, whatsoever! If rational roots exist, however, Theorem 5 does give the form of the rational roots.

and that represents 12 *possible* rational roots. Trial and error leads to *one* that actually is a root, $-\frac{3}{2}$. Synthetic division proves very useful in such a determination. All the roots of $P(x) = 0$, where $P(x) = 2x^3 + 3x^2 - 6x - 9$ are found in the next example.

EXAMPLE 1 Find all roots to $P(x) = 0$, where $P(x) = 2x^3 + 3x^2 - 6x - 9$.

Solution The previous discussion yielded $\pm 1, \pm\frac{1}{2}, \pm 3, \pm\frac{3}{2}, \pm 9, \pm\frac{9}{2}$ as possible rational roots. We try $-\frac{3}{2}$ below:

$$\begin{array}{r|rrrr} -\frac{3}{2} & 2 & 3 & -6 & -9 \\ & & -3 & 0 & 9 \\ \hline & 2 & 0 & -6 & \boxed{0} \end{array}$$

Thus, $-\frac{3}{2}$ is a root. Furthermore, $2x^3 + 3x^2 - 6x - 9 = (x + \frac{3}{2})(2x^2 - 6)$ and other zeros of $P(x)$ are found by examining $2x^2 - 6 = 0$:

$$2x^2 - 6 = 0 \quad \text{or} \quad x = \pm\sqrt{3}$$

Finally, there are three real roots to $2x^3 + 3x^2 - 6x - 9 = 0$; one is rational $(-\frac{3}{2})$ and two are irrational ($\sqrt{3}$ and $-\sqrt{3}$).

EXAMPLE 2 Find all solutions to $x^4 - 3x^3 + 3x - 1 = 0$. (1)

Solution The *possible* rational roots are ± 1. We try 1:

$$\begin{array}{r|rrrrr} 1 & 1 & -3 & 0 & 3 & -1 \\ & & 1 & -2 & -2 & 1 \\ \hline & 1 & -2 & -2 & 1 & \boxed{0} \end{array}$$

Thus, 1 *is* a root. Also, equation (1) can be written as

$$(x - 1)(x^3 - 2x^2 - 2x + 1) = 0.$$

The possible rational roots of $x^3 - 2x^2 - 2x + 1$ are ± 1. We try -1:

$$\begin{array}{r|rrrr} -1 & 1 & -2 & -2 & 1 \\ & & -1 & 3 & -1 \\ \hline & 1 & -3 & 1 & \boxed{0} \end{array}$$

So, equation (1) can be rewritten:

$$(x + 1)(x - 1)(x^2 - 3x + 1) = 0$$

The remaining roots are found by using the quadratic formula on $x^2 - 3x + 1 = 0$:

$$x = \frac{3 \pm \sqrt{9-4}}{2} \quad \text{or} \quad \frac{3+\sqrt{5}}{2}, \frac{3-\sqrt{5}}{2}$$

The four solutions to (1) are $1, -1, \frac{3+\sqrt{5}}{2}$, and $\frac{3-\sqrt{5}}{2}$.

The technique of finding zeros of a polynomial $P(x)$ has been used thus far to obtain the roots of a polynomial equation in a single variable, namely $P(x) = 0$. Next we turn our attention to graphing polynomial functions of the form $y = P(x)$, where it will be seen that the zeros of $P(x)$ are the x-intercepts for the graph of $y = P(x)$.

EXAMPLE 3 Graph $y = x^4 - 3x^3 + 3x - 1$.

Solution The x-intercepts were found in Example 2. They are $(1,0)$, $(-1,0)$, $\left(\frac{3+\sqrt{5}}{2}, 0\right)$, and $\left(\frac{3-\sqrt{5}}{2}, 0\right)$. The y-intercept is clearly $(0,-1)$. Other points are plotted to graph the function below.

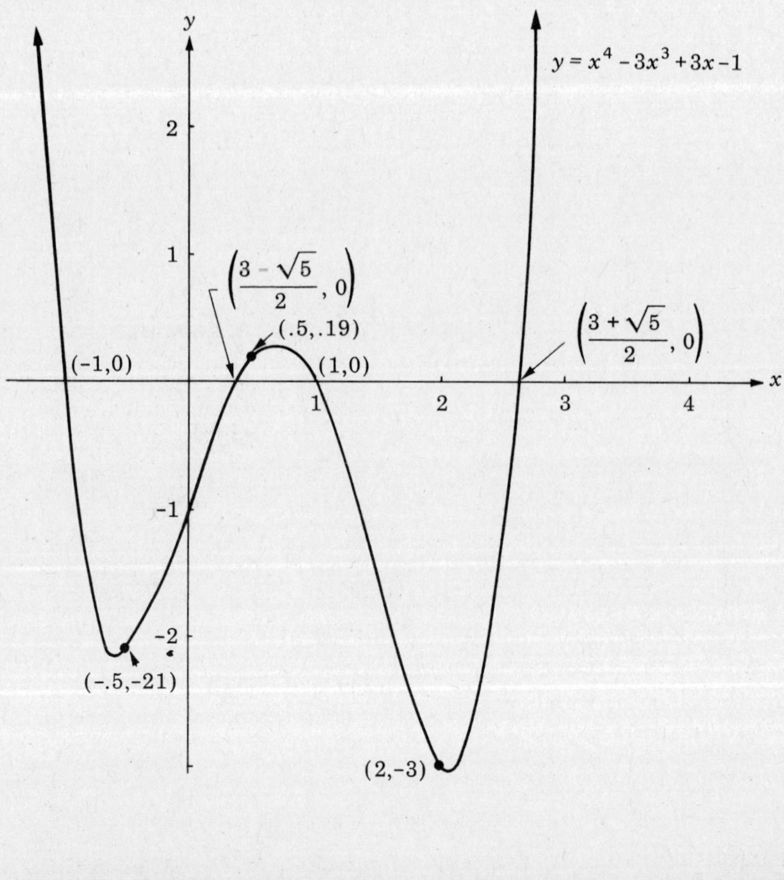

EXAMPLE 4 Graph $y = 2x^3 - 3x^2 - 18x - 8$.

Solution The y-intercept is $(0,-8)$. The x-intercepts are found by solving $2x^3 - 3x^2 - 18x - 8 = 0$. The possible rational roots are $\pm\frac{1}{2}, \pm 1, \pm 2, \pm 4, \pm 8$. We try $-\frac{1}{2}$:

$$\begin{array}{r|rrrr} -\frac{1}{2} & 2 & -3 & -18 & -8 \\ & & -1 & 2 & 8 \\ \hline & 2 & -4 & -16 & \lfloor 0 \end{array}$$

The remaining roots are the solutions to the quadratic equation $2x^2 - 4x - 16 = 0$. They are 4 and -2. Thus the x-intercepts are $(-\frac{1}{2}, 0)$, $(4,0)$, and $(-2,0)$. These are used to graph the cubic function below. Four or five other points are also needed to get a reasonably accurate graph.

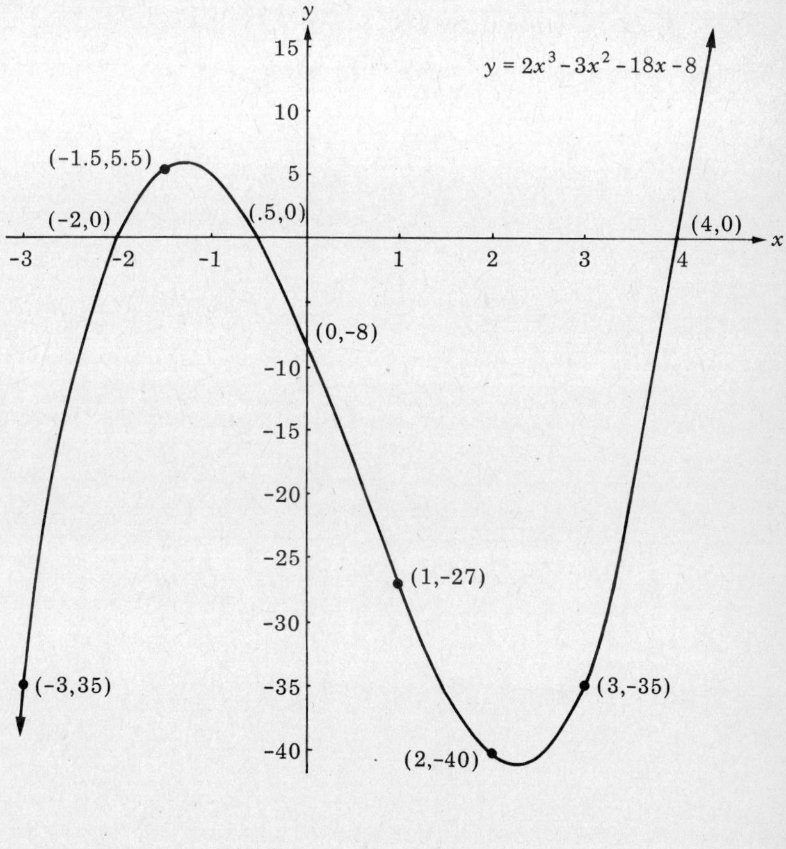

We conclude this section with an example of a polynomial with both rational roots and complex roots

EXAMPLE 5 Find the solutions to $x^4 - 4x^3 + 5x^2 - 4x + 4 = 0$.

Solution The *possible* rational solutions, by Theorem 5 are:

$$\pm 1, \pm 2, \text{ and } \pm 4$$

It is seen below that 2 is a root of multiplicity two:

$$\begin{array}{r|rrrrr} 2 & 1 & -4 & 5 & -4 & 4 \\ & & 2 & -4 & 2 & -4 \\ \hline & 1 & -2 & 1 & -2 & \boxed{0} \end{array}$$

Now, we have shown that

$$x^4 - 4x^3 + 5x^2 - 4x + 4 = (x - 2)(x^3 - 2x^2 + x - 2)$$

and next we examine the polynomial $x^3 - 2x^2 + x - 2$:

$$\begin{array}{r|rrrr} 2 & 1 & -2 & 1 & -2 \\ & & 2 & 0 & 2 \\ \hline & 1 & 0 & 1 & \boxed{0} \end{array}$$

Thus, $x^4 - 4x^3 + 5x^2 - 4x + 4 = (x - 2)^2(x^2 + 1)$ and solutions to $x^4 - 4x^3 + 5x^2 - 4x + 4 = 0$ are 2, 2, i, and $-i$.

EXERCISES FOR SECTION 10.2

In Exercises 1 through 28, a) determine the *possible* rational roots (according to Theorem 5), and b) use synthetic division to find all the roots of each polynomial $P(x) = 0$. (See Examples 1, 2, and 5.)

1. $P(x) = x^3 + x^2 - 4x - 4$
2. $P(x) = x^3 - 4x^2 + x + 6$
3. $P(x) = x^3 - 2x^2 - 11x + 12$
4. $P(x) = x^3 + 3x^2 - 10x - 24$
5. $P(x) = 2x^3 + x^2 - 13x + 6$
6. $P(x) = 3x^3 - 8x^2 - 5x + 6$
7. $P(x) = 2x^4 - 9x^3 - 3x^2 + 34x - 24$
8. $P(x) = 4x^3 + 20x^2 + 19x - 15$
9. $P(x) = 6x^3 - 11x^2 - 24x + 9$
10. $P(x) = 6x^4 - x^3 - 46x^2 + 39x + 18$
11. $P(x) = 32x^3 + 16x^2 - 42x + 9$
12. $P(x) = 320x^3 + 160x^2 - 420x + 90$
13. $P(x) = 2x^3 - 12x - 8$
14. $P(x) = 12x^3 + 16x^2 - 7x - 6$
15. $P(x) = 36x^4 + 36x^3 - x^2 - 9x - 2$
16. $P(x) = x^3 - 4x^2 + 3x + 2$
17. $P(x) = x^3 - 6x - 4$
18. $P(x) = x^3 + 5x^2 + x - 7$
19. $P(x) = x^4 - 2x^3 - 8x^2 + 10x + 15$
20. $P(x) = 2x^4 + 3x^3 - 8x^2 - 9x + 6$
21. $P(x) = 6x^4 - 31x^3 + 19x^2 + 19x + 3$
22. $P(x) = 2x^3 - 5x^2 + 6x - 2$
23. $P(x) = 3x^3 - 13x^2 + 19x - 5$
24. $P(x) = 4x^3 - 11x^2 + 26x - 15$
25. $P(x) = x^4 + x^3 - x^2 + x - 2$
26. $P(x) = x^4 + 3x^3 - 2x^2 - 10x - 12$
27. $P(x) = 4x^4 - 18x^3 + 34x^2 - 6x - 14$
28. $P(x) = x^4 + x^3 - 5x^2 + x - 6$

In Exercises 29 through 39, graph each polynomial function. Be sure to find all intercepts. (See Examples 3 and 4.)

29. $y = x^3 + x^2 - 4x - 4$
30. $y = x^3 - 4x^2 + x + 6$
31. $y = -x^3 - x^2 + 4x + 4$
32. $y = -x^3 + 4x^2 - x - 6$
33. $y = x^4 - 5x^2 + 4$
34. $y = -x^4 + 5x^2 - 4$
35. $y = 2x^4 - 10x^2 + 8$
36. $y = -2x^4 + 10x^2 - 8$
37. $y = 2x^3 - 12x - 8$
38. $y = 2x^4 - 9x^3 - 3x^2 + 34x - 24$
39. $y = x^4 + x^3 - 5x^2 + x - 6$
40. In order to prove Theorem 5, another theorem, called the **Fundamental Theorem of Arithmetic**, is needed:

"Every integer has a unique prime* factorization."

Supply the reasons for each step in the proof of Theorem 5 below:

a) $P\left(\frac{p}{q}\right) = 0$ and $\frac{p}{q}$ is in lowest terms
b) $a_n\left(\frac{p}{q}\right)^n + a_{n-1}\left(\frac{p}{q}\right)^{n-1} + \ldots + a_1\left(\frac{p}{q}\right) + a_0 = 0$
c) $a_np^n + a_{n-1}p^{n-1}q + \ldots + a_1pq^{n-1} + a_0q^n = 0$
d) $a_np^n + a_{n-1}p^{n-1}q + \ldots + a_1pq^{n-1} = -a_0q^n$
e) p is a factor of the left-hand side of d)
f) p is a factor of $-a_0q^n$
g) p is a factor of a_0
h) $a_{n-1}p^{n-1}q + \ldots + a_1pq^{n-1} + a_0q^n = -a_np^n$
i) q is a factor of the left-hand side of h)
j) q is a factor of $-a_np^n$
k) q is a factor of a_n.

For Exercises 41 through 68, express the polynomials of Exercises 1-28 as a product of linear factors.

69. Use the Rational Root Theorem to prove $\sqrt{2}$ is not rational. (*Hint:* examine $P(x) = x^2 - 2$.)
70. In examining the polynomial

$$x^4 - 2x^3 - 5x^2 + 19x - 18$$

a student found 2 was a root by the synthetic division below

$$\begin{array}{r|rrrrr} 2 & 1 & -2 & -5 & 19 & -18 \\ & & 2 & 0 & -10 & 18 \\ \hline & 1 & 0 & -5 & 9 & \boxed{0} \end{array}$$

What can be immediately concluded about the multiplicity of the root 2?

*A prime number is an integer greater than one that has no factors other than one and itself.

10.3 APPROXIMATING IRRATIONAL ZEROS OF POLYNOMIALS

Thus far in this chapter, we have examined various tools that are used to find zeros of polynomials, provided those zeros are rational. In this section we are interested in finding *irrational* roots. Since irrational numbers have unending, nonrepeating decimal expansions, it suffices to approximate such roots.

Consider the polynomial $P(x) = x^3 + x^2 - 4x + 1$. If we apply the Rational Root Theorem, the only possible rational roots of $P(x) = 0$ are ± 1. Synthetic division shows neither is a zero:

$$\begin{array}{r|rrrr} 1 & 1 & 1 & -4 & 1 \\ & & 1 & 2 & -2 \\ \hline & 1 & 2 & -2 & \boxed{-1} \end{array} \qquad \begin{array}{r|rrrr} -1 & 1 & 1 & -4 & 1 \\ & & -1 & 0 & 4 \\ \hline & 1 & 0 & -4 & \boxed{5} \end{array}$$

Thus, $P(1) = -1$ and $P(-1) = 5$; neither +1 nor −1 is a root. If we think of finding a root to $P(x)$ as finding an x-intercept of $y = x^3 + x^2 - 4x + 1$ and if we use the fact that every polynomial function is *continuous* (see Section 4.1), our goal is to find a number between −1 and 1. See the figure below:

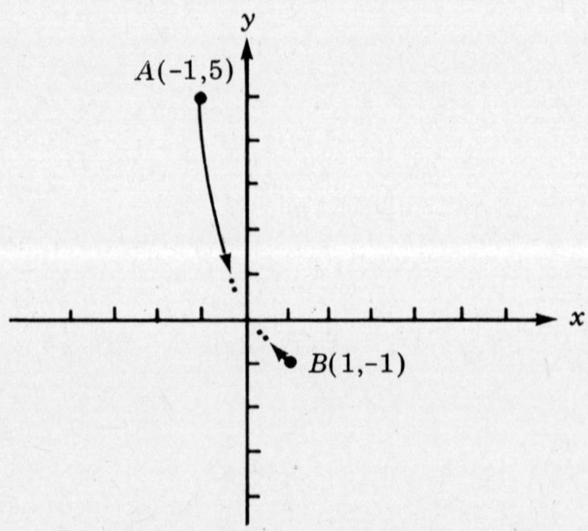

The graph of $y = P(x)$ is continuous so in order to "connect" points A and B we must cross the x-axis. Thus, a root exists *between* −1 and 1. This root is approximated to the nearest tenth by using successive approximations via synthetic division in the next example.

EXAMPLE 1 Find, to the nearest tenth, the real zero of $P(x) = x^3 + x^2 - 4x + 1$ that is between −1 and 1.

Solution Zero is between −1 and 1 and clearly $P(0) = 1$. Since $P(0) > 0$ and $P(1) < 0$, the root must lie between 0 and 1. We try .5:

$$\begin{array}{r|rrrr} .5 & 1 & 1 & -4 & 1 \\ & & .5 & .75 & -1.625 \\ \hline & 1 & 1.5 & -3.25 & -.625 \end{array}$$

Since $P(.5) < 0$, the root must be between 0 and .5. We try .25:

$$\begin{array}{r|rrrr} .25 & 1 & 1 & -4 & 1 \\ & & .25 & .3125 & -.921875 \\ \hline & 1 & 1.25 & -3.6875 & .078125 \end{array}$$

Since $P(.25) > 0$ and $P(.5) < 0$, the root must be between .25 and .5. We try .3:

$$\begin{array}{r|rrrr} .3 & 1 & 1 & -4 & 1 \\ & & .3 & .39 & -1.083 \\ \hline & 1 & 1.3 & -3.61 & -.083 \end{array}$$

Thus, the root is between .25 and .3. To the nearest tenth, $x = .3$.

EXAMPLE 2 Find the real zero of $P(x) = 2x^3 - 4x + 5$ that lies between -2 and -1, to the nearest tenth.

Solution

$$\begin{array}{r|rrrr} -2 & 2 & 0 & -4 & 5 \\ & & -4 & 8 & -8 \\ \hline & 2 & -4 & 4 & -3 \end{array} \qquad \begin{array}{r|rrrr} -1 & 2 & 0 & -4 & 5 \\ & & -2 & 2 & 2 \\ \hline & 2 & -2 & -2 & 7 \end{array}$$

$P(-2) < 0$ and $P(-1) > 0$. We try -1.5:

$$\begin{array}{r|rrrr} -1.5 & 2 & 0 & -4 & 5 \\ & & -3 & 4.5 & -.75 \\ \hline & 2 & -3 & .5 & 4.25 \end{array}$$

$P(-1.5) > 0$ so the root must be between -2 and -1.5. We try -1.9:

$$\begin{array}{r|rrrr} -1.9 & 2 & 0 & -4 & 5 \\ & & -3.8 & 7.22 & -6.118 \\ \hline & 2 & -3.8 & 3.22 & -1.118 \end{array}$$

$P(-1.9) < 0$ and $P(-1.5) > 0$. We try -1.8:

$$\begin{array}{r|rrrr} -1.8 & 2 & 0 & -4 & 5 \\ & & -3.6 & 6.48 & -4.464 \\ \hline & 2 & -3.6 & 2.48 & .536 \end{array}$$

$P(-1.8) > 0$. The root is between -1.9 and -1.8. The last synthetic division, to find $P(-1.85)$ will determine the answer to the nearest tenth:

$$\begin{array}{r|rrrr} -1.85 & 2 & 0 & -4 & 5 \\ & & -3.7 & 6.845 & -5.26325 \\ \hline & 2 & -3.7 & 2.845 & -.26325 \end{array}$$

$P(-1.85) < 0$. Thus, to the nearest tenth, $x = -1.8$.

Although the method above is somewhat useful, it is also tedious. Some of that can be alleviated by high speed computers and calculators. But a drawback in the theory developed thus far is still the lack of knowledge about the *existence* of any real roots. For example, the polynomial $P(x) = x^4 + 1$ has *no* real zeros but none of the theorems explored thus far yields that information. The final theorem that will be useful in determining real zeros of polynomials is **Descartes' Rule of Signs.** It is presented without proof.

THEOREM 6

Descartes' Rule of Signs

Let $P(x)$ be a polynomial with real coefficients. Then:

i) The number of *positive* real zeros of $P(x)$ is either equal to the number of sign variations of $P(x)$'s coefficients or is less than that number by an even integer.

ii) The number of *negative* real zeros of $P(x)$ is either equal to the number of sign variations of $P(-x)$ or is less than that number by an even integer.

To illustrate this theorem consider $P(x) = 2x^3 - 4x + 5$. This polynomial has two variations in sign and from i) above $P(x)$ has either 2 or $2 - 2 = 0$ positive real zeros. Also, $P(-x) = -2x^3 + 4x + 5$ has one change in sign and thus $P(x)$ has one negative real zero. The next example illustrates this further.

EXAMPLE 3 Use Descartes' Rule of Signs to determine the possible number of positive and negative real zeros for each polynomial:

a. $P(x) = x^4 + 5x^3 - 3x^2 + 2x - 1$ b. $Q(x) = x^4 + x + 1$

Solution a. In $P(x)$'s coefficients, there are three sign changes. Thus, the number of positive real zeros of $P(x)$ must be 3 or $3 - 2 = 1$.

$P(-x) = x^4 - 5x^3 - 3x^2 - 2x - 1$ and has one sign change. Hence, $P(x)$ has one negative real root.

b. $Q(x)$ has no sign variations; there are no positive real roots. $Q(-x) = x^4 - x + 1$ and hence $Q(x)$ has either 2 or $2 - 2 = 0$ negative real roots.

The added information of Descartes' Rule of Signs together with the notion of approximating real zeros enables us to graph polynomial functions whose zeros may not be rational, as the next example illustrates.

EXAMPLE 4 Graph $y = x^3 + x^2 - 4x + 1$.

Solution Descartes' Rule of Signs unveils that there will be either two or zero positive x-intercepts (why?) and one negative x-intercept. Since $x = .3$ was an approximate (positive) zero of $x^3 + x^2 - 4x + 1$ (see Example 1), the graph must have three intercepts—two positive and one negative. Repeated synthetic division yields the following information for $P(x) = x^3 + x^2 - 4x + 1$:

x	$P(x) = y$	
−5	−79	
−4	−31	
−3	−5	negative root—between −3 and −2
−2	5	
−1	5	
0	1	positive root—found in Example 1 $\approx .3$
1	−1	positive root—between 1 and 2.
2	5	

The reader should verify that to the nearest tenth, the remaining zeros of $P(x)$ are −2.7 and 1.4. The graph is sketched below:

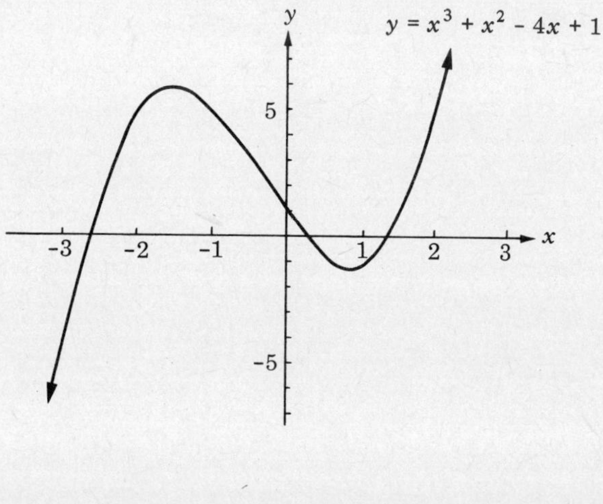

We conclude this section by mentioning certain properties of the real zeros of the general polynomial $P(x) = a_n x^n + a_{n-1}x^{n-1} + \ldots + a_1 x + a_0$ (and hence about the x-intercepts of the graph of $y = P(x)$). First, if n is odd, there will be at least one real zero; if n is even, there may not be any real zeros. Second, complex roots occur in pairs. That is, if $n = 5$ then there will be either 1, 3, or 5 real roots (including possible multiple roots). We depict some cases below. Keep in mind that the list below is not exhaustive (see also Exercises 29-31).

n = 3 cubic polynomial

case I
one real root

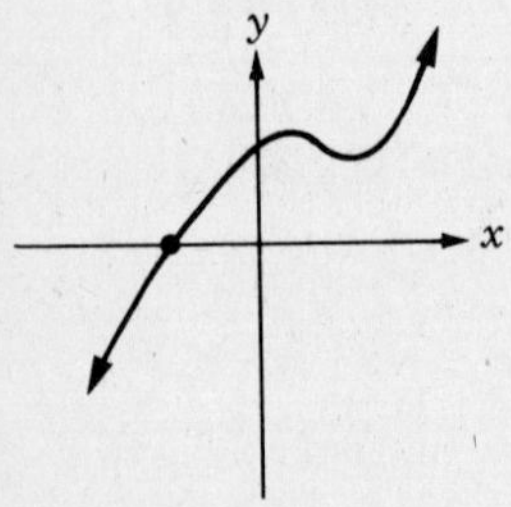

case II
three distinct real roots

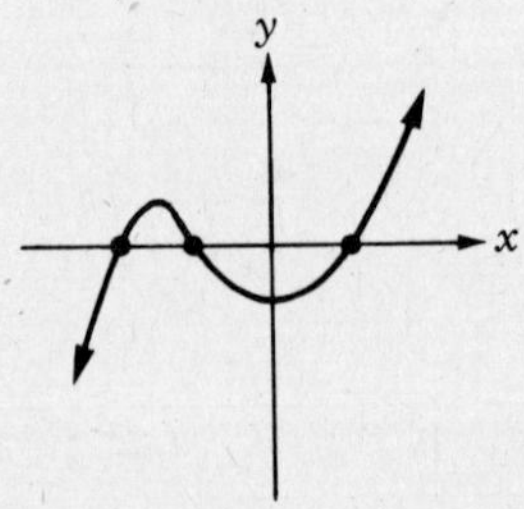

case III
multiple roots

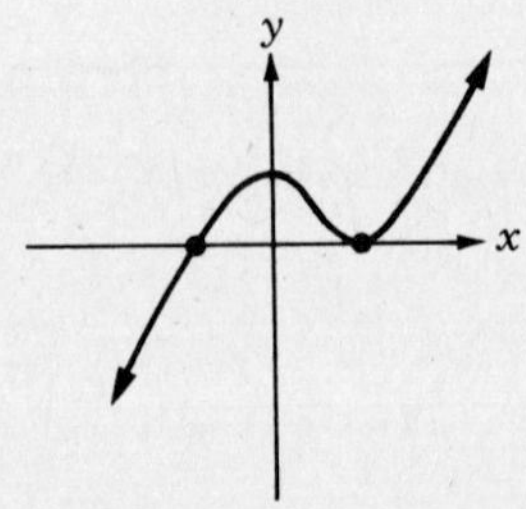

n = 4 quartic polynomial

case I
no real roots

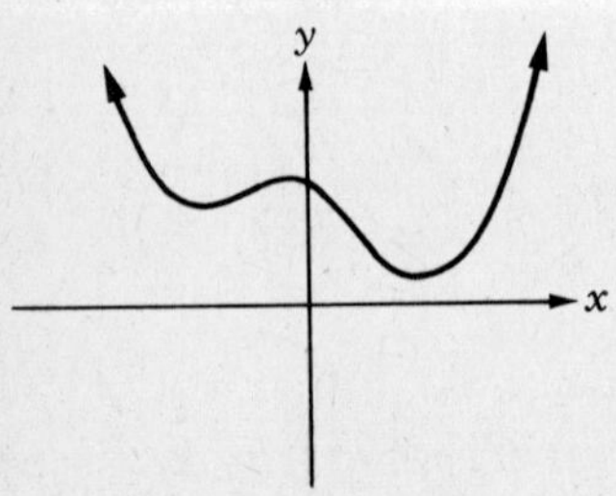

case II
two distinct real roots

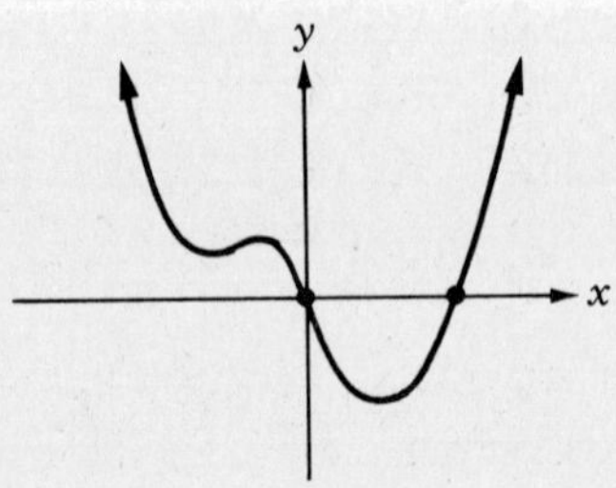

case III
four distinct real roots

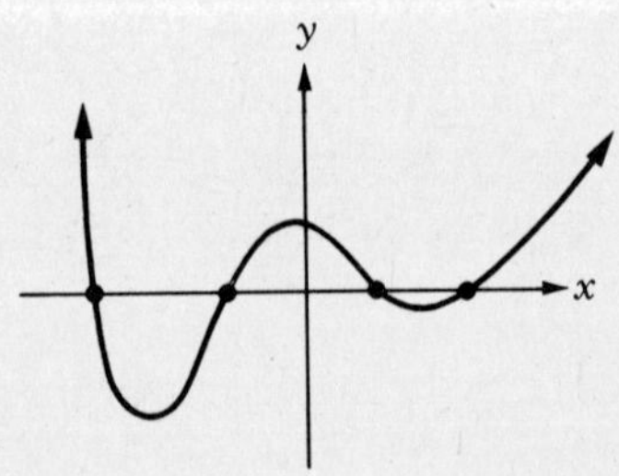

case IV
multiple roots

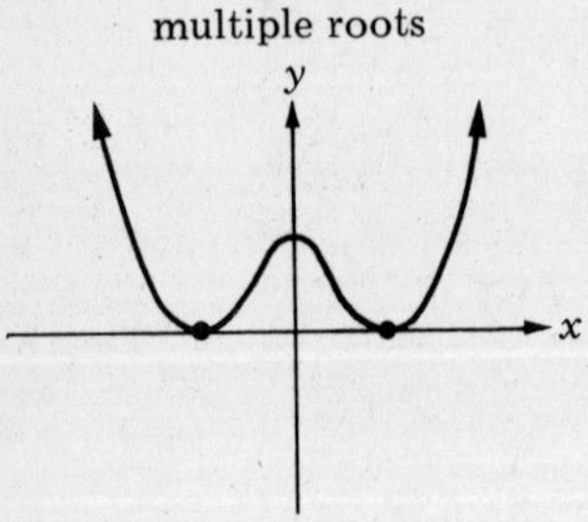

n = 5 quintic polynomial

case I
one real root

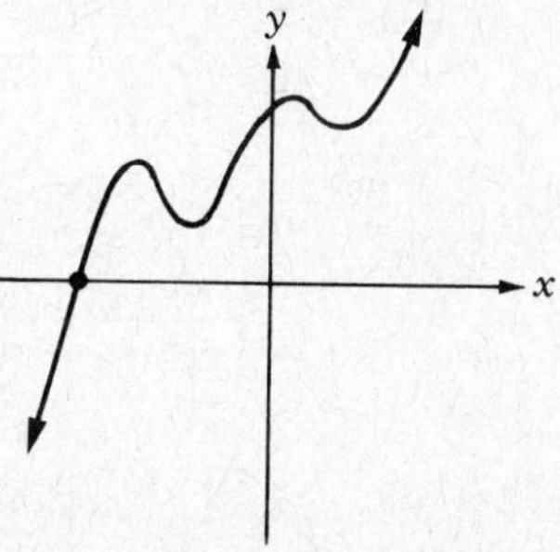

case II
three distinct roots

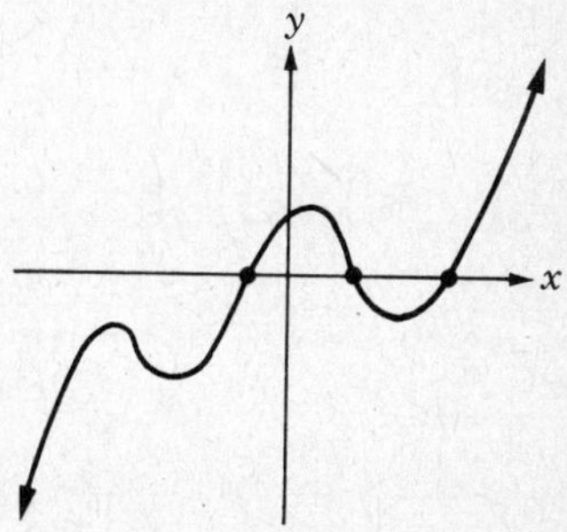

case III
five distinct real roots

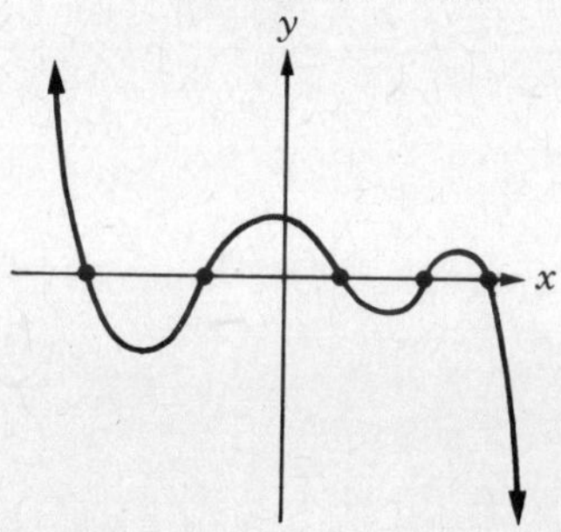

case IV
multiple roots

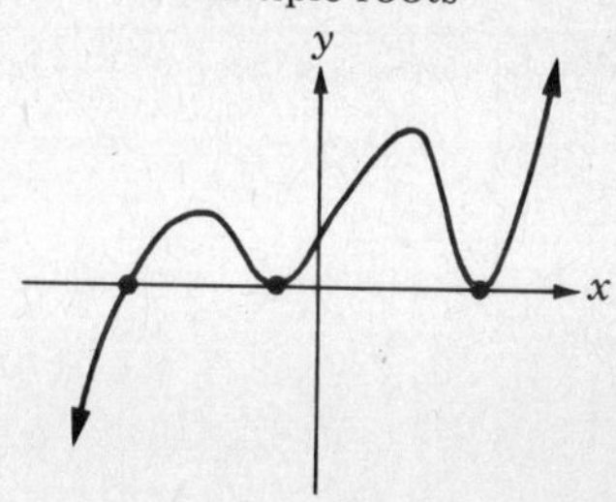

EXERCISES FOR SECTION 10.3

In Exercises 1 through 15, find a root of $P(x) = 0$, to the nearest tenth, in the stated interval. (See Examples 1 and 2.)

1. $P(x) = x^3 - 5x + 1 \; [0,1]$
2. $P(x) = x^3 - 5x + 1 \; [2,3]$
3. $P(x) = x^3 - 5x + 1 \; [-3,-2]$
4. $P(x) = x^3 + 3x^2 - x - 7 \; [1,2]$
5. $P(x) = x^3 - 4x^2 + x - 2 \; [3,4]$
6. $P(x) = x^3 + 10x - 12 \; [1,2]$
7. $P(x) = x^3 - 6x + 2 \; [0,1]$
8. $P(x) = x^3 - 6x + 2 \; [2,3]$
9. $P(x) = x^3 - 6x + 2 \; [-3,-2]$
10. $P(x) = x^3 - 6x^2 + 9x - 2 \; [3,4]$
11. $P(x) = x^3 - 6x^2 + 9x - 2 \; [0,1]$
12. $P(x) = x^4 - 6x^3 + 8x^2 + 2x - 2 \; [.3,.5]$
13. $P(x) = x^4 - 6x^3 + 8x^2 + 2x - 2 \; [-.6,-.5]$
14. $P(x) = x^4 - 6x^3 + 8x^2 + 2x - 2 \; [2.2, 2.3]$
15. $P(x) = x^4 - 6x^3 + 8x^2 + 2x - 2 \; [3.7, 3.8]$

In Exercises 16 through 20, use Descartes' Rule of Signs to determine the possible number of positive and negative real zeros. (See Example 3.)

16. $P(x) = x^3 - 5x + 1$
17. $P(x) = x^3 - 6x^2 + 9x - 2$
18. $P(x) = x^4 + 3x^2 + x + 10$
19. $P(x) = x^4 - 3x^2 + x - 10$
20. $P(x) = x^4 - 3x^3 + x + 10$

In Exercises 21 through 24, prove each statement. In each case, assume a is a positive real number. (See Theorem 6.)

21. If $P(x) = x^n - a$ and n is even, then $P(x)$ has two distinct real zeros.
22. If $P(x) = x^n - a$ and n is odd, then $P(x)$ has one real zero.
23. If $P(x) = x^n + a$ and n is odd, then $P(x)$ has one real zero.
24. If $P(x) = x^n + a$ and n is even, then $P(x)$ has no real zeros.

In Exercises 25 through 28, graph each polynomial function. (See Example 4.)

25. $y = x^3 - 5x + 1$ (See Exercises 1, 2, and 3.)
26. $y = x^3 - 6x + 2$ (See Exercises 7, 8, and 9.)
27. $y = x^4 - 6x^3 + 8x^2 + 2x - 2$ (See Exercises 12–15).
28. $y = x^3 - 5x^2 + 1$
29. The three cases for graphs of a *cubic* polynomial can perhaps be appreciated by graphing the following polynomial functions. Can you see a relationship between the graph and the factored form of the polynomial?
    a. $y = (x + 3)(x^2 + 2x + 5)$ b. $y = (x + 3)(x + 1)(x - 2)$
    c. $y = (x + 1)(x - 1)^2$
30. Proceed, as in Exercise 29, to graph each *quartic* polynomial:
    a. $y = (x^2 + 1)(x^2 + 2)$ b. $y = x(x - 2)(x^2 + x + 2)$
    c. $y = (x - 2)(x - 1)(x + 1)(x + 2)$ d. $y = (x - 2)^2(x + 2)^2$
31. Proceed, as in Exercise 29, to graph each *quintic* polynomial:
    a. $y = (x - 1)(x^2 + 1)(x^2 + 2)$ b. $y = (x + 1)(x - 1)(x - 2)(x^2 + 1)$
    c. $y = x(x + 2)(x - 1)(x - 2)(x - 3)$ d. $y = (x + 2)(x + 1)^2(x - 1)^2$
32. Use the method of Example 1 on the polynomial $P(x) = x^2 - 2$ to find an approximation of $\sqrt{2}$ to the nearest tenth.
33. Two roots of the polynomial $P(x) = x^3 - 6x^2 + 9x - 2$ were *approximated* in Exercises 10 and 11. Express these roots *exactly*. (*Hint:* use the method of the previous section.)

Calculator Exercise

34. The utility of a calculational device in solving the problems of this section is obvious. For instance, the authors, on a programmable calculator (TI 58C), approximated the root found in Example 1 to .273890555. The root in Example 2 was found to be approximately −1.83390585. Each of these approximations was found in well less than a minute! Use a programmable calculator to approximate the roots of the polynomials in Exercises 1-15 to seven decimal places.

10.4 PARTIAL FRACTIONS

In Chapter 1, we saw that the fractions $\dfrac{3}{2x - 1}$ and $\dfrac{7}{x + 3}$ are added as follows:

$$\frac{3}{2x - 1} + \frac{7}{x + 3} = \frac{3(x + 3) + 7(2x - 1)}{(2x - 1)(x + 3)} = \frac{17x + 2}{(2x - 1)(x + 3)}$$

In this section, the *reverse* process, **decomposition**, will be examined. That is, given a fraction like $\dfrac{17x + 2}{(2x - 1)(x + 3)}$, how can we write it as a sum of fractions? This sum will be called the **partial fraction decomposition** and is a useful tool in the study of calculus. It will be studied in four cases for the fraction

$$\frac{P(x)}{Q(x)}$$

where $P(x)$ and $Q(x)$ are polynomials and the degree of $P(x)$ is less than the degree of $Q(x)$.

> **Case I.** The factors of $Q(x)$ are distinct linear factors. For example, $Q(x) = (2x - 1)(x + 2)$.

EXAMPLE 1 Find the partial fraction decomposition of

$$\frac{x+7}{(2x-1)(x+2)}$$

Solution We write

$$\frac{x+7}{(2x-1)(x+2)} = \frac{A}{2x-1} + \frac{B}{x+2} \tag{1}$$

We must find A and B. Multiply both sides of (1) by $(2x - 1)(x + 2)$.

$$x + 7 = A(x + 2) + B(2x - 1) \tag{2}$$

The values for A and B are found by substituting *any* values of x in (2). We choose -2 and $\frac{1}{2}$:

Substitute -2 for x:

$$\begin{aligned} x + 7 &= A(x + 2) + B(2x - 1) \\ 5 &= 0 + B(-5) \\ B &= -1 \end{aligned}$$

Substitute $\frac{1}{2}$ for x:

$$\begin{aligned} x + 7 &= A(x + 2) + B(2x - 1) \\ \tfrac{15}{2} &= \tfrac{5}{2}A \\ 3 &= A \end{aligned}$$

Thus,

$$\frac{x+7}{(2x-1)(x+2)} = \frac{3}{2x-1} - \frac{1}{x+2}$$

> **Case II.** The factors of $Q(x)$ are linear but not all distinct. For example, $Q(x) = (x + 1)^2$.

EXAMPLE 2 Find the partial fraction decomposition of

$$\frac{2x+7}{(x+1)^2}$$

Solution We write

$$\frac{2x+7}{(x+1)^2} = \frac{A}{x+1} + \frac{B}{(x+1)^2}$$

Next, we proceed as in Example 1 to find A and B:

$$2x + 7 = A(x+1) + B$$

Let $x = -1$:

$$5 = B$$

Let $x = 0$:

$$7 = A + B$$
$$7 = A + 5$$
$$A = 2$$

Thus,

$$\frac{2x+7}{(x+1)^2} = \frac{2}{x+1} + \frac{5}{(x+1)^2}$$

The next example illustrates a partial fraction decomposition of a fraction with both multiple and distinct linear factors.

EXAMPLE 3 Find the partial fraction decomposition of

$$\frac{-11x^2 - 3x + 11}{(x+7)(2x+1)^2}$$

Solution

$$\frac{-11x^2 - 3x + 11}{(x+7)(2x+1)^2} = \frac{A}{x+7} + \frac{B}{2x+1} + \frac{C}{(2x+1)^2}$$

$$-11x^2 - 3x + 11 = A(2x+1)^2 + B(x+7)(2x+1) + C(x+7)$$

Let $x = -7$:

$$-507 = 169A$$
$$-3 = A$$

Let $x = -\frac{1}{2}$:

$$\frac{39}{4} = \frac{13}{2}C$$
$$\frac{3}{2} = C$$

Let $x = 0$:

$$11 = A + 7B + 7C$$
$$11 = -3 + 7B + \frac{21}{2}$$
$$\frac{1}{2} = B$$

Thus,

$$\frac{-11x^2 - 3x + 11}{(x+7)(2x+1)^2} = \frac{-3}{x+7} + \frac{\frac{1}{2}}{2x+1} + \frac{\frac{3}{2}}{(2x+1)^2}$$

Case III. The factors of $Q(x)$ are distinct quadratic factors.* For example, $Q(x) = (x^2 + 5)(x^2 + 3)$.

EXAMPLE 4 Find the partial fraction decomposition of

$$\frac{4x^3 + 12x - 2}{(x^2 + 5)(x^2 + 3)}$$

Solution The numerator of each partial fraction will be a linear expression. Otherwise, the method is similar to that of the previous examples:

$$\frac{4x^3 + 12x - 2}{(x^2 + 5)(x^2 + 3)} = \frac{Ax + B}{x^2 + 5} + \frac{Cx + D}{x^2 + 3}$$

$$4x^3 + 12x - 2 = (Ax + B)(x^2 + 3) + (Cx + D)(x^2 + 5)$$

We let $x = 0, -1, 1,$ and 2 and get the following system of equations:

$$\begin{aligned} 3B + 5D &= -2, \text{ using } x = 0 \\ -4A + 4B - 6C + 6D &= -18, \text{ using } x = -1 \\ 4A + 4B + 6C + 6D &= 14, \text{ using } x = 1 \\ 14A + 7B + 18C + 9D &= 54, \text{ using } x = 2 \end{aligned}$$

We leave it to the reader (in Exercise 25) to determine that $A = 4$, $B = 1$, $C = 0$, and $D = -1$. Thus,

$$\frac{4x^3 + 12x - 2}{(x^2 + 5)(x^2 + 3)} = \frac{4x + 1}{x^2 + 5} + \frac{-1}{x^2 + 3}$$

Case IV. $Q(x)$ has repeated quadratic factors. For example, $Q(x) = x(x^2 + 1)^2$.

EXAMPLE 5 Find the partial fraction decomposition of

$$\frac{7x^4 - x^3 + 12x^2 - 3x + 2}{x(x^2 + 1)^2}$$

Solution Notice that the numerator is fourth degree and the denominator is *fifth* degree. We proceed similarly as before:

*In this section we use the term *quadratic factor* to mean a second degree polynomial whose zeros are not real. For example, $x^2 + 1$ is a quadratic factor but $x^2 - 5$ is not, because is can be written as $(x + \sqrt{5})(x - \sqrt{5})$.

$$\frac{7x^4 - x^3 + 12x^2 - 3x + 2}{x(x^2+1)^2} = \frac{A}{x} + \frac{Bx + C}{x^2 + 1} + \frac{Dx + E}{(x^2+1)^2}$$

$$7x^4 - x^3 + 12x^2 - 3x + 2 = A(x^2+1)^2 + (Bx + C)(x)(x^2+1) + (Dx + E)x$$

Letting $x = 0, 1, -1, 2$ and -2 yields:

$$\begin{aligned} A &= 2, \text{ using } x = 0 \\ 4A + 2B + 2C + D + E &= 17, \text{ using } x = 1 \\ 4A + 2B - 2C + D - E &= 25, \text{ using } x = -1 \\ 25A + 20B + 10C + 4D + 2E &= 148, \text{ using } x = 2 \\ 25A + 20B - 10C + 4D - 2E &= 176, \text{ using } x = -2 \end{aligned}$$

The reader is asked to verify (in Exercise 26) that $A = 2, B = 5, C = -1$, $D = 3$, and $E = -2$ so that we have

$$\frac{7x^4 - x^3 + 12x^2 - 3x + 2}{x(x^2+1)^2} = \frac{2}{x} + \frac{5x - 1}{x^2 + 1} + \frac{3x - 2}{(x^2+1)^2}$$

We restate here that the study of calculus depends, in part, upon a knowledge of partial fractions.

EXERCISES FOR SECTION 10.4

In Exercises 1 through 24, find the partial fraction decomposition. (See Examples 1 through 5.)

1. $\dfrac{6x+1}{(2x-3)(x+1)}$
2. $\dfrac{7x-17}{(x-2)(x-3)}$
3. $\dfrac{-x-2}{(2x-1)(x-1)}$
4. $\dfrac{x+2}{(2x-1)(x-1)}$
5. $\dfrac{6x^2+22x+18}{(x+1)(x+2)(x+3)}$
6. $\dfrac{x}{x^2-1}$
7. $\dfrac{5x+2}{(x-1)^2}$
8. $\dfrac{-4x-3}{(2x-3)^2}$
9. $\dfrac{20x^2+96x+100}{x(2x+5)^2}$
10. $\dfrac{x}{x^2-4x+4}$
11. $\dfrac{x^2+5x+1}{x(x+1)^2}$
12. $\dfrac{x^2+5x+1}{x^2(x+1)}$
13. $\dfrac{x^2+5x+1}{x^2(x+1)^2}$
14. $\dfrac{x^2+5x+1}{x(x^2+1)}$
15. $\dfrac{x^2+5x+1}{x^2(x^2+1)}$
16. $\dfrac{x^2-8x+18}{(x-5)^3}$
17. $\dfrac{5x^3+2x^2+14x+4}{(x^2+1)(x^2+x+5)}$
18. $\dfrac{2x^2+3}{(x^2+1)(x^2+2)}$
19. $\dfrac{x^2-2x}{(x^2+9)(x^2+x+7)}$
20. $\dfrac{4x^2-11x-35}{x^3+2x^2+7x}$
21. $\dfrac{4x^4+2x^3+14x^2+5x+12}{x^5+4x^3+4x}$
22. $\dfrac{x^3+7x^2-1}{x^2(x^2+1)^2}$

23. $\dfrac{x^4 + 17x^2 + 82}{(x^2 + 9)^3}$

24. $\dfrac{3}{x^3 - 1}$

25. Verify the coefficients found in Example 4.
26. Verify the coefficients found in Example 5.
27. To save space, we have omitted the checks in each example. a) How would you check Example 1? b) Check each answer in Exercises 1-24.
28. In the examples, the method used to find the coefficients (A, B, C, etc.) was to *substitute* a value for x in the equation. An alternate method, *equating coefficients* can also be used. Supply the reason for the asterisked step for the partial fractions of Example 1:

$$x + 7 = A(x + 2) + B(2x - 1)$$
$$x + 7 = Ax + 2A + 2Bx - B$$
$$x + 7 = (A + 2B)x + 2A - B \quad *$$
$$\left.\begin{aligned} A + 2B &= 1 \\ 2A - B &= 7 \end{aligned}\right\}$$

29. Use the method of equating coefficients (see Exercise 28) to find A, B, and C below:

$$-11x^2 - 3x + 11 = A(2x + 1)^2 + B(x + 7)(2x + 1) + C(x + 7).$$

30. Use the method of equating coefficients (see Exercise 28) to find A, B, C, and D below:

$$4x^3 + 12x - 2 = (Ax + B)(x^2 + 3) + (Cx + D)(x^2 + 5).$$

The remark preceding Example 1 points out that the method of partial fraction decomposition applies to $\dfrac{P(x)}{Q(x)}$ provided the degree of $P(x)$ is less than the degree of $Q(x)$. If that is not the case, one can, by division, obtain a new fraction whose denominator has a greater degree than the numerator:

$$\frac{x^4 - 3x^2 + 7x - 1 \leftarrow P(x)}{x^2 - 1 \leftarrow Q(x)}$$

$$\begin{array}{r} x^2 - 2 \\ x^2 - 1 \,\overline{)\, x^4 - 3x^2 + 7x - 1} \\ \underline{x^4 - x^2 } \\ -2x^2 + 7x - 1 \\ \underline{-2x^2 + 2} \\ 7x - 3 \end{array}$$

Next we write $\dfrac{P(x)}{Q(x)} = x^2 - 2 + \dfrac{7x - 3}{x^2 - 1}$ and perform the decomposition on $\dfrac{7x - 3}{x^2 - 1}$. In Exercises 31 through 34, find the partial fraction decomposition by first using long division.

31. $\dfrac{2x^3 + 5x^2 + 4x + 2}{(x + 1)(x + 2)}$

32. $\dfrac{x^3 - x^2 - 2x - 2}{x^2 - 2x}$

33. $\dfrac{x^4 + x + 5}{x^2 + x}$

34. $\dfrac{x^6 + 6x^2 - 9}{x^2 - 1}$

CHAPTER 10 IN REVIEW

I. Summary

In this chapter we examined the nature and properties of polynomials. In Section 10.1, the Factor and Remainder Theorems helped us determine if a particular number was a zero of a given polynomial. We saw that a root to the polynomial equation $P(x) = 0$, is also the x-intercept of the graph of the polynomial function $y = P(x)$. If c is a zero of $P(x)$, then $P(x)$ has $x - c$ as a factor.

In Section 10.2, the Rational Root Theorem was used to determine whether or not a given polynomial has rational zeros. If a real zero of $P(x)$ is not rational, then it must be irrational and can be approximated using the development of Section 10.3. Also in Section 10.3, the graphs of polynomial functions were explored further, and Descartes' Rule of Signs was used.

Finally, in Section 10.4, partial fraction decomposition, a method important in the study of calculus, was examined.

II. Vocabulary/Symbol True–False Quiz

If the statement is always true, write "TRUE." However, if the statement is false, replace the underlined word or phrase with a different word or phrase that will make the statement true.

1. If $P(x)$ is a polynomial and $P(c) = 0$, then $\underline{x + c}$ is a factor of $P(x)$.
2. If $P(x)$ is a polynomial and $P(c) = 0$, then the point $(c,0)$ is an $\underline{x\text{-intercept}}$ of the graph of $y = P(x)$.
3. If $P(x)$ is a polynomial that has $x - c$ as a factor, then $\underline{P(c) = 0}$.
4. If $P(x)$ is a polynomial, $P(a) < 0$ and $P(b) > 0$, then there must be a number c such that $\underline{P(c) = 0}$.
5. If $P(x)$ is a fifth degree polynomial, then the graph of $y = P(x)$ must have $\underline{\text{at least one}}$ x-intercept.

III. Chapter Test

1. Use synthetic division and the Remainder Theorem to find $P(-2)$ where $P(x) = x^3 + x - 4$.
2. Use synthetic division and the Factor Theorem to determine whether $x + 4$ is a factor of $x^4 + 8x^2 - 16x + 32$.
3. a. List all *possible* rational roots of $P(x) = 0$, where

$$P(x) = 3x^3 + 14x^2 + 7x - 4.$$

   b. Find all solutions to $3x^3 + 14x^2 + 7x - 4 = 0$.

4. Could $-\frac{2}{3}$ possibly be a root of

$$P(x) = 7x^{19} - 5x^{17} + 4x^3 - 19x^2 + 6?$$

Explain.

5. Graph $y = (x - 2)^3$. Be sure to label all intercepts.
6. Graph $y = x^4 - 3x^3 + 3x - 1$. Be sure to label all intercepts.
7. Graph $y = x^3 - 4x^2 + x + 6$. Be sure to label all intercepts.
8. Find a root (to the nearest tenth) of $P(x) = 0$, where

$$P(x) = x^3 - 6x + 2.$$

9. Find a root (to the nearest tenth) of $P(x) = 0$, where

$$P(x) = x^4 - 2.$$

10. Use Descartes' Rule of Signs to determine the possible number of positive and negative roots of $P(x) = x^4 - 5x^3 + x^2 + x - 5$.

In Exercises 11 through 15, find the partial fraction decomposition.

11. $\dfrac{11}{(x - 4)(x + 7)}$

12. $\dfrac{4x - 3}{(x + 1)(2x - 5)}$

13. $\dfrac{4x - 1}{(x + 1)^2}$

14. $\dfrac{4}{x^2 + x + 2}$

15. $\dfrac{2x^2 - x + 3}{(x^2 + 1)^2}$

IV. Student Research Project

Second degree polynomial equations, as we already know, can be solved by means of a formula that involves a radical in terms of the coefficients in the equation. It happens that third and fourth degree polynomial equations can also be solved by formulas that involve radicals in terms of the coefficients of the equations. However, polynomial equations of the fifth degree and higher *cannot* be solved in the same manner.

1. Read pp. 34–36 in *Mathematics: A Humanistic Approach* by G. Joseph Wimbish, Wadsworth Publishing Company, Belmont, California, 1972. Also, read pp. 219–226 in *An Introduction to the History of Mathematics*, 4th ed., by Howard Eves, Holt, Rinehart and Winston, New York, 1976.
2. Solve, by formula, the equation $x^3 - 4x^2 - 9x + 36 = 0$.

CHAPTER 11

Sequences and Series

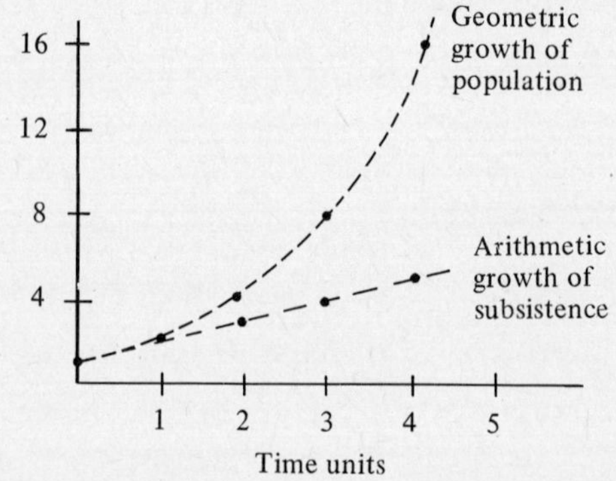

. . . Assuming then, my postulata as granted, I say, that the power of population is indefinitely greater than the power in the earth to produce subsistence for man.

Population, when unchecked, increases in a geometrical ratio. Subsistence increases only in an arithmetical ratio. A slight acquaintance with numbers will shew the immensity of the first power in comparison of the second.

By that law of our nature which makes food necessary to the life of man, the effects of these two unequal powers must be kept equal.

This implies a strong and constantly operating check on population from the difficulty of subsistence. This difficulty must fall some where and must necessarily be severely felt by a large portion of mankind. . . .

Taking the population of the world at any number, a thousand millions, for instance, the human species would increase in the ratio of –1, 2, 4, 8, 16, 32, 64, 128, 256, 512, &c. and subsistence as –1, 2, 3, 4, 5, 6, 7, 8, 9, 10, &c. In two centuries and a quarter, the population would be to the means of subsistence as 512 to 10: in three centuries as 4096 to 13, and in two thousand years the difference would be almost incalculable, though the produce in that time would have increased to an immense extent.

No limits whatever are placed to the productions of the earth; they may increase for ever and be greater than any assignable quantity; yet still the power of population being a power of a superior order, the increase of the human species can only be kept commensurate to the increase of the means of subsistence, by the constant operation of the strong law of necessity acting as a check upon the greater power.

The information about population is from pp. 5 and 9 of Population: The First Essay *by Thomas Robert Malthus with a foreword by Kenneth E. Boulding, The University of Michigan Press (Ann Arbor, 1959).*

11.1 INTRODUCTION

Thus far most of our attention has been directed to functions whose domain is the set of all real numbers or some interval of real numbers. However, in the definitions that follow, note that the domain consists of natural numbers only.

DEFINITION A **finite sequence function** is a function whose domain is a finite subset of the positive integers.

DEFINITION An **infinite sequence function** is a function whose domain is the set of all positive integers.

DEFINITION A **sequence** is the collection of range elements, listed in order, of a sequence function.

EXAMPLE 1 Consider the function $f(n) = n^2 + 2$, whose domain is $\{1,2,3,4\}$.

a. What type of function is f?
b. Determine the sequence formed (or *generated*) by f.

Solution

a. Since the domain of f is a finite subset of the positive integers, f is a finite sequence function.

b. $f(1) = 1^2 + 2 = 3$
$f(2) = 2^2 + 2 = 6$
$f(3) = 3^2 + 2 = 11$
$f(4) = 4^2 + 2 = 18$

The range elements, listed in order, form the finite sequence 3,6,11,18.

EXAMPLE 2 Consider $g(n) = 2n - 3$ with domain $\{1,2,3, \ldots\}$. What sequence is generated by g?

Solution

$$\begin{aligned} g(1) &= 2 \cdot 1 - 3 = -1 \\ g(2) &= 2 \cdot 2 - 3 = 1 \\ g(3) &= 2 \cdot 3 - 3 = 3 \\ g(4) &= 2 \cdot 4 - 3 = 5 \\ &\vdots \qquad\qquad\quad \vdots \\ &\text{etc.} \qquad\quad\ \text{etc.} \end{aligned}$$

Hence, $g(n)$ generates the infinite sequence $\{-1,1,3,5,\ldots\}$.

EXAMPLE 3 Using the sequence function $f(n) = (-1)^{n+1}\dfrac{1}{n^2}$ with domain $D = \{1,2,3,4,5\}$, determine the sequence generated.

Solution

$$\left.\begin{aligned} f(1) &= (-1)^{1+1}\frac{1}{1^2} = 1 \\ f(2) &= (-1)^{2+1}\frac{1}{2^2} = -\frac{1}{4} \\ f(3) &= (-1)^{3+1}\frac{1}{3^2} = \frac{1}{9} \\ f(4) &= (-1)^{4+1}\frac{1}{4^2} = -\frac{1}{16} \\ f(5) &= (-1)^{5+1}\frac{1}{5^2} = \frac{1}{25} \end{aligned}\right\}$$

Notice that term-by-term, the signs alternate. Such a sequence is referred to as **alternating**.

The sequence is $\{1, -\frac{1}{4}, \frac{1}{9}, -\frac{1}{16}, \frac{1}{25}\}$.

Associated with any sequence of numbers is a **series**.

DEFINITION

> A series is an indicated sum of a certain number of terms of a sequence.

As an illustration, let us consider the sequence $2,4,8,16,\ldots$

$$\longrightarrow S_n = 2 + 4 + 8 + \ldots + 2^n.$$

This is a *series* for the sum of the first n terms in the given sequence. A few special cases are:

$$\begin{aligned} S_2 &= 2 + 4 \\ S_3 &= 2 + 4 + 8 \\ S_4 &= 2 + 4 + 8 + 16 \end{aligned}$$

As a shorthand way of writing series we will use **summation notation** or **sigma notation**. The Greek letter Σ (pronounced "sigma") is used to symbolize a sum. For example, $S_n = 2 + 4 + 8 + \ldots + 2^n$ is denoted as

$$\sum_{i=1}^{i=n} 2^i \quad \text{or more briefly as} \quad \sum_{i=1}^{n} 2^i$$

where Σ stands for the words "sum of," i is called the **index of summation** and assumes the values, 1, 2, 3, . . . , n. Furthermore, in this discussion, i has nothing to do with imaginary numbers. The index of summation is called a *dummy variable*, because it does not appear once a series has been written in expanded form.

Again, consider the sequence 2,4,8,16, ... We can write S_6 as:

$$\sum_{i=1}^{6} 2^i \quad \text{or} \quad \sum_{j=1}^{6} 2^j \quad \text{or} \quad \sum_{k=1}^{6} 2^k.$$

The choice of the index letter is unimportant—the structure of the *general term* (2^i, 2^j, 2^k) and the beginning and ending values of the index are important.

EXAMPLE 4 Using summation notation, write the given series:

$$\frac{1}{1} + \frac{1}{3} + \frac{1}{5} + \frac{1}{7} + \ldots + \frac{1}{2n-1}$$

Solution $\frac{1}{1} + \frac{1}{3} + \frac{1}{5} + \frac{1}{7} + \ldots + \frac{1}{2n-1}$ is written as $\sum_{i=1}^{n} \frac{1}{2i-1}$

EXAMPLE 5 Using summation notation, write the given **alternating series**:

$$\frac{2}{4} - \frac{4}{5} + \frac{6}{6} - \frac{8}{7} + \ldots (-1)^{n+1}\frac{2n}{n+3}$$

Solution $\frac{2}{4} - \frac{4}{5} + \frac{6}{6} - \frac{8}{7} + \ldots (-1)^{n+1}\frac{2n}{n+3}$ can be written as

$$\sum_{k=1}^{n} (-1)^{k+1}\frac{2k}{k+3}$$

When a series is written in summation notation and we rewrite it by showing each of its terms, we refer to the latter as the **expanded form**. For instance, the expanded form of $\sum_{j=1}^{4} \frac{1}{3^j}$ is $\frac{1}{3} + \frac{1}{9} + \frac{1}{27} + \frac{1}{81}$.

EXAMPLE 6 Write the series $\sum_{i=1}^{5} \frac{(-1)^i 2}{4i+1}$ in expanded form.

Solution $\sum_{i=1}^{5} \frac{(-1)^i 2}{4i+1} = -\frac{2}{5} + \frac{2}{9} - \frac{2}{13} + \frac{2}{17} - \frac{2}{21}$

EXAMPLE 7 Write the series $\sum_{i=1}^{4} \frac{i}{2^{i+1}}$ in expanded form.

Solution $\sum_{i=1}^{4} \frac{i}{2^{i+1}} = \frac{1}{2^2} + \frac{2}{2^3} + \frac{3}{2^4} + \frac{4}{2^5}$

EXAMPLE 8 Write the series $\sum_{k=2}^{5} \frac{k-1}{2^k}$ in expanded form.

Solution Note that the index value on the bottom of the Σ symbol is 2. Thus the index k assumes the values 2, 3, 4, and 5.

$$\sum_{k=2}^{5} \frac{k-1}{2^k} = \frac{1}{2^2} + \frac{2}{2^3} + \frac{3}{2^4} + \frac{4}{2^5}$$

Examples 7 and 8 serve to demonstrate that

$$\sum_{i=1}^{4} \frac{i}{2^{i+1}} = \sum_{k=2}^{5} \frac{k-1}{2^k} \tag{1}$$

This might not be apparent unless we view the expanded forms of each series. It is possible, however, to start with either summation in (1) and derive the other by performing a *shift of index* as the next example shows.

EXAMPLE 9 Given $\sum_{i=1}^{4} \frac{i}{2^{i+1}}$, perform a shift of index by letting $i = k - 1$, in order to obtain an equivalent summation.

Solution We accomplish this by the substitution $i = k - 1$ in the given expression.

$$\sum_{i=1}^{i=4} \frac{i}{2^{i+1}} =$$

$$\sum_{k-1=1}^{k-1=4} \frac{k-1}{2^{(k-1)+1}} =$$

$$\sum_{k=2}^{k=5} \frac{k-1}{2^k} \quad \text{or}$$

$$\sum_{k=2}^{5} \frac{k-1}{2^k}$$

Sometimes it is necessary to formulate the general term of a sequence or a series. Consider the sequence 2,6,18,54,162, ... Now, in order to formulate the general term it will help if we label the terms as $a_1, a_2, a_3, \ldots$, and then look at how each term is related to the previous term.

$$\begin{array}{ccccc} 2, & 6, & 18, & 54, & 162, \ldots \\ \downarrow & \downarrow & \downarrow & \downarrow & \downarrow \\ a_1 & a_2 & a_3 & a_4 & a_5 \end{array}$$

$$\left.\begin{aligned} a_1 &= 2 = 2 \cdot 3^0 \\ a_2 &= 6 = 2 \cdot 3^1 \\ a_3 &= 18 = 2 \cdot 9 = 2 \cdot 3^2 \\ a_4 &= 54 = 2 \cdot 27 = 2 \cdot 3^3 \\ a_5 &= 162 = 2 \cdot 81 = 2 \cdot 3^4 \end{aligned}\right\}$$

Notice that each term is "2" multiplied by a power of "3." Also the exponent on 3 is one less than the number of the term.

Thus the nth term or general term is

$a_n = 2 \cdot 3^{n-1}$ Note that the exponent *is* one less than the number of the term.

EXAMPLE 10 Formulate the general term for the sequence 8,11,14,17,20, . . .

Solution

$$\left.\begin{aligned} a_1 &= 8 = 2 + 3 \cdot 2 \\ a_2 &= 11 = 2 + 3 \cdot 3 \\ a_3 &= 14 = 2 + 3 \cdot 4 \\ a_4 &= 17 = 2 + 3 \cdot 5 \\ a_5 &= 20 = 2 + 3 \cdot 6 \end{aligned}\right\}$$

Each term is "2" added to "3" times a number that is one more than the number of the term.

Hence

$$a_n = 2 + 3(n + 1) \quad \text{or} \quad a_n = 3n + 5$$

EXERCISES FOR SECTION 11.1

In 1 through 20, use the sequence function, $f(n)$, and domain D, to write the sequence generated by f. (See Examples 1, 2, and 3.)

1. $f(n) = 3n + 2, D = \{1,2,3\}$
2. $f(n) \dfrac{2}{n^2 + 1}, D = \{1,2,3,4\}$
3. $f(n) = 2^{n-1}, D = \{1,2,3, \ldots\}$
4. $f(n) = 2^n, D = \{1,2,3, \ldots\}$
5. $f(n) = 2^{n+1}, D = \{1,2,3, \ldots\}$
6. $f(n) = (-1)^{n+1} 2^n, D = \{1,2,3\}$
7. $f(n) = (-1)^{n+1} \dfrac{n^2}{3}, D = \{1,2,3,4,5\}$
8. $f(n) = \dfrac{n}{n + 1}, D = \{1,2,3,4\}$

9. $f(n) = \frac{n+1}{n+2}$, $D = \{1,2,3,4\}$

10. $f(n) = 2 + (.01)^n$, $D = \{1,2,3, \ldots\}$

11. $f(n) = \frac{1}{n}$, $D = \{1,2,3, \ldots\}$

12. $f(n) = (-1)^{n-1}$, $D = \{1,2,3, \ldots\}$

13. $f(n) = n^2 + 3$, $D = \{1,2,3,4,5\}$

14. $f(n) = n^2 - 3$, $D = \{1,2,3,4,5\}$

15. $f(n) = 2 \cdot 4^{n-1}$, $D = \{1,2,3, \ldots\}$

16. $f(n) = 4 \cdot 2^{n-1}$, $D = \{1,2,3, \ldots\}$

17. $f(n) = \left(\frac{2}{n}\right)^2$, $D = \{1,2,3,4\}$

18. $f(n) = n^n$, $D = \{1,2,3, \ldots\}$

19. $f(n) = \frac{n-1}{2^n}$, $D = \{1,2,3, \ldots\}$

20. $f(n) = 2$, $D = \{1,2,3, \ldots\}$

21. In 1 through 20, above, which sequences are infinite?
22. In 1 through 20, above, which sequences are alternating?

In 23 through 37, write each series in summation notation. (See Examples 4 and 5.)

23. $2 + 4 + 6 + 8 + \ldots + 2n$

24. $\frac{1}{2} + \frac{2}{4} + \frac{3}{8} + \ldots + \frac{n}{2^n}$

25. $\frac{1}{2} - \frac{2}{4} + \frac{3}{8} - \frac{4}{16} + \ldots (-1)^{n+1} \frac{n}{2^n}$

26. $\frac{1}{3} - \frac{1}{9} + \frac{1}{27} - \frac{1}{81} + \ldots (-1)^{n+1} \frac{1}{3^n}$

27. $\frac{2}{2} + \frac{8}{3} + \frac{18}{4} + \ldots + \frac{2n^2}{n+1}$

28. $\frac{1}{1^2} + \frac{1}{2^2} + \frac{1}{3^2} + \frac{1}{4^2} + \frac{1}{5^2}$

29. $\frac{1}{1^2} - \frac{1}{2^2} + \frac{1}{3^2} - \frac{1}{4^2} + \frac{1}{5^2}$

30. $\frac{1}{1^3} + \frac{1}{2^3} + \frac{1}{3^3} + \ldots + \frac{1}{n^3}$

31. $1^2 + 2^2 + 3^2 + 4^2 + \ldots + n^2$

32. $1 + 2 + 3 + 4 + \ldots + n$

33. $1^3 + 2^3 + 3^3 + \ldots + n^3$

34. $1 \cdot 2 + 2 \cdot 3 + 3 \cdot 4 + \ldots + n(n+1)$

35. $\frac{5}{1 \cdot 2} + \frac{5}{2 \cdot 3} + \frac{5}{3 \cdot 4} + \ldots + \frac{5}{n(n+1)}$

36. $2 + 6 + 10 + \ldots + (4n - 2)$

37. $-2 + 0 + 2 + 4 + \ldots + (2n - 4)$

In 38 through 48, write each series in expanded form. (See Examples 6, 7, and 8.)

38. $\sum_{i=1}^{6} \frac{i}{i+4}$

39. $\sum_{i=1}^{n} \frac{i}{i+4}$

40. $\sum_{i=1}^{4} \frac{(-1)^{i-1} i^2}{i+1}$

41. $\sum_{i=1}^{n} \frac{(-1)^{i-1} i^2}{i+1}$

42. $\sum_{k=2}^{6} \frac{k}{k^2+2}$

43. $\sum_{j=1}^{n} (2j+1)(j+1)$

44. $\sum_{i=1}^{n} (-1)^{i+1} \frac{1}{\sqrt{i}}$

45. $\sum_{i=1}^{n} (-1)^i \frac{1}{\sqrt[3]{i}}$

46. $\sum_{i=1}^{4} 3 + (.01)^i$

47. $\sum_{i=1}^{n} \frac{3}{(10)^i}$

48. $\sum_{k=2}^{7} \frac{k+1}{\sqrt{k}}$

In 49 through 53, perform the stated shift of index. (See Example 9.)

49. Rewrite the summation $\sum_{k=2}^{7} \frac{k+1}{\sqrt{k}}$ by letting $k = i + 1$.

50. Rewrite the summation $\sum_{k=2}^{7} \frac{k+1}{\sqrt{k}}$ by letting $k = j - 1$.

51. Write the summation $\sum_{i=1}^{n} \frac{1}{i^2}$ as a summation whose index begins at 2.

52. Write the summation $\sum_{i=2}^{6} \frac{2(i+1)}{3(i+2)(i+3)}$ as a summation whose index begins at 1.

53. Write the summation $\sum_{i=1}^{n} 3 \cdot (\frac{1}{2})^{i-1}$ as a summation whose index begins at 0.

In 54 through 60, formulate the general term for each sequence. (See Example 10.)

54. $\frac{1}{2}, \frac{1}{4}, \frac{1}{8}, \frac{1}{16}, \ldots$

55. $\frac{1}{2}, -\frac{1}{4}, \frac{1}{8}, -\frac{1}{16}, \ldots$

56. 3, 6, 9, 12, 15, . . .

57. 4, 7, 10, 13, 16, . . .

58. $\frac{1}{1}, \frac{1}{5}, \frac{1}{9}, \frac{1}{13}, \ldots$

59. $\frac{2}{2}, \frac{2}{5}, \frac{2}{10}, \frac{2}{17}, \ldots$

60. $a, ar, ar^2, ar^3, ar^4, \ldots$

61. Consider the summation $\sum_{k=1}^{n} ca_k$.

a. Expand the given summation.
b. Remove the common factor c that appears in each term of your answer to part a.
c. Write the expression $a_1 + a_2 + a_3 + \ldots + a_n$ in summation notation.
d. Organize your answers from parts a, b, and c to prove that

$$\sum_{k=1}^{n} ca_k = c \cdot \sum_{k=1}^{n} a_k.$$

62. Consider the summation $\sum_{k=1}^{n} (a_k + b_k)$.

a. Expand the given summation.
b. Rearrange the terms in your answer to part a so that all the a's are grouped together and all the b's are grouped together.
c. Write the expression $a_1 + a_2 + a_3 + \ldots + a_n$ in summation notation.
d. Write the expression $b_1 + b_2 + b_3 + \ldots + b_n$ in summation notation.
e. Organize your answers from parts a, b, c, and d to prove that

$$\sum_{k=1}^{n} (a_k + b_k) = \sum_{k=1}^{n} a_k + \sum_{k=1}^{n} b_k$$

For Exercises 63 through 70, answer TRUE or FALSE.

63. $\sum_{i=1}^{4} i^2 = \sum_{k=2}^{5} (k-1)^2$

64. $\sum_{i=1}^{n} 4 = 4n$

65. $\sum_{i=1}^{n} \left(\frac{1}{i+1} + \frac{1}{i}\right) = \sum_{i=1}^{n} \frac{1}{i+1} + \sum_{i=1}^{n} \frac{1}{i}$

66. $\sum_{k=1}^{n} 3(\frac{1}{2})^k = 3 \sum_{k=1}^{n} (\frac{1}{2})^k$

67. $\sum_{k=1}^{n} 3 + (\frac{1}{2})^k = 3 + \sum_{k=1}^{n} (\frac{1}{2})^k$

68. $\sum_{k=1}^{n} 3 + (\frac{1}{2})^k = 3n + \sum_{k=1}^{n} (\frac{1}{2})^k$

69. $\sum_{j=2}^{4} j^3 = 99$

70. $\sum_{i=1}^{n} a_i b_i = \left(\sum_{i=1}^{n} a_i\right)\left(\sum_{i=1}^{n} b_i\right)$

Computer Exercises

71. An **infinite series** is a series of the form $\sum_{i=1}^{\infty} a_i$. In the study of calculus, it is necessary to determine the nth partial sum, S_n, of the series. For example, if

$$\text{if } S = 1 + \frac{1}{2^2} + \frac{1}{3^2} + \ldots + \frac{1}{n^2} + \ldots = \sum_{i=1}^{\infty} \frac{1}{i^2}$$

then $S_3 = 1 + \frac{1}{2^2} + \frac{1}{3^2} = \frac{49}{36}$.

Write a computer program and use it to find S_5, S_{10}, S_{20}, and S_{25} for each infinite series below.

a. $\sum_{i=1}^{\infty} \frac{1}{i^2}$

b. $1 + \frac{1}{1} + \frac{1}{1 \cdot 2} + \frac{1}{1 \cdot 2 \cdot 3} + \frac{1}{1 \cdot 2 \cdot 3 \cdot 4} + \ldots + \frac{1}{1 \cdot 2 \ldots (n-1)(n)} + \ldots$

c. $\sum_{i=1}^{\infty} (-1)^{i-1} \frac{(0.1)^{i-1}}{i}$

72. A special infinite series, the **harmonic series** is $\sum_{k=1}^{\infty} \frac{1}{k}$. Use a modification of the program of Exercise 71 to find S_{1000}.

11.2 ARITHMETIC SEQUENCES

In this section one special type of sequence known as an **arithmetic sequence** or **arithmetic progression** will be defined and investigated.

DEFINITION

An arithmetic sequence is a sequence in which each term differs from the previous term by the same constant, called the **common difference.**

EXAMPLE 1 Write the first six terms of the arithmetic sequence whose first term is 4 and whose common difference is 3.

Solution To obtain the desired terms we add the common difference to each term in order to obtain the next term. Hence,

$$4, 7, 10, 13, 16, 19$$

EXAMPLE 2 Is the following sequence of four terms $\frac{5}{2}, \frac{3}{2}, \frac{1}{2}, -\frac{1}{2}$ an arithmetic sequence?

Solution We check for a common difference.

$$\left.\begin{aligned} a_2 - a_1 &= \tfrac{3}{2} - \tfrac{5}{2} = -1 \\ a_3 - a_2 &= \tfrac{1}{2} - \tfrac{3}{2} = -1 \\ a_4 - a_3 &= -\tfrac{1}{2} - \tfrac{1}{2} = -1 \end{aligned}\right\} \text{There } is \text{ a common difference. It is } -1.$$

Thus, the given sequence $\frac{5}{2}, \frac{3}{2}, \frac{1}{2}, -\frac{1}{2}$ is an arithmetic sequence.

Later in this section we will use arithmetic sequences to solve certain applied problems. It will be necessary to know how to write the general term of an arithmetic sequence, and also how to compute the sum of the first n terms of an arithmetic sequence. Formulas will now be developed.

By letting d = the common difference of the arithmetic sequence, the first n terms are written as:

$$a_1, a_2, a_3, a_4, \ldots, a_n$$

or

$$\left.\begin{aligned} a_2 &= a_1 + d \\ a_3 &= a_2 + d = a_1 + 2d \\ a_4 &= a_3 + d = a_1 + 3d \\ a_5 &= a_4 + d = a_1 + 4d \end{aligned}\right\} \text{Notice that the coefficient of } d \text{ is one less than the number of the term.}$$

$$\vdots \qquad \vdots$$

Thus, $a_n = a_1 + (n-1)d$

This is a formula for the nth term, or the general term, of any arithmetic sequence. Its use will be demonstrated shortly. Later in this chapter, in the section on mathematical induction, a formal proof will be presented.

THEOREM 1

> If $a_1, a_2, \ldots, a_n, \ldots$ is an arithmetic sequence with common difference d, then the sum of the first n terms is given by
>
> $$S_n = \frac{n}{2}(a_1 + a_n)$$

Proof:

Statement	*Reason*
1. $S_n = a_1 + a_2 + a_3 + \ldots + a_n$	1. Definition of S_n
2. $S_n = a_1 + (a_1 + d) + (a_1 + 2d) + \ldots + (a_n - 2d) + (a_n - d) + a_n$	2. Definition of arithmetic sequence
3. $S_n = a_n + (a_n - d) + (a_n - 2d) + \ldots + (a_1 + 2d) + (a_1 + d) + a_1$	3. Commutative and Associative Properties of Addition.
4. $2S_n = (a_1 + a_n) + (a_1 + a_n) + (a_1 + a_n) + \ldots + (a_1 + a_n)$	4. Adding equations in steps 2 and 3
5. $2S_n = (a_1 + a_n)(1 + 1 + 1 + \ldots + 1)$	5. Distributive Property
6. $2S_n = (a_1 + a_n)n$	6. Addition
7. Thus, $S_n = \frac{n}{2}(a_1 + a_n)$	7. Divide both sides by 2

The formulas are summarized below.

Formulas for an Arithmetic Sequence

The general or nth term: $a_n = a_1 + (n - 1)d$

The sum of the first n terms: $S_n = \frac{n}{2}(a_1 + a_n)$

Some examples involving these formulas will now be presented.

EXAMPLE 3 Consider the arithmetic sequence whose first three terms are $1, -1, -3$.

a. Determine the 19th term.
b. Determine the sum of the first 19 terms.

Solution a. The common difference $d = -2$.

$a_n = a_1 + (n - 1)d$

$a_{19} = 1 + (19 - 1)(-2)$

$a_{19} = -35$

b. $S_n = \frac{n}{2}(a_1 + a_n)$

$S_{19} = \frac{19}{2}[1 + (-35)]$

$S_{19} = -323$

EXAMPLE 4 Determine the first term and the common difference of an arithmetic sequence whose 4th term is 17 and whose 11th term is 52.

Solution

$$a_n = a_1 + (n-1)d \qquad (1)$$

The 4th term is 17, thus $a_4 = 17$. Substituting into equation (1) we obtain:

$$17 = a_1 + (4-1)d \text{ or}$$
$$17 = a_1 + 3d \qquad (2)$$

The 11th term is 52, thus $a_{11} = 52$. Substituting into equation (1) we obtain:

$$52 = a_1 + (11-1)d \text{ or}$$
$$52 = a_1 + 10d \qquad (3)$$

Equations (2) and (3) are a system of 2 linear equations in 2 variables:

$$a_1 + 3d = 17$$
$$a_1 + 10d = 52$$

Solving this system gives

$$a_1 = 2 \text{ and } d = 5$$

EXAMPLE 5 A woman accepts a job as a reporter with a starting salary of $10,500 for the first year, and an increase of $900 per year thereafter.

a. What will her salary be for the 8th year?
b. What will her total earnings be for the first 8 years?

Solution a. The yearly salaries form an arithmetic sequence with $a_1 = 10{,}500$ and $d = 900$. The salary during the 8th year is a_8.

$$a_8 = 10{,}500 + (8-1)(900)$$
$$a_8 = 16{,}800$$

The salary during the 8th year is $16,800.

b. The total for the first 8 years is S_8.

$$S_8 = \tfrac{8}{2}(10{,}500 + 16{,}800)$$
$$S_8 = 109{,}200$$

The total earnings over the first 8 years will be $109,200.

EXAMPLE 6 The owner of a chemical furniture-stripping business charges $2.50 for the first minute plus $1.50 for each minute thereafter for dipping a piece of furniture into the "strip tank." What is the charge for a piece of furniture that remained in the tank for 10 minutes?

Solution We must determine a_{10} for an arithmetic sequence where $a_1 = 2.5$ and $d = 1.5$.

$$a_{10} = 2.5 + (10 - 1)(1.5)$$
$$a_{10} = 16$$

The charge will be \$16.00.

EXERCISES FOR SECTION 11.2

In Exercises 1 through 6, construct the desired arithmetic sequences. (See Example 1.)

1. Construct the first four terms of an arithmetic sequence whose first term is 11 and whose common difference is 4.
2. Construct the first six terms of an arithmetic sequence whose first term is -9 and whose common difference is 6.
3. Construct the first five terms of an arithmetic sequence whose first term is $\frac{2}{5}$ and whose common difference is $\frac{2}{5}$.
4. Construct the first five terms of an arithmetic sequence whose second term is 4 and whose common difference is $\frac{2}{3}$.
5. Construct the first four terms of an arithmetic sequence whose third term is -3 and whose common difference is $\frac{1}{4}$.
6. Construct the first four terms of an arithmetic sequence whose third term is $\frac{1}{4}$ and whose common difference is -3.

In 7 through 12, determine whether or not the given sequence is arithemtic. If so, state the common difference. (See Example 2.)

7. 3,5,8,9,12
8. 1,2,3,4, . . . , 100
9. 15,12,9,6,3,0,−3
10. −1,4,9,14
11. $\frac{1}{2}, \frac{1}{4}, \frac{1}{6}, \frac{1}{8}, \frac{1}{10}, \frac{1}{12}$
12. 2,4,6,8,10,12

In 13 through 22, determine the 15th term, and the sum of the first 15 terms for each of the given arithmetic sequences. (See Example 3.)

13. 3,−1,−5,−9, . . .
14. $\frac{3}{4}, 1, \frac{5}{4}, \frac{6}{4}, \ldots$
15. $-\frac{2}{3}, 1, \frac{8}{3}, \frac{13}{3}, \ldots$
16. 10,4,−2,−8, . . .
17. 0,3,6,9, . . .
18. 2.2, 2.8, 3.4, 4.0, . . .
19. $\frac{7}{4}, \frac{6}{4}, \frac{5}{4}, 1 \ldots$
20. $\log_{10}10000, \log_{10}1000, \log_{10}100, \log_{10}10, \ldots$
21. $x + 2, x, x - 2, x - 4, \ldots$
22. 10, 7, 4, 1, . . .

In 23 through 26, determine the first term and common difference for the described arithmetic sequences. (See Example 4.)

23. The 20th term is 97 and the 1st term is 2.
24. The 3rd term is −2 and the 12th term is −20.
25. The 5th term is 6 and the 13th term is $\frac{50}{3}$.
26. The 2nd term is $-\frac{4}{3}$ and the 4th term is 0.
27. Determine the sum of the first 100 positive integers.
28. Determine the sum of the first 50 positive odd integers.
29. Determine the sum of the first 50 positive even integers.

30. Determine the sum of 1, 2, 3, 4, . . . , n.
31. Determine the sum of the integers 23, 24, . . . , 326.
32. Determine the sum of the integers 4, 2, 0, −2, −4, . . . , −118.

In 33 through 36, determine the first term and common difference of the described arithmetic sequences.

33. $a_{40} = 118, S_{40} = 2{,}380$
34. $a_{100} = \frac{298}{5}, S_{100} = 2{,}990$
35. $a_3 = -2, S_3 = -\frac{24}{9}$
36. $a_{10} = 14, S_{10} = 320$

In 37 through 40, solve each applied problem. (See Examples 5 and 6.)

37. An auditorium has a total of 15 rows with 80 seats in the back row. Proceeding toward the front, each row thereafter has 2 fewer seats than the one behind it.
    a. How many seats are in the front row?
    b. How many seats are in the auditorium?
38. A taxi company charges \$1.00 for the first $\frac{1}{4}$ mile and 25 cents for every $\frac{1}{4}$ mile thereafter. How much is the fare for a 15-mile ride?
39. A scuba diver charges \$2 per minute to work at a depth of 30 feet. For each additional 10 feet below 30 feet, the charge increases by \$1.50 per minute. What is the charge per minute to work at a depth of 100 feet?
40. A carpenter is going to shingle the rear of an A-frame house that has a total of 35 rows of shingles. The bottom row will use 70 shingles, and each row above the bottom has 2 fewer shingles than the one below it.

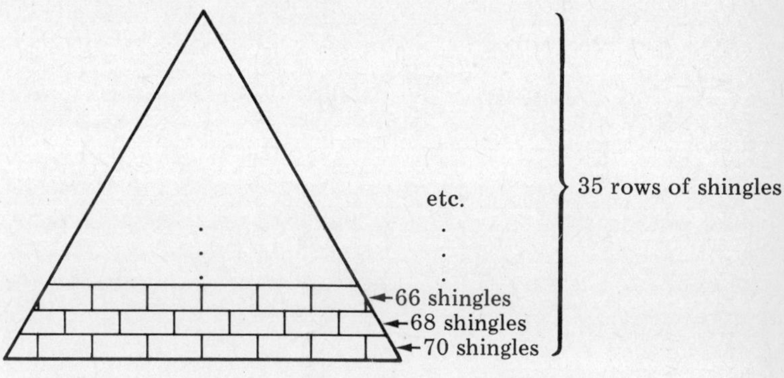

    a. How many shingles will be needed?
    b. If there are 60 shingles per bundle, how many bundles will be needed?
    c. At \$25 a bundle, what is the cost?

Computer Exercise

41. Consider the following BASIC program, written by a student, to solve problems involving arithmetic sequences:

```
1Ø REM    ARITHMETIC SEQUENCES
2Ø PRINT  "WHAT IS THE FIRST TERM"
3Ø INPUT  A1
4Ø PRINT  "WHAT IS THE COMMON DIFFERENCE"
5Ø INPUT  D
6Ø PRINT  "HOW MANY TERMS"
```

```
70  INPUT N
80  L = A1 + (N - 1) * D
90  S = (N / 2) * (A1 + L)
100 PRINT "THE"; N; "TH TERM IS"; L
110 PRINT "THE"; N; "TH PARTIAL SUM IS"; S
120 STOP
```

a. Use the program to find a_n and S_n if $a_1 = -11$, $d = 2$, and $n = 100$.
b. Modify the program so that *each* term in the sequence (from a_1 through a_n) is listed.

11.3 GEOMETRIC SEQUENCES

In this section another type of sequence, known as a **geometric sequence** or **geometric progression**, will be defined and investigated.

DEFINITION A geometric sequence is a sequence in which the ratio of any term after the first to the preceding term is a constant, called the **common ratio.**

EXAMPLE 1 Write the first six terms of the geometric sequence whose first term is 4 and whose common ratio is $\frac{1}{2}$.

Solution To obtain the desired terms we multiply each term by the common ratio to obtain the next term. Hence,

$$4, 2, 1, \tfrac{1}{2}, \tfrac{1}{4}, \tfrac{1}{8}$$

EXAMPLE 2 Is the following sequence of four terms $2, -\frac{4}{3}, \frac{8}{9}, -\frac{16}{27}$ a geometric sequence?

Solution

$$\left.\begin{aligned} a_2 \div a_1 &= -\tfrac{4}{3} \div 2 = -\tfrac{2}{3} \\ a_3 \div a_2 &= \tfrac{8}{9} \div -\tfrac{4}{3} = -\tfrac{2}{3} \\ a_4 \div a_3 &= -\tfrac{16}{27} \div \tfrac{8}{9} = -\tfrac{2}{3} \end{aligned}\right\} \text{There } \textit{is} \text{ a common ratio. It is } -\tfrac{2}{3}.$$

Thus the given sequence $2, -\frac{4}{3}, \frac{8}{9}, -\frac{16}{27}$ is a geometric sequence.

Next we develop formulas for the general term and the sum of the first n terms of a geometric sequence.

Let r = the common ratio of the geometric sequence. The first n terms can be written as:

$$a_1, a_2, a_3, \ldots, a_n$$

However,

$$\left.\begin{aligned} a_2 &= a_1 r \\ a_3 &= a_2 r = a_1 r^2 \\ a_4 &= a_3 r = a_1 r^3 \\ a_5 &= a_4 r = a_1 r^4 \end{aligned}\right\}$$

Notice that the exponent of r is one less than the number of the term.

$$\vdots \qquad \vdots$$

Thus, $a_n = a_1 r^{n-1}$

This is a formula for the nth term or the general term of any geometric sequence. Later in the chapter, in the section on mathematical induction, the reader is asked to prove this formula.

THEOREM 2 If $a_1, a_2, \ldots, a_n$ is a geometric sequence with common ratio r, then the sum of the first n terms is given by

$$S_n = \frac{a_1(1 - r^n)}{1 - r}, \; r \neq 1.$$

Proof:

Statement	*Reason*
1. $S_n = a_1 + a_2 + a_3 + \ldots + a_n$	1. Definition of S_n
2. $S_n = a_1 + a_1 r + a_1 r^2 + a_1 r^3 + \ldots + a_1 r^{n-1}$	2. Definition of geometric sequence
3. $rS_n = a_1 r + a_1 r^2 + a_1 r^3 + a_1 r^4 + \ldots + a_1 r^n$	3. Multiply both sides of step 2 by r
4. $S_n - rS_n = a_1 - a_1 r^n$	4. Subtract step 3 from step 2
5. Hence, $S_n = \dfrac{a_1(1 - r^n)}{1 - r}, \; r \neq 1$	5. Solve for S_n

The formulas are summarized below.

Formulas for a Geometric Sequence*

The general or nth term: $a_n = a_1 r^{n-1}$

The sum of the first n terms: $S_n = \dfrac{a_1(1 - r^n)}{1 - r}$

*Due to the presence of r^{n-1} and r^n in these formulas, a calculator with a y^x key will be helpful in this section.

Some examples involving these formulas will now be presented.

EXAMPLE 3 Consider the geometric sequence whose first three terms are 3, −12, 48.

a. Determine the 9th term.
b. Determine the sum of the first 9 terms.

Solution a. The common ratio is −4.

$$a_n = a_1 r^{n-1}$$
$$a_9 = (3)(-4)^8$$
$$a_9 = 196{,}608$$

b.
$$S_n = \frac{a_1(1-r^n)}{1-r}$$
$$S_9 = \frac{3[1-(-4)^9]}{1-(-4)}$$
$$S_9 = \frac{3(262{,}145)}{5}$$
$$S_9 = 157{,}287$$

EXAMPLE 4 Determine the first term and the common ratio of the geometric sequence whose 3rd term is .027 and whose 7th term is .0002187.

Solution

$$a_n = a_1 r^{n-1} \tag{1}$$

The 3rd term $a_3 = .027$. Substituting into (1) gives:

$$.027 = a_1 r^2. \tag{2}$$

The 7th term $a_7 = .0002187$. Substituting into (1) gives:

$$.0002187 = a_1 r^6 \tag{3}$$

Equations (2) and (3) are now solved simultaneously. Solve (2) for a_1: $a_1 = \frac{.027}{r^2}$. Substitute into (3):

$$\left(\frac{.027}{r^2}\right) r^6 = .0002187$$
$$r^4 = .0081$$
$$r = \pm .3$$

If $r = \pm .3,\ a_1 = \frac{.027}{r^2} = \frac{.027}{(\pm .3)^2} = .3.$

EXAMPLE 5 A mechanic takes a job offering a starting salary of \$8,000 with an increase of 10% per year, based on the previous year's salary.

a. Determine the salary for the 9th year.
b. Determine the amount of earnings for the first 9 years.

Solution a. Each year there is a salary increase of 10% over the previous year. This is equivalent to saying that any year's salary is 110% (or 1.1) of the previous year's salary. Thus the salary for the 9th year is a_9 in the geometric sequence with $a_1 = 8{,}000$ and $r = 1.1$.

$$a_9 = (8{,}000)(1.1)^8$$
$$a_9 = 17{,}148.71$$

The salary for the 9th year is $17,148.71

b. The total earnings for the first 9 years is S_9.

$$S_9 = \frac{(8{,}000)[1-(1.1)^9]}{1-1.1}$$

$$S_9 = \frac{(8{,}000)(-1.3579477)}{-.1}$$

$$S_9 = 108{,}635.82$$

The total earnings for the 9 years is $108,635.82.

EXAMPLE 6 A steam cleaning process can remove 40% of the oil film present on engine parts in 5 minutes. If the initial amount of oil film is D_o, what percent of oil film is left after 35 minutes of cleaning?

Solution The amounts of oil film left after each "block" of 5 minutes form a geometric sequence where $a_1 = D_o$ and $r = .6$. In 35 minutes there are 7 "blocks" of 5 minutes. Hence, we must determine a_7.

$$a_7 = (D_o)(.6)^6$$
$$a_7 = (D_o)(.046656) \approx (4.7\%)(D_o)$$

Thus, after 35 minutes of cleaning, 4.7% of the oil film remains.

In the remaining portion of this section, we will discuss *infinite* geometric sequences. Consider the geometric sequence $\frac{6}{10}, \frac{6}{100}, \frac{6}{1000}, \ldots$ and the associated *infinite series*: $\frac{6}{10} + \frac{6}{100} + \frac{6}{1000} + \ldots$ (4)

This series, although infinite, has a finite sum. Observe:

$\frac{6}{10} + \frac{6}{100} + \frac{6}{1000} + \ldots$

Write each term in decimal form

$.6 + .06 + .006 + \ldots$

Add the terms

$.666\ldots$

$\frac{2}{3}$ ← This is the sum of the infinite series (4).

Not all infinite series have a finite sum. Consider the geometric sequence 2, 4, 8, 16, . . . whose associated infinite series is:

$$2 + 4 + 8 + 16 + \ldots \tag{5}$$

There is no finite sum for the infinite series (5). The terms are increasing in size and the sum keeps increasing beyond bound.

The problem we face now becomes clear—under what conditions will an infinite geometric series have a finite sum? To answer this question, consider the structure of a geometric sequence:

$$a_1, a_1r, a_1r^2, a_1r^3, \ldots, a_1r^{n-1}, \ldots$$

The formula for the sum of the first n terms is

$$S_n = \frac{a_1(1-r^n)}{1-r}$$

We now ask what happens to S_n as n increases without bound*? The answer is found by examining the expression r^n as $n \to \infty$. As long as $-1 < r < 1$ (or equivalently $|r| < 1$), the expression r^n becomes smaller and smaller as $n \to \infty$. Thus, the formula

$$S_n = \frac{a_1(1-r^n)}{1-r} \quad \text{becomes}$$

$$S = \frac{a_1}{1-r} \quad \text{as } n \to \infty$$

The interpretation of this is that an infinite geometric series will have a finite sum, given by S, provided the common ratio r is such that $|r| < 1$.

Formula for the Sum of an Infinite Geometric Series

$$S = \frac{a_1}{1-r}, \text{ provided } |r| < 1$$

EXAMPLE 7 Determine the sum of the infinite geometric series $\frac{1}{3} - \frac{1}{6} + \frac{1}{12} - \frac{1}{24} + \ldots$

Solution The first term $a_1 = \frac{1}{3}$ and the common ratio $r = -\frac{1}{2}$ (notice, $|r| < 1$).

$$S = \frac{a_1}{1-r} = \frac{\frac{1}{3}}{1-(-\frac{1}{2})} = \frac{2}{9}$$

*In mathematics, when a quantity, like n, *increases without bound*, it is customary to write $n \to \infty$ ("as n approaches infinity").

EXAMPLE 8 Determine a quotient of integers that represents the rational number 6.4444 . . . .

Solution First, determine a quotient of integers that represents the repeating part.

$$.4444\ldots = \underbrace{.4 + .04 + .004 + .0004}_{} + \ldots$$

This forms an infinite geometric series with $a_I = .4$ and $r = .1$.

$$S = \frac{a_1}{1-r} = \frac{.4}{1-.1} = \frac{4}{9}$$

Thus $6.444\ldots = 6 + \frac{4}{9} = \frac{58}{9}$.

The answer can be checked by long division.

EXERCISES FOR SECTION 11.3

In Exercises 1 through 6, construct the desired geometric sequences. (See Example 1.)

1. Construct the first five terms of a geometric sequence whose first term is 1 and whose common ratio is $\frac{4}{3}$.
2. Construct the first four terms of a geometric sequence whose first term is $-\frac{1}{3}$ and whose common ratio is $-\frac{1}{2}$.
3. Construct the first six terms of a geometric sequence whose first term is -2 and whose common ratio is $\frac{1}{2}$.
4. Construct the first three terms of a geometric sequence whose first term is $-\frac{1}{2}$ and whose common ratio is $\frac{5}{4}$.
5. Construct the first four terms of a geometric sequence whose third term is -3 and whose common ratio is $\frac{1}{4}$.
6. Construct the first four terms of a geometric sequence whose second term is .3 and whose common ratio is .2.

In 7 through 12, determine whether or not the given sequence is geometric. If so, state the common ratio. (See Example 2.)

7. $2, 4, 6, 8, \ldots$
8. $\frac{1}{2}, \frac{1}{4}, \frac{1}{6}, \frac{1}{8}, \ldots$
9. $-\frac{1}{2}, -\frac{1}{4}, -\frac{1}{8}, -\frac{1}{16}, \ldots$
10. $2, \frac{2}{3}, \frac{2}{9}, \frac{2}{27}, \ldots$
11. $.7, .07, .007, .0007, \ldots$
12. $12, -4, \frac{4}{3}, -\frac{4}{9}, \ldots$

In 13 through 22, determine the specified term, a_n, and sum, S_n, for each of the given geometric sequences. (See Example 3.)

13. $5, 1, \frac{1}{5}, \frac{1}{25}, \ldots$. Determine a_9 and S_9.
14. $\frac{1}{2}, \frac{1}{4}, \frac{1}{8}, \frac{1}{16}, \ldots$. Determine a_8 and S_{10}.
15. $\frac{1}{2}, -\frac{1}{4}, \frac{1}{8}, -\frac{1}{16}, \ldots$. Determine a_8 and S_{10}.
16. $.7, .07, .007, \ldots$. Determine a_{10} and S_{12}.

17. 2, 6, 18, ... Determine a_6 and S_6.
18. $2, -\frac{3}{2}, \frac{9}{8}, \ldots$. Determine a_5 and S_5.
19. $1, \frac{1}{3}, \frac{1}{9}, \ldots$. Determine a_8 and S_8.
20. 1, 3, 9, ... Determine a_8 and S_8.
21. $2, 2x, 2x^2, \ldots$. Determine a_{10} and S_{10}.
22. $\frac{1}{x}, \frac{1}{3x^2}, \frac{1}{9x^3}, \ldots$. Determine a_7 and S_7.

In 23 through 26, determine the first term and common ratio for the described geometric sequences. (See Example 4.)

23. The 2nd term is .09 and the 6th term is .000729.
24. The 3rd term is .004 and the 7th term is .0000064.
25. The 4th term is 24 and the 11th term is 3072.
26. The 1st term is $-\frac{1}{2}$ and the 9th term is $-\frac{19683}{2}$.

In 27 through 30, determine each sum.

27. $\sum_{i=1}^{10} (\frac{1}{2})^i$ 28. $\sum_{i=1}^{10} 2^i$ 29. $\sum_{j=1}^{6} 2(\frac{1}{3})^j$ 30. $\sum_{k=0}^{9} 3(\frac{1}{2})^k$

In 31 through 39, solve each applied problem. (See Examples 5 and 6.)

31. The population of the United States is 200,000,000 and is increasing by 1% each year.
    a. What will the population be in 20 years (assuming the same growth rate)?
    b. What will the population be in 40 years (assuming the same growth rate)?
32. A job pays 1¢ the first day, 2¢ the second day, 4¢ the third day, 8¢ the fourth day, etc.
    a. What is the salary on the 27th day?
    b. What are the total earnings after 27 days?
33. A savings deposit of \$200 is made at a bank offering 6% interest compounded annually. How much money will be on deposit after 20 years?
34. When an automobile is purchased, it begins losing value during each year. This depreciation is expressed as a percent. Suppose an automobile, that originally cost \$6000, depreciates 20% each year. Determine its value after 8 years.
35. In El Salvador the population is increasing 4% each year. If the present population is denoted by P_o, determine El Salvador's population in 10 years.
36. A savings deposit investment of \$1200 earns 8% interest compounded annually. How much money will be on deposit in 3 years?
37. When interest is compounded quarterly, the expressed interest rate is divided by 4 and then using this rate, interest is compounded 4 times a year. Using this method, determine how much money will be on deposit in 3 years if an original deposit of \$1200 earns 8% interest.
38. Oil consumption in the U.S. has been doubling every 14 years. Let the oil consumption now be represented by C_o. What will the consumption be in 70 years (assume the same doubling rate)?

39. An employer offers an employee a choice of two contracts. Each has a starting salary of \$10,000. Contract A offers a raise of \$1,200 each year thereafter. Contract B offers a 10% increase in salary each year over the previous year.
    a. Suppose you plan on working for the employer for 5 years. Determine your total earnings under contracts A and B to see which offers the better deal.
    b. Suppose you plan on working for the employer for 10 years. Determine your total earnings under contracts A and B to see which offers the better deal.
40. The number of bacteria in a certain culture doubles every 24 hours. Let the number of bacteria now be represented by N_o. What will the number of bacteria present in the culture be in a week?

In 41 through 48, determine the sum of each infinite geometric series. (See Example 7.)

41. $\frac{1}{2} + \frac{1}{4} + \frac{1}{8} + \ldots$ 42. $\frac{1}{2} - \frac{1}{4} + \frac{1}{8} - \ldots$ 43. $\frac{2}{3} - \frac{6}{15} + \frac{18}{75} - \ldots$
44. $1 + \frac{9}{10} + \frac{81}{100} + \ldots$ 45. $-1 + \frac{1}{3} - \frac{1}{9} + \ldots$ 46. $-1 - \frac{1}{3} - \frac{1}{9} - \ldots$
47. $\frac{1}{5} + \frac{1}{35} + \frac{1}{245} + \ldots$ 48. $.9 + .09 + .009 + .0009 + \ldots$

In 49 through 56, determine a quotient of integers that represents each rational number. (See Example 8.)

49. .232323 . . . 50. 5.555 . . . 51. 6.555 . . .
52. 1.01232323 . . . 53. 9.1111 . . . 54. 1.999 . . .
55. .27777 . . . 56. .706706 . . .

Computer Exercises

57. Write a computer program to find the nth partial sum of any geometric series given its first two terms.
58. Use your program from Exercise 57 to approximate your answers to the series of Exercises 41–48, using $n = 30$.

11.4 MATHEMATICAL INDUCTION

Mathematical induction is a method of proof particularly suited for proving statements that involve n, where n is a positive integer. However, before developing the method, consider the following formula for adding even integers.

$$2 + 4 + 6 + 8 + \ldots + 2n = n(n + 1)$$

A partial "check" on this formula can be performed as indicated below.

Case $n = 1$:	Case $n = 2$:
$2 = 1(1 + 1)$	$2 + 4 = 2(2 + 1)$
$2 = 2$ ✓	$6 = 6$ ✓

Case $n = 3$:	Case $n = 4$:
$2 + 4 + 6 = 3(3 + 1)$	$2 + 4 + 6 + 8 = 4(4 + 1)$
$12 = 12$ ✓	$20 = 20$ ✓

Case $n = 5$:	Case $n = 6$:
$2 + 4 + 6 + 8 + 10 = 5(5 + 1)$	$2 + 4 + 6 + 8 + 10 + 12 = 6(6 + 1)$
$30 = 30$ ✓	$42 = 42$ ✓

The "checks" above by no means constitute a proof of $2 + 4 + 6 + 8 + \ldots + 2n = n(n + 1)$ for *all* positive integers, n. All we *do* know is that the formula is true for $n = 1, 2, 3, 4, 5, 6$. *Beyond*, $n = 6$ the formula remains unverified. It is possible to continue the checking process for as many positive integral values of n as desired, say $n = 7, 8, 9, 10, \ldots, k$. However, this will *not prove* the formula for any values of $n > k$. What we need is some sort of "chain reaction" so that once the formula is proved for a particular n, the formula will *automatically* follow for the next integer, and the next after that, etc. The method of proof called *mathematical induction* provides us with what we need.

The Principle of Mathematical Induction

Suppose that associated with each positive integer n there is a statement or formula S_n. In order to verify all the statements S_n by the method of mathematical induction:

I. Verify that S_1 is true, and
II. Assume that S_k is true, and then verify that S_{k+1} is true.

Once steps I and II have been carried out, the "chain reaction" previously referred to begins and continues indefinitely.

EXAMPLE 1 Prove $2 + 4 + 6 + 8 + \ldots + 2n = n(n + 1)$.

Solution Proof:

I. We must verify S_1. This is accomplished by letting $n = 1$. We have:

$2 = 1(1 + 1)$ which is true. Thus, S_1 is verified.

II. We assume the truth of S_k and attempt to verify S_{k+1}. By S_k we have:

$2 + 4 + 6 + 8 + \ldots + 2k = k(k + 1)$ ← This is assumed true.

Add $2(k + 1)$ to both sides.

$$2+4+6+8+\ldots+2k+2(k+1)=k(k+1)+2(k+1)$$ Factor the right-hand side

$$2+4+6+8+\ldots+2(k+1)=(k+1)(k+2)$$ ← This is S_{k+1}

We have verified S_{k+1} by assuming the truth of S_k. Hence the proof is complete. Thus $2 + 4 + 6 + 8 + \ldots + 2n = n(n + 1)$.

EXAMPLE 2 Prove $\dfrac{1}{1\cdot 2}+\dfrac{1}{2\cdot 3}+\dfrac{1}{3\cdot 4}+\ldots+\dfrac{1}{n(n+1)}=\dfrac{n}{n+1}$.

Solution Proof:

I. We must verify S_1. Letting $n = 1$ we have:

$$\frac{1}{1(1+1)}=\frac{1}{1+1}$$ which is true. Thus S_1 is verified.

II. We assume the truth of S_k and attempt to verify S_{k+1}. By S_k we have:

$$\frac{1}{1\cdot 2}+\frac{1}{2\cdot 3}+\frac{1}{3\cdot 4}+\ldots+\frac{1}{k(k+1)}=\frac{k}{k+1}$$ ← This is assumed true.

Add $\dfrac{1}{(k+1)(k+2)}$ to both sides.

$$\frac{1}{1\cdot 2}+\frac{1}{2\cdot 3}+\frac{1}{3\cdot 4}+\ldots+\frac{1}{k(k+1)}+\frac{1}{(k+1)(k+2)}$$

$$=\frac{k}{k+1}+\frac{1}{(k+1)(k+2)}$$

$$=\frac{k(k+2)}{(k+1)(k+2)}+\frac{1}{(k+1)(k+2)}$$

$$=\frac{k^2+2k+1}{(k+1)(k+2)}$$

$$=\frac{(k+1)(k+1)}{(k+1)(k+2)}$$

Simplify the right-hand side.

$$\frac{1}{1 \cdot 2} + \frac{1}{2 \cdot 3} + \frac{1}{3 \cdot 4} + \ldots + \frac{1}{(k+1)(k+2)} = \frac{k+1}{k+2}.$$

This is S_{k+1}. We have thus verified S_{k+1} by assuming the truth of S_k. The proof is complete.

Thus, $$\frac{1}{1 \cdot 2} + \frac{1}{2 \cdot 3} + \frac{1}{3 \cdot 4} + \ldots + \frac{1}{n(n+1)} = \frac{n}{n+1}.$$

In the next example we prove that the nth term in an arithmetic sequence is $a_n = a_1 + (n-1)d$.

EXAMPLE 3 Prove that for an arithmetic sequence, the nth term $a_n = a_1 + (n-1)d$, where d is the common difference.

Solution Proof:

I. We must verify S_1. Letting $n = 1$ gives:

$a_1 = a_1 + (1-1)d$ or
$a_1 = a_1$ which is true.
Thus S_1 is verified.

II. Assuming the truth of S_k gives:

$a_k = a_1 + (k-1)d.$ ← This is assumed true.

Add d to both sides.

$$a_k + d = a_1 + (k-1)d + d$$
$$a_{k+1} = a_1 + kd - d + d$$
$$a_{k+1} = a_1 + kd$$ ← This is S_{k+1}.

The proof is complete. Thus $a_n = a_1 + (n-1)d$.

Mathematical induction can also be used when the statements S_n refer to inequalities as the next example shows.

EXAMPLE 4 Prove that if $a > 1$, then $a^n > 1$.

Solution Proof:

I. We must verify S_1. Letting $n = 1$,

$a^1 > 1$ or
$a > 1$, which is true.
Thus S_1 is verified.

II. Assuming the truth of S_k gives:

$$a^k > 1 \leftarrow \text{This is assumed true.}$$

Multiply both sides by a.

$$a^k \cdot a > 1 \cdot a$$
$$a^{k+1} > a$$

Since $a > 1$,

$$a^{k+1} > a > 1$$

So, $a^{k+1} > 1$. ← This is S_{k+1}.

The proof is complete. Hence, if $a > 1$, $a^n > 1$.

EXERCISES FOR SECTION 11.4

In Exercises 1 through 15, prove each formula, where n is any positive integer. (See Examples 1 and 2.)

1. $1 + 2 + 3 + \ldots + n = \dfrac{n(n+1)}{2}$

2. $1 + 4 + 7 + \ldots + (3n - 2) = \dfrac{n(3n-1)}{2}$

3. $1^2 + 2^2 + 3^2 + \ldots + n^2 = \dfrac{n(n+1)(2n+1)}{6}$

4. $1^3 + 2^3 + 3^3 + \ldots + n^3 = \left[\dfrac{n(n+1)}{2}\right]^2$

5. $\dfrac{1}{1 \cdot 2} + \dfrac{1}{2 \cdot 3} + \dfrac{1}{3 \cdot 4} + \ldots + \dfrac{1}{(n-1)n} = \dfrac{n-1}{n}$

6. $\dfrac{1}{2 \cdot 3 \cdot 4} + \dfrac{2}{3 \cdot 4 \cdot 5} + \dfrac{3}{4 \cdot 5 \cdot 6} + \ldots + \dfrac{n}{(n+1)(n+2)(n+3)}$
$= \dfrac{n(n+1)}{4(n+2)(n+3)}$

7. $1 + 3 + 5 + \ldots + (2n - 1) = n^2$

8. $1^2 + 3^2 + 5^2 + \ldots + (2n-1)^2 = \dfrac{n(2n-1)(2n+1)}{3}$

9. $\dfrac{1}{1 \cdot 2 \cdot 3} + \dfrac{1}{2 \cdot 3 \cdot 4} + \dfrac{1}{3 \cdot 4 \cdot 5} + \ldots + \dfrac{1}{n(n+1)(n+2)}$
$= \dfrac{n(n+3)}{4(n+1)(n+2)}$

10. $1 \cdot 2 + 2 \cdot 3 + 3 \cdot 4 + \ldots + n(n+1) = \dfrac{n(n+1)(n+2)}{3}$

11. $1 \cdot 2 \cdot 3 + 2 \cdot 3 \cdot 4 + 3 \cdot 4 \cdot 5 + \ldots + n(n+1)(n+2)$
$= \dfrac{n(n+1)(n+2)(n+3)}{4}$

12. $4 + 4^2 + 4^3 + \ldots + 4^n = \frac{4}{3}(4^n - 1)$
13. $5 + 5^2 + 5^3 + \ldots + 5^n = \frac{5}{4}(5^n - 1)$
14. $x^0 + x^1 + x^2 + \ldots + x^{n-1} = \dfrac{x^n - 1}{x - 1} \quad (x \neq 1)$
15. $\dfrac{1}{2} + \dfrac{1}{4} + \dfrac{1}{8} + \ldots + \dfrac{1}{2^n} = 1 - \dfrac{1}{2^n}$
16. *Derive* the formula given in Example 1 by considering the arithmetic sequence 2, 4, 6, 8, . . . , where $a_1 = 2$ and $d = 2$.
17. *Derive* the formula given in Exercise 1 by considering the arithmetic sequence 1, 2, 3, 4, . . . , where $a_1 = 1$ and $d = 1$.
18. *Derive* the formula given in Exercise 2 by considering the arithmetic sequence 1, 4, 7, 10, . . . , where $a_1 = 1$ and $d = 3$.
19. *Derive* the formula given in Exercise 12 by considering the geometric sequence $4, 4^2, 4^3, \ldots$, where $a_1 = 4$ and $r = 4$.
20. *Derive* the formula given in Exercise 15 by considering the geometric sequence $\frac{1}{2}, \frac{1}{4}, \frac{1}{8}, \ldots$, where $a_1 = \frac{1}{2}$ and $r = \frac{1}{2}$.
21. Given the geometric sequence $a, ar, ar^2, \ldots$, prove, using mathematical induction, that the nth term is $a_n = ar^{n-1}$. (See Example 3.)
22. Prove, using mathematical induction, that if $0 < a < 1$, then $0 < a^n < 1$, where n is any positive integer. (See Example 4.)
23. Prove the law of exponents $(xy)^a = x^a y^a$, where a is any positive integer.
24. Prove the law of exponents $\left(\dfrac{x}{y}\right)^a = \dfrac{x^a}{y^a}$, where a is any positive integer. $(y \neq 0)$.

11.5 THE BINOMIAL THEOREM AND THE BINOMIAL SERIES

There are many situations in the study of algebra, calculus, and probability and statistics where it is necessary to raise a binomial to a power. Consider the expression $(a + b)^n$ where n is a positive integer. Direct multiplication will verify the following facts:

$$(a + b)^1 = a + b$$
$$(a + b)^2 = a^2 + 2ab + b^2$$
$$(a + b)^3 = a^3 + 3a^2b + 3ab^2 + b^3$$
$$(a + b)^4 = a^4 + 4a^3b + 6a^2b^2 + 4ab^3 + b^4$$

We now make the following observations:

1. The first term in the expansion of $(a + b)^n$ is a^n.
2. The last term in the expansion of $(a + b)^n$ is b^n.
3. From left to right, the exponent of a decreases by 1 from term to term, while the exponent of b increases by 1 from term to term (the sum of the exponents of a and b for any term is n).
4. Proceeding from left to right, if the coefficient of any term is multiplied by the exponent of a in that term, and the product is then divided by the *number* of that term, the result is the coefficient of the next term.

The observations just made are *known* to be true for $n = 1, 2, 3, 4$. However, it is possible to prove them for all positive integral values of n by using mathematical induction. We *conjecture* that the following formula is true for all positive integral values of n.

THEOREM 3

The Binomial Theorem

$$(a+b)^n = a^n + na^{n-1}b + \frac{n(n-1)}{2}a^{n-2}b^2 + \frac{n(n-1)(n-2)}{2\cdot 3}a^{n-3}b^3 + \frac{n(n-1)(n-2)(n-3)}{2\cdot 3\cdot 4}a^{n-4}b^4 + \ldots + b^n,$$

where n is a positive integer.

EXAMPLE 1 Use the binomial theorem to expand $(x^2 + 2y)^5$.

Solution Use the binomial theorem with $a = x^2$, $b = 2y$, and $n = 5$.

$$(x^2+2y)^5 = (x^2)^5 + 5(x^2)^4(2y) + \frac{5\cdot 4}{2}(x^2)^3(2y)^2 + \frac{5\cdot 4\cdot 3}{2\cdot 3}(x^2)^2(2y)^3 + \frac{5\cdot 4\cdot 3\cdot 2}{2\cdot 3\cdot 4}(x^2)(2y)^4 + \frac{5\cdot 4\cdot 3\cdot 2\cdot 1}{2\cdot 3\cdot 4\cdot 5}(2y)^5$$

$$(x^2+2y)^5 = x^{10} + 10x^8y + 40x^6y^2 + 80x^4y^3 + 80x^2y^4 + 32y^5$$

Now that the reader has seen how the binomial theorem can be used to expand a binomial with a positive integral power, we turn to more theoretical considerations regarding its structure.

The denominators in the binomial theorem contain products of consecutive positive integers. To write products of this type we use the following definition for the symbol $n!$ (**"n factorial"**).

DEFINITION

$$n! = \begin{cases} n(n-1)(n-2)\ldots(3)(2)(1), & \text{if } n \text{ is a positive integer} \\ 1, & \text{if } n = 0 \end{cases}$$

EXAMPLE 2 Evaluate each expression.

a. $5!$ b. $3!$ c. $\frac{6!}{4!}$ d. $\frac{4!}{0!}$ e. $\frac{5!}{3!\ 2!}$ f. $\frac{3!}{0!\ 3!}$

Solution a. $5! = 5 \cdot 4 \cdot 3 \cdot 2 \cdot 1 = 120$ b. $3! = 3 \cdot 2 \cdot 1 = 6$

c. $\frac{6!}{4!} = \frac{6 \cdot 5 \cdot 4 \cdot 3 \cdot 2 \cdot 1}{4 \cdot 3 \cdot 2 \cdot 1} = 30$ d. $\frac{4!}{0!} = \frac{4 \cdot 3 \cdot 2 \cdot 1}{1} = 24$

e. $\frac{5!}{3!\ 2!} = \frac{5 \cdot 4 \cdot 3 \cdot 2 \cdot 1}{3 \cdot 2 \cdot 1 \cdot 2 \cdot 1} = 10$ f. $\frac{3!}{0!\ 3!} = \frac{3 \cdot 2 \cdot 1}{1 \cdot 3 \cdot 2 \cdot 1} = 1$

Using factorial notation the binomial theorem can now be written as

$$(a+b)^n = a^n + \frac{n}{1!}a^{n-1}b + \frac{n(n-1)}{2!}a^{n-2}b^2 + \frac{n(n-1)(n-2)}{3!}a^{n-3}b^3 + \frac{n(n-1)(n-2)(n-3)}{4!}a^{n-4}b^4 + \ldots + b^n.$$

Next, we state a definition that will enable us to write each coefficient in a more compact form.

DEFINITION The **binomial coefficient,** $\binom{n}{r}$, read as "n on r" is defined to be

$$\binom{n}{r} = \frac{n!}{r!(n-r)!}$$

Observe:

$$\binom{n}{0} = \frac{n!}{0!\ n!} = 1$$

$$\binom{n}{1} = \frac{n!}{1!(n-1)!} = \frac{n\cancel{(n-1)!}}{1\cancel{(n-1)!}} = n$$

$$\binom{n}{2} = \frac{n!}{2!(n-2)!} = \frac{n(n-1)\cancel{(n-2)!}}{2!\cancel{(n-2)!}} = \frac{n(n-1)}{2!}$$

$$\binom{n}{3} = \frac{n!}{3!(n-3)!} = \frac{n(n-1)(n-2)\cancel{(n-3)!}}{3!\cancel{(n-3)!}} = \frac{n(n-1)(n-2)}{3!}$$

$$\binom{n}{n} = \frac{n!}{n!\ 0!} = 1$$

Note: these *are* the coefficients in the binomial theorem.

Hence, we can rewrite the binomial theorem more compactly, as

The Binomial Theorem

$$(a+b)^n = \binom{n}{0}a^n + \binom{n}{1}a^{n-1}b + \binom{n}{2}a^{n-2}b^2 + \binom{n}{3}a^{n-3}b^3 + \ldots + \binom{n}{n}b^n,$$

where n is a positive integer.

EXAMPLE 3 Use the binomial theorem to expand $\left(x - \frac{1}{x}\right)^4$.

Solution Use the binomial theorem with $a = x$, $b = -\frac{1}{x}$, and $n = 4$.

$$\left(x-\frac{1}{x}\right)^4 = \binom{4}{0}(x)^4 + \binom{4}{1}(x)^3\left(-\frac{1}{x}\right) + \binom{4}{2}(x)^2\left(-\frac{1}{x}\right)^2 + \binom{4}{3}(x)\left(-\frac{1}{x}\right)^3 + \binom{4}{4}\left(-\frac{1}{x}\right)^4$$

$$\left(x-\frac{1}{x}\right)^4 = (1)(x^4) + 4x^3\left(-\frac{1}{x}\right) + 6x^2\left(\frac{1}{x^2}\right) + 4x\left(-\frac{1}{x^3}\right) + (1)\left(\frac{1}{x^4}\right)$$

$$\left(x-\frac{1}{x}\right)^4 = x^4 - 4x^2 + 6 - \frac{4}{x^2} + \frac{1}{x^4}$$

Sometimes it is desirable to determine a single specified term in the expansion of $(a + b)^n$. Consequently, the general term, or rth term, will now be determined. A study of the terms in the binomial theorem reveals that each has the structure

$$\binom{n}{\square}a^{n-\square}b^{\square}.$$

Moreover, note that the number in $\square$ is always 1 less than the number of the term. Thus the rth term is:

$$\binom{n}{r-1}a^{n-(r-1)}b^{r-1}$$

EXAMPLE 4 Determine the 7th term in the expansion of $(x - \frac{1}{3}y^2)^{11}$.

Solution Since the rth term is given by

$$\binom{n}{r-1}a^{n-(r-1)}b^{r-1},$$

the 7th term is $\binom{11}{6}(x)^{11-6}\left(-\frac{1}{3}y^2\right)^6$ or

$$\frac{462}{729}x^5y^{12}$$

Thus far in this section it has been assumed that n is a positive integer in the expression $(a + b)^n$. However, if n is a rational number that is *not* a positive integer, the binomial formula generates a series *having no last term*. An infinite series known as the **binomial series** is obtained. We now state without proof:

The Binomial Series

$$(a+b)^n = a^n + na^{n-1}b + \frac{n(n-1)}{2}a^{n-2}b^2 + \frac{n(n-1)(n-2)}{2 \cdot 3}a^{n-3}b^3 + \ldots$$

where n is a rational number that is not a positive integer and $|b| < |a|$.*

The reason for the restriction $|b| < |a|$ is beyond the scope of this book. However, the reader will learn more about this when studying calculus. The binomial series can be used to generate an infinite series** for certain expressions, as the next example shows.

EXAMPLE 5 Use the binomial series to generate an infinite series for $(32 + x)^{1/5}$.

Solution

$$(32+x)^{1/5} = 32^{1/5} + \frac{1}{5}(32)^{-4/5}x + \frac{(\frac{1}{5})(-\frac{4}{5})}{2}(32)^{-9/5}x^2 + \frac{(\frac{1}{5})(-\frac{4}{5})(-\frac{9}{5})}{2 \cdot 3}(32)^{-14/5}x^3 - \ldots,$$

provided $|x| < 32$.

$$(32+x)^{1/5} = 2 + \frac{1}{80}x - \frac{1}{6{,}400}x^2 + \frac{3}{1{,}024{,}000}x^3 - \ldots,$$

provided $|x| < 32$.

*Note that binomial coefficients $\binom{n}{r}$ *cannot* be used to write the binomial series because n is no longer a positive integer.

**Infinite series have a variety of uses in engineering, chemistry, and physics.

EXAMPLE 6 Use the binomial series of the previous example to calculate $\sqrt[5]{36}$ to the nearest hundredth.

Solution $\sqrt[5]{36} = (32 + 4)^{1/5}$. Thus, in the binomial series of the previous example, let $x = 4$. (*Note:* $|4| < 32$)

$$(32 + 4)^{1/5} = 2 + \frac{1}{80}(4) - \frac{1}{6{,}400}(4)^2 + \frac{3}{1{,}024{,}000}(4)^3 - \ldots$$

$$(32 + 4)^{1/5} = 2 + \frac{1}{20} - \frac{1}{400} + \frac{3}{16{,}000} - \ldots$$

$$(32 + 4)^{1/5} = \underbrace{2 + .05 - .0025}_{} + \underbrace{.0001875 - \ldots}_{}$$

These terms have a sum of 2.0475.

These terms have no effect on the second decimal place.

As a result, $\sqrt[5]{36} \approx 2.05$ (to the nearest hundredth).

EXAMPLE 7 Use the binomial series to generate an infinite series for $(1 - x^2)^{-1}$.

Solution

$$(1 - x^2)^{-1} = 1^{-1} + (-1)(1)^{-2}(-x^2) + \frac{(-1)(-2)}{2}(1)^{-3}(-x^2)^2 + \frac{(-1)(-2)(-3)}{2 \cdot 3}(1)^{-4}(-x^2)^3 + \ldots,$$

provided $|x^2| < 1$, or equivalently $x^2 < 1$.

$$(1 - x^2)^{-1} = 1 + x^2 + x^4 + x^6 + \ldots, \text{ provided } x^2 < 1.$$

The result can be expressed as:

$$\frac{1}{1 - x^2} = 1 + x^2 + x^4 + x^6 + \ldots, \text{ provided } x^2 < 1.$$

Incidentally, in Exercise 53, the reader is asked to derive the result of Example 7 using a different method.

EXERCISES FOR SECTION 11.5

In 1 through 10, evaluate each expression. (See Example 2.)

1. $6!$
2. $2!$
3. $\dfrac{6!}{2!}$
4. $\dbinom{6}{2}$
5. $\dfrac{6!}{4!}$
6. $\dbinom{n}{0}$
7. $\dbinom{8}{8}$
8. $\dbinom{100}{99}$
9. $\dbinom{8}{5}$
10. $\dbinom{8}{3}$

In 11 through 25, use the binomial theorem to expand each binomial. (See Examples 1 and 3.)

11. $(x + y)^6$
12. $(x - y)^6$
13. $(a + b)^3$
14. $(a - b)^3$
15. $(-a + b)^3$
16. $(-a - b)^3$
17. $(2x - 3y)^4$
18. $(2x - 3y^2)^4$
19. $(5x + 4)^4$
20. $\left(x - \frac{1}{x^2}\right)^2$
21. $\left(x - \frac{1}{x^2}\right)^4$
22. $(y^3 + x)^3$
23. $(\sqrt{x} + \sqrt{y})^3$
24. $(\sqrt[3]{x} + \sqrt[3]{y})^3$
25. $\left(\frac{1}{x} + \frac{1}{y}\right)^4$
26. Expand $(x + y + z)^4$ by first writing $[(x + y) + z]^4$ and using the binomial theorem with $a = x + y$, $b = z$, and $n = 4$.

In 27 through 30, use the binomial theorem to expand each expression. (See Exercise 26.)

27. $(x - y + 2z)^4$
28. $(2a + 2b + 2c)^2$
29. $(2x - 5y + k)^4$
30. $(x + y + z + w)^3$

In 31 through 40, determine the specified term. (See Example 4.)

31. The 5th term in the expansion of $(a + b)^8$
32. The 4th term in the expansion of $(a - b)^7$
33. The 3rd term in the expansion of $(\sqrt{x} + \sqrt{y})^9$
34. The 4th term in the expansion of $(yx + 6)^5$
35. The 6th term in the expansion of $(2x - 1)^{12}$
36. The 7th term in the expansion of $\left(\frac{1}{y} + 1\right)^{12}$
37. The term involving x^3 in the expansion of $(2x + 1)^6$
38. The term involving y^8 in the expansion of $(y^2 - x)^6$
39. The term involving x in the expansion of $(x^{1/3} - 1)^6$
40. The 4th term in the expansion of $(1 + .02)^{10}$

In 41 through 48, use the binomial series to generate the specified infinite series. (See Examples 5 and 7.)

41. Generate an infinite series for $(1 + x)^{1/3}$. Exhibit the first 4 terms.
42. Generate an infinite series for $(16 - x)^{1/4}$. Exhibit the first 4 terms.
43. Generate an infinite series for $(1 - x)^{-1}$. Exhibit the first 5 terms.
44. Generate an infinite series for $(2 + y)^{1/2}$. Exhibit the first 4 terms.
45. Generate an infinite series for $(1 + x)^{-2}$. Exhibit the first 4 terms.
46. Generate an infinite series for $(1 + x^2)^{2/3}$. Exhibit the first 3 terms.
47. Generate an infinite series for $(4 + x)^{1/2}$. Exhibit the first 4 terms.
48. Generate an infinite series for $(27 - x)^{1/3}$. Exhibit the first 5 terms.

In 49 through 52, calculate as directed. (See Example 6.)

49. Use the answer to Exercise 41 to calculate $\sqrt[3]{1.06}$ to the nearest hundredth.
50. Use the answer to Exercise 42 to calculate $\sqrt[4]{14}$ to the nearest hundredth.
51. Use the answer to Exercise 44 to calculate $\sqrt{2.04}$ to the nearest hundredth.
52. Use the anwer to Exercise 47 to calculate $\sqrt{4.05}$ to the nearest hundredth.
53. Derive the result of Example 7 by considering $1 + x^2 + x^4 + x^6 + \ldots$ to be an infinite geometric series with $a_1 = 1$ and $r = x^2$.

Calculator Exercises

54. Use a calculator to approximate $\sqrt[5]{36}$ to the nearest ten-thousandth, using the method of Example 6.
55. Use a calculator with a "y^x" key to approximate $\sqrt[5]{36}$ directly. Compare with Exercise 54.

CHAPTER 11 IN REVIEW

I. Summary

In this chapter we investigated many aspects of sequences and series. A sequence is the collection of range elements of a sequence function. A series is an indicated sum of a certain number of terms of a sequence.

In Sections 11.2 and 11.3, arithmetic and geometric sequences were studied. Formulas for the general term and the sum of the first n terms were derived. Moreover, we found that when the absolute value of the common ratio is less than 1, an infinite geometric series has a finite sum.

In Section 11.4, the method of proof known as mathematical induction was presented. It is commonly used for proving statements that involve the positive integers, where a case by case type of proof would never end.

The final topic of this chapter was the expansion of $(a + b)^n$. When n is a positive integer the binomial theorem pertains and when n is a rational number that is not a positive integer, then the binomial series is used.

II. Vocabulary/Symbol True-False Quiz

If the statement is true, write "TRUE." However, if the statement is false, replace the underlined word or phrase with a different word or phrase that will make the new statement true.

1. $\sum_{i=1}^{n} (-1)^i \frac{1}{2i+1}$ is an example of an $\underline{\text{alternating series}}$.
2. $\sum_{i=1}^{n} 2i = \underline{n(n+1)}$.
3. The sixth term of any arithmetic sequence is $\underline{a_1 + 5d}$.
4. The 100th term of the geometric sequence $2, 1, \frac{1}{2}, \ldots$ is $\underline{\frac{1}{2^{98}}}$.
5. $\sum_{i=1}^{n} 2i$ is an example of a $\underline{\text{geometric series}}$.
6. $0! = \underline{0}$.
7. $\binom{10}{2} = \underline{45}$.
8. $\frac{1}{1-x^2} = 1 + x^2 + x^4 + x^6 + \ldots$, provided $\underline{x^2 < 1}$.

III. Chapter Test

1. Write the sequence generated by f if the domain of f is the set of positive integers: $f(n) = 2^{n-1}$.
2. a. Determine the twentieth term of the sequence $2, -1, -4, -7, \ldots$.
   b. Determine the sum of the first twenty terms of the sequence.
3. If an arithmetic sequence has a second term equal to -2 and a twentieth term equal to 34, find the first term and common difference.
4. Determine the first term and common difference of an arithmetic sequence with $a_{10} = 29$ and $S_{10} = 155$.
5. Which of the sequences below are geometric?
   a. $2, 4, 12, 48, \ldots$ b. $2, 4, 6, 8, \ldots$
   c. $2, 4, 8, 16, \ldots$
6. Consider the sequence $2, -6, 18, \ldots$
   a. Determine the tenth term.
   b. Determine the sum of the first ten terms.
7. Exhibit a sequence that is *both* arithmetic and geometric.
8. Determine the first term and the common ratio of the geometric sequence whose third term is .02 and whose eighth term is .64.
9. Express .424242 . . . as the quotient of two integers.
10. Find $\sum_{i=1}^{10} 3^i$.
11. Prove $\sum_{i=1}^{n} i^2 = \dfrac{n(n+1)(2n+1)}{6}$.
12. Evaluate:
    a. $4!$ b. $0! + 1! + 2!$ c. $\binom{8}{6}$ d. $\binom{100}{99}$
13. Determine the sixth term in the expansion of $(x - 2y^2)^{10}$.
14. a. Use the binomial series to generate an infinite series for $(4 - x^2)^{-1}$.
    b. For what values of x is the series equal to $(4 - x^2)^{-1}$?

IV. Student Research Project

1. Factorials of numbers get large quickly. For example, $70! > 10^{100}$ and hence is beyond the capacity of most modern calculators. A formula known as **Stirling's approximation** can be used to approximate $n!$ for large n. It is

$$\sqrt{2n\pi}\,(n/e)^n < n! < \sqrt{2n\pi}\,(n/e)^n[1 + 1/(12n - 1)]$$

Use the formula to approximate $60!$.

2. The sequence 1, 1, 2, 3, 5, 8, 13, 21, . . . is known as the **Fibonacci sequence** and has interesting applications in life. Prepare a brief report on this sequence. Consult history of mathematics texts as well as the following sources:

Jacobs, Harold R. *Mathematics: A Human Endeavor*, W. H. Freeman and Co., San Francisco, 1970, pp. 72–78, 290–291.

Newman, James, ed. *The World of Mathematics*, Simon and Schuster, New York, 1956, pp. 718–719.

3. Several interesting infinite series arise in the study of calculus. Some are listed below:

$$e = 1 + 1/1! + 1/2! + \ldots + 1/n! + \ldots \tag{1}$$

$$\ln x = \frac{x-1}{x} + \frac{1}{2}\left(\frac{x-1}{x}\right)^2 + \ldots + \frac{1}{n}\left(\frac{x-1}{x}\right)^n + \ldots, \quad x > \frac{1}{2} \tag{2}$$

$$\cos x = 1 - \frac{x^2}{2!} + \frac{x^4}{4!} - \frac{x^6}{6!} + \ldots \qquad (x \text{ in radians}) \tag{3}$$

a. Approximate e by finding the fifth partial sum of (1).
b. Approximate ln 2 by finding the tenth partial sum of (2). Compare your result with the value for ln 2 from Table II.
c. Approximate cos 45° by finding the fifth partial sum of (3) with $x = \pi/4$. Compare your result with the value for cos 45° from Table III.

APPENDIX **1**

Sets

A **set** is a collection of objects (or elements). In this textbook the sets we work with are sets of numbers. The set of integers between 1 and 5 inclusive, for example, is the collection 1, 2, 3, 4, 5 and braces are often used to denote that set as $\{1,2,3,4,5\}$.

A set can be **finite** or **infinite**. The set of positive integers less than 4, for example, is $\{1,2,3\}$ and thus is finite. The set of positive integers greater than 6 is denoted as $\{7,8,9, \ldots\}$ and is infinite; the three dots $(\ldots)$ indicate the pattern continues indefinitely. To say that 3 is an element of $\{1,2,3,4,5\}$ we write $3 \in \{1,2,3,4,5\}$. However, 3 is not an element of $\{7,8,9,\ldots\}$ and we write $3 \notin \{7,8,9,\ldots\}$.

Not all sets can be described by listing their members. For example, the set H of all real numbers between 0 and 1 cannot be written by listing its elements. We can, however, describe that set by a *rule* and write:

$$H = \{x \in \mathcal{R} \mid 0 < x < 1\} \quad \text{or} \quad H = \{x \mid 0 < x < 1\}$$

This is read "H is the set of all real numbers x such that x is between 0 and 1." This way of describing a set by a rule is called **set-builder notation.**

Two special sets are the set with nothing in it and the set with everything in it. We write $\emptyset$ for the **empty set** (or **null set**)—hence, $\emptyset = \{\ \}$. For any particular discussion involving sets, the set that contains all numbers pertaining to that discussion is called the universal set and is denoted U. For much of the work in this book, $\mathcal{R}$ is considered the universal set.

There are two set operations that we will now define.

DEFINITION

> The **union** of two sets A and B, denoted $A \cup B$, is the set of all elements that are in set A or set B or both.

DEFINITION The **intersection** of two sets A and B, denoted $A \cap B$, is the set of elements that A and B have in common.

EXAMPLE 1 Let $A = \{2,5,7,11\}$ and $B = \{2,4,7,9\}$.

a. Find $A \cup B$ b. Find $A \cap B$

Solution a. $A \cup B = \{2,4,5,7,9,11\}$ b. $A \cap B = \{2,7\}$

EXAMPLE 2 Let $A = \{x \mid x \geqslant 4\}$ and $B = \{x \mid 1 < x < 5\}$

a. Find $A \cup B$ b. Find $A \cap B$

Solution a. $A \cup B = \{x \mid x > 1\}$ b. $A \cap B = \{x \mid 4 \leqslant x < 5\}$

EXERCISES FOR APPENDIX 1

In Exercises 1 through 12, a) find $A \cup B$ and b) find $A \cap B$.

1. $A = \{1,3,5,6,7,8\}$
   $B = \{2,3,5,7,9\}$
2. $A = \{2,4,6,9,10,11\}$
   $B = \{4,11\}$
3. $A = \{1,3,5,7,9\}$
   $B = \{2,4,6,8\}$
4. $A = \{3,4,5,9\}$
   $B = \{1,2,6\}$
5. $A = \{2,9\}$
   $B = \{2,9\}$
6. $A = \{8,4,1\}$
   $B = \{2,5,6,7\}$
7. $A = \{x \mid 0 < x < 1\}$
   $B = \{x \mid x \geqslant 1\}$
8. $A = \{x \mid x \leqslant 2\}$
   $B = \{x \mid x > 2\}$
9. $A = \{x \mid x \leqslant 2\}$
   $B = \{x \mid x \geqslant 2\}$
10. $A = \{x \mid x \text{ is an integer}\}$
    $B = \{x \mid x \text{ is a natural number}\}$
11. $A = \{x \mid x \text{ is a negative integer}\}$
    $B = \{x \mid x \text{ is a positive integer}\}$
12. $A = \{x \mid x \text{ is an integer}\}$
    $B = \{x \mid x \text{ is a rational number}\}$

APPENDIX **2**

The Metric System

The common units of measurement in the metric system are summarized in the following table.

ENGLISH UNITS AND METRIC UNITS				
Length				
1 millimeter (mm)	≈	.04 inches (in)*	or	1 in = 25.4 mm
1 centimeter (cm)	≈	.4 inches (in)	or	1 in = 2.54 cm
1 meter (m)	≈	1.1 yards (yd)	or	1 yd ≈ .92 m
1 kilometer (km)	≈	.62 miles (mi)	or	1 mi ≈ 1.6 km
Mass				
1 gram (g)	≈	.035 ounces (oz)	or	1 oz ≈ 28.3 g
1 kilogram (kg)	≈	2.2 pounds (lb)	or	1 lb ≈ .45 kg
Volume				
1 milliliter (ml)	≈	.03 fluid ounces (fl oz)	or	1 fl oz ≈ 30 ml
1 liter (l)	≈	1.06 quarts (qt)	or	1 qt ≈ .95 l
1 cubic meter (m^3)	≈	1.3 cubic yards (yd^3)	or	1 yd^3 ≈ .76 m^3
Area				
1 square centimeter (cm^2)	≈	.16 square inches (in^2)	or	1 in^2 ≈ 6.5 cm^2
1 square meter (m^2)	≈	1.2 square yards (yd^2)	or	1 yd^2 ≈ 0.8 m^2
1 square kilometer (km^2)	≈	0.4 square miles (mi^2)	or	1 mi^2 ≈ 2.6 km^2

COMMON COMPARISONS WITHIN THE METRIC SYSTEM					
Length			**Mass**		
1 millimeter (mm)	=	$\frac{1}{1000}$ meter (m)	1 milligram (mg)	=	$\frac{1}{1000}$ gram (g)
1 centimeter (cm)	=	$\frac{1}{100}$ meter (m)	1 kilogram (kg)	=	1,000 grams (g)
1 kilometer (km)	=	1,000 meters (m)			
Volume					
1,000 milliliters (ml)	=	1 liter (l)			

*The symbol ≈ means "is approximately equal to." The information in this table is from the United States Department of Commerce, National Bureau of Standards Publication 330, revised 1974.

In the examples that follow, we show how to make English-to-metric and metric-to-English conversions.

EXAMPLE 1 Determine the number of centimeters in 20 inches.

Solution Let x equal the number of centimeters. Set up a proportion (that is equate two equal ratios).

$$\frac{x}{20} = \frac{2.54}{1}$$

This is the ratio of centimeters to inches. This is the ratio of centimeters to 1 inch from the table of units.

$$\frac{x}{20} = \frac{2.54}{1}$$

$$x = 50.8$$

Hence, there are 50.8 cm in 20 in.

EXAMPLE 2 Determine the number of inches in 40 centimeters.

Solution Let x equal the number of inches and set up a proportion.

$$\frac{40}{x} = \frac{2.54}{1}$$

This is the ratio of centimeters to inches This is the ratio of centimeters to 1 inch from the table of units.

$$\frac{40}{x} = \frac{2.54}{1}$$

$$x \approx 15.75$$

Thus, there are about 15.75 inches to 40 cm.

EXAMPLE 3 Determine the number of kilograms in 140 pounds.

Solution Let x equal the number of kilograms and set up a proportion.

$$\frac{x}{140} = \frac{.45}{1}$$

This is the ratio of kilograms to pounds.

This is the ratio of kilograms to 1 pound from the table of units.

$$\frac{x}{140} = \frac{.45}{1}$$

$$x = 63$$

So, there are 63 kilograms in 140 pounds.

EXERCISES FOR APPENDIX 2

1. Determine the number of centimeters in 1 foot.
2. Determine the number of inches in 200 centimeters.
3. Determine the number of kilograms in 120 pounds.
4. Determine the number of grams in 1 pound.
5. Determine the number of milliliters in 3 quarts.
6. Determine the number of square centimeters in a rectangle whose area is 10 in^2.
7. Determine the number of square meters of floor space in a room whose floor dimensions are 18 feet by 12 feet.
8. Determine the number of kilograms in 80 pounds.
9. Determine the number of pounds in 160 kilograms.
10. Determine the number of grams in 15 pounds.
11. Determine the number of liters in a gallon.
12. Determine the number of cubic meters in 5 cubic yards.
13. Determine the number of square meters in 10 square yards.
14. Determine the number of square kilometers in 5.5 square miles.
15. Determine the number of fluid ounces in 100 milliliters.
16. Determine the number of kilometers in 55 miles.
17. Determine the number of miles in 55 kilometers.
18. The formula $F = \frac{9}{5} C + 32$ can be used to convert Celsius temperatures to Fahrenheit temperatures. The formula $C = \frac{5}{9}(F - 32)$ can be used to convert Fahrenheit temperatures to Celsius temperatures. Using the appropriate formula, convert each Celsius temperature below to the corresponding Fahrenheit temperature.

    a. 100° C b. 0° C c. 20° C d. 30° C

    Using the appropriate formula, convert each Fahrenheit temperature given below to the corresponding Celsius temperature.

    e. 90° F f. 32° F g. −14° F h. 58° F i. −40° F

Tables

Table 1 Common Logarithms

X	0	1	2	3	4	5	6	7	8	9
1.0	.0000	.0043	.0086	.0128	.0170	.0212	.0253	.0294	.0334	.0374
1.1	.0414	.0453	.0492	.0531	.0569	.0607	.0645	.0682	.0719	.0755
1.2	.0792	.0828	.0864	.0899	.0934	.0969	.1004	.1038	.1072	.1106
1.3	.1139	.1173	.1206	.1239	.1271	.1303	.1335	.1367	.1399	.1430
1.4	.1461	.1492	.1523	.1553	.1584	.1614	.1644	.1673	.1703	.1732
1.5	.1761	.1790	.1818	.1847	.1875	.1903	.1931	.1959	.1987	.2014
1.6	.2041	.2068	.2095	.2122	.2148	.2175	.2201	.2227	.2253	.2279
1.7	.2304	.2330	.2355	.2380	.2405	.2430	.2455	.2480	.2504	.2529
1.8	.2553	.2577	.2601	.2625	.2648	.2672	.2695	.2718	.2742	.2765
1.9	.2788	.2810	.2833	.2856	.2878	.2900	.2923	.2945	.2967	.2989
2.0	.3010	.3032	.3054	.3075	.3096	.3118	.3139	.3160	.3181	.3201
2.1	.3222	.3243	.3263	.3284	.3304	.3324	.3345	.3365	.3385	.3404
2.2	.3424	.3444	.3464	.3483	.3502	.3522	.3541	.3560	.3579	.3598
2.3	.3617	.3636	.3655	.3674	.3692	.3711	.3729	.3747	.3766	.3784
2.4	.3802	.3820	.3838	.3856	.3874	.3892	.3909	.3927	.3945	.3962
2.5	.3979	.3997	.4014	.4031	.4048	.4065	.4082	.4099	.4116	.4133
2.6	.4150	.4166	.4183	.4200	.4216	.4232	.4249	.4265	.4281	.4298
2.7	.4314	.4330	.4346	.4362	.4378	.4393	.4409	.4425	.4440	.4456
2.8	.4472	.4487	.4502	.4518	.4533	.4548	.4564	.4579	.4594	.4609
2.9	.4624	.4639	.4654	.4669	.4683	.4698	.4713	.4728	.4742	.4757
3.0	.4771	.4786	.4800	.4814	.4829	.4843	.4857	.4871	.4886	.4900
3.1	.4914	.4928	.4942	.4955	.4969	.4983	.4997	.5011	.5024	.5038
3.2	.5051	.5065	.5079	.5092	.5105	.5119	.5132	.5145	.5159	.5172
3.3	.5185	.5198	.5211	.5224	.5237	.5250	.5263	.5276	.5289	.5302
3.4	.5315	.5328	.5340	.5353	.5366	.5378	.5391	.5403	.5416	.5428
3.5	.5441	.5453	.5465	.5478	.5490	.5502	.5514	.5527	.5539	.5551
3.6	.5563	.5575	.5587	.5599	.5611	.5623	.5635	.5647	.5658	.5670
3.7	.5682	.5694	.5705	.5717	.5729	.5740	.5752	.5763	.5775	.5786
3.8	.5798	.5809	.5821	.5832	.5843	.5855	.5866	.5877	.5888	.5899
3.9	.5911	.5922	.5933	.5944	.5955	.5966	.5977	.5988	.5999	.6010
4.0	.6021	.6031	.6042	.6053	.6064	.6075	.6085	.6096	.6107	.6117
4.1	.6128	.6138	.6149	.6159	.6170	.6180	.6191	.6201	.6212	.6222
4.2	.6232	.6243	.6253	.6263	.6274	.6284	.6294	.6304	.6314	.6325
4.3	.6335	.6345	.6355	.6365	.6375	.6385	.6395	.6405	.6415	.6425
4.4	.6435	.6444	.6454	.6464	.6474	.6484	.6493	.6503	.6513	.6522
4.5	.6532	.6542	.6551	.6561	.6571	.6580	.6590	.6599	.6609	.6618
4.6	.6628	.6637	.6646	.6656	.6665	.6675	.6684	.6693	.6702	.6712
4.7	.6721	.6730	.6739	.6749	.6758	.6767	.6776	.6785	.6794	.6803
4.8	.6812	.6821	.6830	.6839	.6848	.6857	.6866	.6875	.6884	.6893
4.9	.6902	.6911	.6920	.6928	.6937	.6946	.6955	.6964	.6972	.6981
5.0	.6990	.6998	.7007	.7016	.7024	.7033	.7042	.7050	.7059	.7067
5.1	.7076	.7084	.7093	.7101	.7110	.7118	.7126	.7135	.7143	.7152
5.2	.7160	.7168	.7177	.7185	.7193	.7202	.7210	.7218	.7226	.7235
5.3	.7243	.7251	.7259	.7267	.7275	.7284	.7292	.7300	.7308	.7316
5.4	.7324	.7332	.7340	.7348	.7356	.7364	.7372	.7380	.7388	.7396

X	0	1	2	3	4	5	6	7	8	9
5.5	.7404	.7412	.7419	.7427	.7435	.7443	.7451	.7459	.7466	.7474
5.6	.7482	.7490	.7497	.7505	.7513	.7520	.7528	.7536	.7543	.7551
5.7	.7559	.7566	.7574	.7582	.7589	.7597	.7604	.7612	.7619	.7627
5.8	.7634	.7642	.7649	.7657	.7664	.7672	.7679	.7686	.7694	.7701
5.9	.7709	.7716	.7723	.7731	.7738	.7745	.7752	.7760	.7767	.7774
6.0	.7782	.7789	.7796	.7803	.7810	.7818	.7825	.7832	.7839	.7846
6.1	.7853	.7860	.7868	.7875	.7882	.7889	.7896	.7903	.7910	.7917
6.2	.7924	.7931	.7938	.7945	.7952	.7959	.7966	.7973	.7980	.7987
6.3	.7993	.8000	.8007	.8014	.8021	.8028	.8035	.8041	.8048	.8055
6.4	.8062	.8069	.8075	.8082	.8089	.8096	.8102	.8109	.8116	.8122
6.5	.8129	.8136	.8142	.8149	.8156	.8162	.8169	.8176	.8182	.8189
6.6	.8195	.8202	.8209	.8215	.8222	.8228	.8235	.8241	.8248	.8254
6.7	.8261	.8267	.8274	.8280	.8287	.8293	.8299	.8306	.8312	.8319
6.8	.8325	.8331	.8338	.8344	.8351	.8357	.8363	.8370	.8376	.8382
6.9	.8388	.8395	.8401	.8407	.8414	.8420	.8426	.8432	.8439	.8445
7.0	.8451	.8457	.8463	.8470	.8476	.8482	.8488	.8494	.8500	.8506
7.1	.8513	.8519	.8525	.8531	.8537	.8543	.8549	.8555	.8561	.8567
7.2	.8573	.8579	.8585	.8591	.8597	.8603	.8609	.8615	.8621	.8627
7.3	.8633	.8639	.8645	.8651	.8657	.8663	.8669	.8675	.8681	.8686
7.4	.8692	.8698	.8704	.8710	.8716	.8722	.8727	.8733	.8739	.8745
7.5	.8751	.8756	.8762	.8768	.8774	.8779	.8785	.8791	.8797	.8802
7.6	.8808	.8814	.8820	.8825	.8831	.8837	.8842	.8848	.8854	.8859
7.7	.8865	.8871	.8876	.8882	.8887	.8893	.8899	.8904	.8910	.8915
7.8	.8921	.8927	.8932	.8938	.8943	.8949	.8954	.8960	.8965	.8971
7.9	.8976	.8982	.8987	.8993	.8998	.9004	.9009	.9015	.9020	.9025
8.0	.9031	.9036	.9042	.9047	.9053	.9058	.9063	.9069	.9074	.9079
8.1	.9085	.9090	.9096	.9101	.9106	.9112	.9117	.9122	.9128	.9133
8.2	.9138	.9143	.9149	.9154	.9159	.9165	.9170	.9175	.9180	.9186
8.3	.9191	.9196	.9201	.9206	.9212	.9217	.9222	.9227	.9232	.9238
8.4	.9243	.9248	.9253	.9258	.9263	.9269	.9274	.9279	.9284	.9289
8.5	.9294	.9299	.9304	.9309	.9315	.9320	.9325	.9330	.9335	.9340
8.6	.9345	.9350	.9355	.9360	.9365	.9370	.9375	.9380	.9385	.9390
8.7	.9395	.9400	.9405	.9410	.9415	.9420	.9425	.9430	.9435	.9440
8.8	.9445	.9450	.9455	.9460	.9465	.9469	.9474	.9479	.9484	.9489
8.9	.9494	.9499	.9504	.9509	.9513	.9518	.9523	.9528	.9533	.9538
9.0	.9542	.9547	.9552	.9557	.9562	.9566	.9571	.9576	.9581	.9586
9.1	.9590	.9595	.9600	.9605	.9609	.9614	.9619	.9624	.9628	.9633
9.2	.9638	.9643	.9647	.9652	.9657	.9661	.9666	.9671	.9675	.9680
9.3	.9685	.9689	.9694	.9699	.9703	.9708	.9713	.9717	.9722	.9727
9.4	.9731	.9736	.9741	.9745	.9750	.9754	.9759	.9763	.9768	.9773
9.5	.9777	.9782	.9786	.9791	.9795	.9800	.9805	.9809	.9814	.9818
9.6	.9823	.9827	.9832	.9836	.9841	.9845	.9850	.9854	.9859	.9863
9.7	.9868	.9872	.9877	.9881	.9886	.9890	.9894	.9899	.9903	.9908
9.8	.9912	.9917	.9921	.9926	.9930	.9934	.9939	.9943	.9948	.9952
9.9	.9956	.9961	.9965	.9969	.9974	.9978	.9983	.9987	.9991	.9996

Table 2 Natural Logarithms

X	0	1	2	3	4	5	6	7	8	9
1.0	0.0000	0.0100	0.0198	0.0296	0.0392	0.0488	0.0583	0.0677	0.0770	0.0862
1.1	0.0953	0.1044	0.1133	0.1222	0.1310	0.1398	0.1484	0.1570	0.1655	0.1740
1.2	0.1823	0.1906	0.1989	0.2070	0.2151	0.2231	0.2311	0.2390	0.2469	0.2546
1.3	0.2624	0.2700	0.2776	0.2852	0.2927	0.3001	0.3075	0.3148	0.3221	0.3293
1.4	0.3365	0.3436	0.3507	0.3577	0.3646	0.3716	0.3784	0.3853	0.3920	0.3988
1.5	0.4055	0.4121	0.4187	0.4253	0.4318	0.4383	0.4447	0.4511	0.4574	0.4637
1.6	0.4700	0.4762	0.4824	0.4886	0.4947	0.5008	0.5068	0.5128	0.5188	0.5247
1.7	0.5306	0.5365	0.5423	0.5481	0.5539	0.5596	0.5653	0.5710	0.5766	0.5822
1.8	0.5878	0.5933	0.5988	0.6043	0.6098	0.6152	0.6206	0.6259	0.6313	0.6366
1.9	0.6419	0.6471	0.6523	0.6575	0.6627	0.6678	0.6729	0.6780	0.6831	0.6881
2.0	0.6931	0.6981	0.7031	0.7080	0.7129	0.7178	0.7227	0.7275	0.7324	0.7372
2.1	0.7419	0.7467	0.7514	0.7561	0.7608	0.7655	0.7701	0.7747	0.7793	0.7839
2.2	0.7885	0.7930	0.7975	0.8020	0.8065	0.8109	0.8154	0.8198	0.8242	0.8286
2.3	0.8329	0.8372	0.8416	0.8459	0.8502	0.8544	0.8587	0.8629	0.8671	0.8713
2.4	0.8755	0.8796	0.8838	0.8879	0.8920	0.8961	0.9002	0.9042	0.9083	0.9123
2.5	0.9163	0.9203	0.9243	0.9282	0.9322	0.9361	0.9400	0.9439	0.9478	0.9517
2.6	0.9555	0.9594	0.9632	0.9670	0.9708	0.9746	0.9783	0.9821	0.9858	0.9895
2.7	0.9933	0.9969	1.0006	1.0043	1.0080	1.0116	1.0152	1.0188	1.0225	1.0260
2.8	1.0296	1.0332	1.0367	1.0403	1.0438	1.0473	1.0508	1.0543	1.0578	1.0613
2.9	1.0647	1.0682	1.0716	1.0750	1.0784	1.0818	1.0852	1.0886	1.0919	1.0953
3.0	1.0986	1.1019	1.1053	1.1086	1.1119	1.1151	1.1184	1.1217	1.1249	1.1282
3.1	1.1314	1.1346	1.1378	1.1410	1.1442	1.1474	1.1506	1.1537	1.1569	1.1600
3.2	1.1632	1.1663	1.1694	1.1725	1.1756	1.1787	1.1817	1.1848	1.1878	1.1909
3.3	1.1939	1.1969	1.2000	1.2030	1.2060	1.2090	1.2119	1.2149	1.2179	1.2208
3.4	1.2238	1.2267	1.2296	1.2326	1.2355	1.2384	1.2413	1.2442	1.2470	1.2499
3.5	1.2528	1.2556	1.2585	1.2613	1.2641	1.2669	1.2698	1.2726	1.2754	1.2782
3.6	1.2809	1.2837	1.2865	1.2892	1.2920	1.2947	1.2975	1.3002	1.3029	1.3056
3.7	1.3083	1.3110	1.3137	1.3164	1.3191	1.3218	1.3244	1.3271	1.3297	1.3324
3.8	1.3350	1.3376	1.3403	1.3429	1.3455	1.3481	1.3507	1.3533	1.3558	1.3584
3.9	1.3610	1.3635	1.3661	1.3686	1.3712	1.3737	1.3762	1.3788	1.3813	1.3838
4.0	1.3863	1.3888	1.3913	1.3938	1.3962	1.3987	1.4012	1.4036	1.4061	1.4085
4.1	1.4110	1.4134	1.4159	1.4183	1.4207	1.4231	1.4255	1.4279	1.4303	1.4327
4.2	1.4351	1.4375	1.4398	1.4422	1.4446	1.4469	1.4493	1.4516	1.4540	1.4563
4.3	1.4586	1.4609	1.4633	1.4656	1.4679	1.4702	1.4725	1.4748	1.4770	1.4793
4.4	1.4816	1.4839	1.4861	1.4884	1.4907	1.4929	1.4951	1.4974	1.4996	1.5019
4.5	1.5041	1.5063	1.5085	1.5107	1.5129	1.5151	1.5173	1.5195	1.5217	1.5239
4.6	1.5261	1.5282	1.5304	1.5326	1.5347	1.5369	1.5390	1.5412	1.5433	1.5454
4.7	1.5476	1.5497	1.5518	1.5539	1.5560	1.5581	1.5602	1.5623	1.5644	1.5665
4.8	1.5686	1.5707	1.5728	1.5748	1.5769	1.5790	1.5810	1.5831	1.5851	1.5872
4.9	1.5892	1.5913	1.5933	1.5953	1.5974	1.5994	1.6014	1.6034	1.6054	1.6074
5.0	1.6094	1.6114	1.6134	1.6154	1.6174	1.6194	1.6214	1.6233	1.6253	1.6273
5.1	1.6292	1.6312	1.6332	1.6351	1.6371	1.6390	1.6409	1.6429	1.6448	1.6467
5.2	1.6487	1.6506	1.6525	1.6544	1.6563	1.6582	1.6601	1.6620	1.6639	1.6658
5.3	1.6677	1.6696	1.6715	1.6734	1.6752	1.6771	1.6790	1.6808	1.6827	1.6845
5.4	1.6864	1.6882	1.6901	1.6919	1.6938	1.6956	1.6974	1.6993	1.7011	1.7029
5.5	1.7047	1.7066	1.7084	1.7102	1.7120	1.7138	1.7156	1.7174	1.7192	1.7210
5.6	1.7228	1.7246	1.7263	1.7281	1.7299	1.7317	1.7334	1.7352	1.7370	1.7387
5.7	1.7405	1.7422	1.7440	1.7457	1.7475	1.7492	1.7509	1.7527	1.7544	1.7561
5.8	1.7579	1.7596	1.7613	1.7630	1.7647	1.7664	1.7681	1.7699	1.7716	1.7733
5.9	1.7750	1.7766	1.7783	1.7800	1.7817	1.7834	1.7851	1.7867	1.7884	1.7901
6.0	1.7918	1.7934	1.7951	1.7967	1.7984	1.8001	1.8017	1.8034	1.8050	1.8066
6.1	1.8083	1.8099	1.8116	1.8132	1.8148	1.8165	1.8181	1.8197	1.8213	1.8229
6.2	1.8245	1.8262	1.8278	1.8294	1.8310	1.8326	1.8342	1.8358	1.8374	1.8390
6.3	1.8405	1.8421	1.8437	1.8453	1.8469	1.8485	1.8500	1.8516	1.8532	1.8547
6.4	1.8563	1.8579	1.8594	1.8610	1.8625	1.8641	1.8656	1.8672	1.8687	1.8703

X	0	1	2	3	4	5	6	7	8	9
6.5	1.8718	1.8733	1.8749	1.8764	1.8779	1.8795	1.8810	1.8825	1.8840	1.8856
6.6	1.8871	1.8886	1.8901	1.8916	1.8931	1.8946	1.8961	1.8976	1.8991	1.9006
6.7	1.9021	1.9036	1.9051	1.9066	1.9081	1.9095	1.9110	1.9125	1.9140	1.9155
6.8	1.9169	1.9184	1.9199	1.9213	1.9228	1.9242	1.9257	1.9272	1.9286	1.9301
6.9	1.9315	1.9330	1.9344	1.9359	1.9373	1.9387	1.9402	1.9416	1.9430	1.9445
7.0	1.9459	1.9473	1.9488	1.9502	1.9516	1.9530	1.9544	1.9559	1.9573	1.9587
7.1	1.9601	1.9615	1.9629	1.9643	1.9657	1.9671	1.9685	1.9699	1.9713	1.9727
7.2	1.9741	1.9755	1.9769	1.9782	1.9796	1.9810	1.9824	1.9838	1.9851	1.9865
7.3	1.9879	1.9892	1.9906	1.9920	1.9933	1.9947	1.9961	1.9974	1.9988	2.0001
7.4	2.0015	2.0028	2.0042	2.0055	2.0069	2.0082	2.0096	2.0109	2.0122	2.0136
7.5	2.0149	2.0162	2.0176	2.0189	2.0202	2.0215	2.0229	2.0242	2.0255	2.0260
7.6	2.0281	2.0295	2.0308	2.0321	2.0334	2.0347	2.0360	2.0373	2.0386	2.0399
7.7	2.0412	2.0425	2.0438	2.0451	2.0464	2.0477	2.0490	2.0503	2.0516	2.0528
7.8	2.0541	2.0554	2.0567	2.0580	2.0592	2.0605	2.0618	2.0631	2.0643	2.0656
7.9	2.0669	2.0681	2.0694	2.0707	2.0719	2.0732	2.0744	2.0757	2.0769	2.0782
8.0	2.0794	2.0807	2.0819	2.0832	2.0844	2.0857	2.0869	2.0882	2.0894	2.0906
8.1	2.0919	2.0931	2.0943	2.0956	2.0968	2.0980	2.0992	2.1005	2.1017	2.1029
8.2	2.1041	2.1054	2.1066	2.1078	2.1090	2.1102	2.1114	2.1126	2.1138	2.1150
8.3	2.1163	2.1175	2.1187	2.1199	2.1211	2.1223	2.1235	2.1247	2.1258	2.1270
8.4	2.1282	2.1294	2.1306	2.1318	2.1330	2.1342	2.1353	2.1365	2.1377	2.1389
8.5	2.1401	2.1412	2.1424	2.1436	2.1448	2.1459	2.1471	2.1483	2.1494	2.1506
8.6	2.1518	2.1529	2.1541	2.1552	2.1564	2.1576	2.1587	2.1599	2.1610	2.1622
8.7	2.1633	2.1645	2.1656	2.1668	2.1679	2.1691	2.1702	2.1713	2.1725	2.1736
8.8	2.1748	2.1759	2.1770	2.1782	2.1793	2.1804	2.1815	2.1827	2.1838	2.1849
8.9	2.1861	2.1872	2.1883	2.1894	2.1905	2.1917	2.1928	2.1939	2.1950	2.1961
9.0	2.1972	2.1983	2.1994	2.2006	2.2017	2.2028	2.2039	2.2050	2.2061	2.2072
9.1	2.2083	2.2094	2.2105	2.2116	2.2127	2.2138	2.2148	2.2159	2.2170	2.2181
9.2	2.2192	2.2203	2.2214	2.2225	2.2235	2.2246	2.2257	2.2268	2.2279	2.2289
9.3	2.2300	2.2311	2.2322	2.2332	2.2343	2.2354	2.2364	2.2375	2.2386	2.2396
9.4	2.2407	2.2418	2.2428	2.2439	2.2450	2.2460	2.2471	2.2481	2.2492	2.2502
9.5	2.2513	2.2523	2.2534	2.2544	2.2555	2.2565	2.2576	2.2586	2.2597	2.2607
9.6	2.2618	2.2628	2.2638	2.2649	2.2659	2.2670	2.2680	2.2690	2.2701	2.2711
9.7	2.2721	2.2732	2.2742	2.2752	2.2762	2.2773	2.2783	2.2793	2.2803	2.2814
9.8	2.2824	2.2834	2.2844	2.2854	2.2865	2.2875	2.2885	2.2895	2.2905	2.2915
9.9	2.2925	2.2935	2.2946	2.2956	2.2966	2.2976	2.2986	2.2996	2.3006	2.3016

ln 10 = 2.3026   ln 20 = 2.9957   ln 50 = 3.9120   ln 100 = 4.6052

Table 3 Trigonometric Values

θ (degrees)	s	sin θ or sin s	csc θ or csc s	cos θ or cos s	sec θ or sec s	tan θ or tan s	cot θ or cot s		
0.0	0.0000	0.0000	UNDEFINED	1.0000	1.0000	0.0000	UNDEFINED	1.5708	90.0
0.1	0.0017	0.0017	572.96	1.0000	1.0000	0.0017	572.96	1.5691	89.9
0.2	0.0035	0.0035	286.48	1.0000	1.0000	0.0035	286.48	1.5673	89.8
0.3	0.0052	0.0052	190.99	1.0000	1.0000	0.0052	190.98	1.5656	89.7
0.4	0.0070	0.0070	143.24	1.0000	1.0000	0.0070	143.24	1.5638	89.6
0.5	0.0087	0.0087	114.59	1.0000	1.0000	0.0087	114.59	1.5621	89.5
0.6	0.0105	0.0105	95.495	0.9999	1.0001	0.0105	95.490	1.5603	89.4
0.7	0.0122	0.0122	81.853	0.9999	1.0001	0.0122	81.847	1.5586	89.3
0.8	0.0140	0.0140	71.622	0.9999	1.0001	0.0140	71.615	1.5568	89.2
0.9	0.0157	0.0157	63.665	0.9999	1.0001	0.0157	63.657	1.5551	89.1
1.0	0.0175	0.0175	57.299	0.9998	1.0002	0.0175	57.290	1.5533	89.0
1.1	0.0192	0.0192	52.090	0.9998	1.0002	0.0192	52.081	1.5516	88.9
1.2	0.0209	0.0209	47.750	0.9998	1.0002	0.0209	47.740	1.5499	88.8
1.3	0.0227	0.0227	44.077	0.9997	1.0003	0.0227	44.066	1.5481	88.7
1.4	0.0244	0.0244	40.930	0.9997	1.0003	0.0244	40.917	1.5464	88.6
1.5	0.0262	0.0262	38.202	0.9997	1.0003	0.0262	38.188	1.5446	88.5
1.6	0.0279	0.0279	35.815	0.9996	1.0004	0.0279	35.801	1.5429	88.4
1.7	0.0297	0.0297	33.708	0.9996	1.0004	0.0297	33.694	1.5411	88.3
1.8	0.0314	0.0314	31.836	0.9995	1.0005	0.0314	31.821	1.5394	88.2
1.9	0.0332	0.0332	30.161	0.9995	1.0006	0.0332	30.145	1.5376	88.1
2.0	0.0349	0.0349	28.654	0.9994	1.0006	0.0349	28.636	1.5359	88.0
2.1	0.0367	0.0366	27.290	0.9993	1.0007	0.0367	27.272	1.5341	87.9
2.2	0.0384	0.0384	26.050	0.9993	1.0007	0.0384	26.031	1.5324	87.8
2.3	0.0401	0.0401	24.918	0.9992	1.0008	0.0402	24.898	1.5307	87.7
2.4	0.0419	0.0419	23.880	0.9991	1.0009	0.0419	23.859	1.5289	87.6
2.5	0.0436	0.0436	22.926	0.9990	1.0010	0.0437	22.904	1.5272	87.5
2.6	0.0454	0.0454	22.044	0.9990	1.0010	0.0454	22.022	1.5254	87.4
2.7	0.0471	0.0471	21.229	0.9989	1.0011	0.0472	21.205	1.5237	87.3
2.8	0.0489	0.0488	20.471	0.9988	1.0012	0.0489	20.446	1.5219	87.2
2.9	0.0506	0.0506	19.766	0.9987	1.0013	0.0507	19.740	1.5202	87.1
3.0	0.0524	0.0523	19.107	0.9986	1.0014	0.0524	19.081	1.5184	87.0
3.1	0.0541	0.0541	18.492	0.9985	1.0015	0.0542	18.464	1.5167	86.9
3.2	0.0559	0.0558	17.914	0.9984	1.0016	0.0559	17.886	1.5149	86.8
3.3	0.0576	0.0576	17.372	0.9983	1.0017	0.0577	17.343	1.5132	86.7
3.4	0.0593	0.0593	16.862	0.9982	1.0018	0.0594	16.832	1.5115	86.6
3.5	0.0611	0.0610	16.380	0.9981	1.0019	0.0612	16.350	1.5097	86.5
3.6	0.0628	0.0628	15.926	0.9980	1.0020	0.0629	15.895	1.5080	86.4
3.7	0.0646	0.0645	15.496	0.9979	1.0021	0.0647	15.464	1.5062	86.3
3.8	0.0663	0.0663	15.089	0.9978	1.0022	0.0664	15.056	1.5045	86.2
3.9	0.0681	0.0680	14.703	0.9977	1.0023	0.0682	14.669	1.5027	86.1
4.0	0.0698	0.0698	14.336	0.9976	1.0024	0.0699	14.301	1.5010	86.0
4.1	0.0716	0.0715	13.987	0.9974	1.0026	0.0717	13.951	1.4992	85.9
4.2	0.0733	0.0732	13.654	0.9973	1.0027	0.0734	13.617	1.4975	85.8
4.3	0.0750	0.0750	13.337	0.9972	1.0028	0.0752	13.300	1.4957	85.7
4.4	0.0768	0.0767	13.035	0.9971	1.0030	0.0769	12.996	1.4940	85.6
		cos θ or cos s	sec θ or sec s	sin θ or sin s	csc θ or csc s	cot θ or cot s	tan θ or tan s	s	θ (degrees)

θ (degrees)	s	sin θ or sin s	csc θ or csc s	cos θ or cos s	sec θ or sec s	tan θ or tan s	cot θ or cot s		
4.5	0.0785	0.0785	12.746	0.9969	1.0031	0.0787	12.706	1.4923	85.5
4.6	0.0803	0.0802	12.469	0.9968	1.0032	0.0805	12.429	1.4905	85.4
4.7	0.0820	0.0819	12.204	0.9966	1.0034	0.0822	12.163	1.4888	85.3
4.8	0.0838	0.0837	11.951	0.9965	1.0035	0.0840	11.909	1.4870	85.2
4.9	0.0855	0.0854	11.707	0.9963	1.0037	0.0857	11.665	1.4853	85.1
5.0	0.0873	0.0872	11.474	0.9962	1.0038	0.0875	11.430	1.4835	85.0
5.1	0.0890	0.0889	11.249	0.9960	1.0040	0.0892	11.205	1.4818	84.9
5.2	0.0908	0.0906	11.034	0.9959	1.0041	0.0910	10.988	1.4800	84.8
5.3	0.0925	0.0924	10.826	0.9957	1.0043	0.0928	10.780	1.4783	84.7
5.4	0.0942	0.0941	10.626	0.9956	1.0045	0.0945	10.579	1.4765	84.6
5.5	0.0960	0.0958	10.433	0.9954	1.0046	0.0963	10.385	1.4748	84.5
5.6	0.0977	0.0976	10.248	0.9952	1.0048	0.0981	10.199	1.4731	84.4
5.7	0.0995	0.0993	10.068	0.9951	1.0050	0.0998	10.019	1.4713	84.3
5.8	0.1012	0.1011	9.8955	0.9949	1.0051	0.1016	9.8448	1.4696	84.2
5.9	0.1030	0.1028	9.7283	0.9947	1.0053	0.1033	9.6768	1.4678	84.1
6.0	0.1047	0.1045	9.5668	0.9945	1.0055	0.1051	9.5144	1.4661	84.0
6.1	0.1065	0.1063	9.4105	0.9943	1.0057	0.1069	9.3572	1.4643	83.9
6.2	0.1082	0.1080	9.2593	0.9942	1.0059	0.1086	9.2052	1.4626	83.8
6.3	0.1100	0.1097	9.1129	0.9940	1.0061	0.1104	9.0579	1.4608	83.7
6.4	0.1117	0.1115	8.9711	0.9938	1.0063	0.1122	8.9152	1.4591	83.6
6.5	0.1134	0.1132	8.8337	0.9936	1.0065	0.1139	8.7769	1.4573	83.5
6.6	0.1152	0.1149	8.7004	0.9934	1.0067	0.1157	8.6428	1.4556	83.4
6.7	0.1169	0.1167	8.5711	0.9932	1.0069	0.1175	8.5126	1.4539	83.3
6.8	0.1187	0.1184	8.4457	0.9930	1.0071	0.1192	8.3863	1.4521	83.2
6.9	0.1204	0.1201	8.3238	0.9928	1.0073	0.1210	8.2636	1.4504	83.1
7.0	0.1222	0.1219	8.2055	0.9925	1.0075	0.1228	8.1444	1.4486	83.0
7.1	0.1239	0.1236	8.0905	0.9923	1.0077	0.1246	8.0285	1.4469	82.9
7.2	0.1257	0.1253	7.9787	0.9921	1.0079	0.1263	7.9158	1.4451	82.8
7.3	0.1274	0.1271	7.8700	0.9919	1.0082	0.1281	7.8062	1.4434	82.7
7.4	0.1292	0.1288	7.7642	0.9917	1.0084	0.1299	7.6996	1.4416	82.6
7.5	0.1309	0.1305	7.6613	0.9914	1.0086	0.1317	7.5958	1.4399	82.5
7.6	0.1326	0.1323	7.5611	0.9912	1.0089	0.1334	7.4947	1.4382	82.4
7.7	0.1344	0.1340	7.4635	0.9910	1.0091	0.1352	7.3962	1.4364	82.3
7.8	0.1361	0.1357	7.3684	0.9907	1.0093	0.1370	7.3002	1.4347	82.2
7.9	0.1379	0.1374	7.2757	0.9905	1.0096	0.1388	7.2066	1.4329	82.1
8.0	0.1396	0.1392	7.1853	0.9903	1.0098	0.1405	7.1154	1.4312	82.0
8.1	0.1414	0.1409	7.0972	0.9900	1.0101	0.1423	7.0264	1.4294	81.9
8.2	0.1431	0.1426	7.0112	0.9898	1.0103	0.1441	6.9395	1.4277	81.8
8.3	0.1449	0.1444	6.9273	0.9895	1.0106	0.1459	6.8548	1.4259	81.7
8.4	0.1466	0.1461	6.8454	0.9893	1.0108	0.1477	6.7720	1.4242	81.6
8.5	0.1484	0.1478	6.7655	0.9890	1.0111	0.1495	6.6912	1.4224	81.5
8.6	0.1501	0.1495	6.6874	0.9888	1.0114	0.1512	6.6122	1.4207	81.4
8.7	0.1518	0.1513	6.6111	0.9885	1.0116	0.1530	6.5350	1.4190	81.3
8.8	0.1536	0.1530	6.5366	0.9882	1.0119	0.1548	6.4596	1.4172	81.2
8.9	0.1553	0.1547	6.4637	0.9880	1.0122	0.1566	6.3859	1.4155	81.1
		cos θ or cos s	sec θ or sec s	sin θ or sin s	csc θ or csc s	cot θ or cot s	tan θ or tan s	s	θ (degrees)

θ (degrees)	s	sin θ or sin s	csc θ or csc s	cos θ or cos s	sec θ or sec s	tan θ or tan s	cot θ or cot s		
9.0	0.1571	0.1564	6.3925	0.9877	1.0125	0.1584	6.3138	1.4137	81.0
9.1	0.1588	0.1582	6.3228	0.9874	1.0127	0.1602	6.2432	1.4120	80.9
9.2	0.1606	0.1599	6.2546	0.9871	1.0130	0.1620	6.1742	1.4102	80.8
9.3	0.1623	0.1616	6.1880	0.9869	1.0133	0.1638	6.1066	1.4085	80.7
9.4	0.1641	0.1633	6.1227	0.9866	1.0136	0.1655	6.0405	1.4067	80.6
9.5	0.1658	0.1650	6.0589	0.9863	1.0139	0.1673	5.9758	1.4050	80.5
9.6	0.1676	0.1668	5.9963	0.9860	1.0142	0.1691	5.9124	1.4032	80.4
9.7	0.1693	0.1685	5.9351	0.9857	1.0145	0.1709	5.8502	1.4015	80.3
9.8	0.1710	0.1702	5.8751	0.9854	1.0148	0.1727	5.7894	1.3998	80.2
9.9	0.1728	0.1719	5.8164	0.9851	1.0151	0.1745	5.7297	1.3980	80.1
10.0	0.1745	0.1736	5.7588	0.9848	1.0154	0.1763	5.6713	1.3963	80.0
10.1	0.1763	0.1754	5.7023	0.9845	1.0157	0.1781	5.6140	1.3945	79.9
10.2	0.1780	0.1771	5.6470	0.9842	1.0161	0.1799	5.5578	1.3928	79.8
10.3	0.1798	0.1788	5.5928	0.9839	1.0164	0.1817	5.5026	1.3910	79.7
10.4	0.1815	0.1805	5.5396	0.9836	1.0167	0.1835	5.4486	1.3893	79.6
10.5	0.1833	0.1822	5.4874	0.9833	1.0170	0.1853	5.3955	1.3875	79.5
10.6	0.1850	0.1840	5.4362	0.9829	1.0174	0.1871	5.3435	1.3858	79.4
10.7	0.1868	0.1857	5.3860	0.9826	1.0177	0.1890	5.2924	1.3840	79.3
10.8	0.1885	0.1874	5.3367	0.9823	1.0180	0.1908	5.2422	1.3823	79.2
10.9	0.1902	0.1891	5.2883	0.9820	1.0184	0.1926	5.1929	1.3806	79.1
11.0	0.1920	0.1908	5.2408	0.9816	1.0187	0.1944	5.1446	1.3788	79.0
11.1	0.1937	0.1925	5.1942	0.9813	1.0191	0.1962	5.0970	1.3771	78.9
11.2	0.1955	0.1942	5.1484	0.9810	1.0194	0.1980	5.0504	1.3753	78.8
11.3	0.1972	0.1959	5.1034	0.9806	1.0198	0.1998	5.0045	1.3736	78.7
11.4	0.1990	0.1977	5.0593	0.9803	1.0201	0.2016	4.9595	1.3718	78.6
11.5	0.2007	0.1994	5.0159	0.9799	1.0205	0.2035	4.9152	1.3701	78.5
11.6	0.2025	0.2011	4.9732	0.9796	1.0209	0.2053	4.8716	1.3683	78.4
11.7	0.2042	0.2028	4.9313	0.9792	1.0212	0.2071	4.8288	1.3666	78.3
11.8	0.2059	0.2045	4.8901	0.9789	1.0216	0.2089	4.7867	1.3648	78.2
11.9	0.2077	0.2062	4.8496	0.9785	1.0220	0.2107	4.7453	1.3631	78.1
12.0	0.2094	0.2079	4.8097	0.9781	1.0223	0.2126	4.7046	1.3614	78.0
12.1	0.2112	0.2096	4.7706	0.9778	1.0227	0.2144	4.6646	1.3596	77.9
12.2	0.2129	0.2113	4.7321	0.9774	1.0231	0.2162	4.6252	1.3579	77.8
12.3	0.2147	0.2130	4.6942	0.9770	1.0235	0.2180	4.5864	1.3561	77.7
12.4	0.2164	0.2147	4.6569	0.9767	1.0239	0.2199	4.5483	1.3544	77.6
12.5	0.2182	0.2164	4.6202	0.9763	1.0243	0.2217	4.5107	1.3526	77.5
12.6	0.2199	0.2181	4.5841	0.9759	1.0247	0.2235	4.4737	1.3509	77.4
12.7	0.2217	0.2198	4.5486	0.9755	1.0251	0.2254	4.4374	1.3491	77.3
12.8	0.2234	0.2215	4.5137	0.9751	1.0255	0.2272	4.4015	1.3474	77.2
12.9	0.2251	0.2232	4.4793	0.9748	1.0259	0.2290	4.3662	1.3456	77.1
13.0	0.2269	0.2250	4.4454	0.9744	1.0263	0.2309	4.3315	1.3439	77.0
13.1	0.2286	0.2267	4.4121	0.9740	1.0267	0.2327	4.2972	1.3422	76.9
13.2	0.2304	0.2284	4.3792	0.9736	1.0271	0.2345	4.2635	1.3404	76.8
13.3	0.2321	0.2300	4.3469	0.9732	1.0276	0.2364	4.2303	1.3387	76.7
13.4	0.2339	0.2317	4.3150	0.9728	1.0280	0.2382	4.1976	1.3369	76.6
		cos θ or cos s	sec θ or sec s	sin θ or sin s	csc θ or csc s	cot θ or cot s	tan θ or tan s	s	θ (degrees)

θ (degrees)	s	sin θ or sin s	csc θ or csc s	cos θ or cos s	sec θ or sec s	tan θ or tan s	cot θ or cot s		
13.5	0.2356	0.2334	4.2837	0.9724	1.0284	0.2401	4.1653	1.3352	76.5
13.6	0.2374	0.2351	4.2527	0.9720	1.0288	0.2419	4.1335	1.3334	76.4
13.7	0.2391	0.2368	4.2223	0.9715	1.0293	0.2438	4.1022	1.3317	76.3
13.8	0.2409	0.2385	4.1923	0.9711	1.0297	0.2456	4.0713	1.3299	76.2
13.9	0.2426	0.2402	4.1627	0.9707	1.0302	0.2475	4.0408	1.3282	76.1
14.0	0.2443	0.2419	4.1336	0.9703	1.0306	0.2493	4.0108	1.3265	76.0
14.1	0.2461	0.2436	4.1048	0.9699	1.0311	0.2512	3.9812	1.3247	75.9
14.2	0.2478	0.2453	4.0765	0.9694	1.0315	0.2530	3.9520	1.3230	75.8
14.3	0.2496	0.2470	4.0486	0.9690	1.0320	0.2549	3.9232	1.3212	75.7
14.4	0.2513	0.2487	4.0211	0.9686	1.0324	0.2568	3.8947	1.3195	75.6
14.5	0.2531	0.2504	3.9939	0.9681	1.0329	0.2586	3.8667	1.3177	75.5
14.6	0.2548	0.2521	3.9672	0.9677	1.0334	0.2605	3.8391	1.3160	75.4
14.7	0.2566	0.2538	3.9408	0.9673	1.0338	0.2623	3.8118	1.3142	75.3
14.8	0.2583	0.2554	3.9147	0.9668	1.0343	0.2642	3.7849	1.3125	75.2
14.9	0.2601	0.2571	3.8890	0.9664	1.0348	0.2661	3.7583	1.3107	75.1
15.0	0.2618	0.2588	3.8637	0.9659	1.0353	0.2679	3.7321	1.3090	75.0
15.1	0.2635	0.2605	3.8387	0.9655	1.0358	0.2698	3.7062	1.3073	74.9
15.2	0.2653	0.2622	3.8140	0.9650	1.0363	0.2717	3.6806	1.3055	74.8
15.3	0.2670	0.2639	3.7897	0.9646	1.0367	0.2736	3.6554	1.3038	74.7
15.4	0.2688	0.2656	3.7657	0.9641	1.0372	0.2754	3.6305	1.3020	74.6
15.5	0.2705	0.2672	3.7420	0.9636	1.0377	0.2773	3.6059	1.3003	74.5
15.6	0.2723	0.2689	3.7186	0.9632	1.0382	0.2792	3.5816	1.2985	74.4
15.7	0.2740	0.2706	3.6955	0.9627	1.0388	0.2811	3.5576	1.2968	74.3
15.8	0.2758	0.2723	3.6727	0.9622	1.0393	0.2830	3.5339	1.2950	74.2
15.9	0.2775	0.2740	3.6502	0.9617	1.0398	0.2849	3.5105	1.2933	74.1
16.0	0.2793	0.2756	3.6280	0.9613	1.0403	0.2867	3.4874	1.2915	74.0
16.1	0.2810	0.2773	3.6060	0.9608	1.0408	0.2886	3.4646	1.2898	73.9
16.2	0.2827	0.2790	3.5843	0.9603	1.0413	0.2905	3.4420	1.2881	73.8
16.3	0.2845	0.2807	3.5629	0.9598	1.0419	0.2924	3.4197	1.2863	73.7
16.4	0.2862	0.2823	3.5418	0.9593	1.0424	0.2943	3.3977	1.2846	73.6
16.5	0.2880	0.2840	3.5209	0.9588	1.0429	0.2962	3.3759	1.2828	73.5
16.6	0.2897	0.2857	3.5003	0.9583	1.0435	0.2981	3.3544	1.2811	73.4
16.7	0.2915	0.2874	3.4800	0.9578	1.0440	0.3000	3.3332	1.2793	73.3
16.8	0.2932	0.2890	3.4598	0.9573	1.0446	0.3019	3.3122	1.2776	73.2
16.9	0.2950	0.2907	3.4399	0.9568	1.0451	0.3038	3.2914	1.2758	73.1
17.0	0.2967	0.2924	3.4203	0.9563	1.0457	0.3057	3.2709	1.2741	73.0
17.1	0.2985	0.2940	3.4009	0.9558	1.0463	0.3076	3.2506	1.2723	72.9
17.2	0.3002	0.2957	3.3817	0.9553	1.0468	0.3096	3.2305	1.2706	72.8
17.3	0.3019	0.2974	3.3628	0.9548	1.0474	0.3115	3.2106	1.2689	72.7
17.4	0.3037	0.2990	3.3440	0.9542	1.0480	0.3134	3.1910	1.2671	72.6
17.5	0.3054	0.3007	3.3255	0.9537	1.0485	0.3153	3.1716	1.2654	72.5
17.6	0.3072	0.3024	3.3072	0.9532	1.0491	0.3172	3.1524	1.2636	72.4
17.7	0.3089	0.3040	3.2891	0.9527	1.0497	0.3191	3.1334	1.2619	72.3
17.8	0.3107	0.3057	3.2712	0.9521	1.0503	0.3211	3.1146	1.2601	72.2
17.9	0.3124	0.3074	3.2536	0.9516	1.0509	0.3230	3.0961	1.2584	72.1
		cos θ or cos s	sec θ or sec s	sin θ or sin s	csc θ or csc s	cot θ or cot s	tan θ or tan s	s	θ (degrees)

θ (degrees)	s	sin θ or sin s	csc θ or csc s	cos θ or cos s	sec θ or sec s	tan θ or tan s	cot θ or cot s		
18.0	0.3142	0.3090	3.2361	0.9511	1.0515	0.3249	3.0777	1.2566	72.0
18.1	0.3159	0.3107	3.2188	0.9505	1.0521	0.3269	3.0595	1.2549	71.9
18.2	0.3176	0.3123	3.2017	0.9500	1.0527	0.3288	3.0415	1.2531	71.8
18.3	0.3194	0.3140	3.1848	0.9494	1.0533	0.3307	3.0237	1.2514	71.7
18.4	0.3211	0.3156	3.1681	0.9489	1.0539	0.3327	3.0061	1.2497	71.6
18.5	0.3229	0.3173	3.1515	0.9483	1.0545	0.3346	2.9887	1.2479	71.5
18.6	0.3246	0.3190	3.1352	0.9478	1.0551	0.3365	2.9714	1.2462	71.4
18.7	0.3264	0.3206	3.1190	0.9472	1.0557	0.3385	2.9544	1.2444	71.3
18.8	0.3281	0.3223	3.1030	0.9466	1.0564	0.3404	2.9375	1.2427	71.2
18.9	0.3299	0.3239	3.0872	0.9461	1.0570	0.3424	2.9208	1.2409	71.1
19.0	0.3316	0.3256	3.0716	0.9455	1.0576	0.3443	2.9042	1.2392	71.0
19.1	0.3334	0.3272	3.0561	0.9449	1.0583	0.3463	2.8878	1.2374	70.9
19.2	0.3351	0.3289	3.0407	0.9444	1.0589	0.3482	2.8716	1.2357	70.8
19.3	0.3368	0.3305	3.0256	0.9438	1.0595	0.3502	2.8556	1.2339	70.7
19.4	0.3386	0.3322	3.0106	0.9432	1.0602	0.3522	2.8397	1.2322	70.6
19.5	0.3403	0.3338	2.9957	0.9426	1.0608	0.3541	2.8239	1.2305	70.5
19.6	0.3421	0.3355	2.9811	0.9421	1.0615	0.3561	2.8083	1.2287	70.4
19.7	0.3438	0.3371	2.9665	0.9415	1.0622	0.3581	2.7929	1.2270	70.3
19.8	0.3456	0.3387	2.9521	0.9409	1.0628	0.3600	2.7776	1.2252	70.2
19.9	0.3473	0.3404	2.9379	0.9403	1.0635	0.3620	2.7625	1.2235	70.1
20.0	0.3491	0.3420	2.9238	0.9397	1.0642	0.3640	2.7475	1.2217	70.0
20.1	0.3508	0.3437	2.9099	0.9391	1.0649	0.3659	2.7326	1.2200	69.9
20.2	0.3526	0.3453	2.8960	0.9385	1.0655	0.3679	2.7179	1.2182	69.8
20.3	0.3543	0.3469	2.8824	0.9379	1.0662	0.3699	2.7034	1.2165	69.7
20.4	0.3560	0.3486	2.8688	0.9373	1.0669	0.3719	2.6889	1.2147	69.6
20.5	0.3578	0.3502	2.8555	0.9367	1.0676	0.3739	2.6746	1.2130	69.5
20.6	0.3595	0.3518	2.8422	0.9361	1.0683	0.3759	2.6605	1.2113	69.4
20.7	0.3613	0.3535	2.8291	0.9354	1.0690	0.3779	2.6464	1.2095	69.3
20.8	0.3630	0.3551	2.8161	0.9348	1.0697	0.3799	2.6325	1.2078	69.2
20.9	0.3648	0.3567	2.8032	0.9342	1.0704	0.3819	2.6187	1.2060	69.1
21.0	0.3665	0.3584	2.7904	0.9336	1.0711	0.3839	2.6051	1.2043	69.0
21.1	0.3683	0.3600	2.7778	0.9330	1.0719	0.3859	2.5916	1.2025	68.9
21.2	0.3700	0.3616	2.7653	0.9323	1.0726	0.3879	2.5782	1.2008	68.8
21.3	0.3718	0.3633	2.7529	0.9317	1.0733	0.3899	2.5649	1.1990	68.7
21.4	0.3735	0.3649	2.7407	0.9311	1.0740	0.3919	2.5517	1.1973	68.6
21.5	0.3752	0.3665	2.7285	0.9304	1.0748	0.3939	2.5386	1.1956	68.5
21.6	0.3770	0.3681	2.7165	0.9298	1.0755	0.3959	2.5257	1.1938	68.4
21.7	0.3787	0.3697	2.7046	0.9291	1.0763	0.3979	2.5129	1.1921	68.3
21.8	0.3805	0.3714	2.6927	0.9285	1.0770	0.4000	2.5002	1.1903	68.2
21.9	0.3822	0.3730	2.6811	0.9278	1.0778	0.4020	2.4876	1.1886	68.1
22.0	0.3840	0.3746	2.6695	0.9272	1.0785	0.4040	2.4751	1.1868	68.0
22.1	0.3857	0.3762	2.6580	0.9265	1.0793	0.4061	2.4627	1.1851	67.9
22.2	0.3875	0.3778	2.6466	0.9259	1.0801	0.4081	2.4504	1.1833	67.8
22.3	0.3892	0.3795	2.6354	0.9252	1.0808	0.4101	2.4383	1.1816	67.7
22.4	0.3910	0.3811	2.6242	0.9245	1.0816	0.4122	2.4262	1.1798	67.6
		cos θ or cos s	sec θ or sec s	sin θ or sin s	csc θ or csc s	cot θ or cot s	tan θ or tan s	s	θ (degrees)

θ (degrees)	s	sin θ or sin s	csc θ or csc s	cos θ or cos s	sec θ or sec s	tan θ or tan s	cot θ or cot s		
22.5	0.3927	0.3827	2.6131	0.9239	1.0824	0.4142	2.4142	1.1781	67.5
22.6	0.3944	0.3843	2.6022	0.9232	1.0832	0.4163	2.4023	1.1764	67.4
22.7	0.3962	0.3859	2.5913	0.9225	1.0840	0.4183	2.3906	1.1746	67.3
22.8	0.3979	0.3875	2.5805	0.9219	1.0848	0.4204	2.3789	1.1729	67.2
22.9	0.3997	0.3891	2.5699	0.9212	1.0856	0.4224	2.3673	1.1711	67.1
23.0	0.4014	0.3907	2.5593	0.9205	1.0864	0.4245	2.3559	1.1694	67.0
23.1	0.4032	0.3923	2.5488	0.9198	1.0872	0.4265	2.3445	1.1676	66.9
23.2	0.4049	0.3939	2.5384	0.9191	1.0880	0.4286	2.3332	1.1659	66.8
23.3	0.4067	0.3955	2.5282	0.9184	1.0888	0.4307	2.3220	1.1641	66.7
23.4	0.4084	0.3971	2.5180	0.9178	1.0896	0.4327	2.3109	1.1624	66.6
23.5	0.4102	0.3987	2.5078	0.9171	1.0904	0.4348	2.2998	1.1606	66.5
23.6	0.4119	0.4003	2.4978	0.9164	1.0913	0.4369	2.2889	1.1589	66.4
23.7	0.4136	0.4019	2.4879	0.9157	1.0921	0.4390	2.2781	1.1572	66.3
23.8	0.4154	0.4035	2.4780	0.9150	1.0929	0.4411	2.2673	1.1554	66.2
23.9	0.4171	0.4051	2.4683	0.9143	1.0938	0.4431	2.2566	1.1537	66.1
24.0	0.4189	0.4067	2.4586	0.9135	1.0946	0.4452	2.2460	1.1519	66.0
24.1	0.4206	0.4083	2.4490	0.9128	1.0955	0.4473	2.2355	1.1502	65.9
24.2	0.4224	0.4099	2.4395	0.9121	1.0963	0.4494	2.2251	1.1484	65.8
24.3	0.4241	0.4115	2.4301	0.9114	1.0972	0.4515	2.2148	1.1467	65.7
24.4	0.4259	0.4131	2.4207	0.9107	1.0981	0.4536	2.2045	1.1449	65.6
24.5	0.4276	0.4147	2.4114	0.9100	1.0989	0.4557	2.1943	1.1432	65.5
24.6	0.4294	0.4163	2.4022	0.9092	1.0998	0.4578	2.1842	1.1414	65.4
24.7	0.4311	0.4179	2.3931	0.9085	1.1007	0.4599	2.1742	1.1397	65.3
24.8	0.4328	0.4195	2.3841	0.9078	1.1016	0.4621	2.1642	1.1380	65.2
24.9	0.4346	0.4210	2.3751	0.9070	1.1025	0.4642	2.1543	1.1362	65.1
25.0	0.4363	0.4226	2.3662	0.9063	1.1034	0.4663	2.1445	1.1345	65.0
25.1	0.4381	0.4242	2.3574	0.9056	1.1043	0.4684	2.1348	1.1327	64.9
25.2	0.4398	0.4258	2.3486	0.9048	1.1052	0.4706	2.1251	1.1310	64.8
25.3	0.4416	0.4274	2.3400	0.9041	1.1061	0.4727	2.1155	1.1292	64.7
25.4	0.4433	0.4289	2.3314	0.9033	1.1070	0.4748	2.1060	1.1275	64.6
25.5	0.4451	0.4305	2.3228	0.9026	1.1079	0.4770	2.0965	1.1257	64.5
25.6	0.4468	0.4321	2.3144	0.9018	1.1089	0.4791	2.0872	1.1240	64.4
25.7	0.4485	0.4337	2.3060	0.9011	1.1098	0.4813	2.0778	1.1222	64.3
25.8	0.4503	0.4352	2.2976	0.9003	1.1107	0.4834	2.0686	1.1205	64.2
25.9	0.4520	0.4368	2.2894	0.8996	1.1117	0.4856	2.0594	1.1188	64.1
26.0	0.4538	0.4384	2.2812	0.8988	1.1126	0.4877	2.0503	1.1170	64.0
26.1	0.4555	0.4399	2.2730	0.8980	1.1136	0.4899	2.0413	1.1153	63.9
26.2	0.4573	0.4415	2.2650	0.8973	1.1145	0.4921	2.0323	1.1135	63.8
26.3	0.4590	0.4431	2.2570	0.8965	1.1155	0.4942	2.0233	1.1118	63.7
26.4	0.4608	0.4446	2.2490	0.8957	1.1164	0.4964	2.0145	1.1100	63.6
26.5	0.4625	0.4462	2.2412	0.8949	1.1174	0.4986	2.0057	1.1083	63.5
26.6	0.4643	0.4478	2.2333	0.8942	1.1184	0.5008	1.9970	1.1065	63.4
26.7	0.4660	0.4493	2.2256	0.8934	1.1194	0.5029	1.9883	1.1048	63.3
26.8	0.4677	0.4509	2.2179	0.8926	1.1203	0.5051	1.9797	1.1030	63.2
26.9	0.4695	0.4524	2.2103	0.8918	1.1213	0.5073	1.9711	1.1013	63.1
		cos θ or cos s	sec θ or sec s	sin θ or sin s	csc θ or csc s	cot θ or cot s	tan θ or tan s	s	θ (degrees)

θ (degrees)	s	sin θ or sin s	csc θ or csc s	cos θ or cos s	sec θ or sec s	tan θ or tan s	cot θ or cot s		
27.0	0.4712	0.4540	2.2027	0.8910	1.1223	0.5095	1.9626	1.0996	63.0
27.1	0.4730	0.4555	2.1952	0.8902	1.1233	0.5117	1.9542	1.0978	62.9
27.2	0.4747	0.4571	2.1877	0.8894	1.1243	0.5139	1.9458	1.0961	62.8
27.3	0.4765	0.4586	2.1803	0.8886	1.1253	0.5161	1.9375	1.0943	62.7
27.4	0.4782	0.4602	2.1730	0.8878	1.1264	0.5184	1.9292	1.0926	62.6
27.5	0.4800	0.4617	2.1657	0.8870	1.1274	0.5206	1.9210	1.0908	62.5
27.6	0.4817	0.4633	2.1584	0.8862	1.1284	0.5228	1.9128	1.0891	62.4
27.7	0.4835	0.4648	2.1513	0.8854	1.1294	0.5250	1.9047	1.0873	62.3
27.8	0.4852	0.4664	2.1441	0.8846	1.1305	0.5272	1.8967	1.0856	62.2
27.9	0.4869	0.4679	2.1371	0.8838	1.1315	0.5295	1.8887	1.0838	62.1
28.0	0.4887	0.4695	2.1301	0.8829	1.1326	0.5317	1.8807	1.0821	62.0
28.1	0.4904	0.4710	2.1231	0.8821	1.1336	0.5339	1.8728	1.0804	61.9
28.2	0.4922	0.4726	2.1162	0.8813	1.1347	0.5362	1.8650	1.0786	61.8
28.3	0.4939	0.4741	2.1093	0.8805	1.1357	0.5384	1.8572	1.0769	61.7
28.4	0.4957	0.4756	2.1025	0.8796	1.1368	0.5407	1.8495	1.0751	61.6
28.5	0.4974	0.4772	2.0957	0.8788	1.1379	0.5430	1.8418	1.0734	61.5
28.6	0.4992	0.4787	2.0890	0.8780	1.1390	0.5452	1.8341	1.0716	61.4
28.7	0.5009	0.4802	2.0824	0.8771	1.1401	0.5475	1.8265	1.0699	61.3
28.8	0.5027	0.4818	2.0758	0.8763	1.1412	0.5498	1.8190	1.0681	61.2
28.9	0.5044	0.4833	2.0692	0.8755	1.1423	0.5520	1.8115	1.0664	61.1
29.0	0.5061	0.4848	2.0627	0.8746	1.1434	0.5543	1.8040	1.0647	61.0
29.1	0.5079	0.4863	2.0562	0.8738	1.1445	0.5566	1.7966	1.0629	60.9
29.2	0.5096	0.4879	2.0498	0.8729	1.1456	0.5589	1.7893	1.0612	60.8
29.3	0.5114	0.4894	2.0434	0.8721	1.1467	0.5612	1.7820	1.0594	60.7
29.4	0.5131	0.4909	2.0371	0.8712	1.1478	0.5635	1.7747	1.0577	60.6
29.5	0.5149	0.4924	2.0308	0.8704	1.1490	0.5658	1.7675	1.0559	60.5
29.6	0.5166	0.4939	2.0245	0.8695	1.1501	0.5681	1.7603	1.0542	60.4
29.7	0.5184	0.4955	2.0183	0.8686	1.1512	0.5704	1.7532	1.0524	60.3
29.8	0.5201	0.4970	2.0122	0.8678	1.1524	0.5727	1.7461	1.0507	60.2
29.9	0.5219	0.4985	2.0061	0.8669	1.1535	0.5750	1.7391	1.0489	60.1
30.0	0.5236	0.5000	2.0000	0.8660	1.1547	0.5773	1.7321	1.0472	60.0
30.1	0.5253	0.5015	1.9940	0.8652	1.1559	0.5797	1.7251	1.0455	59.9
30.2	0.5271	0.5030	1.9880	0.8643	1.1570	0.5820	1.7182	1.0437	59.8
30.3	0.5288	0.5045	1.9821	0.8634	1.1582	0.5844	1.7113	1.0420	59.7
30.4	0.5306	0.5060	1.9762	0.8625	1.1594	0.5867	1.7045	1.0402	59.6
30.5	0.5323	0.5075	1.9703	0.8616	1.1606	0.5890	1.6977	1.0385	59.5
30.6	0.5341	0.5090	1.9645	0.8607	1.1618	0.5914	1.6909	1.0367	59.4
30.7	0.5358	0.5105	1.9587	0.8599	1.1630	0.5938	1.6842	1.0350	59.3
30.8	0.5376	0.5120	1.9530	0.8590	1.1642	0.5961	1.6775	1.0332	59.2
30.9	0.5393	0.5135	1.9473	0.8581	1.1654	0.5985	1.6709	1.0315	59.1
31.0	0.5411	0.5150	1.9416	0.8572	1.1666	0.6009	1.6643	1.0297	59.0
31.1	0.5428	0.5165	1.9360	0.8563	1.1679	0.6032	1.6577	1.0280	58.9
31.2	0.5445	0.5180	1.9304	0.8554	1.1691	0.6056	1.6512	1.0263	58.8
31.3	0.5463	0.5195	1.9249	0.8545	1.1703	0.6080	1.6447	1.0245	58.7
31.4	0.5480	0.5210	1.9194	0.8536	1.1716	0.6104	1.6383	1.0228	58.6
		cos θ or cos s	sec θ or sec s	sin θ or sin s	csc θ or csc s	cot θ or cot s	tan θ or tan s	s	θ (degrees)

θ (degrees)	s	sin θ or sin s	csc θ or csc s	cos θ or cos s	sec θ or sec s	tan θ or tan s	cot θ or cot s		
31.5	0.5498	0.5225	1.9139	0.8526	1.1728	0.6128	1.6319	1.0210	58.5
31.6	0.5515	0.5240	1.9084	0.8517	1.1741	0.6152	1.6255	1.0193	58.4
31.7	0.5533	0.5255	1.9031	0.8508	1.1753	0.6176	1.6191	1.0175	58.3
31.8	0.5550	0.5270	1.8977	0.8499	1.1766	0.6200	1.6128	1.0158	58.2
31.9	0.5568	0.5284	1.8924	0.8490	1.1779	0.6224	1.6066	1.0140	58.1
32.0	0.5585	0.5299	1.8871	0.8480	1.1792	0.6249	1.6003	1.0123	58.0
32.1	0.5603	0.5314	1.8818	0.8471	1.1805	0.6273	1.5941	1.0105	57.9
32.2	0.5620	0.5329	1.8766	0.8462	1.1818	0.6297	1.5880	1.0088	57.8
32.3	0.5637	0.5344	1.8714	0.8453	1.1831	0.6322	1.5818	1.0071	57.7
32.4	0.5655	0.5358	1.8663	0.8443	1.1844	0.6346	1.5757	1.0053	57.6
32.5	0.5672	0.5373	1.8612	0.8434	1.1857	0.6371	1.5697	1.0036	57.5
32.6	0.5690	0.5388	1.8561	0.8425	1.1870	0.6395	1.5637	1.0018	57.4
32.7	0.5707	0.5402	1.8510	0.8415	1.1883	0.6420	1.5577	1.0001	57.3
32.8	0.5725	0.5417	1.8460	0.8406	1.1897	0.6445	1.5517	0.9983	57.2
32.9	0.5742	0.5432	1.8410	0.8396	1.1910	0.6469	1.5458	0.9966	57.1
33.0	0.5760	0.5446	1.8361	0.8387	1.1924	0.6494	1.5399	0.9948	57.0
33.1	0.5777	0.5461	1.8312	0.8377	1.1937	0.6519	1.5340	0.9931	56.9
33.2	0.5794	0.5476	1.8263	0.8368	1.1951	0.6544	1.5282	0.9913	56.8
33.3	0.5812	0.5490	1.8214	0.8358	1.1964	0.6569	1.5224	0.9896	56.7
33.4	0.5829	0.5505	1.8166	0.8348	1.1978	0.6594	1.5166	0.9879	56.6
33.5	0.5847	0.5519	1.8118	0.8339	1.1992	0.6619	1.5108	0.9861	56.5
33.6	0.5864	0.5534	1.8070	0.8329	1.2006	0.6644	1.5051	0.9844	56.4
33.7	0.5882	0.5548	1.8023	0.8320	1.2020	0.6669	1.4994	0.9826	56.3
33.8	0.5899	0.5563	1.7976	0.8310	1.2034	0.6694	1.4938	0.9809	56.2
33.9	0.5917	0.5577	1.7929	0.8300	1.2048	0.6720	1.4882	0.9791	56.1
34.0	0.5934	0.5592	1.7883	0.8290	1.2062	0.6745	1.4826	0.9774	56.0
34.1	0.5952	0.5606	1.7837	0.8281	1.2076	0.6771	1.4770	0.9756	55.9
34.2	0.5969	0.5621	1.7791	0.8271	1.2091	0.6796	1.4715	0.9739	55.8
34.3	0.5986	0.5635	1.7745	0.8261	1.2105	0.6822	1.4659	0.9721	55.7
34.4	0.6004	0.5650	1.7700	0.8251	1.2120	0.6847	1.4605	0.9704	55.6
34.5	0.6021	0.5664	1.7655	0.8241	1.2134	0.6873	1.4550	0.9687	55.5
34.6	0.6039	0.5678	1.7610	0.8231	1.2149	0.6899	1.4496	0.9669	55.4
34.7	0.6056	0.5693	1.7566	0.8221	1.2163	0.6924	1.4442	0.9652	55.3
34.8	0.6074	0.5707	1.7522	0.8211	1.2178	0.6950	1.4388	0.9634	55.2
34.9	0.6091	0.5721	1.7478	0.8202	1.2193	0.6976	1.4335	0.9617	55.1
35.0	0.6109	0.5736	1.7434	0.8192	1.2208	0.7002	1.4281	0.9599	55.0
35.1	0.6126	0.5750	1.7391	0.8181	1.2223	0.7028	1.4229	0.9582	54.9
35.2	0.6144	0.5764	1.7348	0.8171	1.2238	0.7054	1.4176	0.9564	54.8
35.3	0.6161	0.5779	1.7305	0.8161	1.2253	0.7080	1.4124	0.9547	54.7
35.4	0.6178	0.5793	1.7263	0.8151	1.2268	0.7107	1.4071	0.9529	54.6
35.5	0.6196	0.5807	1.7221	0.8141	1.2283	0.7133	1.4019	0.9512	54.5
35.6	0.6213	0.5821	1.7179	0.8131	1.2299	0.7159	1.3968	0.9495	54.4
35.7	0.6231	0.5835	1.7137	0.8121	1.2314	0.7186	1.3916	0.9477	54.3
35.8	0.6248	0.5850	1.7095	0.8111	1.2329	0.7212	1.3865	0.9460	54.2
35.9	0.6266	0.5864	1.7054	0.8100	1.2345	0.7239	1.3814	0.9442	54.1
		cos θ or cos s	sec θ or sec s	sin θ or sin s	csc θ or csc s	cot θ or cot s	tan θ or tan s	s	θ (degrees)

θ (degrees)	s	sin θ or sin s	csc θ or csc s	cos θ or cos s	sec θ or sec s	tan θ or tan s	cot θ or cot s		
36.0	0.6283	0.5878	1.7013	0.8090	1.2361	0.7265	1.3764	0.9425	54.0
36.1	0.6301	0.5892	1.6972	0.8080	1.2376	0.7292	1.3713	0.9407	53.9
36.2	0.6318	0.5906	1.6932	0.8070	1.2392	0.7319	1.3663	0.9390	53.8
36.3	0.6336	0.5920	1.6892	0.8059	1.2408	0.7346	1.3613	0.9372	53.7
36.4	0.6353	0.5934	1.6852	0.8049	1.2424	0.7373	1.3564	0.9355	53.6
36.5	0.6370	0.5948	1.6812	0.8039	1.2440	0.7400	1.3514	0.9338	53.5
36.6	0.6388	0.5962	1.6772	0.8028	1.2456	0.7427	1.3465	0.9320	53.4
36.7	0.6405	0.5976	1.6733	0.8018	1.2472	0.7454	1.3416	0.9303	53.3
36.8	0.6423	0.5990	1.6694	0.8007	1.2489	0.7481	1.3367	0.9285	53.2
36.9	0.6440	0.6004	1.6655	0.7997	1.2505	0.7508	1.3319	0.9268	53.1
37.0	0.6458	0.6018	1.6616	0.7986	1.2521	0.7536	1.3270	0.9250	53.0
37.1	0.6475	0.6032	1.6578	0.7976	1.2538	0.7563	1.3222	0.9233	52.9
37.2	0.6493	0.6046	1.6540	0.7965	1.2554	0.7590	1.3175	0.9215	52.8
37.3	0.6510	0.6060	1.6502	0.7955	1.2571	0.7618	1.3127	0.9198	52.7
37.4	0.6528	0.6074	1.6464	0.7944	1.2588	0.7646	1.3079	0.9180	52.6
37.5	0.6545	0.6088	1.6427	0.7934	1.2605	0.7673	1.3032	0.9163	52.5
37.6	0.6562	0.6101	1.6390	0.7923	1.2622	0.7701	1.2985	0.9146	52.4
37.7	0.6580	0.6115	1.6353	0.7912	1.2639	0.7729	1.2938	0.9128	52.3
37.8	0.6597	0.6129	1.6316	0.7902	1.2656	0.7757	1.2892	0.9111	52.2
37.9	0.6615	0.6143	1.6279	0.7891	1.2673	0.7785	1.2846	0.9093	52.1
38.0	0.6632	0.6157	1.6243	0.7880	1.2690	0.7813	1.2799	0.9076	52.0
38.1	0.6650	0.6170	1.6207	0.7869	1.2708	0.7841	1.2753	0.9058	51.9
38.2	0.6667	0.6184	1.6171	0.7859	1.2725	0.7869	1.2708	0.9041	51.8
38.3	0.6685	0.6198	1.6135	0.7848	1.2742	0.7898	1.2662	0.9023	51.7
38.4	0.6702	0.6211	1.6099	0.7837	1.2760	0.7926	1.2617	0.9006	51.6
38.5	0.6720	0.6225	1.6064	0.7826	1.2778	0.7954	1.2572	0.8988	51.5
38.6	0.6737	0.6239	1.6029	0.7815	1.2796	0.7983	1.2527	0.8971	51.4
38.7	0.6754	0.6252	1.5994	0.7804	1.2813	0.8012	1.2482	0.8954	51.3
38.8	0.6772	0.6266	1.5959	0.7793	1.2831	0.8040	1.2438	0.8936	51.2
38.9	0.6789	0.6280	1.5925	0.7782	1.2849	0.8069	1.2393	0.8919	51.1
39.0	0.6807	0.6293	1.5890	0.7771	1.2868	0.8098	1.2349	0.8901	51.0
39.1	0.6824	0.6307	1.5856	0.7760	1.2886	0.8127	1.2305	0.8884	50.9
39.2	0.6842	0.6320	1.5822	0.7749	1.2904	0.8156	1.2261	0.8866	50.8
39.3	0.6859	0.6334	1.5788	0.7738	1.2923	0.8185	1.2218	0.8849	50.7
39.4	0.6877	0.6347	1.5755	0.7727	1.2941	0.8214	1.2174	0.8831	50.6
39.5	0.6894	0.6361	1.5721	0.7716	1.2960	0.8243	1.2131	0.8814	50.5
39.6	0.6911	0.6374	1.5688	0.7705	1.2978	0.8273	1.2088	0.8796	50.4
39.7	0.6929	0.6388	1.5655	0.7694	1.2997	0.8302	1.2045	0.8779	50.3
39.8	0.6946	0.6401	1.5622	0.7683	1.3016	0.8332	1.2002	0.8762	50.2
39.9	0.6964	0.6414	1.5590	0.7672	1.3035	0.8361	1.1960	0.8744	50.1
40.0	0.6981	0.6428	1.5557	0.7660	1.3054	0.8391	1.1918	0.8727	50.0
40.1	0.6999	0.6441	1.5525	0.7649	1.3073	0.8421	1.1875	0.8709	49.9
40.2	0.7016	0.6455	1.5493	0.7638	1.3092	0.8451	1.1833	0.8692	49.8
40.3	0.7034	0.6468	1.5461	0.7627	1.3112	0.8481	1.1792	0.8674	49.7
40.4	0.7051	0.6481	1.5429	0.7615	1.3131	0.8511	1.1750	0.8657	49.6
		cos θ or cos s	sec θ or sec s	sin θ or sin s	csc θ or csc s	cot θ or cot s	tan θ or tan s	s	θ (degrees)

θ (degrees)	s	sin θ or sin s	csc θ or csc s	cos θ or cos s	sec θ or sec s	tan θ or tan s	cot θ or cot s		
40.5	0.7069	0.6494	1.5398	0.7604	1.3151	0.8541	1.1709	0.8639	49.5
40.6	0.7086	0.6508	1.5366	0.7593	1.3171	0.8571	1.1667	0.8622	49.4
40.7	0.7103	0.6521	1.5335	0.7581	1.3190	0.8601	1.1626	0.8604	49.3
40.8	0.7121	0.6534	1.5304	0.7570	1.3210	0.8632	1.1585	0.8587	49.2
40.9	0.7138	0.6547	1.5273	0.7559	1.3230	0.8662	1.1544	0.8570	49.1
41.0	0.7156	0.6561	1.5243	0.7547	1.3250	0.8693	1.1504	0.8552	49.0
41.1	0.7173	0.6574	1.5212	0.7536	1.3270	0.8724	1.1463	0.8535	48.9
41.2	0.7191	0.6587	1.5182	0.7524	1.3291	0.8754	1.1423	0.8517	48.8
41.3	0.7208	0.6600	1.5151	0.7513	1.3311	0.8785	1.1383	0.8500	48.7
41.4	0.7226	0.6613	1.5121	0.7501	1.3331	0.8816	1.1343	0.8482	48.6
41.5	0.7243	0.6626	1.5092	0.7490	1.3352	0.8847	1.1303	0.8465	48.5
41.6	0.7261	0.6639	1.5062	0.7478	1.3373	0.8878	1.1263	0.8447	48.4
41.7	0.7278	0.6652	1.5032	0.7466	1.3393	0.8910	1.1224	0.8430	48.3
41.8	0.7295	0.6665	1.5003	0.7455	1.3414	0.8941	1.1184	0.8412	48.2
41.9	0.7313	0.6678	1.4974	0.7443	1.3435	0.8972	1.1145	0.8395	48.1
42.0	0.7330	0.6691	1.4945	0.7431	1.3456	0.9004	1.1106	0.8378	48.0
42.1	0.7348	0.6704	1.4916	0.7420	1.3478	0.9036	1.1067	0.8360	47.9
42.2	0.7365	0.6717	1.4887	0.7408	1.3499	0.9067	1.1028	0.8343	47.8
42.3	0.7383	0.6730	1.4859	0.7396	1.3520	0.9099	1.0990	0.8325	47.7
42.4	0.7400	0.6743	1.4830	0.7385	1.3542	0.9131	1.0951	0.8308	47.6
42.5	0.7418	0.6756	1.4802	0.7373	1.3563	0.9163	1.0913	0.8290	47.5
42.6	0.7435	0.6769	1.4774	0.7361	1.3585	0.9195	1.0875	0.8273	47.4
42.7	0.7453	0.6782	1.4746	0.7349	1.3607	0.9228	1.0837	0.8255	47.3
42.8	0.7470	0.6794	1.4718	0.7337	1.3629	0.9260	1.0799	0.8238	47.2
42.9	0.7487	0.6807	1.4690	0.7325	1.3651	0.9293	1.0761	0.8221	47.1
43.0	0.7505	0.6820	1.4663	0.7314	1.3673	0.9325	1.0724	0.8203	47.0
43.1	0.7522	0.6833	1.4635	0.7302	1.3696	0.9358	1.0686	0.8186	46.9
43.2	0.7540	0.6845	1.4608	0.7290	1.3718	0.9391	1.0649	0.8168	46.8
43.3	0.7557	0.6858	1.4581	0.7278	1.3741	0.9424	1.0612	0.8151	46.7
43.4	0.7575	0.6871	1.4554	0.7266	1.3763	0.9457	1.0575	0.8133	46.6
43.5	0.7592	0.6884	1.4527	0.7254	1.3786	0.9490	1.0538	0.8116	46.5
43.6	0.7610	0.6896	1.4501	0.7242	1.3809	0.9523	1.0501	0.8098	46.4
43.7	0.7627	0.6909	1.4474	0.7230	1.3832	0.9556	1.0464	0.8081	46.3
43.8	0.7645	0.6921	1.4448	0.7218	1.3855	0.9590	1.0428	0.8063	46.2
43.9	0.7662	0.6934	1.4422	0.7206	1.3878	0.9623	1.0392	0.8046	46.1
44.0	0.7679	0.6947	1.4396	0.7193	1.3902	0.9657	1.0355	0.8029	46.0
44.1	0.7697	0.6959	1.4370	0.7181	1.3925	0.9691	1.0319	0.8011	45.9
44.2	0.7714	0.6972	1.4344	0.7169	1.3949	0.9725	1.0283	0.7994	45.8
44.3	0.7732	0.6984	1.4318	0.7157	1.3972	0.9759	1.0247	0.7976	45.7
44.4	0.7749	0.6997	1.4293	0.7145	1.3996	0.9793	1.0212	0.7959	45.6
44.5	0.7767	0.7009	1.4267	0.7133	1.4020	0.9827	1.0176	0.7941	45.5
44.6	0.7784	0.7022	1.4242	0.7120	1.4044	0.9861	1.0141	0.7924	45.4
44.7	0.7802	0.7034	1.4217	0.7108	1.4069	0.9896	1.0105	0.7906	45.3
44.8	0.7819	0.7046	1.4192	0.7096	1.4093	0.9930	1.0070	0.7889	45.2
44.9	0.7837	0.7059	1.4167	0.7083	1.4118	0.9965	1.0035	0.7871	45.1
45.0	0.7854	0.7071	1.4142	0.7071	1.4142	1.0000	1.0000	0.7854	45.0
		cos θ or cos s	sec θ or sec s	sin θ or sin s	csc θ or csc s	cot θ or cot s	tan θ or tan s	s	θ (degrees)

Answers to Selected Problems

Odd-Numbered Exercises
Vocabulary True–False Quizzes
Chapter Tests
Appendices

CHAPTER 1

Exercises ■ 1.1

1. True 3. False 5. True 7. False 9. True 11. True 13. False 15. True 17. True 19. a. Identity Property for Multiplication b. Distributive Property c. Definition of Subtraction d. Commutative Property for Multiplication e. Distributive Property f. Adding 6 and −2 g. Commutative Property for Multiplication and Definition of Subtraction 21. a. Definition of Subtraction b. Associative Property for Addition c. Commutative Property for Addition d. Associative Property for Addition e. Distributive Property and Adding 3 and 6 f. Commutative Property for Addition g. Associative Property for Addition h. Inverse Property for Addition i. Identity Property for Addition 23. $2x + 11$ 25. $7x + 2$ 27. 5 29. $2w$ 31. $7x_1 - 16x_2$ 33. $-x^2 + 20x + 9$ 35. $8{y_1}^2 + 2y_1y_2 - {y_2}^2$ 37. True, Distributive Property 39. False, try $a = 2, b = 3, c = 4$ 41. False, try $a = 1, b = 6, c = 2$ 43. True, Definition of Division 45. False, try $a = 6, b = 2, c = 3, d = 1$ 47. False, try $a = 6, b = 3, c = 5$ 49. False, try $x = 4$ 51. True 53. True 55. a. 26 b. 15.3333 c. division by zero

Exercises ■ 1.2

1. $2 \leqslant 8$ 3. $-7 > -10$ 5. $7 > -3$ 7. $8 \leqslant |-24|$ 9. $\frac{22}{7} > \pi$ 11. $2 \geqslant 0$ 13. $-2 < 0$ 15. $4 \geqslant -4$ 17. $-100 < 2$ 19. $1.6 < 1.63$ 21. 10 23. 5 25. 25 27. 2 29. 3 31. $x = \pm 10$ 33. $x = \pm 1$ 35. $x = \pm\pi$ 37. $|x| = 3$ 39. $|x| = 6$ 41. $|x| = 2\pi$ 43. 3 45. 4 47. 20 49. −4 or 8 51. −4 or 8 53. 0 57. True 59. a. $S = .120428$ b. oops

Exercises ■ 1.3

1. x^{12} 3. $x^{10}y^5$ 5. x^7 7. $\frac{1}{x^7}$ 9. x^3 11. x^3 13. y^2 15. $\frac{6}{z^2}$ 17. −1 19. $\frac{1}{x^{24}}$ 21. $\frac{1}{x+y}$ 23. $\frac{y^4}{x^2}$ 25. 1 27. 1 29. $2a^5$ 31. $\frac{1}{pq^2}$ 33. $\frac{x^2y^6}{27}$ 35. $\frac{1}{x^6}$ 37. $\frac{9}{16}$ 39. $2x^3y$ 41. 6 43. $2.39 \cdot 10^{10}$ 45. $4 \cdot 10^{-1}$ 47. $9.3 \cdot 10^{-8}$ 49. $2.76 \cdot 10^{13}$ 51. $2 \cdot 10^9$ 53. $1.28 \cdot 10^{39}$ 55. $2.90 \cdot 10^{23}$ 57. a. 1) Definition of x^a 2) Definition of P^b where $P = x^a$ 3) b factors of x^a is the same as ab factors of x b. 1) Exponent Law #1 2) Law #5 about Quotients 3) Simplification of $\frac{x^b}{x^b}$ 4) Identity Property for Multiplication c. 1) Definition of P^a where $P = \frac{x}{y}$ 2) Law #5 about Quotients 3) Definition of x^a and y^a

Exercises ■ 1.4

1. yes 3. no 5. yes 7. no 9. yes 11. $x^3 + x^2$ 13. $5y^3 - 3y^2 + 2y + 2$
15. $x^5 + 6x^4 - x^3 - x + 7$ 17. $5y^3 - 11y^2 - 12y$ 19. $2x^3 - 6x^2 + x + 6$ 21. $x^5 - 3x^4$
23. $-6x^5 - 2x^4$ 25. $ac - ad - bc + bd$ 27. $8x^2 + 6x - 5$ 29. $x^2 - 1$ 31. $x^2 + xy - 2y - 4$
33. $ad + ae + af + bd + be + bf + cd + ce + cf$ 35. $3x^3 - 14x^2 + 9x - 4$ 37. $x^4 - 4$ 39. $x^3 - 13x + 12$
41. $x^3 + 3x^2 + 3x + 1$ 43. $8y^3 + 12y^2 + 6y + 1$ 45. $8x^3 + 36x^2 + 54x + 27$ 47. $x^4 + 4x^3 + 6x^2 + 4x + 1$
49. $x^5 + 5x^4 + 10x^3 + 10x^2 + 5x + 1$ 51. $x^3 - 3x^2 + \frac{3}{x} - \frac{1}{x^3}$ 53. $8x^3 + 12x + \frac{6}{x} + \frac{1}{x^3}$ 55. $2x^6 - x^4 + 4$
57. $x^4 + \frac{2}{3}x + \frac{2}{3x}$ 59. $1 - \frac{3}{w} + \frac{1}{w^3}$ 61. $3 + \frac{1}{xy}$ 63. $x - 5 + \frac{4}{x^2}$ 65. 1 67. 0 69. 2
71. 3 73. $\frac{2x^3 - 3x^2 + 2x - 1}{x - 1} = 2x^2 - x + 1$ 75. $\frac{x^4 + 0x^3 - x^2 + 20x + 1}{x + 3} = x^3 - 3x^2 + 8x - 4 + \frac{13}{x + 3}$
77. $\frac{x^2 - 7x + 12}{x - 5} = x - 2 + \frac{2}{x - 5}$ 79. $\frac{x^3 - 1}{x - 1} = x^2 + x + 1$ 81. $\frac{2x^3 + 2x^2 - 3x + 4}{x + 2} = 2x^2 - 2x + 1 + \frac{2}{x + 2}$
83. $\frac{5x^3 - 14x^2 + 4x - 21}{x - 3} = 5x^2 + x + 7$ 85. $\frac{2x^5 + x^4 - 14x^3 + x + 2}{x + 3} = 2x^4 - 5x^3 + x^2 - 3x + 10 + \frac{-28}{x + 3}$
89. False, try $a = 3$, $b = 4$ 91. False, try $a = 3$, $b = 4$ 93. True

Exercises ■ 1.5

1. $x^2(x^2 + 2)$ 3. $y(x^2 - x + y)$ 5. $2xy(z + 1)$ 7. $-2xz(xz + y - y^2)$ 9. $3a^2b^4(a^2 + b^2)$
11. $(a + b)(x + y)$ 13. $(a + b)(x - y)$ 15. $(c + a)(2x + 3y)$ 17. $(x - 3)(x - 4)$ 19. $(6x + 1)(x + 1)$
21. $(2x + 3y)(x - y)$ 23. $2(x + 2y)(4x - 5y)$ 25. $(3x - 4)(x - 5)$ 27. $(x - 7)(x - 8)$
29. $4(x + y)(x - 5y)$ 31. $x(x + 2)(x - 6)$ 33. $x^3(2x + 1)(3x - 2)$ 35. $xy(3x + y)(3x + 2y)$
37. $2y(y + 1)^2$ 39. $(2x + 3)^2$ 41. $(x - y)(x + y)$ 43. $(x - y)(x^2 + xy + y^2)$
45. $(x - y)(x^2 + xy + y^2)(x + y)(x^2 - xy + y^2)$ 47. $(x - 1)(x^2 + x + 1)$ 49. $x(8x - 1)^2$
51. $(x - 2y)(x + 2y)(x^2 + 4y^2)$ 53. $(2x + 1)(4x^2 - 2x + 1)$ 55. $(5x + y)(25x^2 - 5xy + y^2)$
57. $(x - 1)(x + 1)(x^2 + 3)$ 59. $(x + y)(a + 1)^2$ 61. $10(x^2 + y^2)$ 63. $(x - 1)(x + 1)(x^2 + 1)(x^4 + 1)$
65. $(a^2 - a - 1)(a^2 + a + 1)$

Exercises ■ 1.6

3. -1 5. $\frac{x + 4}{x + 2}$ 7. $\frac{x^2 - 3x + 9}{x + 4}$ 9. $\frac{x - 2}{2x + 3}$ 11. $\frac{2(x - 1)}{(x + 2)(x + 1)}$ 13. $-\frac{(x + y)^2}{2}$
15. $\frac{2 + 7x}{y^2(4 + x)}$ 17. $\frac{x}{y}$ 19. $\frac{x}{2y(x + 1)}$ 21. $\frac{(a + b)^2}{x - y}$ 23. $\left(\frac{x - 6}{x - 3}\right)^2$ 25. $\frac{x}{x - y}$
27. $\frac{(a - b)(a - 4b)}{a + b}$ 29. $\frac{w^4}{(r^2 + w^2)(r - w)}$ 31. $\frac{2z^3 + 4y}{xy^2z^3}$ 33. $\frac{yz + xz + xy}{xyz}$ 35. $\frac{x^2 - x - 8}{(x - 3)(x + 4)}$
37. $\frac{2}{x - 1}$ 39. $\frac{x^2 + x + 1}{(x + 1)^2(x + 2)}$ 41. $\frac{a^2 - a - 1}{(a - 1)^3}$ 43. $\frac{a^2 + a - 1}{a^2(a + 1)}$ 45. $\frac{k}{k + 2}$
47. $\frac{5a^2 + 14ab + 10b^2}{(2a + 3b)(a + 2b)(a + b)}$ 49. $\frac{3a - 3b - 3a^2 - 18ab - 27b^2}{(a + 3b)^2(a - 3b)}$ 51. $\frac{x + 1}{x^2}$ 53. $\frac{4x + 2}{4x - 1}$
55. $\frac{x^3}{x - 1}$ 57. $\frac{r^3 + r}{r^3 + 1}$ 59. $\frac{1}{x + y}$ 61. False, try $p = 2$, $q = 2$
63. True, addition of fractions with equal denominators

Exercises ■ 1.7

1. 5 3. 3 5. 2 7. 1 9. $2\sqrt{3}$ 11. $2\sqrt[3]{2}$ 13. $2\sqrt[5]{4}$ 15. $2x^2y^2z^3\sqrt{2y}$
17. $\frac{\sqrt{35xy}}{7y}$ 19. $\frac{\sqrt{10xy}}{2xy^2}$ 21. $\sqrt[4]{6}$ 23. x 25. $\sqrt{2}$ 27. $a^2b\sqrt{2}$ 29. $xy^2\sqrt{2}$
31. $2x\sqrt{2}$ 33. $2xy^2z^4\sqrt{xz}$ 35. $5a^3b^3\sqrt{2b}$ 37. $wz\sqrt[3]{9w}$ 39. $x + y$ 41. $(x + y)\sqrt{x + y}$

43. xy 45. $x^2y^3\sqrt{xy}$ 47. $abc^3\sqrt[4]{a^3c}$ 49. $x^2y^3\sqrt[3]{y}$ 51. $3x\sqrt{x}$ 53. $1-5y\sqrt{x}$

55. $5\sqrt{a}+\sqrt{b}$ 57. $2-2\sqrt{x}$ 59. a. $-2\sqrt{2}-2\sqrt{3}$ b. $\dfrac{4x(\sqrt{x}-\sqrt{y})}{x-y}$ c. $4+\sqrt{15}$

61. $\dfrac{-1}{x(x+h)\left(\dfrac{1}{\sqrt{x+h}}+\dfrac{1}{\sqrt{x}}\right)}$ 63. False, try $a = 9$, $b = 4$ 65. True, rule of radicals

Exercises ■ 1.8

1. 3 3. 2 5. 4 7. 16 9. 2 11. 8 13. 64 15. 3 17. $\frac{1}{9}$ 19. $\frac{1}{3125}$ 21. $x^{2/3}$
23. a^2 25. x^6y 27. y 29. $36x^{4/3}y^{1/2}$ 31. $x^{1/8}$ 33. $x-y\sqrt{x}$ 35. $\frac{a}{b}\sqrt{b}$
37. $\dfrac{z^2}{x^{1/2}y^{2/3}}$ 39. $\dfrac{1}{x^{3/2}y^3}$ 41. $x^{5/6}$ 43. $\dfrac{1}{(xy)^{1/12}}$ 45. $\sqrt[4]{x}$ 47. $4a^2b^{3/4}$ 49. $(x+y)^{5/6}$
51. $3\cdot 10^{-3}$ 53. $2\cdot 10^{-1}$ 55. $(x-\sqrt{5})(x+\sqrt{5})$ 57. $(\sqrt{a}-b)(\sqrt{a}+b)$ 59. $(\sqrt{x}-2y)(\sqrt{x}+2y)$
61. $x+2\sqrt{xy}+y$ 63. $a-b$ 65. $x+10x^{1/2}+25$ 67. $x-2+1/x$ 69. $(x-1-\sqrt{5})(x-1+\sqrt{5})$
71. $(x-3)(x+1)$ 73. $(x+2-\sqrt{2})(x+2+\sqrt{2})$ 75. $x(x+4)$ 77. $(y-\frac{5}{2}-\sqrt{6})(y-\frac{5}{2}+\sqrt{6})$
79. .6810 81. 1.2599 83. 31.6227 85. 27.3313 87. 1.5848 89. 2.0239

Exercises ■ 1.9

1. −1 3. 1 5. −1 7. 1 9. −1 11. $6i$ 13. $10i$ 15. $2i\sqrt{2}$ 17. $2i\sqrt{5}$ 19. $2i\sqrt{7}$
21. $x=\frac{1}{2}$, $y=-4$ 23. $x=3$, $y=2$ 25. $x=2$, $y=1$ 27. $1+i$ 29. $0+i$ 31. $0+0i$ 33. $8-2i$
35. $8+4i$ 37. $-1+2i$ 39. $7+6i$ 41. $\frac{9}{5}-\frac{2}{5}i$ 43. $5-12i$ 45. $-12+0i$ 47. $-2+2i$
49. x^2+y^2 51. 7 53. True 55. False 57. True 59. True

Vocabulary/Symbol True-False Quiz

1. integer 2. True 3. 1 4. True 5. $|x|$ 6. True 7. x^2+6x+9 8. True 9. True
10. $\dfrac{d+2b}{bd}$

Chapter Test

1. $8.4\cdot 10^{-4}$ 2. a. 11 b. −11 3. a. $x-5$ b. $x^2+2x+14$ 4. $2x^3+7x^2-4x$
5. $2x^2-3x+6$ 6. $2y^5+3y^4+1$ 7. a. 53 b. 1 8. $(x+3)(x+4)$
9. $(t-1)(t+1)(t^2+1)(t^4+1)$ 10. $3y^2(x-1)(x+1)$ 11. $(2x+3)^2$ 12. $(2x-3y)(2x+3y)$
13. $-\dfrac{3x^2-3x-16}{(x-3)(x+3)(x-4)}$ 14. $\dfrac{y-x}{y+xy^2}$ 15. $Q(x)=3x^2-3x+2$, $R(x)=2$ 16. $-2\sqrt{2x}$ 17. $\dfrac{x^2y^3}{z^6}$
18. −1 19. $14+5i$ 20. $\frac{3}{5}-\frac{4}{5}i$

CHAPTER 2

Exercises ■ 2.1

1. $x=5$ 3. $x=5$ 5. $y=-3$ 7. $x=\frac{15}{2}$ 9. $x=\frac{3}{2}$ 11. $x=-1$ 13. $s=\frac{7}{2}$ 15. $x=0$
17. $x=10$ 19. $x=\frac{6}{17}$ 21. $y=3$ 23. no solution 25. identity 27. no solution 29. $x=\frac{22}{8}$
31. $x=2$ 33. $t=-\frac{7}{2}$ 35. $x=7$ 37. $y=14.875$ 39. $x=-2$ 41. $x=15$ or $x=-7$
43. $x=\frac{13}{2}$ or $x=\frac{3}{2}$ 45. $x=\frac{3}{2}$ or $x=\frac{11}{2}$ 47. no solution 49. True 51. True 53. a. −1.66667
b. −3.33335 c. 2.70588

Exercises ■ 2.2

1. \$200 3. \$15.39 5. 6 7. a. \$10,000 at 11% and \$20,000 at 5.5% b. \$20,000 at 11% and \$10,000 at 5.5% 9. Andy's rate is 11 km/hr and Katie's rate is 21 km/hr. 11. 3:30 p.m. the same day 13. 200 grams 15. Add .1 gallon of water and .1 gallon of pure antifreeze. 17. 9:30 a.m. that day 19. \$16.11 21. 1 ten, 4 fives, and 12 ones 23. 50 mi/hr and 70 mi/hr 25. 20 nines, 40 thirteens, 40 fifteens 27. 21 mi/hr 29. 495 gallons of milk and 15 gallons of cream 31. $1\frac{2}{3}$ gallons of 60% insecticide and $3\frac{1}{3}$ gallons of water 33. $33\frac{1}{3}$ ml 35. a. 50.7 pounds per square inch b. $D = \dfrac{P - 14.7}{.45}$ c. 189.56 ft 37. $w = \dfrac{A}{l}$ 39. $r = \dfrac{d}{t}$ 41. $d = \dfrac{C}{\pi}$ 43. $t = \dfrac{v_2 - v_1}{a}$

45. $n = \dfrac{l - a}{d} + 1$ 47. $y = \dfrac{c - ax}{b}$ 49. a. $a = \dfrac{18 - h}{1.1}$ b. 9 years 1 month (approx.)

51. a. $q = \dfrac{L}{.943}$ b. 2.12 53. a. $F = \dfrac{9C + 160}{5}$ b. $C = \dfrac{5F - 160}{9}$ c. $212°F$ d. $-23\frac{1}{3}°C$

55. a. 32 seconds b. $v = 32{,}768s$ c. 32,768 d. 163,840 57. a. $P = 30a + 50j$ b. \$700 59. $K = 12(295x + 345y)$

Exercises ■ 2.3

1. $3 < 9$ 3. $-2 < 1$ 5. $|-3| < |-5|$ 7. $[-2,\infty)$ 9. $(-\infty,1)$ 11. $(-\infty,0]$ 13. $(-1,\infty)$ 15. $[-2,\infty)$ 17. $(4,\infty)$ 19. $(-\frac{13}{3},\frac{113}{3}]$ 21. $[-\frac{1}{3},3)$ 23. $(3,6)$ 25. $[-5.2, 5.85)$ 27. $(-\infty,-1] \cup [3,\infty)$ 29. $(-\infty,-5) \cup (9,\infty)$ 31. $(-2,7)$ 33. $(-\infty,-\frac{28}{3}] \cup [4,\infty)$ 35. $(-\infty,-6] \cup [-2,\infty)$ 37. $(-\frac{1}{4},\frac{5}{12})$ 39. $(-2,4)$ 41. no solution 43. $\mathcal{R}$ 45. no solution 47. $x > 200$ 49. $-19\frac{4}{9}° < C < -5°$ 51. a. 3) The associative, distributive, and commutative laws used to simplify $(b + c) - (a + c)$. 4) Statements 2 and 3 5) Definition of "<." b. 2) Given 3) Definition of "<." 4) Product of two positive real numbers is positive. 5) Distributive Law 6) Statements 4 and 5. 7) Definition of "<." c. 3) Distributive Law 4) Statements 2 and 3. 5) Given 6) Statements 4 and 5. 7) Definition of "<."

Exercises ■ 2.4

1. -2 and 9 3. -1 and -12 5. -2 and 12 7. $-\frac{5}{2}$ (double root) 9. -1 and $\frac{9}{2}$ 11. 2 and 3 13. $\pm 2i$ 15. -3 and $\frac{1}{2}$ 17. $\frac{3}{2}$ and $\frac{5}{2}$ 19. $-\frac{1}{4}$ and -4 21. $-\frac{1}{3}$ and 1 23. $-\frac{1}{3}$ and -1

25. $1 \pm i$ 27. $\dfrac{3 \pm \sqrt{17}}{2}$ 29. $3 \pm \frac{1}{2}i$ 31. $\frac{3}{2}$ (double root) 33. $\dfrac{7 \pm \sqrt{29}}{10}$ 35. $-\sqrt{2}$ and $-2\sqrt{2}$

37. $\dfrac{3 \pm \sqrt{97}}{2}$ 39. $-\sqrt{2}$ (double root) 41. $\dfrac{7 \pm \sqrt{1 + 4y}}{2}$ 43. $\dfrac{8 \pm \sqrt{64 - s}}{4}$ 45. $\dfrac{-3 \pm \sqrt{8y - 23}}{4}$

49. 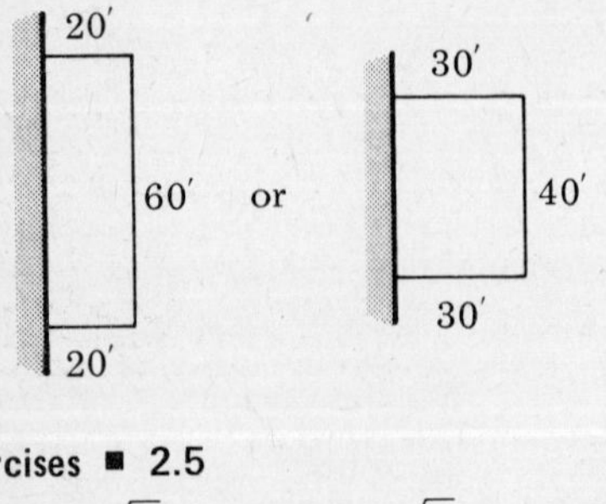

51. a. 256 ft b. 2.5 seconds c. .5 seconds and 4.5 seconds

53. $s = \dfrac{177 \pm \sqrt{168C - 35199}}{84}$ 57. 1.86603, .133973

59. This is not a quadratic equation.

Exercises ■ 2.5

1. $\pm 2, \pm\sqrt{5}$ 3. $\pm 2, \pm\sqrt{2}$ 5. $0, \pm\sqrt{2}, \pm\sqrt{10}$ 7. $\pm 2, \pm\dfrac{\sqrt{3}}{3}$ 9. $\pm 1, \pm\frac{1}{5}\sqrt{55}$ 11. $\frac{1}{9}$ 13. 1, 8

15. 1, 64 17. 0, 8, 27 19. $-\frac{1}{4}, \frac{2}{5}$ 21. $-\frac{27}{8}, -8$ 23. $\frac{15}{2}$ 25. no solution 27. -1 29. no solution 31. $-\frac{11}{5}$ 33. -2 35. $\frac{400}{3}$ 37. no solution 39. 8 41. $-1, -9\frac{3}{4}$ 43. $-\frac{3}{4}$

45. $\frac{14}{3}$ 47. a. 50 b. \$60 c. \$95 49. $A = 1 - \frac{m^2}{4d^2}$ 51. a. \$180.18 b. $F = \frac{n(n+1)}{k(k+1)}C$

c. $k = \frac{-1 + \sqrt{1 + \frac{4Cn(n+1)}{F}}}{2}$

Exercises ■ 2.6

1. $-2 < x < 5$ 3. $-2 \leqslant x \leqslant 5$ 5. $x \leqslant 2$ or $x \geqslant 4$ 7. $-1 < x < \frac{5}{2}$ 9. $x > 6$ or $x < -\frac{8}{3}$
11. $x \geqslant \frac{7}{9}$ or $x \leqslant -1$ 13. $0 < x < 1$ 15. $x \geqslant 4$ or $x \leqslant -\frac{3}{2}$ 17. $-3 < x < -\frac{1}{4}$ 19. $x \leqslant -\frac{8}{3}$ or $x \geqslant \frac{1}{4}$
21. $y < -\sqrt{3}$ or $y > \sqrt{3}$ 23. $x \leqslant 3 - \sqrt{3}$ or $x \geqslant 3 + \sqrt{3}$ 25. $-1 - \sqrt{11} \leqslant x \leqslant -1 + \sqrt{11}$
27. $x > 0$ or $x < 0$ 29. $x > \frac{5}{2}$ or $x < -\frac{5}{2}$ 31. $x \leqslant -4 \cup 1 \leqslant x \leqslant 3$ 33. $x \leqslant -2 \cup -\frac{1}{2} \leqslant x \leqslant 5$
35. $1 < x < 2 \cup x > 3$ 37. $-1 < x < 2$ 39. $-1 < x \leqslant 2$ 41. $-2 \leqslant x \leqslant -1 \cup x > 4$
43. $x < -6 \cup 1 < x < 2$ 45. $x < 0 \cup \frac{1}{8} \leqslant x < \frac{1}{2} \cup x \geqslant 1$ 47. $x < -4 \cup -\frac{3}{2} < x < 0$ 49. \$1.80 (at least)

Vocabulary/Symbol True-False Quiz

1. True 2. True 3. True 4. True 5. $\{-9, 7\}$ 6. $ac > bc$ 7. no 8. True
9. $b^2 - 4ac$ 10. True

Chapter Test

1. $x = 2$ 2. $x = -\frac{5}{7}$ 3. $x = \frac{c-b}{a}$ 4. 600 pounds 5. $-\frac{1}{2}, 7$ 6. $-\frac{1}{3}$ (double root)
7. $\frac{-1 \pm \sqrt{7}}{3}$ 8. 6, 9 9. $\pm\sqrt{3}, \pm\sqrt{5}$ 10. $-\frac{5}{4}, 2$ 11. $(-\infty, 2)$ 12. $(-\infty, -\frac{5}{3}]$ 13. $(4, 10)$
14. $(-\infty, -5) \cup (6, \infty)$ 15. $[-\frac{7}{6}, \frac{9}{6}]$ 16. $(-\infty, 3] \cup [4, \infty)$ 17. $(-\frac{7}{2}, \frac{1}{3})$
18. $\left(-\infty, \frac{-7-\sqrt{57}}{4}\right] \cup \left[\frac{-7+\sqrt{57}}{4}, \infty\right)$ 19. $(-3, 1) \cup (5, \infty)$ 20. $(-\infty, -\frac{11}{3}] \cup [1, 4)$

CHAPTER 3

Exercises ■ 3.1

1.

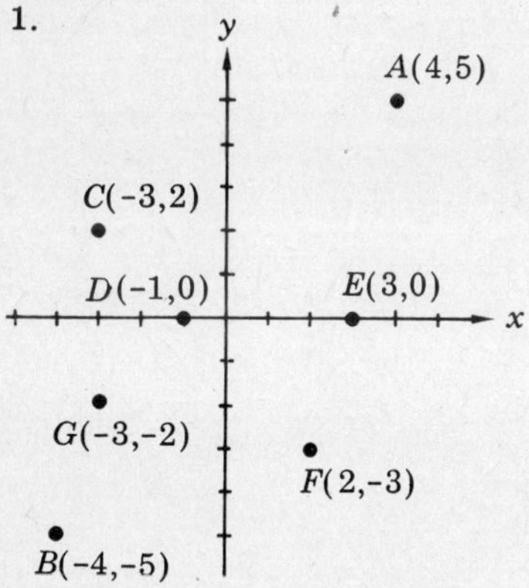

3. $A(-5,-4)$, $B(2,3)$, $C(-3,5)$, $D(4,-2)$, $E(-2,-2)$
5. 13 7. 10 9. $2\sqrt{2}$ 11. $4\sqrt{5}$ 13. 2π 15. True

17. a.

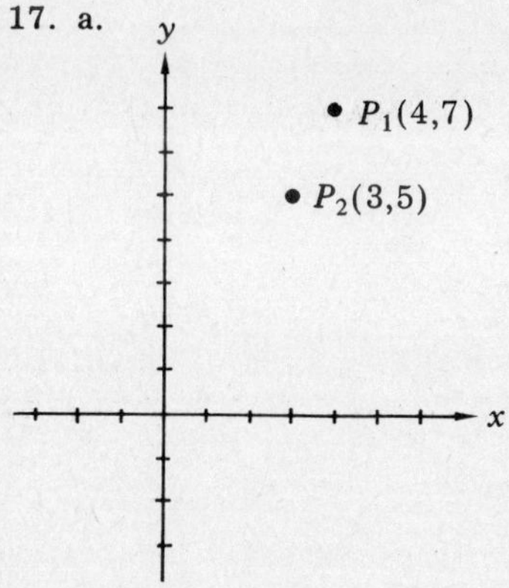

b. $(\frac{7}{2}, 6)$

19. a.

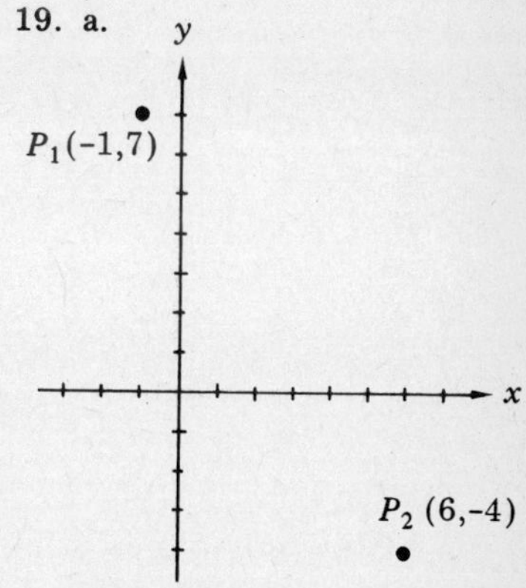

b. $(\frac{5}{2}, \frac{3}{2})$

21. a.

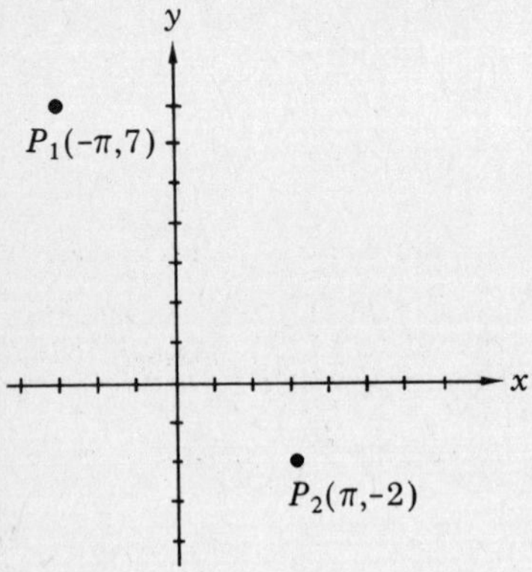

b. $(0,\frac{5}{2})$

29. 1 or 13 31. −1 or 39 33. $\frac{4}{7}$

Exercises ■ 3.2

1. a. $\{-3,-1,1,3,5\}$ b. $\{-1,0,2\}$ c. It is a function. 3. a. $\{-4,-2,0,2,4\}$ b. $\{-4,-2,0,2,4\}$ c. It isn't a function. 5. a. $\mathcal{R}$ b. $\{y|y \leqslant 3\}$ c. It is a function. 7. a. $\mathcal{R}$ b. $\{y|y \leqslant 1\}$ c. It is a function. 9. a. $\{x|x = \pm 3, -2 \leqslant x \leqslant 2\}$ b. $\{y|y = \pm 3, -2 \leqslant y \leqslant 2\}$ c. It isn't a function. 11. a. $\{0,6\}$ b. $\{0,6\}$ c. It is a function.
13. a. $\{\pm\sqrt{2}\}$ b. $\{0\}$ c. It is a function. 15. a. $\{3\}$ b. $\{12\}$ c. It is a function.
17. a. $\{12\}$ b. $\{3,6,9,11\}$ c. It isn't a function. 19. a. $\{8,10,12\}$ b. $\{1,2\}$ c. It is a function. 21. a. $\{-5,3,\pi\}$ b. $\{1,2,3,4\}$ c. It isn't a function.

23. a.

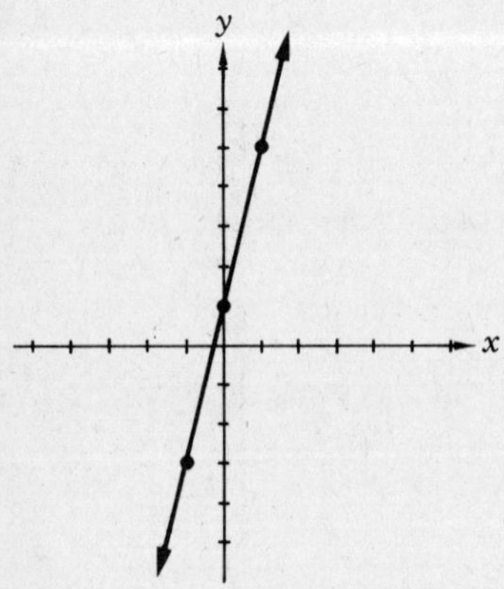

b. $\mathcal{R}$ c. $\mathcal{R}$ d. is a function

25. a.

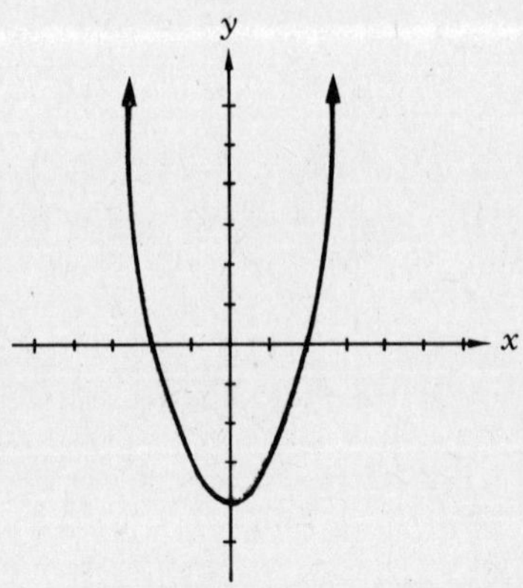

b. $\mathcal{R}$ c. $\{y|y \geqslant -4\}$ d. is a function

27. a. 1 b. −5 c. 5 d. 2 29. a. $-\frac{1}{2}$ b. −2 c. $\frac{1}{2}$ d. $\frac{1}{2}$

31. a. −1 b. $-\frac{2}{5}$ c. undefined d. $\dfrac{-2}{(x+h-2)(x-2)}$ 33. a. $\sqrt{3}$ b. 0 c. $\sqrt{5}$
d. $\dfrac{1}{\sqrt{x+h+3}+\sqrt{x+3}}$ 35. a. b b. $-3m+b$ c. $2m+b$ d. m 37. a. −3 b. $-\frac{1}{3}$, 3

39. a. −4 b. $\dfrac{1 \pm \sqrt{33}}{4}$ 41. a. b b. $-\dfrac{b}{m}$ 43. a. $\dfrac{1}{(3-x)^2}$ b. $-\dfrac{1}{x^2}+3$ 45. a. 2 b. 8

47. a. $g \circ f(x) = 280 + 14x$ b. $S = 280 + 14x$ c. $S = g \circ f(x)$ 49. Is not one-to-one.
51. Is one-to-one. 53. Is not one-to-one. 55. Is one-to-one. 57. Is not one-to-one.

Exercises ▪ 3.3

1. a.

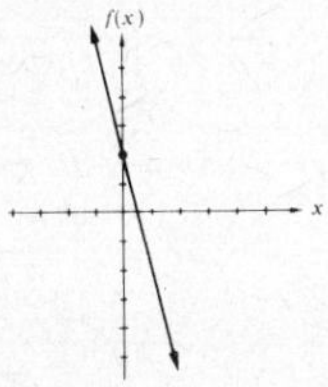

b. $m = -3$, $b = 2$
c. Domain $\mathcal{R}$, Range $\mathcal{R}$

3. a.

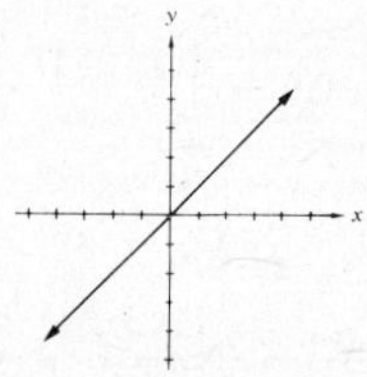

b. $m = 1$, $b = 0$
c. Domain $\mathcal{R}$, Range $\mathcal{R}$

5. a.

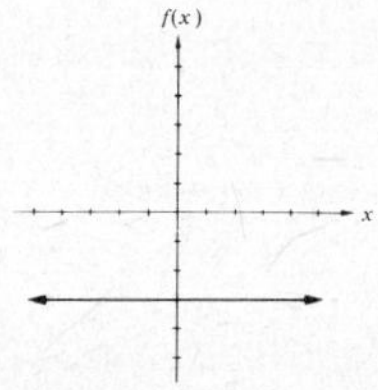

b. $m = 0$, $b = -3$
c. Domain $\mathcal{R}$, Range $\{-3\}$

7. a.

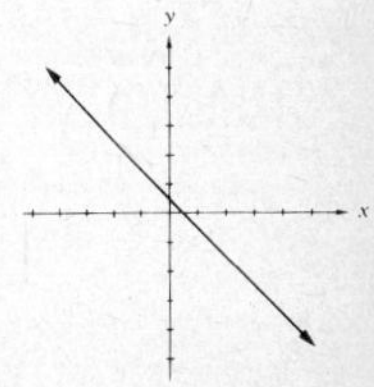

b. $m = -1$, $b = \frac{1}{2}$
c. Domain $\mathcal{R}$, Range $\mathcal{R}$

9. a.

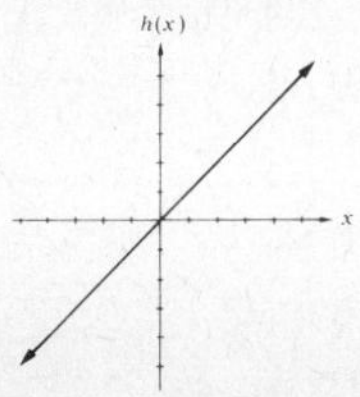

b. $m = 1$, $b = 0$
c. Domain $\mathcal{R}$, Range $\mathcal{R}$

11. a.

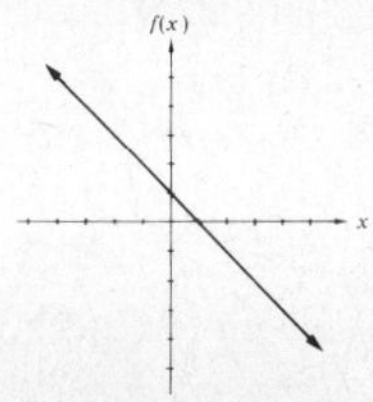

b. $m = -1$, $b = 1$
c. Domain $\mathcal{R}$, Range $\mathcal{R}$

13. -3 15. $\frac{5}{2}$ 17. $-\frac{3}{2}$ 19. $-2\sqrt{3}$
21. undefined 23. $y = -2x + 4$ 25. $y = 3x$
27. $y = x + 7$ 29. $x + y = -7$ 31. $y = -x - 7$
33. $y = 3x - 1$ 35. $y = 4$ 37. $x - y - 6 = 0$
39. $y = 8$ 41. $y = 3$ 43. a. $x + 2y = 13$
b. $y = \frac{3}{2}x$

Exercises ▪ 3.4

1. a. opens up
b. $(0,-6)$, $(2,0)$, $(-3,0)$
c. $(-\frac{1}{2}, -\frac{25}{4})$
d.

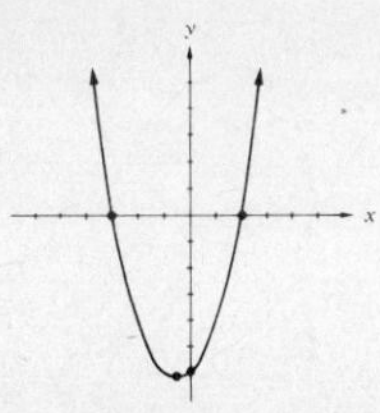

e. Domain $\mathcal{R}$
Range $\{y \mid y \geqslant -\frac{25}{4}\}$

3. a. opens down
b. $(0,6)$, $(2,0)$, $(-3,0)$
c. $(-\frac{1}{2}, \frac{25}{4})$
d.

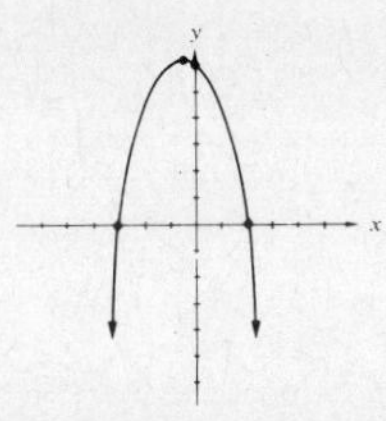

e. Domain $\mathcal{R}$
Range $\{y \mid y \leqslant \frac{25}{4}\}$

5. a. opens up
b. $(0,-8)$, $(2,0)$, $(-4,0)$
c. $(-1,-9)$
d.

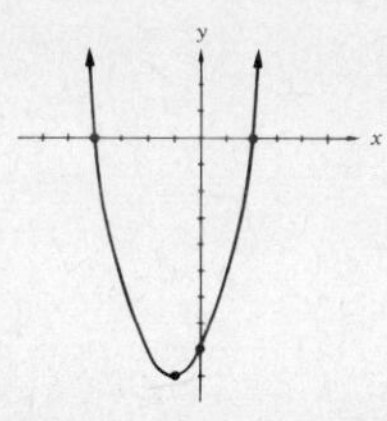

e. Domain $\mathcal{R}$
Range $\{y \mid y \geqslant -9\}$

7. a. opens up
b. $(0,2)$
c. $(1,1)$
d.

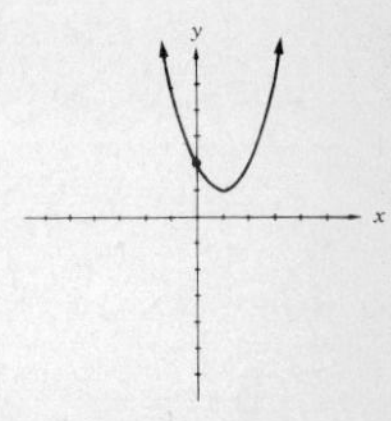

e. Domain $\mathcal{R}$
Range $\{y \mid y \geqslant 1\}$

9. a. opens down
b. $(0,8)$, $(-\frac{4}{3},0)$, $(2,0)$
c. $(\frac{1}{3},\frac{25}{3})$
d.

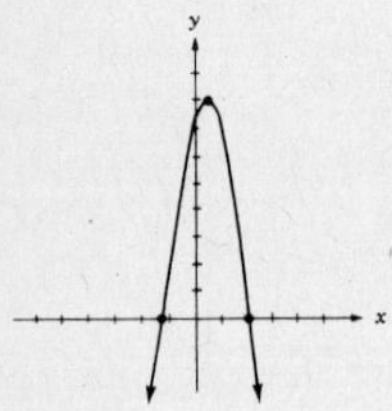

e. Domain $\mathcal{R}$
Range $\{y \mid y \leqslant \frac{25}{3}\}$

11. a. opens up
b. $(0,8)$, $(\frac{2}{3},0)$, $(4,0)$
c. $(\frac{7}{3},-\frac{25}{3})$
d.

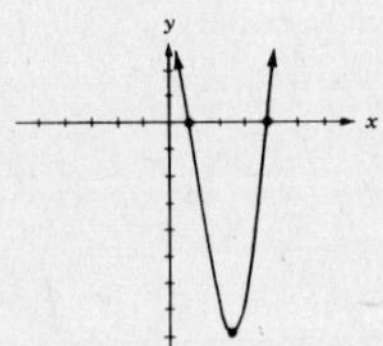

e. Domain $\mathcal{R}$
Range $\{y \mid y \geqslant -\frac{25}{3}\}$

13. a. opens up
b. $(0,4)$
c. $(0,4)$
d.

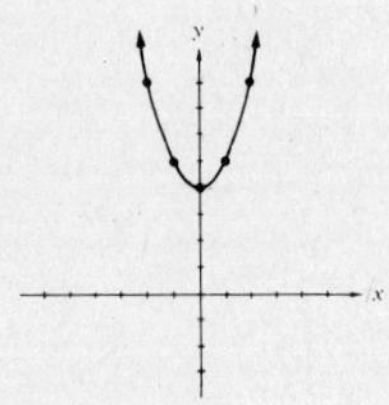

e. Domain $\mathcal{R}$
Range $\{y \mid y \geqslant 4\}$

15. a. opens down
b. $(0,4)$, $(2,0)$, $(-2,0)$
c. $(0,4)$
d.

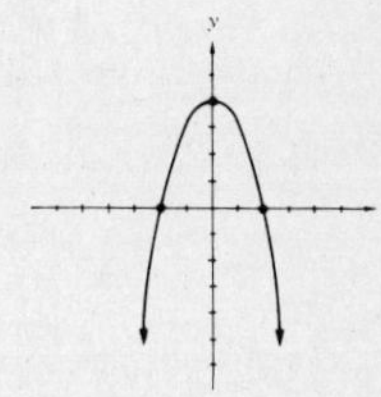

e. Domain $\mathcal{R}$
Range $\{y \mid y \leqslant 4\}$

17. a. opens up
b. $(0,37)$
c. $(3,1)$
d.

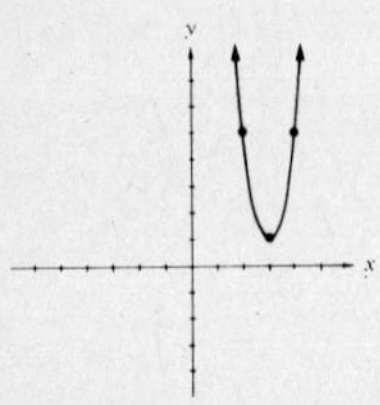

e. Domain $\mathcal{R}$
Range $\{y \mid y \geqslant 1\}$

19. a. opens up
b. $(0,1)$, $(\frac{1}{2},0)$
c. $(\frac{1}{2},0)$
d.

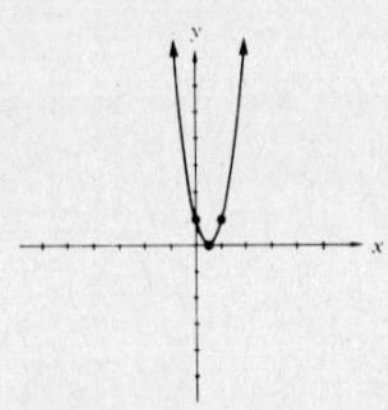

e. Domain $\mathcal{R}$
Range $\{y \mid y \geqslant 0\}$

21. a. opens down
b. $(0,-8)$, $(-2,0)$
c. $(-2,0)$
d.

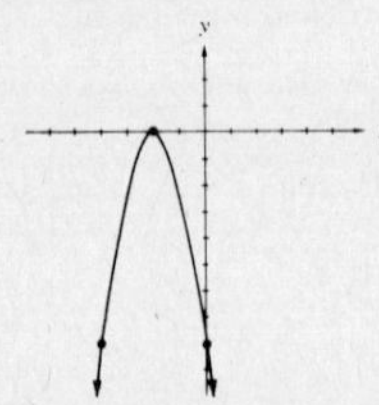

e. Domain $\mathcal{R}$
Range $\{y \mid y \leqslant 0\}$

23. a. opens up
b. $(0,9)$, $(\frac{3}{2},0)$
c. $(\frac{3}{2},0)$
d.

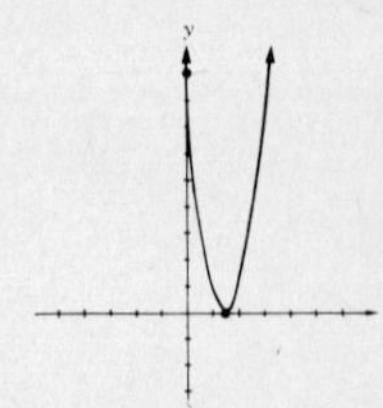

e. Domain $\mathcal{R}$
Range $\{y \mid y \geqslant 0\}$

25. a. opens up
b. $(0,1)$, $(-1,0)$
c. $(-1,0)$
d.

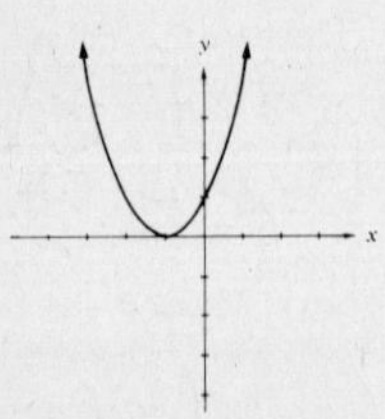

e. Domain $\mathcal{R}$
Range $\{y \mid y \geqslant 0\}$

27. a. opens up
b. $(0,5)$
c. $(\frac{1}{2},4)$
d.

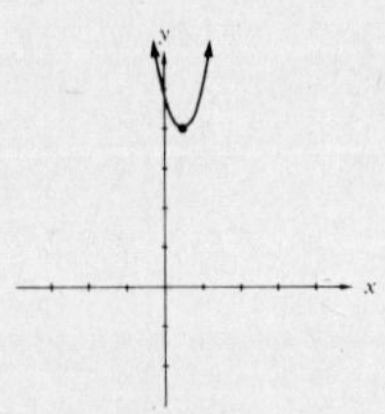

e. Domain $\mathcal{R}$
Range $\{y \mid y \geqslant 4\}$

Exercises ■ 3.5

1. a.

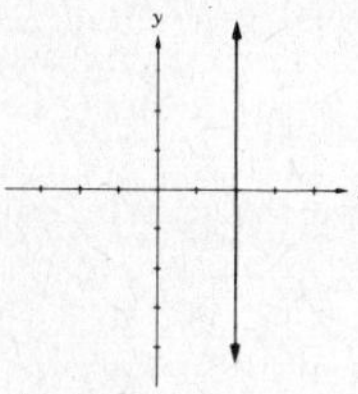

b. Domain {2}
Range $\mathcal{R}$

3. a.

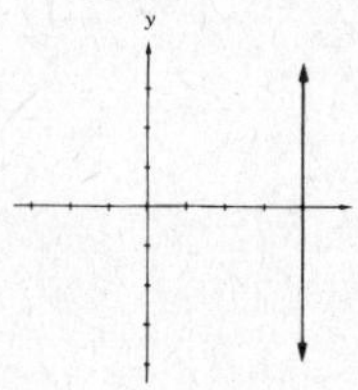

b. Domain {4}
Range $\mathcal{R}$

5. a.

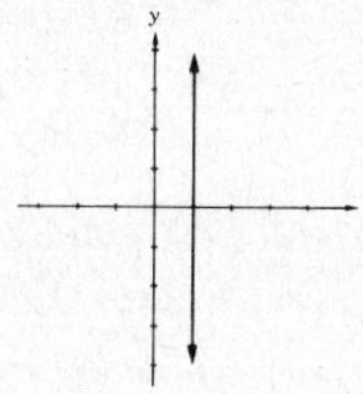

b. Domain {1}
Range $\mathcal{R}$

7. a.

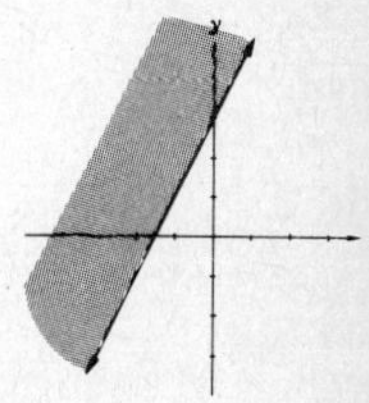

b. Domain $\mathcal{R}$
Range $\mathcal{R}$

9. a.

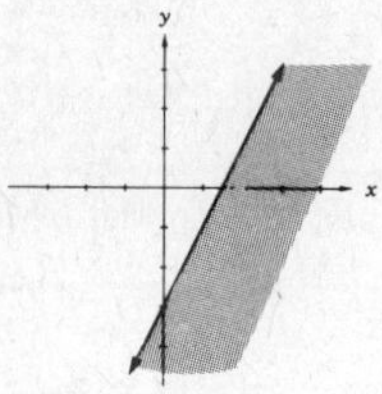

b. Domain $\mathcal{R}$
Range $\mathcal{R}$

11. a.

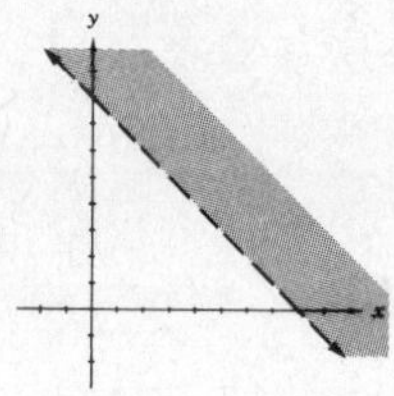

b. Domain $\mathcal{R}$
Range $\mathcal{R}$

13. a.

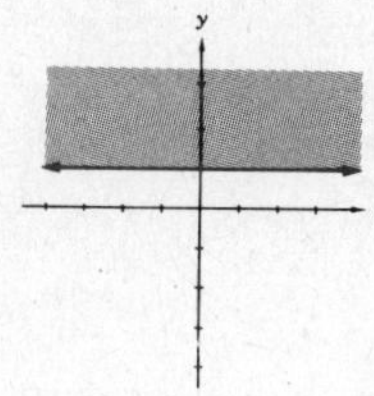

b. Domain $\mathcal{R}$
Range $\{y|y \geqslant 1\}$

15. a.

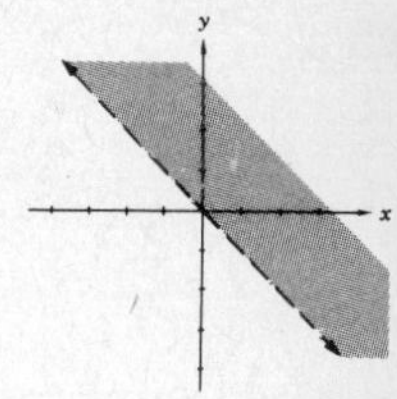

b. Domain $\mathcal{R}$
Range $\mathcal{R}$

17. a.

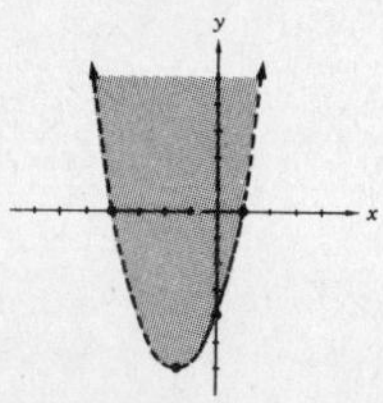

b. Domain $\mathcal{R}$
Range $\{y|y > -\frac{25}{4}\}$

19. a.

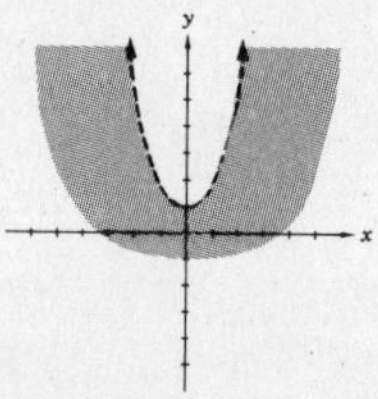

b. Domain $\mathcal{R}$
Range $\mathcal{R}$

21. a.

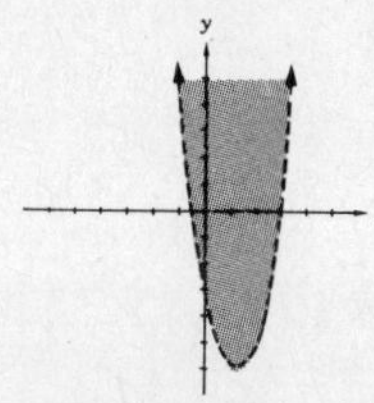

b. Domain $\mathcal{R}$
Range $\{y|y > -\frac{49}{8}\}$

23. a.

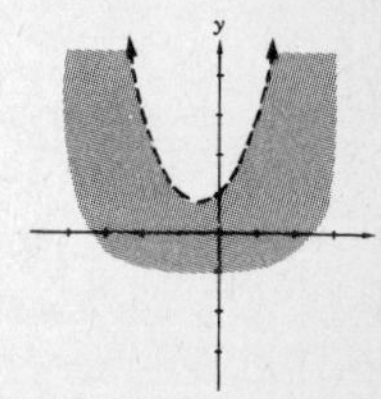

b. Domain $\mathcal{R}$
Range $\mathcal{R}$

25. a. opens right
b. (−8,0), (0,2), (0,−4)
c. (−9,−1)
d.

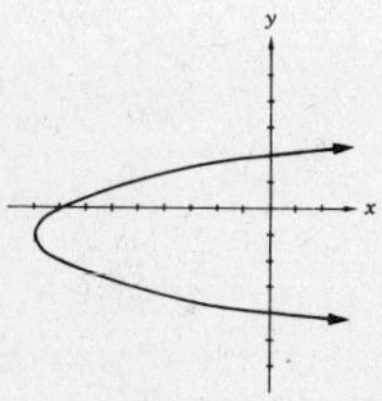

e. Domain $\{x|x \geqslant -9\}$
Range $\mathcal{R}$

27. a. opens left
b. (−5,0), (0,5), (0,1)
c. (4,3)
d.

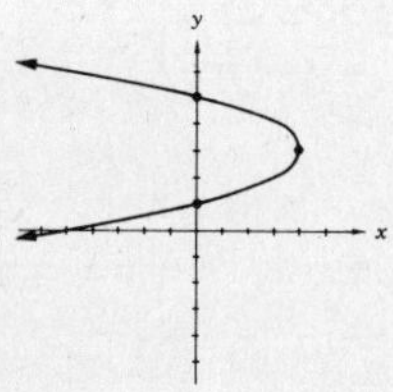

e. Domain $\{x|x \leqslant 4\}$
Range $\mathcal{R}$

29. a. opens left
b. (4,0), (0,2), (0,−2)
c. (4,0)
d.

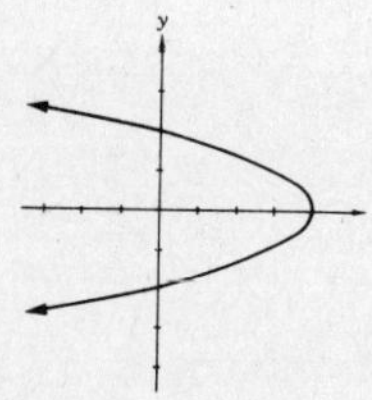

e. Domain $\{x|x \leqslant 4\}$
Range $\mathcal{R}$

31. a. opens right
b. (1,0), $(0,\frac{1}{2})$
c. $(0,\frac{1}{2})$
d.

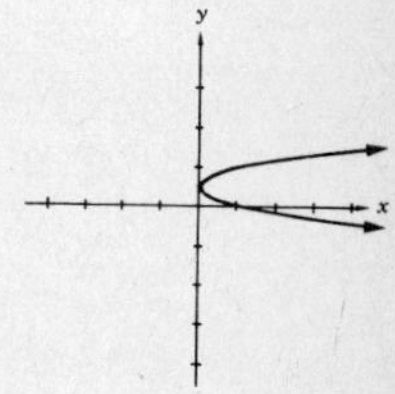

e. Domain $\{x|x \geqslant 0\}$
Range $\mathcal{R}$

33. a. opens left
b. (2,0), (0,2), (0,−2)
c. (2,0)
d.

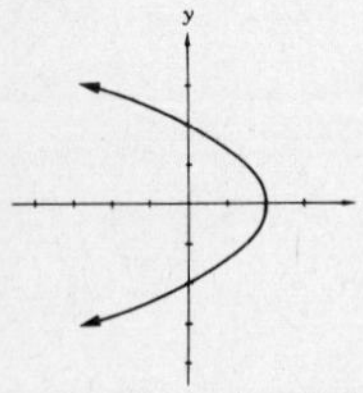

e. Domain $\{x | x \leqslant 2\}$
Range $\mathcal{R}$

35.

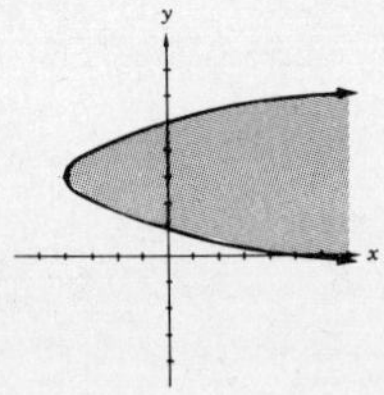

37.

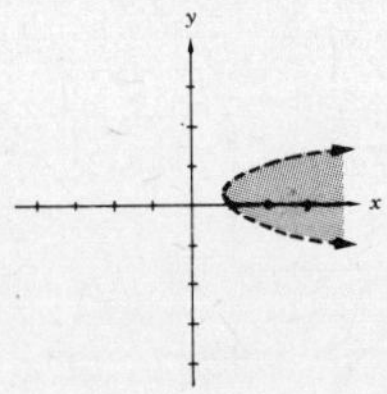

39.

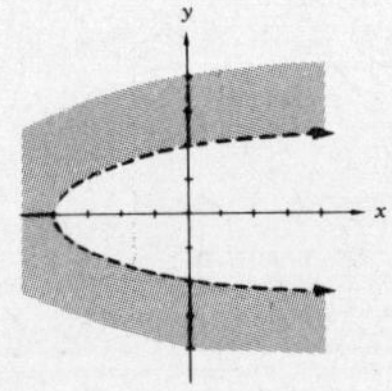

41.

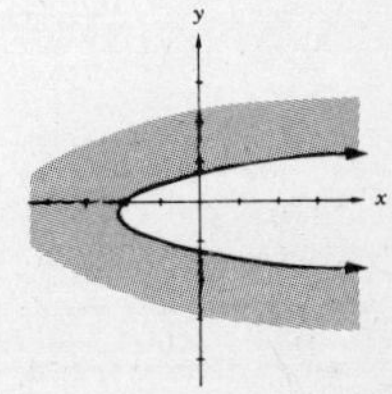

Exercises ■ 3.6

1. a. $R^{-1} = \{(2,1), (4,3)\}$
b.

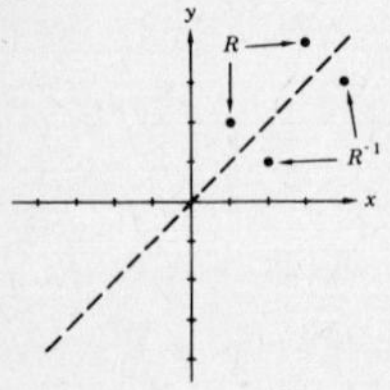

c. R and R^{-1} are both functions.

3. a. $R^{-1} = \{(5,-1), (7,-1)\}$
b.

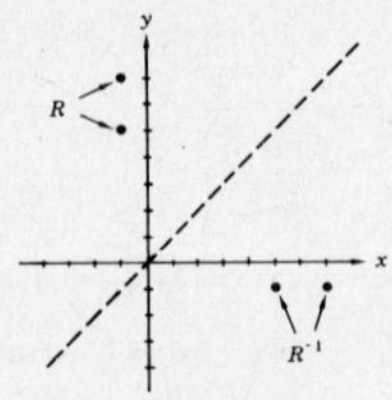

c. R is not a function. R^{-1} is a function.

5. a. $y = x - 3$
b.

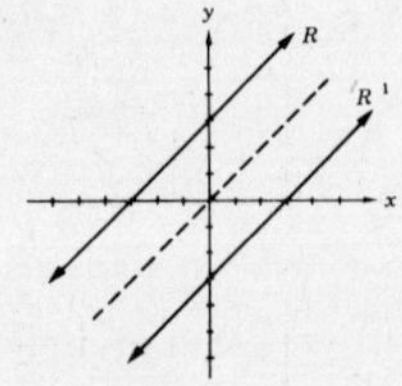

c. R and R^{-1} are both functions.

7. a. $y = \frac{1}{2}x - \frac{3}{2}$
b.

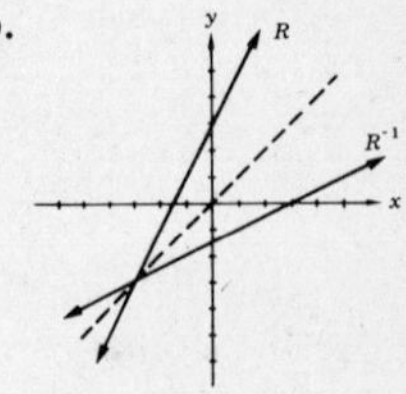

c. R and R^{-1} are both functions.

9. a. $x + y = 2$
b.

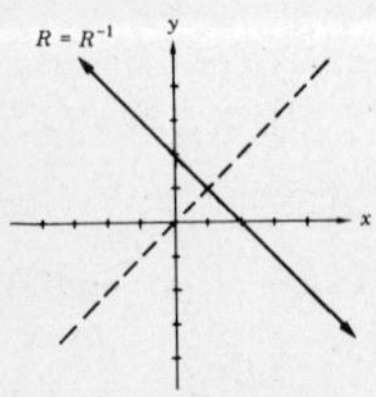

c. R and R^{-1} are both functions.

11. a. $x = 2$
b.

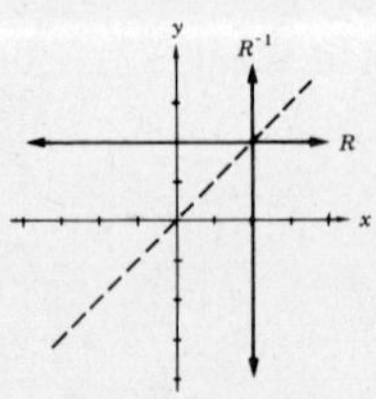

c. R is a function. R^{-1} is not a function.

13. a. $x = y^2 + 1$ or $y = \pm\sqrt{x - 1}$
b.

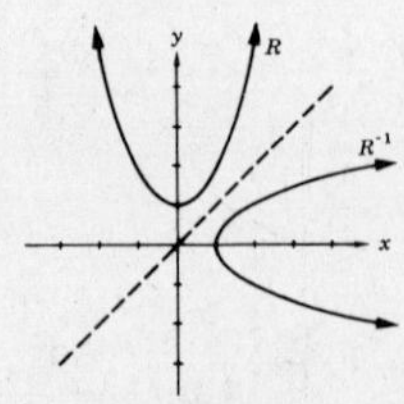

c. R is a function. R^{-1} is not a function.

15. a. $y = x^2 - 2$
b.

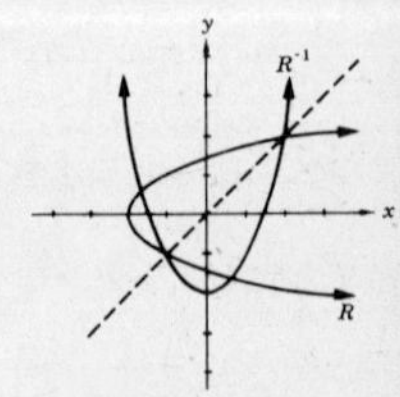

c. R is not a function. R^{-1} is a function.

17. a. $x = y^2 + 4y + 3$ or $y = -2 \pm \sqrt{1 + x}$
b.

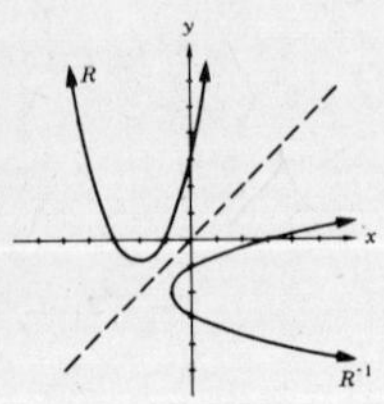

c. R is a function. R^{-1} is not a function.

19. a. $y = x^2 + x - 6$
b.

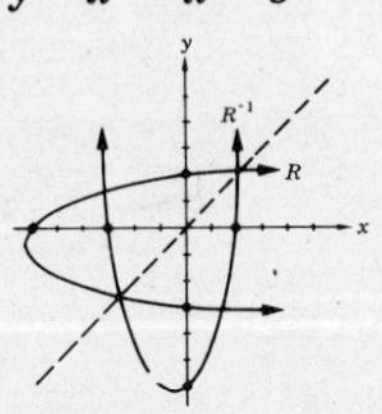

c. R is not a function. R^{-1} is a function.

21. a. $x = 9 - y^2$ or $y = \pm\sqrt{9 - x}$
b.

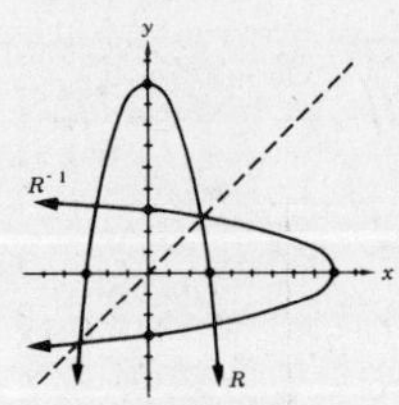

c. R is a function. R^{-1} is not a function.

23. a. $x = (y + 1)^3$ or $y = \sqrt[3]{x} - 1$
b.

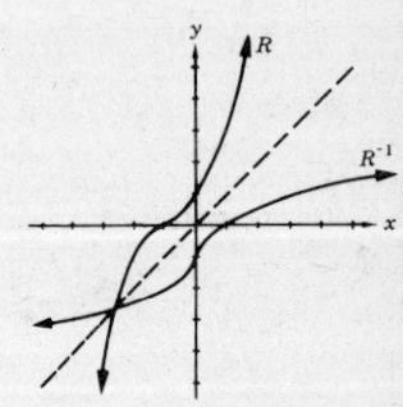

c. R and R^{-1} are both functions.

25. a. $f^{-1}(x) = \frac{1}{2}x + \frac{5}{2}$ 27. a. $f^{-1}(x) = 6x - 7$ 29. a. $f^{-1}(x) = x^{-1/3}$

33. a. $f^{-1}(x) = \dfrac{x}{m} - \dfrac{b}{m}$ b. $m \neq 0$

Exercises ■ 3.7

1. a.

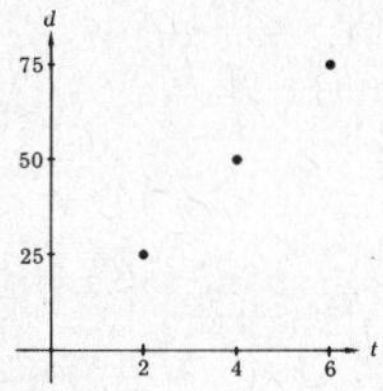

b. $d = \frac{25}{2}t$

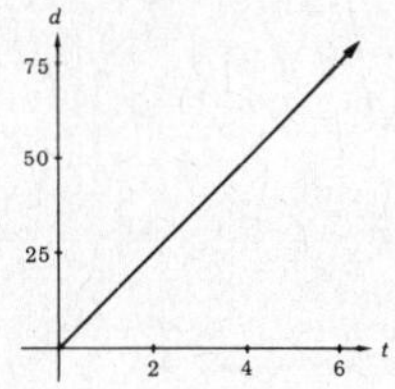

c. 87.5 m

3. a.

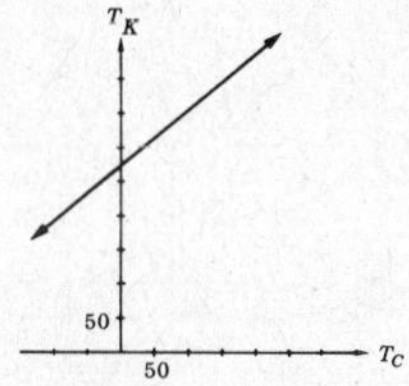

b. $293^\circ K$

5. a.

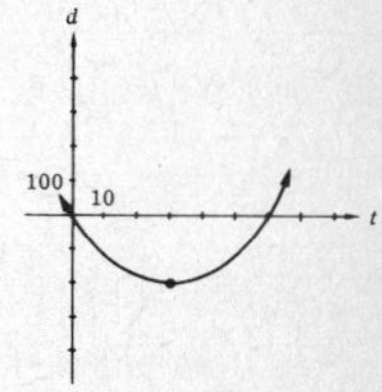

b. The portion in the 4th quadrant.
c. 200 ft *below* the surface.
d. 60 seconds

7. a. $A = x(450 - x)$ or $A = -x^2 + 450x$

b.

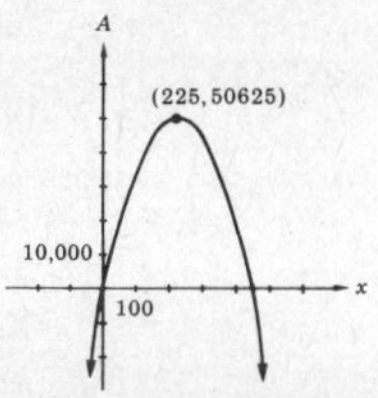

c. The portion in the first quadrant.
d. 225 by 225

9. a.

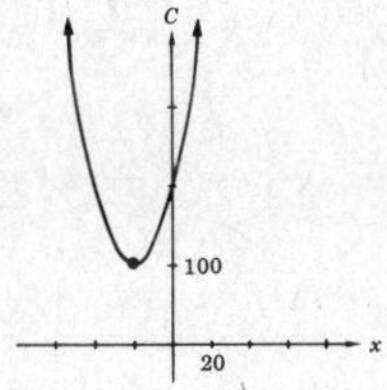

b. The portion in the first quadrant.

11. $w = kz^2$ 13. $z = kx^3y^2$ 15. $F = k\dfrac{Mm}{r^2}$

17. a. $P = ks$ b. $k = 4$ 19. a. $Q = \dfrac{k}{r^2}$
b. $k = .032$ c. $Q = \dfrac{.032}{r^2}$ d. .00355556
e. .00088889

21. a. $T = k\sqrt{L}$ b. T doubles, if length quadruples.
c. T increases by a factor of 1.0488, if length increases by a factor of 1.1. 23. a. $E = kv^3$
b. E increases by a factor of 8 if velocity doubles.
c. E increases by a factor of 27 if velocity triples.

Vocabulary/Symbol True-False Quiz

1. $\sqrt{61}$ 2. True 3. True 4. undefined 5. downward and has vertex (0,4) 6. (3,0) and (4,0)
7. True 8. True 9. True 10. True

Chapter Test

1. A function is a relation such that each domain element corresponds to exactly one range element. 2. 5
3. a. yes b. Domain $\{x|x \geqslant -3\}$, Range $\{y|y \geqslant 0\}$ 4. a. -6 b. $1 - 10x - 5h$ 5. $y = 2x - 7$
6. -1 7. $y = x + 7$

8.

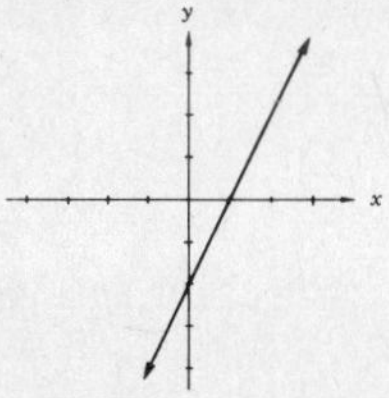

9.

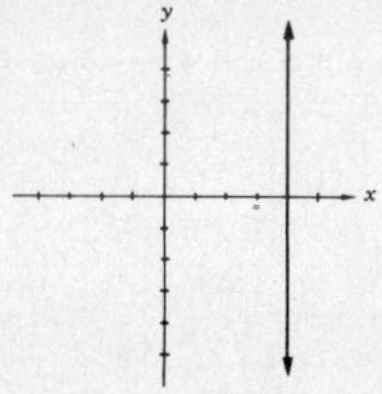

10.

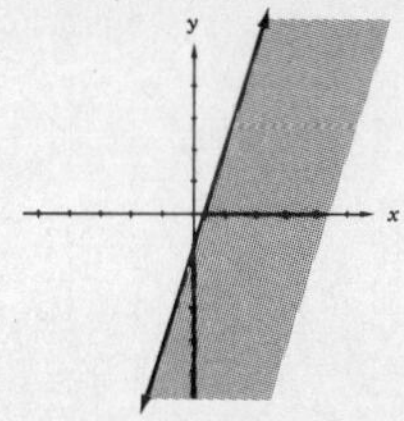

11. a. opens up; intercepts are (0,2), (−1,0); vertex (−1,0)

b.

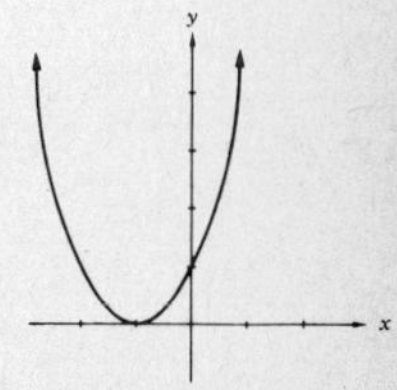

c. Domain $\mathcal{R}$
Range $\{y|y \geqslant 0\}$

12. a. 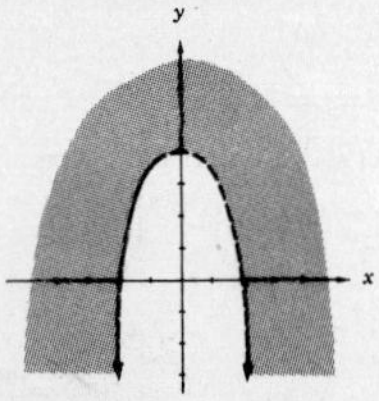

b. Domain $\mathcal{R}$
Range $\mathcal{R}$

13. a. 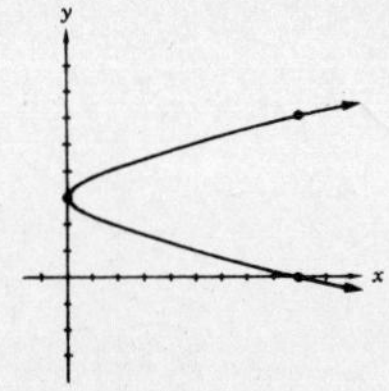

b. Domain $\{x | x \geqslant 0\}$
Range $\mathcal{R}$

14. a. $f^{-1}(x) = \frac{1}{2}x - \frac{3}{2}$

b. 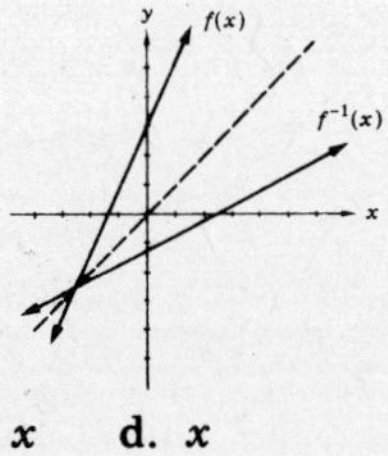

c. x d. x

15. a. $g^{-1}(x) = \pm\sqrt{x - 1}$

b. no

16. a. 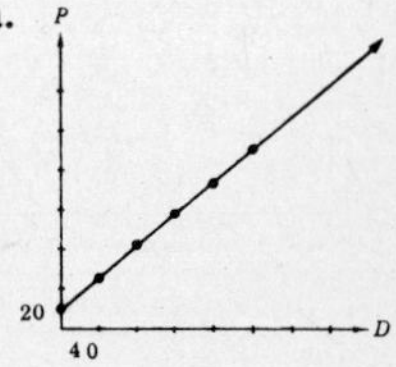

b. 95.7 lb/in^2

CHAPTER 4

Exercises ■ 4.1

1. a. 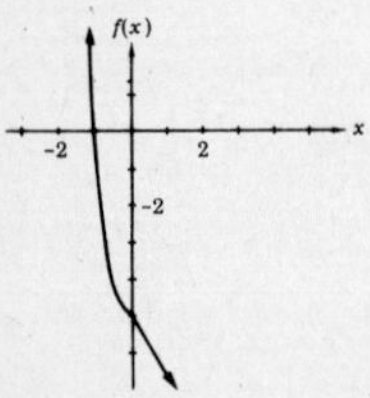

b. continuous

3. a. 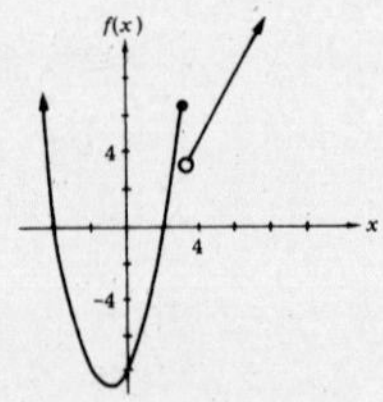

b. discontinuous at $x = 3$

5. a. 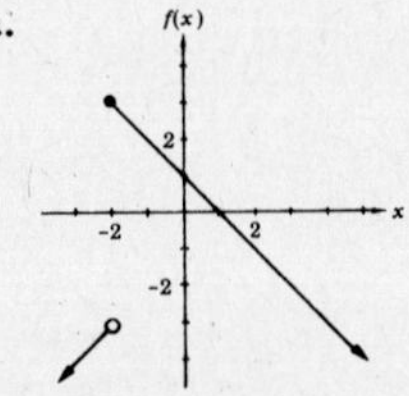

b. discontinuous at $x = -2$

7. a. 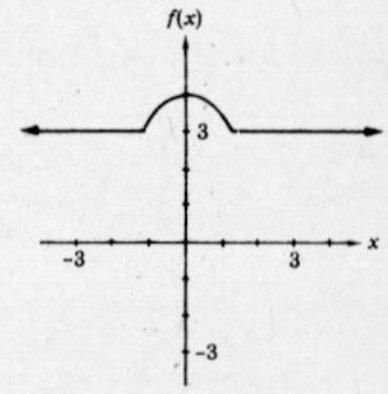

b. continuous

9. a. 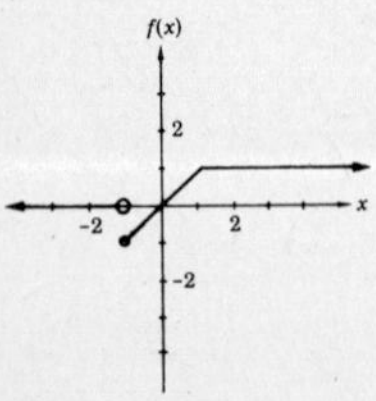

b. discontinuous at $x = -1$

11. a. 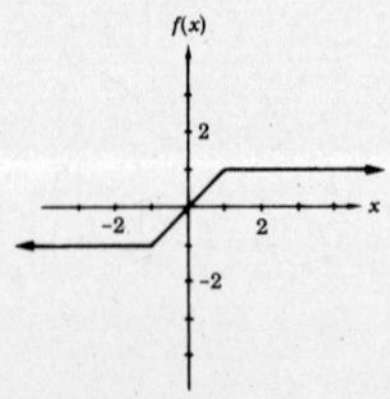

b. continuous

13. a. 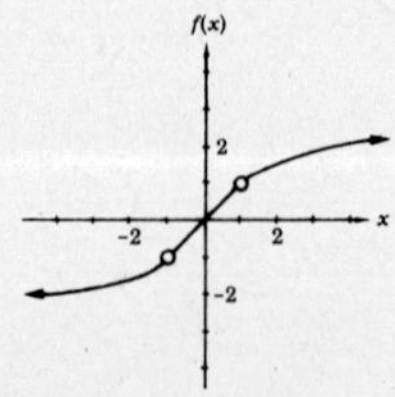

b. discontinuous at $x = \pm 1$

15. a. 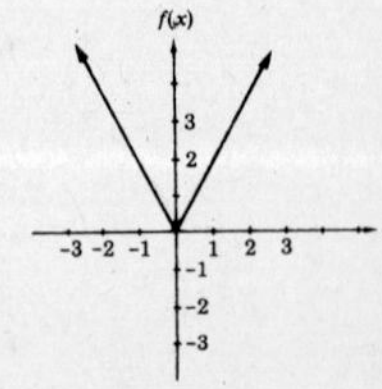

b. continuous

17. a. 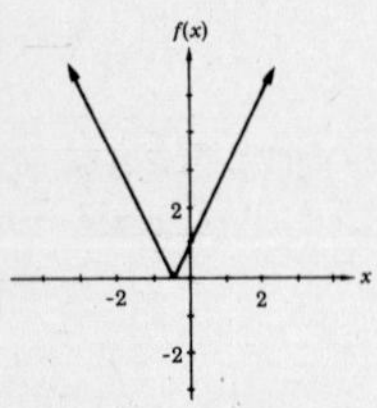

b. continuous

19. a. 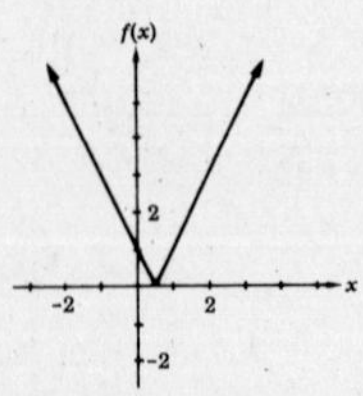

b. continuous

21. a. 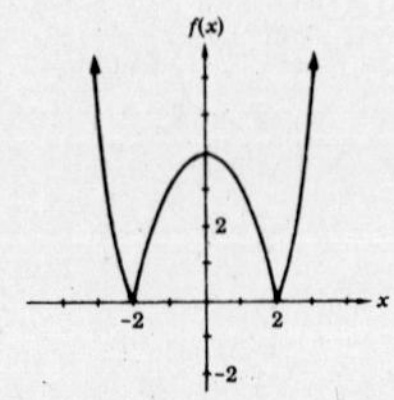

b. continuous

23. a. 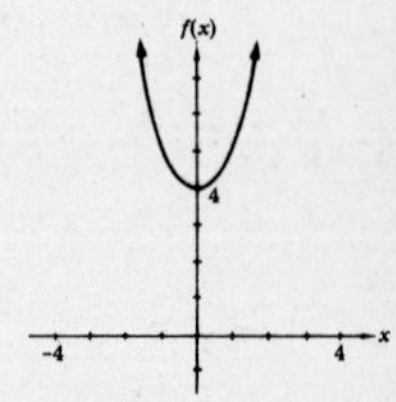

b. continuous

25. a. 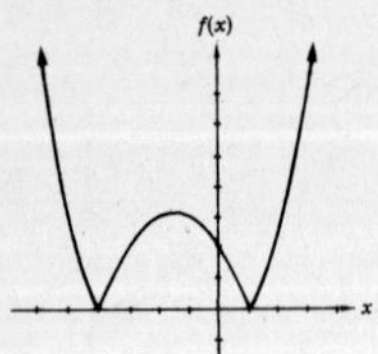

b. continuous

27. a. $\mathcal{R}$ b. $\{y | 3 \leqslant y \leqslant 4\}$

29. a. $\mathcal{R}$ b. $\{y | y \leqslant -1 \text{ or } 0 \leqslant y \leqslant 1\}$

31. 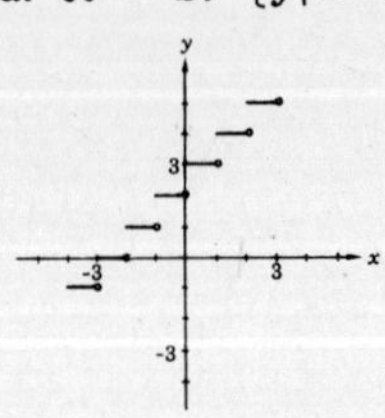

33. a. 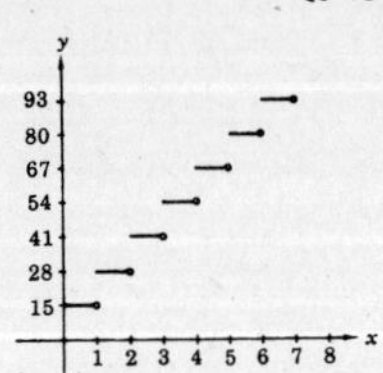

b. Domain $\{x | x \geqslant 0\}$
Range $\{y | y = 15 + 13n, n = 0, 1, 2, \ldots\}$

35. 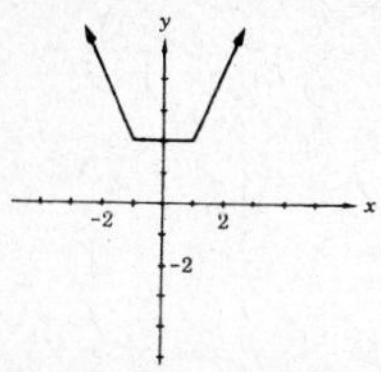

37. 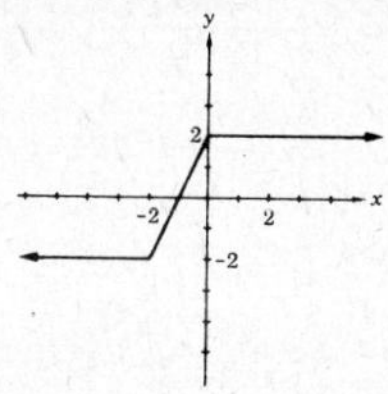

Exercises ■ 4.2

1. y-axis 3. origin 5. origin 7. y-axis 9. origin 11. y-axis 13. neither 15. neither
17. a. none b. $y = 0$ 19. a. $x = 5, x = -2$ b. $y = 0$ 21. a. $x = 2, x = -2$ b. $y = 1$
23. a. $x = 2, x = -2$ b. $y = -1$ 25. a. $x = 2, x = -2$ b. $y = 0$ 27. a. $x = 9$ b. $y = 1$
29. a. $x = 0$ b. $y = 1$ 31. y-axis, origin 33. a. yes b. yes c. yes 35. True 37. yes

Exercises ■ 4.3

1. 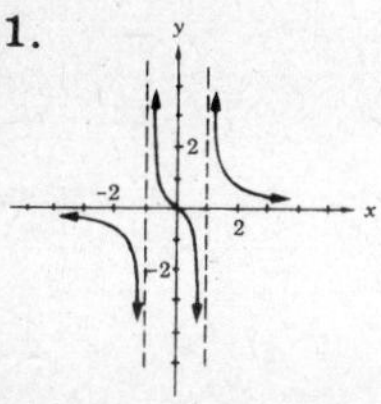

3. 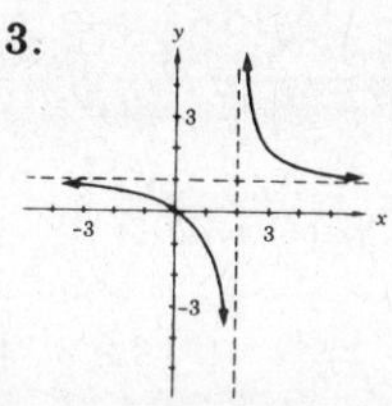

5. 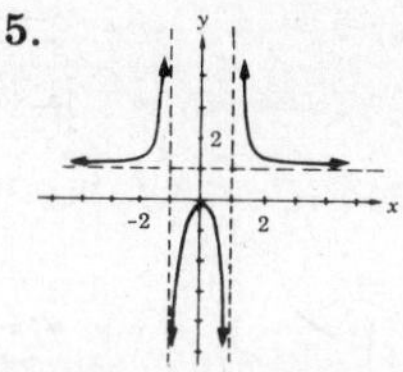

7. 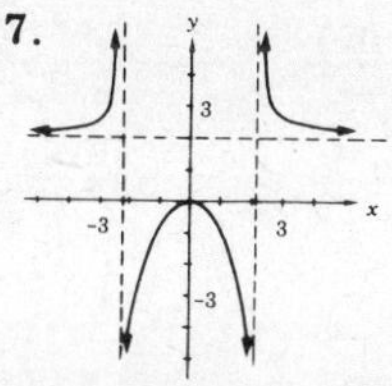

9. 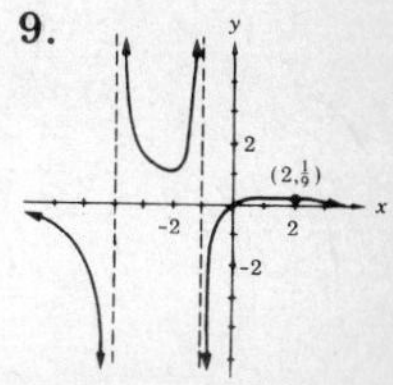

11. 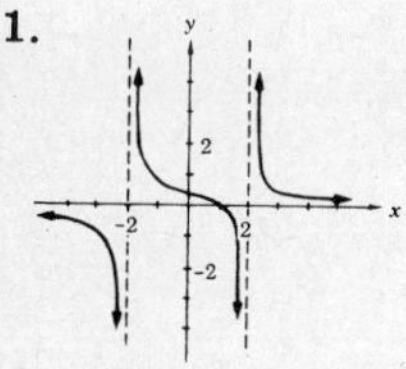

13. 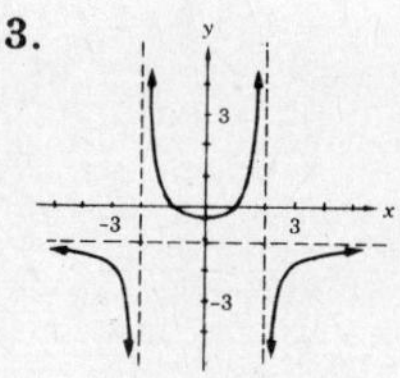

15. 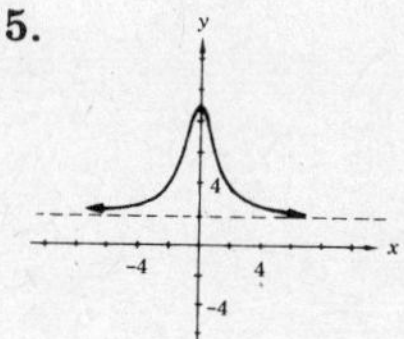

17. 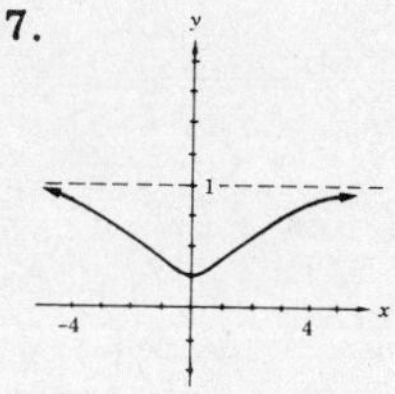

19. 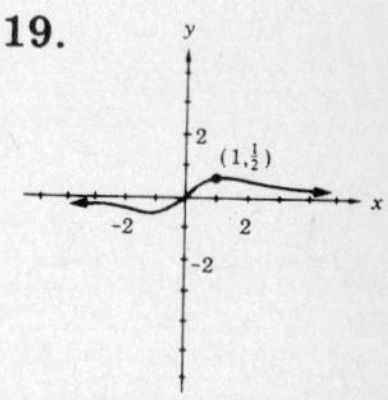

21. 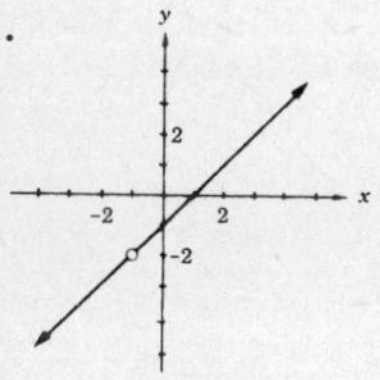

23. 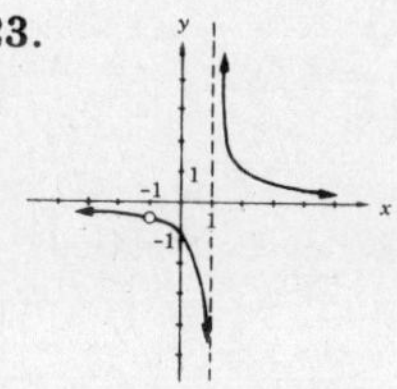

25. a. $\{y \mid y \leqslant 0 \text{ or } y > 1\}$
b. $\{y \mid y \geqslant 0\}$
c. $\{y \mid y \leqslant -\frac{4}{9} \text{ or } y > 0\}$

27. $\{y \mid y \neq -4\}$

29. 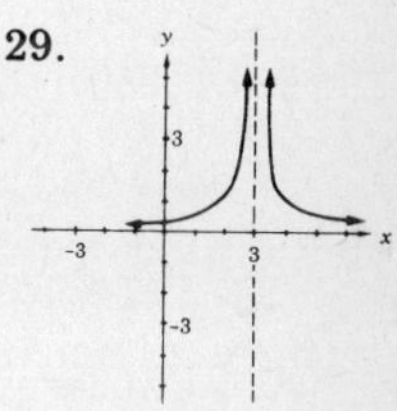

Exercises ■ 4.4

1. a. (3,2) b. 1
c. 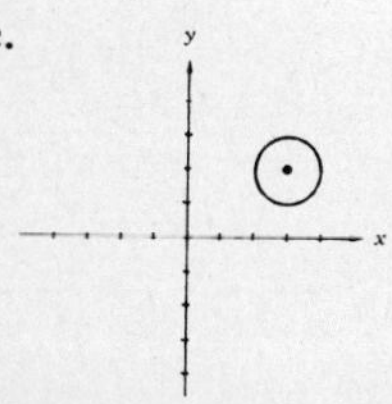
d. Domain $\{x \mid 2 \leqslant x \leqslant 4\}$
Range $\{y \mid 1 \leqslant y \leqslant 3\}$

3. a. (1,0) b. 3
c. 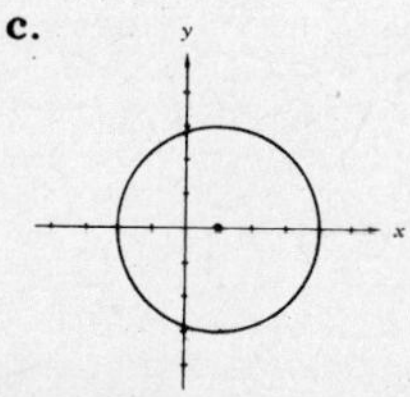
d. Domain $\{x \mid -2 \leqslant x \leqslant 4\}$
Range $\{y \mid -3 \leqslant y \leqslant 3\}$

5. a. (1,−2) b. 3
c. 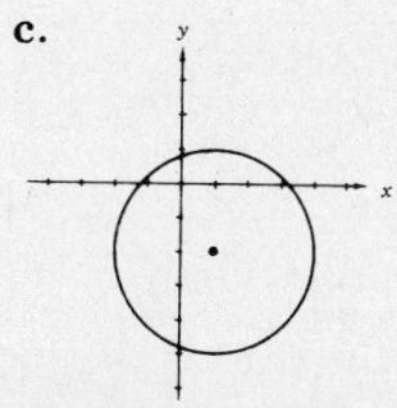
d. Domain $\{x \mid -2 \leqslant x \leqslant 4\}$
Range $\{y \mid -5 \leqslant y \leqslant 1\}$

7. a. (0,0) b. $\sqrt{3}$
c. 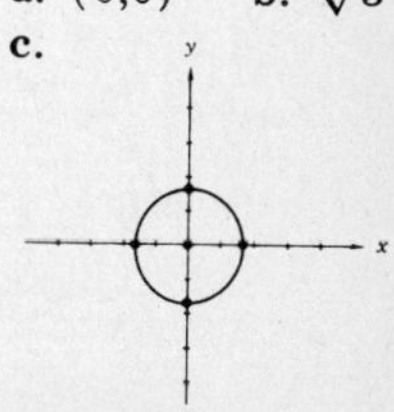
d. Domain $\{x \mid -\sqrt{3} \leqslant x \leqslant \sqrt{3}\}$
Range $\{y \mid -\sqrt{3} \leqslant y \leqslant \sqrt{3}\}$

9. a. (±4,0), (0,±5)
b.
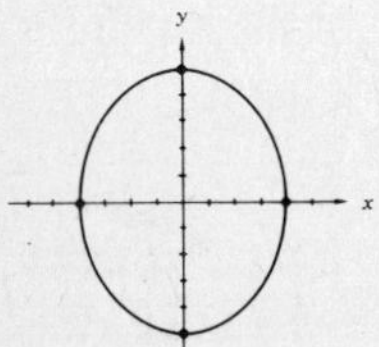
c. Domain
$\{x|-4 \leqslant x \leqslant 4\}$
Range
$\{y|-5 \leqslant y \leqslant 5\}$

11. a. (±2,0), (0,±7)
b.
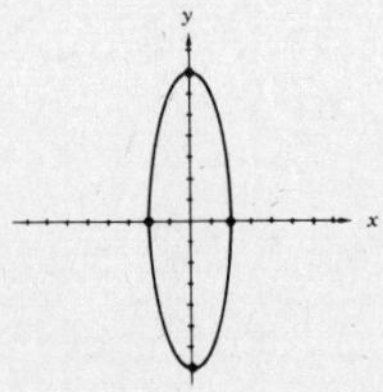
c. Domain
$\{x|-2 \leqslant x \leqslant 2\}$
Range
$\{y|-7 \leqslant y \leqslant 7\}$

13. a. (±3,0), (0,±1)
b.
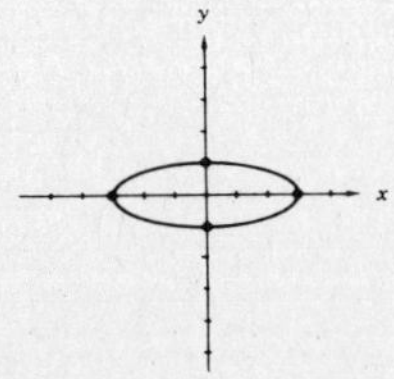
c. Domain
$\{x|-3 \leqslant x \leqslant 3\}$
Range
$\{y|-1 \leqslant y \leqslant 1\}$

15. a. $(\pm 2,0), (0,\pm\sqrt{2})$
b.
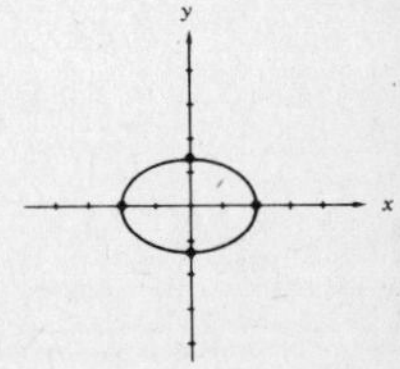
c. Domain
$\{x|-2 \leqslant x \leqslant 2\}$
Range
$\{y|-\sqrt{2} \leqslant y \leqslant \sqrt{2}\}$

17. a. (0,1) b. $y = -1$
c.
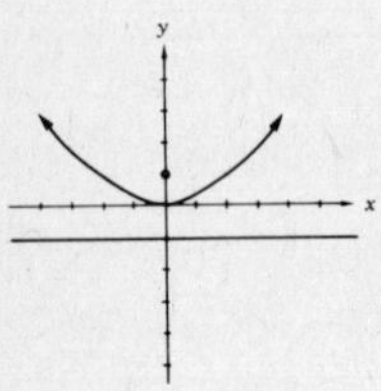

19. a. (1,0) b. $x = -1$
c.
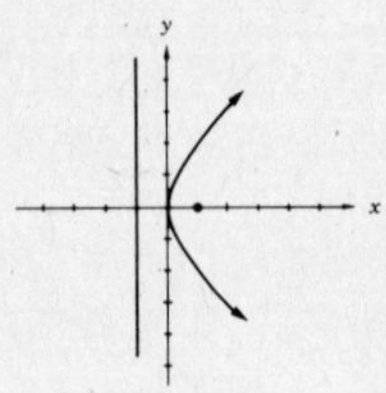

21. a. $(0,-\frac{1}{4})$ b. $y = \frac{1}{4}$
c.
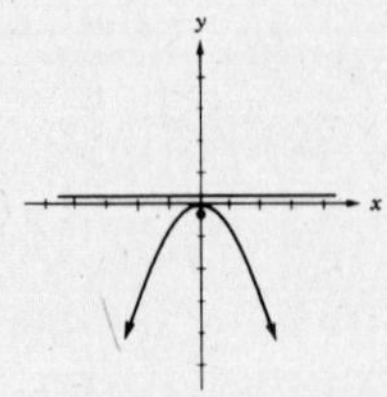

23. a. $(0,\frac{1}{2})$ b. $y = -\frac{1}{2}$
c.
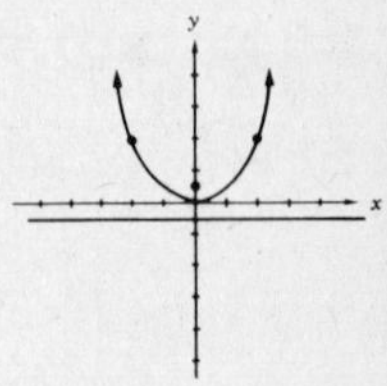

25. a. (±3,0)
b. $y = \pm\frac{2}{3}x$
c.
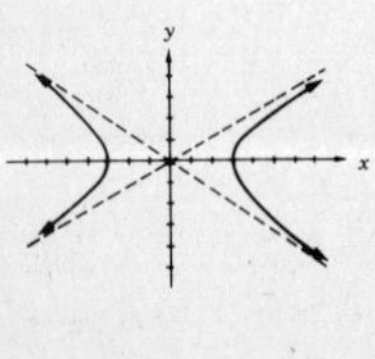
d. Domain
$\{x|x \leqslant -3 \text{ or } x \geqslant 3\}$
Range $\mathcal{R}$

27. a. (±1,0)
b. $y = \pm x$
c.
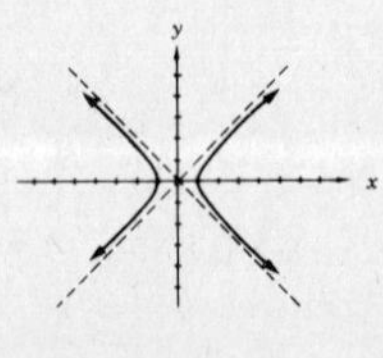
d. Domain
$\{x|x \leqslant -1 \text{ or } x \geqslant 1\}$
Range $\mathcal{R}$

29. a. (0,±3)
b. $y = \pm 3x$
c.
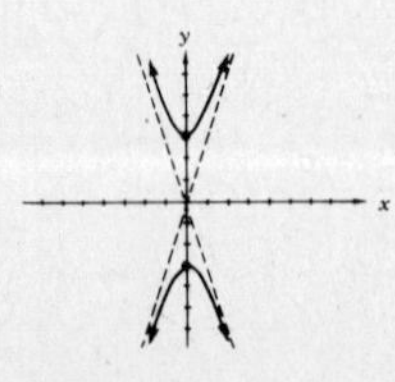
d. Domain $\mathcal{R}$
Range
$\{y|y \leqslant -3 \text{ or } y \geqslant 3\}$

31. a. (±4,0)
b. $y = \pm\frac{3}{4}x$
c.
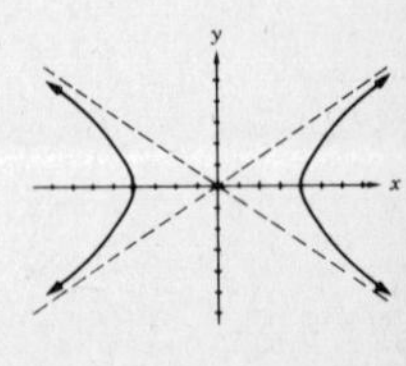
d. Domain
$\{x|x \leqslant -4 \text{ or } x \geqslant 4\}$
Range $\mathcal{R}$

33. $(x-2)^2 + (y-3)^2 = 5$; circle 35. $y = \frac{3}{2}x^2 - 3$; parabola 37. $x^2 + y^2 = 2$; circle

43. b.

x	$\frac{3}{2}\sqrt{x^2-4}$	$\frac{3}{2}x$
10	14.696938	15
20	29.849623	30
30	44.899889	45
100	149.97	150
1000	1499.997	1500
1,000,000	1,500,000	1,500,000

Vocabulary/Symbol True-False Quiz

1. True 2. True 3. True 4. True 5. True 6. $x = 2$ 7. values of $x > 4$ or $x < -4$
8. $(x-1)^2 + (y-2)^2 = 9$ 9. True 10. $y = \pm\frac{1}{2}x$

Chapter Test

1. a. 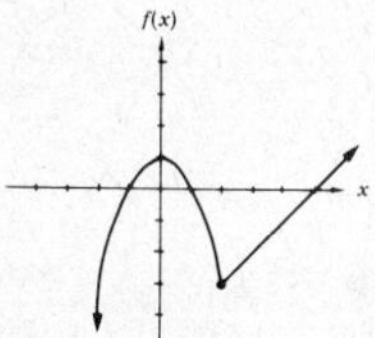

b. yes

2. 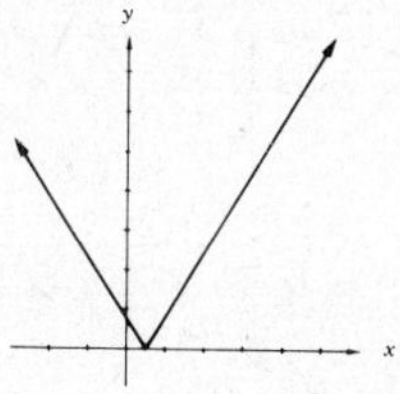

3. Domain $\mathcal{R}$
Range $\{y|y \geqslant 0\}$

4. a. (0,0) b. origin
c. $x = \pm 4$, $y = 0$
d. Domain $\{x|x \neq \pm 4\}$
Range $\mathcal{R}$
e. $x < -4$: $f(x)$ below x-axis
$-4 < x < 0$: $f(x)$ above x-axis
$0 < x < 4$; $f(x)$ below x-axis
$x > 4$: $f(x)$ above x-axis
f. 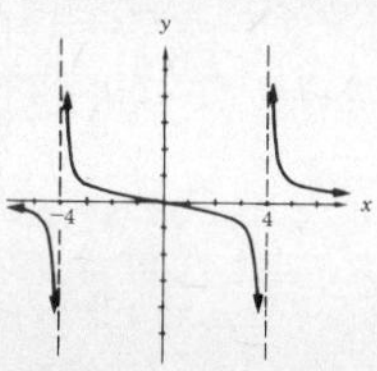

5. a. $(0, -\frac{1}{16})$ b. y-axis
c. $x = \pm 4$, $y = 0$
d. Domain $\{x|x \neq \pm 4\}$
Range $\{y|y > 0 \text{ or } y \leqslant -\frac{1}{16}\}$
e. $x < -4$: $f(x)$ above x-axis
$-4 < x < 4$: $f(x)$ below x-axis
$x > 4$: $f(x)$ above x-axis
f. 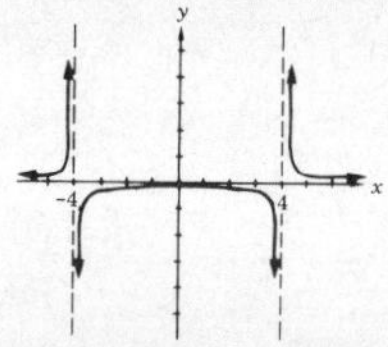

6. a. (0,0) b. y-axis
c. $x = \pm 4$, $y = 1$
d. Domain $\{x|x \neq \pm 4\}$
Range $\{y|y > 1 \text{ or } y \leqslant 0\}$
e. $x < -4$: $f(x)$ above x-axis
$-4 < x < 4$: $f(x)$ below x-axis
$x > 4$: $f(x)$ above x-axis
f. 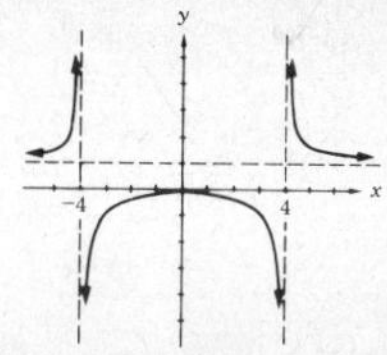

7. a. (0,0) b. origin
c. $y = 0$
d. Domain $\mathcal{R}$
Range $\{y|-\frac{1}{8} \leqslant y \leqslant \frac{1}{8}\}$
e. $x < 0$: $f(x)$ below x-axis
$x > 0$: $f(x)$ above x-axis
f. 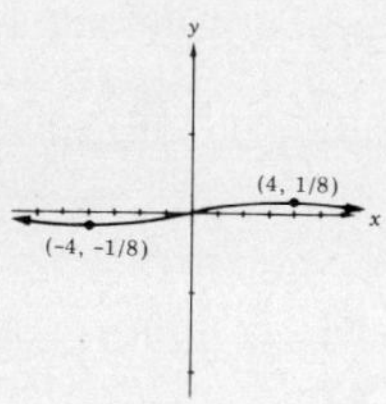

8. a. $(4,0)$, $(0,\frac{1}{3})$
b. none apparent
c. $x = -4$, $x = 3$, $y = 0$
d. Domain $\{x|x \neq -4 \text{ or } x \neq 3\}$
Range $\left\{y|y \leqslant \dfrac{9 - 4\sqrt{2}}{49} \text{ or } y \geqslant \dfrac{9 + 4\sqrt{2}}{49}\right\}$
e. $x < -4$: $f(x)$ below x-axis
$-4 < x < 3$: $f(x)$ above x-axis
$3 < x < 4$: $f(x)$ below x-axis
$x > 4$: $f(x)$ above x-axis
f. 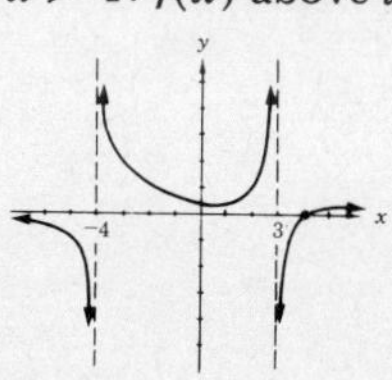

9. a. $(0, -\frac{1}{3})$
b. none apparent
c. $x = 3$, $y = 0$
d. Domain $\{x|x \neq -4 \text{ or } x \neq 3\}$
Range $\{y|y \neq 0 \text{ or } y \neq -\frac{1}{7}\}$
e. $x < 3$: $f(x)$ below x-axis
$x > 3$: $f(x)$ above x-axis
f. 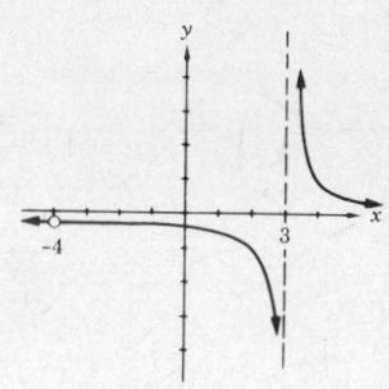

10. 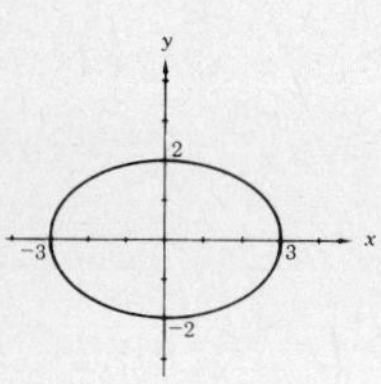

11. 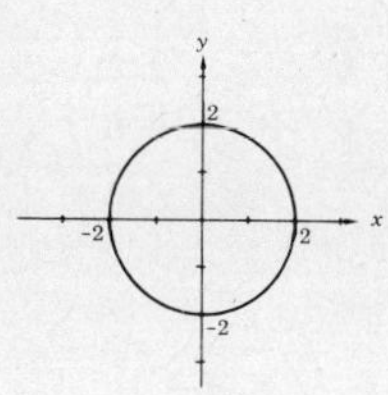

12.

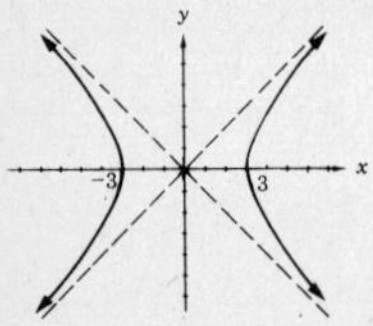

13.

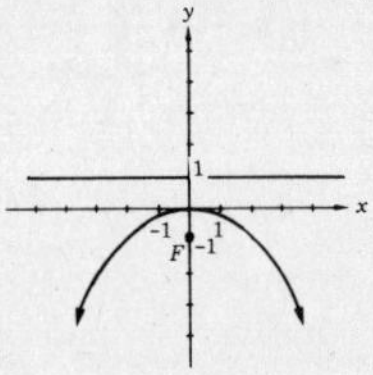

14.

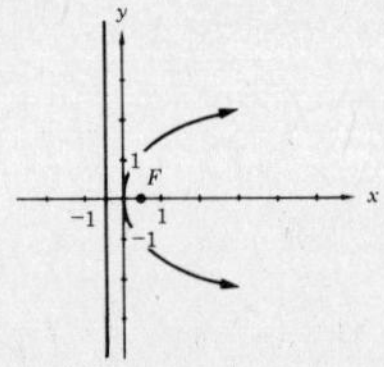

15.

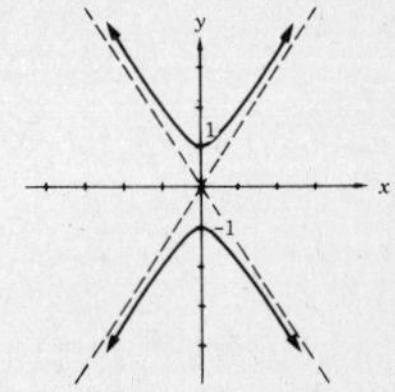

CHAPTER 5

Exercises ■ 5.1

1. 1 **3.** 512 **5.** 2 **7.** $\frac{1}{8}$ **9.** $\sqrt{2}$

11. a.

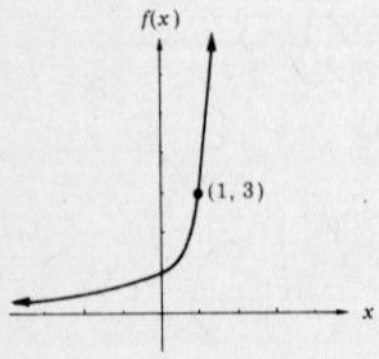

b. Domain $x \in \mathcal{R}$
Range $y > 0$
c. as $x \to -\infty$, $y \to 0$
as $x \to +\infty$, $y \to +\infty$

13. a.

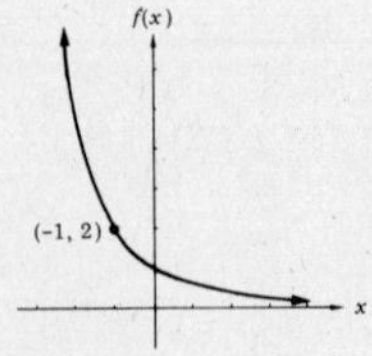

b. Domain $x \in \mathcal{R}$
Range $y > 0$
c. as $x \to -\infty$, $y \to +\infty$
as $x \to +\infty$, $y \to 0$

15. a.

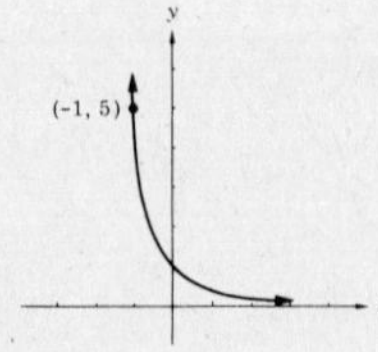

b. Domain $x \in \mathcal{R}$
Range $y > 0$
c. as $x \to -\infty$, $y \to +\infty$
as $x \to +\infty$, $y \to 0$

17. a.

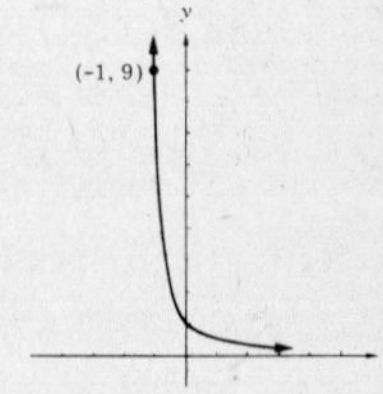

b. Domain $x \in \mathcal{R}$
Range $y > 0$
c. as $x \to -\infty$, $y \to +\infty$
as $x \to +\infty$, $y \to 0$

19. a.

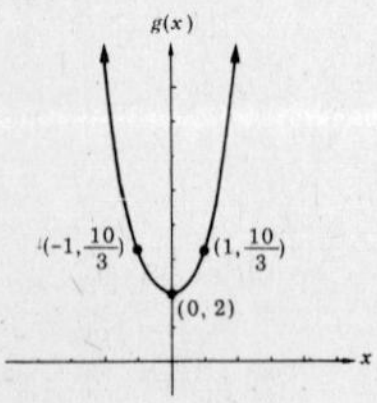

b. Domain $x \in \mathcal{R}$
Range $y \geqslant 2$
c. as $x \to -\infty$, $y \to +\infty$
as $x \to +\infty$, $y \to +\infty$

21. a.

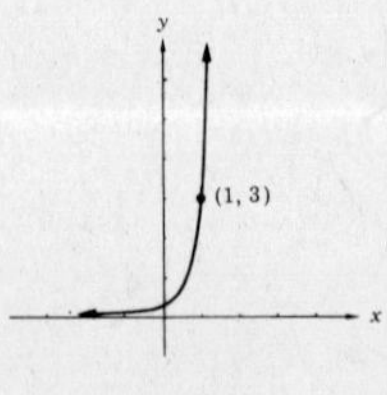

b. Domain $x \in \mathcal{R}$
Range $y > 0$
a. as $x \to -\infty$, $y \to 0$
as $x \to +\infty$, $y \to +\infty$

23. a.

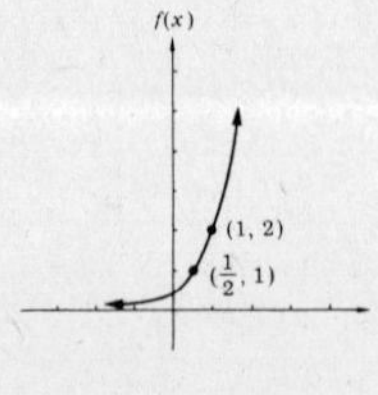

b. Domain $x \in \mathcal{R}$
Range $y > 0$
c. as $x \to -\infty$, $y \to 0$
as $x \to +\infty$, $y \to +\infty$

25. a.

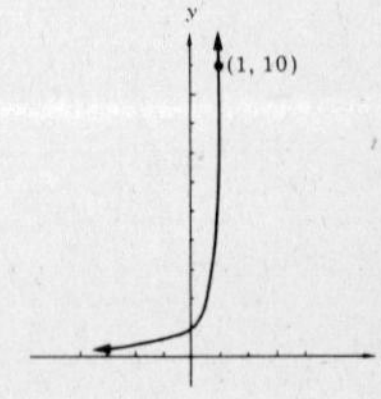

b. Domain $x \in \mathcal{R}$
Range $y > 0$
c. as $x \to -\infty$, $y \to 0$
as $x \to +\infty$, $y \to +\infty$

27. a.

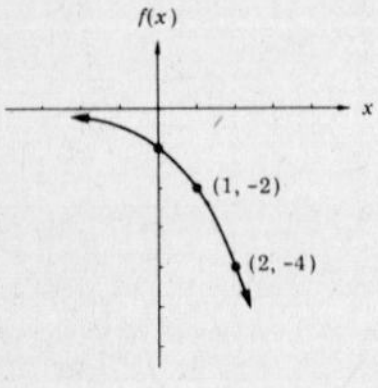

b. Domain $x \in \mathcal{R}$
Range $y < 0$
c. as $x \to -\infty$, $y \to 0$
as $x \to +\infty$, $y \to -\infty$

29. a.

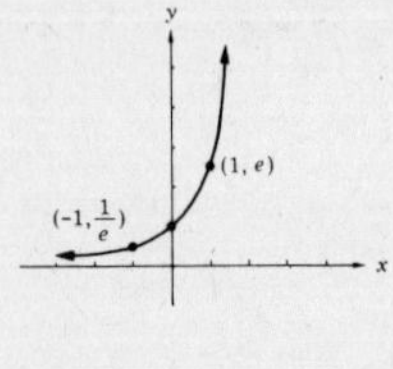

b. Domain $x \in \mathcal{R}$
Range $y > 0$
c. as $x \to -\infty$, $y \to 0$
as $x \to +\infty$, $y \to +\infty$

31. 5 **33.** −2 **35.** 0 **37.** 0 **39.** 2

45.

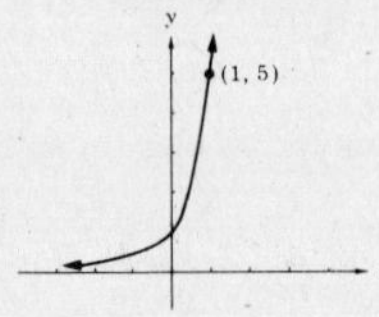

47.

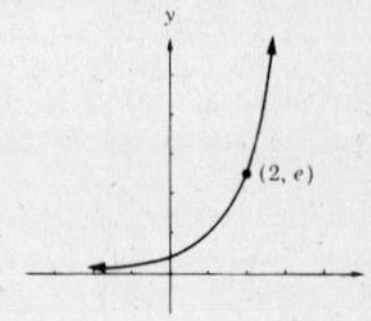

49.
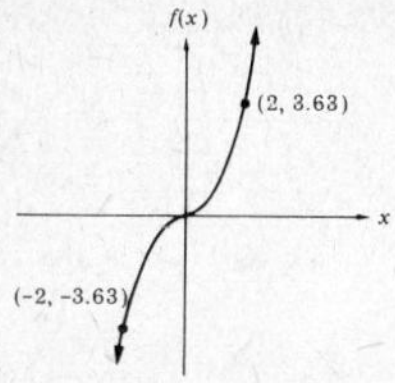

51.
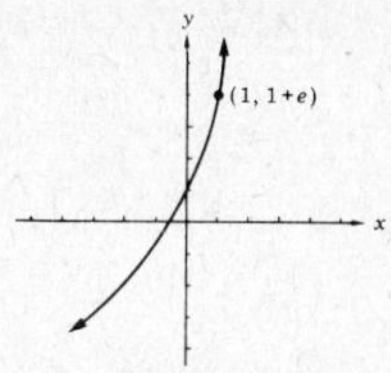

53.
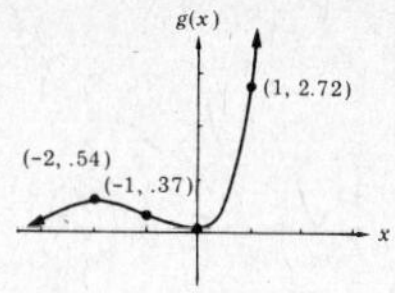

55. e^{π}

Exercises ■ 5.2

1. $\log_2 32 = 5$ **3.** $\log_9 3 = \frac{1}{2}$ **5.** $\log_{.1} .01 = 2$ **7.** $\log_7 49 = 2$ **9.** $\log_{10} 1000 = 3$ **11.** $2^5 = 32$
13. $(\frac{1}{5})^Q = 5$ **15.** $10^3 = 1000$ **17.** $(b)^P = x - y$ **19.** $e^M = 7$

21. a.
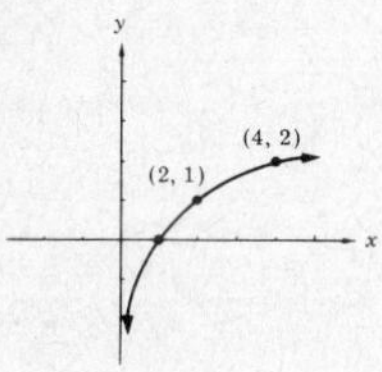

b. Domain $x > 0$
Range $y \in \mathcal{R}$
c. as $x \to 0^+$, $y \to -\infty$
as $x \to +\infty$, $y \to +\infty$

23. a.
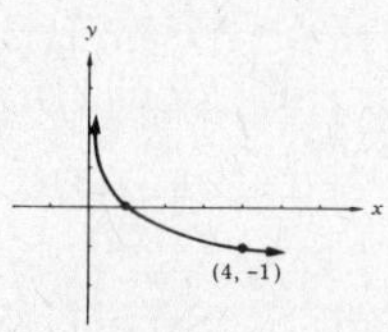

b. Domain $x > 0$
Range $y \in \mathcal{R}$
c. as $x \to 0^+$, $y \to +\infty$
as $x \to +\infty$, $y \to -\infty$

25. a.
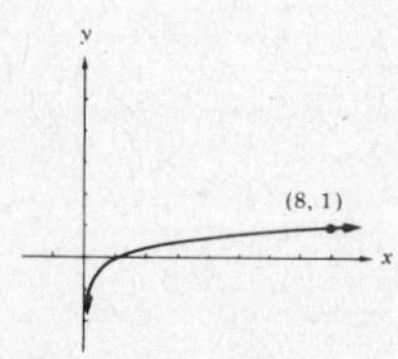

b. Domain $x > 0$
Range $y \in \mathcal{R}$
c. as $x \to 0^+$, $y \to -\infty$
as $x \to +\infty$, $y \to +\infty$

27. a.
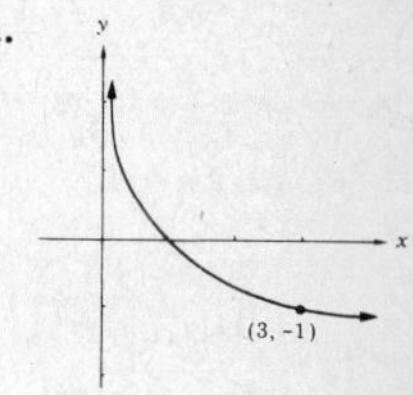

b. Domain $x > 0$
Range $y \in \mathcal{R}$
c. as $x \to 0^+$, $y \to +\infty$
as $x \to +\infty$, $y \to -\infty$

29. a.
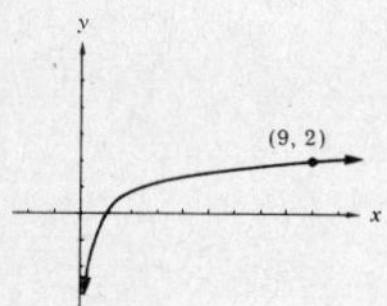

b. Domain $x > 0$
Range $y \in \mathcal{R}$
c. as $x \to 0^+$, $y \to -\infty$
as $x \to +\infty$, $y \to +\infty$

31.
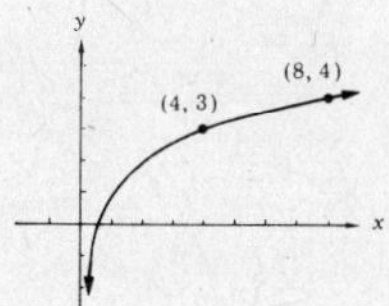

33.
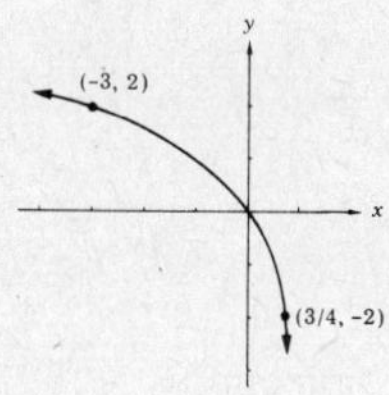

35.
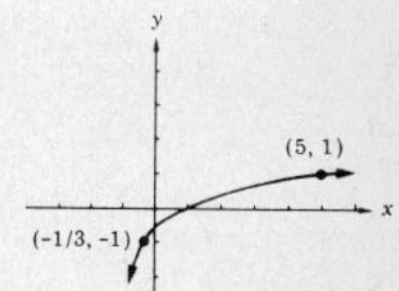

37.
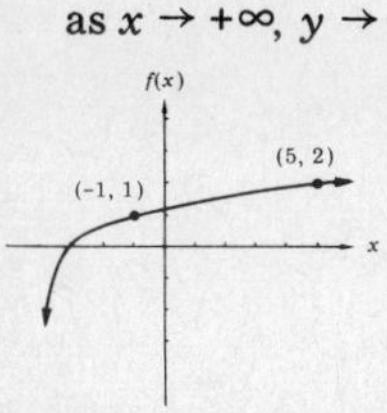

39.
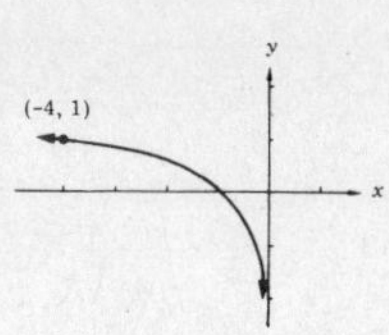

41.
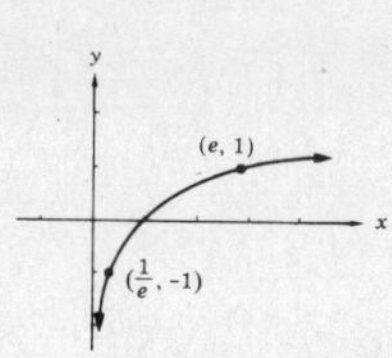

43. 0 **45.** .8293
47. .4418 **49.** .0969
51. .8893

59.
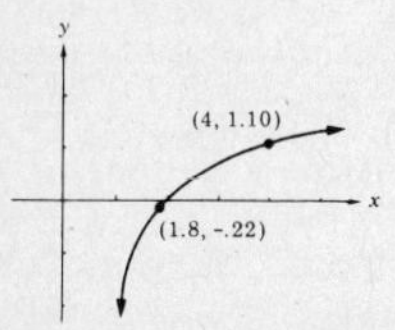

61.
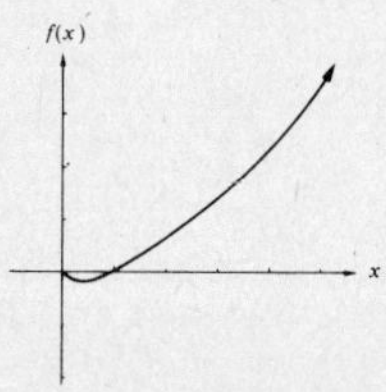

63.
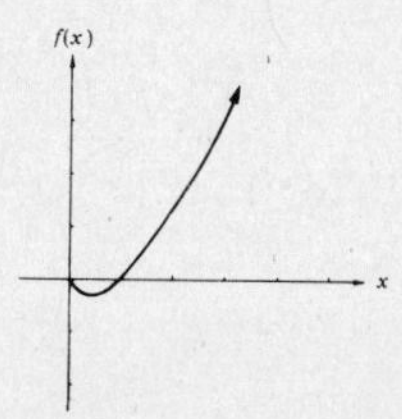

65.
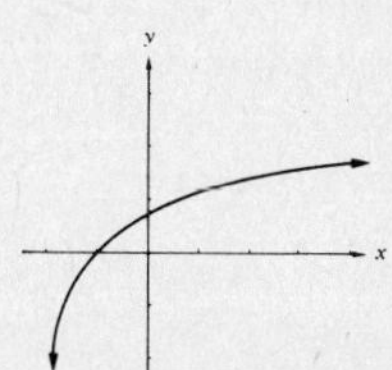

67.
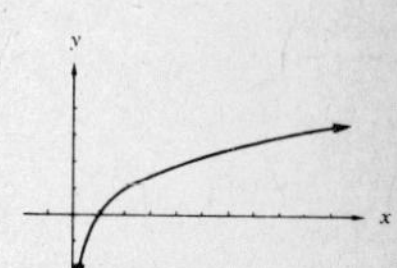

Exercises ■ 5.3

3. $1 + \log_2 7$ 5. $\log_b \pi + 2\log_b r$ 7. cannot be expanded 9. $\log_8 x - 2\log_8 y$ 11. $\frac{1}{3}\log_b(x+h) - \frac{1}{3}\log_b x$
13. $-\log_b a$ 15. $(x+1)\log_e x$ 17. $\log_{10}2 + \log_{10}(x-y)$ 19. $\frac{2}{3}\log_5 2 + \frac{2}{3}\log_5 x + \frac{1}{3}\log_5 y$
21. $\log_2 \frac{3}{t}$ 23. $\log_b \frac{\sqrt{x}}{10y}$ 25. cannot be written as a single logarithm 27. $\log_3 \frac{x}{yz}$
29. cannot be written as a single logarithm 31. $\log_6 \sqrt[3]{3x}$ 33. $\log_8 2$
35. cannot be written as a single logarithm 37. $\ln \frac{x}{2}$ 39. $\log_2 7$ 41. 1.40 43. 1.19 45. .83
47. 2.11 49. 3.77 51. 7.96 53. .65 55. 12.17 57. .63 59. .78 61. .28 63. 8.17
65. 2.30 67. 6.90 69. ±31.62 73. 2.30 75. 6.91

Exercises ■ 5.4

1. .91855453 3. 8.9185545 5. .9185545 − 8 7. .30319606 9. .8733206 − 1 11. 1.8733206
13. 1.447158 15. 6 17. 845,000 19. .845 21. .306 23. .000381 25. 10,100
27. 9.7383052 29. $2.8763 \cdot 10^{36}$ 31. 9.7762048 33. 10.488088 35. .37014285 37. 4.8009919
39. −2.5355105 41. .04571406 43. .99995657 45. 2.8660509 47. 791.40721
49. $5.5335 \cdot 10^{-10}$ 51. .00002183 53. 24.507556 55. 101.62487 57. no answer

Exercises ■ 5.5

1. 1.3862944 3. 5.9914645 5. 8.3710107 7. 4.3228073 9. −.14003 11. −.29169009
13. 5.4638318 15. −5.381699 17. 1.7749524 19. 230 21. 9.0403691 23. 3.9499385
25. 19.999955 27. 7.9248231 29. .30071269 31. .65044405 33. .08802803 35. .00025621
37. 7.7399869 39. 10.000149 41. 1.9036539 43. .97826739 45. $1.0995 \cdot 10^{12}$
47. 1.9520737 49. 1096.6332 51. no answer

Exercises ■ 5.6

1. 2.5113916 3. −.150515 5. .27751627 7. −1.3624876 9. 2.7095113 11. ±1.1413085
13. −2.113012 15. −2.4849067 17. .81767844 19. 2.7095113 21. −1.5 23. 0
25. 2.6234637 27. .80226081 29. ±1.2999402 31. $x = \ln(y + \sqrt{y^2+1})$ 33. $x = \frac{1}{2}\ln\left(\frac{y+1}{y-1}\right)$
35. 98 37. $e^2 - 2$ 39. $0, \frac{1}{2}$ 41. 25.148669 43. 3 45. e^e 47. .0001 49. 22

Exercises ■ 5.7

1. a. \$431.79 b. 9.000646 years (in reality, 10 years) 3. a. \$441.61 b. 8.75069 years (in reality, 9 years) 5. a. \$445.11 b. 8.66433 years 7. 27.465307 years 9. 11,440 years 11. a. 6 b. $1.58 \cdot 10^7 A_0$ 13. a. $2.7 \cdot 10^8$ b. 110 years 15. a. 819,200 b. $\frac{2}{3}$ hour

17. a. $K = .0016653$

b. $N = \dfrac{45}{3 + 12e^{-.02497956t}}$

c. 6.07 billion

d.

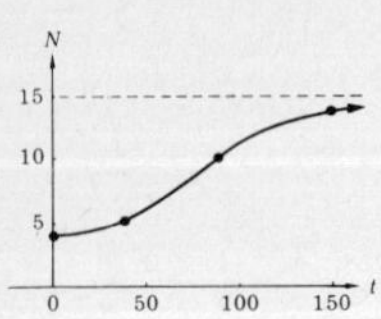

19. a.

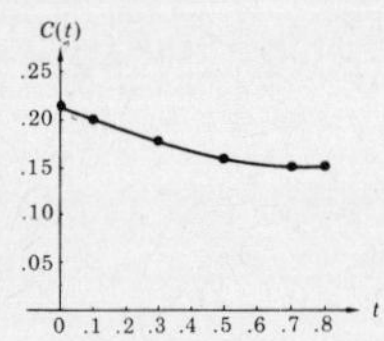

b. −39.95%

21. a. 7 b. 6.26 c. $2 \cdot 10^{-8}$ moles per liter

Vocabulary/Symbol True-False Quiz

1. $b > 0, b \neq 1$ 2. True 3. True 4. True 5. $\{x|x > 3\}$ 6. True 7. True 8. True 9. True 10. True

Chapter Test

1. 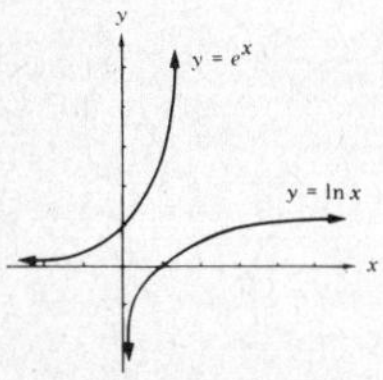

2. a. 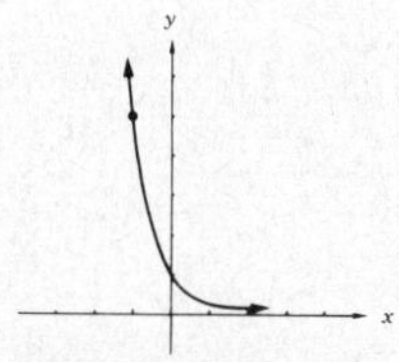

b. Domain $x \in \mathbb{R}$
Range $y > 0$

3. a. 1.7715568 b. .26573946 c. −.7212464
4. $10^y = x^2 - 1$ 5. 2.5053176 6. $\log_3 \frac{xy}{z}$
7. .01229986 8. $\frac{2}{3}$ 9. $5^{\sqrt{3}}$ 10. .39079374
11. 0, .63092975 12. 3, 5 13. $z = \ln(y + \sqrt{y^2 - 1})$
14. 29.638584 15. a. \$1349.66
b. 26.836906 years (in reality, 26 years and 302 days)

CHAPTER 6

Exercises ■ 6.1

1. $\frac{\pi}{4}$ 3. $\frac{3\pi}{4}$ 5. $\frac{5\pi}{4}$ 7. $\frac{7\pi}{4}$ 9. $\frac{\pi}{6}$ 11. $\frac{2\pi}{3}$ 13. $\frac{4\pi}{3}$ 15. $\frac{5\pi}{9}$ 17. $\frac{7\pi}{9}$ 19. 135° 21. 225° 23. 150° 25. 15° 27. 75° 29. 120° 31. 210° 33. 180° 35. 360° 37. 114.6° 39. 343.8° 41. 229.2° 43. 57.3° 45. 143.2° 47. 3.14 in 49. 5π cm 51. $\frac{9\pi}{2}$ cm 53. 16 in 55. 5 in 57. $\frac{\pi}{3}$ in^2 59. $\frac{5}{36}$ in^2 61. $\frac{5}{2}$ in^2 63. 76° 42′ 65. 76°43′12″ 67. 30.239° 69. 10°7′30″ 71. 20° 73. 19°30′ 75. 6° 77. 90° 79. 45° 81. 139°39′50″

Exercises ■ 6.2

1. 1 and $\sqrt{2}$ 3. 3 and $3\sqrt{2}$ 5. $2\sqrt{2}$ and $2\sqrt{2}$ 7. $\sqrt{2}$ and $\sqrt{2}$ 9. $\frac{3}{2}\sqrt{2}$ and $\frac{3}{2}\sqrt{2}$ 11. 2 and $2\sqrt{3}$ 13. $\frac{4}{3}\sqrt{3}$ and $\frac{8}{3}\sqrt{3}$ 15. $\frac{3}{2}$ and $\frac{3}{2}\sqrt{3}$ 17. $2x$ and $x\sqrt{3}$ 19. $\frac{a}{d}$ 21. $\frac{p}{r}$ 23. $\frac{y}{x}$ 25. $\frac{a}{c}$ 27. $\frac{z}{y}$ 29. $\sin 90° = 1$, $\cos 90° = 0$, $\tan 90°$ = undefined, $\cot 90° = 0$, $\sec 90°$ = undefined, $\csc 90° = 1$ 31. .133986 33. 2.366201 35. .984807 37. .920504 39. .176326 41. 1 43. 5.23 cm 45. $\measuredangle B = 50°$, $a = 1.7$, $c = 2.6$ 47. $\measuredangle A = 40°$, $a = 3.9$, $b = 4.6$ 49. $g = 5$, $\measuredangle D = 36.9°$, $\measuredangle F = 53.1°$ 51. $\measuredangle C = 68.8°$, $d = 8.3$, $c = 7.7$ 53. $\cos\theta = \frac{4}{5}$, $\tan\theta = \frac{3}{4}$, $\cot\theta = \frac{4}{3}$, $\sec\theta = \frac{5}{4}$, $\csc\theta = \frac{5}{3}$ 55. $\sin\theta = \frac{3}{\sqrt{34}}$, $\cos\theta = \frac{5}{\sqrt{34}}$, $\cot\theta = \frac{5}{3}$, $\sec\theta = \frac{\sqrt{34}}{5}$, $\csc\theta = \frac{\sqrt{34}}{3}$ 57. $\cos\theta = \sqrt{1-y^2}$, $\tan\theta = \frac{y}{\sqrt{1-y^2}}$, $\cot\theta = \frac{\sqrt{1-y^2}}{y}$, $\sec\theta = \frac{1}{\sqrt{1-y^2}}$, $\csc\theta = \frac{1}{y}$ 59. $\sin\theta = \frac{1}{\sqrt{a^2+1}}$, $\cos\theta = \frac{a}{\sqrt{a^2+1}}$, $\tan\theta = \frac{1}{a}$, $\sec\theta = \frac{\sqrt{a^2+1}}{a}$, $\csc\theta = \sqrt{a^2+1}$ 61. .6773 63. .7357 65. 1.6963 67. 1.1608 69. .0131 71. .6772613 73. .73574258 75. 1.6963088 77. 1.1608309 79. .0130896

Exercises ■ 6.3

1. $s = -\frac{\pi}{4}$, $x = \frac{\sqrt{2}}{2}$, $y = -\frac{\sqrt{2}}{2}$ 3. $s = -\frac{3\pi}{4}$, $x = -\frac{\sqrt{2}}{2}$, $y = -\frac{\sqrt{2}}{2}$ 5. $s = -\frac{5\pi}{4}$, $x = -\frac{\sqrt{2}}{2}$, $y = \frac{\sqrt{2}}{2}$ 7. $s = -\frac{7\pi}{4}$, $x = \frac{\sqrt{2}}{2}$, $y = \frac{\sqrt{2}}{2}$

9. a.

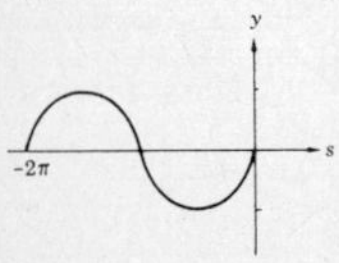

b.

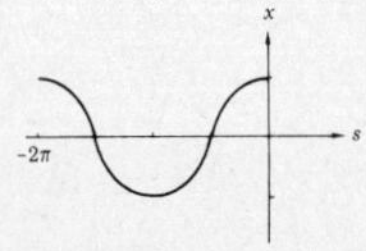

13. a.

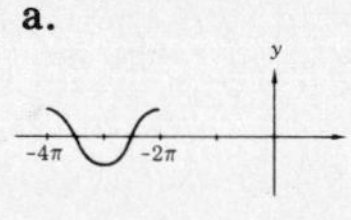

b. 1

15. a.

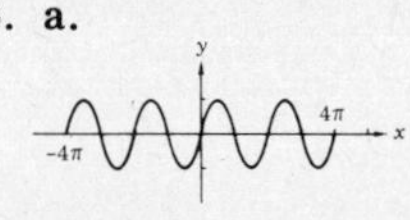

b. 4

17. a.

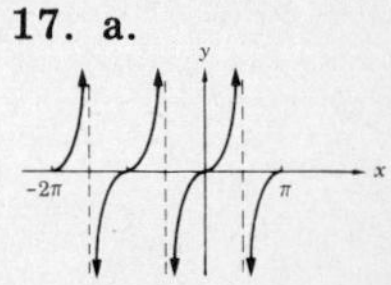

b. 3

19. a.

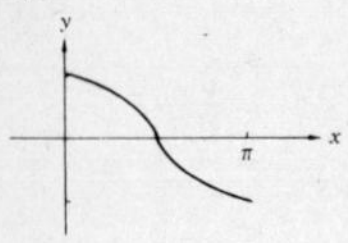

b. $\frac{1}{2}$

21. $x = \frac{k\pi}{2}$; k is any integer

23. $x = \frac{(2k+1)\pi}{6}$; k is any integer

25. $x = \frac{k\pi}{3}$; k is any integer

27. $w = \frac{k\pi}{6}$; k is any integer

29. $y = \frac{\pi}{2}\left(k - \frac{1}{2}\right)$; k is any integer

31. $x = \frac{\pi}{2}(k+1)$; k is any integer

33. $x = \frac{\pi}{2}\left(k - \frac{1}{2}\right)$; k is any integer

35. $x = \frac{\pi}{2}\left(k + \frac{3}{2}\right)$; k is any integer

37. $k = \frac{\pi}{2}(k+1)$; k is any integer

39. $\theta = \frac{\pi}{4}\left(k - \frac{3}{2}\right)$; k is any integer

41.

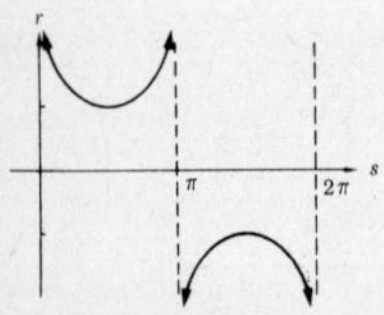

43.

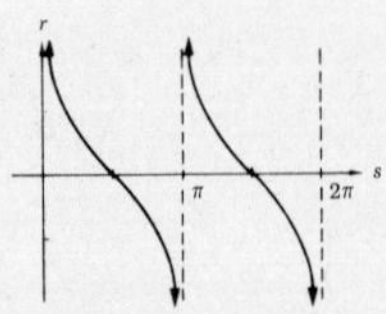

Exercises ■ 6.4

1. $\sin\theta = \frac{3\sqrt{10}}{10}$, $\cos\theta = \frac{\sqrt{10}}{10}$, $\tan\theta = 3$, $\csc\theta = \frac{\sqrt{10}}{3}$, $\sec\theta = \sqrt{10}$, $\cot\theta = \frac{1}{3}$

3. $\sin\theta = \frac{3\sqrt{10}}{10}$, $\cos\theta = \frac{\sqrt{10}}{10}$, $\tan\theta = 3$, $\csc\theta = \frac{\sqrt{10}}{3}$, $\sec\theta = \sqrt{10}$, $\cot\theta = \frac{1}{3}$

5. $\sin\theta = \frac{9\sqrt{82}}{82}$, $\cos\theta = \frac{\sqrt{82}}{82}$, $\tan\theta = 9$, $\csc\theta = \frac{\sqrt{82}}{9}$, $\sec\theta = \sqrt{82}$, $\cot\theta = \frac{1}{9}$

7. $\sin\theta = -1$, $\cos\theta = 0$, $\tan\theta =$ undefined, $\csc\theta = -1$, $\sec\theta =$ undefined, $\cot\theta = 0$

9. $\sin\theta = \frac{\sqrt{10}}{10}$, $\cos\theta = \frac{3\sqrt{10}}{10}$, $\tan\theta = \frac{1}{3}$, $\csc\theta = \sqrt{10}$, $\sec\theta = \frac{\sqrt{10}}{3}$, $\cot\theta = 3$

11. $\sin\theta = \frac{3\sqrt{10}}{10}$, $\cos\theta = \frac{-\sqrt{10}}{10}$, $\tan\theta = -3$, $\csc\theta = \frac{\sqrt{10}}{3}$, $\sec\theta = -\sqrt{10}$, $\cot\theta = -\frac{1}{3}$

13. $\sin\theta = \frac{2\sqrt{5}}{5}$, $\cos\theta = -\frac{\sqrt{5}}{5}$, $\tan\theta = -2$, $\csc\theta = \frac{\sqrt{5}}{2}$, $\sec\theta = -\sqrt{5}$, $\cot\theta = -\frac{1}{2}$

15. $\sin\theta = \frac{3\sqrt{10}}{10}$, $\cos\theta = \frac{-\sqrt{10}}{10}$, $\tan\theta = -3$, $\csc\theta = \frac{\sqrt{10}}{3}$, $\sec\theta = -\sqrt{10}$, $\cot\theta = -\frac{1}{3}$

17. $\sin\theta = \frac{5\sqrt{26}}{26}$, $\cos\theta = -\frac{\sqrt{26}}{26}$, $\tan\theta = -5$, $\csc\theta = \frac{\sqrt{26}}{5}$, $\sec\theta = -\sqrt{26}$, $\cot\theta = -\frac{1}{5}$

19. $\sin\theta = -\frac{3}{5}$, $\cos\theta = -\frac{4}{5}$, $\tan\theta = \frac{3}{4}$, $\csc\theta = -\frac{5}{3}$, $\sec\theta = -\frac{5}{4}$, $\cot\theta = \frac{4}{3}$

21. $\sin 0^\circ = 0$, $\cos 0^\circ = 1$, $\tan 0^\circ = 0$, $\csc 0^\circ =$ undefined, $\sec 0^\circ = 1$, $\cot 0^\circ =$ undefined

23. $\sin 360^\circ = 0$, $\cos 360^\circ = 1$, $\tan 360^\circ = 0$, $\csc 360^\circ =$ undefined, $\sec 360^\circ = 1$, $\cot 360^\circ =$ undefined

25. $\sin(180^\circ) = 0$, $\cos(180^\circ) = -1$, $\tan(180^\circ) = 0$, $\csc(180^\circ) =$ undefined, $\sec(180^\circ) = -1$, $\cot(180^\circ) =$ undefined

27. $\sin 540^\circ = 0$, $\cos 540^\circ = -1$, $\tan 540^\circ = 0$, $\csc 540^\circ =$ undefined, $\sec 540^\circ = -1$, $\cot 540^\circ =$ undefined

29. $\sin 720^\circ = 0$, $\cos 720^\circ = 1$, $\tan 720^\circ = 0$, $\csc 720^\circ =$ undefined, $\sec 720^\circ = 1$, $\cot 720^\circ =$ undefined

31. $\sin(-270^\circ) = 1$, $\cos(-270^\circ) = 0$, $\tan(-270^\circ) =$ undefined, $\csc(-270^\circ) = 1$, $\sec(-270^\circ) =$ undefined, $\cot(-270^\circ) = 0$

33. $\sin 810^\circ = 1$, $\cos 810^\circ = 0$, $\tan 810^\circ =$ undefined, $\csc 810^\circ = 1$, $\sec 810^\circ =$ undefined, $\cot 810^\circ = 0$

35. $\sin(-360^\circ) = 0$, $\cos(-360^\circ) = 1$, $\tan(-360^\circ) = 0$, $\csc(-360^\circ) =$ undefined, $\sec(-360^\circ) = 1$, $\cot(-360^\circ) =$ undefined
37. 20° 39. 30° 41. 80° 43. 80° 45. 80° 47. 1.2516 49. .8584 51. .8584 53. .2832
55. .1616
59. $\sin 68^\circ = .9272$, $\cos 68^\circ = .3746$, $\tan 68^\circ = 2.4751$, $\csc 68^\circ = 1.0785$, $\sec 68^\circ = 2.6695$, $\cot 68^\circ = .4040$
61. $\sin 3.8397 = -.6428$, $\cos 3.8397 = -.7660$, $\tan 3.8397 = .8391$, $\csc 3.8397 = -1.5557$, $\sec 3.8397 = -1.3054$, $\cot 3.8397 = 1.1918$
63. $\sin 144.6^\circ = .5793$, $\cos 144.6^\circ = -.8151$, $\tan 144.6^\circ = -.7051$, $\csc 144.6^\circ = 1.7263$, $\sec 144.6^\circ = -1.2268$, $\cot 144.6^\circ = -1.4071$
65. $\sin 1.5708 = 1.0000$, $\cos 1.5708 = .0000$, $\tan 1.5708 =$ undefined, $\csc 1.5708 = 1.0000$, $\sec 1.5708 =$ undefined, $\cot 1.5708 = .0000$
67. $\sin(-105^\circ) = -.9659$, $\cos(-105^\circ) = -.2588$, $\tan(-105^\circ) = 3.7321$, $\csc(-105^\circ) = -1.0353$, $\sec(-105^\circ) = -3.8637$, $\cot(-105^\circ) = .2679$,
69. $\sin(-3) = -.1409$, $\cos(-3) = -.9900$, $\tan(-3) = .1423$, $\csc(-3) = -7.0972$, $\sec(-3) = -1.0101$, $\cot(-3) = 7.0264$
71. $\sin 212^\circ = -.5299$, $\cos 212^\circ = -.8480$, $\tan 212^\circ = .6249$, $\csc 212^\circ = -1.8871$, $\sec 212^\circ = -1.1792$, $\cot 212^\circ = 1.6003$

73. 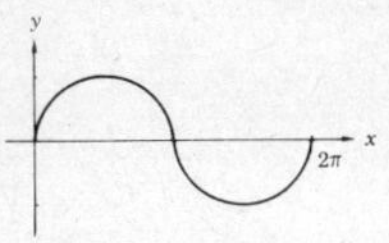

75. 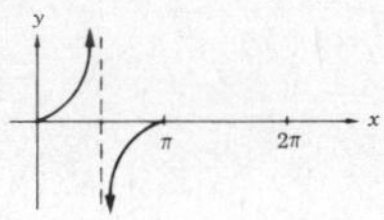

77. 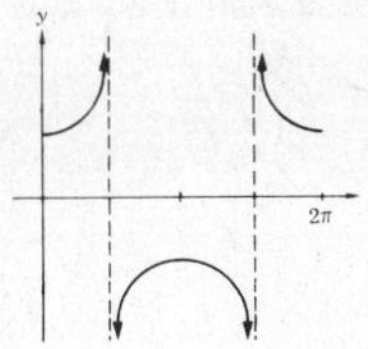

Exercises ■ 6.5

1. 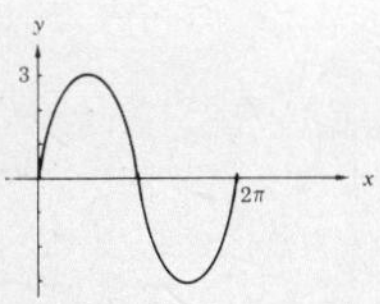

3. 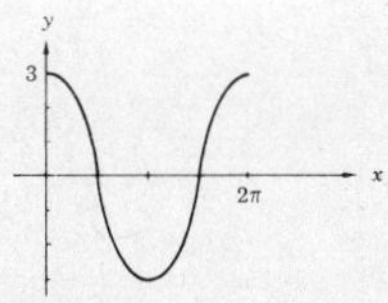

5. 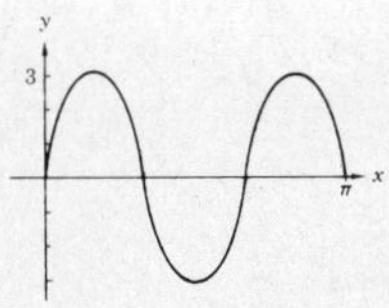

7. 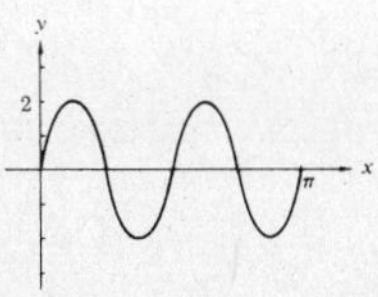

9. 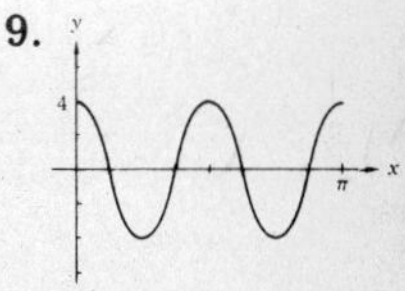

11. 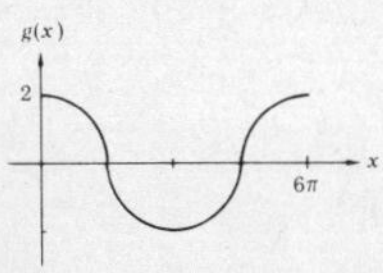

13. 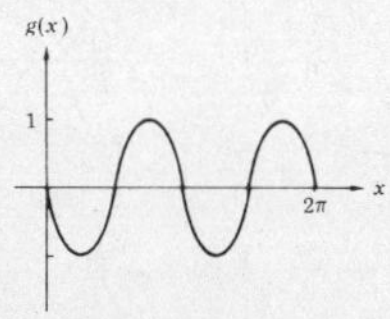

15. 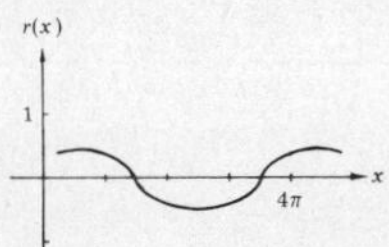

17. 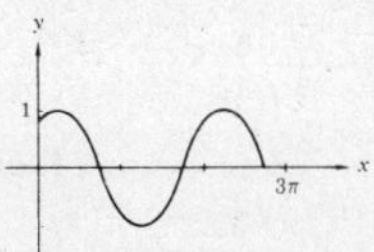

19. 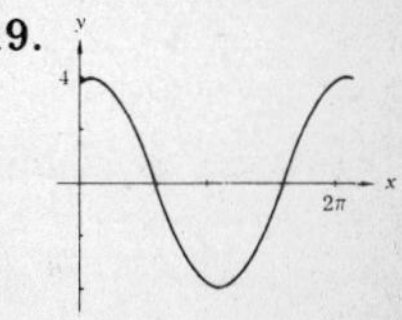

21. 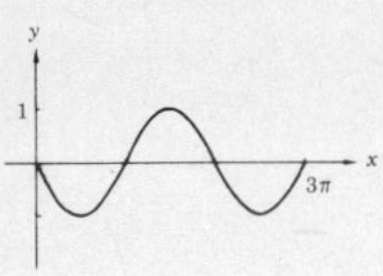

23. 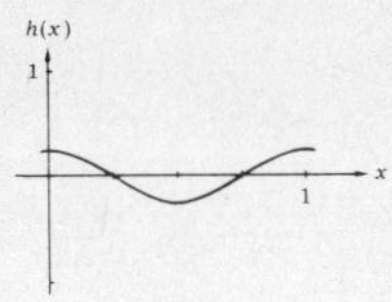

25. 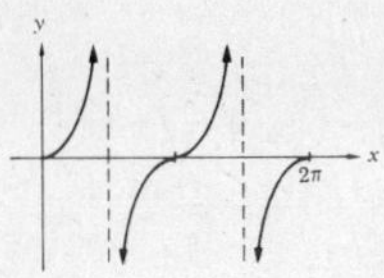

27. 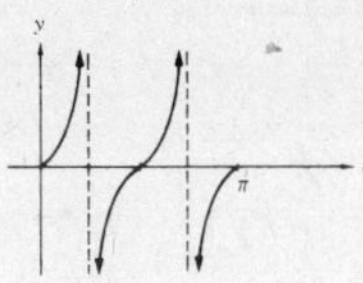

29. 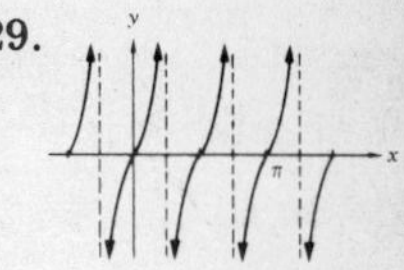

31. 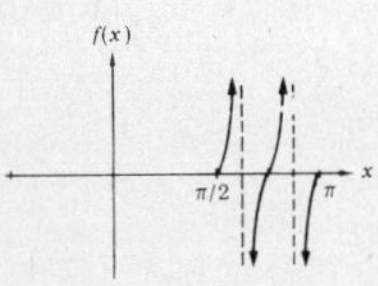

33. 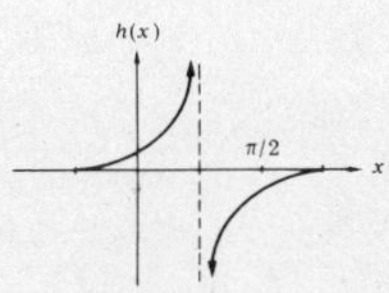

35. 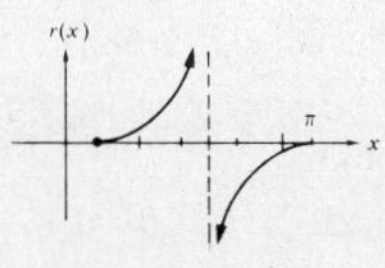

37. 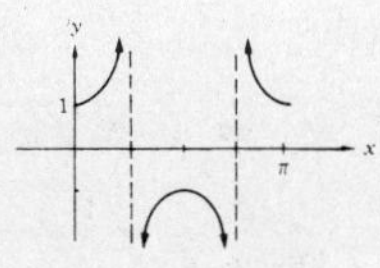

39. 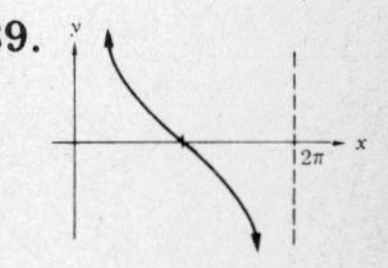

41. 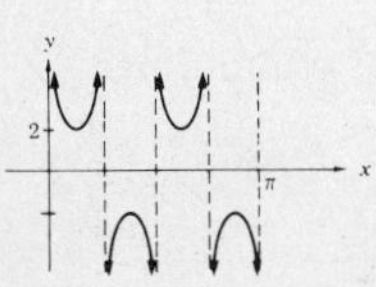

43. 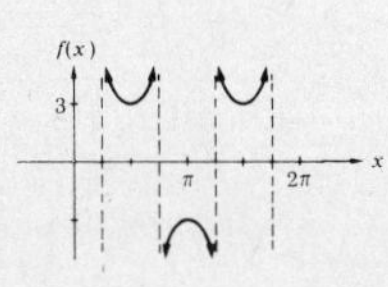

45. 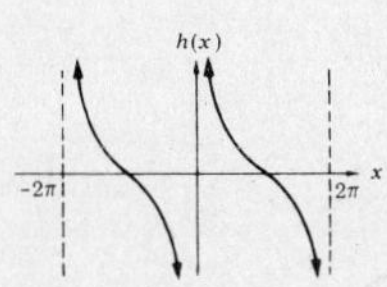

47. 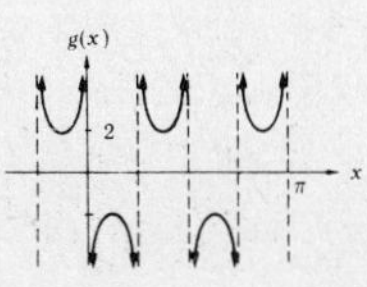

49. 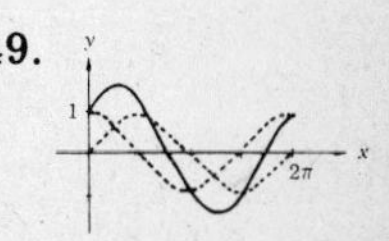

51. 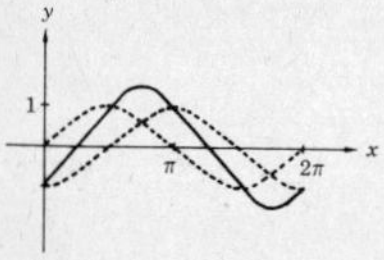

53. 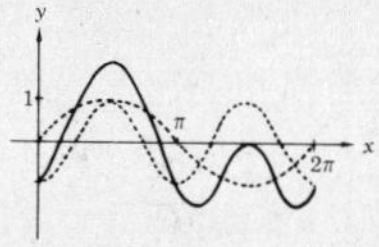

55. 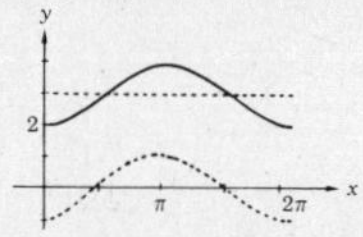

57. 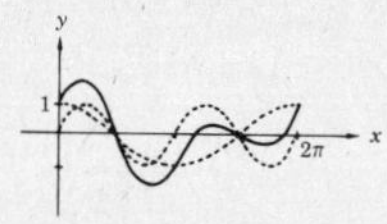

61. 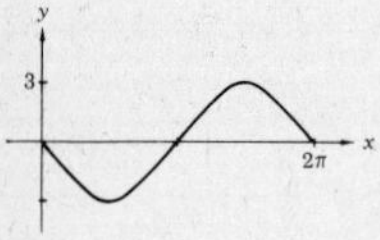

63. 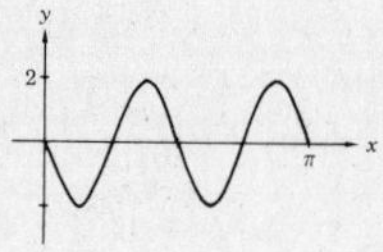

65. 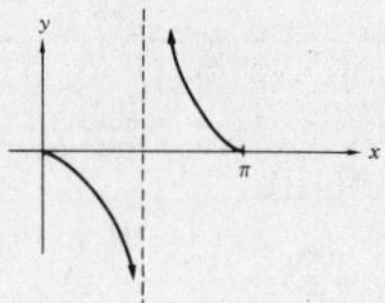

67. 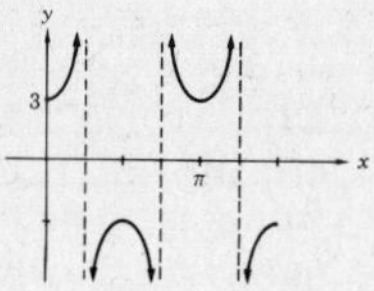

69. 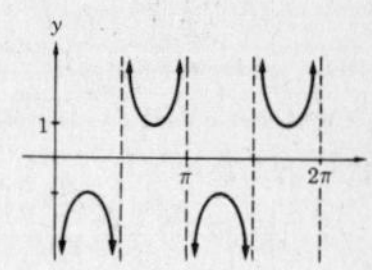

71. 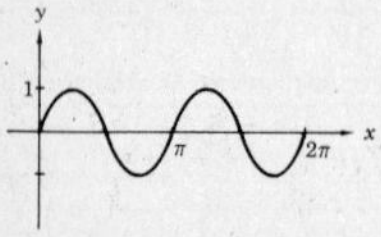

73. 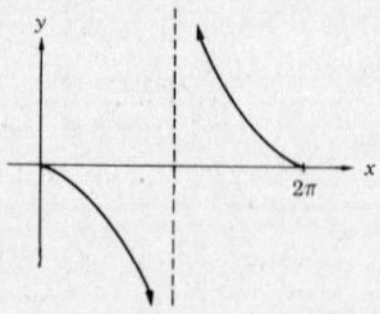

75. 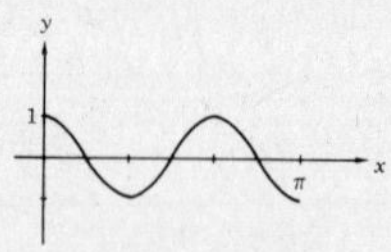

77. $f \circ g(x) = (\sin x)^2$
79. $f \circ g(x) = \tan \sqrt{x}$
81. $f \circ g(x) = (2 \cos x + 1)^{1/2}$
83. 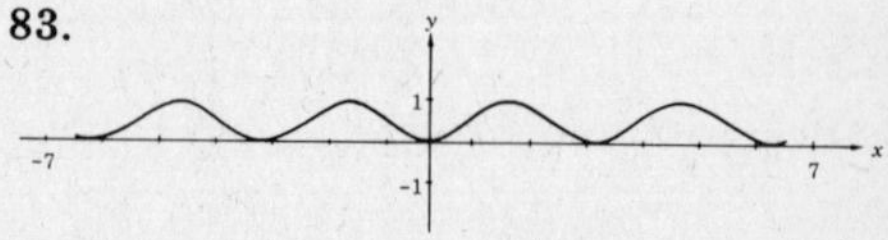

Exercises ■ 6.6

1. 97.4 ft
3. 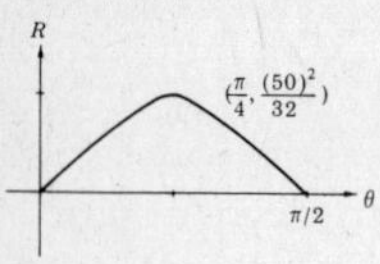

5. 14.8 m 7. a. 5.64 m b. 3.39 m 9. a. 9.85 ft b. 66.0° c. 41.2°
11. a. 2304.94 ft b. 31.0° 15. 38,893 ft 17. .21 miles 19. c. 35,308.77 ft^2
21. 25.14 cm 23. $\measuredangle A = 69.4°$ and $\measuredangle B = 20.6°$

25. 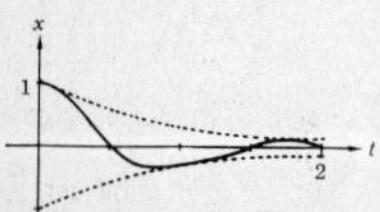

27. 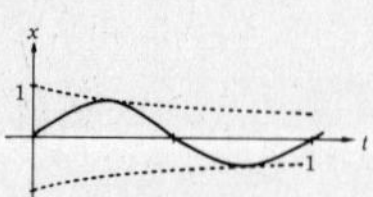

Vocabulary/Symbol True-False Quiz

1. True 2. $\frac{\pi}{10}$ ft^2 3. True 4. 45 in and 77.9 in 5. True 6. True 7. True 8. True
9. True 10. $-\frac{4}{3}$

Chapter Test

1. a. $\frac{\pi}{4}$ b. $\frac{25}{6}\pi$ c. $-\frac{\pi}{2}$ 2. a. 225° b. 45° c. 114.59° 3. a. 100.5888° b. 12°45′36″
4. sin 60° = .8660, cos 60° = .5000, tan 60° = 1.7320, csc 60° = 1.1547, sec 60° = 2.0000, cot 60° = .5773
5. $\measuredangle Q = 49°$, $q = 6.90$, $r = 9.15$ 6. a. .6794 b. .3223 c. 95.490
7. $x = \frac{k\pi}{5}$, k is any integer 8. $x = \frac{\pi}{2}\left(k - \frac{1}{2}\right)$, k is any integer 9. $x = \frac{\pi}{2}(k + 1)$, k is any integer
10. a. .4848 b. −.8746 c. −3.7321 d. 5.1446 e. −5.7588 f. undefined
11. a. .9093 b. −.9900 c. 1.1578 d. −.2958 e. 1.0415 f. 3.5789
12. $\sin\theta = -\frac{12}{13}$, $\cos\theta = \frac{5}{13}$, $\tan\theta = -\frac{12}{5}$, $\csc\theta = -\frac{13}{12}$, $\sec\theta = \frac{13}{5}$, $\cot\theta = -\frac{5}{12}$
13. 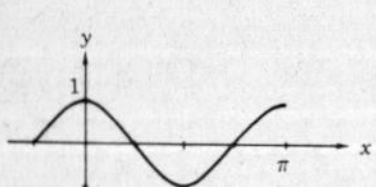
Amplitude = 1
Period = π
Phase shift = $-\frac{\pi}{4}$

14. 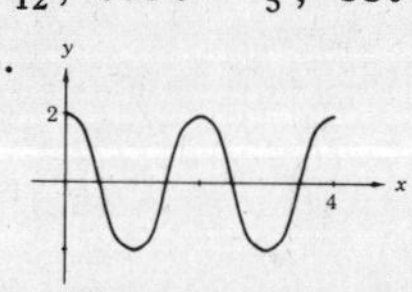
Amplitude = 2
Period = 2
Phase shift = 0

15. 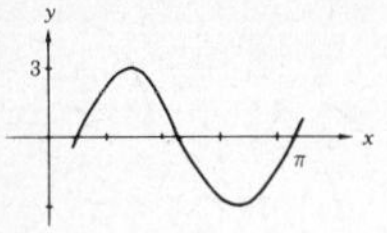

Amplitude = 3
Period = π
Phase shift = $\frac{\pi}{8}$

16. 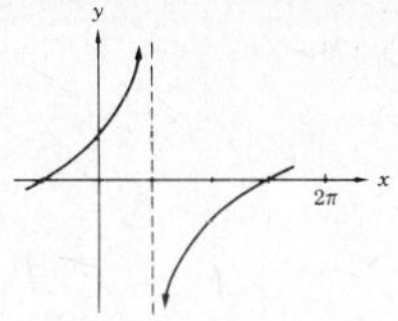

Period = 2π
Phase shift = $-\frac{\pi}{2}$

17. 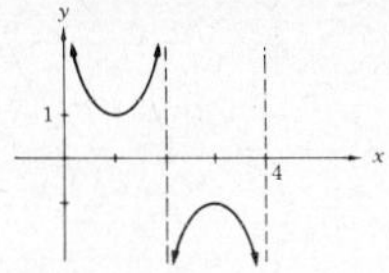

Period = 4
Phase shift = 0

18. 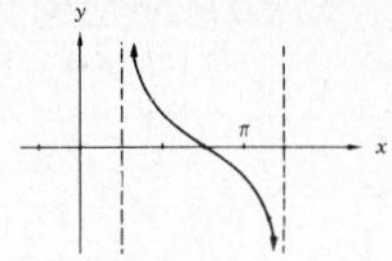

Period = π
Phase shift = $\frac{\pi}{4}$

19. 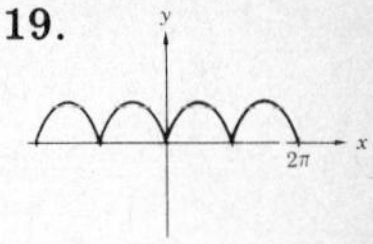

20. a. 1402.17 m b. 47.9°

CHAPTER 7

Exercises ■ 7.1

1. $\measuredangle C = 105°$, $b = 8.9$ cm, $c = 15.0$ cm 3. $\measuredangle A = 48°$, $c = 3.6$, $a = 3.1$ 5. $\measuredangle B = 24.2°$, $\measuredangle C = 135.8°$, $c = 40.8$ or $\measuredangle B = 155.8°$, $\measuredangle C = 4.2°$, $c = 4.3$ 7. no solution 9. $\measuredangle C = 57.8°$, $\measuredangle A = 12.2°$, $a = 2.2$
11. $\measuredangle C = 66.8°$, $\measuredangle A = 63.2°$, $a = 5.8$ or $\measuredangle C = 113.2°$, $\measuredangle A = 16.8°$, $a = 1.9$ 13. $\measuredangle C = 17.5°$, $\measuredangle A = 12.5°$, $a = 7.2$
15. $\measuredangle B = 56.4°$, $\measuredangle C = 93.6°$, $c = 12.0$ or $\measuredangle B = 123.6°$, $\measuredangle C = 26.4°$, $c = 5.3$ 17. $\measuredangle B = 30°$, $\measuredangle C = 60°$, $c = 6.9$
19. no solution 21. $\measuredangle B = 88.9°$, $\measuredangle C = 53.1°$, $a = 3.1$ cm 23. $\measuredangle A = \measuredangle B = 60°$, $c = 5$ in
25. $\measuredangle A = 22.3°$, $\measuredangle B = 49.5°$, $\measuredangle C = 108.2°$ 27. no solution 29. $\measuredangle A = \measuredangle C = 87.1°$, $\measuredangle B = 5.8°$
31. $\measuredangle A = \measuredangle B = 75.5°$, $\measuredangle C = 29°$ 33. $\measuredangle A = 22.3°$, $\measuredangle C = 27.7°$, $b = 8.9$ m 35. $\measuredangle A = 22.6°$, $\measuredangle B = 90°$, $\measuredangle C = 67.4°$
37. $\measuredangle A = 29.9°$, $\measuredangle C = 65.1°$, $c = 18.2$ 39. $\measuredangle A = \measuredangle B = 40°$, $c = 9.2$ cm 41. a. 9.2 m b. 70.9° c. 3.8 m d. 13 m 43. a. 9.8 m b. 67.4° c. 3.9 m d. 13.7 m 47. a. 2439.1 m b. 1175.1 m 51. The sum of *any* two sides must be greater than the third side. 53. no solution
55. $\measuredangle B = 24.787°$, $\measuredangle C = 135.272°$, $c = 41.310$ or $\measuredangle B = 155.213°$, $\measuredangle C = 4.846°$, $c = 4.959$

Exercises ■ 7.2

3. $\frac{\sqrt{5}}{2}$ 5. $\frac{\sqrt{5}}{3}$ 7. $\frac{-3\sqrt{5}}{5}$ 9. $\frac{-\sqrt{6}}{12}$ 11. $\frac{2\sqrt{2}}{3}$ 13. $\sqrt{1-Q^2}$ 15. $\sqrt{\frac{K^2}{4}-1}$ 29. $\frac{\sqrt{6}-\sqrt{2}}{4}$
31. $\frac{\sqrt{2}}{2}$ 33. $\frac{\sqrt{2}-\sqrt{6}}{4}$ 35. $\frac{1}{2}$ 45. $\frac{4\sqrt{5}}{9}$ 47. $\frac{4\sqrt{5}}{9}$ 49. $-\frac{7}{8}$ 51. $-\frac{\sqrt{15}}{8}$
53. $\cos 3u = 4\cos^3 u - 3\cos u$ 55. $\cos 5u = 16\cos^5 u - 20\cos^3 u + 5\cos u$
57. $\sin 4u = 4(1 - 2\sin^2 u)\sin u(1 - \sin^2 u)^{1/2}$
59. $-\frac{\sqrt{6}}{4}$ 61. $\frac{\sqrt{3}}{3}$ 63. $-\frac{\sqrt{2}}{2}$ 65. $\frac{\sqrt{30}}{10}$ 67. $\frac{\sqrt{10}}{10}$ 69. $\frac{2\sqrt{5}}{5}$ 73. $\cos\sqrt{1-2x}$
79. $\frac{1}{2}\cos 9x + \frac{1}{2}\cos x$ 81. $2\sin 7s + 2\sin 3s$ 87. $2\sin\frac{5x}{2}\cos\frac{x}{2}$ 89. $2\cos 4x\cos 2x$

Exercises ■ 7.3

47. $\frac{3\pi}{2} + 2k\pi$ 49. $\frac{\pi}{3} + 2k\pi$ 51. $\frac{\pi}{6} + 2k\pi, \frac{5\pi}{6} + 2k\pi$ 53. $\frac{\pi}{4} + 2k\pi, \frac{5\pi}{4} + 2k\pi$ 55. $\frac{\pi}{2} + 2k\pi$
57. $\frac{4\pi}{3} + 2k\pi, \frac{5\pi}{3} + 2k\pi$ 59. $\frac{3\pi}{4} + k\pi, (2k+1)\pi$ 61. $\frac{\pi}{4} + 2k\pi, \frac{5\pi}{4} + 2k\pi, \frac{3\pi}{4} + 2k\pi, \frac{7\pi}{4} + 2k\pi$ 63. $k\pi$
65. $\frac{\pi}{6} + 2k\pi, \frac{5\pi}{6} + 2k\pi, \frac{3\pi}{2} + 2k\pi$ 67. $\frac{k\pi}{2}$ 69. no solution 71. 30°, 330°, 73. 10°, 50°, 70°, 110°, 130°, 170°, 190°, 230°, 250°, 290°, 310°, 350° 75. $22\frac{1}{2}°, 112\frac{1}{2}°$ 77. 210°, 330° 79. 180°
81. 210°, 270°, 330°

Exercises ■ 7.4

1. $r = \sin t$ 3. $r = \tan \frac{2}{3}t$ 5. $r = 3 \csc \dfrac{t-2}{8}$ 7. $r = 3 \sin \dfrac{1-t}{4} - 2$ 9. $r = \dfrac{1}{\tan^{-1}\left(\dfrac{1-t}{4}\right)}$
11. $r = \cot[\cot(t+1)]$

13. 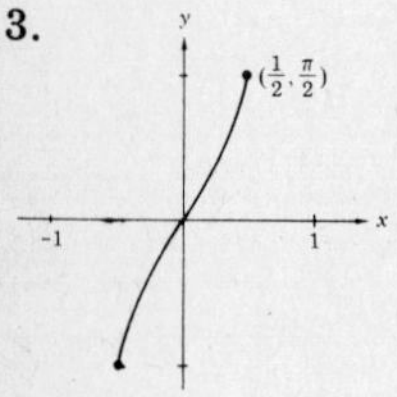

15. 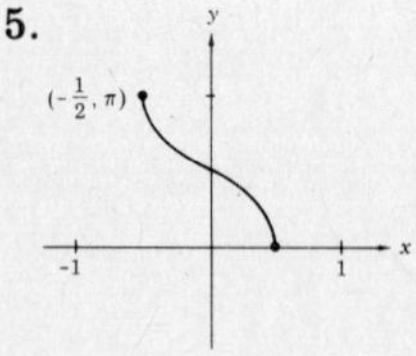

17. 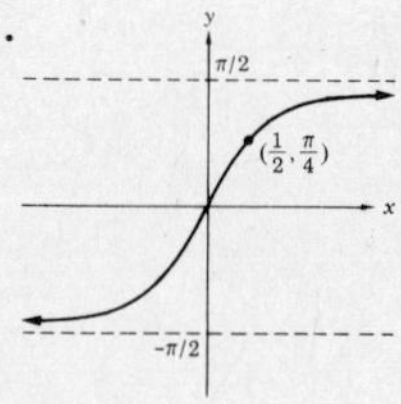

19. 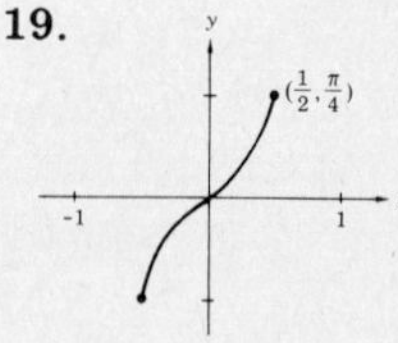

21. 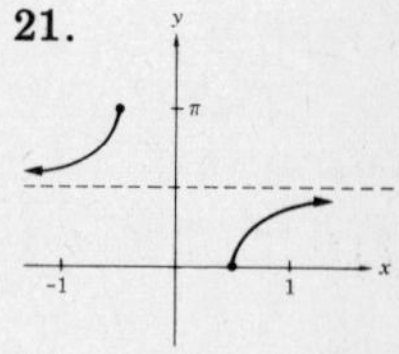

23. 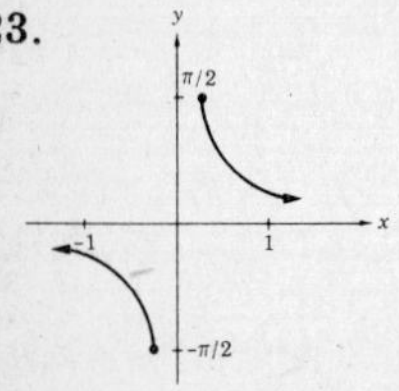

25. $\dfrac{\pi}{4}$ 27. $-\dfrac{\pi}{3}$ 29. $\dfrac{\pi}{4}$ 31. $-\dfrac{\pi}{3}$ 33. $\dfrac{\pi}{3}$ 35. 1.1990587 37. −1.1990587
39. 1 41. $\dfrac{\sqrt{2}-\sqrt{6}}{4}$ 43. $\dfrac{3-4\sqrt{3}}{10}$ 45. $\sqrt{1-4x^2}$ 47. $\sqrt{1-x^2}$
49. $\pm\dfrac{x}{\sqrt{1-x^2}}$ 51. $\sqrt{1-4x^2}$ 53. 1.12631927, 5.0199926 55. no solution
57. $3.4128564 + 2k\pi$, $6.0119215 + 2k\pi$ 63. −.20807342, .54030231, 9.9500417, 99.995, 999.9995, 10,000 65. .43643578 67. 12.63644 69. −.94310108

Exercises ■ 7.5

1. (2,380°), (−2,200°) 3. (2,200°), (2,560°) 5. $\left(3,-\dfrac{3\pi}{2}\right)$, $\left(-3,-\dfrac{\pi}{2}\right)$ 7. $\left(-4,-\dfrac{\pi}{4}\right)$, $\left(4,-\dfrac{5\pi}{4}\right)$
9. $\left(5,-\dfrac{\pi}{3}\right)$, $\left(-5,\dfrac{2\pi}{3}\right)$

11. 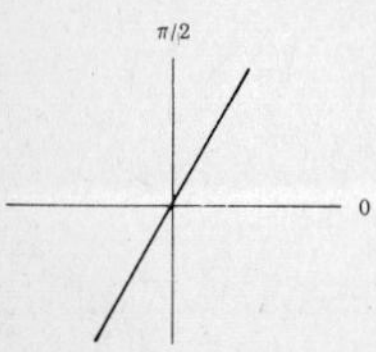

13. 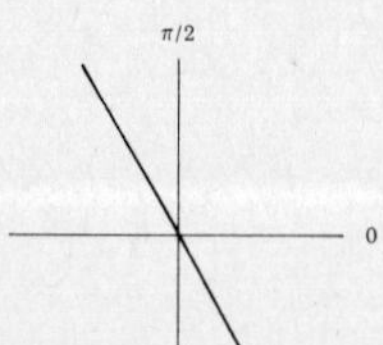

15. 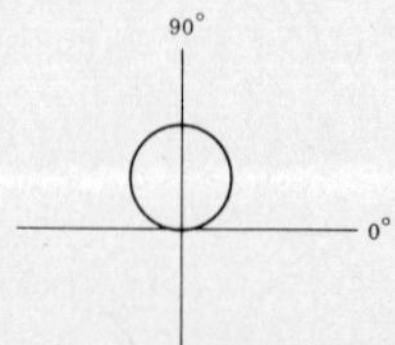

17. 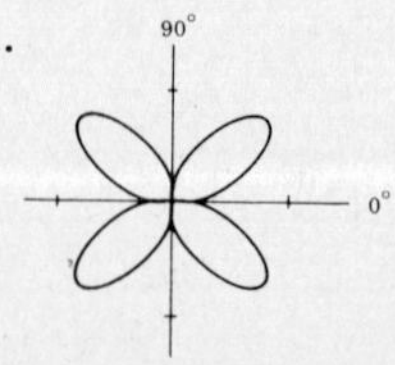

19. 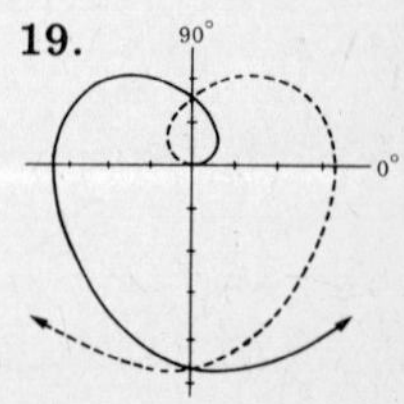

21. 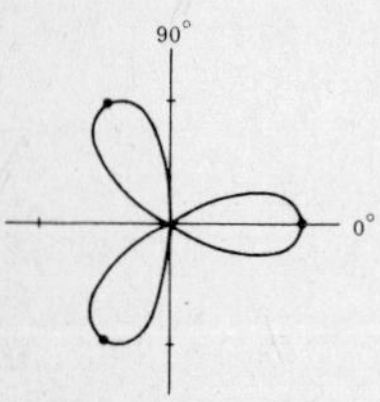

23. 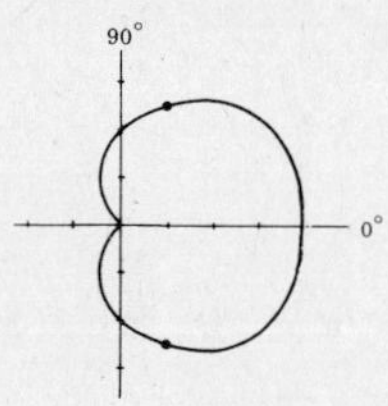

25. 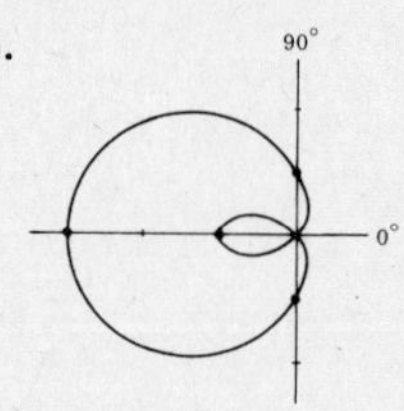

27. 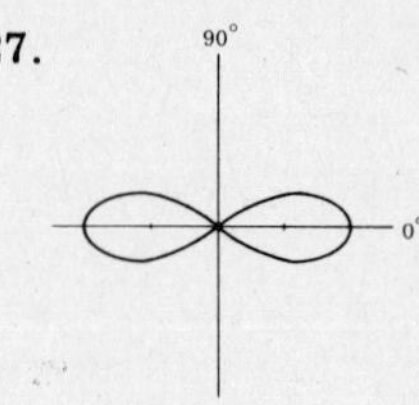

29. 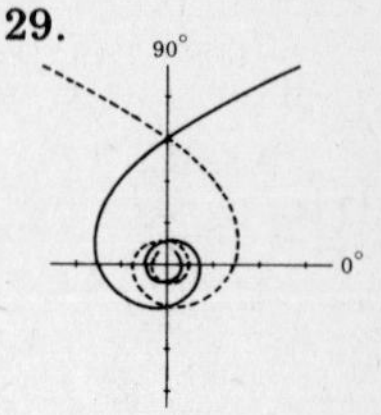

31. $x^2 + y^2 - y = 0$ 33. $(x^2 + y^2)^{3/2} = x^2 - y^2$ 35. $\sqrt{x^2 + y^2} = 3$ or $x^2 + y^2 = 9$
37. $\sqrt{x^2 + y^2} = 2 - \dfrac{x}{\sqrt{x^2 + y^2}}$ 39. $\sqrt{x^2 + y^2} = \tan^{-1} \dfrac{y}{x}$ 41. $r = 3$ 43. $r^2 \sin 2\theta = 9$
45. $3r \cos\theta + 2r \sin\theta = 1$ 47. $r = 2\cos\theta + 4\sin\theta$ 49. $r \sin\theta = 4$

Exercises ■ 7.6

1. A and B 3. C and D 23. a. −2i + 2j b. 2i − 2j c. $2\sqrt{2}$ 25. a. −2i + 4j b. 2i − 4j c. $2\sqrt{5}$ 27. a. −i − 2j b. i + 2j c. $\sqrt{5}$ 29. a. −i + j b. i − j c. $\sqrt{2}$ 31. a. −5i − 5j b. 5i + 5j c. $5\sqrt{2}$ 33. a. −2i − 2j b. 2i + 2j c. $2\sqrt{2}$ 35. a. 2i + 4j b. 0i + 0j c. −i − 2j 37. a. −i + 4j b. i − 4j c. 2i − 8j 39. a. 3i − 3j b. −i + 3j c. −3i + 6j
41. a. 3i + 4j b. −i − 6j c. −3i − 11j 43. a. 7i − j b. −3i + j c. −8i + 2j 45. a. 3i + 3j

b. $-5\mathbf{i} - \mathbf{j}$ c. $-9\mathbf{i} - 3\mathbf{j}$ 47. a. $6\mathbf{i} - 5\mathbf{j}$ b. $0\mathbf{i} + \mathbf{j}$ c. $-3\mathbf{i} + 4\mathbf{j}$ 49. a. $0\mathbf{i} + 7\mathbf{j}$ b. $-4\mathbf{i} + 11\mathbf{j}$ c. $-6\mathbf{i} + 13\mathbf{j}$ 51. yes 53. yes 55. no 57. yes 59. yes 61. 24.2° 63. 100.3° 65. 90° 67. 71.6° 69. 56.3°

Exercises ■ 7.7

1. $\sqrt{2}\text{ cis }\frac{\pi}{4}$ 3. $\sqrt{2}\text{ cis }\frac{3\pi}{4}$ 5. $\text{cis }\frac{\pi}{3}$ 7. $6\text{ cis }\frac{2\pi}{3}$ 9. $4\text{ cis }\frac{3\pi}{2}$ 11. $6\sqrt{2}\text{ cis }\frac{\pi}{4}$

13. $5\text{ cis}(\tan^{-1}\frac{4}{3}) \approx 5\text{ cis }53.1^\circ$ 15. $2\sqrt{5}\text{ cis}(\tan^{-1}\frac{1}{2}) \approx 2\sqrt{5}\text{ cis }26.6^\circ$ 17. $\frac{3\sqrt{2}}{2} + \frac{3\sqrt{2}}{2}i$

19. $\sqrt{2} - \sqrt{2}\,i$ 21. $-\frac{\sqrt{2}}{2} - \frac{\sqrt{2}}{2}i$ 23. $\frac{3}{2} - \frac{3\sqrt{3}}{2}i$ 25. $1.8794 + .6840i$ 27. $4i$ 29. -2

31. $\text{cis }2\pi$ 33. $\text{cis }\frac{5\pi}{12}$ 35. $\text{cis}\left(-\frac{\pi}{4}\right)$ 37. $6\text{ cis }\frac{\pi}{2}$ 39. $\text{cis }\frac{\pi}{6}$ 41. $6\text{ cis }0^\circ$ 43. $10\text{ cis }\frac{3\pi}{2}$

45. 16 47. -2^{10} 49. $-1.4641 + 5.4641i$ 51. $-128 - 128i$ 53. -2^{50} 55. $78.24 + 20.96i$

57. 1 59. $3, -\frac{3}{2} + \frac{3\sqrt{3}}{2}i, -\frac{3}{2} - \frac{3\sqrt{3}}{2}i$ 61. $\frac{\sqrt{2}}{2} + \frac{\sqrt{2}}{2}i, -\frac{\sqrt{2}}{2} - \frac{\sqrt{2}}{2}i$

63. $\text{cis}\left(\frac{\pi}{4} + \frac{k\pi}{2}\right), k = 0,1,2,3$ 65. $2^{1/3}\text{ cis}(110^\circ + 120^\circ k), k = 0,1,2$ 67. $\sqrt[6]{2}\text{ cis}\left(\frac{\pi}{12} + \frac{2k\pi}{3}\right), k = 0,1,2$

69. $\text{cis }\frac{(2k+1)\pi}{3}, k = 0,1,2$ 71. $\text{cis }\frac{2k\pi}{5}, k = 0,1,2,3,4$ 73. $2^{1/4}\text{ cis}\left(\frac{\pi}{3} + \frac{k\pi}{2}\right), k = 0,1,2,3$

79. $\cos 3\theta = 4\cos^3\theta - 3\cos\theta, \sin 3\theta = 3\sin\theta - 4\sin^3\theta$

Vocabulary/Symbol True-False Quiz

1. True 2. True 3. $\pm\frac{4}{5}$ 4. True 5. True 6. True 7. True 8. True 9. (0,0) to (3,−1) 10. True

Chapter Test

1. $\measuredangle C = 110^\circ, a = 2.74, b = 2.13$ 2. $\measuredangle A = 136.2^\circ, \measuredangle B = 19.8^\circ, a = 20.43$ 3. $\measuredangle A = 28.9^\circ, \measuredangle B = 104.5^\circ, \measuredangle C = 46.6^\circ$ 4. $a = 4.97, \measuredangle B = 54.8^\circ, \measuredangle C = 101.2^\circ$ 5. a. $-\frac{1}{5}\sqrt{21}$ b. $\sqrt{.7}$ c. $\frac{1}{2}\sqrt{21}$ d. $-\sqrt{.3}$ 12. $\frac{\pi}{3} + 2k\pi, \frac{2\pi}{3} + 2k\pi$, where k is any integer 13. $\frac{\pi}{3} + 2k\pi, \frac{5\pi}{3} + 2k\pi$, where k is any integer 14. $\frac{\pi}{4} + k\pi, \frac{3\pi}{4} + k\pi$, where k is any integer 15. $\frac{\pi}{6} + 2k\pi, \frac{5\pi}{6} + 2k\pi, (\sin^{-1}\frac{1}{3}) + 2k\pi, (\pi - \sin^{-1}\frac{1}{3}) + 2k\pi$, k is any integer 16. a. $\frac{\pi}{4}$ b. $\frac{3\pi}{4}$ c. $-\frac{\pi}{2}$ d. $-\frac{\pi}{3}$ 17. $\frac{x}{\sqrt{1-x^2}}$

18.

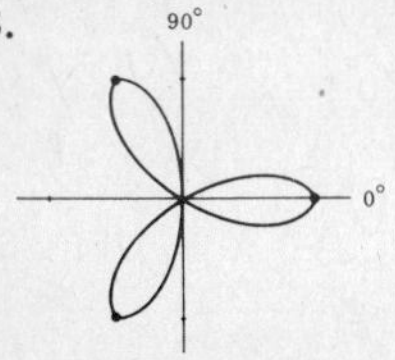

19.

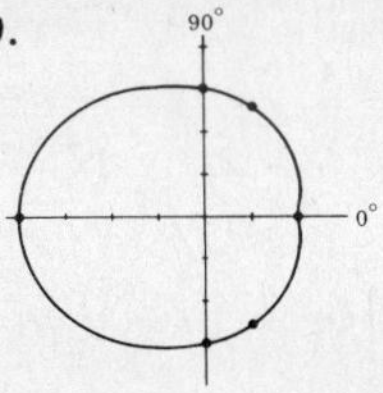

20. a. $\sqrt{13}$ b. 5 c. $5\mathbf{i} - \mathbf{j}$ d. $\sqrt{26}$ e. $-\mathbf{i} + 7\mathbf{j}$ f. $17\mathbf{j}$ g. -6 21. 120.6° 22. 16 23. $3\text{ cis }\frac{2k\pi}{3}, k = 0,1,2$ 24. $\text{cis}\left(\frac{\pi}{4} + \frac{k\pi}{2}\right), k = 0,1,2,3$ 25. $\text{cis}\left(\frac{\pi}{3} + \frac{2k\pi}{3}\right), k = 0,1,2$

CHAPTER 8

Exercises ■ 8.1

1. 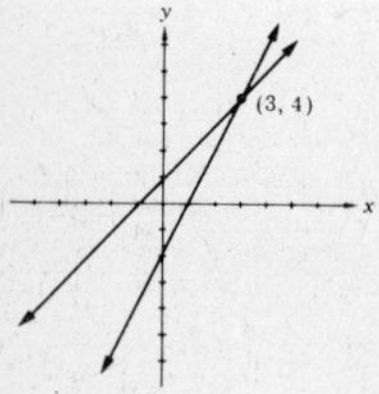

3. 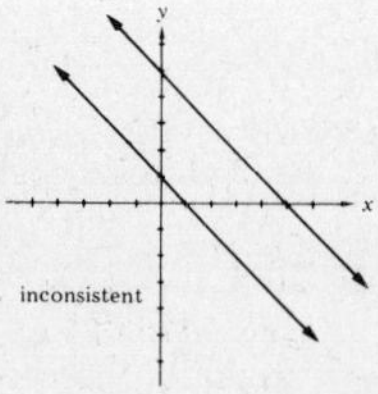

5. 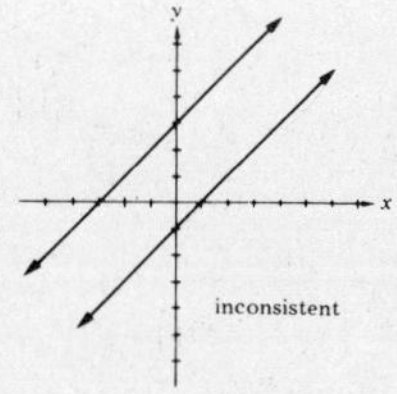

7. 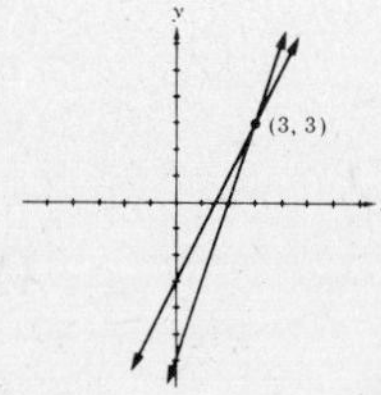

9. 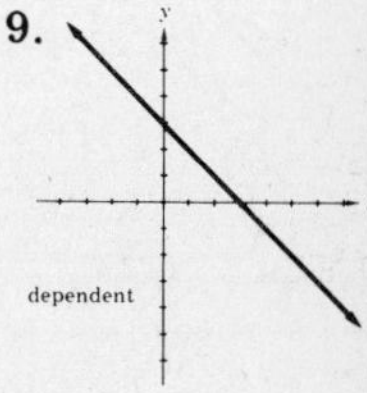

11. 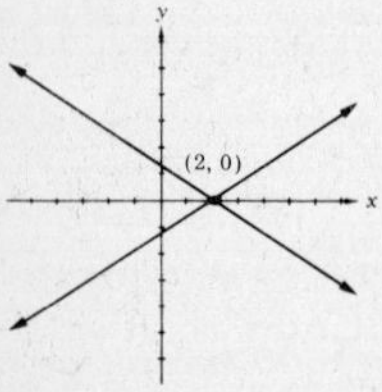

13. (3,4) 15. inconsistent 17. (1,2) 19. (3,2) 21. (5,−4) 23. (−4,1) 25. $(\frac{1}{2},-\frac{7}{3})$ 27. inconsistent 29. dependent 31. (−1,2) 33. $\frac{7}{2}$ and $\frac{17}{2}$ 35. 4 quarters and 26 half-dollars 37. \$3 for adults and \$2.50 for children 39. The information forms a dependent system. 41. $s = \frac{1}{2}$, $t = -\frac{1}{3}$ 43. $s = 4$, $t = 1$ 45. In the variables $\frac{1}{r}$ and $\frac{1}{s}$ this is a dependent system. 47. $m = 2, n = \sqrt{2}$ or $m = 2, n = -\sqrt{2}$ or $m = -2, n = \sqrt{2}$ or $m = -2, n = -\sqrt{2}$ 49. (7.6176,1.2464) 51. (−.4018,.2363) 53. (.9051,.9133)

Exercises ■ 8.2

1. (1,2,3) 3. (3,1,−2) 5. (2,4,3) 7. (−1,−1,3) 9. (−1,−2,6) 11. $(0,-1,\frac{1}{2})$ 13. (3,2,−2) 15. (0,0,0) 17. $(\frac{1}{2},\frac{1}{3},\frac{1}{4})$ 19. (1,3,2) 21. $(\frac{1}{2},-1,\frac{7}{12})$ 23. $(1,\frac{1}{2},-\frac{1}{2})$ 25. $(-\frac{7}{2},8,\frac{17}{2})$ 31. 10 one-dollar bills, 5 two-dollar bills, 15 five-dollar bills 33. balloons cost 50¢ each, pennants cost 75¢ each, hats cost \$2 each 35. $y = 2x^2 - x - 3$

Exercises ■ 8.3

1. (4,0), (−2,−6) 3. (6,0), (−6,0) 5. $(\frac{12}{5},\frac{12}{5}), (-\frac{12}{5},-\frac{12}{5})$ 7. (1,3), (−1,3), (1,−3), (−1,−3) 9. $\left(\frac{\sqrt{2}}{2},\frac{\sqrt{2}}{2}\right), \left(-\frac{\sqrt{2}}{2},-\frac{\sqrt{2}}{2}\right)$ 11. (2,1), (−1,4) 13. no solution 15. (2,1), (−2,1), (1,4), (−1,4) 17. $(3,-2), (-\frac{2}{3},9)$ 19. (2,−4) 21. $(e,1)$ 23. (84,3) 25. (1,5) 27. (1,9) 29. $(1,0), (2,\log_3 2)$

31. 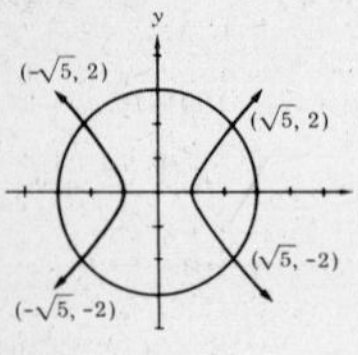

33. 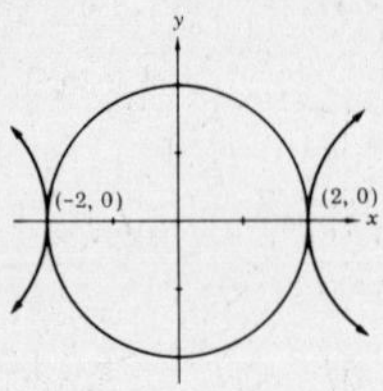

35. 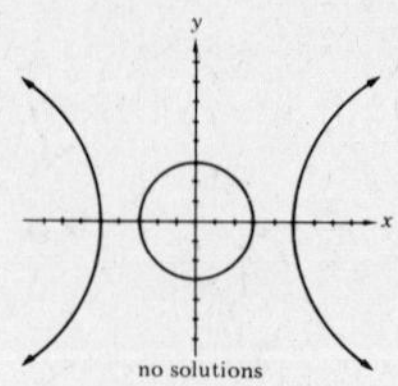

37. −6 and 12 39. $2+\sqrt{2}$ and $2-\sqrt{2}$ 41. $\frac{1}{2}$ and $\frac{2}{3}$ 43. 1.8 and 8 45. $(\frac{3}{2},\frac{1}{2}), (-\frac{3}{2},-\frac{1}{2}), (1,\frac{1}{2}), (-1,-\frac{1}{2})$, $\left(\frac{3\sqrt{11}}{11},\frac{\sqrt{11}}{11}\right), \left(-\frac{3\sqrt{11}}{11},-\frac{\sqrt{11}}{11}\right)$, $\left(\frac{2\sqrt{11}}{11},\frac{\sqrt{11}}{11}\right), \left(-\frac{2\sqrt{11}}{11},-\frac{\sqrt{11}}{11}\right)$

Exercises ■ 8.4

1. 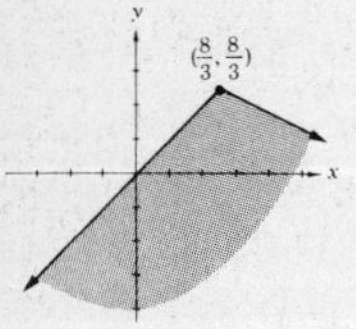

3. 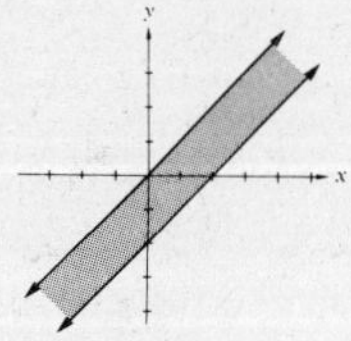

5. 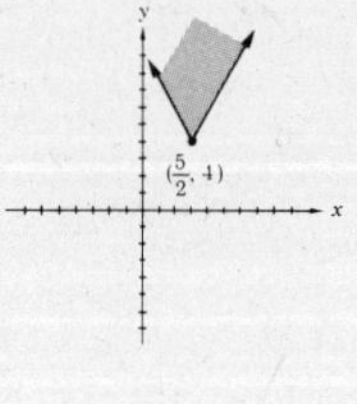

7. 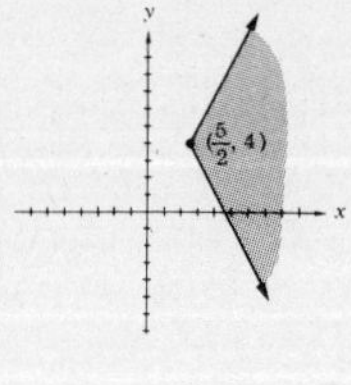

9. 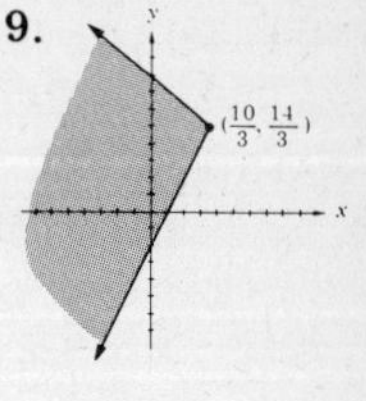

11. 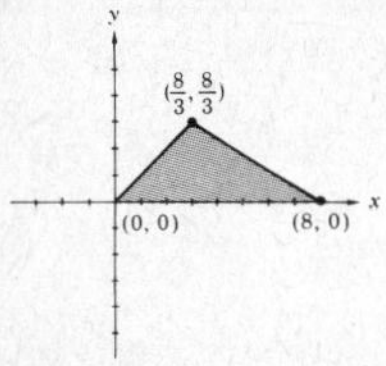

13. 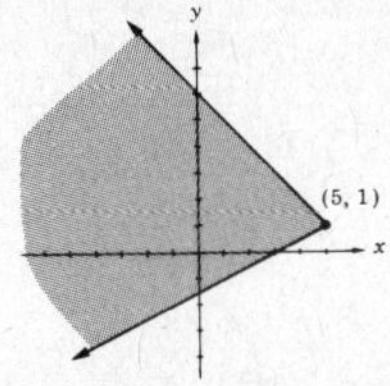

15. 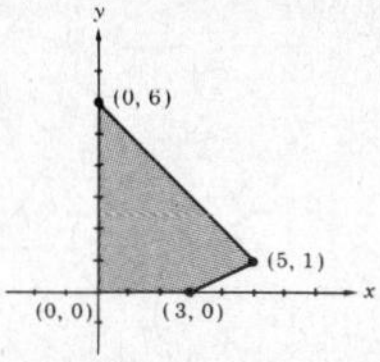

17. 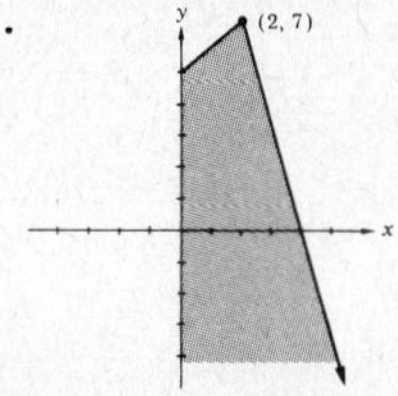

19. 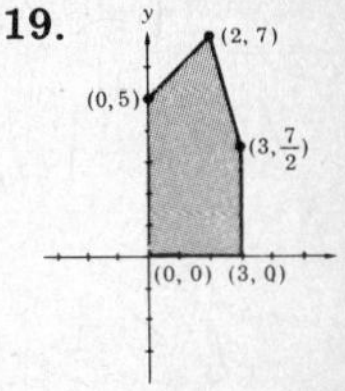

21. 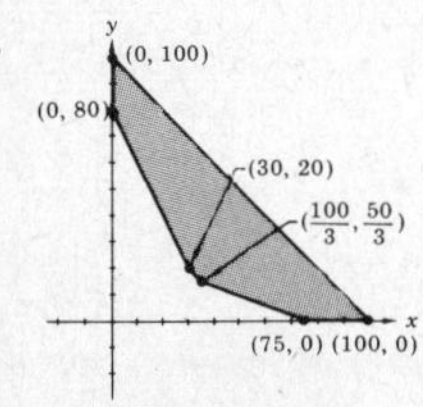

23. 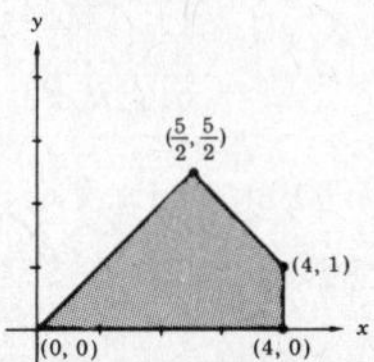

25. F's maximum is 90 when $x = 4$ and $y = 5$.
27. G's minimum is -40 when $x = 6$ and $y = 2$.
29. $10x - y$ achieves a maximum of 60 at (6,0).
31. $x + y$ achieves a maximum of 12 at (6,6).
33. $5x + y$ achieves a maximum of 40 at (8,0).
35. a. $x \leqslant 40, y \leqslant 60, x + y \leqslant 80, x \geqslant 0, y \geqslant 0$

b. 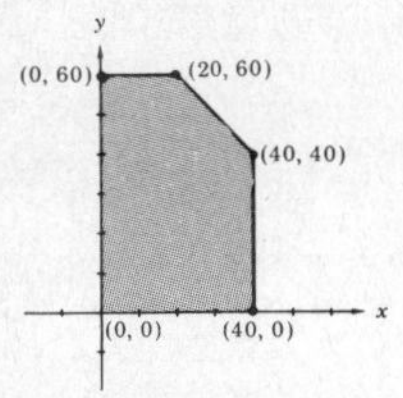

c. The company should produce 20 block sets and 60 toy trains to maximize profit.
37. The company should produce 30 pianos and 5 organs to realize a profit of \$6500.
39. Joseph's costs are zero when he produces nothing!
43. Robert's minimum consumption of 612.5 calories is obtained with $\frac{5}{2}$ servings of sardines and $\frac{5}{6}$ servings of milk.

Vocabulary/Symbol True-False Quiz

1. True 2. True 3. True 4. True 5. True 6. True 7. constraints 8. objective function

Chapter Test

1. 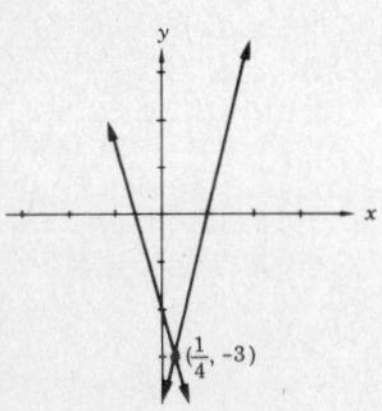

2. 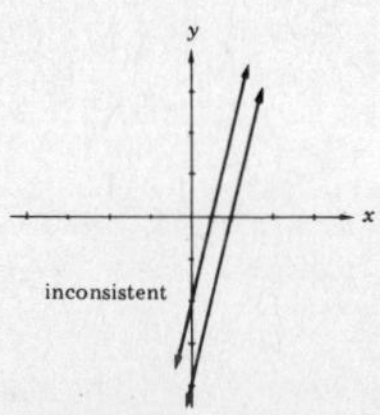

3. 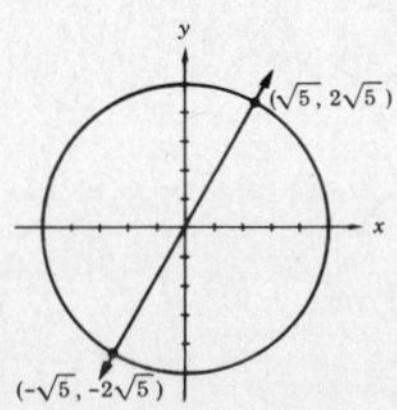

4. 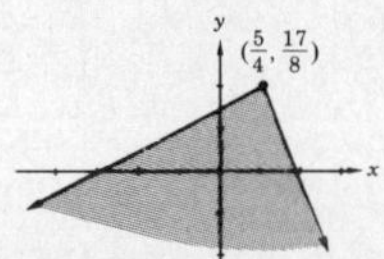

5. 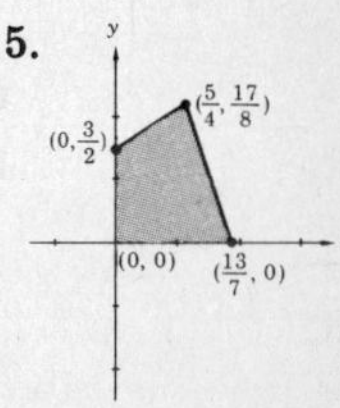

6. 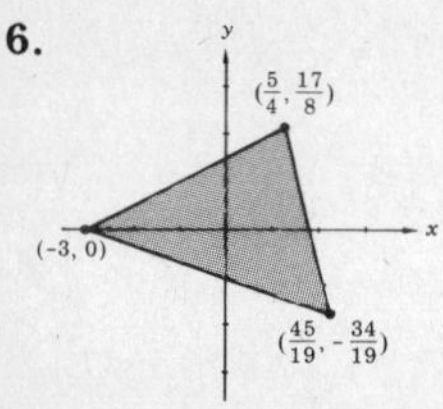

7. $(-\frac{35}{23}, \frac{100}{23})$ 8. dependent 9. (3,2,1) 10. (0,0,−2) 11. (4,−3), (−3,4)
12. (1,−4) 13. $\frac{11}{2}$ and $-\frac{1}{2}$ 14. 5 pennies, 5 nickels, 20 dimes
15. C achieves its minimum value of zero at (0,0).

CHAPTER 9

Exercises ■ 9.1

1. $x = 5, y = 3$ 3. $x = 3, y = 2$ 5. $x_1 = 1, x_2 = 2, x_3 = 4$ 7. $x = \frac{1}{2}, y = \frac{1}{3}, z = 0$ 9. $x = 0, y = -1, z = \frac{1}{2}$
11. inconsistent 13. dependent 15. dependent 17. $x_1 = 3, x_2 = -6, x_3 = .4$
19. $x_1 = 1, x_2 = 2, x_3 = 3, x_4 = 4$ 21. $x_1 = 4, x_2 = 3, x_3 = -2, x_4 = -5$ 23. $x_1 = 1, x_2 = -1, x_3 = 2, x_4 = 3$
25. $x_1 = 2, x_2 = 3, x_3 = 4, x_4 = 6$

27. a. $x_1 = \dfrac{a_{22}b_1 - a_{12}b_2}{a_{11}a_{22} - a_{12}a_{21}}$, $x_2 = \dfrac{a_{11}b_2 - a_{21}b_1}{a_{11}a_{22} - a_{12}a_{21}}$ b. $a_{11}a_{22} - a_{12}a_{21} \neq 0$ 29. a. $x_1 = 2, x_2 = -\frac{1}{3}, x_3 = -1$
b. $x_1 = -\frac{3}{2}, x_2 = \frac{1}{2}, x_3 = 1$ c. $x_1 = 2, x_2 = -\frac{2}{3}, x_3 = -1$ 31. dolls: \$8, train sets: \$25, tricycles: \$15
33. $x_1 = 3, x_2 = -1$ 35. $x_1 = 7.61757, x_2 = 1.246385$ 37. $x_1 = -1.67364, x_2 = -.990237$
39. $x_1 = .5858954, x_2 = .4783816$

Exercises ■ 9.2

1. $\begin{bmatrix} 1 & 6 & 12 \\ 2 & 4 & 7 \end{bmatrix}$ 3. $\begin{bmatrix} 0 & 0 \\ 0 & 2 \\ 0 & 4 \end{bmatrix}$ 5. $\begin{bmatrix} 0 & 5 & 10 \\ 15 & 20 & 25 \end{bmatrix}$ 7. $\begin{bmatrix} \frac{1}{2} & 3 & 6 \\ 1 & 2 & \frac{7}{2} \end{bmatrix}$ 9. $\begin{bmatrix} 2 & 11 & 22 \\ 1 & 4 & 9 \end{bmatrix}$ 11. undefined

13. $\mathbf{A} \bullet \mathbf{B} = \begin{bmatrix} 2 & -2 \\ 5 & -16 \end{bmatrix}$ $\mathbf{B} \bullet \mathbf{A} = \begin{bmatrix} -6 & 10 \\ 7 & -8 \end{bmatrix}$ 15. $\mathbf{A} \bullet \mathbf{B} = \begin{bmatrix} 0 & 0 \\ 30 & 20 \end{bmatrix}$ $\mathbf{B} \bullet \mathbf{A} = \begin{bmatrix} 0 & -35 \\ 0 & 20 \end{bmatrix}$

17. $\mathbf{A} \bullet \mathbf{B} = \begin{bmatrix} 1 & 8 \\ 5 & -10 \end{bmatrix}$ $\mathbf{B} \bullet \mathbf{A} = \begin{bmatrix} 1 & 8 \\ 5 & -10 \end{bmatrix}$ 19. $\mathbf{A} \bullet \mathbf{B} = \begin{bmatrix} 3 & 2 & 1 \\ 2 & 2 & 1 \\ 1 & 1 & 1 \end{bmatrix}$ $\mathbf{B} \bullet \mathbf{A} = \begin{bmatrix} 1 & 1 & 1 \\ 1 & 2 & 2 \\ 1 & 2 & 3 \end{bmatrix}$

21. $\mathbf{A} \bullet \mathbf{B} = \begin{bmatrix} 1 & -6 & 5 \\ 0 & 16 & -12 \\ 8 & 36 & -23 \end{bmatrix}$ $\mathbf{B} \bullet \mathbf{A} = \begin{bmatrix} -7 & 1 \\ -24 & 1 \end{bmatrix}$ 23. $\mathbf{A} \bullet \mathbf{B} = \begin{bmatrix} -.05 & .2 & 2.8 \\ 1.59 & 2.36 & -23.64 \end{bmatrix}$ $\mathbf{B} \bullet \mathbf{A}$ is undefined

25. $\mathbf{A} \bullet \mathbf{B} = \mathbf{B} \bullet \mathbf{A} = \begin{bmatrix} 1 & 2 & 3 & 4 \\ 5 & 6 & 7 & 8 \\ 9 & 10 & 11 & 12 \\ 13 & 14 & 15 & 16 \end{bmatrix}$ 31. a. $[3 \quad 15 \quad 30]$ b. $\begin{bmatrix} 24 \\ -4 \\ 8 \end{bmatrix}$ c. 21 d. undefined

33. Four A panels, four B panels, four C panels, and one D panel at lumber yard 3.

35. a. $\begin{bmatrix} 1 & 4 \\ 2 & 5 \\ 3 & 6 \end{bmatrix}$ b. $\begin{bmatrix} 5 \\ 7 \\ -6 \end{bmatrix}$ c. $[1 \quad 2 \quad \frac{1}{2}]$ d. $\begin{bmatrix} 1 & 0 & 0 & 0 \\ 0 & 1 & 0 & 0 \\ 0 & 0 & 1 & 0 \\ 0 & 0 & 0 & 1 \end{bmatrix}$ e. $\begin{bmatrix} \frac{1}{2} & 1 & \frac{1}{2} \\ \frac{2}{5} & 0 & \frac{1}{3} \\ \frac{1}{10} & 0 & \frac{1}{6} \end{bmatrix}$ 37. 2A 39. $\mathbf{A} \bullet \mathbf{C}$

41. $\mathbf{A} + \mathbf{B} = \begin{bmatrix} 0 & 9 & 15 \\ 11 & 8 & -6 \end{bmatrix}$ $2\mathbf{A} - \mathbf{B} = \begin{bmatrix} 3 & -3 & 6 \\ -8 & 7 & 12 \end{bmatrix}$ $\mathbf{A} \bullet \mathbf{C} = \begin{bmatrix} 47 & -15 & 2 \\ -25 & 74 & 5 \end{bmatrix}$ $\mathbf{C} \bullet \mathbf{D} = \begin{bmatrix} -17 & 10 & 36 \\ -34 & 21 & 70 \\ 43 & 60 & -16 \end{bmatrix}$

Exercises ■ 9.3

3. no 5. no 7. yes 9. yes 11. $\begin{bmatrix} 1 & 0 \\ -2 & 1 \end{bmatrix}$ 13. $\begin{bmatrix} -1 & -1 \\ 1 & -1 \end{bmatrix}$ 15. $\begin{bmatrix} -4 & 7 \\ 3 & -5 \end{bmatrix}$

17. $\begin{bmatrix} \frac{1}{8} & \frac{5}{4} & 0 \\ \frac{3}{8} & \frac{5}{8} & \frac{1}{8} \\ \frac{3}{4} & \frac{1}{4} & \frac{1}{4} \end{bmatrix}$ 19. $\begin{bmatrix} \frac{17}{39} & -\frac{2}{39} & -\frac{7}{39} \\ -\frac{9}{39} & \frac{24}{39} & \frac{6}{39} \\ -\frac{8}{39} & \frac{17}{39} & \frac{1}{39} \end{bmatrix}$ 21. a. singular b. nonsingular c. singular

23. $x_1 = 1, x_2 = 2, x_3 = 3$ 25. $x_1 = 1, x_2 = 10, x_3 = -5$ 27. $x_1 = -1, x_2 = -7, x_3 = 9$
29. $x_1 = -1, x_2 = \frac{1}{2}, x_3 = \frac{1}{2}$ 31. $x_1 = 0, x_2 = 0, x_3 = 5, x_4 = 6$ 33. $x_1 = -1, x_2 = -2, x_3 = -3, x_4 = 6$
37. a. $x_1 = -4.76837\text{E} - 06$, $x_2 = -9.53674\text{E} - 06$, $x_3 = 5$, $x_4 = 5.99999$

Exercises ■ 9.4

1. −3 3. −3 5. 30 7. −35 9. 0 11. $M_{12} = -6, A_{12} = 6$ 13. $M_{21} = -6, A_{21} = 6$
15. $M_{23} = -6, A_{23} = 6$ 17. −211 19. −581 21. a. 192 b. 192 c. 192 29. −2684 31. 0

Exercises ■ 9.5

1. $x = 1, y = 1$ 3. $x = -\frac{1}{2}, y = \frac{1}{2}$ 5. $x = -\frac{1}{4}, y = \frac{3}{4}$ 7. $x_1 = 0, x_2 = 0$ 9. $x_1 = 4, x_2 = 10$
11. $x = -4, y = 5, z = 11$ 13. $x = 0, y = 0, z = 7$ 15. $x = -2, y = -4, z = -3$ 17. $x_1 = 0, x_2 = 6, x_3 = 2$
19. $x_1 = .1, x_2 = .2, x_3 = -.3$ 21. a. $|\mathbf{D}| = 0$, $|\mathbf{D}_{x_1}| = 2$, $|\mathbf{D}_{x_2}| = 1$, $|\mathbf{D}_{x_3}| = -3$

Vocabulary/Symbol True-False Quiz

1. True 2. True 3. 15 4. their corresponding components are equal 5. True
6. The number of columns of **A** must equal the number of rows of **B**. 7. True 8. True 9. True
10. True

Chapter Test

1. $x_1 = 3, x_2 = -2$ 2. $x_1 = -\frac{2}{23}, x_2 = \frac{74}{23}$ 3. dependent 4. $x_1 = 4, x_2 = \frac{1}{2}, x_3 = -6$ 5. dependent

6. $x_1 = 1, x_2 = 2, x_3 = 3, x_4 = -2$ 7. $\begin{bmatrix} \frac{1}{4} & \frac{3}{4} \\ \frac{7}{4} & \frac{13}{4} \\ \frac{13}{4} & \frac{23}{4} \end{bmatrix}$ 8. $\begin{bmatrix} 66 & 49 \\ 55 & 45 \\ -2 & 17 \end{bmatrix}$ 9. $\begin{bmatrix} 0 & 49 \\ -8 & 2 \\ 7 & 35 \end{bmatrix}$ 10. $\begin{bmatrix} 7 & 23 & -11 \\ 5 & 20 & -10 \\ -1 & -9 & 30 \end{bmatrix}$

11. $\begin{bmatrix} -5 & -3 \\ \frac{1}{2} & 5 \end{bmatrix}$ 12. $\begin{bmatrix} 9 & -11 \\ -4 & 5 \end{bmatrix}$ 13. $\begin{bmatrix} \frac{1}{5} & -\frac{1}{10} \\ \frac{4}{15} & \frac{1}{30} \end{bmatrix}$ 14. $\begin{bmatrix} .8 & -4.4 & .2 \\ -.2 & 1.6 & .2 \\ 0 & 1 & 0 \end{bmatrix}$ 15. $\begin{bmatrix} \frac{1}{4} & 0 & 0 & 0 \\ 0 & \frac{1}{4} & 0 & 0 \\ 0 & 0 & \frac{1}{4} & 0 \\ 0 & 0 & 0 & \frac{1}{4} \end{bmatrix}$

16. a. $\begin{bmatrix} 1 & -3 \\ -8 & 6 \end{bmatrix}$ b. $x = -1, y = -7$ 17. 0 18. -5 19. -15 20. $x_1 = 1, x_2 = 0, x_3 = 1$

CHAPTER 10

Exercises ■ 10.1

1. -2 3. -6 5. -3 7. 435 9. 0 11. 152 13. 3 15. $-\frac{7}{9}$ 17. True
19. True 21. No 23. Yes 25. No 27. Yes 29. No 31. $-3, -4, 2$ 33. $\sqrt{2}i, -\sqrt{2}i, 3, 4$
35. $1, -1, \frac{\sqrt{2}}{2}i, -\frac{\sqrt{2}}{2}i$ 37. n must be even 41. $.4, i, -i$

Exercises ■ 10.2

1. a. $\pm1, \pm2, \pm4$ b. $-1, 2, -2$ 3. a. $\pm1, \pm2, \pm3, \pm4, \pm6, \pm12$ b. $-3, 1, 4$ 5. a. $\pm\frac{1}{2}, \pm1, \pm\frac{3}{2}, \pm2, \pm3, \pm6$
b. $-3, \frac{1}{2}, 2$ 7. a. $\pm1, \pm2, \pm3, \pm4, \pm6, \pm8, \pm12, \pm24, \pm\frac{1}{2}, \pm\frac{3}{2}$ b. $-2, 1, 4, \frac{3}{2}$
9. a. $\pm1, \pm3, \pm9, \pm\frac{3}{2}, \pm\frac{1}{2}, \pm\frac{9}{2}, \pm\frac{1}{3}, \pm\frac{1}{6}$ b. $-\frac{3}{2}, \frac{1}{3}, 3$ 11. a. $\pm1, \pm\frac{1}{2}, \pm\frac{1}{4}, \pm\frac{1}{8}, \pm\frac{1}{16}, \pm\frac{1}{32}, \pm3, \pm\frac{3}{2}, \pm\frac{3}{4}, \pm\frac{3}{8}, \pm\frac{3}{16}, \pm\frac{3}{32}$,
$\pm9, \pm\frac{9}{2}, \pm\frac{9}{4}, \pm\frac{9}{8}, \pm\frac{9}{16}, \pm\frac{9}{32}$ b. $-\frac{3}{2}, \frac{1}{4}, \frac{3}{4}$ 13. a. $\pm1, \pm2, \pm4, \pm8, \pm\frac{1}{2}$ b. $-2, 1-\sqrt{3}, 1+\sqrt{3}$
15. a. $\pm1, \pm\frac{1}{2}, \pm\frac{1}{3}, \pm\frac{1}{4}, \pm\frac{1}{6}, \pm\frac{1}{9}, \pm\frac{1}{12}, \pm\frac{1}{18}, \pm\frac{1}{36}, \pm2, \pm\frac{2}{3}, \pm\frac{2}{9}$ b. $-\frac{2}{3}, -\frac{1}{2}, -\frac{1}{3}, \frac{1}{2}$ 17. a. $\pm1, \pm2, \pm4$
b. $-2\ 1-\sqrt{3}, 1+\sqrt{3}$ 19. a. $\pm1, \pm3, \pm5, \pm15$ b. $-1, 3, -\sqrt{5}, \sqrt{5}$ 21. a. $\pm1, \pm3, \pm\frac{1}{2}, \pm\frac{3}{2}, \pm\frac{1}{6}, \pm\frac{1}{3}$
b. $-\frac{1}{3}, \frac{3}{2}, 2-\sqrt{5}, 2+\sqrt{5}$ 23. a. $\pm1, \pm5, \pm\frac{1}{3}, \pm\frac{5}{3}$ b. $\frac{1}{3}, 2+i, 2-i$ 25. a. $\pm1, \pm2$ b. $-2, 1, i, -i$
27. a. $\pm1, \pm2, \pm7, \pm14, \pm\frac{1}{2}, \pm\frac{7}{2}, \pm\frac{1}{4}, \pm\frac{7}{4}$ b. $-\frac{1}{2}, 1, 2-\sqrt{3}i, 2+\sqrt{3}i$

29. 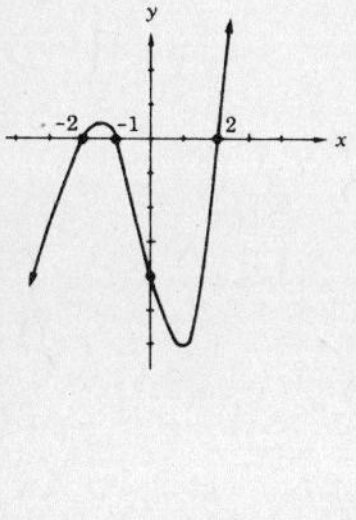

31. 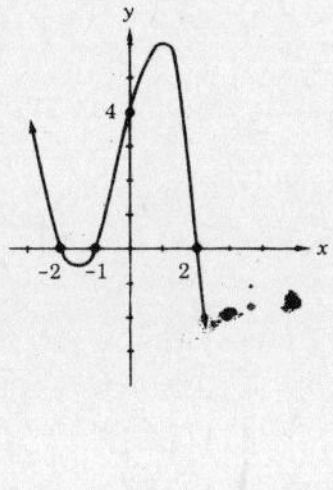

33. 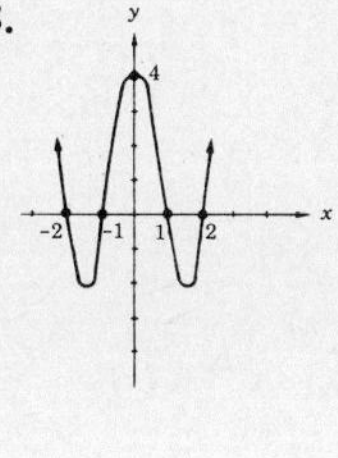

35. 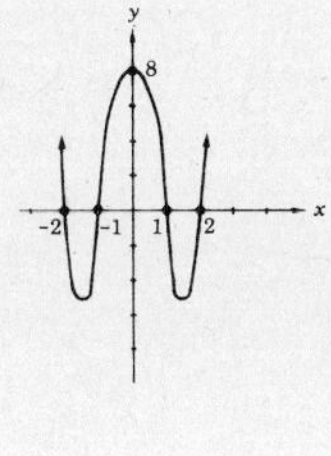

37. 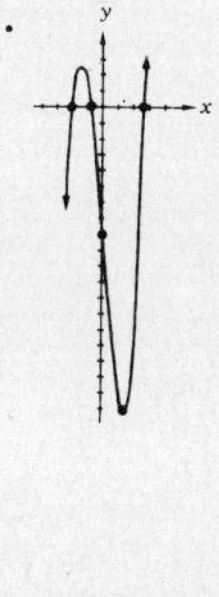

39. 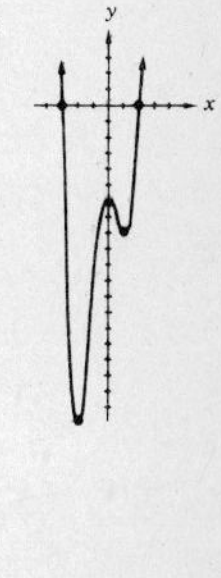

41. $(x+1)(x-2)(x+2)$ 43. $(x+3)(x-1)(x-7)$ 45. $(x+3)(2x-1)(x-2)$
47. $(x+2)(2x-3)(x-1)(x-4)$ 49. $(2x+3)(3x-1)(x-3)$ 51. $(2x+3)(4x-1)(4x-3)$
53. $(x+2)(x-1+\sqrt{3})(x-1-\sqrt{3})$ 55. $(3x+2)(2x+1)(3x+1)(2x-1)$ 57. $(x+2)(x-1+\sqrt{3})(x-1-\sqrt{3})$
59. $(x+1)(x-3)(x+\sqrt{5})(x-\sqrt{5})$ 61. $(3x+1)(2x-3)(x-2+\sqrt{5})(x-2-\sqrt{5})$
63. $(3x-1)(x-2-i)(x-2+i)$ 65. $(x+2)(x-1)(x-i)(x+i)$ 67. $(2x+1)(x-1)(x-2-\sqrt{3}i)(x-2+\sqrt{3}i)$

Exercises ■ 10.3

1. .2 3. −2.3 5. 3.9 7. .3 9. −2.6 11. .3 13. −.5 15. 3.8 17. pos, 3 or 1 neg, 0
19. pos, 3 or 1 neg, 1

25.
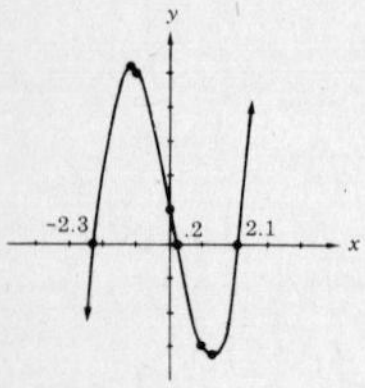

27.
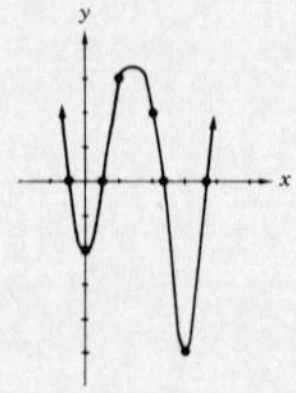

29. a.
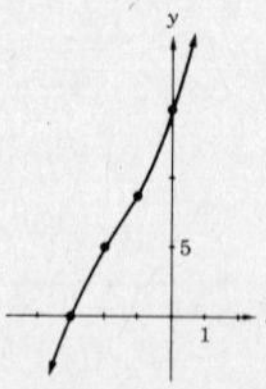

b.
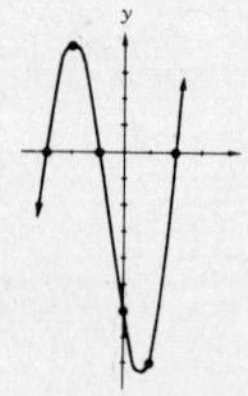

c.
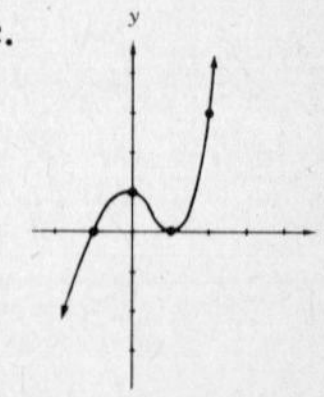

31. a.

b.
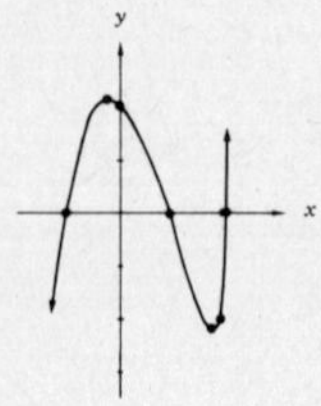

c.
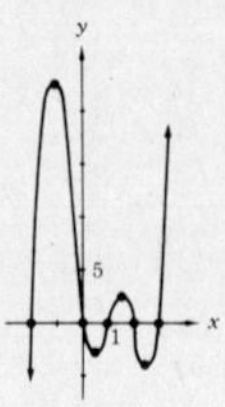

d.
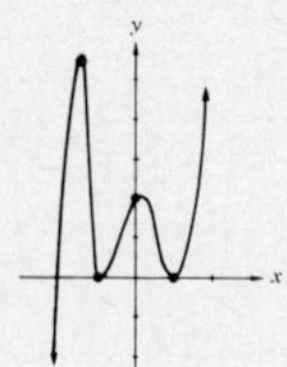

33. $2 \pm \sqrt{3}$

Exercises ■ 10.4

1. $\frac{4}{2x-3} + \frac{1}{x+1}$ 3. $\frac{5}{2x-1} - \frac{3}{x-1}$ 5. $\frac{1}{x+1} + \frac{2}{x+2} + \frac{3}{x+3}$ 7. $\frac{5}{x-1} + \frac{7}{(x-1)^2}$

9. $\frac{4}{x} + \frac{7}{2x+5} - \frac{19}{(2x+5)^2}$ 11. $\frac{1}{x} + \frac{3}{(x+1)^2}$ 13. $\frac{3}{x} + \frac{1}{x^2} - \frac{3}{x+1} - \frac{3}{(x+1)^2}$ 15. $\frac{5}{x} + \frac{1}{x^2} + \frac{-5x}{x^2+1}$

17. $\frac{2x+1}{x^2+1} + \frac{3x-1}{x^2+x+5}$ 19. $\frac{x}{x^2+9} - \frac{x}{x^2+x+7}$ 21. $\frac{3}{x} + \frac{x+2}{x^2+2} + \frac{1}{(x^2+2)^2}$

23. $\frac{1}{x^2+9} - \frac{1}{(x^2+9)^2} + \frac{10}{(x^2+9)^3}$ 29. $A=-3, B=.5, C=1.5$ 31. $2x-1+\frac{1}{x+1}+\frac{2}{x+2}$

33. $x^2-x+1+\frac{5}{x}-\frac{5}{x+1}$

Vocabulary/Symbol True-False Quiz

1. $x-c$ 2. True 3. True 4. True 5. True

Chapter Test

1. −14 2. No 3. a. $\pm1, \pm2, \pm4, \pm\frac{1}{3}, \pm\frac{2}{3}, \pm\frac{4}{3}$ b. $\frac{1}{3}, -1, -4$ 4. No—contradicts Rational Root Theorem

5.
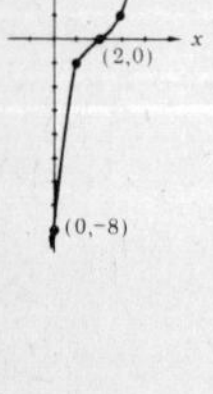

6.
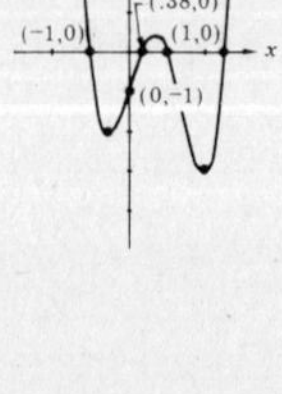

7.
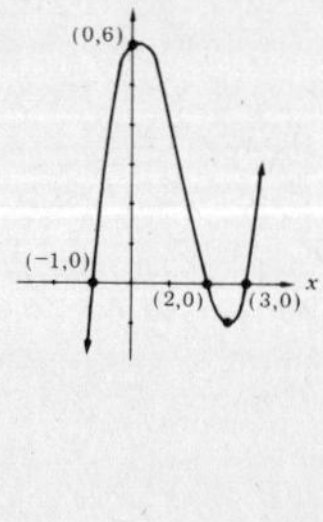

8. .3, 2.3, −2.6 9. ±1.2
10. pos, 3 or 1 neg, 1
11. $\frac{1}{x-4} + \frac{-1}{x+7}$ 12. $\frac{1}{x+1} + \frac{2}{2x-5}$
13. $\frac{4}{x+1} - \frac{5}{(x+1)^2}$
14. $\frac{-4}{x+2} + \frac{4}{x+1}$ 15. $\frac{2}{x^2+1} + \frac{-x+1}{(x^2+1)^2}$

CHAPTER 11

Exercises ■ 11.1

1. 5, 8, 11 3. 1, 2, 4, 8, . . . 5. 4, 8, 16, . . . 7. $\frac{1}{3}, -\frac{4}{3}, 3, -\frac{16}{3}, \frac{25}{3}$ 9. $\frac{2}{3}, \frac{3}{4}, \frac{4}{5}, \frac{5}{6}$ 11. $1, \frac{1}{2}, \frac{1}{3}, \ldots$
13. 4, 7, 12, 19, 28 15. 2, 8, 32, 128, . . . 17. $4, 1, \frac{4}{9}, \frac{1}{4}$ 19. $0, \frac{1}{4}, \frac{1}{4}, \frac{3}{16}, \ldots$
21. 3, 4, 5, 10, 11, 12, 15, 16, 18, 19, 20 23. $\sum_{k=1}^{n} 2k$ 25. $\sum_{k=1}^{n} (-1)^{k+1} \frac{k}{2^k}$ 27. $\sum_{k=1}^{n} \frac{2k^2}{k+1}$
29. $\sum_{k=1}^{5} \frac{(-1)^{k+1}}{k^2}$ 31. $\sum_{k=1}^{n} k^2$ 33. $\sum_{k=1}^{n} k^3$ 35. $\sum_{k=1}^{n} \frac{5}{k(k+1)}$ 37. $\sum_{k=1}^{n} (2k-4)$
39. $\frac{1}{5} + \frac{2}{6} + \frac{3}{7} + \ldots + \frac{n}{n+4}$ 41. $\frac{1}{2} - \frac{4}{3} + \frac{9}{4} - \ldots + \frac{(-1)^{n-1}n^2}{n+1}$
43. $3 \cdot 2 + 5 \cdot 3 + 7 \cdot 4 + \ldots + (2n+1)(n+1)$ 45. $-1 + \frac{1}{\sqrt[3]{2}} - \frac{1}{\sqrt[3]{3}} + \ldots + (-1)^n \frac{1}{\sqrt[3]{n}}$
47. $.3 + .03 + .003 + \ldots + \frac{3}{10^n}$ 49. $\sum_{i=1}^{6} \frac{i+2}{\sqrt{i+1}}$ 51. $\sum_{k=2}^{n+1} \frac{1}{(k-1)^2}$ 53. $\sum_{k=0}^{n-1} 3 \cdot (\frac{1}{2})^k$
55. $(-1)^{n+1}(\frac{1}{2})^n$ 57. $3n+1$ 59. $\frac{2}{n^2+1}$ 61. a. $ca_1 + ca_2 + ca_3 + \ldots + ca_n$ b. $c(a_1 + a_2 + \ldots + a_n)$
c. $\sum_{k=1}^{n} a_k$ 63. True 65. True 67. False 69. True 71. a. $S_5 \approx 1.46361$, $S_{10} \approx 1.54977$, $S_{20} \approx 1.59616$, $S_{25} \approx 1.60572$ b. $S_5 \approx 2.7083333$, $S_{10} \approx 2.71828153$, $S_{20} \approx 2.718281828$, $S_{25} \approx e$
c. $S_5 \approx .953103$ $S_{10} \approx S_{20} \approx S_{25} \approx .953102$

Exercises ■ 11.2

1. 11, 15, 19, 23 3. $\frac{2}{5}, \frac{4}{5}, \frac{6}{5}, \frac{8}{5}, \frac{10}{5}$ 5. $-3\frac{1}{2}, -3\frac{1}{4}, -3, -2\frac{3}{4}$ 7. No 9. Yes, −3 11. No
13. $a_{15} = -53$, $S_{15} = -375$ 15. $a_{15} = 22\frac{2}{3}$, $S_{15} = 165$ 17. $a_{15} = 42$, $S_{15} = 315$ 19. $a_{15} = -1.75$, $S_{15} = 0$
21. $a_{15} = x - 26$, $S_{15} = 15x - 180$ 23. $a_1 = 2$, $d = 5$ 25. $a_1 = \frac{2}{3}$, $d = \frac{4}{3}$ 27. 5050 29. 2550
31. 53,048 33. $a_1 = 1$, $d = 3$ 35. $a_1 = \frac{2}{9}$, $d = -\frac{10}{9}$ 37. a. 52 b. 990 39. \$12.50
41. a. $a_{100} = 189$, $S_{100} = 8900$

Exercises ■ 11.3

1. $1, \frac{4}{3}, \frac{16}{9}, \frac{64}{27}, \frac{256}{81}$ 3. $-2, -1, -\frac{1}{2}, -\frac{1}{4}, -\frac{1}{8}, -\frac{1}{16}$ 5. $-48, -12, -3, -\frac{3}{4}$ 7. No 9. Yes, $r = \frac{1}{2}$
11. Yes, $r = .1$ 13. $a_9 = .0000128$, $S_9 = 6.25$ 15. $a_8 = -\frac{1}{256}$, $S_{10} = .333008$ 17. $a_6 = 486$, $S_6 = 728$
19. $a_8 = .00045725$, $S_8 = 1.49977$ 21. $a_{10} = 2x^9$, $S_{10} = \frac{2(1-x^{10})}{1-x}$ 23. $a_1 = .3$, $r = .3$ 25. $a_1 = 3$, $r = 2$
27. .999023 29. .998628 31. a. 241,621,790 b. 294,824,502 33. \$605.12 35. $1.4233 \cdot P_o$
37. \$1521.89 39. a. A = \$62,000, B = \$61,051 b. A = \$154,000, B = \$159,374.25 41. 1 43. $\frac{5}{12}$
45. $-\frac{3}{4}$ 47. $\frac{7}{30}$ 49. $\frac{23}{99}$ 51. $\frac{59}{9}$ 53. $\frac{82}{9}$ 55. $\frac{5}{18}$

Exercises ■ 11.5

1. 720 3. 360 5. 30 7. 1 9. 56 11. $x^6 + 6x^5y + 15x^4y^2 + 20x^3y^3 + 15x^2y^4 + 6xy^5 + y^6$
13. $a^3 + 3a^2b + 3ab^2 + b^3$ 15. $-a^3 + 3a^2b - 3ab^2 + b^3$ 17. $16x^4 - 96x^3y + 216x^2y^2 - 216xy^3 + 81y^4$
19. $625x^4 + 2000x^3 + 2400x^2 + 1280x + 256$ 21. $x^4 - 4x + \frac{6}{x^2} - \frac{4}{x^5} + \frac{1}{x^8}$
23. $x\sqrt{x} + 3x\sqrt{y} + 3y\sqrt{x} + y\sqrt{y}$ 25. $\frac{1}{x^4} + \frac{4}{x^3y} + \frac{6}{x^2y^2} + \frac{4}{xy^3} + \frac{1}{y^4}$
27. $x^4 + y^4 + 16z^4 - 4x^3y + 8x^3z - 4xy^3 - 8y^3z - 24x^2yz + 6x^2y^2 + 24x^2z^2 + 24xy^2z - 48xyz^2 + 32xz^3 + 24y^2z^2$
29. $16x^4 + 625y^4 + k^4 - 160x^3y + 32x^3k - 1000xy^3 - 500y^3k - 240x^2yk + 600x^2y^2 + 24x^2k^2 + 60xy^2k - 120xyk^2 + 8xk^3 + 150y^2k^2$ 31. $70a^4b^4$ 33. $36xy^3\sqrt{x}$ 35. $-101{,}376x^7$ 37. $160x^3$ 39. $-20x$

41. $1+\frac{1}{3}x-\frac{1}{9}x^2+\frac{5}{81}x^3-\ldots\ |x|<1$ 43. $1+x+x^2+x^3+x^4+\ldots\ |x|<1$
45. $1-2x+3x^2-4x^3+\ldots\ |x|<1$ 47. $2+\frac{1}{4}x-\frac{1}{64}x^2+\frac{3}{512}x^3-\ldots\ |x|<4$ 49. 1.02 51. 1.43
55. 2.047672511

Vocabulary/Symbol True-False Quiz

1. True 2. True 3. True 4. True 5. arithmetic 6. 1 7. True 8. True

Chapter Test

1. 1, 2, 4, 8, . . . 2. a. −55 b. −530 3. $a_1=-4,\ d=2$ 4. $a_1=2,\ d=3$ 5. c 6. a. −39,366
b. 29,525 7. any constant sequence: $a, a, a, \ldots$ 8. $a_1=.005,\ r=2$ 9. $\frac{14}{33}$ 10. 88,572
12. a. 24 b. 4 c. 28 d. 100 13. $-6720x^4y^{10}$ 14. $\frac{1}{4}+\frac{1}{16}x^2+\frac{1}{64}x^4+\ldots\ |x|<1$

APPENDIX 1

Exercises

1. a. {1,2,3,5,6,7,8,9} b. {3,5,7} 3. a. {1,2,3,4,5,6,7,8,9} b. ∅ 5. a. {2,9} b. {2,9}
7. a. $\{x \mid x>0\}$ b. ∅ 9. a. $\mathcal{R}$ b. $\{x \mid x=2\}$ 11. a. {x | x is any integer except 0} b. ∅

APPENDIX 2

Exercises

1. 30.48 cm 3. 54.55 kg 5. 2830.2 ml 7. 20 m^2 9. 352 lb 11. 3.77 l 13. 8.33 m^2
15. 3 fl oz 17. 34.1 mi

Index

Abscissa, 128
Absolute value, 11, 93-94, 190, 385
  of a complex number, 385
  inequalities, 93-94
Additive identity, 4
Additive inverse, 4
Amplitude, 300
Angle
  central, 277
  coterminal, 308
  degree measurement, 274, 276
  of depression, 332
  of elevation, 328
  interior, 284
  quadrantal, 308
  radian measurement, 274, 276
  reference, 310
  right, 282
  standard position, 306
Arc length, 278
Argand diagram, 384
Arithmetic sequence, 500-504
  formulas for, 502
Associative properties, 4
Asymptote, 199-200
  of a hyperbola, 217-218
  oblique, 203
Augmented matrix, 428
Axes, 127
Axioms (of real numbers), 4
Axis of symmetry, 156

Base (of an exponential expression), 16
BASIC (programming language), 10, 24, 55, 75, 106, 132, 145, 246, 315, 375, 443, 450, 458, 462, 505, 513
Basic solutions, 358
Binomial
  coefficient, 520
  expression, 25
  series, 522-523
  theorem, 519-522

Cardioid, 374
Carlyle, method of, 186
Cartesian
  coordinate system, 126-128
  product, 126
Central angle, 277
Characteristic, 247
Circle, 211
Circular functions, 300
Closure properties, 4
Coefficient
  defined, 17
  matrix of, 447
Cofactor, 452
Common denominator, 42-43
Common logarithm, 241
Commutative properties, 4
Completing the square, 58
Complex
  fractions, 44-46
  number, 60-65
  plane, 385
Composition of functions, 140
Conditional equations, 70
Congruent triangles, 282
Conic sections, 210-219
  degenerate, 221
Continuous function, 189
Convex bounded set, 416
Coordinates, 128
Correspondence diagram, 135
Cosecant, 285
Cosine, 285
Cotangent, 285
Coterminal, 308
Counterexample, 7
Counting numbers (see natural numbers)
Cramer's Rule, 458-461

Damped harmonic oscillation, 330
Decomposition, 484
Degenerate conics, 221
Degree, angle measurement, 274-276
DeMoivre's Theorem, 384-390
Dependent system, 396, 400
Depression, angle of, 332
Descartes' Rule of Signs, 480-483
Determinant, 451-461
  Cramer's Rule, 458-461
  n-by-*n*, 453
  two-by-two, 451
Direct variation, 177
Discontinuous function, 189
Discriminant, 99
Distance formula, 128-130
Distributive property, 4

Division
  defined, 5
  Form I, 29
  Form II, 30
  laws, 8
  by zero, 7, 54
Domain, 133
Dot product, 381
Double-angle formulas, 349

Elevation, angle of, 328
Elimination method, 398-400
Ellipse, 212
Empty set, 528
Equal vectors, 376
Equations
  conditional, 70
  equivalent, 70
  fractional, 109-110
  identity, 70
  linear, 70-74
  literal, 82-84
  quadratic, 97-104
  quadratic in form, 106-109
  radical, 110-112
  solution set of, 70
Equilateral triangle, 282
Equivalent equations, 70
Even function, 203
Exponential
  equations, 230-232
  function, 226
  growth and decay, 228
Exponents
  integral, 16-22
  laws of, 17, 19
  negative, 19
  rational, 55-58
  and scientific notation, 20-22
  zero, 18
Extraneous solutions, 109, 259

Factor Theorem, 468
Factoring, 34-38
  common factor, 35
  difference of two cubes, 37
  difference of two squares, 37
  GCF, 35
  sum of two cubes, 37
Feasible solution, 417
Fibonacci sequence, 527
Finite sequence function, 493
Foil, 26
Formulas, 83-84, 502, 507
Fractional
  equations, 109-110
  expressions, 39-46
Frictionless harmonic oscillation, 330
Function
  circular, 300
  composite, 140
  continuous, 189
  discontinuous, 190
  even, 203
  exponential, 226
  identity, 145
  inverse of a, 168
  linear, 145-151
  logarithmic, 234
  notation, 137-138
  odd, 203
  one-to-one, 144
  piecewise, 188-192
  rational, 195, 204-208
  trigonometric, 299-304
  vertical line test for, 137
  zero of, 138
Fundamental Theorem of
  algebra, 469
  arithmetic, 477

Gaussian elimination method, 431-433
GCF (greatest common factor), 35
Geometric sequence, 506-511
  formulas for, 507
  infinite, 510
General linear equation, 72
Generalized *n*-by-*n* linear system, 430
Greatest integer function, 191

Half-angle formulas, 350
Half-plane, 161
Harmonic series, 500
Homogeneous system, 434
Horizontal asymptote, 200
Hyperbola, 217
Hypotenuse, 282

Identity
  function, 145
  fundamental, 346
  matrix, 444
  properties, 4
  trigonometric, 307, 346-351
Imaginary numbers, 60-61
Inconsistent system, 396, 400
Indeterminant expressions, 7
Index of
  radical, 49
  summation, 495
Inequality
  linear in two variables, 161
  systems of, 414-416 (see also order relations)
Infinite sequence function, 493
Integral exponents, 16-22
Intercepts, 154
Interest, 77
Interpolation, 249
Intersection of sets, 529
Interval notation, 91-94
Inverse
  of a matrix, 444-449
  property of real numbers, 4
  relation, 168-173
  of a trigonometric function, 363-368
  variation, 177
Irrational
  numbers, 3
  roots, 478-483
Isosceles triangle, 282

Joint variation, 177

Law(s) of
  cosines, 341-344
  exponents, 17, 19
  logarithms, 241
  sines, 339-341
LCM (least common multiple), 42
Leading coefficient, 25
Lemniscate, 375
Less than ($<$), 10
Limaçon, 375
Linear
  equations, 70-74
  functions, 145-151
  inequalities, 161
  interpolation, 249
  programming, 416-420
Literal equations, 82-84
Logarithm
  characteristic of, 247
  equations, 258-259
  functions, 234
  laws, 241
  mantissa of, 247

Mantissa, 247
Map (of a relation), 135
Markup, 75
Mathematical induction, 513-517
Matrices, 427-465
  addition of, 436
  algebra of, 436-440
  augmented, 428
  coefficient, 447
  determinant of, 451-456
  equality of, 436
  identity, 444
  inverse of, 444-449
  multiplication of, 439-440
  nonsingular, 450
  reduced, 429
  scalar multiplication of, 437
  singular, 450
  transpose of, 442
Metric system, 530-532
Midpoint formula, 130
Minor, 452
Modulus, 385
Monomials, 17
Multiplicative identity, 4
Multiplicative inverse, 4
Multiplicity of roots, 98

Natural
  exponential function, 230
  logarithm, 237, 252-255
  numbers, 2
Nonsingular matrices, 450
*n*th roots of unity, 390
Number line, 10
Number systems, 2-3
  axioms of, 4
  complex, 60-65
  imaginary, 60-61
  integers, 2
  irrational numbers, 3
  natural numbers, 2
  number line, 10
  rational numbers, 2
  real numbers, 3

Objective function, 417
Oblique asymptote, 203, 217
Odd function, 203
One-to-one function, 144, 153
Optimal feasible solution, 417
Order relations, 11-14, 89-94
  and absolute value, 93-94
Ordered pairs, 127
Ordinate, 128
Origin, 11, 126
  symmetry with respect to, 197

Parabola, 155, 214
  vertex of, 155
Partial fraction decomposition, 484-488
Period (of a trigonometric function), 300
Phase shift, 319
Pi, 68, 275
Piecewise function, 188-192
Point-slope form, 150
Polar coordinates, 369-374
Pole, 370
  tangents at, 372
Polynomials, 24-38, 467, 482-484
  cubic, 482, 484
  degree of, 25
  factoring, 34-38
  leading coefficient, 25
  multiplication of, 26
  notation, 28
  *n*th degree, 467
  quartic, 482, 484
  quintic, 483, 484
Principle of Exponentiation, 53
Principle of Mathematical Induction, 514
Principle *n*th root, 48
Products, special, 27
Proportionality constant, 177
Pythagorean Theorem, 129, 283, 289

Quadrantal angles, 308
Quadrants, 128
Quadratic
  equations, 97-104
  formula, 99
  functions, 153-159
  inequality, 114-118, 164
Quartic polynomial, 482, 484
Quintic polynomial, 483, 484

Radian measurement of angles, 275, 276
Radical equation, 110-112
Radical expressions, 48-53
  index of, 49
Radius vector, 306
Range, 133
Rational exponents, 55-58
Rational functions, 204-208
Rational numbers, 2
Rational Root Theorem, 472
Rationalizing the denominator, 50, 54
Rationalizing the numerator, 55
Real number(s)
  axioms of, 4
  line, 10
Reciprocal identities, 307
Rectangular coordinate system, 126
Rectilinear motion, 78
Reduced matrix, 429
Reference angle, 310
Relation, 132-135
  domain and range of, 132
  mapped, 135
Remainder Theorem, 467
Resultant vector, 377
Revenue equation, 176
Right triangle, 282
Roots
  irrational, 478-483
  multiple, 98
  of polynomials, 467
  principal, 48
  rational, 472
  of unity, 390
Rose, 373, 374
Row transformations, 428

Scalar, 375
Scalar product, 381
Scale ratio, 174
Scientific notation, 20-22
Secant, 285
Sequence, 493-511
  alternating, 494
  arithmetic, 500-504
  function, 493
  geometric, 506-511
Series, 494-511
  alternating, 495
  harmonic, 500
Set-builder notation, 528
Sets, 528-529
Sign graphs, 115
Sign intervals, 201
Similar triangles, 283
Simultaneous solutions, 396

Sine, 285
Singular matrices, 450
Slope, 149
Slope-intercept form, 150
Spiral, 374
Standard position of an angle, 306
Standard trigonometric function, 299, 302
Stirling's formula, 526
Subscripts, 7
Substitution method, 397-398
Subtraction, 4
Summation notation, 495-497
Supply equations, 119
Symmetry
  origin, 197
  x-axis, 203
  y-axis, 196
Synthetic division, 31-32, 478-480
Systems of equations, 396-433
  elimination method, 398-400
  generalized *n*-by-*n* system, 430
  homogeneous, 434
  matrix solutions, 427-433
  nonlinear, 409-412
  substitution method, 397-398
  three-by-three linear systems, 403-406
  two-by-two linear systems, 396-401
Systems of inequalities, 414-416

Tangent, 285
Tangent function, 302
Theorems, 7
Transformations, 398, 427
  row, 428
Triangle inequality, 15
Triangles
  congruent, 282
  equilateral, 282
  isosceles, 282
  right, 282
  similar, 283
  solving, 288-290
Trigonometric equations, 357-360
  basic solutions, 358
Trigonometric function, 299-304, 363-368
  inverse, 363-368
Trigonometric identities, 307, 346-351
  addition formulas, 348
  double-angle formulas, 349
  fundamental, 346
  half-angle, 350
Trigonometric ratios, 285
Trinomials, 25
Two-point form, 153

Union of sets, 528
Unit circle, 221, 294
Unit vectors, 379

Variation, 177-179
Vectors, 375-382
  addition of, 376, 377
  algebra of, 379
  dot product, 381
  equal, 376
  multiplication by scalar, 378
  resultant, 377
  scalar product, 381
  unit, 379
Vertex
  form for a quadratic, 160
  of a parabola, 155
Vertical asymptote, 199
Vertical line test, 137, 155, 299

x-axis, 127
  symmetry with respect to, 203
x-intercept, 138

y-axis, 127
  symmetry with respect to, 196
y-intercept, 154

Zero
  division by, 7
  exponent, 18
  of a function, 138
  laws of, 7
  of a polynomial, 467